T0231198

Phosphorus Biogeochemistry *in* SUBTROPICAL ECOSYSTEMS

Phosphorus Biogeochemistry *in* SUBTROPICAL ECOSYSTEMS

EDITED BY

K. R. Reddy
Graduate Research Professor
Wetland Biogeochemistry Laboratory
Soil and Water Science Department
Institute of Food and Agriculture Sciences
University of Florida, Gainesville, FL

G. A. O'Connor
Professor
Soil and Water Science Department
Institute of Food and Agriculture Sciences
University of Florida, Gainesville, FL

C. L. Schelske
Carl S. Swisher Professor
Department of Fisheries and Aquatic Sciences
Institute of Food and Agriculture Sciences
University of Florida, Gainesville, FL

CRC Press is an imprint of the
Taylor & Francis Group, an informa business

Reprinted 2010 by CRC Press
CRC Press
6000 Broken Sound Parkway, NW
Suite 300, Boca Raton, FL 33487
270 Madison Avenue
New York, NY 10016
2 Park Square, Milton Park
Abingdon, Oxon OX14 4RN, UK

Library of Congress Cataloging-in-Publication Data

Phosphorus biogeochemistry of subtropical ecosystems / edited by K.R. Reddy, G.A. O'Connor, C.L. Schelske.
p. cm.
Based on papers from a symposium held in Clearwater, Fla., July 14–16, 1997.
Includes bibliographical references and index.
ISBN 1-56670-331-X
1. Phosphorus cycle (Biogeochemistry)--Florida--Congresses. 2. Aquatic ecology--Florida--Congresses. 3. Wetland ecology--Florida--Congresses. 4. Soil ecology--Florida--Congresses. I. Reddy, K. R. (K. Ramish) II. O'Connor, George A., 1944– III. Schelske, Claire L.
QH344.P5 1999
577'.14—dc21 98-50246
CIP

International Standard Book Number 1-56670-331-X
Library of Congress Card Number 98-50246

Preface

Phosphorus is one of the major nutrients limiting the productivity of terrestrial, wetland, and aquatic ecosystems. Since many of these ecosystems are hydrologically linked, the quality of water leaving one ecosystem can significantly affect the quality of water of another ecosystem. Phosphorus has been identified as one of the major nutrients affecting many of these ecosystems. For example, Florida's ecosystems are sensitive to anthropogenic nutrient loads, and many aquatic systems are currently in either eutrophic or hypereutrophic condition. Over the last decade, several research projects were conducted on Florida's ecosystems by scientists from state and federal agencies and private industry to address water quality issues and to develop best management practices to control nutrient loads. Similar studies are conducted in other subtropical ecosystems. These interdisciplinary projects involved studies on interactive physical, biological, geological, and chemical processes regulating the fate and transport of phosphorus in soils/sediments and water components of terrestrial, wetland, and aquatic ecosystems (e.g., St. Johns River Basin, Oklawaha River Basin, Lake Apopka, Lake Okeechobee Basin, Lake Okeechobee, and the Everglades). Biochemical processes have been studied at various spatial and temporal scales, ranging from small (microbial cell, particle level, or laboratory scale) to relatively large (field-plot or ecosystem level) scales. Conclusions derived from studies conducted at different scales are frequently influenced by disciplinary bias. Thus, although important research has been conducted, linkage or synthesis to address the common targeted goal (to effectively use this research information to solve the problem) is lacking. The chapters in this book provide such a linkage.

The chapters are divided into the following specific groups:

1. Ecological analysis and global issues
2. Biogeochemical transformations
3. Biogeochemical responses
4. Transport processes
5. Phosphorus management
6. Synthesis

Although Florida's ecosystems are used as a case example, results presented are discussed in the context of global issues. A five-member panel evaluated the results of various studies reported and developed a synthesis paper that included recommendations and future directions for research and management.

About the Editors

Dr. K. Ramesh Reddy is the Graduate Research Professor of Wetland Biogeochemistry in the Soil and Water Science Department, University of Florida. Dr. Reddy and his coworkers conduct research on biogeochemical cycles in wetlands and aquatic ecosystems. He is the author or coauthor of more than 240 scientific papers. Major research interests are spatial and temporal variations in biogeochemical processes regulating the fate of nutrients and other contaminants in wetlands and aquatic systems. Dr. Reddy conducted research on several subtropical ecosystems of Florida, including Lake Apopka, Lake Okeechobee, Everglades, Okeechobee Drainage Basin, Indian River Lagoon, and others. He is a Fellow of the Soil Sciences Society of America and the American Society of Agronomy; Chairman (1992) Div. A-5 (Environmental Quality), American Society of Agronomy; Chairman (1994) Div. S-10 (Wetland Soils), Science Society of America; University of Florida Research Award, 1990, 1991–92; Edward Deevey Jr. Award, 1998.

Dr. George A. O'Connor is Professor of Soil Chemistry in the Soil and Water Science Department at the University of Florida. He has researched the fate and transport of various solutes in soils and plants for more than 25 years. For the last decade or so, he has studied the reactions of nonhazardous wastes in soils to determine safe levels of waste applications to soils. His work on biosolids-borne pollutants resulted in his appointment to the Peer Review Committee for the national (EPA) rule regulating land application of sewage sludge. He was a member of the Technical Advisory Committee for the rewrite of Florida's residuals reuse policy. Current research efforts focus on the bioavailability and leachability of biosolids-borne oxyanions (Mo and P) in Florida soils. He is a Fellow of the Soil Science Society of America and the American Society of Agronomy, and was Chair of the Soil and Water Science Department at the University of Florida from 1990 to 1994. He is coauthor of the textbook *Soil Chemistry.*

Dr. Claire L. Schelske is the Eminent Scholar Chair and the Carl S. Swisher Professor of Water Resources, Department of Fisheries and Aquatic Sciences, University of Florida. Prof. Schelske's research is concerned with factors affecting the biological productivity of lakes, particularly how nutrient enrichment affects water quality and produces undesirable effects in lakes. One of the related management questions arising from this research is, "Has nutrient enrichment changed historically?" Because long-term records from lake sampling are not available, research has been directed toward studying historical changes from analysis of the sediment record in lakes. Dr. Schelske and his colleagues have conducted research on the Laurentian Great Lakes, the St. Johns River, and Florida lakes including Lake Apopka and Lake Okeechobee. He was Secretary and President of the American Society of Limnology and Oceanography and is a National Representative of the International Association of Theoretical and Applied Limnology.

Contributors

F.J. Aldridge
Department of Fisheries and Aquatic Sciences
University of Florida
Gainesville, FL

Martin T. Auer
Department of Civil and Environmental Engineering
Michigan Technological University
Houghton, MI

Nicholas G. Aumen
Ecosystem Restoration Department
South Florida Water Management District
West Palm Beach, FL

Lawrence E. Battoe
St. Johns River Water Management District
Department of Water Resources
Palatka, FL

A.B. (Del) Bottcher
Soil and Water Engineering Technology, Inc.
Gainesville, FL

Joseph N. Boyer
Southeast Environmental Research Program
Florida International University
Miami, FL

Patrick L. Brezonik
Department of Civil Engineering and Water Resource Center
University of Minnesota
Minneapolis, MN

Jerry Brooks
Florida Department of Environmental Protection
Tallahassee, FL

Kenneth L. Campbell
Agricultural and Biological Engineering Department
University of Florida
Gainesville, FL

John C. Capece
Southwest Florida Research and Education Center
University of Florida
Immokalee, FL

Steven C. Chapra
Civil, Environmental, Architectural Engineering Department
University of Colorado
Boulder, CO

M.J. Chimney
South Florida Water Management District
West Palm Beach, FL

T. Chua
Wetland Biogeochemistry Lab
University of Florida
Soil and Water Science Department
Gainesville, FL

Michael F. Coveney
St. Johns River Water Management District
Department of Water Resources
Palatka, FL

Thomas A. DeBusk
DB Environmental Laboratories
Rockledge, FL

W.F. DeBusk
Soil and Water Science Department
University of Florida
Gainesville, FL

Forrest E. Dierberg
DB Environmental Laboratories, Inc.
Rockledge, FL

H. Carl Fitz
Everglades Systems Research Division
South Florida Water Management District
West Palm Beach, FL

T. Fontaine
South Florida Water Management District
West Palm Beach, FL

D.A. Graetz
Soil and Water Science Department
University of Florida
Gainesville, FL

Willie Harris
Soil and Water Science Department
University of Florida
Gainesville, FL

Karl E. Havens
Ecosystem Restoration Department
South Florida Water Management District
West Palm Beach, FL

S.R. Humphrey
Department of Natural Science
University of Florida
Gainesville, FL

Wade Hurt
Soil and Water Science Department
University of Florida
Gainesville, FL

Forrest T. Izuno
Agricultural and Biological Engineering
University of Florida
Institute of Food and Agricultural Sciences
Everglades Research and Education Center
Belle Glade, FL

R. Thomas James
Ecosystem Restoration Department
South Florida Water Management District
West Palm Beach, FL

Kang-Ren Jin
Ecosystem Restoration Department
South Florida Water Management District
West Palm Beach, FL

Ronald D. Jones
Southeast Environmental Research Program
Florida International University
Miami, FL

Robert H. Kadlec
Wetland Management Services
Chelsea, MI

W.F. Kenney
Department of Fisheries and Aquatic Sciences
University of Florida
Gainesville, FL

T.C. Kosier
South Florida Water Management District
West Palm Beach, FL

T.J. Logan
School of Natural Resource
Ohio State University
Columbus, OH

E. Lowe
St. Johns River Water Management District
Palatka, FL

T. MacVicar
MacVicar Federico Lab Inc.
West Palm Beach, FL

Paul V. McCormick
Everglades Systems Research Division
South Florida Water Management District
West Palm Beach, FL

S.L. Miao
Everglades Systems Research Division
South Florida Water Management District
West Palm Beach, FL

M.Z. Moustafa
Ecosystem Restoration Department
South Florida Water Management District
West Palm Beach, FL

V.D. Nair
Soil and Water Science Department
University of Florida
Institute of Food and Agricultural Sciences
Gainesville, FL

Susan Newman
Everglades Systems Research Division
South Florida Water Management District
West Palm Beach, FL

G.A. O'Connor
Soil and Water Science Department
University of Florida
Gainesville, FL

Curtis D. Pollman
Tetratech Inc.
Gainesville, FL

K.R. Reddy
Soil and Water Science Department
University of Florida
Gainesville, FL

R.D. Rhue
Soil and Water Science Department
University of Florida
Gainesville, FL

Curtis J. Richardson
Duke University Wetland Center
Nicholas School of the Environment
Duke University
Durham, NC

J. Stephen Robinson
Department of Soil Science
University of Reading
Reading, Berkshire, U.K.

Barry H. Rosen
Planning Department
South Florida Water Management District
West Palm Beach, FL

C.L. Schelske
Department of Fisheries and Aquatic Sciences
University of Florida
Gainesville, FL

Leonard J. Scinto
Southeastern Environmental Research Program
Florida International University
University Park
Miami, FL

Andrew N. Sharpley
USDA-ARSl
Pasture Systems and Watershed Management Research Lab
University Park, PA

Y. Peter Sheng
Coastal and Oceanographic Engineering Department
University of Florida
Gainesville, FL

Fred H. Sklar
Everglades Systems Research Division
South Florida Water Management District
West Palm Beach, FL

Alan D. Steinman
Ecosystem Restoration Department
South Florida Water Management District
West Palm Beach, FL

David L. Stites
St. Johns River Water Management District
Department of Water Resources
Palatka, FL

Terry K. Tremwel
St. Johns Water Management District
Palatka, FL

William W. Walker, Jr.
Environmental Engineer
Concord, MA

Robert G. Wetzel
Department of Biological Sciences
University of Alabama
Tuscaloosa, AL

Paul J. Whalen
Everglades Regulation Division
Regulation Department
South Florida Water Management District
West Palm Beach, FL

J.R. White
Wetland Biogeochemistry Lab
University of Florida
Soil and Water Science Department
Gainesville, FL

A.Wright
Wetland Biogeochemistry Lab
Soil and Water Science Department
University of Florida
Gainesville, FL

Joyce Zhang
Ecosystem Restoration Department
South Florida Water Management District
West Palm Beach, FL

Acknowledgments

The financial assistance provided by the sponsors of the symposium from which this book is derived made it possible to invite several experts, support conference operations, and compile this book. The editors wish to acknowledge the financial support of the University of Florida (Institute of Food and Agricultural Sciences [IFAS], Center for Natural Resources, and Soil and Water Science Department), the St. Johns River Water Management District (SJRWMD), and the South Florida Water Management District (SFWMD).

Compiling this volume is the result of the efforts of a number of people. We wish to acknowledge the efforts of program committee members (Dr. K.L. Campbell, University of Florida; Tom Fontaine, SFWMD; and Ed Low, SJRWMD). The editors also were the members of the program committee, with K.R. Reddy serving as chair.

We also acknowledge the assistance of several graduate students of the University of Florida Wetland Biogeochemistry Laboratory, and Dr. V.D. Nair of the University of Florida Soil and Water Science Department, during the symposium. We especially want to thank the authors who chose to prepare manuscripts in accordance with reviewers' requirements and to share their technical knowledge and experience.

Special appreciation goes to Ms. Pam Marlin, who played a key role in various stages of the symposium and preparation of this book. We also thank the following scientists for providing critical review of the manuscripts: R. Brown, K.L. Campbell, J. Capece, D.L. Childers, M. Coveney, E.M. D'Angelo, W.F. DeBusk, F.T. Dierberg, T. Fontaine, W.D. Graham, L.G. Goldsborough, S. Humphrey, W.G. Harris, R.H. Kadlec, L. Keenan, M.S. Koch, E. Lowe, T. MacVicar, J.E. Margurger, B.L. McNeal, V.D. Nair, S. Newman, A. Ogram, H. Pant, E.J. Phlips, S.C. Reed, J.A. Robbins, and R.R. Twilley.

We acknowledge the IFAS Office of Conferences and Institutes and its staff for organizing the nontechnical portion of the symposium. The cover design for the book was provided by Donald W. Poucher and Katrina Vitkus, IFAS-UF Educational Media and Services.

Any opinions, findings, conclusions, or recommendations expressed in this publication do not necessarily reflect the views of the University of Florida or the editors of this book.

K.R. Reddy
G.A. O'Connor
C.L. Schelske

Contents

VI PHOSPHORUS MANAGEMENT

Section I
Introduction

Phosphorus Biogeochemistry of Subtropical Ecosystems: Florida as a Case Example

Symposium Overview and Synthesis

K.R. Reddy, G.A. O'Connor, and C.L. Schelske

Nonpoint source pollution, especially from agricultural sources, has been identified as the major source of nutrients to, and degrader of, wetlands and aquatic systems (USEPA, 1996). For example, in subtropical ecosystems of Florida, agricultural drainage water discharged from the Everglades Agricultural Area (EAA) into the Water Conservation Areas (WCAs) of the Everglades has shifted oligotrophic wetland areas into eutrophic conditions. Similarly, nutrient loads from the dairy industry on Lake Okeechobee seem to have shifted the lake toward hypereutrophic conditions. Increased nutrient loads from agricultural and urban activities are adversely affecting Florida Bay and other estuaries in Florida. Similar examples can be cited for other areas of the U.S.

Many uplands, wetlands, and aquatic systems are hydrologically linked, and the quality of water leaving one ecosystem can significantly impact the water quality of a receiving ecosystem. Thus, management practices implemented in one ecosystem impact adjacent ecosystems, and holistic, coordinated management of ecosystems is needed to improve water quality.

Biogeochemistry is an interdisciplinary science that includes the study of interactive biological, geological, and chemical processes regulating the fate and transport of nutrients and contaminants in soil, water, and atmospheric components of an ecosystem. Biogeochemical processes are studied at various spatial and temporal scales, from small (microbial cell, particle level, or laboratory scale) to relatively large (field-plot or watershed level) scales. Conclusions derived from studies conducted at different scales are frequently influenced by disciplinary bias. Thus, although important research has been conducted, linkage or synthesis to address the common targeted goal (to effectively use this research information to solve problems) is lacking. Because of the differences in scientific approaches used by scientists working in upland, wetland, and aquatic systems, similarities among these systems are often ignored. These problems must be resolved by more active communication between the disciplines.

This introduction provides an overview and synthesis of the issues discussed at a symposium held during July 14 through 16, 1997, in Clearwater, Florida. The purpose of the symposium was to provide a necessary forum for the synthesis and interpretation of the current status of P biogeochemistry research in upland, wetland, and aquatic ecosystems, with Florida as a case example. Main issues related to P biogeochemistry discussed at the symposium were

1. What are the major sources of excess P supply to the environment?
2. What is the capacity of soils to retain P?
3. What processes govern P transfer from one ecosystem to another?
4. How can we assess the recovery of an ecosystem from P loads?

We summarize the results in the following major topics: sources of P, transformations, transport processes, management options, and future directions.

1.1 SOURCES

Phosphorus inputs through atmospheric deposition and weathering of natural minerals can maintain wetlands and aquatic systems under eutrophic conditions and can significantly impact trophic conditions of oligotrophic wetlands and aquatic systems. Often, P inputs from these sources receive less attention because of the difficulties in accurate quantification (Brezonik and Pollman, 1999). Estimates of atmospheric P deposition range from 20 to 80 mg P/m^2 year (Redfield, 1998) and, in Florida, represent the second largest P input, accounting for approximately 10% of the total measured imports (Reddy et al., 1999a). The relative contribution of natural sources, compared to imports through fertilizers, feeds, and organic solids, should be quantified when determining P budgets and in developing management strategies to restore these disturbed ecosystems.

Phosphorus is imported anthropogenically into upland ecosystems through fertilizers, organic wastes, wastewater, and animal feeds. Some of this P is exported out of the system in the product produced, and some is exported to adjacent aquatic systems, but the majority remains within the system. Long histories of P-rich material application elevate soil P levels in many ecosystems, resulting in residual P sufficient to meet the P requirements of crops grown on these soils. Application of organic wastes based on crop nitrogen needs usually applies P in excess of crop needs. Soils contaminated with P as a result of intensive use, or abandoned agricultural lands with long history of fertilization, can continue to release P for years after farming activities on these lands have ceased.

Fertilizers are a major source of P added to upland ecosystems. Best management practices and improved recognition of residual P levels in soils have decreased the overall use of fertilizer P (Sharpley, 1999). It is critical to determine the sustainable P application rates necessary to maintain the balance between production agriculture and sound water quality. We must also develop methodologies to manage these ecosystems for maximum P retention while maintaining acceptable water quality.

To address these issues, we need a clear understanding of the transformations regulating P availability and the stability of stored P in different ecosystems.

1.2 BIOGEOCHEMICAL TRANSFORMATIONS

Unlike C and N, much of the P added stays within an ecosystem in a variety of forms. Inorganic soil P forms are better characterized than organic P forms. Methodologies to characterize soil P forms are operationally defined but provide indications of the bioavailability and stability of stored P in soils and sediments (Graetz and Nair, 1999; Newman and Robinson, 1999). However, more refined methodologies are needed to characterize soil P, especially organic P forms. Soil P undergoes various transformations as it cycles through inorganic P pools (associated with Fe, Al, Ca, and Mg minerals) and organic P pools (through plants, animals, microbes, and soil organic matter). Predicting these transformations requires detailed understanding of the soil/sediment characteristics, chemical equilibria, kinetics of organic and inorganic P transformations, and the ways P availability is affected under various management strategies.

Chemical processes affecting P concentration in porewaters of soils and sediments are sorption and precipitation that incorporate P into the solid phase (Rhue and Harris, 1999). Phosphate sorption by soils and sediments is frequently related to the amount of amorphous and poorly crystalline forms of Fe and Al present (Rhue and Harris, 1999; Richardson, 1999). In highly reactive soils and sediments, P sorption can be hysteretic, suggesting irreversible binding. However, Rhue and Harris (1999) suggest that the hysteresis effect may be due to kinetics of adsorption and desorption. At high P activities, this effect may be compounded by precipitation reactions. Phosphorus solubility is regulated by both pH and Eh. Although the effect of these regulators is clearly demonstrated on the fate of inorganic P, their interactive effects with biological processes are poorly understood. Soils or sediments loaded with P tend to lose their capacity to buffer porewater P concentration, increasing the equilibrium P concentration (EPC) in soil *or* sediment porewater. At present, we do not have a good understanding of the processes that regulate the P sorption capacity of soils and sediments, especially in relation to surface and subsurface water quality.

In wetland soils and aquatic sediments, organic P typically constitutes >50% of total P, whereas in upland soils it can vary from 10 to 90% (Sharpley, 1999; Newman and Robinson, 1999). Organic P accumulation in soils and sediments is dictated by biological productivity and accumulation of detrital material. Conventional chemical fractionation schemes provide information on the lability of organic P form and their bioavailability but offer little insight on the forms of P. Organic P forms can be characterized using ^{31}P-NMR (nuclear magnetic resonance) methods (Newman and Robinson, 1999), but NMR methods are subject to various interferences and need further refinement.

Microbial metabolism influences organic P breakdown by direct biochemical mobilization of extracellular enzymes and by altering physicochemical environments in the soil (ionic composition, Eh, pH, and other conditions) (Wetzel, 1999). Organic P cycling in soils and sediments is limited by the availability of bioavailable inorganic

P. Approximately 70% of organic P mineralization is performed by microbes, showing the importance of microbial processes in actively cycling organic P (Wetzel, 1999).

1.3 BIOGEOCHEMICAL RESPONSES

Wetlands and lakes receive P discharged from adjacent drainage basins. Often, the amount of P leaving uplands is small in relation to the overall P budget, but sufficient to cause nutrient imbalances and alter the trophic status of the receiving water bodies. Wetlands and lakes differ in how they respond to P loading. Many lakes tend to function as continuously stirred tank reactors, where added P is uniformly mixed in the water column, and impacts are observed throughout the lake. However, large lakes (such as Lake Okeechobee) may not function as stirred tank reactors but may exhibit distinct zones in water column chemistry and trophic conditions (Schelske et al., 1999). In addition, phytoplankton or microbe turnover occurs in hours or days, compared to scales of weeks for physical mixing and circulation in large lakes. Wetlands tend to function more as plug-flow reactors, where P added is rapidly removed by microbes, periphyton, vegetation, and sorption, and precipitation reactions in areas proximal to the point of discharge. These conditions create distinct gradients in P content in the water and soil, similar to those observed in the Everglades Water Conservation Areas (DeBusk et al., 1994; Newman et al., 1997).

In freshwater wetlands, periphyton (attached algae and bacteria) respond rapidly to increased P loading (McCormick and Scinto, 1999) by increasing productivity and growth rates and shifting community composition. This suggests that periphyton may be useful as sensitive indicators of environmental impacts.

The detrital layer and periphyton mats host a wide range of microbial populations involved in nutrient cycling that are affected by P loading. Microbial populations, microbial biomass and respiration, organic N mineralization, nitrification, denitrification and N_2 fixation in detrital and soil layers of the Water Conservation Area 2a (WCA-2a) of the Everglades are affected by P loading (Reddy et al., 1999b). Overall, P loading increases the size of the microbial pool and organic soil accretion rates and increases the release of inorganic N and P into the water column.

Increased nutrient availability, either from external or internal sources, can influence the growth rates and composition of macrophytes. In the Everglades, changes in macrophyte structure (shift from saw grass to cattails) and functions were related to P availability (Miao and DeBusk, 1999). Phosphorus loading to WCA-2a of the Everglades increased macrophyte production, photosynthesis, and decomposition, increasing rates of organic matter accumulation (Miao and DeBusk, 1999).

The trophic status of lakes is usually determined by measuring total N, total P, and chlorophyll-*a* concentrations of the water column. However, much of the total N and P is not bioavailable, so measuring these parameters inaccurately indicates the nutrient-limiting status of the lake (Schelske et al., 1999). When bioassays showed N - limited biomass production in hypereutrophic lakes, TN:TP ratios falsely predicted that P was the limiting nutrient. Predictions based on bioavailable N and P (dissolved inorganic N and soluble reactive P) may more reliably identify the limiting nutrient status of the lake (Schelske et al., 1999).

1.4 TRANSPORT PROCESSES

The amount of P leaving uplands is affected by a number of factors including hydrology, soil type, land use, and management factors. Often, many P outputs are not adequately characterized because of the complexity associated with hydrologic processes in uplands. Phosphorus is transported in dissolved P (DP) and particulate P (PP) forms. Phosphorus transfer from one ecosystem to another requires knowledge of biogeochemistry and hydrology as related to fate and transport.

Many of Florida's uplands soils have had their low P retention capacities exceeded by high P loadings with wastes and commercial fertilizers, resulting in movement of soluble P in surface and subsurface drainage (Campbell and Capace, 1999). Surface flow is the dominant pathway in areas with flat topography (<0.3% slope), whereas P transport through subsurface flow is greatest in areas with surface slopes of >0.3%. Flatwood watersheds with sandy surface horizons (common in Florida) have high infiltration rates, resulting in P transport through subsurface flow. For many ecosystems, the amounts and forms of P leaving the watershed are not quantified, but outputs have been quantified for the Okeechobee Basin and the Everglades Agricultural Area (Bottcher et al., 1999; Izuno and Whalen, 1999).

Phosphorus movement in wetlands occurs via vertical and horizontal movement of water and is influenced by topography, hydraulic conductivity of soils, and spatial vegetational patterns. In certain ecosystems, vertical transport moves P to groundwater. The dominant flow path, however, is surface flow, which occurs as sheet *or* plug flow (Kadlec, 1999). Slow moving waters generally occupy a larger fraction of the wetland surface. Low velocities, coupled with resistance created by vegetation, can result in short-circuiting flow.

Wetlands function as major sinks for suspended particles. Low flow rates and vegetation resistances enhance particle settling and associated particulate P. This results in accretion of organic matter and the building of new soil material. Phosphorus transport due to sediment resuspension and bioturbation is minimal (Kadlec, 1999). Vertical transport (exchange between soil and overlying water column) occurs through infiltration, diffusion, and transpiration, although diffusive flux of P is generally slow compared to other vertical transport processes (Kadlec, 1999). Influences of biotic processes (e.g., bioturbation, plant uptake and transpiration, and root zone decomposition) on mobilization of P are poorly understood and must be addressed in future research. The role of wetlands in P retention cannot be accurately determined until all hydrologic flow paths are adequately quantified.

Phosphorus movement between sediments and overlying water column in lakes is well described and thoroughly researched (e.g., Brezonik and Pollman, 1999; Chapra and Auer, 1999; Sheng, 1999). Hydrodynamic processes, including wind-driven circulation, wind-induced bottom stress and sediment resuspension, advection, turbulent mixing, and deposition significantly influence P exchange between sediments and overlying water columns (Sheng, 1999). In shallow lakes and estuaries, currents and waves cause significant resuspension and vertical mixing of sediment, resulting in release of P to the water column. In deep aquatic systems, vertical stratification can occur, with low levels of dissolved oxygen at the sediment-water interface. A detailed discussion on the importance of hydrodynamic processes on P mobilization is presented by Brezonik and Pollman (1999) and Sheng (1999).

1.5 PHOSPHORUS MANAGEMENT

Phosphorus management for sound water quality is a critical issue where long-term application of P has increased soil P levels. In many agricultural lands, soil test values predict little or no response to fertilizers P by crops grown on these soils. Upland ecosystems of Florida are often dominated by Flatwood soils, where P management is a major challenge. These soils generally have poor P retention capacities, and water bodies typically associated with these uplands are very sensitive to P inputs. Accordingly, P control strategies have been implemented, including Best Management Practices (BMPs), control of P imports (Bottcher et al., 1999; Izuno and Whalen, 1999), purchase of agricultural lands adjacent to sensitive water bodies, reducing fertilizer recommendations, and incorporating buffer strips and wetlands in the ecosystem. Despite BMP development, implementation by farmers is often problematic, requiring financial incentives and education and outreach programs. Sometimes, legislative action forces implementation of BMPs and P control strategies. For example, a 25% total P load reduction is legislatively mandated for P discharges from the Everglades Agricultural Area into the Everglades WCAs (Izuno and Whalen, 1999).

Soils enriched with P can continue to release P long after the P inputs (e.g., farming activities) cease. Some of these areas are being acquired by state and federal agencies and converted into wetlands. On-site management strategies must be developed to immobilize soluble P in such areas to reduce P transport into adjacent water bodies.

Phosphorus management strategies implemented in uplands also affect adjacent wetlands and lakes. Buffer wetlands intercept P discharges from uplands and reduce P loads to sensitive water bodies. In Florida, wetlands have been constructed adjacent to lakes (Lake Apopka) and natural wetlands (the Everglades). The constructed buffer wetland adjacent to Water Conservation Area-1 of the Everglades is capable of removing approximately 80% of inflow P and producing an expected outflow total P levels of <50 μg L^{-1} at low loads (1.55 ± 0.12 g m^{-2} y^{-1}) (Moustafa et al., 1999). Lower effluent P requirements necessitate greater treatment area requirements. Greater P loads (e.g., 10 g m^{-2} y^{-1}) increase the effluent total P concentration on the order of 1 mg L^{-1} (Richardson, 1999). Empirical relationships suggest that P loading increases should be limited to 1 g m^{-2} y^{-1} to limit effluent P concentrations < 40 μg L^{-1} (Richardson, 1999). Others, however, question establishing one critical loading rate based on these empirical relationships without considering site-specific conditions and uncertainties associated with these predictions. Management strategies to enhance P removal effectiveness of wetlands, including vegetation harvesting, substrates (soils and limestone), and addition of chemical amendments can help to meet stringent effluent P criteria (DeBusk and Dierberg, 1999), but their cost effectiveness is questionable.

Nonpoint sources of P often dominate the eutrophication of aquatic systems, including lakes, and P control strategies are needed to reduce overall P load. Such P load reductions on two shallow lakes [Lake Apopka (125 km^2) and Lake

Okeechobee (1,732 km^2)] were effective in stabilizing water column total P concentration (Battoe et al., 1999; Steinman et al., 1999). However, not all lakes will respond similarly to the reduction of P loads. Lakes in areas with high background P levels, or those with internal P reserves in sediments, can require decades to respond to load reductions (Brezonik and Pollman, 1999). For many lakes, historic water quality data are not available. Paleolimnological studies (including analysis of diatom microfossils and other proxies) and trophic/nutrient models may provide direction to establish limits for management decisions (Brenner et al., 1993; Schelske et al., 1999).

1.6 SYNTHESIS

Predicting the fate and transport of P within and between ecosystems requires knowledge about and evaluation of, the "whole" system. Experimental and monitoring approaches, however, frequently provide only site-specific information of P impacts on a given ecosystem. Furthermore, these latter approaches are often expensive and time consuming. Empirical and mechanistic modeling approaches offer an alternative way of synthesizing experimental data and predicting long-term performance and water quality. For example, spatial decision support systems, using integrated GIS and hydrologic/water quality models, are available for decision makers and regulators to evaluate systems at a watershed scale (Campbell, 1999). Similarly, Fritz and Sklar (1999) developed a landscape model that incorporates ecosystem hydrology, chemistry, and biology to evaluate various management alternatives in the Everglades wetlands. Both approaches are useful tools and offer the potential to evaluate the system at a landscape level. Kadlec and Walker (1999) developed a simple management model to evaluate P effects on wetlands that uses a mass balance approach. The model has been calibrated using the data from the Everglades wetlands and is capable of predicting effluent total P concentration for site-specific conditions.

Several models are also available that describe P dynamics in lakes (Chapra and Auer, 1999; Sheng, 1999). The most common and simple model, often used by managers, is the P budget model developed by Vollenweider (1968). Empirical models, based on relationship between chlorophyll-*a* and total P concentration, can describe trophic status of lakes. Mechanistic models developed for temperate deep lakes with long residence times, however, may not be directly applicable to shallow lakes in subtropical environments such as Florida.

The predictive capability of models depends on availability of proper (input) experimental data and model validation. Without validation, models are little more than academic exercises. Model development and validation should continuously involve experimentalists, and seek conceptual evaluation of model parameters and data needs.

1.7 CONCLUSIONS AND FUTURE DIRECTIONS

Results presented in this book clearly demonstrate the significance of biogeochemical transformations in P transfer within and across ecosystems. Although the studies reported focus on Florida ecosystems, insights gained can be used to manage P in other ecosystems. Implementation of BMPs, purchase of land adjacent to sensitive water bodies, creating buffer wetlands, and in situ P management are some of the positive approaches used as P control strategies. Despite numerous studies over the past half century (especially in uplands and lakes), major data gaps exists in our understanding of P (fate and transport). At present, data exist for various transformations and transport processes for site-specific conditions, but we lack the appropriate framework to integrate these data for application to other sites. Scientists working on uplands, wetlands, and lakes often use individual approaches in measuring processes at laboratory or field scale, without due consideration of the established approaches in other disciplines. Often, we focus on the differences among disciplines and fail to recognize the similarities. We need to develop a more integrated approach to address issues at the ecosystem level. Biogeochemistry of an ecosystem is studied at different spatial scales, i.e., fundamental process level (molecular or particle level) to landscape level. Major data gaps exist at both extremes of spatial scale. At the fundamental level, we need to strengthen our understanding of P biogeochemistry at a molecular level, especially the biotic and abiotic interactions in water, soils, and sediments. Similarly, we need refined tools to integrate the experimental data to evaluate ecosystems at various spatial scales for more realistic predictions of P behavior in soils and sediments in relation to water quality. Future research should focus on developing easily measurable ecological indicators for use by managers and regulators to evaluate the recovery of disturbed ecosystem, and to set realistic limits for restoration. However, these indicators should be supported by basic science and thoroughly evaluated in reference to natural variations in undisturbed ecosystems.

REFERENCES

Battoe, L.E., M.F. Coveney, E.F. Lower, and D.L. Stites. 1999. The role of phosphorus reduction and export in the restoration of Lake Apopka. Chapter 22, this book.

Bottcher, A.B., T.K. Tremwel, and K.L. Campbell. 1999. Phosphorus management in flatwood (Spodosols) soils. Chapter 17, this book.

Brenner, M., T.J. Whitmore, M.S. Flannery, and M.W. Binford. 1993. Paleolimnological methods for defining target conditions in lake restoration: Florida case studies. Lake and Reserv. Manage. 7:209–217.

Brezonik, P. and C.D. Pollman, 1999. Phosphorus chemistry and cycling in Florida lakes: Global issues and local perspectives. Chapter 3, this book.

Campbell, K.L. and J.C. Capace. 1999. Hydrologic processes influencing phosphorus transport. Chapter 14, this book.

Chapra, S.C. and M.T. Auer. 1999. Management models to evaluate phosphorus loads in lakes. Chapter 28, this book.

DeBusk, W.F., K.R. Reddy, M.S. Koch, and Y. Wang. 1994. Spatial distribution of soil nutrients in a northern everglades marsh: Water Conservation Area 2A. Soil Sci. Soc. Am. J. 58:543–552.

DeBusk, T.A. and F.E. Dierberg. 1999. Techniques for optimizing phosphorus removal in treatment wetlands. Chapter 20, this book.

Fritz, H.C. and F.H. Sklar. 1999. Ecosystem analysis of phosphorus impacts and altered hydrology in the Everglades: A landscape modeling approach. Chapter 26, this book.

Graetz, D.A. and V.D. Nair. 1999. Inorganic forms of phosphorus in soils and sediments. Chapter 6, this book.

Izuno, F.T. and P.J. Whalen. 1999. Phosphorus management in organic soils. Chapter 18, this book.

Kadlec, R.H. 1999. Transport of phosphorus in wetlands. Chapter 15, this book.

Kadlec, R.H. and W.W. Walker. 1999. Management models to evaluate phosphorus impacts on wetlands. Chapter 27, this book.

Miao, S. and W.F. DeBusk. 1999. Effects of phosphorus enrichment on structure and function of sawgrass and cattail communities in Florida wetlands. Chapter 11, this book.

McCormick, P.V. and L.J. Scinto. 1999. Influence of phosphorus loading on wetlands periphyton assemblages: A case study from the Everglades. Chapter 12, this book.

Moustafa, M.Z. and M.J. Chimney. 1999. Phosphorus retention by the Everglades nutrient removal project: An Everglades stormwater treatment area. Chapter 21, this book.

Newman, S., K.R. Reddy, W.F. DeBusk, Y. Wang, G. Shih, and M.M. Fisher. 1997. Spatial distribution of soil nutrients in a Northern Everglades Marsh: Water Conservation Area 1. Soil Sci. Soc. Am. J. 61:1275–1283.

Newman S. and J.S. Robinson. 1999. Forms of organic phosphorus in water, soils and sediments. Chapter 8, this book.

Reddy, K.R., E. Lowe, and T. Fontaine. 1999. Phosphorus in Florida's ecosystems: Analysis of current issues. Chapter 4, this book.

Reddy, K.R., J.R. White, and T. Chua. 1999. Influence of phosphorus loading on microbial processes in soil and water column of wetlands. Chapter 10, this book.

Redfield, G.W. 1998. Quantifying atmospheric deposition of phosphorus: A conceptual model and literature review for environmental management. South Florida Water Management District. Techn. Publ. WRE#60 pp. 35 West Palm Beach, Fl.

Richardson, C.J. 1999. The role of wetlands in storage, release and cycling of phosphorus on the landscape: A 25 year retrospective. Chapter 2, this book.

Rhue, R.D. and W.G. Harris. 1999. Phosphorus sorption/desorption reactions in soils and sediments. Chapter 7, this book.

Schelske, C.L., F.J. Aldridge, and W.F. Kenney. 1999. Assessing nutrient limitation and trophic state in Florida lakes. Chapter 13, this book.

Sharpley, A.N. 1999. Global issues of phosphorus in terrestrial ecosystems. Chapter 1, this book.

Sheng, Y.P. 1999. Effects of hydrodynamic processes on phosphorus distribution in aquatic ecosystems. Chapter 16, this book.

Steinman, A.D., K.E. Havens, N.G. Aumen, R.T. James, K.R. Jin, J. Zhang, and B.H. Rosen. 1999. Phosphorus in Lake Okeechobee: Sources, sinks and strategies. Chapter 23, this book.

Wetzel, R.G. 1999. Organic phosphorus mineralization in soils and sediments. Chapter 9, this book.

Vollenweider, 1968. Scientific fundamentals of the eutrophication of lakes and flowing waters, with particular reference to nitrogen and phosphorus as factors in eutrophication. Tech. Rep. DAS/CSI/68.27, Environmental Directorate, Organization for Economic Cooperation and Development (OECD), Paris, 154 pp.

Section II

Ecological Analysis and Global Issues

buildup of soil P above levels required for crops, prompting concern of unacceptable P losses in areas vulnerable to runoff and erosion.

1.2 INTRODUCTION

Phosphorus (P) is an essential element for plant growth and its input has long been recognized as necessary to maintain profitable crop production. Phosphorus inputs can also increase the biological productivity of surface waters. Although nitrogen (N) and carbon (C) are essential to the growth of aquatic biota, most attention has focused on P inputs, because of the difficulty in controlling the exchange of N and C between the atmosphere and water, and fixation of atmospheric N by some blue–green algae. Thus, P is often the limiting element and its control is of prime importance in reducing the accelerated eutrophication of fresh waters. As salinity increases, N generally becomes the element controlling aquatic productivity (Thomann and Mueller, 1987).

Advanced eutrophication of surface water leads to problems with its use for fisheries, recreation, industry, or drinking, because of the increased growth of undesirable algae and aquatic weeds and oxygen shortages caused by their senescence and decomposition. Also, periodic surface blooms of cyanobacteria occur in drinking water supplies throughout the world, and may pose a serious health hazard to livestock and humans (Kotak et al., 1993; Lawton and Codd, 1991; Martin and Cooke, 1994).

Since the late 1960s, point sources of water pollution have been reduced, because of their relative ease of identification and control. Even so, water quality problems remain and, as further point source control becomes less cost-effective, attention is now being directed towards the contribution of agricultural nonpoint sources to P in surface waters. In recent reports to Congress, U.S. Environmental Protection Agency has identified agricultural nonpoint source pollution as the major source of stream and lake contamination that prevents attainment of the water quality goals identified in the Clean Water Act (USEPA, 1988). Specifically, eutrophication has been identified as the critical problem in those surface waters having impaired water quality in the USA. Agriculture has been estimated to be a major source of nutrients in these waters (USEPA, 1996).

In more and more areas, the potential for P loss in runoff has been increased by the continual land application of fertilizer and/or manure from intensive livestock operations (McFarland and Hauck, 1995; Sharpley et al., 1997). In most cases, these operations have concentrated in certain geographical regions, and the infrastructure and economy of many rural communities now depend on the operations. As a result, global issues facing the management of P in terrestrial ecosystems are now focusing on:

- Balancing inputs and outputs of P in agricultural systems at regional and local scales
- Identifying defensible threshold levels of soil P above which the potential enrichment of runoff P becomes unacceptable
- Targeting critical source areas of P at a watershed scale for cost-effective remediation.

1.3 BACKGROUND

A brief description of P cycling in terrestrial ecosystems, in terms of P inputs and outputs is given as background for the main global issues facing agriculture. These processes are outlined in Fig. 1.1. No attempt has been made to quantify these inputs and outputs on this general figure, as the amounts of P, particularly those input via manure, can vary greatly from region to region. The reasons for these variations and imbalances will be discussed in detail in a later section.

1.3.1 Phosphorus Inputs

Rainfall

Although the amount of P input in rainfall is not agronomically important compared to fertilizer and manure inputs, it has been shown to be significant in the P cycle of oligotrophic lakes. In a study of Lake Michigan, Murphy and Doskey (1975) reported a 30-fold greater total P (TP) concentration in rainfall than in lake water. In fact, the input of P in rainfall may contribute a significant proportion of the TP input in large lakes. For example, Elder (1975) estimated that rainfall P could account for up to 50% of the P entering Lake Superior. About 25–50% of the TP in rainfall is soluble and is directly available to organisms in the lake (Peters, 1977). Schindler and Nighswander (1970) attributed most of the enrichment of Clear Lake, Ontario, to rainfall, and similar observations have been made for several Wisconsin lakes by Lee (1973).

Fertilizers

Phosphorus fertilizers have had enormous impacts on agricultural production worldwide, and are essential inputs in the conversion of virgin and impoverished land to agricultural use. Large increases in crop yields are obtained when inorganic and organic P fertilizers are applied to P-deficient soil, and maintaining adequate soil P fertility is considered essential to make optimum the use of other plant nutrients, particularly N.

The manufacture of inorganic P fertilizers expanded dramatically after the World War II in response to the demand for increased agricultural output. In developed countries, large processing plants were built to convert imported rock phosphate into a variety of water-soluble and partially water-soluble P fertilizer products. Basic slag, a by-product from the steel industry, also became widely used. As a consequence, soil P fertility was improved and, in many areas, is now at a level at which the need for P fertilizer is less. Further, there is increasing awareness among farmers that P is a comparatively expensive plant nutrient, and that its application needs to be cost effective. As a result, world production of phosphate rock decreased by about 10% from 1990 to 1992 (Louis, 1993).

Feeds

On a national basis, inputs of P in feed to agricultural systems are generally less than those input in fertilizer (Table 1.1). The 2.5-fold greater input of P in feed than

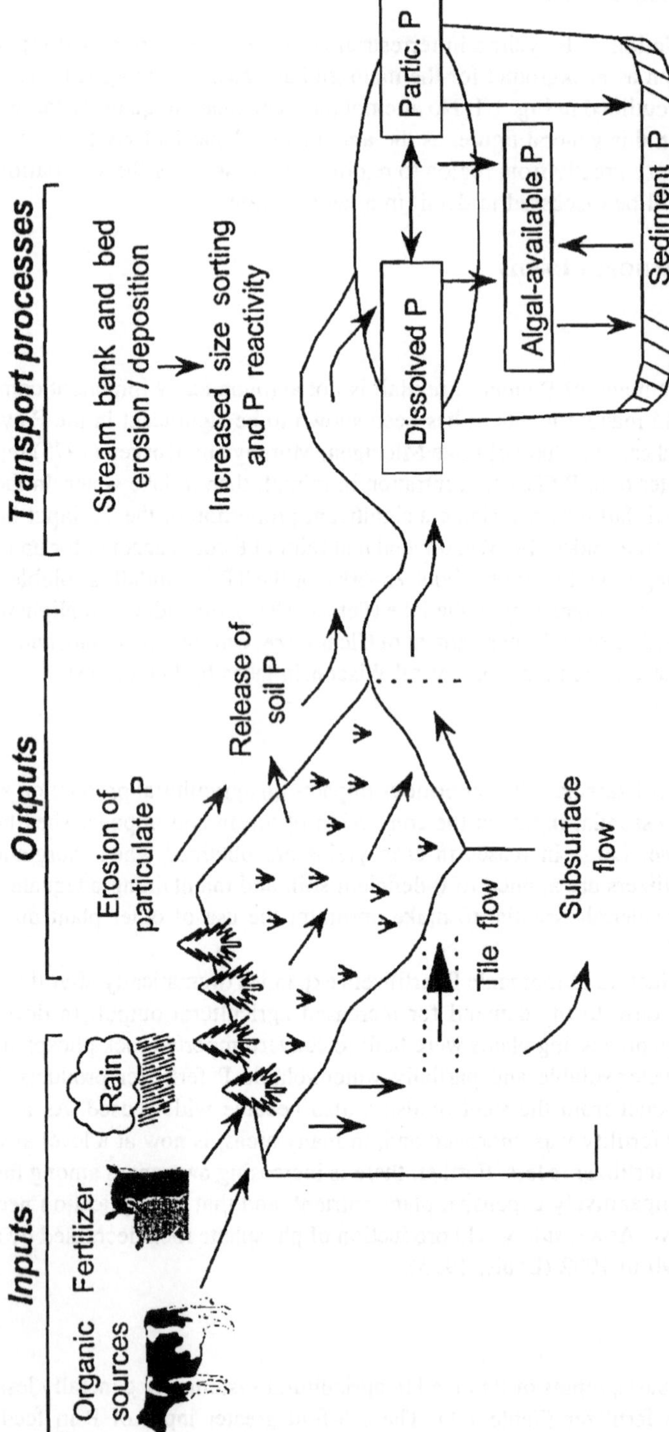

FIGURE 1.1 The inputs, outputs, and processes during transport of P to surface waters in agricultural ecosystems.

fertilizer for the Netherlands results from the fact that confined livestock operations comprise most of the agricultural production there. In the U.S., inputs of P in feed vary at local and regional scales, and as a function of livestock density and production.

The maintenance of a satisfactory P content in livestock feeds through crop fertilization and/or supplementation with inorganic P minerals is required to optimize animal performance. Reduced liveweight gains and lower milk yields have been observed in beef and dairy cows on diets containing inadequate P (Bass et al., 1981; Kincaid et al., 1981). On the other hand, diets containing P in excess of dietary requirements do not improve animal performance and increase the amount of P excreted (Brodison et al., 1989; Morse et al., 1992). In the Netherlands, the concentration of P in manure decreased temporarily during the World War II when concentrates and fertilizers were less available. Reductions in concentrate P contents are now being implemented to help similarly reduce the amounts of P excreted (Wadman et al., 1987).

Organic Sources

Primary sources of organic P include livestock manures and sewage sludge. Bone meal (or flour) and various miscellaneous by-products produced from the sea are minor sources. The P contents and compositions of fresh livestock manures vary widely with the type and age of the animal, the type of bedding material used, and the composition and digestibility of the diet, particularly where mineral supplements (di–calcium phosphate) are used (Table 1.2). Haynes and Williams (1992) found that the proportion of inorganic P (mostly tri–calcium phosphate) in sheep feces increased with increasing rates of superphosphate applied to pasture. Upon storage, the inorganic P component in feces is further increased by microbial mineralization of the organic P fraction (Gerritse and Vriesma, 1984). In stored manures, inorganic P represents 70 to 90% of the total P content and is largely composed of calcium phosphates of varying solubility. The P contained in livestock manures is more slowly available than in water-soluble P fertilizers, but is generally considered to become totally available in the longer term (Sharpley and Sisak, 1997).

Sewage sludge is considered a useful source of P for agricultural crops, but its P content and potential plant availability can also vary with the sludge source and method of treatment. Sewage sludge P contents can be larger where P is chemically precipitated at sewage treatment works before effluent discharge. Phosphorus contained in anaerobically digested sludge largely occurs as inorganic precipitates of Ca, Fe, or Al (Hue, 1995). Sludges produced after only primary or secondary treatment have a higher content of organic P (Hinedi et al., 1989). As with manures, sewage sludge P is generally considered to be about 50% available in the first year of application, although tertiary treatment of sludges tends to reduce plant available P (MAFF, 1994; Zhang et al., 1990).

1.3.2 Soil Phosphorus

Soil P exists in inorganic and organic forms (Fig. 1.2). In most agricultural soils, 50 to 75% of the P is inorganic, although this fraction can vary from 10 to 90%.

TABLE 1.1
Phosphorus Balance and Efficiency of Plant and Animal Uptake of P for the U.S. and Several European Countries (data for U.S. adapted from National Research Council [1993] and for European countries from Isermann, 1991)

		Input		Output			Efficiency of		
	Area in Agriculture	Fertilizer	Feed	Animal	Plant	Surplus	Plant uptake	Animal uptake	Total uptake
	10^6 ha	kg P ha^{-1} yr^{-1}					%		
E. Germany	6.2	25	6	3	1	27	59	10	11
W. Germany	12.0	27	10	10	3	24	76	34	35
Ireland	5.7	11	1	3	1	8	72	22	30
Netherlands	2.3	18	44	17	5	40	69	24	38
Switzerland	1.1	22	11	6	4	23	91	18	28
United Kingdom	18.5	9	3	2	1	9	55	18	25
United States	394.7	39	5	13	5	26	56	15	33

TABLE 1.2
Average P, N, and K Contents (dry weight basis) of animal manures (data adapted from Eck and Stewart, 1995, Gilbertson et al., 1979, and Hue, 1995)

Animal	Nitrogen	Phosphorus	Potassium
	g kg^{-1}		
Animal manure			
Beef	32.5	9.6	20.8
Dairy	39.6	6.7	31.6
Poultry layers	49.0	20.8	20.8
Broiler litter	40.0	16.9	19.0
Sheep	44.4	10.3	30.5
Swine	76.2	17.6	26.2
Turkeys	59.6	16.5	19.4
Sewage sludge	33.0	23.0	3.0

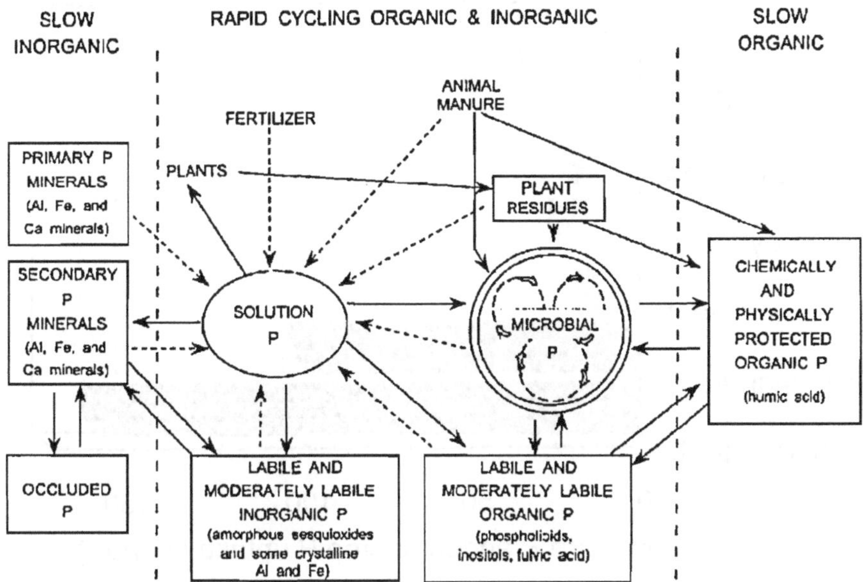

FIGURE 1.2 The soil P cycle: its components and measurable fractions (adapted from Stewart and Sharpley, 1987).

Inorganic P forms are dominated by hydrous sesquioxides, amorphous, and crystalline Al and Fe compounds in acidic, noncalcareous soils and by Ca compounds in alkaline, calcareous soils. Organic P forms include relatively labile phospholipids, nucleic acids, inositols, and fulvic acids, while more resistant forms are comprised of humic acids. The lability of these forms of P is operationally defined as the extent to which extractants of increasing acidity or alkalinity, applied sequentially, can dissolve soil P.

After application, P is either taken up by the crop or immobilized in soil by precipitation or adsorption reactions (Frossard et al., 1995; Syers and Curtin, 1989). In general, Al and Fe surfaces control these reactions in acidic soils, while Ca dominates in neutral or basic environments. After the initial reaction, there is a gradual adsorption of added P, rendering an increasing proportion of soil P unavailable for plant uptake (Fig. 1.2). Small amounts of added P may also be incorporated into organic P (McLaughlin et al., 1988) that may become resistant to mineralization by complexation with Al and Fe (Tate, 1984).

The repeated application of P in excess of crop utilization increases soil test P (Table 1.3). In many areas, this has resulted from the application of manures at rates designed to meet crop N requirements. As illustrated in Fig. 1.3, amounts of P added in "average" dairy manure (20 Mg ha^{-1}) and poultry litter (10 Mg ha^{-1}) applications are considerably greater than that removed in harvested crops. The result is an accumulation of soil P. Once built up, the decline in available soil P, however, is slow after further applications are stopped (Table 1.4). McCollum (1991) estimated that without further P additions, 16 to 18 years of corn (*Zea mays* L.) or soybean

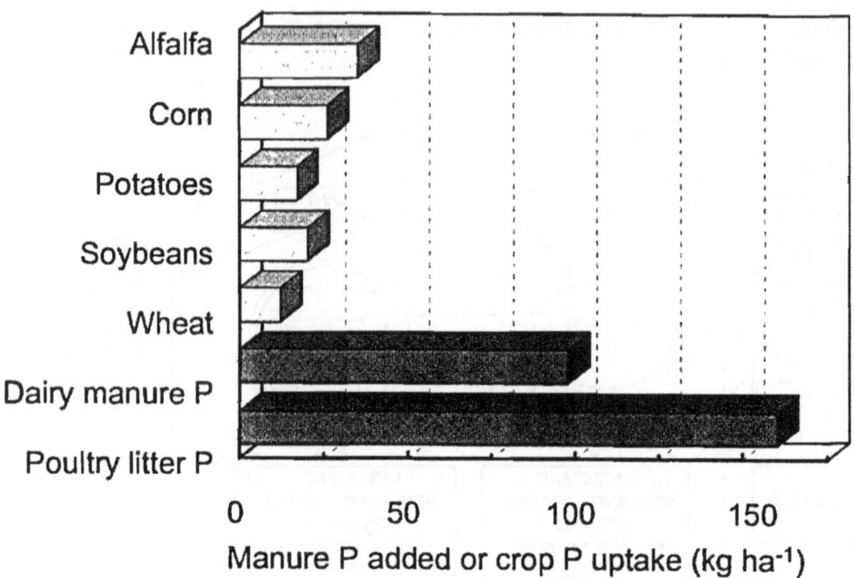

FIGURE 1.3 Phosphorus added in dairy manure (20 Mg ha^{-1}) and poultry litter (8 Mg ha^{-1}) is considerably greater than that removed in the harvested crop (hg ha^{-1}) when manure is added to provide about 300 kg N ha^{-1}.

[*Glycine max* (L.) Merr.] production would be needed to deplete soil test P (Mehlich-3 P, Mehlich, 1984) in a Portsmouth fine sandy loam from 100 mg P kg^{-1} to the agronomic threshold level of 20 mg P kg^{-1}. Several authors have found the rate of decrease in available soil P is related to the soil P sorption capacity (Aquino and Hanson, 1984; Sharpley 1996).

Organic P plays a critical role in soil P cycling, with microbial P fluxes a key intermediary both within and among organic and inorganic pools (Fig. 1.2). Brookes et al. (1984) found biomass P fluxes in grassland soils (23 kg P ha^{-1} yr^{-1}) were much greater than P uptake by the grass (12 kg P ha^{-1} yr^{-1}).

1.3.3 Phosphorus Outputs

Agricultural produce

The output of P in plant and livestock products for the USA and several European countries is given in Table 1.1. Although the magnitude of these outputs varies within and among countries, the relative proportions of P uptake in plant and livestock products are similar (Table 1.1). In spite of the relatively efficient recovery of P in crop production of 56 to 91%, the recovery in livestock production is only 10 to 34%, so that total P recovery by agriculture is only 11 to 38% (Table 1.1). Thus, the efficiency of P recovery in agriculture is dominated by livestock production, as 76 to 94% of the total crop production is fed to animals. Clearly, agricultural systems that include confined livestock operations supported by off-farm feeds can determine

TABLE 1.3
Available Soil P of Soil Treated with Fertilizer or Manure for Several Years vs. Untreated Soil

Soil	Crop	Added P	Time	Available soil P			Ref. and location
				Method	Untreated	Treated	
		kg ha^{-1} yr^{-1}			mg kg^{-1}		
Fertilizer							
Raub, sil	Mixed	22	25	Bray I	18	24	Barber, 1979, Indiana
		54	25		18	71	
Portsmouth, fsl	Mixed veg.	20	9	Mehlich I	18	73	Cox et al., 1981, NC
Batcombe, cl		27	19	Olsen	16	44	and Rothamsted
Richfield, scl	Mixed veg.	20	14	Bray I	12	54	Hooker et al., 1983, KS
		40	14		12	56	
Pullman, cl	Sorghum	56	8	Bray I	15	76	Sharpley et al., 1984, TX
Keith, sil	Wheat	11	6	Bray I	22	31	McCallister et al., 1987, NE
		33	6		24	47	
Rosebud, sil	Wheat	11	6	Bray I	10	28	
		33	6		10	48	
Beef manure							
Lethbridge, cl	Barley	160	11	Bray I	22	424	Chang et al., 1991, Alberta
		320	11		22	736	
		480	11		22	893	
Pullman, cl	Sorghum	90	8	Bray I	15	63	Sharpley et al., 1993, TX
		273	8		15	230	
		840	5		15	270	
Poultry litter							
Cahaba, vfsl	Grass	130	12	Bray I	5	216	Sharpley et al., 1993, OK
Ruston, fsl		100	12		12	342	
Stigler, sl		35	35		14	239	
Sandsones	Grass		10	Mehlich I	30	230	Kingery et al., 1994, AL
Swine manure							
Norfolk, l	Grass	109	11	Mehlich I	80	235	King et al., 1990, NC
		218	11		80	310	
		437	11		80	450	
Captina, sl	Grass	101	9	Bray I	5	121	Sharpley et al., 1991, OK
Sallisaw, l	Grass	81	15		6	147	
Stigler, sl	Wheat	37	9		15	82	
Cecil, sl	Grass	160	3	Melich I	19	45	Reddy et al., 1980, NC
		320	3		19	100	

the overall efficiency of P recycling in agriculture and thereby, the magnitude of P surpluses or potential soil accumulations.

The efficiency of P uptake by plants depends on several complex and interrelated edaphic, management, and environmental factors (Sharpley et al., 1992). Even so, P removal at average crop yields varies from crop to crop (Table 1.5). This presents a potential to enhance soil test P reductions, through selective cropping (Kelling, 1991; Pierzynski and Logan, 1993).

Transport in runoff

Amounts of P transported from agricultural watersheds are mainly a function of hydrology, in terms of when and where surface runoff occurs, soil P content, and amount and placement of P added as fertilizer or manure. This assumes that, in most cases, P export from watersheds occurs in surface rather than subsurface flow. In some regions, notably Florida, western Australia, and The Netherlands, most P is transported in subsurface drainage.

TABLE 1.4
The Decrease in Available Soil P after P Applications Are Stopped

Soil*	Crop	Time	Available soil P				Ref. and location
			Method	Initial	Final	Treated	
		yr		mg kg^{-1}		mg kg^{-1} yr^{-1}	
Thurlow, l	Small grains	9	Olsen	13	4	1.0	Cambell, 1965, MT
		9		20	4	1.8	
Georgeville, scl	Small grains	9	Mehlich I	60	6	6.0	Cox et al., NC and Saskatchewan
		7		3	1	0.1	
Haverhill, cl	Wheat/fallow	7	Olsen	7	2	0.6	
		14		40	25	1.1	
		14		74	33	2.9	
		14		134	69	4.6	
Portsmouth, fsl	Small grains	8	Mehlich I	23	18	0.6	
		9		54	26	3.1	
Sceptre, c	Wheat/fallow	8	Olsen	45	18	3.4	
		8		67	18	6.1	
		8		147	40	13.4	
Williams, l	Wheat/barley	16	Olsen	26	8	1.1	Halverson and Black, 1985, MT
		16		45	14	1.9	
Richfield, scl	Corn	8	Bray-1	12	8	0.5	Hooker et al., 1983, KS
		8		22	14	1.0	
Carroll, cl	Wheat/flax	8	Olsen	71	10	7.6	Spratt et al., 1980, Manitoba
		8		135	23	14.0	
		8		222	50	21.5	
Waskada, l	Wheat/flax	8	Olsen	48	9	4.9	
		8		88	23	8.1	
		8		200	49	18.9	Wager et al., 1986, Manitoba
Waskada, cl	Wheat/flax	8	Bray	140	50	11.3	
		8		320	80	30.0	

*fsl = fine sandy loam, l = loam, scl = silty clay loam, cl = clay loam, c = clay.

Phosphorus is transported in dissolved (DP) and particulate (PP) forms (Fig. 1.1). Particulate P includes P sorbed by soil particles and organic matter eroded during runoff, and constitutes the major proportion (60 to 90%) of P transported from cultivated land. As a result, processes determining soil erosion also determine PP transport. Runoff from grass or forest land carries little sediment, and is, therefore, generally dominated by DP. The transport of DP in runoff is initiated by the desorption, dissolution, and extraction of P from soil and plant material (Fig. 1.1). These processes occur as a portion of rainfall interacts with a thin layer of surface soil (1 to 5 cm) before leaving the field as runoff (Sharpley, 1985).

Whereas DP is, for the most part, immediately available for biological uptake (Nurnberg and Peters, 1984), PP can provide a long-term source of P for aquatic biota (Carignan and Kalff, 1980). The algal-availability of PP can vary from 10 to 90% depending on the nature of the eroding soil (Sharpley, 1993). Together DP and algal-available PP constitute algal-available P or P available for uptake by aquatic biota (Fig. 1.1).

Increases in P loss in surface runoff have been measured after fertilizer and/or manure applications (Table 1.6). Phosphorus losses are influenced by the rate, time, and method of fertilizer and manure application, form of P applied, amount and time of rainfall after application, and vegetative cover. As expected, larger additions of P increased P losses and in most cases, the proportion of added P losses in runoff. In a couple of studies, P applications reduced P loss in surface runoff, probably

TABLE 1.5
Mean P Concentration, Yield, and P Removal for Selected Crops (adapted from Pierzynski and Logan, 1993)

Crop	P concentration	Yield	P removal
	G P kg^{-1}	Mg ha^{-1}	kg P ha^{-1}
Row Crops—Grain			
Barley	3.4	2.2	7.4
Corn	2.8	9.4	25.8
Oat	3.4	2.9	9.7
Rice	2.4	4.0	9.7
Sorghum	3.2	3.8	12.3
Soybean	5.8	3.4	19.6
Wheat	4.5	2.7	12.3
Forage Crops			
Alfalfa	2.5	13.4	34.7
Coastal Bermuda	3.2	22.4	70.6
Corn silage	0.6	67.2	39.2
Fescue	4.4	8.9	32.2
Johnsongrass	3.5	26.9	93.0
Orchardgrass	3.2	10.1	31.4
Red clover	2.5	9.0	22.4
Sudangrass	2.5	6.7	16.8
Vegetable Crops			
Carrot	0.3	50.0	14.6
Onion	0.5	41.4	19.6
Potato	0.4	6.3	16.9
Sugar cane	0.2	224.0	49.3
Tomato	0.4	67.2	26.9

because of reduced runoff and erosion associated with increased protective vegetative cover afforded by fertilization (Table 1.6).

Generally, the concentration and loss of P in subsurface flow is small because of sorption of P by P-deficient subsoils (Table 1.7). Subsurface flow includes tile drainage and natural subsurface flow: tile drainage is percolating water intercepted by artificial drainage systems, such as tile or mole drains. Such drains accelerate drainage (and P) movement into streams. In general, P concentrations and losses in natural subsurface flow are lower than in tile drainage (Table 1.7) because of the longer contact time between subsoil and natural subsurface flow than tile drainage, enhancing DP removal. Increased sorption of P from percolating water accounted for lower TP losses from 1.0 (0.50 kg ha^{-1} yr^{-1}) than 0.6 m (1.07 kg ha^{-1} yr^{-1}) deep tiles draining a Brookston clay soil under alfalfa (Culley et al., 1983).

Greater P losses can occur in acid organic or peaty soils, where the adsorption affinity and capacity for P is low because of the predominantly negative charged surfaces and the complexing of Al and Fe by organic matter, than in most mineral soils (Miller, 1979; White and Thomas, 1981). Recent studies have shown P adsorption is independent of organic matter content of several sandy Coastal Plains soils,

TABLE 1.6
Effect of Fertilizer and Manure Application on P Loss in Surface Runoff

Land use	P added	Phosphorus loss: Dissolved	Total	Percent*	Ref. and location
	kg ha^{-1} yr^{-1}				
Fertilizer					
Grass	0	0.02	0.22		McColl et al., 1977, New Zealand
	75	0.04	0.33	0.01	
No-till corn	0	0.70	2.00		McDowell & McGregor, 1984, MS
	30	0.80	1.80	−0.7	
Conventional corn	0	0.10	13.89		
	30	0.20	17.70	12.7	
Wheat	0	0.20	1.60		Nicolaichuk & Read, 1978, Saskatchewan
	54	1.20	4.10	4.6	
Bahiagrass	0	0.88	1.29		Rechcigl et al., 1990, FL
	12	1.10	1.15	−1.2	
	48	2.36	2.87	3.3	
Grass	0	0.50	1.17		Sharpley & Syers, 1979a, New Zealand
	50	2.80	5.54	8.7	
Dairy manure†					
Alfalfa	0	0.10	0.10		Young & Mutchler, 1976, MN
–spring	21	1.90	3.70	17.1	
–fall	55	4.80	7.40	13.3	
Corn	0	0.10	0.20		
–spring	21	0.20	0.60	1.9	
–fall	55	1.00	1.60	2.5	
Poultry litter					
Grass	0	0.01	0.91		Heathman et al., 1995, OK
	140	0.10	2.96	1.5	
Fescue	0	0.00	0.00		Edwards & Daniel, 1993c, AR
	54	1.20	1.20	2.2	
	108	2.40	2.70	2.5	
	215	4.70	5.80	2.7	
Fallow	0	0.10	0.40		Westerman et al., 1983, NC
	83	1.72	5.61	6.3	
	165	4.71	15.58	8.2	
Poultry manure					
Grass	0	0.00	0.10		Edwards & Daniel, 1992, AR
	76	1.10	2.10	2.6	
	304	4.30	9.70	3.2	
Grass	0	0.01	0.40		Westerman et al., 1983, NC
	47	0.21	5.00	9.8	
	95	1.40	12.4	12.6	
Swine manure					
Fescue	9	0.00	0.00		Edwards & Daniel, 1993a, AR
	19	1.50	1.50	7.9	
	38	4.80	4.80	12.6	

*Percent P applied lost in runoff.

†Manure applied in either spring or fall.

TABLE 1.7
Effect of Fertilizer P Application on the Loss of P in Subsurface Flow

	P applied	Concentration		Amount			
		Dissolved	Total	Dissolved	Total		
Land use	kg ha^{-1} yr^{-1}	mgL^{-1}		kg ha^{-1} yr^{-1}		Percent	Reference
Alfalfa	0	0.180	–	0.12	–	–	Bolton et al., 1970, Canada
(tile drainage)	29	0.210	–	0.19	–	–	
Continuous corn	40	0.007	–	0.03	–	–	Burwell et al., 1977, IA
	66	0.009	–	0.04	–	–	
Terraced corn	67	0.028	–	0.17	–	–	
Bromegrass	40	0.005	–	0.03	–	–	
Continuous corn	0	0.020	0.120	0.13	0.42		Culley et al., 1983, Canada
(tile drainage)	30	0.110	0.470	0.20	0.62	0.06	
Bluegrass sod	0	0.02	0.17	0.06	0.15	–	
(tile drainage)	30	1.01	4.30	0.16	0.37	0.7	
Oats	0	0.02	0.11	1.10	0.29		
(tile drainage)	30	0.42	1.52	0.20	0.50	0.7	
Alfalfa	0	0.02	0.13	0.12	0.32		
(tile drainage)	30	0.37	1.40	0.20	0.51	0.6	
Corn	17	0.02	0.06	0.01	0.03	–	Hanway & Laflen, 1974, IA
(tile drainage)	42	0.00	0.00	0.00	0.00	–	
	44	0.01	0.03	0.01	0.05	–	
Grass	0	0.02	0.04	0.04	0.48		Sharpley & Syers, 1979b,
	50	0.03	0.05	0.12	0.19	–7.2	New Zealand
Grass	0	0.06	0.13	0.08	0.17		
(tile drainage)	50	0.19	0.35	0.44	0.81	1.3	

but that P was more readily desorbed from horizons (Bh) containing Al-organic matter complexes (Zhou et al., 1997). For further information on P biogeochemistry in wetlands, excellent reviews by Reddy et al. (1995), Richardson (1985), Richardson and Marshall (1986), and others in this issue are recommended to the readers. Phosphorus is also susceptible to movement through sandy soils with low P sorption capacities, in soils that have become waterlogged, leading to conversion of Fe (III) to Fe (II) and the mineralization of organic P, and with preferential flow through macropores and earthworm holes (Sharpley et al., 1979).

Transformations between dissolved and particulate forms of P such as sorption/desorption and deposition/resuspension occur during stream flow (Fig. 1.1). On arrival at the receiving lake, further exchange of P at the sediment–water interface influences algal availability of P. These transformations in P availability should be considered in assessing the potential biological impact of P from agricultural ecosystems.

1.4 REGIONAL IMBALANCES OF PHOSPHORUS IN AGRICULTURE

The rapid growth and intensification of the livestock industry in certain areas of the U.S. and Europe, has created national and regional imbalances in system inputs and outputs of P (Table 1.1). Averaged for these developed countries, only 30% of the fertilizer and feed input was output in crop and livestock produce, resulting in an annual P surplus of 22 kg ha^{-1}. In the United Kingdom, for example, a national P balance sheet for 1993 showed annual P surpluses between external P inputs and outputs of about 9 kg ha^{-1}, when averaged over the total utilizable agricultural land area (Wither, 1997). For all countries except The Netherlands, fertilizer P inputs were similar to or greater than P surpluses (Table 1.1). This simple assessment highlights the potential for regional imbalances of P and uneven redistribution of manures within these countries, as fertilizer P inputs appear to be needed in many regions of these same counties to maintain adequate soil P fertility.

Prior to World War II, farming communities tended to be self-sufficient in that enough feed was produced locally and recycled to meet livestock requirements. As a result, a sustainable food chain tended to exist (Fig. 1.4). After World War II, increased fertilizer use in crop production has consolidated farming systems, creating specialized crop and livestock operations that efficiently coexist in different regions of the country (Fig. 1.4).

For example, the ratio of livestock to crop income for farms in the Midwest Cornbelt has decreased since 1939 (Lanyon, 1995). Over the same period, there has been a dramatic increase in these ratios for several Eastern U.S. states with concentrated livestock operations (Fig. 1.4). In these states, livestock sales exceed crop income. In 1995 over half the corn grain produced in the Cornbelt was exported as animal feed, while states in the Southeast imported 83% of their grain for confined livestock operations (Lanyon and Thompson, 1996). In fact, less than 30% of the grain produced on farms today is fed on farms where it is grown (USDA, 1989).

The addition of more P than removed in crop harvest for several years can increase soil test P (Fig. 1.5). As manure applications are routinely based on their

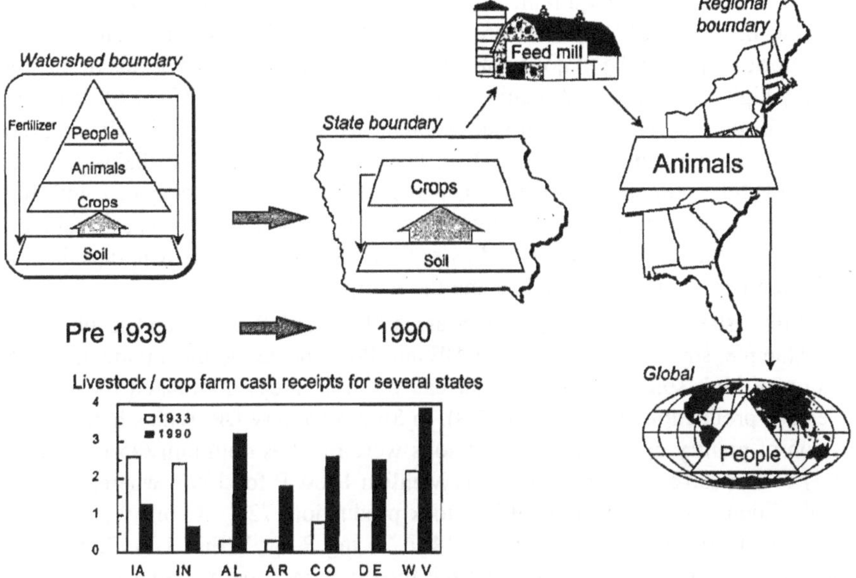

FIGURE 1.4 Relationship between changes in the spatial distribution of farming practices and the ratio of livestock/crop receipts for pre-1939 and 1990 (adapted from Lanyon, 1995; and Lanyan and Thompson, 1996).

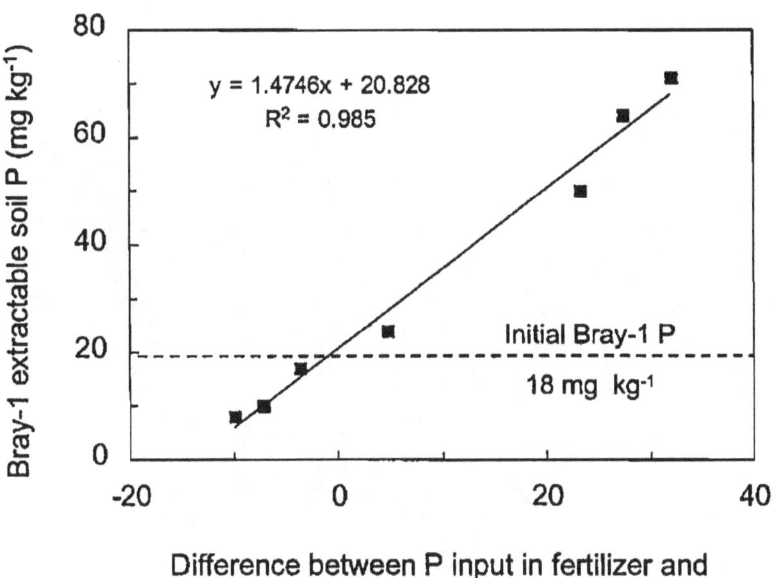

FIGURE 1.5 Bray-1 extractable P content of the surface (0 to 15 cm) of a Raub silt loam in Indiana as a function of the difference between P input as fertilizer and export in harvested crop over a 25-year period (adapted from Barber, 1979).

N content and crop N requirement, the main result of this imbalance has been an increase in soil test P in these areas (Fig. 1.6). In 1989, several state soil test laboratories reported the majority of soils analyzed had soil test P levels in the high or very high categories, that require little or no P fertilization (Fig. 1.6). It is clear that high soil P levels are a regional problem, with the majority of soils in several states testing medium or low (Fig. 1.6). For example, most Great Plains soils still require P for optimum crop yields. Unfortunately, problems associated with high soil P are aggravated by the fact that many of these soils are located in lake-rich states and near sensitive water bodies such as the Great Lakes, Chesapeake, and Delaware Bays, Lake Okeechobee and the Everglades (Fig. 1.6).

Within states and regions, distinct areas of general P deficit and surplus exist. For example, soil test summaries for DE and PA indicate the magnitude and localization of high soil test P levels that can occur in areas dominated by intensive livestock production (Figs. 1.7 and 1.8). In Sussex County DE, with a high concentration of poultry operations, 87% of soils were rated as optimum (25 to 50 mg P kg^{-1}) or excessive (>50 mg P kg^{-1}) as Mehlich I test P for 1995, whereas in New Castle County, with only limited livestock production, 72% of soils were rated as low (0 to 13 mg P kg^{-1}) or medium (14 to 25 mg P kg^{-1}) (Fig. 1.7). Similarly, in Lancaster County PA, dominated by the livestock and poultry industries, 77% of soils were rated as optimum (51 to 75 mg P kg^{-1}) or excessive (>76 mg P kg^{-1}) as Mehlich-3 test P for 1996, whereas nearby Adams County with orchard and crop production, was dominated (70%) by low (<30 mg P kg^{-1}) and medium (31 to 50 mg P kg^{-1}) soil test P (Fig. 1.8).

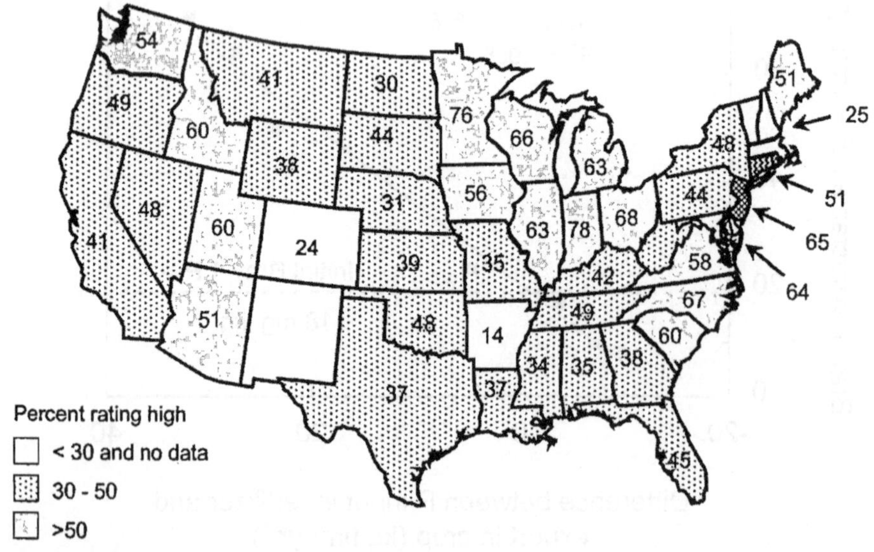

FIGURE 1.6 Percent of soil samples testing high or above for P in 1989. Highlighted states have 50% or greater of soil samples testing in the high or above range (adapted from Sharpley et al., 1994).

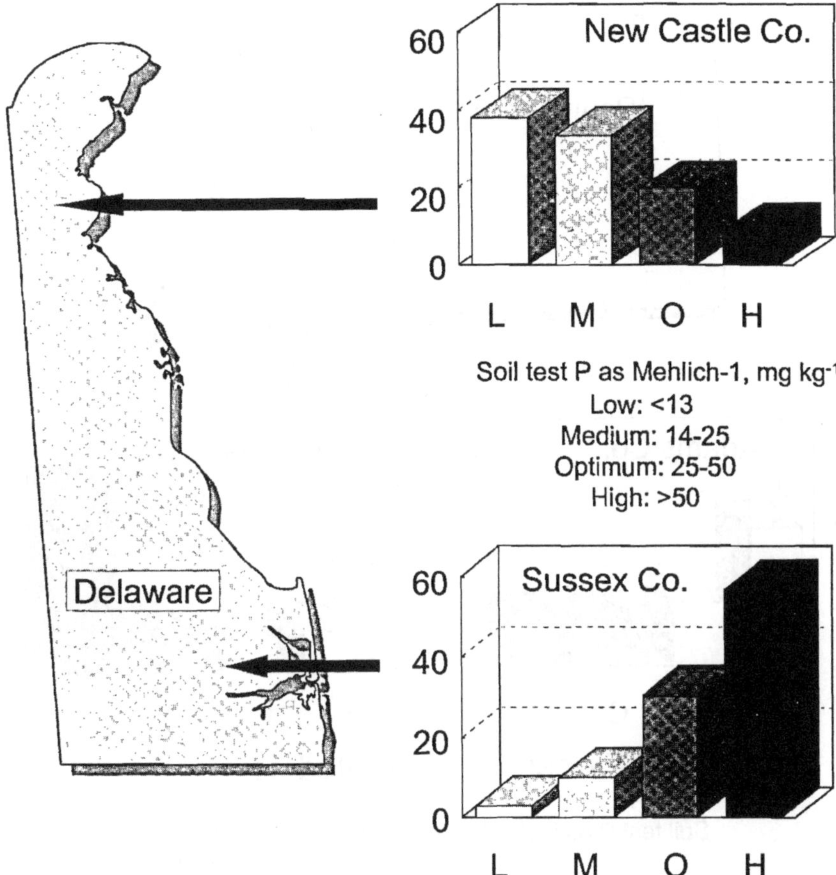

FIGURE 1.7 Percent of soils rated as low, medium, optimum, and high from 1995 soil test summaries for New Castle (little animal production) and Sussex counties (high concentration of poultry operations), Delaware with intensive grain and vegetable production (adapted from Sims, 1997).

Clearly, the transfer of feed components, such as corn grain and soybean oil meal, among regions creates a P imbalance, as manures are rarely transported more than 10 miles from where they are produced. However, mandatory transport of manure from surplus areas to nearby farms where the nutrients are needed faces several significant obstacles. First, it must be shown that the current location is unsuitable, based on soil properties, crop nutrient requirements, hydrology, etc. From European experiences this may be difficult to justify scientifically because of the large temporal and spatial variability in the factors controlling N and P mobility in soils and transport to ground or surface waters. Second, in many areas there is no clearly defined legal basis for requiring farmers in one physiographic area to perform management practices that are not required on neighboring farms. The greatest success with redistribution of manure nutrients is likely to occur when the general

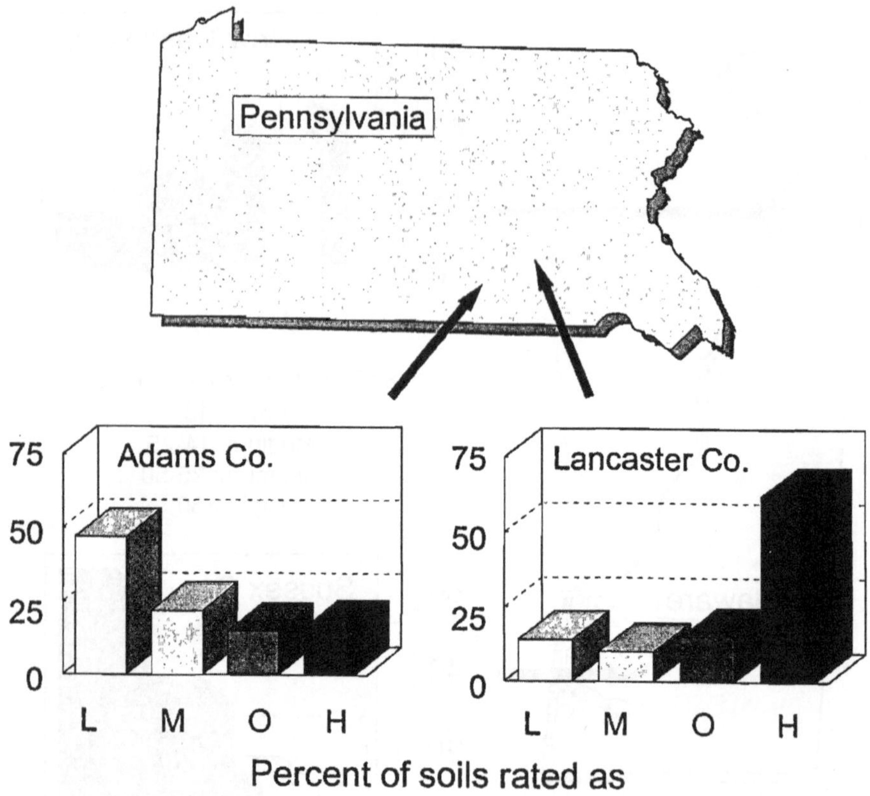

FIGURE 1.8 Percent of soils rated as low, medium, optimum, and high from 1996 soil test summaries for Adams (little animal production) and Lancaster counties (high concentration of livestock and poultry operations), Pennsylvania.

goals set by a national (or state) government are supported by consumers, local governments, the farm community, and the livestock industry involved. This may initially require cost-sharing to facilitate an overall assessment of the situation and the subsequent transport of manures from one area to another.

Manipulation of dietary P intake by animals may help reduce the surplus of P (input–output) in livestock operations. Morse et al. (1992) recorded a 23% reduction in excretion of P in feces and a 17% reduction in total P excretion when the daily P intake of dairy cows was reduced from 82 to 60 g day^{-1}. Increasing the dietary P intake from 82 to 112 g day^{-1} increased excretion of P in feces by 49% and total P excretion by 37%. Thus, there is a clear indication that amounts of excreted P can be reduced by carefully matching dietary P inputs to ruminant animal requirements, especially as P intakes above minimum dietary requirements do not seem to confer any milk yield advantage. In fact, Mahan and Howes (1995) found that balancing supplemental P to dietary intake requirements of the animal would reduce P use by 15%.

Enzyme additives for livestock feed that will increase the efficiency of P uptake during digestion are also being tested. Development of such enzymes that would be cost-effective in terms of livestock weight gain may reduce the P content of manure. One example is the use of phytase, an enzyme that enhances the efficiency of P recovery from phytin in grains fed to poultry (Cromwell et al., 1993; Kornegay et al., 1996). It is now common to supplement poultry feed with mineral forms of P because of the low digestibility of phytin, which contributes to P enrichment of poultry manures and litters.

In the Northeast U.S., attempts are being made to introduce rotational grazing into dairy operations to reduce the amount of feed imported onto the farm (Ford, 1994). Use of intensively grazed pasture has the potential to increase dairy farm profits, provide labor savings and, as environmental concerns become greater, reduce off-farm inputs of P. The increased income is mainly because of decreased feed and fertilizer costs, as manure was used to supplement fertilizer P. Intensively grazed pasture is not "the answer" for all dairy farms and its success depends to a large extent on the existing crop mix and above average managerial ability, particularly pasture management.

1.5 THRESHOLD SOIL PHOSPHORUS LEVELS

One of the main impacts of regional imbalances of P on farm management and policy has been to force many states to consider the development of recommendations for P applications and watershed management based on the potential for P loss in runoff, as well as crop P requirements. A major difficulty in the development of these recommendations has been the identification of threshold levels of soil P that are high enough to raise concerns about the potential for unacceptable levels of P loss in runoff. Examples from several states are given in Table 1.8. Establishing these levels is often a highly controversial process for two reasons. First, the data base relating soil test P to runoff P is limited to a few soils and crops, and there is often a reluctance to rely upon data of this type generated in other regions. Second, the economic implications of establishing soil test P levels that may limit manure applications are significant. In many areas dominated by animal-based agriculture, there simply is no economically viable alternative to land application. Because of these factors, those most affected by these soil test P limits are vigorously challenging their scientific basis. Clearly, there is a need to assess the validity of the use of soil test P values as indicators of P loss in runoff.

Several studies have reported that the loss of DP in runoff depends on the soil P content of surface soil. For example, a highly significant, linear relationship was obtained between the DP concentration of runoff and soil P content (Mehlich 3) of surface soil (5 cm) from cropped and grassed watersheds in Arkansas, Oklahoma, and New Zealand (Fig. 1.9). A similar dependence of the DP concentration of runoff on Bray-1 P was found by Romkens and Nelson (1974) for a Russell silt loam in

TABLE 1.8
Threshold Soil Test P Values and P Management Recommendations (adapted from Sharpley et al., 1996)

State	Critical value (mg kg^{-1})	Soil test P method	Management recommendations for water quality protection
Arkansas	150	Mehlich 3	*At or above 150 mg kg^{-1} soil P:* Apply no more P, provide buffers next to streams, overseed pastures with legumes to aid P removal, and provide constant soil cover to minimize erosion.
Delaware	120	Mehlich 3	Above 120 *mg kg^{-1} soil P:* Apply no more P until soil P is significantly reduced.
Ohio	150	Bray-1	Above 150 *mg kg^{-1} soil P:* Reduce erosion and reduce or eliminate P additions.
Oklahoma	130	Mehlich 3	*30 to 130 mg kg^{-1} soil P:* Half P rate on >8% slopes. *130 to 200 mg kg^{-1} soil P:* Half P rate on all soils and institute practices to reduce runoff and erosion. *Above 200 mg kg^{-1} soil P:* P rate not to exceed crop removal.
Michigan	75	Bray-1	*Above 75 mg kg^{-1} soil P:* P application not to exceed crop removal. *Above 150 mg kg^{-1} soil P:* Apply no P from any source.
Texas	200	Texas A&M	*Above 200 mg kg^{-1} soil P:* P addition not to exceed crop removal.
Wisconsin	75	Bray-1	*Above 75 mg kg^{-1} soil P:* Rotate to P demanding crops and reduce P additions. *Above 150 mg kg^{-1} soil P:* Discontinue P applications.

Illinois (r^2 = 0.81) and on water extractable soil P (r^2 = 0.61) of 17 Mississippi watersheds by Schreiber (1988) and 11 Oklahoma watersheds by Olness. (1975) (r^2 = 0.88).

These and similar studies related DP in runoff to soil P determined by traditional soil test methods for estimating plant availability of soil P. While they show promise in describing the relationship between the level of runoff and soil P, they are limited for several reasons. First, while DP is an important water quality parameter, it only represents the dissolved portion of runoff P readily available for aquatic plant growth. It does not represent sorbed P that can become available through desorption. To overcome this limitation, an approach using iron–oxide impregnated strips of filter paper has been developed to estimate the algal-available P in runoff (Sharpley, 1993). Acting as a P sink, Fe–oxide strips may have a stronger theoretical basis than chemical extraction in estimating algal-available P.

Using simulated rainfall (2.54 cm hr^{-1} for 30 min), Sharpley (1995) found Fe–oxide soil P was related to the algal-available P concentration of runoff from 10

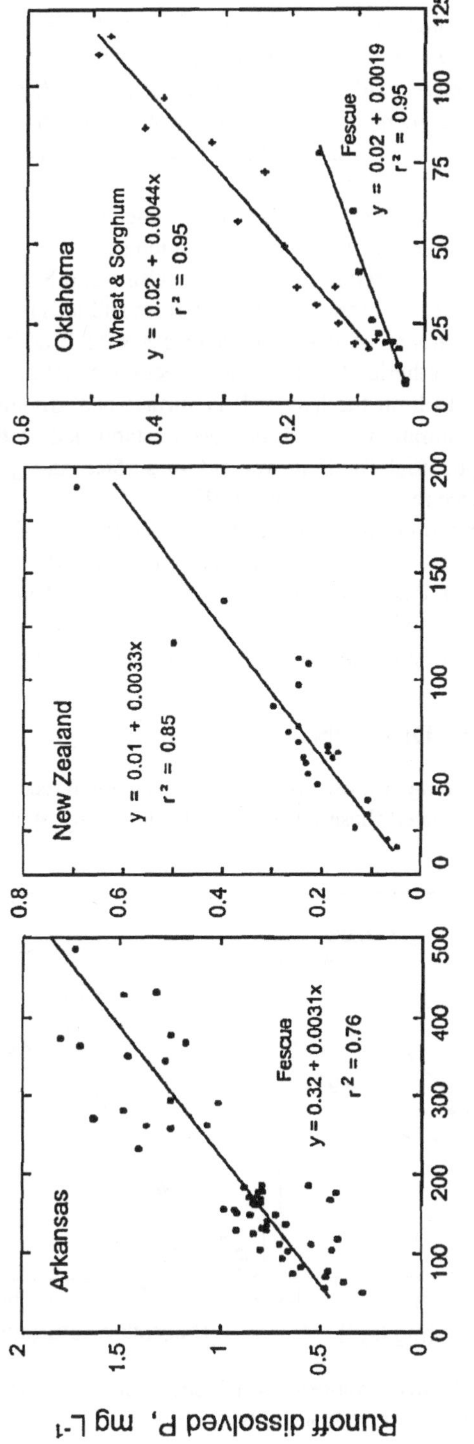

FIGURE 1.9 Effect of soil test P (Mehlich 3) on the dissolved P concentration of runoff from watersheds in Arkansas, New Zealand, and Oklahoma.

Oklahoma soils ranging from sandy loam to clay in texture (Fig. 1.10). Thus, for a given soil P level, the concentration of P maintained in runoff will be influenced by soil type, because of differences in P buffering capacity between soils caused by varying levels of clay, Fe and Al oxides, carbonates and organic matter. For example, a surface soil Fe–oxide P content of 200 mg kg^{-1} would support a runoff algal-available P concentration of 0.53 mg L^{-1} for the clay but 1.65 mg L^{-1} for the sandy loam (Fig. 1.10).

Another approach, developed in the Netherlands to determine the potential for DP movement in drainage water, estimates soil P sorption saturation as the percentage of P sorption capacity as extractable soil P (Breeuwsma and Silva, 1992). This approach is based on the fact that more P is released from soil to runoff or leaching water as P saturation or amount of P sorbed increases with P additions. Soil P saturation is used in the Netherlands where farm recommendations for manure management are designed to limit the loss of P in surface and ground waters. For Dutch soils, a critical P saturation of 25% has been established as the threshold value above which the potential for P movement in surface and ground waters becomes unacceptable (Breeuwsma and Silva, 1992).

When the P sorption saturation of the Oklahoma soils was calculated using Fe–oxide P as extractable soil P, a single relationship described the dependence of algal-available P in runoff on P saturation for all soils (Fig. 1.10). Thus, P saturation better describes the effect of soil type in the differential release of soil P to runoff and potential for P loss in runoff than traditional soil test P measures.

1.6 TARGETING REMEDIATION

Threshold soil P levels are being proposed to guide P management recommendations. In most cases, agencies that need these levels, hope to uniformly apply a threshold

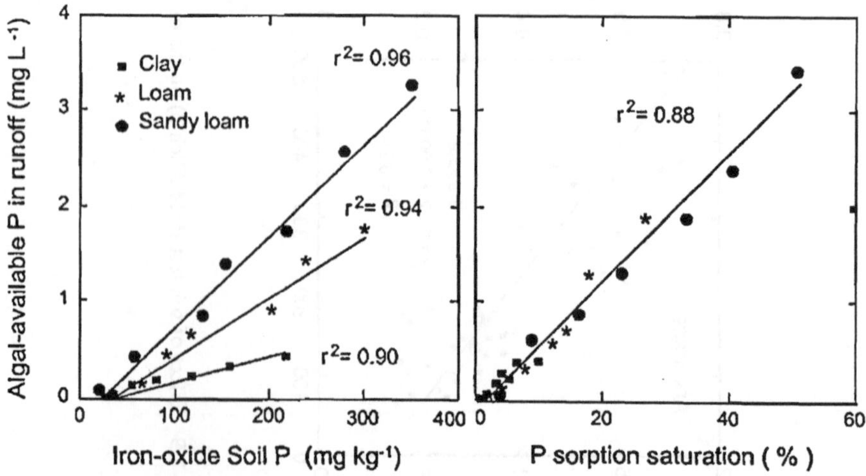

FIGURE 1.10 Relationship between iron-oxide strip P and P sorption saturation and algal-available P concentration of runoff for several soil types.

value to areas and states under their domain. However, it is often too simplistic to use threshold soil P levels as the sole criterion to guide P management and P applications. In fact, these values will have little meaning unless they are used in conjunction with an estimate of a site's potential for runoff and erosion. This is based on the general observation that 90% of annual algal-available P export from watersheds occurs from only 10% of the land area during a relatively few large storms (Pionke et al., 1997). For example, more than 75% of annual runoff from watersheds in Ohio (Edwards and Owens, 1991) and Oklahoma (Smith et al., 1991) occurred in one or two severe storms events. These events contributed over 90% of annual TP export (0.2 and 5.0 kg ha^{-1} yr^{-1}, respectively).

Loss of P from the land surface to the stream is controlled primarily by the interaction of the P "source" factors (functions of soil, crop, and management) with its "transport" factors (runoff, erosion, and channel processes). Attempts to reduce P loss from agricultural land uses has followed an evolutionary process. Initially, the problem was documented (Omernik, 1977; Ryden et al., 1973) and the important soil P-runoff water interactions assessed (Nelson and Logan, 1980; Taylor and Kunishi, 1971). As our understanding increased, high soil P levels were perceived to be most critical, and efforts were directed toward minimizing build-up and availability of surface soil P through the use of soil test P recommendations to guide fertilizer and manure applications (Sims, 1993; Sharpley et al., 1996). However, this has not always accomplished the desired or expected P loss reductions (Sharpley et al., 1994). As a result, we are now addressing the interaction between this source term, as manifested by soil P levels, and the transport mechanisms of runoff and erosion (Gburek et al., 1996; Pote et al., 1996). Preventing P loss has taken on the added dimension of defining, targeting, and remediating source-areas of P that combine high soil P levels with high erosion and surface runoff potentials.

There is a substantial body of work and understanding related to the dynamic interactions between soil P and runoff water that control P levels in runoff at point or plot scales (Ryden et al., 1973; Sharpley et al., 1994). Extension of this knowledge to multifield or watershed scales becomes problematic, but critical to the issue of P management, when spatially variable P sources, sinks, and transport processes are linked by the watershed-scale flow system. Further, a comprehensive P management strategy must address down-gradient water quality impacts because this is where the impact of land management will be assessed. Such a strategy must link effects at the local scale (i.e., the field where specific management practices are implemented), with the scale of the logical management unit (i.e., the farm), and with the larger scales at which impacts are evaluated (i.e., the watershed).

Without using this source area perspective to target application of P fertility, surface runoff and erosion control technology, conventionally applied remediations, may not produce the desired results and may prove to be an inefficient and not a cost-effective approach to the problem (Heatwole et al., 1987; Prato and Wu, 1991). As we become better at identifying the hydrologic controls on P transport and specific source areas of runoff, we must incorporate this knowledge into our land use and management recommendations for reduction of P loss from agricultural watersheds.

The importance of the linkage between runoff producing variable source area and areas of high soil P in determining P export from watersheds has been clearly demonstrated for a 26 ha watershed in east central Pennsylvania, by Zollweg et al. (1995). Phosphorus export from the agricultural watershed is dominated by storm flow that contributes approximately 90% of annual algal-available P export (Pionke and Kunishi, 1992). Zollweg et al. (1995) integrated a variable source area storm runoff model (Soil-Moisture-based Runoff Model, Zollweg, 1994) with algorithms to predict runoff P from soil P where runoff occurs (Sharpley and Smith, 1989). Runoff of fairly uniform depth was generated in the near-stream areas where soil P was lowest. Thus, P loss in runoff from these areas tends to be low. Loss is greater from the upper regions of the watershed where storm runoff originated at slope breaks and in areas of converging subsurface flow within cropped fields with high soil P contents (Fig. 1.11, Zollweg et al., 1995). Overall, P loss tended to reflect soil P content in areas where runoff occurred demonstrating the need to delineate critical source areas of P export and where remediation efforts may be best directed.

In cooperation with several research scientists, NRCS has developed a simple P index as a screening tool for use by field staffs, watershed planners, and farmers to rank the vulnerability of fields as sources of P loss in runoff (Lemunyon and Gilbert, 1993). The index accounts for and ranks transport and source factors controlling P loss in runoff and sites where the risk of P movement is expected to be higher than that of others. In an attempt to address P issues, NRCS is recommending that by 1998 the index be used in development of waste management and nutrient utilization plans for farms with confined animal operations. The intent of the effort is to identify sources of P, target more effective remedial strategies, and prevent further build-up of soil P, while maintaining the viability of crop and livestock production.

1.7 CONCLUSIONS

Issues facing agronomic and environmental P management in terrestrial ecosystems are similar in most developed countries. Specialized farming systems between and within these countries has tended to dismantle natural P cycles, resulting in an imbalanced flow of P from areas of fertilizer manufacture and grain production to areas of intensive crop and livestock operations. As a result, localized areas of high soil P content occur less than 100 miles from areas of low soil P fertility. In undeveloped countries, however, socio-economic constraints generally limit P use such that many soils are still deficient with respect to crop production.

Many farm management plans for P assume that if erosion is controlled, so will P losses. These measures address P control primarily through soil conservation measures. Less attention has been directed to P cycling in terrestrial ecosystems and the source management of P at field, farm, or watersheds scales. As a result, there has been a general increase in soil P contents in localized areas of intensive crop and livestock production, with increased losses of P in surface runoff and drainage water more frequently noted.

Although the relationship between soil and runoff P has not been quantified over wide areas, it is clear that the potential for P loss in runoff and thereby, accelerated eutrophication, increases as soil P accumulates. Unfortunately soil P decline via crop removal is slow, levels will remain elevated for several years after application

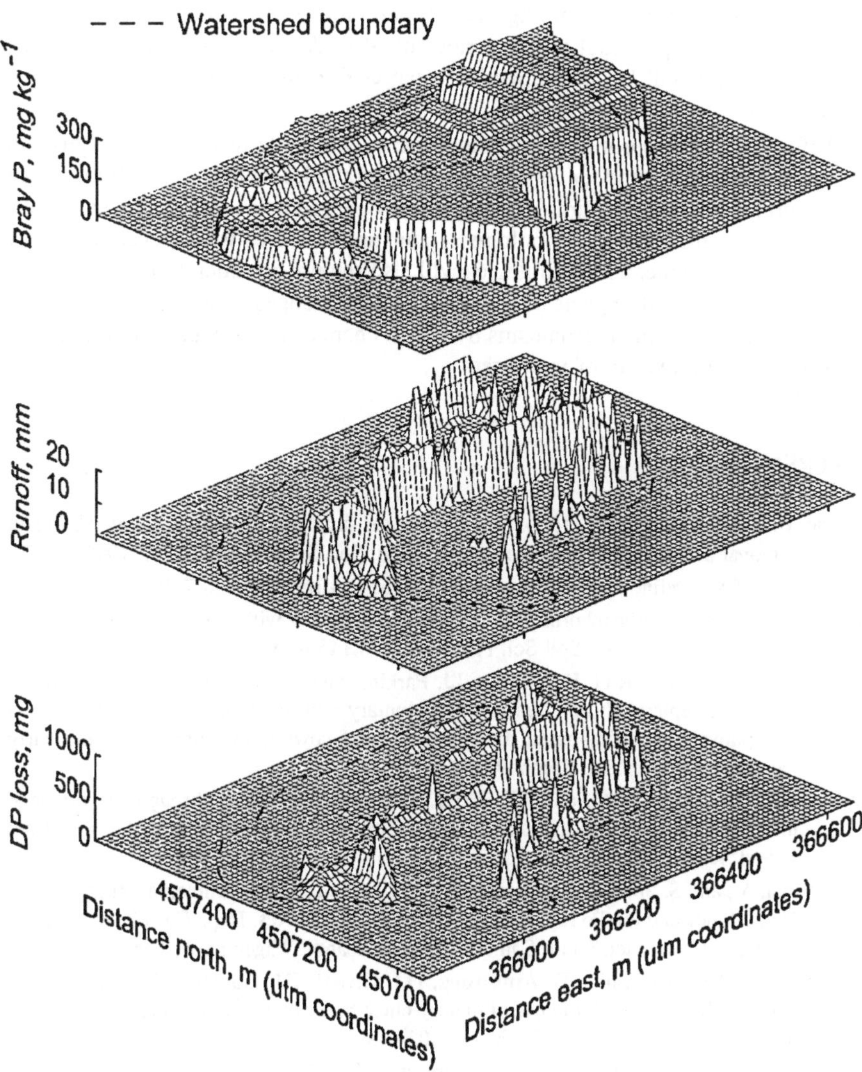

FIGURE 1.11 Simulation of runoff depth, soil P (Bray-1 P) and dissolved P loss from an eastern Pennsylvania watershed during an April 1992 storm (data adapted from Zollweg et al., 1995).

have ceased. To a certain extent, these concerns have not been addressed because managing agricultural inputs and outputs is often much more costly and restrictive to a farmer than general N management. As a result, N continues to drive manure management, and exacerbates the build-up of soil P.

Future programs should reinforce the fact that all fields do not contribute equally to P export from watersheds. In general, about 90% of P export is coming from only 10% of the watershed during only a relatively few large storms. Although soil P content is important in determining the concentration of P in runoff, it is likely that runoff and erosion potential will override soil P in determining P export. Clearly, remedial systems will be most effective if targeted to hydrologically active source areas of runoff in a watershed.

Efforts to increase our understanding of P cycling in terrestrial ecosystems and develop technically sound, defensible remedial strategies that minimize P loss from agricultural land, will require interdisciplinary research involving soil scientists, hydrologists, agronomists, limnologists, and animal scientists. Development of guidelines to implement such strategies will also require consideration of the socio-economic and political impacts of any management changes on both rural and urban communities, and of the mechanisms by which change can be achieved in a diverse and dispersed community of land users.

REFERENCES

Aquino, B.F., and R.G. Hanson. 1984. Soil phosphorus supplying capacity evaluated by plant removal and available phosphorus extraction. Soil Sci. Soc. Am. J. 48:1091–1096.

Bailey, R.G. 1983. Delineation of ecosystem regions. Environ. Managt. 7:365–373.

Barber, S.A. 1979. Soil phosphorus after 25 years of cropping with five rates of phosphorus application. Commun. Soil Sci. Plant Anal. 10:1459–1468.

Bass, J.M., G. Fishwick, R.G. Heminway, J.J. Parkins, and N.S. Ritchie. 1981. The effects of supplementary phosphorus on the voluntary consumption and digestibility of a low phosphorus straw-based diet given to beef cows during pregnancy and early lactation. J. Agric. Sci. Camb. 97: 365–372.

Bolton, E.F., J.W. Aylesworth, and F.R. Hove. 1970. Nutrient losses through tile drainage under three cropping systems and two fertility levels on a Brookston clay soil. Can. J. Soil Sci. 50:272–279.

Breeuwsma, A., and S. Silva. 1994. Phosphorus fertilization and environmental effects in The Netherlands and the Po region (Italy). Rep. 57. Agric. Res. Dep. The Winand Staring Centre for Integrated Land, Soil and Water Research. Wageningen, The Netherlands.

Brodison, J.A., E.A. Goodall, J.D. Armstrong, D.I. Givens, F.J. Gordon, W.J. McCaughey, and J.R. Todd. 1989. Influence of dietary phosphorus on the performance of lactating dairy cattle. J. Agric. Sci. Camb. 112:303–306.

Brookes, P.C., D.S. Powlson, and D.S. Jenkinson. 1984. Phosphorus in the soil microbial biomass. Soil Biol. Biochem. 16:169–175.

Burwell, R.E., G.E. Schuman, H.G. Heinemann, and R.G. Spomer. 1977. Nitrogen and phosphorus movement from agricultural watersheds. J. Soil and Water Conserv. 32:226–230.

Campbell, R.E. 1965. Phosphorus fertilizer residual effects on irrigated crops in rotation. Soil Sci. Soc. Am. Proc. 29:67–70.

Carignan, R., and J. Kalff. 1980. Phosphorus sources for aquatic weeds: Water or sediments? Science 207:987–989.

Chang, C., T.G. Sommerfeldt, and T. Entz. 1991. Soil chemistry after eleven annual applications of cattle feedlot manure. J. Environ. Qual. 20:475–480.

Cox, F.R., E.J. Kamprath, and R.E. McCollum. 1981. A descriptive model of soil test nutrient levels following fertilization. Soil Sci. Soc. Am. J. 45:529–532.

Cromwell, G.L., T.S. Stahly, R.D. Coffey, H.J. Moneque, and J.H. Randolph. 1993. Efficacy of phytase in improving the bioavailability of phosphorus in soybean meal and corn-soybean diets for pigs. Animal Sci. 71:1831–1840.

Culley, J.L.B., E.F. Bolton, and V. Bernyk. 1983. Suspended solids and phosphorus loads from a clay soil: I. Plot studies. J. Environ. Qual. 12:493–498.

Eck, H.V., and B.A. Stewart. 1995. Manure. p. 169–198. *In* J.E. Rechcigl (ed.), Soil amendments and environmental quality. CRC Press, Boca Raton, FL.

Edwards, D.R., and T.C. Daniel. 1992. Potential runoff quality effects of poultry manure slurry applied to fescue plots. Trans. Am. Soc. Agric. Eng. 35:1827–1832.

Edwards, D.R., and T.C. Daniel. 1993a. Runoff quality impacts of swine manure applied to fescue plots. Trans. Am. Soc. Agric. Eng. 36:81–80.

Edwards, W.M., and L.B. Owens. 1991. Large storm effects on total soil erosion. J. Soil Water Conserv. 46:75–77.

Elder, F.C. 1975. International Joint Commission Program for Atmospheric Loading of the Upper Great Lakes. Second Interagency Committee on Marine Science and Engineering Conference on the Great Lakes, Argonne, IL.

Ford, S.A. 1994. Economics of pasture systems. Dairy Economics Fact Sheet No 1. Pennsylvania State Univ. College of Agric. Sci., Cooperative Extension Service, University Park, PA. 9 pp.

Frossard, E., M. Brossard, M.J. Hedley, and A. Metherell. 1995. Reactions controlling the cycling of P in soils. p. 107–147. *In:* H. Tiessen (ed.), Phosphorus in the global environment. J. Wiley and Sons Ltd., New York, NY.

Gburek, W.J., and W.R. Heald. 1974. Soluble phosphate output of an agricultural watershed in Pennsylvania. Water Resour. Res. 10:113–118.

Gburek, W.J., A.N. Sharpley, and H.B. Pionke. 1996. Identification of critical sources for phosphorus export from agricultural catchments. p. 263–282. *In*: M.G. Anderson and S.M. Brooks (eds.), Advances in hillslope processes. J. Wiley, Chichester, England.

Gerritse, R.G., and R. Vriesma. 1984. Phosphate distribution in animal waste slurries. J. Agric. Sci. Camb. 102:159–161.

Gilbertson, C.B., F.A. Norstadt, A.C. Mathers, R.F. Holt, A.P. Barnett, T.M. McCalla, C.A. Onstad, and R.A. Young. 1979. Animal waste utilization on cropland and pastureland. USDA Utilization Research Report No. 6. 135 pp.

Halvorson, A.D., and A.L. Black. 1985. Fertilizer phosphorus recovery after seventeen years of dryland cropping. Soil Sci. Soc. Am. J. 49:933–937.

Hanway, J.J., and J.M. Laflen. 1974. Plant nutrient losses from tile outlet terraces. J. Environ. Qual. 7:208–212.

Haynes, R.J., and P.H. Williams. 1992. Long term effect of superphosphate on accumulation of soil phosphorus and exchangeable cations on a grazed, irrigated pasture site. Plant and Soil 42:123–133.

Heathman, G.C., A.N. Sharpley, S.J. Smith, and J.S. Robinson. 1995. Poultry litter application and water quality in Oklahoma. Fert. Res. 40:165–173.

Heatwole, C.D., A.B. Bottcher, and L.B. Baldwin. 1987. Modeling cost-effectiveness of agricultural nonpoint pollution abatement programs in two Florida basins. Water Res. Bull. 23:127–131.

Hinedi, Z.R., A.C. Chang, and R.W.K. Lee. 1989. Characterization of phosphorus in sewage sludge extracts using phosphorus-31 nuclear magnetic resonance spectroscopy. J. Environ. Qual. 18:323–329.

Hooker, M.L., R.E. Gwin, G.M. Herron, and P. Gallagher. 1983. Effects of long-term annual applications of N and P on corn grain yields and soil chemical properties. Agron. J. 75:94–99.

Hue, N.V. 1995. Sewage sludge. p. 199–247. *In:* J.E. Rechcigl (ed.), Soil amendments and environmental quality. CRC Press, Boca Raton, FL.

Isermann, K. 1990. Share of agriculture in nitrogen and phosphorus emissions into the surface waters of Western Europe against the background of their eutrophication. Fert. Res. 26:253–269.

Kelling, K.A., E.E. Schulte, L.G. Bundy, S.M. Combs, and J.B. Peters. 1991. Soil test recommendations for field, vegetable, and fruit crops. Univ. of Wisconsin Coop. Ext. Serv. Bull. A2809.

Kincaid, R.L., J.K. Hilliers, and J.D. Cronrath. 1981. Calcium and phosphorus supplementation of rations for lactating dairy cows. J. Dairy Sci. 64: 754–758.

King, L.D., J.C. Burns, and P.W. Westerman. 1990. Long-term swine lagoon effluent applications on Coastal Bermudagrass: II. Effects on nutrient accumulations in soil. J. Environ. Qual. 19:756–760.

Kingery, W.L., C.W. Wood, D.P. Delaney, J.C. Williams, and G.L. Mullins. 1994. Impact of long-term land application of broiler litter on environmentally related soil properties. J. Environ. Qual. 23:139–147.

Kornegay, E.T., D.M. Denbow, Z. Yi, and V. Ravindran. 1996. Response of broilers to graded levels of microbial phytase added to maize-soybean-meal-based diets containing three levels of nonphytase phosphorus. British J. Nutrition 75:839–852.

Kotak, B.G., S.L. Kenefick, D.L. Fritz, C.G. Rousseaux, E.E. Prepas, and S.E. Hrudey. 1993. Occurrence and toxicological evaluation of cyanobacterial toxins in Alberta lakes and farm dugouts. Water Res. 27:495–506.

Lanyon, L.E. 1995. Does nitrogen cycle?: Changes in the spatial dynamics of nitrogen with industrial fixation. J. Prod. Agric. 8:70–78.

Lanyon, L.E., and P.B. Thompson. 1996. Changing emphasis of farm production. p. 15–23. *In*: M. Salis and J. Popow (eds.), Animal agriculture and the environment: Nutrients, pathogens, and community relations. Northeast Regional Agricultural Engineering Service, Ithaca, New York.

Lawton, L.A., and G.A. Codd. 1991. Cyanobacterial (blue-green algae) toxins and their significance in UK and European waters. J. Inst. Wat. Environ. Managt. 5:460–465.

Lee, G.F. 1973. Role of phosphorus in eutrophication and diffuse source control. Water Res. 7:111–128.

Lemunyon, J.L., and R.G. Gilbert. 1993. Concept and need for a phosphorus assessment tool. J. Prod. Agric. 6: 483–486.

Louis, P.L. 1993. Availability of fertilizer raw materials. Proc. Fert. Soc. No. 336. The Fertilizer Society, Peterborough, England, 13.

MAFF. 1994. Fertilizer recommendations for agricultural and horticultural crops. Min. Agric. Fish. Food. Ref. Book 209. HMSO, London, England.

Mahan, D.C., and D. Howes. 1995. Environmental aspects with particular emphasis on phosphorus, selenium, and chromium in livestock feed. p. 75–91. *In*: Anon (ed), 13th Annual Pacific Northwest Animal Nutrition Conference, Portland, OR.

Martin, A., and G.D. Cooke. 1994. Health risks in eutrophic water supplies. Lake Line 14:24–26.

McCallister, D.L., C.A. Shapiro, W.R. Raun, F.N. Anderson, G.W. Rhem, O.P. Engelstadt, M.O. Russelle, and R.A. Olson. 1987. Rate of phosphorus and potassium buildup/decline with fertilization for corn and wheat on Nebraska Mollisols. Soil Sci. Soc. Am. J. 51:1646–1652.

McColl, R.H.S., E. White, and A.R. Gibson. 1977. Phosphorus and nitrate runoff in hill pasture and forest catchments, Taita, New Zealand. N.Z. J. Mar. Freshwater Res. 11:729–744.

McCollum, R.E. 1991. Buildup and decline in soil phosphorus: 30-year trends on a Typic Umprabuult. Agron. J. 83:77–85.

McDowell, L.L., and K.C. McGregor. 1984. Plant nutrient losses in runoff from conservation tillage corn. Soil Tillage Res. 4:79–91.

McFarland, A., and L. Hauck. 1995. Livestock and the environment: Scientific underpinnings for policy analysis. Report No. 1, Texas Inst. for Applied Environ. Res., Stephenville, TX. Published by Tarleton State Univ., Stephenville, TX. 140 pp.

Miller, M.H. 1979. Contribution of nitrogen and phosphorus to subsurface drainage water form intensively cropped mineral and organic soils in Ontario. J. Environ. Qual. 8:42–48.

Morse, D., H.H. Head, C.J. Wilcox, H.H. van Horn, C.D. Hissem, and B. Harris, Jr. 1992. Effects of concentration of dietary phosphorus on amount and route of excretion. J. Dairy Sci. 75:3039–3045.

Murphy, T.J., and P.V. Doskey. 1975. Inputs of phosphorus from precipitation to Lake Michigan. U.S. EPA Report No. 600/3-75-005. Duluth, MN.

National Research Council. 1993. Soil and water quality: An agenda for agriculture. National Academy Press, Washington, DC.

Nelson, D.W., and T.J. Logan. 1980. Chemical processes and transport of phosphorus. p. 65–91. *In*: F.W. Schaller and G.W. Bailey (eds.), Agricultural management and water quality. Iowa State Univ. Press, Ames, IA.

Nicholaichuk, W., and D.W.L. Read. 1978. Nutrient runoff from fertilized and unfertilized fields in western Canada. J. Environ. Qual. 7:542–544.

Nurnberg, G.K., and R.H. Peters. 1984. Biological availability of soluble reactive phosphorus in anoxic and oxic freshwaters. Can. J. Fish. Aquat. Sci. 41:757–765.

Olness, A.E., S.J. Smith, E.D. Rhoades, and R.G. Menzel. 1975. Nutrient and sediment discharge from agricultural watersheds in Oklahoma. J. Environ. Qual. 4:331–336.

Omernik, J.M. 1977. Nonpoint source–stream nutrient level relationships: A nationwide study. EPA-600/3-77-105. Corvallis, OR.

Peters, R.H. 1977. Availability of atmospheric orthophosphate. J. Fish. Res. Board. Can. 34:918–924.

Pierzynski, G.M., and T.J. Logan. 1993. Crop, soil, and management effects on phosphorus soil test levels. J. Prod. Agric. 6:513–520.

Pionke, H.B., W.J. Gburek, A.N. Sharpley, and J.A. Zollweg. 1997. Hydrologic and chemical controls on phosphorus loss from catchments. *In* H. Tunney (ed.), Phosphorus loss to water from agriculture. CAB Press, Cambridge, England. (In press).

Pionke, H.B., and H.H. Kunishi. 1992. Phosphorus status and content on suspended sediments and streamflow. Soil Sci. 153:452–462.

Potash and Phosphate Institute. 1990. Soil test summaries: Phosphorus, potassium, and pH. Better Crops 74:16–19.

Pote, D.H., T.C. Daniel, A.N. Sharpley, P.A. Moore, D.R. Edwards, and D.J. Nichols. 1996. Relating extractable phosphorus to phosphorus losses in runoff. Soil Sci. Soc. Am. J. 60:855–859.

Prato, T., and S. Wu. 1991. Erosion, sediment, and economic effects of conservation compliance in an agricultural watershed. J. Soil Water Conserv. 46:211–214.

Rechcigl, J.E., C.G. Payne, A.D. Bottcher, and P.S. Porter. 1990. Development of fertilizer practices for beef cattle pastures to minimize nutrient loss in runoff. South Florida Water Management District Report, South Florida Water Management District, West Palm Beach, FL.

Reddy, K.R., O.A. Diaz, L.J. Scinto, and M. Agami. 1995. Phosphorus dynamics in selected wetlands and streams of the Lake Okeechobee Basin. Ecol. Eng. 5:183–207.

Reddy, K.R., M.R. Overcash, R. Kahleel, and P.W. Westerman. 1980. Phosphorus absorption-desorption characteristics of two soils used for disposal of animal manures. J. Environ. Qual. 9:86–92.

Richardson, C.J. 1985. Mechanisms controlling phosphorus retention capacity in freshwater wetlands. Science (Washington, D.C.) 228:1424–1427.

Richardson, C.J., and P.E. Marshall. 1985. Processes controlling movement, storage, and export of phosphorus in a fen peatland. Ecol. Monogr. 56:279–302.

Romkens, M.J.M., and D.W. Nelson. 1974. Phosphorus relationships in runoff from fertilized soil. J. Environ. Qual. 3:10–13.

Ryden, J.C., J.K. Syers, and R.F. Harris. 1973. Phosphorus in runoff and streams. Adv. Agron. 25:1–45.

Schindler, D.W., and J.E. Nighswander. 1970. Nutrient supply and primary production in Clear Lake, eastern Ontario. J. Fish. Res. Board, Can. 27:260–262.

Schreiber, J.D. 1988. Estimating soluble phosphorus (PO_4–P) in agricultural runoff. J. Miss. Acad. Sci. 33:1–15.

Sharpley, A.N. 1985. Depth of surface soil-runoff interaction as affected by rainfall, soil slope and management. Soil Sci. Soc. Am. J. 49:1010–1015.

Sharpley, A.N. 1993. Assessing phosphorus bioavailability in agricultural soils and runoff. Fert. Res. 36:259–272.

Sharpley, A.N. 1995. Dependence of runoff phosphorus on soil phosphorus. J. Environ. Qual. 24:920–926.

Sharpley, A.N. 1996. Residual availability of phosphorus in manured soils. Soil Sci. Soc. Am. J. 60:1459–1466.

Sharpley, A.N., and I. Sisak. 1997. Differential availability of manure and inorganic sources of phosphorus in soil. Soil Sci. Soc. Am. J. 61:1503–1508.

Sharpley, A.N., and S.J. Smith 1989. Prediction of soluble phosphorus transport in agricultural runoff. J. Environ. Qual. 18:313–316.

Sharpley, A.N., and J.K. Syers. 1979a. Phosphorus inputs into a stream draining an agricultural watershed: II. Amounts and relative significance of runoff types. Water, Air and Soil Pollut. 11:417–428.

Sharpley, A.N., and J.K. Syers. 1979b. Loss of nitrogen and phosphorus in tile drainage as influenced by urea application and grazing animals. N.Z. J. Agric. Res. 22:127–131.

Sharpley, A.N., S.J. Smith, and W.R. Bain. 1993. Nitrogen and phosphorus fate from long-term poultry litter applications to Oklahoma soils. Soil Sci. Soc. Am. J. 57:1131–1137.

Sharpley, A.N., J.K. Syers, and J.A. Springett. 1979. Effect of surface-casting earthworms on the transport of phosphorus and nitrogen in surface runoff from pasture. Soil Biol. Biochem. 11:459–462.

Sharpley, A.N., T.C. Daniel, J.T. Sims, and D.H. Pote. 1996. Determining environmentally sound soil phosphorus levels. J. Soil Water Conserv. 51:160–166.

Sharpley, A.N., J.J. Meisinger, J. Power, and D. Suarez. 1992. Root extraction of nutrients associated with long-term soil management. Adv. Soil Sci. 19:151–217.

Sharpley, A.N., S.J. Smith, B.A. Stewart, and A.C. Mathers. 1984. Forms of phosphorus in soil receiving cattle feedlot waste. J. Environ. Qual. 13:211–215.

Sharpley, A.N., J.J. Meisinger, A. Breeuwsma, T. Sims, T.C. Daniel, and J.S. Schepers. 1997. Impacts of animal manure management on ground and surface water quality. p. 173–242. *In*: J. Hatfield (ed.), Effective management of animal waste as a soil resource. Lewis Publishers, Boca Raton, FL.

Sharpley, A.N., B.J. Carter, B.J. Wagner, S.J. Smith, E.L. Cole, and G.A. Sample. 1991. Impact of long-term swine and poultry manure applications on soil and water resources in eastern Oklahoma. Okla. State Univ., Tech. Bull. T169, 51 pp.

Sharpley, A.N., S.C. Chapra, R. Wedepohl, J.T. Sims, T.C. Daniel, and K.R. Reddy. 1994. Managing agricultural phosphorus for protection of surface waters: Issues and options. J. Environ. Qual. 23:437–451.

Sims, J.T. 1993. Environmental soil testing for phosphorus. J. Prod. Agric. 6:501–507.

Sims, J.T. 1997. Animal waste management: Agricultural and environmental issues in the United States and Western Europe. *In*: Animal waste management: National and international perspectives. Canadian Soil Sci. Soc., Lethbridge, Alberta, Canada. 20 pp.

Smith, S.J., A.N. Sharpley, J.R. Williams, W.A. Berg, and G.A. Coleman. 1991. Sediment-nutrient transport during severe storms. p. 48–55. *In*: S.S. Fan and Y.H. Kuo (eds.), Fifth Interagency Sedimentation Conf. March 1991, Las Vegas, NV. Federal Energy Regulatory Commission, Washington, DC.

Sommers, L.E. 1977. Chemical composition of sewage sludges and analysis of their potential use as fertilizers. J. Environ. Qual. 6:225–232.

Sommers, L.E., D.W. Nelson, and K.J. Yost. 1976. Variable nature of chemical composition of sewage sludges. J. Environ. Qual. 5:303–306.

Spratt, E.D., F.G. Warder, L.D. Bailey, and D.W.L. Read. 1980. Measurement of fertilizer phosphorus residues and its utilization. Soil Sci. Soc. Am. J. 44:1200–1204.

Stewart, J.W.B., and A.N. Sharpley. 1987. Controls on dynamics of soil and fertilizer phosphorus and sulfur. p. 101–121. *In*: R.F. Follett, J.W.B. Stewart, and C.V. Cole (eds.), Soil fertility and organic matter as critical components of production. SSSA Spec. Pub. 19, Am. Soc. Agron., Madison, WI.

Syers, J.K., and D. Curtin. 1989. Inorganic reactions controlling phosphorus cycling. p. 17–29. *In*: H. Tiessen (ed.), Phosphorus cycles in terrestrial and aquatic ecosystems. UNDP, Pub. by Saskatchewan Inst. Pedology, Saskatoon, Canada.

Tate, K.R. 1984. The biological transformation of P in soil. Plant and Soil 76:245–256.

Taylor, A.W., and H.M. Kunishi. 1971. Phosphate equilibria on stream sediment and soil in a watershed draining an agricultural region. J. Agric. Food Chem. 19:827–831.

Thomann, R.V., and J.A. Mueller. 1987. Principles of surface water quality modeling and control. 644 pp. Harper Collins Publ. Inc., New York, NY.

U.S. Environmental Protection Agency. 1988. Quality criteria for water. EPA 440/5-86-001. USEPA, Office of Water Regulations and Standards. U.S. Govt. Print. Office (PB81-226759), Washington, DC.

U.S. Department of Agriculture. 1989. Fact book of agriculture. Misc. Publ. No. 1063, Office of Public Affairs, Washington, DC. pp. 17.

U.S. Environmental Protection Agency. 1996. Environmental indicators of water quality in the United States. EPA 841-R-96-002. USEPA, Office of Water (4503F), U.S. Govt. Printing Office, Washington, DC. 25 pp.

Wadman, W.P., C.M.J. Sluijsmans, and L.C.N. De Ła Lande Cremer. 1987. Value of animal manures: changes in perception. p. *In*: H.G. Van der Meer (ed.), Animal Manure on Grassland and Fodder crops, Martinus Nijhoff Publishers, Dordrecht, The Netherlands.

Wagar, B.I., J.W.B. Stewart, and J.L. Henry. 1986. Comparison of single large broadcast and small annual seed-placed phosphorus treatments on yield and phosphorus and zinc content of wheat on chernozemic soils. Can. J. Soil Sci. 66:237–248.

Westerman, P.W., T.L. Donnely, and M.R. Overcash. 1983. Erosion of soil and poultry manure—a laboratory study. Trans. ASAE 26:1070–1078, 1084.

White, R.E., and G.W. Thomas. 1981. Hydrolysis of aluminum on weakly acidic organic exchangers: implications for phosphorus adsorption. Fert. Res. 2:159–167.

Wither, P.J.A. 1997. Phosphorus cycling in United Kingdom agriculture and implications for water quality. *In*: Anon (ed), Phosphates in water and losses from agricultural land. Soc. Chemical Industry, London, England.

Young, R.A., and C.K. Mutchler. 1976. Pollution potential of manure spread on frozen ground. J. Environ. Qual. 5:174–179.

Zhang, L.M., J.L. Morel, and E. Frossard. 1990. Phosphorus availability in sewage sludge. p. 32. *In*: Proc. 1st ESA Congress, Paris. A. Scaife (ed.). European Society of Agronomy, Paris, France.

Zhou, M., R.D. Rhue, and W.G. Harris. 1997. Phosphorus sorption characteristics of Bh and Bt horizons from sandy Coastal Plain soils. Soil Sci. Soc. Am. J. 61:1364–1369.

Zollweg, J.A. 1994. Effective use of geographic information systems for rainfall-runoff modeling. Ph.D. Dissertation, Cornell Univ., Ithaca, NY.

Zollweg, J.A., W.J. Gburek, A.N. Sharpley, and H.B. Pionke. 1995. GIS-based delineation of source areas of phosphorus within northeastern agricultural watersheds. *In*: Proc. IAHS Symposium on Modeling and Management of Sustainable Basin-scale Water Resource Systems (in press).

2 The Role of Wetlands in Storage, Release, and Cycling of Phosphorus on the Landscape: A 25-Year Retrospective

Curtis J. Richardson

2.1 ABSTRACT

A brief historical overview of early wetland research that became the basis of our understanding of P cycling and storage in wetlands is reviewed. Next, a synopsis of the mechanisms controlling P retention in wetlands is assessed and presented for a number of wetland ecosystems. A cross-sectional statistical model developed from the EPA national wetland database is refined as a predictive model for assessing P retention capacity in wetlands. Finally, this model is tested and calibrated with data from a site in the Everglades of south Florida to estimate P assimilative capacity for this freshwater wetland. An analysis of the North American Wetland database suggests that P concentration in wetland effluent increases exponentially after a P loading threshold is exceeded. The data showed a clear pattern of low phosphorus output concentration when the total phosphorus mass loading rate was less that 1.0 g m^{-2} yr^{-1} but some variation among sites exists. For some wetlands the critical loading rate may fall closer to the lower end of the P threshold zone (0.4 g m^{-2} yr^{-1}) if changes in ecosystem properties are a concern. The annual outflow P median

1-56670-331-X/99/$0.00+$.50

concentration was 40μg/L with very low variation. This suggests that wetlands have a threshold for P that, if exceeded, may result in a moving front of elevated P concentrations within the wetlands until the P loading reaches the threshold. All wetland types have the capacity to store some P without significant changes in downstream P concentrations. As a general rule, the "One-Gram Assimilative Capacity Rule" for P loadings may be useful within freshwater wetlands if long-term storage of P and low P effluent concentrations are to be maintained. A site-specific test analysis of the assimilative capacity rule confirmed this for the Everglades. The analysis suggested that to permanently store P and prevent downstream movement of elevated P concentrations, background P loadings within the wetland must be near or below 1.0 g $m^{-2} yr^{-1}$. Data analysis also suggests that wetlands can continue to retain additional P as P loadings increase, but at the cost of increasing P outflow concentrations. Lower P retention efficiencies exist at higher P loadings. Finally, the P assimilative capacity concept for most wetlands has not been tested, but current analysis suggests that the "one-gram rule" may be a good first approximation for most freshwater wetlands.

2.2 INTRODUCTION

Where did the often accepted ecological concept that wetlands function on the landscape as efficient nutrient sinks (i.e., nutrient retention areas) come from? Some literature suggests that wetlands can function as either a nutrient source or sink, depending on the nutrient in question or the season of the year. (Godfrey et al., 1985; Richardson, 1985; Kadlec and Knight, 1996) Is the sink concept for phosphorus (P) retention flawed by being based simply on empirical input/output studies for systems that have not yet reached saturation? Do we understand the mechanisms controlling P retention capacity in both natural and constructed wetlands to the degree that we can predict recycling rates and storage capacity for P if we know loading rates? Do wetlands have a threshold for P, and, if exceeded, does it cause a "moving front of P" within the ecosystem? Finally, what capacity, if any, do wetlands have to store P on a long-term basis without a significant change in the wetlands ecosystem structure and function? These questions, for the most part, have remained only partially answered for wetland ecosystems.

Phosphorus releases are important because nonpoint P inputs into downstream water remains the major eutrophication problem within many watersheds. An understanding and quantification of the mechanisms responsible for the retention and processing of phosphorus in wetland ecosystems is, thus, essential to improving water quality on the landscape.

In this article, I first provide a brief historical overview of early wetland research that became the basis of our understanding of P cycling and storage in wetlands. Next, a synopsis of the mechanisms controlling P retention in wetlands is presented. A cross-sectional model (Reckhow and Qian, 1994) developed from the EPA national wetland database is refined as a predictive model for P retention capacity in wetlands. Finally, this model is tested and calibrated with data from a site in the Everglades of south Florida to estimate P assimilative capacity for this freshwater wetland.

2.3 HISTORICAL OVERVIEW

It was nearly twenty five years ago that I was asked by John and Bob Kadlec to join a research team at the University of Michigan to investigate using wetlands to remove nutrients (primarily nitrogen and phosphorus) from wastewater. The concept that wetlands were "Mother Nature's kidneys" had led a number of scientists in the U.S. (Tourbier and Pierson, 1976; Tilton et al., 1976; Good et al., 1978) to hypothesize that wetlands were efficient nutrient sinks. Their ideas were based to some degree on the early plant work of Siedel (1955, 1961) at the Max Planck Institute in Plön. In her initial studies she had used *Scirpus lacustris* to treat phenol and livestock wastewaters (Vymazal et al., 1998). It was an exciting area of research during the 1970s. Data on nutrient cycling and retention in wetlands in the United States was almost nonexistent, and the National Science Foundation was willing to fund applied research through a program known as RANN (Research Applied to National Needs). We began to test the nutrient sink concept at the Houghton Lake peatlands in the north central section of Michigan's lower peninsula at microcosm, mesocosm, and field scales.

2.3.1 Early Mesocosm and Pilot-Scale Studies

Research from Houghton Lake was presented at a national symposium in Ann Arbor, Michigan during May of 1976, as was research by a number of leading scientists including H. T. Odum and K. C. Ewel (the effects of adding nutrients to natural and constructed wetlands in Florida cypress domes), D. F. Whigham and R. L. Simpson (Maryland tidal marshes), R. E. Turner and J. W. Day (Louisiana coastal marshes), and M. M. Small and F. L. Spangler (constructed wetlands). A review of those proceedings edited by D. L. Tilton and others, unfortunately never widely distributed, revealed two opposing views of the capacity of wetlands to store and remove nutrients (Tilton et al., 1976). Macrophytes were either deemed to be the key to successful wetland treatment, or they were relegated to only a supportive role in the process. Moreover, two different criteria for success were presented:

1. black box input/output analysis
2. process studies focusing on controlling mechanisms

Not surprisingly, these perspectives and methods still exist today.

Primarily, most engineers, hydrologists, and biological researchers at the symposium presented simple input/output studies and were optimistic about the potential of utilizing wetlands for wastewater treatment. This was based, in part, on findings that consistently indicated biological oxidation demand (BOD) removal rates from 77 to 91%, high potential for nitrogen removal via denitrification, and phosphorus storage up to 99% (Tilton et al., 1976).

In the 1970s the mechanistic approach of studying nutrient cycles in wetlands was in its infancy. Many of the studies concluded that they were not sure why, how, or for what period their systems could store or process nutrients or, in other cases, why they unpredictably exported nutrients (Tilton et al., 1976). Almost all the papers ended with a statement for the need for more research. However, the euphoria that a simple, cost-effective nutrient retention system was available using wetlands spread rapidly through the consulting world, as well as the Environmental Protection Agency (EPA). Before clearly understanding if wetlands removed nutrients or how long these systems would work, some governmental agencies, including EPA, were pushing in the late 1970s for a policy to utilize wetlands for wastewater treatment (Bastian and Reed, 1979).

There was no doubt that many of these wetlands were processing and storing some nutrients, but data on phosphorus retention efficiency ranged from 10% to 90% (Tilton et al., 1976; Richardson and Nichols, 1985). However, these studies were based on data for only a few years of treatment from a variety of natural and constructed systems, some of which were not even constructed to function as wetlands. Most importantly, the mechanisms controlling nutrient removal and storage for these ecosystems were, for the most part, unknown. Also, few researchers realized that most of the purification and nutrient retention was carried out in the substrata of the system and not by the plants. An exception was Spangler et al. (1976). They had reported that only 5% of the P went into harvestable plant tissue in their artificial marshes and, in fact, nonvegetated gravel controls often removed as much P as the vegetated systems. Researchers on Michigan fens also reported that macrophytes showed no significant uptake of N or P in the field in the first year under increased nutrient loading conditions; most of the P additions accumulated in the litter and soils (Richardson et al., 1976). These and other studies noted that outputs of nutrients increased significantly during the winter when the plants and other biotic components died back (Godfrey et al., 1985). Early investigations showed that wetland nutrient retention differed greatly among wetland types and varied by element. Output concentrations changed seasonally, and the processes were significantly under biological control.

Phosphorus output concentrations and retention efficiency also showed no consistent pattern, nor were the controlling mechanisms known. A review of the limnology and soils literature available at the time, along with several studies in European peatlands, clearly showed that Al, Fe and Ca controlled phosphate chemistry in water and soil. Moreover, anaerobic conditions resulted in the release of P in acid lake waters due to the conversion of insoluble ferric phosphate to soluble ferrous phosphate (Rigler, 1956; Hayes and Phillips, 1958; Fox and Kamprath, 1971; Wetzel, 1983). Patrick and Khalid (1974) also showed this process occurred in wetland soil systems. Unfortunately, little of this process-level information was utilized in designing wetlands for P removal, with the result that a large number of natural and constructed wetlands failed, especially in terms of phosphorus retention.

The good news is that many of the early success findings on nutrient retention at the Michigan symposium have since been reconfirmed for many elements in numerous wetland systems (Godfrey et al., 1985; Richardson and Nichols, 1985; Reddy and Smith, 1987; Hammer, 1989; Kadlec and Knight, 1996). Nitrogen outputs are not generally a problem due to high denitrification rates and storage in wetland systems (Bowden, 1987). Sediment retention and BOD reductions are highly efficient in most wetlands (Richardson, 1989; Richardson and Nichols, 1985; Kadlec and Knight, 1996). However, the problem with large variations in P storage and release data from wetlands continued into the 1990s. This was ignored, misunderstood, or of little interest to wastewater engineers who were primarily concerned with N, BOD, and suspended solids removal.

The variations in P field data suggested that the only way to understand P retention capacity and the mechanism controlling P cycling was to carry out uniform microcosm and field level studies on many different wetlands. Earlier studies at the Houghton Lake peatland had shown that microbial uptake and soil adsorption capacity controlled the amount of P made available for plant growth (Richardson and Marshall, 1986). Permanent long-term storage of P was in the peat as either chemically adsorbed forms or as recalcitrant organic P. Comparative studies determining phosphorus retention capacity of wetlands from soil chemistry variables (Al, Fe, Ca, Mg, pH,) and organic matter content revealed that P adsorption capacity could be predicted from the amorphous acid oxalate-aluminum and iron content in acidic soils, and that mineral soil wetland types were more efficient than peatlands in retaining P (Richardson, 1985). Moreover, wetlands, especially peatlands, loaded with wastewater became P-saturated in a few years and exported P.

In contrast, terrestrial upland sites with higher amorphous Fe and Al content in the soil had almost unlimited P storage capacity as compared to wetlands. A major conclusion from this research was that "the selection of natural freshwater wetlands with peat soils for efficient processing and storage of P, especially at high concentrations, is thus unwarranted" (Richardson, 1985) This was often incorrectly interpreted to mean that peatlands do not retain P. The correct interpretation is that they do store P, but at a very low rate. However, quantification of the mechanisms controlling P storage and release as well as an analysis of how these mechanisms affect P retention capacity among different wetland types is needed to understand the wide variations reported for P storage and releases.

2.4 CONTROLLING MECHANISMS

Maximizing long-term high-efficiency retention of P requires understanding the most efficient pathways which move P from the water column into a permanent sink. The major P storage pool in natural wetlands is the soil/litter compartment (Richardson and Marshall, 1986; Verhoeven, 1986; Richardson, 1989). Soil P primarily occurs in the +5 valence state (oxidized), because all lower oxidation states are thermodynamically unstable and readily oxidized to PO_4^{3-} even in reduced wetland soils (Lindsay, 1979). There are no changes in P valence during the biotic assimilation of inorganic P or during decomposition of organic P by microorganisms. Phosphorus has only a minor gaseous phase (phosphine), therefore, this ion has only a liquid/solid biogeochemical cycle controlling storage (Devai and Delaune, 1995).

Phosphorus that enters the wetland water column is rapidly absorbed by bacteria, periphyton, and floating aquatic plants (Fig. 2.1). Phosphorus is also adsorbed on soil or sediment particles if sufficient Al, Fe, Ca, and Mg are present. Which ion is more active in sequestering P depends on the pH of the system and the amount of the ion present. In acidic soil, inorganic P is adsorbed on hydrous oxides of Fe and Al, and P may precipitate as insoluble Fe–phosphates (Fe–P) and Al–P (Fig. 2.1). Precipitation as insoluble Ca–P or Mg–P is the dominant transformation at pH's greater than 8.0. Radioisotope P studies have shown that 10 to 20% of the P is

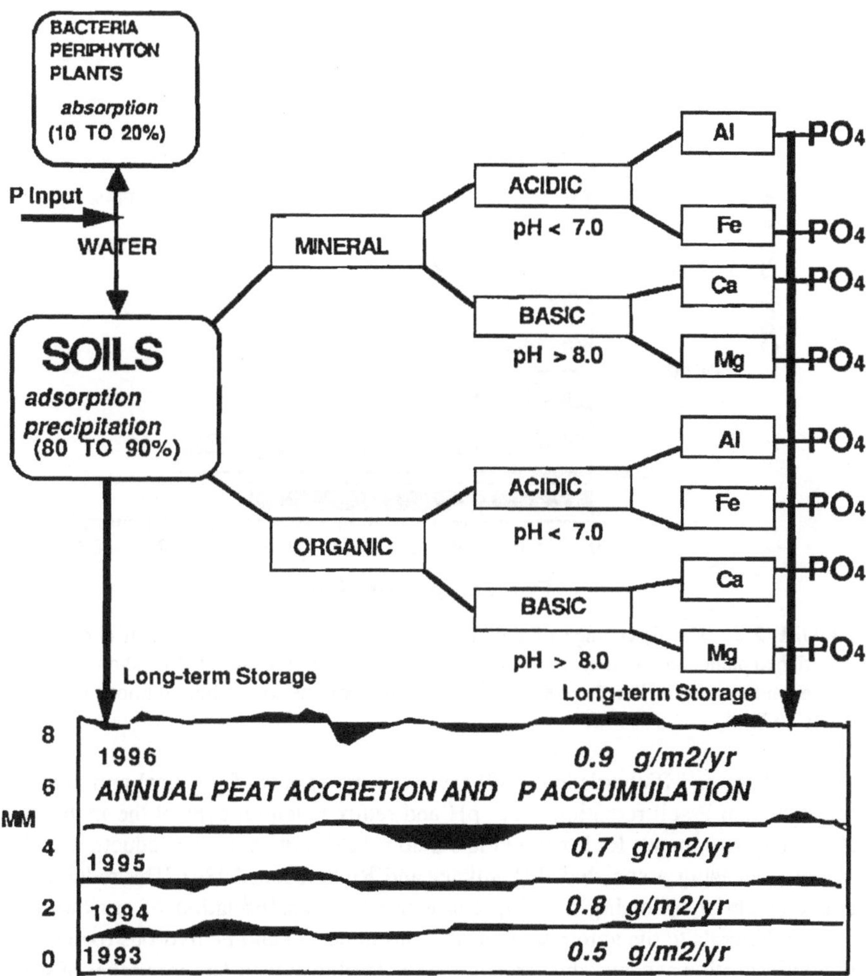

FIGURE 2.1 Phosphorus control model for wetlands. P inputs to the wetland surface are initially controlled by biotic uptake (absorption), chemical adsorption, and precipitation. Long-term P storage is via peat/sediment accumulation.

controlled by the biotic uptake initially (Davis, 1982; Richardson, 1985). Specifically, Davis found that ^{32}P added to the Everglades water column was sequestered in the soils (50 to 60%) and litter compartments (10 to 30%), depending on the P enrichment status of the communities (Fig. 2.2). Macrophytes controlled less than 10% and periphyton around 20% of the radioactive P inputs after one week (Fig. 2.2). The unenriched periphyton and soils sequestered more ^{32}P than they did at the enriched sites. The Michigan mesocosm study also found 80 to 90% of added P was located in the soils compartment after one or two weeks (Richardson, 1985). Thus, quantification of the rate of P storage in wetlands soils is needed to assess P availability and retention in wetlands.

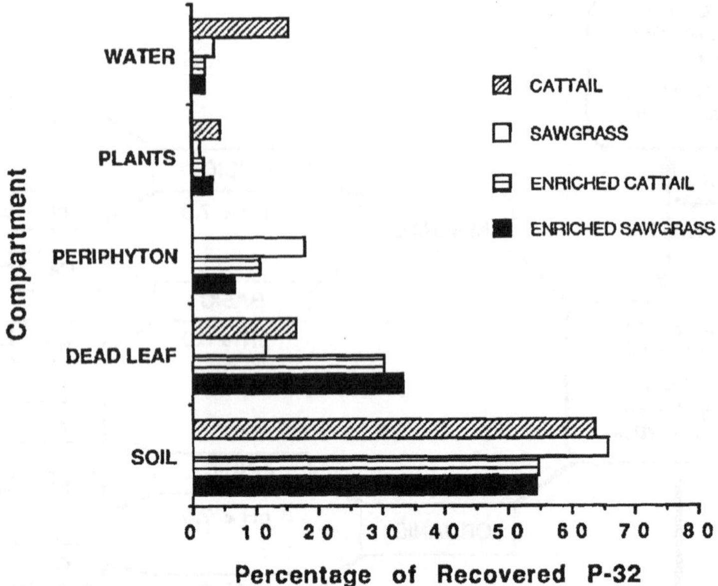

FIGURE 2.2 The percentage of radioactive phosphorus (^{32}P) recovered in enriched and unenriched plant communities in WCA-2A of the northern Everglades. Values shown represent the percentage of the ^{32}P recovered in each compartment one week after additions (data from Davis 1982).

Faulkner and Richardson (1989) reviewed the effects of physical and chemical properties on P soil retention. Both pH and redox potential control the mobility of P in the environment. Redox potentials below +120 mv cause the reduction of Fe^{3+} to Fe^{2+}, releasing associated P (Faulkner and Richardson, 1989). The reduction of redox potential following flooding can cause the transformation of crystalline Al and Fe minerals to the amorphous form. Amorphous Al and Fe hydrous oxides have higher P sorption capacity than crystalline oxides due to their larger number of singly coordinated surface hydroxyl ions (Reddy and Smith, 1987). Recent work by Petrovic and Kastelan-Macan (1996) suggests that the binding of inorganic P can involve the formation of a complex association of Ca, Mg, Fe and Al (pH dependent) bound to humic substances (Fig. 2.3). Humic substances can act as bridges between humic macromolecules and phosphate ions. The other possibility is ligand exchange of the metal complex and the formation of insoluble metal–phosphates. In addition Petrovic and Kastelan-Macan have shown in desorption experiments, high concentrations of fulvic acids compete for sorption sites on mineral particles. Such fulvics can increase P desorption by 10 to 20% (Kastelan-Macan and Petrovic, 1996). These findings have important consequences for storage and release of P in peat-based wetlands that have increased P additions.

A theoretical hierarchy of the processes that control P retention in wetlands founded on earlier research findings, and the literature is shown in Table 2.1 (Reddy and Smith, 1987; Richardson and Craft, 1993; Richardson et al., 1997). The magnitude of the processes is presented in decreasing order of importance. Research

A. Phosphate binding onto solid HS-Me complex

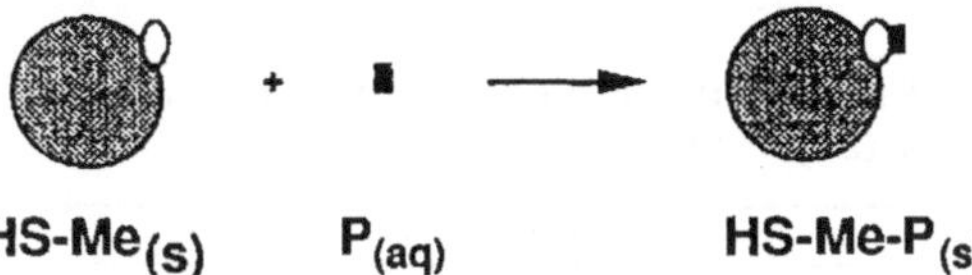

B. Ligand exchange and precipitation of insoluble phosphate

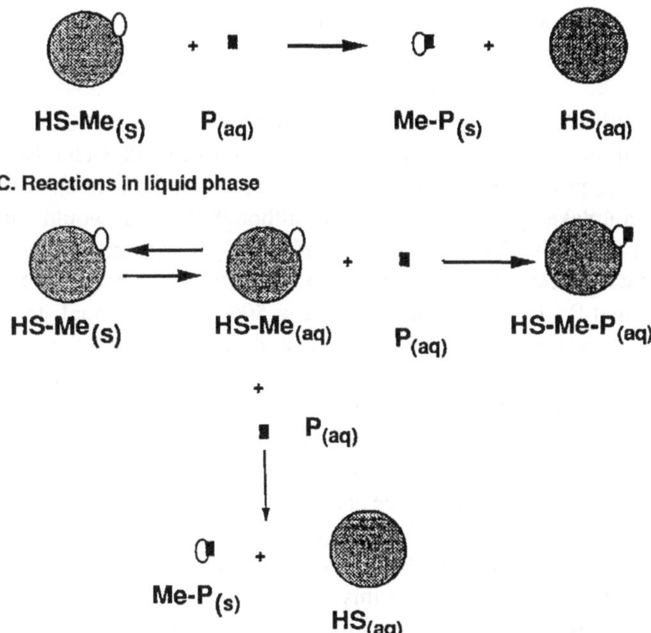

FIGURE 2.3 Potential mechanisms for the binding of phosphate ions onto humic complexes, dissolved or solid. Humic substances (HS), phosphorus (P), and metals (Me) are represented in both solid (s) and liquid (aq) phases. Possible ligand exchange processes are also shown that result in the formation of insoluble metal-phosphates (adapted with permission from Petrovic and Kastelan-Macan 1996).

findings suggest that peat/soil accretion is the dominant long-term (LT) mechanism responsible for P storage in wetlands (Table 2.1, Fig. 2.1). In annually flooded wetlands (e.g., an alluvial swamp in Illinois) annual P storage in sediments reached nearly 3.6 g m^{-2} yr^{-1} because of massive inputs of sediment during flooding conditions (Mitsch et al., 1979). Output of P was 0.34 g P m^{-2} yr^{-1}, which is only 10% of input, but a high rate of loss compared to an average output of 0.024 g P m^{-2} yr^{-1} in upland ecosystems (Richardson, 1985). The peat accumulation rate in peatlands is very slow, with the world average of accretion being only 1 to 2 mm per year (Richardson and Nichols, 1985; Craft and Richardson, 1993a). Concomitant with peat buildup is P storage with values averaging 0.005 to 0.024 g P m^{-2} yr^{-1} for

TABLE 2.1
Mechanisms Controlling Long-Term (LT) and Short-Term (ST) P Storage in Wetlands

Mechanism	Magnitude	Rate
Peat/soil accretion (LT)	high	very slow
Soil adsorption(LT)	low/moderate	moderate
Precipitation (LT)	moderate	fast
Plant uptake (ST)	low/moderate	slow
Detritus sorption (ST)	low	fast
Microbial uptake (ST)	very low	very fast

unfertilized wetlands (Richardson, 1985). In nutrient-enriched wetlands long-term P accretion can reach nearly 1.0 g $m^{-2}\,yr^{-1}$ (Fig. 2.1) (Craft and Richardson, 1993b).

The chemical processes of soil adsorption and precipitation are considered more important than uptake by plants and detritus, although the rates would vary considerably among wetlands (Richardson and Craft, 1993). For example, the amount of P stored by each process summarized for the Houghton Lake peatland in Michigan demonstrates that P adsorption can reach nearly 50% of total annual storage (Table 2.2) (Richardson and Marshall, 1986). A general comparison of the amount of P adsorption among wetland soil types demonstrates that wetland systems vary greatly in terms of the amount of P that can be removed by this process (Fig. 2.4). The systems with the highest adsorption capacity have been shown to have mineral soils, which contain the highest amounts of Fe or Al (Richardson, 1985). The pocosin peat soils have the lowest amount of Fe and Al and the lowest P adsorption potential. The high P adsorption response of the Everglades soil is related to its high Ca content (Richardson and Vaithiyanathan, 1995). Assessment of desorption rates are also required to determine net storage via this mechanism.

Microbial uptake is very fast, but the magnitude (amount stored) is very low. For example, only 12 to 13% of the P in a northern peat-based fen was estimated to be stored in microbial biomass (Table 2.2) (Richardson and Marshall, 1986). In contrast, recent work in the Everglades has shown that microbes control as much as 35% of the P that is stored and cycled in the unenriched soils of the Everglades (Table 2.3) (Richardson et al., 1997). Of importance, and seldom recognized, is the amount of P that can be sequestered by the algal component of wetlands, especially areas with open water. Both the Everglades and Houghton Lake studies indicated as much as 20 to 25% of the added P was stored annually by these organisms (Fig. 2.2, Table 2.2).

The effects of P enrichment on the processes controlling storage is nicely demonstrated by a study in the Florida Everglades. The percentage of P controlled by long-term and short-term processes in soil varied considerably in unenriched and enriched plant communities (Table 2.3). In the enriched area the P soil concentration in the top 5 cm was (1500 mg kg^{-1}, nearly 3 times that found in the unenriched zone (Richardson et al., 1997). Total P and major refractory fractions followed the loading and concentration gradients. The refractory fractions made up 80% of the P stored in the enriched zone, but only 58% in the unenriched area (Table 2.3). Long-term

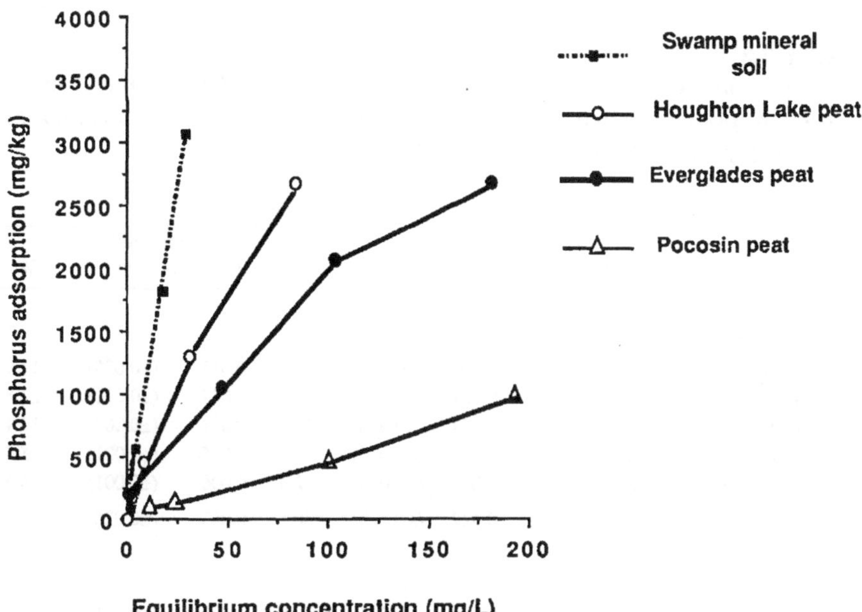

FIGURE 2.4 Isotherms of P adsorption capacity in both mineral and peat based wetlands are related to the equilibrium concentration of PO_4-P remaining in solution. The amount of PO_4-P added in solution (as KH_2PO_4) ranged from 16 to 260 mg/L. (Swamp, Pocosin, and Houghton Lake data are from Richardson 1985. Everglades data are from Richardson and Vaithiyanathan 1995.)

TABLE 2.2
Short-Term and Long-Term Phosphorus Storage Capacity (Houghton Lake, Michigan)

	Range (g m^{-2} yr^{-1})	% of total
Microbial	0.5–1.0	12–13
Algal	1.0	12–25
Sedge	1.0–2.5	25–30
Soil adsorption	1.5–3.8	38–46
Total (short-term)	4.0–8.3	
Long-term storage capacity (field estimates, 5 years)		
0.92 ± 0.15 g m^{-2} yr^{-1}		

storage was mainly as humic organic P and insoluble P. Iron and Al controlled little of the P, whereas >10% was controlled by Ca precipitation in both the enriched and unenriched sites. In the unenriched portion of the Everglades, labile fractions comprise (42% of the stored P, but only 20% in the enriched zone. Thus, labile P fractions make up a lower proportion of total P in the enriched zone versus the unenriched zone. The total amount of P, however, is much less in the unenriched zone.

TABLE 2.3
Average annual accretion rate of P fractions since 1964 (^{137}Cs peak) in enriched and unenriched areas along a gradient in WCA-2A of the Everglades (Richardson et al., 1997)

	Accretion rate of P (g m^{-2} y^{-1})					
Dominant plant communities	**Enriched (cattail)**[a]		**%**	**Unenriched (saw grass/slough)**		**%**
P fractions						
Refractory (long-term storage)						
Insoluble organic P	0.099	(±0.029)	18.6	0.013	(±0.005)	11.9
Humic organic P	0.221	(±0.065)	41.6	0.027	(±0.011)	24.8
Ca bound P	0.074	(±0.022)	13.9	0.011	(±0.004)	10.1
Occluded Fe/Al-bound inorg. P	0.020	(±0.005)	3.6	0.006	(±0.001)	5.5
Surface Fe/Al-bound inorg. P	0.013	(±0.003)	2.4	0.006	(±0.001)	5.5
Subtotal	0.427	(±0.120)	80.8	0.063	(±0.020)	57.8
Labile (short-term storage)						
Bicarbonate extr. org. P	0.021	(±0.006)	3.8	0.003	(±0.001)	2.8
Microbial biomass P	0.069	(±0.014)	13.0	0.038	(±0.010)	34.8
Exchangeable inorg. P	0.017	(±0.003)	3.1	0.005	(±0.001)	4.6
Subtotal	0.107	(±0.020)	19.9	0.046	(±0.010)	4.6
Total	0.530	(±0.140)	100.0	0.110	(±0.030)	100.0

[a] Accretion rate is calculated by summing the P content of each form down to the depth corresponding to the ^{137}Cs peak and dividing by 26 y. The enriched areas are represented by the average rates at four stations. All accretion rates were significantly greater ($P < 0.05$ using t tests) in the enriched area compared with the unenriched area. Standard deviations are in parentheses. The percentage of each P fraction is also shown.

Using Cs-137 dating techniques, Craft and Richardson (1993b) estimated the average annual storage rate of each form of P in the enriched and unenriched areas over the past 26 years (Table 2.3). This revealed two important findings. First, microbial absorption accounted for 35% of the P sequestered in the unenriched zone, but this was reduced to 13% in the enriched cattail zone (Qualls and Richardson, 1995). Maintaining P availability in the unenriched zone is apparently important to organisms since this area of the Everglades has been shown to be P-limited (Craft and Richardson, 1993b). Second, the average long-term total refractory P sequestered in the cattail zone (0.43 g m^{-2} yr^{-1}) was nearly 8 times that stored in the unenriched area, but well below 1.0 g m^{-2} yr^{-1}. This study shows that P enrichment greatly increases long-term total soil P storage up to an average of 0.53 g m^{-2} yr^{-1}, but not all of it is refractory. Moreover, P additions can significantly reduce the importance of short-term microbial P cycling in wetlands. Collectively, these findings suggest that long-term storage of P in a wetland can be related to the P accretion rate for each wetland.

2.5 GENERAL MODEL FOR P LOADINGS–EFFLUENT RELATIONSHIPS UTILIZING THE NORTH AMERICAN WETLAND DATA BASE (NAWDB)

Qian and Richardson (1997) analyzed data from a large number of wetlands throughout the United States to assess the effects of different P loadings on long-term storage rates and effluent concentration patterns. The North American Wetlands for Water Quality Treatment Database (NAWDB) is a United States Environmental Protection Agency effort to collect and summarize the effectiveness of using wetlands, both constructed and natural, as a low-cost alternative for removing pollutants (Knight et al., 1994). Data represent over 125 sites, in some cases covering nearly 25 years of analysis. An early release of the database was made available in 1992 (Kadlec and Newman, 1992). Reckhow and Qian (1994) and Qian (1995) used data from this early release to develop statistical models for predicting output P concentrations.

Cross-sectional data sets are appealing because they include a range of wetland responses to nutrient inputs. This means that empirical evidence discovered from these data are likely to have a relatively broad inference base. The NAWDB includes input and output phosphorus concentrations (P_{in} and P_{out}, in mg/L), hydraulic loading rate (q_s, in cm/day), treatment area (in hectares), and areal input and output P mass loading rates (L_{in} and L_{out}, in g m^{-2} yr^{-1}). Unfortunately, most investigators at the wetland sites used the same water flow rate for input and output. This results in a self-correlation problem and a misleadingly high correlation when plotting loading rates against output concentrations (Reckhow and Qian, 1994). To overcome this problem, numerous relationships were tested with the data set, and a simple model was developed relating areal P loading rate (g m^{-2} yr^{-1}) versus P output concentrations (Fig. 2.5) (Richardson et al., 1997). This figure suggests that effluent output P concentrations remain nearly constant as long as annual inputs are below 1 g m^{-2} yr^{-1} P. As P loading increases from 1 to 5 g m^{-2} yr^{-1} into a wetland, P effluent concentrations increase exponentially and can reach 5000 µg · L^{-1} at 10 g m^{-2} yr^{-1} of P mass loading. P outputs continue to increase as P loadings increase from 10 to 1000 g m^{-2} yr^{-1}. This figure suggests that a significant increase of P effluent output concentration occurs with loading increases above some threshold. Additional P is stored in the wetlands with increased loading, but retention efficiency is reduced. The graph also indicates that loadings below some assimilative P capacity (i.e., a value near mean long-term P storage capacity noted earlier) for the wetlands results in no significant increases in downstream P concentrations. Statistical analysis of the NAWDB reveals that when P loadings to a wetland are kept below 1 g m^{-2} yr^{-1}, P output concentrations remain not only constant but extremely low median of ≅ 40 µg · L^{-1} (Fig. 2.6) (Richardson et al., 1997). However, analysis at this scale reveals that P effluent output ranges from 10 to near 100 µg · L.

A piecewise linear model proposed earlier by Reckhow and Qian (1994) was used to do exploratory analysis and to prepare a nonparametric regression model fit with the data (Fig. 2.7). In the figure, the solid line is the piecewise model, and the dotted line represents a nonparametric regression model. The figure shows that the piecewise linear model is a reasonably good representation of the relationship. The piecewise model may reflect the main mechanism of phosphorus retention in wet-

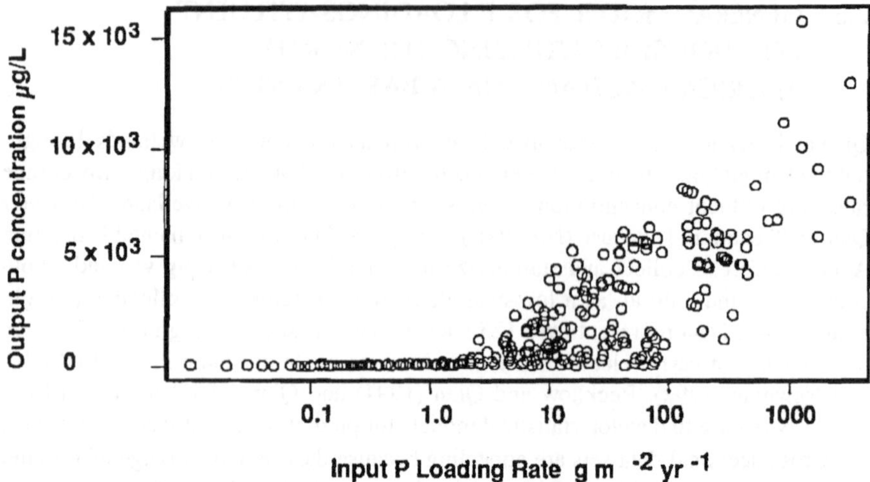

FIGURE 2.5 Effects of input total phosphorus (P) loadings (g m^{-2} yr^{-1}) on P output concentrations (µg · L^{-1} P) for the original North American wetland database (NAWDB). Total sites are 126, with data collected over several years, n = 317 (Knight et al., 1994).

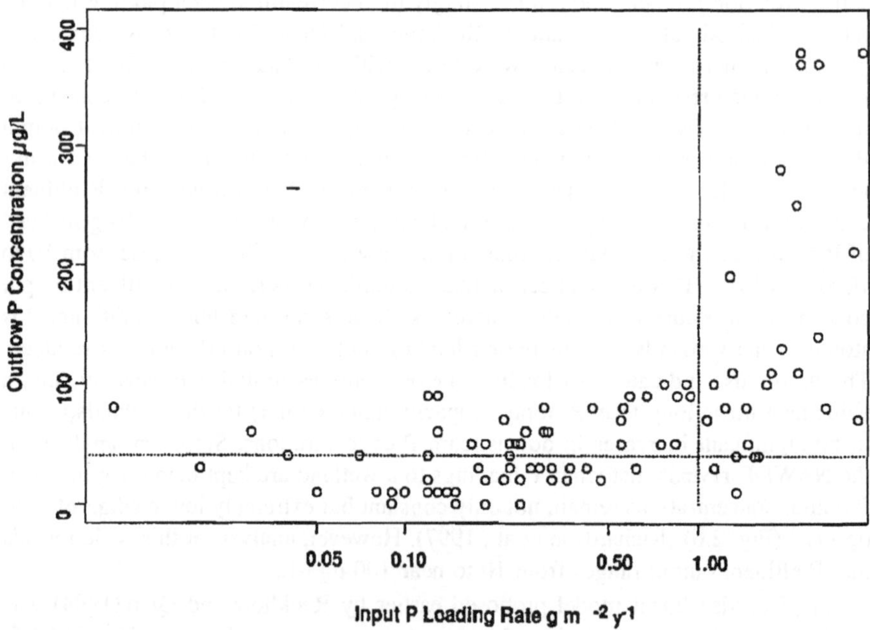

FIGURE 2.6 An expanded plot of input P loading rates (g m^{-2} yr^{-1}) vs. wetland outflow P concentrations (µg · L^{-1} P) (data from Knight et al., 1993). The median value of 40 µg/L TP is shown as a dashed horizontal line and was calculated on effluent values only below 1 g m^{-2} yr^{-1} (from Richardson et al., 1997, with permission).

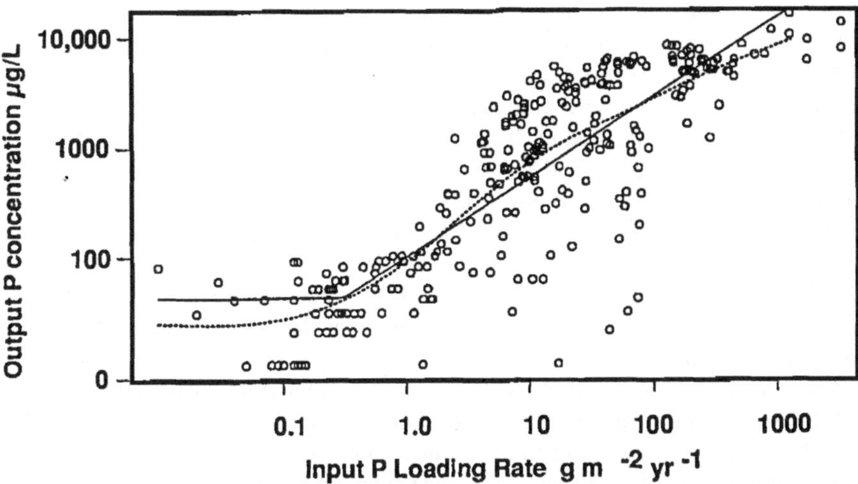

FIGURE 2.7 Effects of input total phosphorus (P) loadings (g m^{-2} yr^{-1}) on P output concentrations (µg · L^{-1} P) for the North American wetland database (NAWDB) (Knight et al., 1994). A piecewise linear model (solid line) and a nonparametric regression model (dashed line) are fitted to the data of the NAWDB (models adapted from Reckhow and Qian, 1994). The change points in data with both models are below 1 g m^{-2} yr^{-1} and can be used to predict the P output relationships from the loading data. The estimated 95% confidence interval of the change point is 0.4 to 1.4 g m^{-2} yr^{-1}.

lands, i.e., annual P storage capacity based on the buildup of peat and soil sediments. When the loading rate of phosphorus to the wetland is below the annual P storage rate, both models predict low output values with little change in the water column P concentrations. When the storage rate is exceeded, phosphorus is apparently available to the water column, and greater concentrations of P are found in the outflow waters. The change point in the models may represent the P assimilative capacity of a wetland. The variations around the change point suggest that the assimilative capacity is a site-specific quantity. The value found around the inflection point from all these wetlands indicates a value that ranges from 0.4 to 1.4 g m^{-2} yr^{-1} (Qian, 1995; Richardson et al., 1997). From the analysis of Reckhow and Qian (1994) and Richardson et al. (1997), a conservative estimate of long-term P storage capacity was estimated to be between a high of 1.4 g m^{-2} yr^{-1} to 0.4 g m^{-2} yr^{-1}. This suggests that 1 g should be considered an upper conservative value for P loading assimilative capacity for wetland ecosystems not designed specifically to treat wastewater.

Long-term storage (defined here as wetland P assimilative capacity) is apparently indicative of a threshold loading rate that maintains a stable outflow of low phosphorus concentrations. When the phosphorus loading rate of a wetland is below the P assimilation capacity, most of the added phosphorus will be stored in the peat or sediment, and no significant increase in the water column concentration occurs (Fig. 2.1, Table 2.2). When the loading rate is above the P assimilative capacity, only part of the added phosphorus can be stored via soil accretion, and the remaining P remains in the water column; hence the P concentration will increase downstream. Finally, the high variation in concentrations in wetland effluent at the higher loadings in the NAWDB is largely due to the differences in the wetland systems (e.g., age of a constructed wetland, soil type, climate, and type of wastewater) (Fig. 2.5). Collectively the cross-sectional studies (Figs. 2.5 through 2.7) and mechanistic studies (Tables 2.2 and 2.3, Figs. 2.2 and 2.4) suggest a P assimilation capacity which ranges from 0.4 to 1.4 g m^{-2} yr^{-1}. Moreover, P mass loadings at or below 1.0 g m^{-2} yr^{-1} result in a mean downstream P release of 42.8 μg/L^{-1} ± 7.3 (±1 SE). However, some variations in the P effluent can be expected to occur at even these low P loadings depending on the wetland type, season, size of the wetland, age of the system, etc. This suggests that each wetland should be assessed if the actual range of P effluent outputs are important.

These findings (Figs. 2.5 through 2.7) led to the "One-Gram Assimilative Capacity Rule," the interpretation being that phosphorus loadings into freshwater wetlands above 1.0 g m^{-2} yr^{-1} will result in a significant increase in P concentration above median baseline outputs (>40 μg · L^{-1}), once short-term uptake processes are saturated (Richardson et al., 1997). Analyses in this paper suggest that the processes that control P retention capacity for wetlands are saturated at loadings above 1.0 g m^{-2} yr^{-1} when the short-term processes are overloaded after the "aging phenomena" (Kadlec, 1985). This rule does not hold if P is added in conjunction with Fe, Ca or Al in the wastewater entering the wetland since continuous P precipitation will occur (Cooke et al., 1992).

2.6 A TEST OF THE ONE-GRAM RULE

More wetlands were included in the 1994 release of the NAWDB database, with better estimates of loading rates than the earlier data set (Knight et al., 1994). Qian and Richardson applied a Bayesian change point detection method on a piecewise linear model of the data and developed a procedure for estimating assimilative capacity when site-specific data are available (Qian and Richardson, 1997). This method was applied to data from Water Conservation Area 2A (WCA-2A) in the northern Everglades (Fig. 2.8) to develop a single-wetland "change-point" curve. The solid line in the figure is the probability of overloading WCA-2A at a given P loading rate, and the dotted line is the nonparametrically fitted model of P outflow concentration as a function of P loading rate. These results suggest that WCA-2A will have mean effluent releases near 20 μg/L P in areas with inputs below 1.0 g m^{-2} yr^{-1}, and the probability of overloading the system is near 10%. At loadings in excess of 1.0 g m^{-2} yr^{-1}, the effluent P concentration doubles, and the probability of overloading the system is nearly 100%. Overloading of P results in the downstream movement of P in the water column once short-term uptake processes are surpassed and long-term storage capacity is exceeded (Richardson and Craft, 1993).

To further assess the relationship of P loadings to assimilative capacity, we collected and analyzed data along a P gradient in WCA-2A of the northern Everglades (Craft and Richardson, 1993a and b, Qualls and Richardson, 1995; Richardson et al., 1997). These field data measurements further supported the one-gram rule since this ecosystem fully assimilated P loadings within the wetlands. It is important

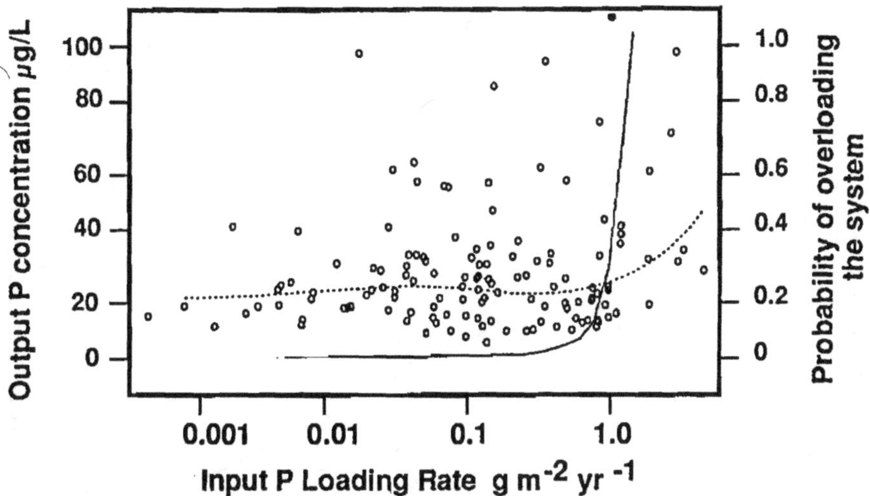

FIGURE 2.8 Outflow P concentrations as a function of P loading for WCA-2A based on the use of a nonparametrically fitted model (dashed line). The probability of overloading WCA- 2A at a given loading rate (solid line) is based on the Bayesian change point detection method following Qian and Richardson, 1997. The estimated 95% confidence interval of the change point for WCA-2A is 0.5 to 1.5 g m^{-2} yr^{-1}.

to note that the P loadings (Fig. 2.9) at each distance were calculated independently using a first order model developed by Walker (1995). Inherent in this model are the assumptions of a settling rate constant (K) under conditions of steady invariant inflow (Kadlec, 1994). These loadings were then compared to measured soil, water and plant community responses to denote changes in the ecosystem. Increased P concentrations and community changes occur above the one-gram loading zone. Few changes are found below the threshold zone. No increase in P above baseline concentrations to downstream waters occurred below the lowest part of the threshold zone of loading, i.e., below 0.4 g m^{-2} yr^{-1} (Fig. 2.9). The highly enriched area (zone 0 to 2 km), which received the highest average P loadings (2.7 g m^{-2} yr^{-1}), maintained water column concentrations > 100 µg · L^{-1} P (Fig. 2.9). The moderately impacted zone (2 to 5 km) received loadings at or above 1 g m^{-2} yr^{-1} and had water column TP concentrations from 50 to 100 µg · L^{-1} P. It was not until loadings were below 1.0 g m^{-2} yr^{-1} (> 5 km) into the nonimpacted zone that TP concentrations fell below 50 µg · L^{-1} P, near the median output value (40 µg · L^{-1} P) reported for all wetlands in the NAWDB. The final output from the non-enriched zone was similar to the 20 µg L^{-1} predicted from the nonparametric site specific model (Fig. 2.8).

The relationships and models presented in this cross-sectional analysis were successfully used to develop a general model of the P assimilative capacity for freshwater wetlands. If low output P concentrations are a goal for natural or constructed wetlands, then loadings must apparently be reduced to 1 g m^{-2} yr^{-1} or lower within the wetland. However, the Bayesian change point method should be used with a piecewise linear model to develop specific parameters for each wetland type.

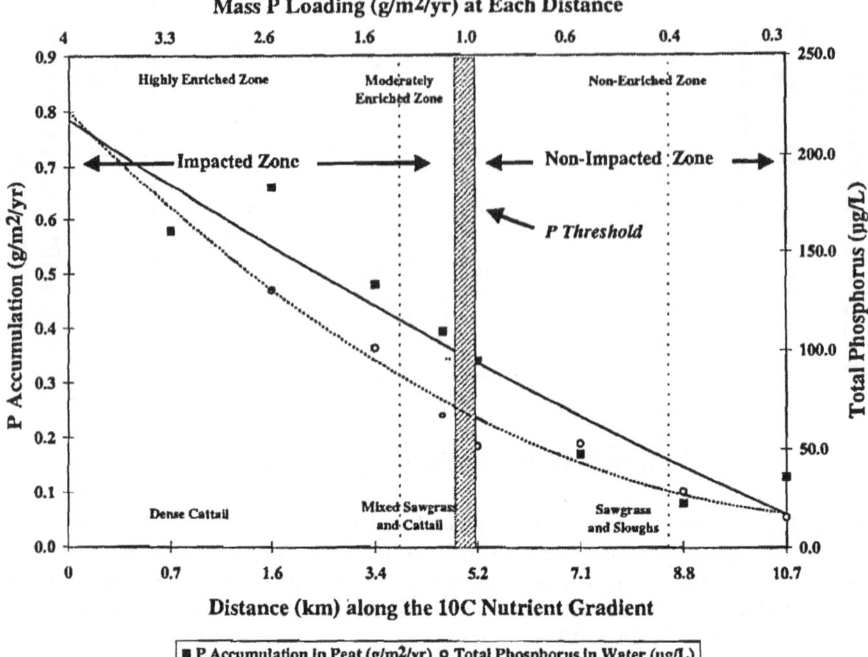

FIGURE 2.9 The relationship between surface water total phosphorus and mean soil P accumulations along an eutrophication gradient in WCA-2A in the central Everglades of Florida. Estimated phosphorus mass loadings over the past 26 years are shown for set distances from the Hillsboro Canal and are calculated following a P retention model developed for the Everglades wetland (Walker, 1995). The model was run successively to calculate the amount of P stored in each zone, and the remaining P was used as the load to the downstream zone. The degree of enrichment and amount of impact are shown for each zone and are based on the amount of P loadings, elevated water column P concentrations, and changes in plant community structure (Richardson et al., 1995; Craft and Richardson, 1997). The "one-gram threshold," along with the 95% confidence intervals (1.4 to 0.4 g m^{-2} yr^{-1}), are shown by the shaded area and dashed lines, respectively.

It has been suggested that the central tendency of the cross-sectional data set of the NAWDB cannot be used to draw general conclusions about general ecosystem behavior (R.H. Kadlec, personal communication). This is true if it is used as the final model to estimate particular parameters for specific sites. The exact change point is a case of such a parameter. However, it is clear that wetlands that can reduce P loadings to 1.0 g m^{-2} yr^{-1} or lower will have few if any ecosystem changes downstream caused by P additions. The effects of increased hydrologic inputs will have to be considered.

Finally, the goals of optimizing maximum P retention capacity for a wetland is not compatible with the P assimilation capacity concept unless a large area of wetland is available and an impacted zone is acceptable within the highest load zone. For example, the impacted zones are found in the areas receiving 4.0 to 1.0 g m^{-2} yr^{-1}

of P loadings in the Everglades (Fig. 2.9). The sizing of the wetlands necessary to remove P loadings can also be estimated from the model of the relationship of mass P storage inputs to water column P inputs (Richardson et al., 1997; Qian and Richardson, 1997). This is

$$A = \frac{C_i Q}{PL_i}$$

where A = total P storage area (m^2), C = P input concentrations ($\mu g \cdot L^{-1}$), and PL = P loading inputs ($g\ m^{-2}\ yr^{-1}$).

This concept has been further developed in a paper by Lowe and Keenan (1997).

2.7 CONCLUSIONS

I have attempted to shed some light on the questions posed in the beginning of the paper. The sink concept for P is not flawed but rather is often misrepresented by the data sets used. The permanent storage rate of P must be based on the P assimilative capacity (long-term storage) of the wetland being tested, not on short-term uptake values. Research findings seem to support a proposed hierarchy of controlling mechanisms, with peat/soil accretion accounting for most of the long-term P storage in wetlands. The long term (nonrecycling) forms of P that control permanent storage of phosphorus in alkaline wetlands are organic P (50 to 70%) and calcium phosphate (15 to 20%). In acid mineral–soil wetlands, Fe and Al adsorption control P retention chemistry. Organic P storage dominates alkaline fen systems; however, microbial organisms and algae, not macrophytes, control the short-term uptake of P in open water areas. Importantly, the role of each mechanism–either as a contributor to short-term or long-term storage of P in wetlands–is now known. These findings, however, lead to unanswered questions regarding the rates of recycling for each mechanism.

An analysis of the North American Wetland database suggests that P concentrations in wetland effluent increase exponentially after a P loading threshold is exceeded. The data showed a clear pattern of low phosphorus output concentration when the total phosphorus mass loading rate was less that 1 $g\ m^{-2}\ yr^{-1}$, but some variation among sites exists. For some wetlands the critical loading rate may fall closer to the lower end of the P threshold zone (0.4 $g\ m^{-2}\ yr^{-1}$) if changes in ecosystem properties are a concern. The annual outflow P median concentration was 40 μg/L with very low variation. This suggests that wetlands have a threshold for P that, if exceeded, may result in a moving front of elevated P concentrations within the wetlands until the P loading reaches the threshold. All wetland types have the capacity to store some P without significant changes in downstream P concentrations.

As a general rule, the "One-Gram Assimilative Capacity Rule" for P loadings may be useful within freshwater wetlands if long-term storage of P and low P effluent concentrations are to be maintained. A site-specific test analysis of the assimilative capacity rule confirmed this for the Everglades. The analysis suggested that to permanently store P and prevent downstream movement of elevated P concentrations, background P loadings within the wetland must be below 1 g m^{-2} yr^{-1}. No significant changes in plant community structure and macroinvertebrates were found below this threshold loading zone (Richardson et al., 1998). Data analysis also suggests that wetlands can continue to retain additional P as P loadings increase, but at the cost of increasing P outflow concentrations. Lower P retention efficiencies exist at higher P loadings. Finally, the P assimilative capacity concept for most wetlands has not been tested, but current analysis suggests that the "one-gram rule" may be a first approximation for most freshwater wetlands.

2.8 ACKNOWLEDGMENTS

I would like to thank all my former wetlands students for the privilege of working with them and for their invaluable help with the phosphorus research. Funding was provided by the Everglades Agricultural Area Environmental Protection District of Florida. We thank W. Willis, C. Best, P. Heine, J. Rice, and D. Heath for lab and statistical analysis. I would also like to thank Dr. C. Craft and Dr. R. G. Qualls for use of their data. Dr. S. Qian helped greatly with the original modeling efforts and provided Figs. 2.7 and 2.8. Appreciation is extended to the South Florida Water Management District for field access to the Everglades.

REFERENCES

Bastian, R.K. and S.C. Reed. 1979. Aquaculture systems for wastewater treatment: seminar proceedings and engineering assessment. MCD-67. EPA 430/90-80-006. Washington, D.C.

Bowden, W.B. 1987. The biogeochemistry of nitrogen in freshwater wetlands. Biogeochemistry 4:313–348.

Cooke, J.G., L. Stub, and N. Mora. 1992. Fractionation of phosphorus in the sediment of a wetland after a decade of receiving sewage effluent. Journal of Environmental Quality 21:726–732.

Craft, C.B., and C.J. Richardson. 1993a. Peat accretion and N, P, and organic C accumulation in nutrient enriched and unenriched Everglades peatlands. Ecological Applications 3:446–458.

Craft, C.B., and C.J. Richardson. 1993b. Peat accretion and phosphorus accumulation along a eutrophication gradient in the northern Everglades. Biogeochemistry 22:133–156.

Craft, C.B. and C.J. Richardson. 1997. Relationships between soil nutrients and plant species composition in Everglades peatlands. Journal of Environmental Quality 26:224–232.

Davis, S.M. 1982. Patterns of radiophosphorus accumulation in the Everglades after its introduction into surface water. Tech. Pub. 82-2. South Florida Water Management District. West Palm Beach.

Devai, I., and R.D. Delaune. 1995. Evidence for phosphine production and emission from Louisiana and Florida marsh soils. Organic Geochemistry 23:277–279.

Faulkner, S.P., and C.J. Richardson. 1989. Physical and chemical characteristics of freshwater wetland soils. p. 41–72. *In* D.A. Hammer (ed.) Constructed wetlands for wastewater treatment: municipal, industrial, and agricultural. Lewis Publishers, Chelsea, Michigan.

Fox, R.L., and E.J. Kamprath. 1971. Adsorption and leaching of P in acid organic soils and high organic matter sand. Soil Science Society of America 35:154–156.

Godfrey, P.J., E.R. Kaynor, S. Pelczarski, and J. Benforado, J. (ed.) 1985. Ecological considerations in wetlands treatment of municipal wastewaters. Van Nostrand Reinhold, New York.

Good, R.E., D.F. Whigham, and R.L. Simpson (ed.) 1978. Freshwater wetlands. Academic Press, New York.

Hammer, D.A. (ed.) 1989. Constructed wetlands for wastewater treatment: municipal, industrial, and agricultural. Lewis Publishers, Chelsea, Michigan.

Hayes, F.R., and J.E. Phillips. 1958. Lake water and sediment. IV. Radiophosphorus equilibrium with mud, plants, and bacteria under oxidized and reduced conditions. Limnology and Oceanography 3:459–475.

Kadlec, R.H. 1985. Aging phenomena in wastewater wetlands. p. 338–350. *In* P.J. Godfrey, E.R. Kaynor, S. Pelczonski, and J. Benforado (ed.) Ecological considerations in wetlands treatment of municipal wastewater, Van Nostrand Reinhold, New York.

Kadlec, R.H. 1994. Phosphorus uptake in Florida marshes. Water Science and Technology 30:225–234.

Kadlec, R.H., and R.L. Knight. 1996. Treatment wetlands. Lewis, Boca Raton, Florida.

Kadlec, R.H., and S. Newman. 1992. Phosphorus removal in wetland treatment areas: principal and data report, Everglades stormwater treatment area design support. South Florida Water Management District, West Palm Beach.

Kastelan-Macan, M., and M. Petrovic. 1996. The role of fulvic acids in phosphorus sorption and release from mineral particles. Wet. Sci. Tech. 34:259–265.

Knight, R.L., R.H. Kadlec, S.C. Reed, R.W. Ruble, J.D. Waterman, and D.S. Brown. 1994. North American wetlands for water quality treatment database. EPA/600/C-94/002. Washington, D.C.

Knight, R.L., R.W. Rubles, R.H. Kadlec, and S.C. Reed. 1993. Wetlands for wastewater treatment performance data base. p. 35–58. *In* G. Moshiri (ed.) Constructed wetlands for water quality improvement. CRC Press, Boca Raton, Florida.

Lindsay, A.L. 1979. Chemical equilibria in soils. John Wiley and Sons, New York.

Lowe, E.F. and L.W. Keenan. 1997. Managing phosphorus based cultural eutrophication in wetlands: A conceptual approach. Ecological Engineering 9:109–118.

Mitsch, W.J., C.L. Dorge, and J.R. Weimhoff. 1979. Ecosystem dynamics and a phosphorus budget of an alluvial cypress swamp in southern Illinois. Ecology 60:1116–1124.

Patrick, W.H., Jr., and R.A. Khalid. 1974. Phosphate release and sorption by soils and sediments: effect of aerobic and anaerobic conditions. Science 186:53–55.

Petrovic, M., and M. Kastelan-Macan, M. 1996. The uptake of inorganic phosphorus by insoluble metal-humic complexes. Wet. Sci. Tech. 34:253–258.

Qian, S.S. 1995. A nonparametric Bayesian model of phosphorus retention in wetlands. Ph.D. dissertation. Duke University, Nicholas School of the Environment, Durham, North Carolina.

Qian, S.S., and C.J. Richardson. 1997. Estimating the long-term phosphorus accretion rate in the Everglades: A Bayesian approach with risk assessment. Water Resources Research 33:1681–1688.

Qualls, R G., and C.J. Richardson. 1995. Forms of soil phosphorus along a nutrient enrichment gradient in the northern Everglades. Soil Science 160:183–198.

Reckhow, K.H., and S.S. Qian. 1994. Modeling phosphorus trapping in wetlands using generalized additive models. Water Resources Research 30:3105–3114.

Reddy, K.R., and W.H. Smith (ed.) 1987. Aquatic plants for water treatment and resource recovery. Magnolia Publishing, Orlando, Florida.

Richardson, C.J. 1985. Mechanisms controlling phosphorus retention capacity in freshwater wetlands. Science 228:1424–1427.

Richardson, C.J. 1989. Freshwater wetlands: transformers, filters, or sinks. p. 25–46. *In* R.R. Sharitz and J.W. Gibbons (ed.) Freshwater wetlands and wildlife. Report CONF-8888603101, DOE Symposium Series No. 61, USDOS Office of Scientific and Technical Information, Oak Ridge, Tennessee.

Richardson, C.J., and C.B. Craft. 1993. Effective phosphorus retention in wetlands–fact or fiction? p. 271–282. *In* C.B. Moshiri (ed.) Constructed wetlands for water quality improvement. Lewis Publishing, Boca Raton, Florida.

Richardson, C.J., C.B. Craft, R.G. Qualls, J. Stevenson, P. Vaithiyanathan. 1995. Annual report: effects of phosphorus and hydroperiod alterations on ecosystem structure and function in the Everglades. Duke Wetland Center Publication 95-05. Nicholas School of the Environment, Duke University, Durham, North Carolina.

Richardson, C.J., J.A. Kadlec, W.A. Wentz, J.P.M. Chamie, and R.H. Kadlec. 1976. Background ecology and the effects of nutrient additions on a central Michigan wetland. p. 34–74. *In* M.W. Lefor, W.C. Kennard, and T.B. Helfgott (ed.) Proceedings: Third Wetlands Conference. Institute of Water Resources No. 26, Storrs, Connecticut.

Richardson, C.J., and P.E. Marshall. 1986. Processes controlling movement, storage, and export of phosphorus in a fen peatland. Ecological Monographs 56:279–302.

Richardson, C.J. and D.S. Nichols. 1985. Ecological analysis of wastewater management criteria in wetland ecosystems. p. 351–391. *In* P.J. Godfrey, E.R. Kaynor, S. Pelczonski, and J. Benforado (ed.) Ecological considerations in wetlands treatment of municipal wastewater, Van Nostrand Reinhold, New York.

Richardson, C.J., S.S. Qian, and C.B. Craft. 1998. Phosphorus assimilative capacity in freshwater wetlands: a new paradigm. Environmental Science and Technology, (submitted).

Richardson, C.J., S.S. Qian, C.B. Craft, and R.G. Qualls. 1997. Predictive models for phosphorus retention in wetlands. Wetlands Ecology and Management 4:159–175.

Richardson, C.J., and P. Vaithiyanathan. 1995. Phosphorus sorption characteristics of the Everglades soils along an eutrophication gradient. Soil Science Society of America, 59:1782–1788.

Rigler, F.H. 1956. A tracer study of the phosphorus cycle in lake water. Ecology 37: 550–.

Siedel, K. 1955. Die Flechtbinse *Scirpus lacustris*. p. 37–52. *In* Ökologie, Morphologie und Entwicklung, ihre Stellung bei den Volkern und ihre wirtschaftliche Bedeuting. Sweizerbart'sche Verlgsbuchhandlung, Stuttgart.

Siedel, K. 1961. Zur Problematik der Keim- und Pflanzengewasser. Verh. Internat. Verein. Limnol. 14:1035–1039.

Spangler, F.L., C.W. Fetter, Jr., and W.E. Sloey. 1976. Artificial and natural marshes as wastewater treatment systems in Wisconsin. p. 215–240. *In* D.L. Tilton, R.H. Kadlec, and C.J. Richardson (ed.) Freshwater wetlands and sewage effluent disposal, a national symposium. The University of Michigan, Ann Arbor.

Tilton, D.L., R.H. Kadlec, and C.J. Richardson. 1976. Freshwater wetlands and sewage effluent disposal, a national symposium. The University of Michigan, Ann Arbor.

Tourbier, J. and R. W. Pierson, Jr. 1976. Biological Control of Water Pollution. University of Pennsylvania Press, Philadelphia.

Verhoeven, J.T.A. 1986. Nutrient dynamics in minerotrophic peat mires. Aquatic Botany 25:117–137.

Vymazal, J., H. Brix, P.F. Cooper, M.B. Green, and R. Haberl (ed.) 1998. Constructed wetlands for wastewater treatment in Europe. Backhuys Publishers, Leiden.

Walker, W.W., Jr. 1995. Design basis for Everglades stormwater treatment areas. Water Resources Bulletin 31:671–685.

Wetzel, R.G. 1983. Limnology. Saunders, New York.

3 Phosphorus Chemistry and Cycling in Florida Lakes: Global Issues and Local Perspectives

Patrick L. Brezonik and Curtis D. Pollman

3.1 ABSTRACT

Substantial progress has been made over the past 40 years in understanding the chemical and biological cycling of phosphorus (P) in lakes, a consequence of its key role in controlling lake productivity. This chapter reviews the progress, primarily in the context of studies on Florida lakes, and describes major issues that remain to be addressed. At the landscape level, P concentrations and export rates can be predicted as a function of land use conditions in many parts of the country, but the accuracy of such predictions for Florida lakes is still crude because of their complicated hydrology and poorly defined catchment boundaries. Atmospheric loadings contribute significantly to the P budgets of Florida lakes, even large, shallow

1-56670-331-X/99/$0.00+$.50

eutrophic ones like Lake Apopka. Much of this loading is from particulate dry fall rather than wet precipitation. Because the atmospheric transport scale for P-containing particles is small (order of tens to hundreds of meters), land-based precipitation samplers are likely to overestimate atmospheric P loading to large lakes.

Mild winters and long seasons for biological growth in Florida lakes dampen the seasonal patterns of P concentrations that occur typically in temperate lakes. Colored natural organic matter (humic substances), high concentrations of which are found in many Florida lakes, significantly affects the relationship between chlorophyll and total P levels, and further work is needed on this subject. In contrast, P cycling processes are rather insensitive to pH, at least at the whole ecosystem level, within the wide range of pH (~4 to 9) found in Florida lakes.

P cycling in lakes cannot be fully understood without considering the sedimentary P cycle and sediment–water interactions. Recent studies have shown that sediment P accumulation rates can be determined even in such shallow lakes as Okeechobee. P accumulation in this lake has increased by a factor of four during the 20th century, suggesting that its external P loading has in creased substantially over that time period. P sorption to sediments often fits a simple Langmuir isotherm, but more complicated models that consider electrostatic interactions and nonuniform binding site energies are needed in some cases. Finally, internal loading of P from sediments is a critical factor affecting the recovery of shallow eutrophic lakes once external loading is controlled. A model that explicitly considers internal loading shows that this process is likely to maintain hypereutrophic conditions in Lake Apopka for many decades after external sources are controlled, unless additional steps are taken to remove or isolate the P-rich sediments from the water column.

3.2 INTRODUCTION

Phosphorus is one of the key elements affecting ecosystem productivity and perhaps the key nutrient in most freshwater ecosystems. As such, aquatic scientists from many disciplines have studied its chemistry, cycling processes, effects on ecosystem productivity, and interactions with the cycles of other nutrient elements for many decades. Phosphorus biogeochemistry is a mature field of study, the main features of which are now relatively well understood. In certain respects, current studies on phosphorus can be regarded as "filling in the details." From a practical perspective, however, many of these details are crucial for effective water quality management.

In our view, the "global issues" pertaining to phosphorus (P) in aquatic systems are mostly local issues. Political scientists often say that "all politics is local." So too can it be argued that water quality issues are all (or at least mostly) local. Because scientific interest in P in aquatic systems is linked to its impacts on water quality, it can be argued that phosphorus issues are largely local. Consequently, our approach in this chapter on current issues for P cycling in aquatic systems is both local and personal. The issues are described in the context of advances made in understanding P cycling through studies conducted mostly in Florida over the past few decades, many involving our research groups. Nearly all of these studies were initiated as a result of concerns about the quality of Florida's lakes. Because conditions in lakes depend strongly on inputs and conditions of their surrounding watersheds, we begin with a discussion of landscape-level issues and then turn to P cycling in the water column of lakes. The P cycle in lakes is not limited to the water column, however. Bottom sediments play a crucial role in regulating water column concentrations, and the last part of the chapter discusses the role of sediments as long-term repositories for P and as reservoirs (i.e., potential sources) of P for continued internal loading of lake waters.

3.3 LANDSCAPE LEVEL

3.3.1 EXPORT COEFFICIENTS

Surface water ecosystems are a reflection of the landscapes in which they occur, and we cannot fully understand what is happening in a lake or stream, either its structure or the way it functions, outside the context of its surrounding watershed. Huge amounts of data have been collected over the past few decades on P levels and loadings in the nation's rivers and streams, and these data have enabled water quality scientists to make broad generalizations regarding land-use/nutrient relationships (e.g., Dillon and Kirchner, 1981; Omernik, 1977). For example, numerous studies have shown that both runoff concentrations and areal export rates (e.g., kg of P exported per km^2 of land in a given use category per year) are much higher for P and N forms from agricultural lands than from forested watersheds. Urban watersheds, which have the highest fraction of impervious surfaces among the major land-use categories, tend to have still higher values than agricultural landscapes because impervious areas produce rapid and high amounts of surface runoff and relatively lower amounts of infiltration. This situation promotes P export. In contrast, water that reaches streams via infiltration and subsurface flow tends to have lower P concentrations because phosphorus generally is not mobile in soils.

For a variety of reasons (described below), Florida is a difficult region in which to measure or predict nutrient export from watersheds. The available literature indicates that landscape/land-use conditions explain only part of the variations in P and N concentrations and loads in tributaries draining Florida watersheds. Export coefficients (e.g., kg P lost per hectare per year) for a given land-use type may range over an order of magnitude (Baker et al., 1981). Application of the land-use export-coefficient approach thus leads to large uncertainties in estimates of P and N loads for a given watershed.

There are considerable differences between the land-use/export relationships for P and those for N. For example, in analyzing tributary data for nutrients from 41 Florida watersheds from the EPA's National Eutrophication Survey, Baker et al. (1981) found that drainage area alone explained a high fraction of the variance in total nitrogen (TN) loading ($r^2 = 0.84$) and did almost as well as a regression that included several land uses (Table 3.1). In contrast, drainage area by itself explained only 21% of the variance in total phosphorus (TP) loading. Multiple regressions involving several land-use characteristics explained much higher proportions of the variance, but even the best predictive relationship for TP loading left 30% of the variance unexplained (Table 3.1). The scatter in Fig. 3.1 indicates that literature-based estimates of TP loadings from Florida watersheds have limited accuracy.

TABLE 3.1
Predictive Equations for Nutrient Loading Based on 41 Florida Watersheds from the National Eutrophication Survey[a]

$TP_L = 0.48(DA) + 71.20$	$r^2 = 0.21$
$TP_L = 6.0(Urb) + 1.2(CrPas) - 1.4(RA) + 1.705$	$r^2 = 0.71$
$TN_L = 5.1(DA) + 82.35$	$r^2 = 0.84$
$TN_L = 4.2(CrPas) + 16.1(NFWet) + 16.0(WA) + 15145$	$r^2 = 0.87$

[a] TP_L and TN_L = total P and N loading (kg/yr); DA = drainage area (km^2); land-cover types (all in ha): Urb = urban area, CrPas = crop and pasture land, NFWet = nonforested wetland, RA = range land, WA = open water. From Baker et al. (1981).

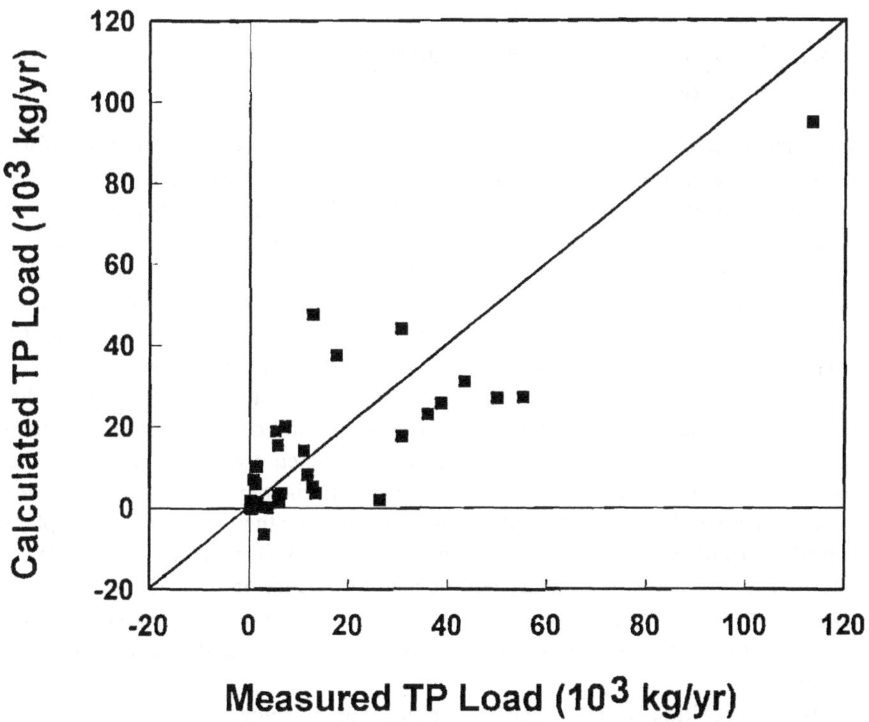

FIGURE 3.1 Predicted versus measured phosphorus export from Florida watersheds based on data from the National Eutrophication Survey and the regression relationship shown in the figure (from Baker et al., 1981).

Compounding the difficulties in estimating P loads to Florida lakes is the ill-defined hydrology of these systems. Florida's terrain tends to be flat (especially so in the southern part of the state), making delineation of watershed boundaries inexact. Also, soils are very sandy in most of Florida so that well defined stream tributaries do not occur in the catchments of many small lakes. About two-thirds of the lakes in the state are seepage lakes, so defined by having no distinct surface water inflow and outflow, and many of these lakes receive appreciable quantities of solutes from shallow groundwater inflow (Pollman et al., 1991; Pollman, submitted). This makes accurate measurement of both water budgets and P loads difficult tasks.

Moreover, although an impressive and commendable amount of nutrient-related research has been conducted on landscape-lake relationships in Florida over the past two decades, much of it has been focused on a few large "problem lakes", especially Okeechobee and Apopka. The drainage basins of these lakes are not wholly representative of the catchments of numerous smaller lakes in Florida that also suffer from an over-abundance of nutrients. Further studies are needed to refine land-use-based P and N loading rates, develop predictive models for nutrient export from Florida landscapes, and quantify landscape-lake quality relationships for these systems.

3.3.2 Importance of Atmospheric Deposition as a Source of Phosphorus

Wet atmospheric deposition is well known as an important source of nitrogen forms (ammonium and nitrate), and long-term data on these contributions are available nationally from the National Atmospheric Deposition Program's network (http://nadp.nrel.colostate.edu/NADP/nadphome). The importance of rainfall as an N source is not surprising because the nitrogen cycle includes important volatile N forms (NH_3, NO_x), for which the atmosphere is a major reservoir. In contrast, no significant volatile species are involved in the global P cycle. The P content of atmospheric deposition is thought to arise primarily from wind erosion of soil and dust particles, and consequently, P levels in wet deposition are presumed to be low. Data to assess nationwide trends in P levels of rain and dry fall are comparatively sparse. Partly because of extensive studies conducted by the authors and their colleagues on nutrients in rainfall, Florida is among the regions that are exceptions to this statement; Lake Tahoe (California–Nevada) is another site where atmospheric contributions of P and N have been studied extensively (Jassby et al., 1994, 1995). The first large-scale study on nutrient deposition in Florida involved collection of bulk precipitation at 24 sites (Fig. 3.2) and wet-only and dry-only deposition at four sites for a two-year period in the late 1970s (Brezonik et al., 1983a; Hendry et al., 1981).

Bulk deposition of TP varied by a more than a factor of six among the sites (17 to 111 mg m^{-2} yr^{-1}), and although the quantitative significance of bulk deposition as a measure of wet plus dry deposition is uncertain (cf. McDowell et al., 1997), areal deposition rates clearly were related to land-use conditions, with agricultural areas having the highest TP deposition (Table 3.2). In contrast to TN, which is associated mostly with wet deposition (on average, 63%), sites with wet/dry collectors all had low deposition rates for TP in wet-only precipitation, and dry fall inferred from the dry collectors accounted for >80% of the atmospheric TP deposition (Table 3.3).

The fact that such a large fraction of atmospheric P loading appears to come from dry deposition poses special problems in measuring or estimating atmospheric

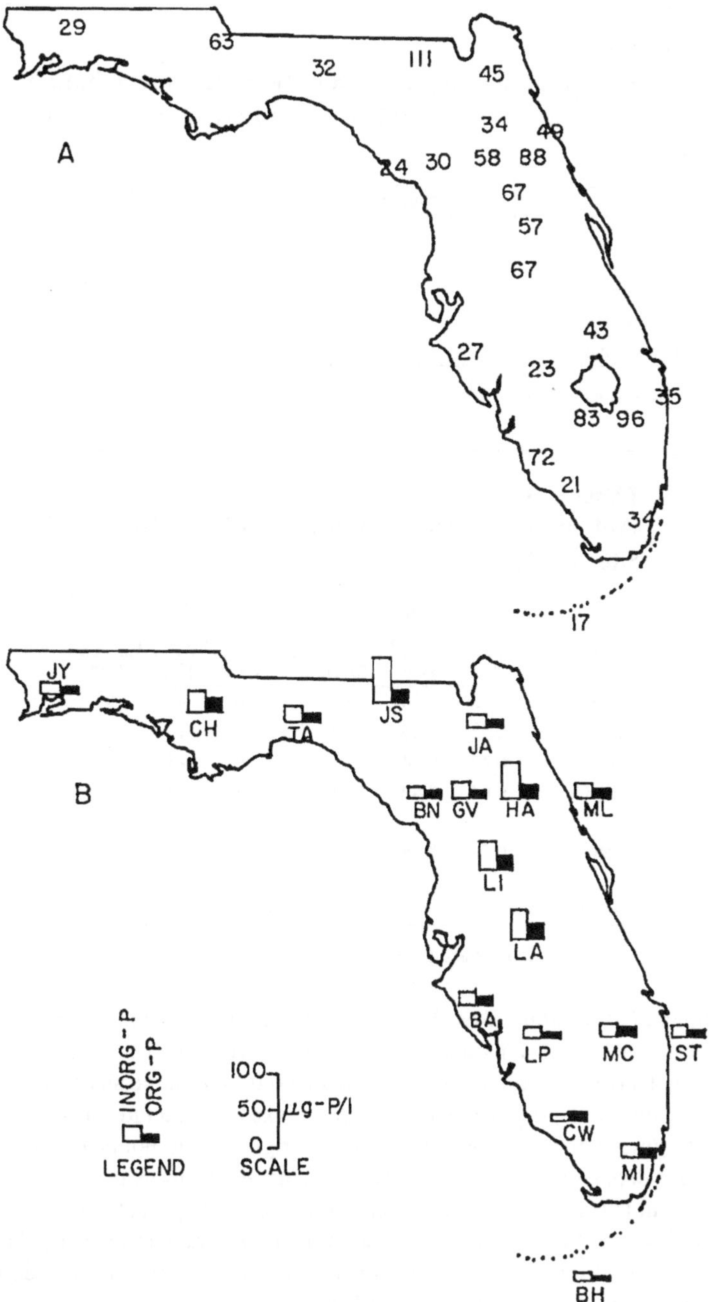

FIGURE 3.2 (A) Total P deposition (mg P m^{-2} yr^{-1}) by bulk precipitation across Florida (May 1978 to April 1979); (B) Volume-weighted average concentrations (g/L) of inorganic and organic P in bulk precipitation. For clarity, not all sites are shown. Statewide means shown in legend (from Brezonik et al., 1983).

TABLE 3.2
Annual Deposition Rates of TN and TP in Bulk Precipitation in Florida, Grouped According to Land Use at Collection Site[a]

Land use	Number of sites	TN g m^{-2} yr^{-1}	TP g m^{-2} yr^{-1}
Coastal	4	0.58	0.031
Urban	5	0.76	0.050
Rural (nonagricultural)	7	0.62	0.027
Rural (agricultural)	8	0.88	0.066
Statewide average	24	0.75	0.051

[a] From Brezonik et al., 1983a

TABLE 3.3
Wet and Dry Deposition of TP and TN at Four Sites in Florida[a]

	TN g m^{-2} yr^{-1}			TP mg m^{-2} yr^{-1}		
Site	Wet	Dry	Total	Wet	Dry	Total
Gainesville	0.63	0.28	0.91	16	42	58
Cedar Key	0.52	0.25	0.77	6	18	24
Apopka	0.47	0.34	0.81	9	48	57
Belle Glade	0.64	0.49	1.13	12	84	96
Mean	0.57	0.34	0.91	11	48	59
Percent wet			63			19

[a] From Brezonik et al., 1983a.

contributions of P to large lakes. It is reasonable to assume that much of the P in dry fall is derived from soils and related particles (this idea is supported by the variations in deposition with land-use), which tend to have large sizes and settle out rapidly. In addition, the dry deposition signal to lakes also includes insects, spiders, and plant fragments that, because of their local origin, would not be considered part of the net dry deposition load to terrestrial systems. Phosphorus in dry fall and bulk precipitation thus is mostly of local origin (transport scale probably on the order of tens to hundreds of meters, short-range rather than long-range transport). For small lakes, deposition of locally derived organic particles can overwhelm wet deposition and the deposition of fine aerosols derived from more regional sources (Cole et al., 1990). According to these authors, this local signal declines exponentially with distance from shore within a distance of some 25 m; thus for large lakes like Okeechobee and Apopka, a bulk deposition collector located on land likely will seriously over-estimate dry fall contributions. Accurate loading studies on such lakes thus would seem to require deposition samplers located on rafts on a transect extending toward the middle of the lake.

Recent data from the Florida Atmospheric Mercury Study (FAMS), which included nine stations located throughout Florida, suggests the important role of coarse particles in atmospheric P loadings (Pollman and Landing, in prep.). Wet and bulk deposition samples were collected aboard 14.5 m high towers in an effort to minimize effects of local particle reentrainment and deposition on measured rainfall chemistry. Total P was included for analysis in addition to mercury (Hg), the primary analyte of concern. Figure 3.3 compares replicate measurements of TP in bulk and wet deposition, and Hg in wet deposition. When compared to the precision of replicate Hg measurements, the precision of TP measurements in wet deposition is quite poor (pooled relative standard deviation = 12.5 and 37.9%, respectively). The relatively high precision of wet deposition Hg measurements is a consequence of washout of aerosol Hg and reactive gaseous Hg, which concentrates on small particles through adsorption and likely reflects long-range and regional scale transport. Conversely, the rather poor precision of wet deposition TP likely reflects highly variable inputs of coarse particles on a very small spatial scale (e.g., insects or insect parts). This variability becomes exacerbated for replicate bulk deposition measurements of TP (pooled relative standard deviation = 70.8%).

Clearly, dry deposition of P to lakes can be important, and because the process is dominated by local scale phenomena, the relative importance to lakes is scale-dependent. Because of the inherent problems in using and interpreting nutrient data collected from bulk collectors, their use is discouraged (McDowell et al., 1997). Wet-only collectors, which are covered except during precipitation events, are less prone to contamination and are widely used in atmospheric deposition studies. However, because the data in Table 3.3 show that P deposition is associated with dry deposition, exclusive use of wet-only collectors will underestimate atmospheric P contributions. Thus, methods need to be developed to characterize P in airborne particulates and aerosols, including size distribution, deposition velocities, and reactivity. McDowell et al. (1997) recommend direct aerosol measurements to infer dry deposition rates, coupled with careful wet deposition measurements that avoid contamination by bird droppings and insects, to develop estimates of total atmospheric loadings of P.

Although contamination of bulk deposition samples by birds or insects cannot always be avoided, sufficient data are available from past studies to place reasonable upper limits on concentrations representing uncontaminated deposition samples so that contaminated samples can be eliminated from the data base. A review of the data collected by Brezonik et al. (1983) on Florida rainfall leads to the conclusion that TP concentrations in bulk precipitation greater than about 75 µg/L are unlikely to occur except when samples are contaminated by bird droppings or dead insects. For wet-only samples, much lower concentrations are expected, and values greater than ~15 µg/L are likely to reflect contamination. Pollman and Landing (in prep.) conducted a similar analysis using replicate precision to identify compromised samples. Analysis of the cumulative distributions of the screened rainfall samples indicated that wet and bulk samples above 10 and 106 µg/L respectively were likely to occur only 10% of the time.

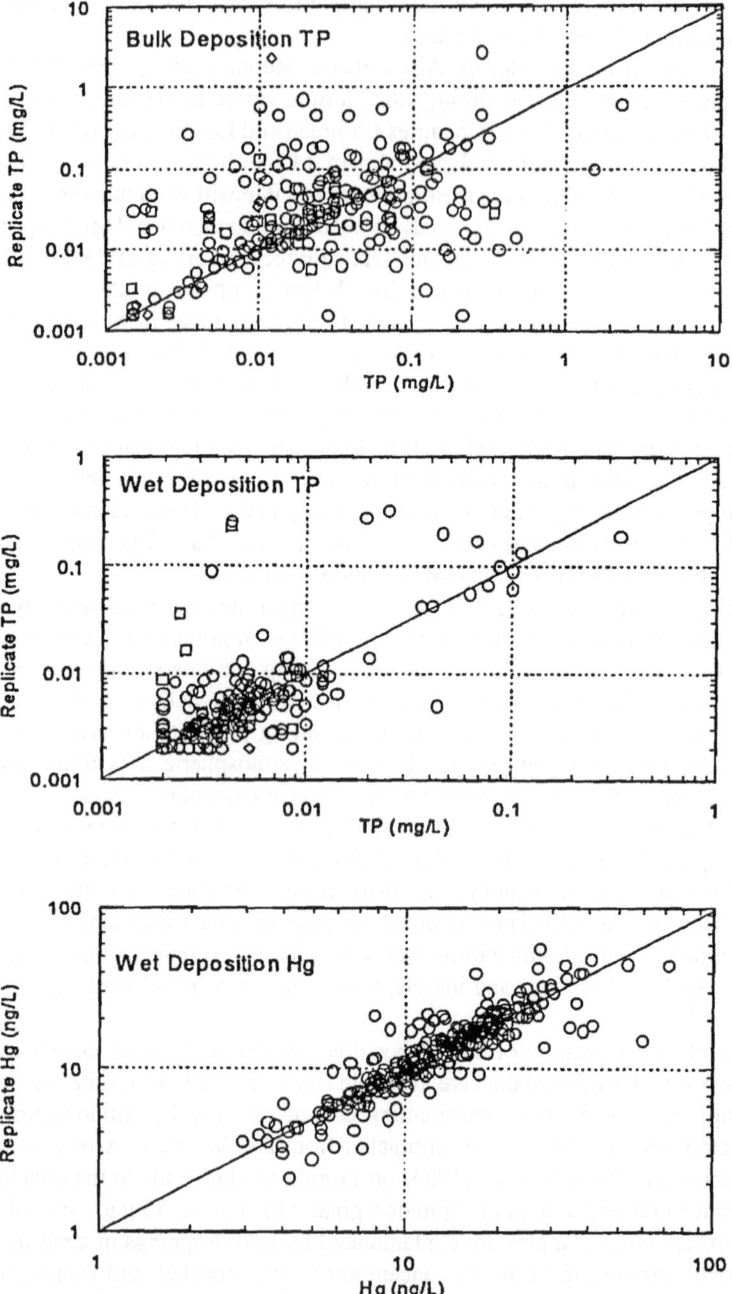

FIGURE 3.3 Comparison of replicate TP and Hg concentration measurements in field replicates collected at FAMS network sites in Florida. Top: replicate TP concentrations in bulk deposition. Center: replicate TP concentrations in wet deposition. Bottom: replicate Hg concentrations in wet deposition (Pollman and Landing, unpublished).

The importance of not including contaminated samples in the data base used to compute atmospheric P loadings is not just an issue of academic science. Atmospheric deposition makes important contributions to the nutrient (N and P) budgets of many lakes (e.g., Jassby et al. 1995). Atmospheric deposition contributes ~44–58 mg P m^{-2} yr^{-1} to McCloud Lake, a highly oligotrophic and acidic water body in north-central Florida (~25 miles east of Gainesville), and this represents nearly all of the P loading to this pristine seepage lake (Ogburn, 1984). The lake has no tributary streams; its uninhabited catchment is small and wholly undeveloped; and its soils are very sandy, leading to a lack of surface runoff.

Atmospheric contributions can be important even for lakes at the other end of the trophic continuum. On an areal basis, atmospheric P loading to hypereutrophic Lake Apopka is about the same as that for McCloud Lake (~46 to 57 mg P m^{-2} yr^{-1}). The lower value is based on unpublished data of the St. Johns River Water Management District (SJWMD) for 1989 to 1992, censored to remove outliers suspected to reflect bird contamination. The upper value is based on results from a wet/dry collector maintained near the lake in 1978 to 79 (Hendry et al., 1981). Because the lake has a large surface area (124 km^2), the areal rate represents a total contribution of ~5750 to 7070 kg P per year. These estimates may be high because of potential bias from use of land-based collectors (described above). However, if we accept the estimates at face value, they represent ~28 to 33% of the total P loading that the SJWMD has estimated for the lake.

Atmospheric loadings also constitute a significant fraction of the "allowable" loadings developed by eutrophication scientists and used by regulatory agencies in developing management and protection plans for lakes. If atmospheric contributions of P estimated for a lake are too high because the data base includes contaminated samples, an agency may set an erroneously low level of loading allowed from "potentially controllable sources," such as agricultural and urban runoff. In the case of McCloud Lake, atmospheric P loading is about half of the upper loading limits proposed for oligotrophic lakes; i.e.,100 mg m^{-2} yr^{-1} (Vollenweider, 1975) and 130 mg m^{-2} yr^{-1} specifically for Florida lakes (Baker et al., 1981). In the case of Lake Apopka, the atmospheric loading rates given above are about two-thirds of the total P loading from sources considered to be uncontrollable and more than one-third of the total loading limit of ~0.13 g m^{-2} yr^{-1} proposed for the lake by the SJWMD in 1995 (Lowe and Battoe, unpublished).

3.3.3 Regional Patterns in Surface Water Phosphorus Levels

Concern about surface water quality in recent decades has resulted in the collection of a huge amount of data on the distribution of P and related trophic state variables in lakes across the country. In some cases, these data are sufficient to construct regional scale maps showing patterns of P abundance in lakes. For example, Kiilsgaard et al. (1996) developed such a map for lakes in the northeastern United States based on data from 2893 lakes (about 13% of the water bodies > 1 ha) in the region. Subregions on their map are colored according to the modal value of TP (μg/L) of lakes in the delineated region. In that associated histograms for many subregions show a wide range of TP values for lakes within them, such maps are a simplification of landscape trends. Nonetheless, a central tendency is exhibited in most cases. Similar maps showing patterns of surface water TP levels are available for the upper Great Lakes states (Omernik et al., 1989). Such maps can be useful as management tools—in understanding large-scale trends, posing questions about causes of those trends, and focusing land management activities.

3.4 PHOSPHORUS CYCLING IN THE WATER COLUMN OF LAKES

Limnology began as an observational science in the late 19th century (NRC, 1996). By the mid 20th century, extensive data were available on some (northern) lake districts to characterize seasonal trends and regional patterns, and limnology progressed at least partly into an experimental science. Limnological studies on Florida lakes lagged well behind those in northern states. As recently as the 1960s, Florida was almost "unknown territory" with regard to basic trophic state conditions and nutrient chemistry/cycling in its lakes. Nevertheless, many northern limnologists believed that Florida lakes were high in P, probably because the state was well known for phosphate mining, and there was a lack of appreciation for the spatial variability of hydrologic and geologic conditions in the state. Florida lakes also were thought by many limnologists to be eutrophic (because of the supposed high P levels and a perception, also erroneous, that all the state's lakes were very shallow). This section describes several issues related to P levels and cycling in Florida lakes based primarily on studies by the senior author. Most of the studies were conducted in relation to two major lake problems of the past quarter century, eutrophication and acidification.

3.4.1 Phosphorus-Trophic State Relationships: Influence of Color

Regional studies conducted in the late 1960s and early 1970s in response to public concerns about eutrophication showed the generalizations cited above to be false. A study on 55 lakes in north-central Florida (Shannon and Brezonik, 1972) found a great diversity of lake types, in terms of morphometry, basic water chemistry, and trophic conditions, just in that part of the state. Total P concentrations exhibit a wide range in this region, from highly oligotrophic values of a few μg/L for "sand-hill" lakes in the Trail Ridge region ~40 to 60 km east of Gainesville to hypereutrophic values in the hundreds of μg/L for shallow lakes impacted by waste water effluents and agricultural activities scattered around the region. Dissolved organic color was found to be an important chemical factor characterizing the lakes. A cluster analysis using six basic chemical variables (pH, alkalinity, acidity, color, conductivity, and calcium) grouped the lakes into four types primarily reflecting color and pH conditions.

Further cluster analyses on the lakes using seven indicators of trophic conditions grouped the low-color lakes into three classes that were readily interpretable in terms of the well known oligotrophic, mesotrophic, and eutrophic classes (Fig. 3.4a). In

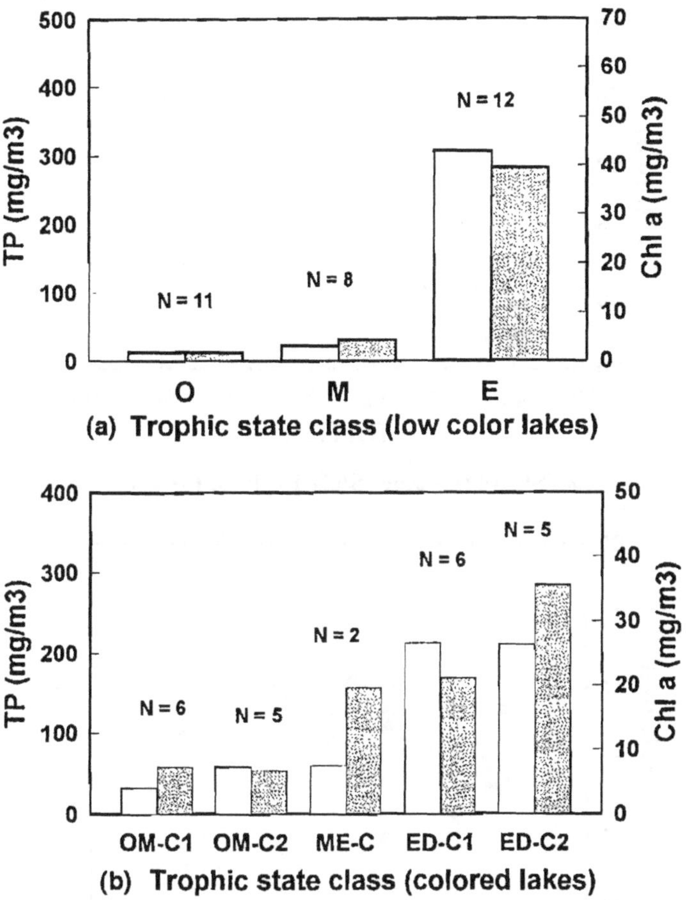

FIGURE 3.4 (a) Mean TP (open bars) and chlorophyll *a* (shaded bars) in three trophic classes of 31 low-color lakes (generally < 50 Pt-Co units); (b) same variables for five classes of 24 high-color lakes (generally > 50 Pt-Co units). N = number of lakes in a class; classes delineated by Shannon and Brezonik (1972) by cluster analysis based on six trophic state indicators.

contrast, the colored lakes formed five groups that were difficult to relate to classic trophic categories and yielded more confusing patterns of TP and chlorophyll values (Fig. 3.4b). Differences between low-color and highly colored lakes relative to their responses to increased nutrient concentrations can be seen in plots of Carlson's (1977) trophic state index (TSI) based on chlorophyll versus his TSI based on TP (Fig. 3.5). The relationship for colored lakes exhibits greater scatter and a lower slope (smaller increase in net chlorophyll production per unit increase in TP) than lakes with lower color. Other studies have shown that chlorophyll–TP relationships differ in river impoundments, compared with natural lakes, apparently because nonalgal turbidity, which affects light penetration, typically is much higher in impoundments than in lakes (e.g., Canfield and Bachman, 1981; Jones and Knowl-

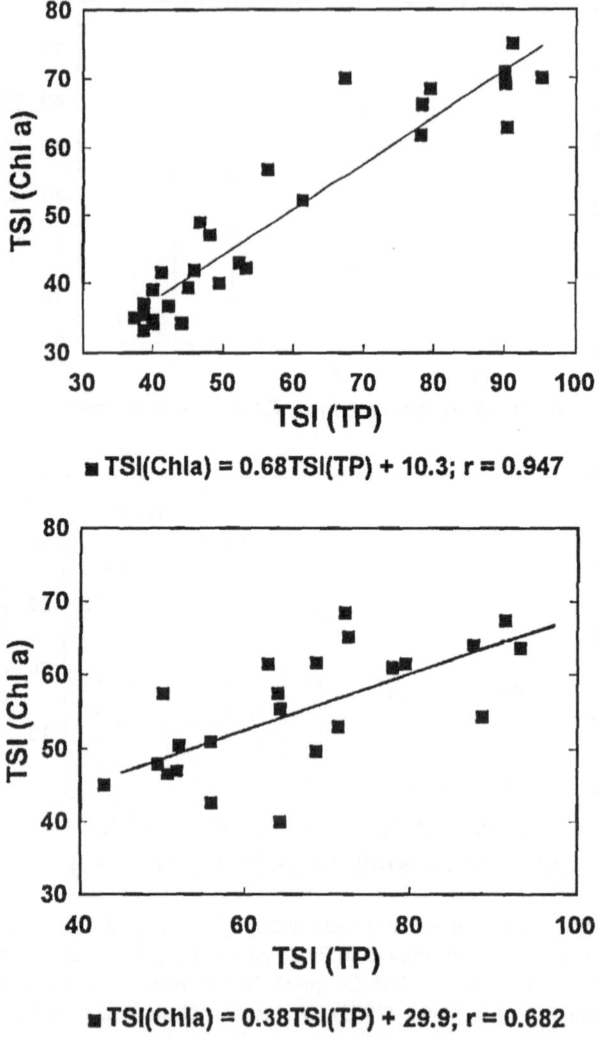

FIGURE 3.5 Carlson trophic state index based on chlorophyll (TSI Chla) versus index based on total P (TSI TP). Top: low-color lakes in Fig. 3.4; bottom: high-color lakes in Fig. 3.4.

ton, 1993). In summary, many Florida lakes exhibit high levels of organic (humic) color, and this apparently affects the way a lake responds to phosphorus relative to other trophic state conditions. Further studies are needed to better define trophic response-nutrient loading relationships in colored lakes.

3.4.2 Effects of pH on Phosphorus Cycling

Another finding of early surveys of Florida lakes was the presence of many acidic (or at least very low alkalinity), soft-water lakes in the Trail Ridge area of the north-central part of the state. This finding also contradicted a common view that Florida's lakes had hard water, a misunderstanding based on the well known fact that the peninsula is underlain by massive deposits of limestone. Soft-water lakes in the Trail Ridge area are fairly isolated (hydrologically) from the underlying limestone Floridan Aquifer. Studies in the late 1970s to early 1980s showed that many of these lakes are vulnerable to acidification by atmospheric deposition (e.g., Brezonik et al., 1980; Hendry and Brezonik, 1984), and this led to studies on the potential impacts of acidification on nutrient chemistry and cycling.

Part of the impetus for these studies was the "oligotrophication hypothesis" of Grahn et al. (1974), who proposed that acidification inhibited P cycling and thus caused a decline in lake productivity. A two-year survey of acid-sensitive lakes in north-central Florida across gradients of pH and P (Brezonik et al., 1984) showed many changes in the composition of the zooplankton and benthic invertebrate communities that were attributable to variations in pH, but no evidence was found for a specific effect of pH on trophic conditions, either on nutrient concentrations themselves or on planktonic responses (e.g., chlorophyll levels) to a given nutrient concentration.

Inferences from this survey were supported by experimental studies on P cycling in McCloud Lake (Ogburn, 1984; Ogburn and Brezonik, 1986), which demonstrated that pH affects P cycling processes only at very low pH, values that are unlikely to result from atmospheric deposition. In one study, the pH of pelagic enclosures in the lake was lowered from an already low ambient pH of 4.7 to values of 4.1 and 3.6, and the enclosures all were spiked with orthophosphate. No differences were observed in the rate of uptake of inorganic phosphate by plankton in the enclosures (Fig. 3.6), but rates of mineralization of particulate P and dissolved organic P back to SRP (soluble reactive phosphate, approximately equal to orthophosphate; see section on P forms) were inhibited at the lowest pH. In contrast, the amount of inorganic phosphate released from decomposing macrophytes (*Eleocharis* sp.) in laboratory microcosms over a 7 month period was independent of pH in the range 3.7 to 5.5, and initial release rates actually were higher at the lowest pH (Ogburn et al., 1987).

Additional studies as part of an experimental acidification of Little Rock Lake (LRL), Wisconsin (Brezonik et al., 1993) also support the conclusion that gross features of P cycling in lakes are insensitive to pH, although effects were observed on some processes. The north basin of LRL, a highly dilute, low alkalinity seepage lake in a small, forested catchment, was acidified in three two-year increments of ~0.5 pH units from an initial value of 6.1 successively to 5.6, 5.1, and 4.7 (target values on an annual average basis). Acidification was achieved by repeated additions of technical grade H_2SO_4. The south basin of the lake was maintained as a reference, and the two basins were separated by a dacron-reinforced PVC curtain at a narrow constriction between the basins. Detailed monitoring of the basins showed no evidence of an acidification effect on average, minimum, and maximum concentrations or seasonal distributions of SRP and TP in the epilimnion of the treatment basin at pH 5.6 and 5.1 (Fig. 3.7). A small increase in the summer mean for epilimnetic SRP was noted at pH 4.7, along with a general increase in the variability of SRP (Sampson et al., 1994). The near lack of response is despite laboratory studies that showed sorption of SRP by LRL sediments to be highly pH-dependent (Detenbeck and Brezonik, 1991b), with sorption increasing as pH decreased.

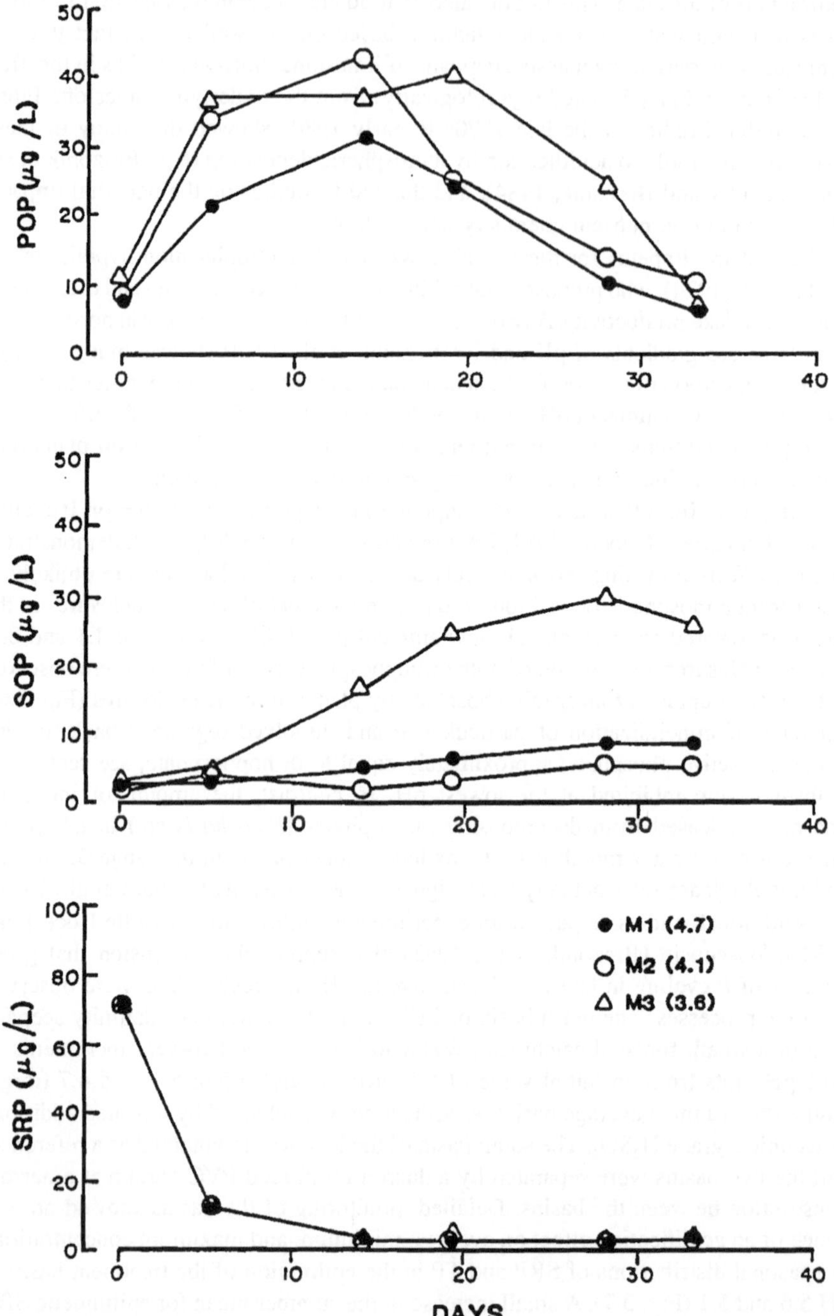

FIGURE 3.6 Changes in concentrations of P forms in McCloud Lake pelagic enclosures maintained at three pH values (shown in parentheses) after spiking with orthophosphate on day 0. Top: particulate organic P; center: soluble organic P; bottom: SRP (from Ogburn, 1983).

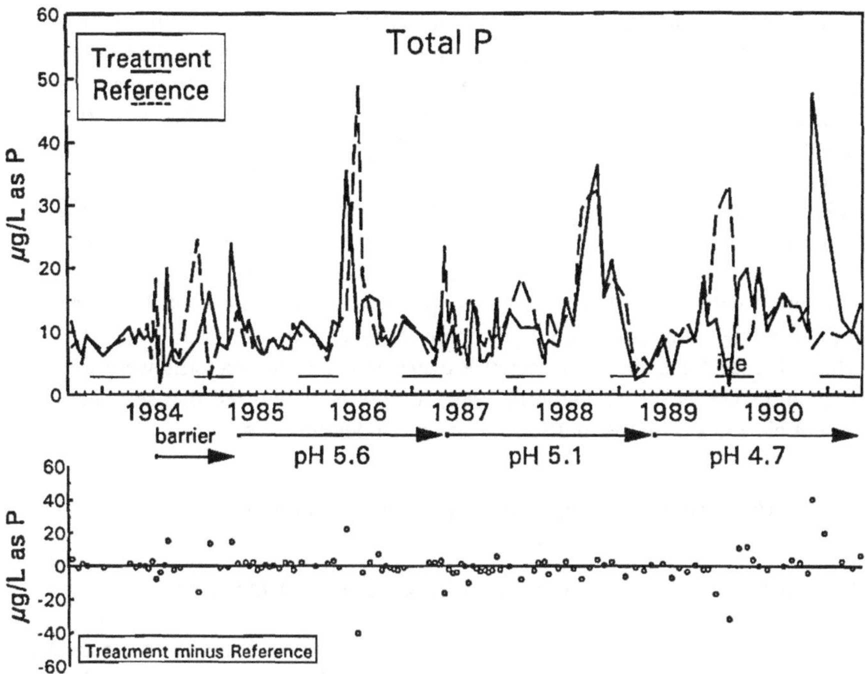

FIGURE 3.7 Top: concentrations of TP in treatment and reference basins of Little Rock Lake, WI over course of acidification experiment. Treatment periods shown below plot. Bottom: treatment basin TP minus reference basin TP versus time (source: C. Sampson and P. Brezonik, unpublished data).

One apparent effect of acidification on water column P was a steady increase in the maximum SRP concentration in the hypolimnion of the treatment basin, from 70 µg/L in 1984 (the year before acidification began) to about 380 µg/L in 1990, the last year at pH 4.7 (Sampson and Brezonik, unpublished data). Because the south basin is too shallow to stratify, it does not serve as a reference for comparison in this case. However, the trend probably reflects interactions among the phosphorus, iron and sulfur cycles of the basin. Acidification increased sulfate concentrations in the lake (a natural consequence of adding sulfuric acid) from about 2.5 mg/L (26 µ*M*) to 7.1 mg/L (74 µ*M*). This led to increased sulfide and dissolved iron concentrations in the hypolimnion. The increased hypolimnetic SRP levels may reflect release of sorbed P from ferric oxyhydroxide as it underwent reductive dissolution. Because the hypolimnion of the treatment basin comprises only a small fraction (~6%) of the total basin volume, increased hypolimnetic SRP levels did not result in measurable increases in surface water SRP.

3.4.3 Seasonal Patterns of Phosphorus in Florida Lakes

An early observation from lake surveys conducted in the 1960s and 1970s is that Florida lakes generally do not follow the classic seasonal pattern of temperate lakes (peak concentrations early in spring, a rapid decline as a result of spring algal blooms, and low inorganic P levels during summer). Instead, seasonal patterns are much more stochastic, both within and between years, as shown for a four-year sequence of SRP and TP data on lakes in the Ocklawaha chain (Fig. 3.8). Mild, short winters and the absence of ice cover allow virtually year-round plant growth; minimum water temperatures even in northern Florida lakes seldom drop below 8 to 9° C. The lack of seasonal patterns is most pronounced in the most eutrophic and shallowest of these lakes. A correlogram for chlorophyll *a* levels (Fig. 3.9) shows almost no evidence of a seasonal pattern in Lake Apopka, but an increasing negative autocorrelation with a lag of six months is apparent for the somewhat deeper and less eutrophic lakes downstream. Oligotrophic lakes such as McCloud Lake also show little evidence for a simple seasonal P cycle, but seasonal patterns were found for silica and some N forms (Ogburn, 1984). It now is recognized that seasonal cycles are not always simple for nutrients even in temperate lakes.

3.4.4 Phosphorus Forms and Turnover Times in Lakes

Phosphorus Turnover: Seasonal and Mass Balance Approaches

In early studies of nutrient cycling in lakes, limnologists focused on seasonal patterns, under the implicit assumption that the rate of nutrient turnover in the water column was defined by the seasonal cycle (i.e., the late winter maximum and summer minimum). Indirect evidence from mass balance studies can be used to demonstrate that P cycling in Florida lakes is more rapid than implied by seasonal variations in pool sizes. Ogburn (1984) measured sedimentation of particulate P in McCloud Lake using cylindrical sediment traps suspended in the water column and found rates of 10 to 48 mg m^{-2} mo^{-1} (mean = 31 mg m^{-2} mo^{-1}). Mass sediment fluxes were three orders of magnitude higher (12 to 66 g m^{-2} mo^{-1}; mean of 37 g m^{-2} mo^{-1}). Extrapolated to an annual basis, the values become 380 mg m^{-2} yr^{-1} (P) and 430 g m^{-2} yr^{-1} (total sedimentation). In contrast, atmospheric deposition (wet plus dry) contributes only about 44 to 58 mg P m^{-2} yr^{-1} to the surface of McCloud Lake. For reasons discussed earlier, atmospheric deposition is thought to account for most of the P loading to the lake. On this basis, the annual P loss from the water column of the lake by settling is about 6 to 8 times greater than the annual external P input to the lake. This high rate can be maintained only by substantial internal recycling. If the lake is considered on an average, steady-state basis, its total P reservoir turns over in about 0.12 to 0.16 years (~45 to 60 days). Although the data on which these calculations are based are limited and the estimates are only approximate, they illustrate the dynamic character of P cycling in lakes.

Phosphorus Forms in Lakes

The above calculations lumped all the P in the water column into one pool, and the turnover time assumes all P forms in the water are equally reactive. This is a gross simplification of reality; there are many distinct pools of P in aquatic systems.

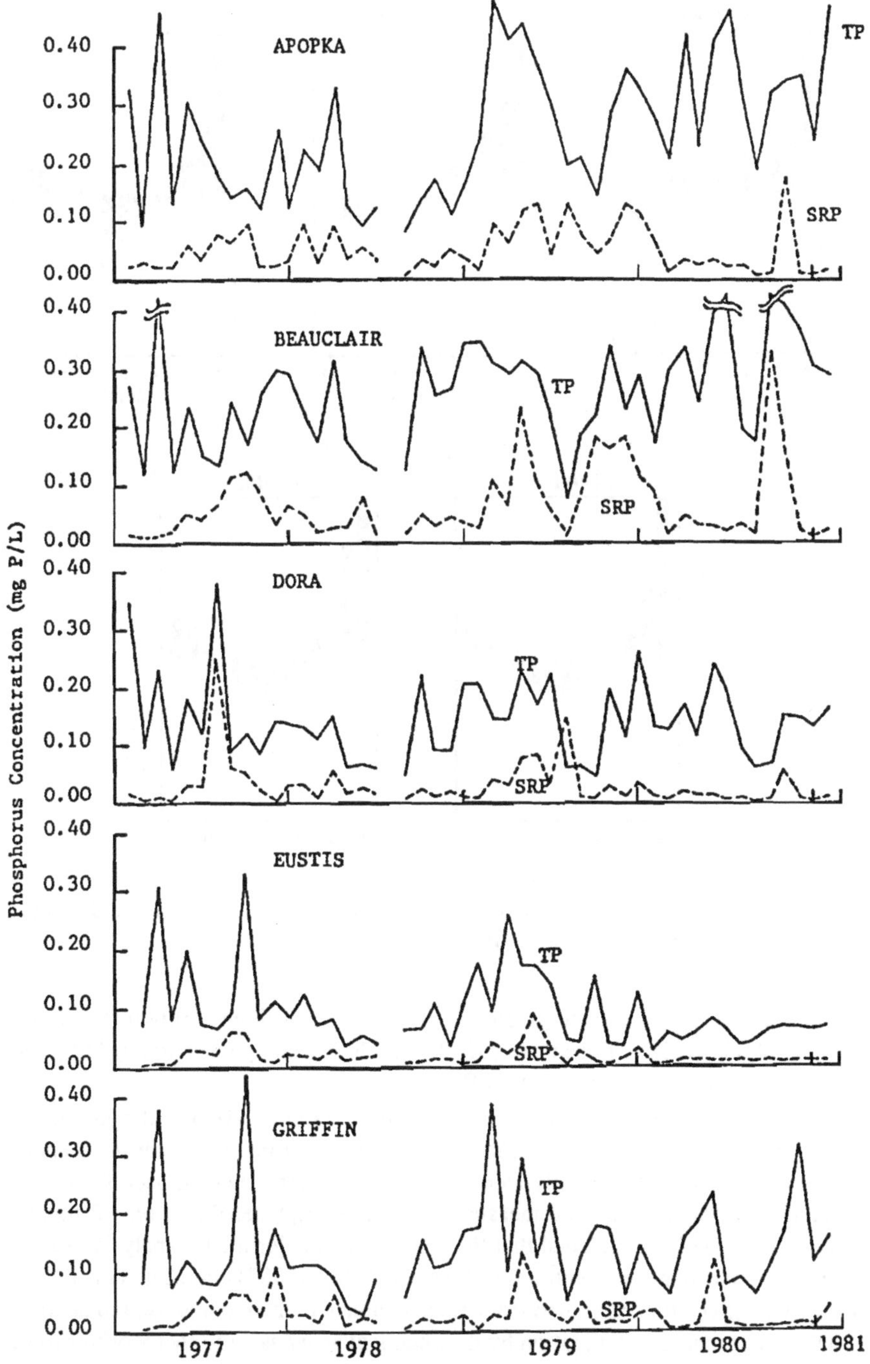

FIGURE 3.8 Mean monthly concentrations of TP and SRP in five lakes of the Ocklawaha chain over the period January, 1977 to January, 1981 (from Brezonik et al., 1981).

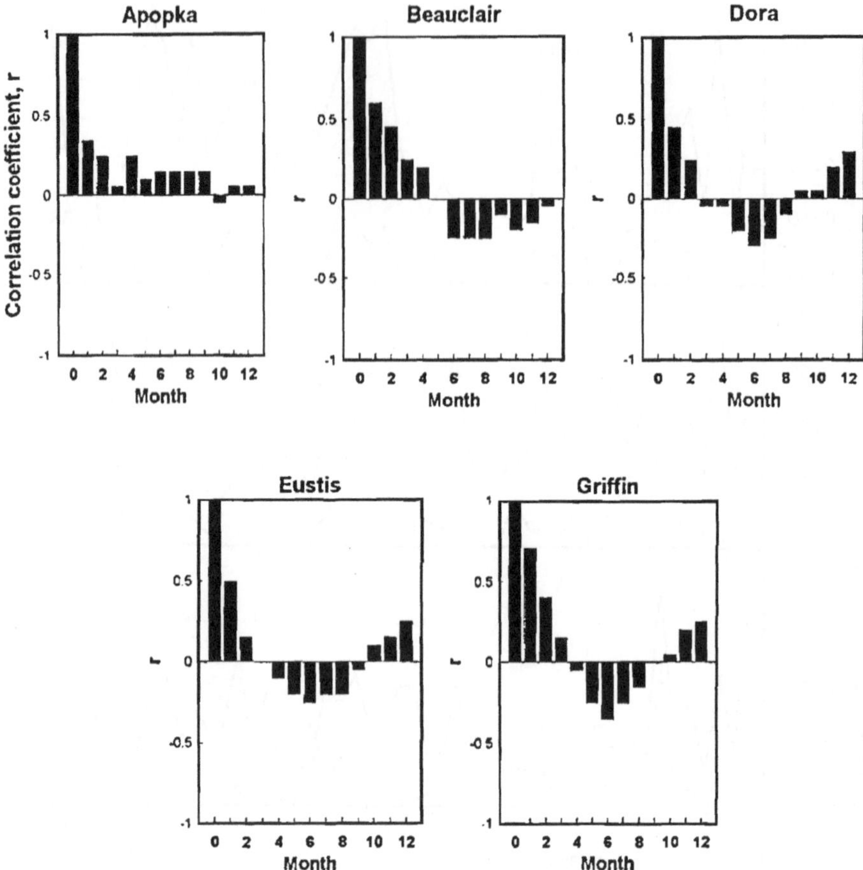

FIGURE 3.9 Correlograms for chlorophyll *a* concentrations in five lakes of the Oklawaha chain for lag period of 1 to 12 months (redrawn from Preston, 1983).

Routine analytical measurements commonly distinguish three forms in the water column: soluble reactive phosphate (SRP), soluble organic phosphorus (SOP), and suspended (particulate) phosphorus (PP).

SRP is closely identified with orthophosphate ($H_2PO_4^-$, HPO_4^{2-}, depending on pH), but small amounts of labile dissolved organic P forms may hydrolyze under the acidic conditions of the molybdenum-blue colorimetric method used for orthophosphate analysis and thus be measured as orthophosphate. Based on ^{32}P uptake studies, Rigler (1968) concluded that the conventional colorimetric analysis overestimates orthophosphate concentrations. Lean and White (1983) argued that the inconsistency in Rigler's data stemmed from an inadequacy in the method used to analyze the data rather than overestimation by the chemical analysis, but other workers have reported similar hydrolysis problems (e.g., see Broberg and Pettersson, 1988). Results of the analysis usually are reported in an "operationally-defined" context as soluble reactive phosphate (SRP).

Early studies by Rigler (1964) concluded that the soluble organic P fraction constitutes about 18% of the TP across a wide range of trophic conditions in temperate lakes. However, a few examples for Florida lakes indicates there is substantial variability in the concentration of SOP and the fraction of TP it represents. Ogburn (1984) found that SOP was very low (~1 to 3 μg/L) in oligotrophic McCloud Lake and constituted about 15 to 20% of the TP. In contrast, SOP in Lake Okeechobee in October 1980 ranged from 13 to 109 μg/L (mean = 42 μg/L) and represented 15 to 84% of TP (Brezonik et al., 1983b).

The composition of the SOP fraction is still poorly characterized, but it is thought to be diverse in terms of structure and size (ranging from simple organic phosphate compounds to macromolecules in the colloidal size range), bioavailability to aquatic microorganisms (bacteria and algae), and lability toward hydrolysis. Investigation of these characteristics has been of interest to aquatic scientists for decades, and there is a large literature on this subject (e.g., Broberg and Persson, 1988; Boström et al. 1988). The limited success that has been achieved in understanding the nature and reactivity of SOP reflects more on the complexity of this fraction than the intensity of the efforts.

Many investigators have used size-exclusion chromatography to evaluate the molecular size distribution of SOP forms in lake water. Lean (1973) demonstrated that uptake of SRP by aquatic bacteria and algae results in rapid formation of a small amount of a labile, low molecular weight (MW) SOP compound (~250 daltons), which subsequently becomes associated with a high-MW colloid [$>5 \times 10^6$)]. The latter material was thought to represent the majority of the nonparticulate P in the water. Minear (1972) found that up to 20% of the SOP recovered from algal cultures and lake water was high MW material (>50,000 daltons) consisting primarily of DNA fragments. Eisenreich and Armstrong (1977) found that inositol phosphate (IP) esters were a component of the SOP in Lake Mendota, Wisconsin. Inositol (hexahydroxycyclohexane) is widely distributed in plants and animals, and formation of IP esters is a mechanism of P storage in cells. A significant fraction of the IP esters in Lake Mendota was associated with high MW organic matter, and they accounted for about 26% of the TP in the water. Based on size separations using Sephadex G-25, Brezonik et al. (1983b) found that the SOP in Lake Okeechobee was predominantly high MW material ($> \sim 10^4$), and the ratio of high:low MW SOP decreased from the littoral zone to the pelagic zone. Qualitative evidence for the presence of IP esters in SOP from Lake Okeechobee was found using a technique developed by Eisenreich and Armstrong (1977).

Another approach to characterizing SOP has been to expose the fraction to various phosphate-hydrolyzing enzymes (phosphatases). Herbes et al. (1975) were among the first to use this approach. They found that up to 50% of the SOP in water from two Michigan lakes was hydrolyzable by the enzyme phytase, but none was hydrolyzable by alkaline phosphatase and phosphodiesterase. (Phytase hydrolyzes IP esters but is not specific for this substrate.) In contrast to the above findings, Franko and Heath (1979) found low-MW SOP compounds that were hydrolyzable by alkaline phosphatase in two eutrophic Ohio lakes. Alkaline phosphatase (APase) hydrolyzes phosphomonoesters to release ortho P. Many other investigators have found evidence for both APase and APase-hydrolyzable SOP in lakes (e.g., Jansson et al. 1988; Cotner and Wetzel, 1991; Newman et al., 1994). In general, synthesis of APase is induced in algae and bacteria in response to low orthophosphate concentrations. Some of the enzyme activity in lakes is associated with particles (i.e., cells, including cell surfaces), but activity also is found in the water itself, indicating that cells excrete the enzyme into their surrounding growth medium (Jansson et al. 1988). Newman and Reddy (1992) found that APase in sediments is associated mainly with sediment particles rather than the interstitial water. APase activity was highest in the top few cm of sediment in Lake Apopka, and sediment resuspension led to short-term increases in APase activity in the water column. Sediment redox status affected APase activity, with significantly lower activity occurring under anoxic conditions in Lake Apopka sediments (Newman and Reddy, 1993). However, short-term depletion of dissolved oxygen (few hours) did not affect APase activity in the water column.

A suite of phosphate-hydrolyzing enzymes was used to characterize SOP from eight locations in Lake Okeechobee (Brezonik et al., 1983b). Half the locations, all in littoral areas, showed small increases in SRP (~1 to 7 μg/L) when incubated with APase. In contrast, none of the sites showed evidence of SOP hydrolyzable by phosphodiesterase, but small, positive responses were found when samples were concentrated ten-fold. Phytase induced the release of SRP at two littoral stations. Only 5 to 10% of the SOP was hydrolyzed in standard 24-h incubations, but much longer incubations (14 days) resulted in as much as 90% hydrolysis in a few cases.

The particulate phosphorus (PP) fraction in lakes is even more heterogeneous than the SOP fraction. PP may include inorganic forms, e.g phosphate sorbed onto clay suspended in the water and suspended precipitates of phosphate with such cations as Ca, Al, and Fe. Inorganic PP probably is of minor importance in most Florida lakes, however; the PP fraction is comprised mostly of organic forms: detrital P from dead and decomposing cells, bacteria (free-living and attached to detritus), phytoplankton, and various zooplankton classes (from protozoa to crustaceans). The phytopankton and zooplankton categories are themselves highly diverse in size and complexity of the individual cells or organisms. Of course, if one is interested in studying P dynamics on a lake ecosystem basis, uptake, storage, and release by benthos, much larger organisms (fish, macrophytes), and sediments also must be measured.

Detailed discussion of P forms in lake sediments is beyond the scope of this chapter, but it is pertinent to note that a wide variety of mineral and organic forms of varying bioavailability are found in lake sediments, including phosphate loosely associated with ion exchange sites, sorbed onto or occluded in iron and aluminum hydroxides, and bound in Ca minerals such as apatite $Ca_5(PO_4)_3X$, where X = F or OH). The amounts of P in these categories within sediments usually are determined by sequential extraction methods adapted from the field of soil chemistry (e.g., Williams et al. 1971; Pettersson et al. 1988). Information on the chemical forms and distribution of phosphorus in sediments is needed to understand and predict the magnitude and mechanisms of P cycling between sediments and the overlying water column; such information has been the focus of several recent studies on Lakes Apopka and Okeechobee (e.g., Olila et al., 1995; Brezonik and Engstrom, 1998).

Phosphorus Turnover: Isotope Tracer Approaches

Isotope tracer studies on aquatic systems dating back to the 1940s for P and 1960s for N demonstrate that nutrient forms cycle much more rapidly than implied by seasonal variations in concentrations. Concentrations present in a given compartment at any time thus are a balance between dynamic source and sink processes. Hutchinson and Bowen (1947) described an early effort to study P dynamics at the lake ecosystem level by adding 10 mCi of ^{32}P to Linsley Pond, a small lake in Connecticut. Hayes and Phillips (1958) used ^{32}P in microcosms and estimated the following turn over times for components of lake systems: ~1 week for total P in the water column, 0.3 days for phytoplankton and bacterial cells, and 1 day for zooplankton. Prepas (1983) reported that orthophosphate turnover times in Canadian lakes were highly variable but generally rapid, ranging from 17 min to 17 h in spring and 2 min to 36 h in summer. Short turnover times for ortho P (i.e., $<$ 1h) are considered to indicate that P is limiting for planktonic growth (Lean et al., 1983).

Isotope tracer studies have been used only to a limited extent to quantify the dynamics of P cycling in Florida lakes, and definitive information on cycling rates is not available. However, the warm temperatures that occur for most of the year should promote rapid microbial kinetics and P cycling, and it is reasonable to assume that turnover of P pools in Florida lakes is rapid. Ogburn (1984) used ^{32}P to study inorganic P uptake and release by seston (plankton) in pelagic enclosures at McCloud Lake. Enclosures were maintained at pH values of 4.6 [ambient], 4.1 and 3.7, but no consistent trends attributable to pH were found. A wide range was found for uptake rates (0.6 to 12 g L^{-1} h^{-1}; median = 3.3 g L^{-1} h^{-1}) and turn over times of inorganic P in the lake (0.4 to 10 h; median = 1.2 h) during August and September, 1982. Similar data were obtained on P release rates by seston that had been labeled with ^{32}P in preliminary incubations with ^{32}P-enriched orthophosphate, separated from the solution containing ^{32}P, and resuspended in ^{32}P-free enclosure water. Rates of P release by seston during the same period as the uptake measurements ranged from 0.5 to 10.2 g L^{-1} h^{-1} (median = 1.7 g L^{-1} h^{-1}). Calculated turnover times for seston P were quite short, ranging from 0.5 to 14 h (median = 2.5 h).

The development of more detailed information on P cycling dynamics in aquatic (planktonic) food webs has been hindered until recently by several factors, including: (i) lack of effective ways to separate seston into ecologically meaningful compartments (beyond a few simple size classes) and (ii) inadequate models and tools by which to analyze tracer kinetic data. This is an active area of research, however, and several detailed studies on Scandinavian lakes have been published recently on this topic (e.g, Lyche et al., 1996; Vadstein et al., 1995; Salonen et al., 1994).

In summary, despite the efforts made over the past few decades to understand the nature and cycling dynamics of key P pools in Florida lakes, our knowledge is still meager. More efforts need to be made to apply "state-of-the-science" analytical methods, isotope-tracer procedures, and advanced modeling techniques to these questions in various types of Florida lakes.

3.5 ROLE OF BOTTOM SEDIMENTS IN THE PHOSPHORUS CYCLE OF LAKES

Limnologists have recognized for many decades that bottom sediments play two crucial roles in the processing of substances in lakes. They act as: 1. permanent repositories (or net sinks) for materials, recording the history of past lake and watershed conditions in sediment strata, 2. reservoirs of substances (e.g., nutrients, metals, pollutants) available for recycling into the water column. Several studies are described below that exemplify advances made in understanding both aspects of P biogeochemistry in Florida lake sediments over the past two decades.

3.5.1 Sediments as Repositories: Historical Phosphorus Loading in Lake Okeechobee

Concern has been expressed for close to 30 years about the eutrophication of Lake Okeechobee and the possibility that its ecological quality is declining. Many studies have been undertaken to measure trends in the lake's water quality and nutrient inputs (e.g., see Kratzer and Brezonik, 1984; Janus et al., 1990; Aumen, 1995; James et al., 1995ab), but important issues remain unresolved, including the role of external P loading in controlling trophic state conditions. Although there is disagreement about the efficacy of nutrient control for a lake some believe to be naturally eutrophic, the South Florida Water Management District (SFWMD), which is responsible for managing the lake, targeted a 40% reduction in external P inputs to reverse a perceived long-term decline in water quality. A key question in this debate concerns the extent to which P loading to the lake has changed in the approximately 100 years since European settlement and development of the lake's drainage basin began.

Brezonik and Engstrom (1998) used a paleolimnological approach (i.e., sediment stratigraphic analyses) to answer this question by examining trends in P accumulation in dated sediment cores and quantifying whole-lake P accumulation rates as a proxy for external nutrient inputs. They used this approach because water quality monitoring data do not extend back to predisturbance conditions (or even to the period before 1969). The short monitoring record prevents scientists and managers from knowing the extent to which the lake has been changed by drainage basin disturbances. Without such knowledge, there is little scientific basis for selecting a target for loading reduction (Brenner et al., 1993). For example, if the lake received high P loadings under predisturbance conditions, as suggested by Canfield and Hoyer (1988), it would be unrealistic to establish oligotrophic or mesotrophic loading goals.

It was not certain at the outset of the paleolimnological study whether such methods could be used on Lake Okeechobee because the lake is so shallow and has a large wind fetch (> 50 km). Resuspension of sediments during major events such as hurricanes or even by summer squalls and weather fronts was a serious concern. If sediment resuspension occurred to deep enough layers, it would blur the deposition record to the extent that strata could not be dated and temporal trends could not be inferred. P accumulation rates in depositional zone sediments of the lake were measured on 11 mud-zone cores and two peat-zone core dated by ^{210}Pb. Although difficulties were encountered in interpreting ^{210}Pb data from some sites, reliable dating of sediments from the mud zone of the lake was found to be feasible.

Sediment accumulation rates in the mud zone increased during the present century by an average of about twofold (Fig. 3.10); moreover, accumulation of organic sediments in the lake during presettlement time apparently was much slower than during the past century. Concentrations of all forms of sedimentary P, but especially nonapatite inorganic-P and organic-P (which are potentially more bioavailable than apatite-P), also increased on average by a factor of two since presettlement times, and most of the increase occurred after 1940. Annual P accumulation rates in the lake's sediments thus increased about fourfold during the 1900s (Fig. 3.10), with most of the increase occurring in the past 40 to 50 years. Recent accumulation rate of sedimentary P in the lake (~850 mg m^{-2} yr^{-1}, roughly during the 1980s) agrees within a factor of 1.5 with the lake net P retention (570 mg P m^{-2} yr^{-1}; 14-year average, assuming uniform deposition only in the lake's mud zone) calculated from published input-output mass balances (Janus et al., 1990). The difference between these two estimates could reflect uncertainties in the input-output budgets, inaccuracies in the whole lake accumulation rates calculated from the sediment cores, or both.

Overall, this study demonstrated the applicability of paleolimnological approaches to studying nutrient-related questions relevant to lake management. It is

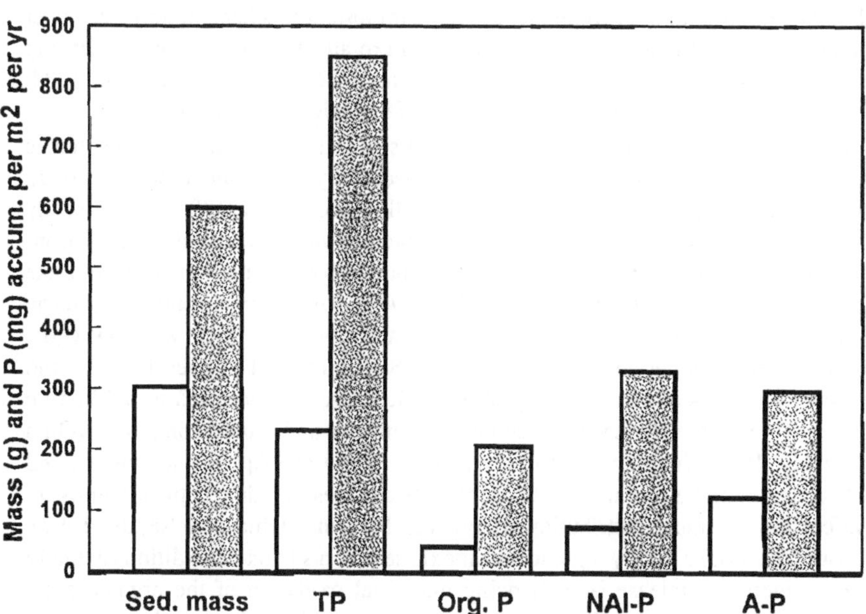

FIGURE 3.10 Mean accumulation rates of sediment, TP, and three forms of sedimentary P (organic P, nonapatite inorganic P, apatite P) in mud zone of Lake Okeechobee for two time periods: before1910 (open bars) and 1980s (shaded bars) (data from Brezonik and Engstrom, 1998).

clear from this work that sediment dating is possible even for rather shallow lakes (which are common in Florida) and that direct measurement of P accumulation in lake sediments by stratigraphic methods complements indirect calculations based on lake mass balance approaches. For Lake Okeechobee, the findings verify that human impacts on the lake have been substantial, that recent trends are a continuation of past deterioration, and that the SFWMD's reduction goals are modest relative to the natural state of the lake.

3.5.2 Sediments as Temporary Reservoirs for Phosphorus: Role of Internal Loading in Maintaining Eutrophic Conditions

Introduction

Recycling of P from sediments to the water column of lakes can occur by at least five mechanisms: diffusion of dissolved P across the sediment-water interface; redox-driven dissolution/ desorption; wind-and-current-driven sediment resuspension, bioturbation by benthic invertebrates and fish; and plant uptake with subsequent excretion into the water column or release into the water column when the plants die. The quantitative importance of these mechanisms varies widely among lakes, depending on physical, chemical and biological conditions in the water column and sediments. Even within a given lake, the most important mechanism may vary over time and location. Some attention has been paid to all of these mechanisms during recent decades (e.g., Davis et al., 1975; Lerman, 1978; van der Loeff et al., 1984; Boström et al. 1988; Granéli and Solander, 1988; Andersson et al. 1988).

Factors affecting fluxes of P from sediments of Lakes Apopka and Okeechobee have been studied extensively during the past decade by Reddy and coworkers (e.g., Moore et al. 1991; Moore and Reddy, 1994; Olila and Reddy, 1995, 1997; Reddy et al. 1996; Van Rees et al., 1996). Redox potential and pH, which control iron-phosphate interactions, are thought to be important in controlling P fluxes from Lake Okeechobee sediments (Moore and Reddy, 1994). Under oxic conditions, ferric phosphate was found to control P solubility, and porewater SRP was low (~0.1 mg/L). Under reducing conditions, porewater SRP increased to over 1 mg/L, and porewater levels increased markedly at low pH. However, decreases in pH had no effect on P release from Lake Okeechobee sediments under oxic conditions (Olila and Reddy, 1995). In contrast, iron levels are low in Lake Apopka sediments, and its P occurs mainly as organic forms. Moderate changes in redox status are unlikely to affect the chemical distribution of organic P forms (Olila and Reddy, 1997). Exposure of Apopka highly organic and anoxic sediments to oxic conditions actually could accelerate P release by enhancing microbial oxidation of the organic sediments. A diffusive flux of 2.7 mg SRP m^{-2} d^{-1} was found for Lake Apopka sediments in quiescent laboratory incubations (Moore et al., 1991). This is sufficient to increase lake water SRP by about 500 µg/L per year. Of course, conditions in the lake are not quiescent, and the above value probably represents a lower bound on annual P loading from the sediment to the water column.

Although wind-driven sediment resuspension is unlikely to be a major mechanism for P recycling in small, moderately deep lakes, it is potentially an important recycling mechanism in large, shallow lakes, such as Apopka and Okeechobee. Reddy et al. (1996) concluded that P fluxes from sediment resuspension are higher than diffusive flux contributions in Lake Apopka. Resuspended sediments can act as P sources in two ways: 1. mixing of sediment pore water, which may have high P concentrations, into the water column, and 2. desorption of phosphate from sediment surfaces. The former mechanism is quantitatively unimportant unless the depth of sediment that is resuspended is much larger than typically occurs (usually a few mm or at most a few cm in severe wind events). Desorption of phosphate from resuspended sediment presupposes the presence of sorption sites on sediment surfaces to which phosphate is bound. This implies that the binding is reversible and can be described as an equilibrium process.

Phosphorus Sorption Models

Phosphorus sorption by lake sediments (and by soils) commonly is modeled as a reversible binding process by the classic Langmuir isotherm (Table 3.4). This model assumes 1. single-layer coverage by solutes (or sorbate) onto sorbing surfaces (the sorbent) and 2. a uniform binding energy for all sites on the sorbent; i.e., all binding sites are assumed to have the same affinity for sorbate. Pollman (1983) found that the Langmuir model adequately described P sorption by sediments from Lake Apopka and some sediments from Lake Okeechobee (Fig. 3.11), but a sediment with high organic content and another with a high clay-silt content did not fit the model (Table 3.5). These sediments and Lake Apopka sediment yielded the highest release of SRP when the sediments were mixed with phosphate-free water buffered to pH 8.3. Olila and Reddy (1993) also found that P sorption fit a Langmuir isotherm for mud, sand and littoral sediments from Lake Okeechobee, but sorption by peat sediments did not (but did fit the Freundlich equation). Highly organic Lake Apopka sediments showed no sorption tendency at low soluble P concentrations (< 1.5 mg/L) and fit a linear adsorption (partition) coefficient at high SRP levels.

Given the strict assumptions of the Langmuir model relative to the known complexity of natural sorbents, lack of fit is not surprising, and the fact that sorption data often do fit this model may be the greater surprise. Use of the Langmuir model to fit P sediment sorption data should be regarded as a practical, data-fitting procedure rather than a mechanistic approach. Indeed, Detenbeck and Brezonik (1991a) found that fit of P sorption data for sediments from Little Rock Lake to a Langmuir plot and its linearized form, C/X versus C (C = solution-phase P concentration at equilibrium; X = amount of P sorbed per mass of sediment; Fig. 3.12a,b) was only apparent and may represent a "spurious self-correlation" (Kenney, 1983), i.e., a correlation that results from a plot of a variable (C) versus a function of the same variable (X/C) when the common variable (C) has a much larger range than the unique variable (X). A plot of X/C versus X (Fig. 3.12c), which also is linear if the Langmuir model applies, is not linear, and the Langmuir model thus does not fit these data. Better fits were obtained (Fig. 3.12d,e) with more complicated models (Table 3.4) that portray site binding energy as a normal distribution (Perdue and

TABLE 3.4
Equations of Models Used for P Sorption

Langmuir Model

$$X = \frac{X_m KC}{1 + KC}$$

X = amount of P sorbed per mass of sorbent (μg/g)
X_m = sorption capacity for P (μg/g)
K = constant related to binding energy (L/μg)
C = equilibrium P concentration in solution

Assumptions: monolayer coverage, homogeneous surface, uniform binding energy for all sites.

Normal Distribution Model (Perdue and Lytle, 1983)

$$\theta = \frac{X}{X_m} = \frac{1}{\sigma\sqrt{2\pi}}\int_{-\infty}^{\infty}\frac{C10^{\log K}}{1 + C10\log K}\frac{\exp[-0.5(\mu - \log K)^2]}{\sigma^2}d\log K$$

θ = macroscopic binding parameter defined by summing over all i sites.

$$\theta = \frac{X}{X_m} = \frac{\sum X_i}{\sum X_{mi}} = \frac{X_{mi}K_iC}{1 + K_iC}$$

Assumption: binding sites have continuous, normal distribution of binding energies characterized by mean logK = μ and standard deviation = σ; binding constant K (relating to binding energy) is normally distributed in "logK space."

Barrow's (1983) Electrostatic Model

$$\theta = \frac{1}{\sigma\sqrt{2\pi}}\int_{\Psi_{a0}-5\sigma}^{\Psi_{a0}+5\sigma}\frac{(\alpha_i\gamma C)K_i\exp(-z_iF\Psi_a/RT)}{1 + (\alpha_i\gamma C)K_i\exp(-z_iF\Psi_a/RT)}\exp[(\overline{\Psi}_{a0} - \Psi_{a0})/\sigma]^2 d\Psi_{a0}$$

ψ_a = electrostatic potential (V) in plane of sorption; ψ_{a0} = initial ψ_a (in absence of sorbed P; ψ_{a0} = mean electrostatic potential (V); σ = standard deviation of ψ_{a0}; α_2 = ionization fraction, γ = activity coefficient, and z_i = charge on HPO_4^{2-}; K_i = intrinsic binding constant (M^{-1}); F = Faraday's constant, R = gas constant; T = temperature (K).

Predicts binding of ions to charged surfaces. Assumption: homogeneous surface (same intrinsic binding energy for all sites); actual binding energy for site depends on net surface charge, which depends on extent of binding of ions.

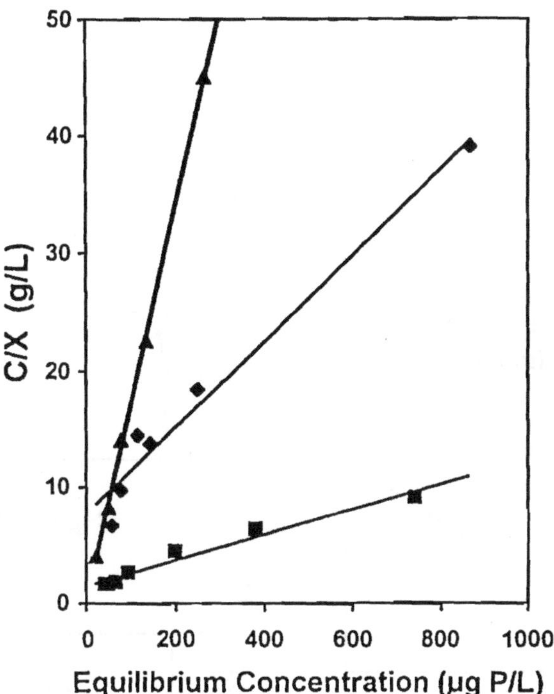

FIGURE 3.11 Langmuir plots for P sorption by two organic-rich mud-zone sediments: (diamonds) station O 10 [north end of lake; volatile solids (VS) = 12.7%] and (squares) station O 11 [center of lake; VS = 17.2%]; and (triangles) a silty-sandy sediment O 13 [NW of Observation Shoal] from Lake Okeechobee (redrawn from Pollman, 1983).

TABLE 3.5
Summary of Langmuir Sorption Parameters for Lake Okeechobee and Lake Apopka Sediments[a]

Sediment sample[b]	NAI-P[c] µg/g	Γ_m µg P/g $10^3 \times$ L/µg	b	r[d]	p[d]
O 10	12	27.4	4.30	0.982	<0.01
O 11	35	91.0	7.50	0.982	<0.01
O 13	6	6.5	20.80	0.970	<0.01
O 14	96	220.0	1.74	0.719	NS
A 5	108	108.0	34.60	0.999	<0.01

[a] From Pollman, 1983.

[b] Data not included for one other Okeechobee site (O 7) because of poor fit to model.

[c] NAI-P = nonapatite inorganic P content of sediment sample, a measure of the maximum P initially sorbed to the sediment.

[d] r = correlation coefficient for fit to Langmuir model, p = significance level.

Lytle, 1983) or account for electrostatic interactions on the charged sediment surfaces (Barow, 1983).

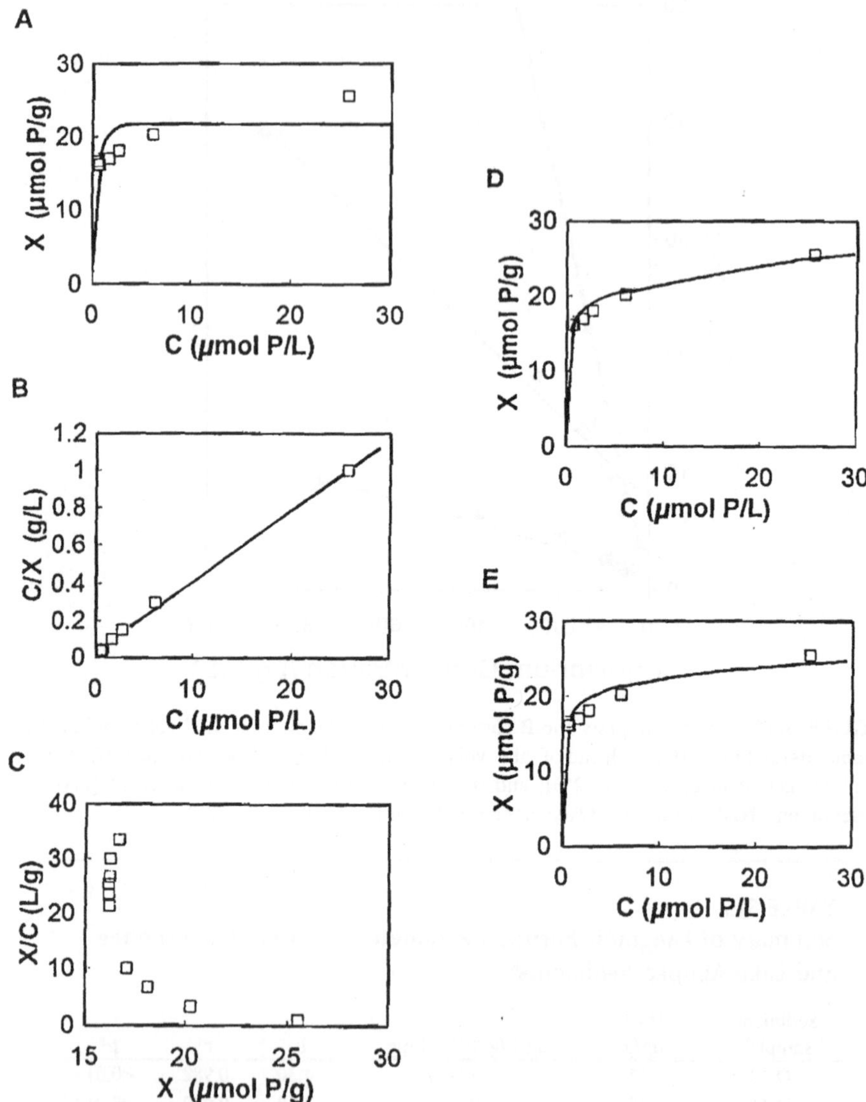

FIGURE 3.12 Equilibrium sorption data for sediment from Little Rock Lake, WI at pH 4.5 plotted according to (A) Langmuir model, (B) linearized transformation of Langmuir model, (C) alternative linearization of Langmuir model, (D) Perdue-Lytle normal distribution model, and (E) Barrow's electrostatic model (redrawn from Detenbeck and Brezonik, 1991a).

EPC

Regarding the role of sediment resuspension in recycling P to the water column, the first issue that must be addressed is whether the sediment will act as a source of P (i.e., desorb P to the water) or a sink (i.e., sorb P from the water). In a simple way, this question can be answered by comparing water column P concentrations with sediment EPC values. The EPC (equilibrium phosphorus concentration) is the aqueous P concentration at which no net sorption or desorption occurs when a sediment is suspended in a water sample. If P sorption for a sediment follows the Langmuir model, it can be shown that the EPC for the sediment depends on K, the binding constant of the Langmuir model, and on the fraction of binding sites on the sediment occupied by P (Pollman, 1983). EPCs can be determined experimentally from a plot of the measured amount of sorption or desorption versus initial aqueous concentration of SRP (Fig. 3.13). As the summary in Table 3.6 indicates, EPC values for sediments from Lakes Apopka and Okeechobee are higher than SRP values in the lakes' water column. The EPC values for Lake Okeechobee in Table 3.6 are generally higher than the range of 4 to 84 μg/L reported by Olila and Reddy (1993), but at least some of the differences may be attributed to different sampling locations. For example, their lowest values are for littoral sediments that were not sampled by Pollman (1983). The exact value of EPC for Lake Apopka sediment is difficult to determine because the sediment is saturated with phosphate, but in any case, the EPC is much greater than the lake's SRP concentration. Olila and Reddy (1993) reported an EPC of ~3.8 mg/L for Lake Apopka sediments. Clearly, these sediments serve as P sources when they are resuspended in lake water.

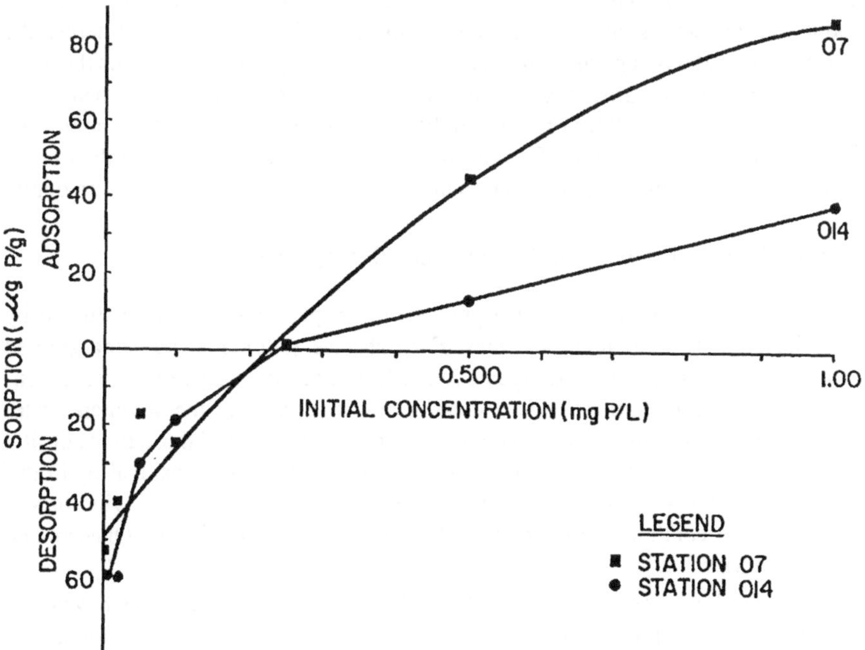

FIGURE 3.13 SRP sorption and desorption as function of initial SRP concentration for two sediments from Lake Okeechobee. EPC values (SRP concentrations at no net sorption): 225 μg/L for site O 7 (highly organic mud-zone site in NE end of lake; VS = 35.1%) and 235 g/L for site O 14 (highly organic site in South Bay; VS = 42.7%) (from Pollman, 1983).

TABLE 3.6
Measured and Calculated Equilibrium Phosphorus Concentrations (EPC) at pH 8.3 for Sediments from Lakes Okeechobee and Apopka[a]

		EPC	
Station	Φ_0[b]	Measured, μg/L	Calculated,[c] μg/L
O 07	—	225	—
O 10	0.438	217	181
O 11	0.385	80	83
O 13	0.923	520	577
O 14	0.420	230	417
A 05	0.996	470–2000	7200

[a] From Pollman, 1983.
[b] Initial fractional surface coverage by Pi.
[c] From Languir model.

Typical dissolved inorganic phosphorus profiles in sediment porewater show increasing concentrations below the sediment-water interface, grading from concentrations at the interface equivalent to or approximating the water column to a subsurface maximum (Berner, 1980). This gradient reflects the balance between production (decomposition of accreting organic P) and losses (transport across the interface coupled with upward diffusion). If we assume equilibrium prevails throughout the depth of the surficial sediments, EPC will increase with depth below the interface as well. The effect of resuspending surficial sediments is thus nonlinear; *ceteris paribus,* resuspension of 2 mm of sediment should release more than twice as much P than 1 mm sediment because the weighted EPC will increase with depth.

Internal Phosphorus Loading by Sediment Resuspension

The physical forces required to resuspend sediment particles have been studied by hydraulic engineers for many decades. For cohesive sediments (i.e., fine-grained sediments containing clays and organic matter), it generally is agreed that resuspension does not occur until a critical shear stress is reached at the sediment-water interface. Hydrodynamic equations relate this stress to wind speed through the orbital wave action produced by wind stress. A complicated, three-dimensional hydrodynamic model that evaluates sediment resuspension as a function of wind-driven turbulence in a continuous-time domain has been developed for Lake Okeechobee (Sheng, 1998).

A simpler but nonetheless physically realistic model (Fig. 3.14) was developed a decade earlier (Pollman, 1983; Brezonik et al., 1983b) to predict the amount of sediment resuspension as a function of wind speed, fetch, and sediment characteristics. The model was coupled with sorption-desorption equilibria to assess net effects of resuspension on water column SRP levels. Model runs over a range of wind speeds (Table 3.7) indicated that sediment resuspension should have relatively small effects on water column SRP in Lake Okeechobee. In contrast, the model

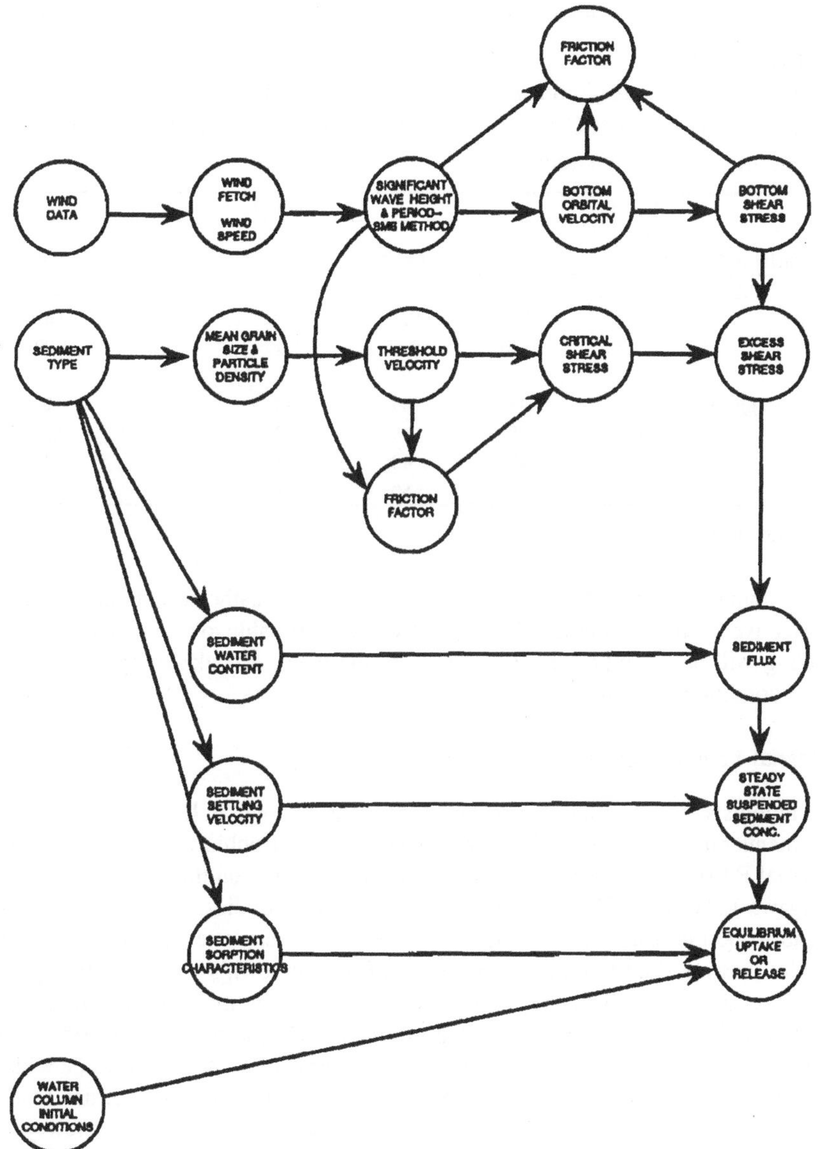

FIGURE 3.14 Schematic of sediment resuspension-nutrient release model (redrawn from Pollman, 1983).

predicted an SRP increase in Lake Apopka of 60 to 70% at wind speeds of only 15 mph (6.7 m/s) and a near doubling at 25 mph (11 m/s). The large increases predicted for Lake Apopka reflect the very shallow nature of the lake (mean depth < 2 m), highly flocculent character (and low density) of the sediments, and fact that the sediments essentially are saturated with P. Wind speeds in the range of 15 to 25 mph

TABLE 3.7
Predicted Changes in Suspended Solids and Soluble Reactive Phosphate at Stations in Lakes Okeechobee and Apopka Caused by Wind-Induced Sediment Resuspension[a]

Variable	Wind speed, m/s	Station				
		O 10	O 11	O 13	A 5	A 8
SS (g/L)	6.7	0.38	—	0.13	0.63	0.76
	8.9	0.59	0.31	0.90	0.76	0.87
	11.2	0.76	0.45	1.64	0.92	1.00
SRP (μg/L)	0	45	45	45	50	50
	6.7	48	45	47	81	85
	8.9	49	48	48	86	90
	11.2	57	50	48	93	96

[a] From Pollman, 1983.

occur commonly over the lake as a result of frontal weather events in winter and convective storms in summer. Consequently, it is reasonable to conclude that internal loading will be an important source of P to the water of Lake Apopka so long as these conditions remain in effect.

Internal Loading and Lake Recovery

Limnologists first recognized in the 1970s that internal P loading from bottom sediments could delay the recovery of lakes from eutrophic conditions after their external P loadings have been controlled. A classic study on this topic was done on Shagawa Lake, near Ely, Minnesota. After extensive studies, internal loading was found to be the cause for a delay in the lake's recovery after a tertiary treatment system was installed to remove phosphorus from the town's waste water (e.g., see Larsen et al., 1981). Subsequent studies with a dynamic mass balance model that accounted for P loading from enriched sediments (Chapra and Canale, 1991) indicated complete recovery of the lake would take many decades. In contrast, a widely-used simpler model that does not include an explicit term for internal loading (Vollenweider, 1975) predicted recovery would occur in a matter of a few years. This prediction is not supported by monitoring data from the 1980s.

The importance of internal P loading in affecting and sometimes controlling lake recovery rates now is widely accepted by limnologists, and guidelines for conducting diagnostic studies on eutrophic lakes to design restoration and management programs recommend that the importance of this process be measured. If internal loading is found to be significant, management practices (such as in-lake alum treatment or sediment dredging) typically are included in the restoration plan.

A recent review of the topic (Welch and Cooke, 1995) concluded that recovery is delayed by internal loading to a greater extent in large, shallow, unstratified lakes than in deeper lakes that stratify and develop anoxic bottom waters, even though it is widely accepted that P release from sediments is more rapid under anoxic conditions. At least two factors help to explain this apparent contradiction. First, the volume of water available to dilute the internal load is smaller (per unit area of sediment) in shallow lakes than in deep lakes. Second, in stratified lakes a thermal barrier (the metalimnion) separates the internally-loaded hypolimnion from the euphotic zone in the epilimnion during the critical summer growth period, but in unstratified lakes, P is released directly into the zone where algae are growing.

A long-term study was conducted on 10 shallow lakes in western Europe to evaluate the importance of internal loading in delaying their recovery after external loads to the lakes were reduced (Sas, 1989). Data from this study (Table 3.8) indicate that recovery can be delayed for many years in such lakes. In fact, only two lakes exhibited recovery (in the sense of having a much lower internal P load) within the 10-year study period: Schlachtensee (Germany) and Veluwemeer (Netherlands). Special circumstances apply to both lakes, and neither represents a case of rapid recovery in a shallow, unstratified lake as a result simply of decreasing its external nutrient loading rate. Schlachtensee has a small surface area (< 40 ha) and is relatively deep (mean depth of 4.6 m) for its size. As a result, it stratifies thermally and does not belong in the category of lakes being discussed here. Veluwemeer also received other restoration measures.

TABLE 3.8
Trends in Internal P Loading in Shallow Lakes Following External Load Reductions[a]

Lake	Mean depth, m	% load reduction	Net release		Years after
			Before	After	
Søbygaard (DK)	1.0	83	21.1	52.5 (−13)	4(8)
Hylke Sø (DK)	7.4	40	15.5	19.7	5
Norrviken (SK)	5.4	70	6.7	6.5	10
Neagh (GB)	8.9	10	4.4	4.4	12
Glum Sø (DK)	1.8	70	36.7	20.2	5
Trummen (SK)	2.0	80	20.0	20.0	10
Alderfen Broad (GB)	0.6	96	4.7	3.5	7
Vallentuna (SK)	2.7	90	18.0	9.0	10
Schlachtensee (G)	4.6	90	17.2	1.9	4
Veluwemeer (ND)	1.3	52	3.9	0.4	2

[a] Summarized from Welch and Cooke, 1995.

The above findings have important implications for the restoration of shallow eutrophic lakes in Florida (and elsewhere) and also have important implications regarding the use of P mass balance models to predict recovery rates of shallow lakes. The conventional Vollenweider (1975) model predicts in-lake P as a simple function of external loading, hydraulic loss through the lake's outlet, and net sedimentation,

$$d[P]_l/dt = L_{ext}\frac{A}{V} - \frac{[P]_l}{V}(Q_{out} + \sigma_p V) \tag{1}$$

where $[P]_l$ is the in-lake P concentration (g m^{-3}); L_{ext} is the total external P loading rate on a lake-area basis (g m^{-2} yr^{-1}), A is lake area (m^2), V is lake volume (m^3), Q_{out} is the hydraulic outflow (m^3 yr^{-1}), and σ_P is the first-order P sedimentation coefficient (yr^{-1}). σ_P represents the net of two opposing processes: sedimentation by all mechanisms, including gravitational settling of algae and detrital particles, and internal loading (release from sediments) by all the processes described earlier. Neither of these processes is likely to remain constant if external loading is changed significantly. Consequently, this model is not appropriate to predict recovery of lakes after their external loads are reduced.

A better model would decompose σ_P into two terms, thus treating internal loading explicitly:

$$\sigma = \frac{v_P}{z} - \frac{L_{int}A}{[P]_l V} \tag{2}$$

where v_P is the gross sedimentation velocity (m yr^{-1}); z is the mean depth of the lake (m); and L_{int} is the rate of internal loading (g m^{-2} yr^{-1}). L_{int} is a function of the sediment characteristics and can be considered initially as simply a linear function of the P content of the active sediment layer. Substituting Eq. 2 into Eq. 1 now yields a model that defines sedimentation and sediment release of P as distinct processes:

$$(d[P]_l/dt = (L_{ext} + L_{int}))\frac{A}{V} - P_l\frac{(Q_{out} + v_p A)}{V} \tag{3}$$

Because internal loading is incorporated explicitly in Eq. 3, different time-dependent behavior is predicted by the two models when they are used to simulate the temporal dynamics of recovery of perturbed systems. We illustrate this different behavior for Lake Apopka (Fig. 3.15). Dramatically different rates of recovery are obtained depending on which model is used. The Vollenweider model predicts rapid recovery: TP declines from a starting concentration of 236 μg/L to < 24 μg/L within about three years in response to a decrease in external P loading by a factor of ten. In contrast, the internal loading model shows that the lake continues to have high TP concentrations (> 100 μg/L) more than 40 years after external loading is decreased by the same factor. On the basis of this exercise, it appears that internal P loading, which is widely recognized as an important process in shallow, hypereutrophic lakes, may cause a substantial delay in the recovery of water quality in Lake Apopka.

It should be noted that Lake Apopka and also Lake Okeechobee represent more extreme conditions of shallowness, as quantified by the Osgood Index, than any of the lakes listed in Table 3.8. This index, the ratio of a lake's mean depth (in m) to the square root of its surface area (in km^2), is related to the likelihood that the lake will develop thermal stratification and is an inverse measure of the extent of its water column that is well mixed (Osgood, 1988; Welch and Cooke, 1995). The index also

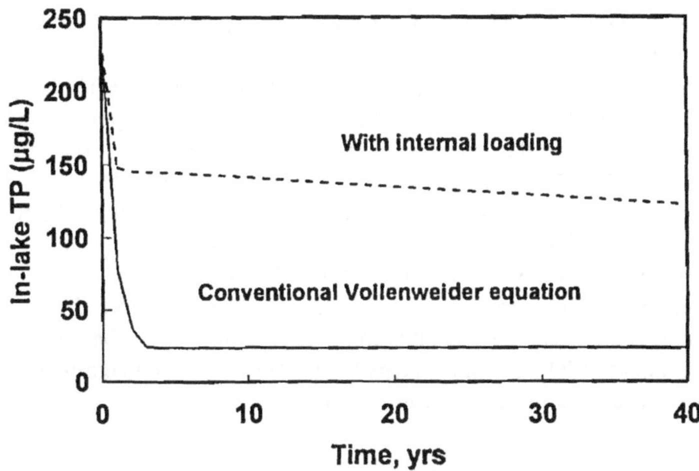

FIGURE 3.15 Predicted response of lake water TP concentrations in Lake Apopka to a step decrease of 90% in external P loading L_{ext} from 0.152 mg P m^{-2} da^{-1} to 0.0152 mg P m^{-2} da^{-1} at t = 0 using the conventional Vollenweider mixed-reactor model (Eq. 1) and a similar model that explicitly considers internal P loading (Eq. 3). Values of model terms and parameters: A = 1.21×10^8 m^2; V = 2.125×10^8 m^3; Q_{out} = 7.051×10^7 m^3 yr^{-1}; t_w = 3.01 yr; σ_P = 1.054 yr^{-1} (based on current loadings, current $[P]_l$ = 236 µg/L, and assumed steady-state); v_P = 7.98 m/yr (computed from assumed P internal loading rate); and assumed initial internal loading rate = 4.0 mg P m^{-2} da^{-1}. Internal loading model predicts $[P]_l$ = 130 µg/L at 40 yr. If assumed initial internal loading rate is 2.0 mg P m^{-2} da^{-1}, predicted $[P]_l$ at t = 40 yr is 122 µg/L. If initial internal loading is 1.0 mg P m^{-2} da^{-1}, predicted $[P]_l$ at t = 40 yr is 107 µg/L.

is likely to be inversely correlated with the influence a lake's sediment has on water column conditions. Low values indicate shallow conditions and/or large surface areas, both of which enhance the influence of wind on resuspension of fine-grained or flocculent (organic) sediments. Values of the index for Lakes Apopka and Okeechobee, 0.16 and 0.06 respectively, are lower than the values for any of the lakes in Table 3.8.

It must be emphasized that the fact that internal P loading is important in a lake does not mean that a decrease in external P loading has no effect on P levels in the lake. For lakes that have nutrient cycles strongly influenced by internal loading, response to external perturbations will follow two stages: an initial, short-term response governed by the apparent P residence time in the water column, and a second, much slower response governed by the P residence time in the surficial sediment layer in contact with the water column. The latter residence time is a function of 1. the magnitude of the available sediment pool (both in terms of lability of the P and depth of interaction), and 2. the rate at which sediments accrete and ultimately are lost to the system via deep burial. Ultimately, the sediment pool will reequilibrate to progressively lower rates of gross sedimentation and changing sedimentary P characteristics as the lake trophic state improves, and L_{int} will decline accordingly. Depending upon the relative magnitude of internal loading, the initial response of a lake to decreases in external loading may be relatively large or inconsequential.

The results in Fig. 3.15 predict that Lake Apopka would respond rapidly to a decrease in external loading, but the magnitude of this rapid response would not be sufficient to move the lake from its hypereutrophic state. The initial decline in P (from 236 μg/L to ~150 μg/L) occurs on a time scale of about one year and reflects direct "flushing" effects resulting from decreased external loading. The resulting P concentration is still at a level associated with undesirable levels of chlorophyll and poor water clarity. The results also suggest that the time scale for diminution of the lake's internal P loading is such that decades would pass before the lake's water quality improved substantially, a situation that probably would be unacceptable to the public, unless in-lake treatment steps to directly lessen the magnitude of internal loading were included in the restoration program.

3.6 ACKNOWLEDGMENTS

The senior author has been fortunate to have worked with many bright and energetic graduate students during his career, and many of the accomplishments described in this paper are due to their efforts. Review of the manuscript by Lorin Hatch and his assistance with references is appreciated. Funding for our past studies on P cycling was provided by several Florida agencies, especially the Department of Environmental Protection and South Florida Water Management District, the University of Florida Water Resources Research Center, and federal agencies, especially the U.S. Environmental Protection Agency, whose support is gratefully acknowledged.

REFERENCES

Aumen, N. G. 1995. The history of human impacts, lake management, and limnological research on Lake Okeechobee, Florida (USA). *Archiv. Hydrobiol.* Adv. in Limnol., 45: 1–16.

Baker, L. A., P. L. Brezonik, and C. R. Kratzer. 1981. Nutrient loading-trophic state relationships in Florida lakes. Rept. No. 56, Water Resources Res. Center, Univ. Florida, Gainesville, 126 p.

Barrow, N. J. 1983. A mechanistic model for describing sorption and desorption of phosphate by soil. *J. Soil Sci.* 43: 733–750.

Berner, R. A. 1980. Early diagenesis: a theoretical approach. Princeton University Press, Princeton, NJ.

Boström, B., G. Persson, and B. Broberg. 1988. Bioavailability of different phosphorus forms in freshwater systems. *Hydrobiol.* 170: 133–156.

Boström, B., J. M. Andersen, S. Fleischer, an M. Jansson. 1988. Exchange of phosphorus across the sediment-water interface. *Hydrobiol.* 170: 229–244.

Brenner, M., T. J. Whitmore, M. S. Flannery, and M. W. Binford. 1993. Paleolimnological methods for defining target conditions for lake restoration: Florida case studies. *Lake Reserv. Manag.* 7: 209–217.

Brezonik, P. L. and D. R. Engstrom. 1998. Modern and historic accumulation rates of phosphorus in Lake Okeechobee, Florida. *J. Paleolimnol.* (in press).

Brezonik, P. L. and E. E. Shannon. 1971. Trophic states of Florida lakes. Univ. of Florida, Water Resources Research Center, Gainesville, Publ. 13, 105 p.

Brezonik, P. L., E. C. Blancher, II, V. B. Myers, C. Hilty, M. Leslie, C. R. Kratzer, G. D. Marbury, B. Snyder, T. L. Crisman, and J. J. Messer. 1979. Factors affecting primary production in Lake Okeechobee, Florida. Rept. 07-79-01, Dept. Envir. Eng. Sci., Univ. of Florida, Gainesville, 300 p.

Brezonik, P. L., T. L. Crisman, and S. Preston. 1981. Limnological studies on Lake Apopka and the Ocklawaha lakes. 4. Water quality in 1980 and a summary of four-year trends. Rept. to Florida Dept. Environ. Regulation, Tallahassee. Rept. 07-81-01, Dept. Environ. Eng. Sci., Univ. of Florida, Gainesville.

Brezonik, P. L., T. L. Crisman, and R. L. Schulze. 1984. Planktonic communities in Florida softwater lakes of varying pH. *Canad. J. Fish. Aquat. Sci.* 41: 46–56.

Brezonik, P. L., J. G. Eaton, T. M. Frost, P. J. Garrison, T. K. Kratz, C. E. Mach, J. H. McCormick, J. A. Perry, W. A. Rose, C. J. Sampson, B. C. L. Shelley, W. A. Swenson, and K. E. Webster. 1993. Experimental acidification of Little Rock Lake, Wisconsin: chemical and biological changes over the pH range 6.1 to 4.7. *Canad. J. Fish. Aquat. Sci.* 50: 1101–1121.

Brezonik, P.L., C.D. Hendry and E.S. Edgerton. 1980. Acid rainfall and sulfate deposition in Florida. *Science* 208: 1027–1029.

Brezonik, P. L., C. D. Hendry, E. S. Edgerton, R. Schulze, and T. L. Crisman. 1983a. Acidity, nutrients, and minerals in atmospheric precipitation over Florida: Deposition patterns, mechanisms and ecological effects. EPA-600/S3-83-004 U.S. EPA. Corvallis, OR. (NTIS document PB 83-165 837).

Brezonik, P. L., W. C. Huber, C. D. Pollman, M. Blosser, S. Zoltewicz, C. Miles, and L. Lane. 1983b. Evaluation and modeling of internal nutrient cycling processes in Lake Okeechobee, Florida. Rept. to SFWMD. Dept. Environ. Eng. Sci., Univ. of Florida, Gainesville, 327 p.

Broberg, O. and G. Persson. 1988. Particulate and dissolved phosphorus forms in freshwater: composition and analysis. Hydrobiol. 170: 61–90.

Broberg, O. and K. Pettersson. 1988. Analytical determination of orthophosphate in water. Hydrobiol. 170: 45–60.

Canfield, D. E. and R. W. Bachmann. 1981. Prediction of total phosphorus concentrations, chlorophyll a, and Secchi depths in natural and artificial lakes. Can. J. Fish. Aquat. Sci. 38: 414–423.

Canfield, D. E. and M. V. Hoyer, 1988. The eutrophication of Lake Okeechobee. Lake Reserv. Manag. 4: 91–99.

Carlson, R. E. 1977.A trophic state index for lakes. Limnol. Oceanogr. 22: 361–369.

Chapra, S. C. and R. P. Canale. 1991. Long-term phenomenological model of phosphorus and oxygen for stratified lakes. Water Research 25: 707–715.

Cole, J. J., N. F. Caraco, and G. E. Likens. 1990. Short-range atmospheric transport: a significant source of phosphorus to an oligotrophic lake. Limnol. Oceanogr. 35: 1230–1237.

Cotner, J. B. and R. G. Wetzel. 1991. Bacterial phosphatases from different habitats in a small, hardwater lake, pp. 187–205 in R. J. Chróst [Ed.], Microbial enzymes in aquatic environments. Springer-Verlag, New York.

Davis, R. B., D. L. Thurlow, and F. E. Brewster. 1975. Effects of burrowing tubificids on the exchange of phosphorus between lake sediment and overlying water. Verh. int. Ver. Limnol. 19: 382–394.

Detenbeck, N. E. and P. L. Brezonik. 1991a. Phosphorus sorption by lake sediments. 1. Comparison of equilibrium models. Environ. Sci. Technol. 25: 395–403.

Detenbeck, N. E. and P. L. Brezonik. 1991b. Phosphorus sorption by lake sediments. 2. Effects of pH and other solution variables. Environ. Sci. Technol. 25: 403–409.

Dillon, P. J. and W. B. Kirchner. 1981. The effects of geology and land use on the export of phosphorus from watersheds. Water Research 9: 135–148.

Eisenreich, S. J. and D. E. Armstrong. 1977. Chromatographic investigations of inositol phosphate esters in lake waters. Environ. Sci. Technol. 11: 497–501.

Franko, D. A. and R. T. Heath. 1979. Functionally distinct classes of complex phosphorus compounds in lake water. Limnol. Oceanogr. 24: 463–473.

Grahn, O., H. Hultberg, and L. Landner. 1974. Oligotrophication: a self accelerating process in lakes subjected to excessive supply of acid substances. Ambio 3: 93–94.

Granéli, W. and D. Solander. 1988. Influence of aquatic macrophytes on phosphorus cycling in lakes. Hydrobiol. 170: 245–266.

Hansen, P. S., E. J. Phlips, and F. J. Aldridge. 1997. The effects of sediment resuspension on phosphorus available for algal growth in a shallow subtropical lake, Lake Okeechobee. Lake Reserv. Manag. 13: 154–159.

Hayes, F. R. and J. E. Phillips. 1958. Lake water and sediment, IV. Radiophosphorus equilibrium with mud, plants, and bacteria under oxidized and reduced conditions. Limnol. Oceanogr. 3: 459–475.

Hendry, C. D., E. S. Edgerton, and P. L. Brezonik. 1981. Atmospheric deposition of nitrogen and phosphorus in Florida, pp. 199–206 in: Atmospheric Pollutants in Natural Waters, S.J. Eisenreich (Ed.), Ann Arbor Sci. Publ., Ann Arbor, MI.

Hendry, C. D. and P. L. Brezonik. 1984. Chemical composition of softwater Florida lakes and their sensitivity to acid precipitation. Water Resources Bull. 20: 75–86.

Herbes, S. E., H. E. Allen, and K. H. Mancy. 1975. Enzymatic characterization of soluble organic phosphorus in lake water. Science 187: 432–434.

Hutchinson, G. E. and V. T. Bowen. 1947. A direct demonstration of the phosphorus cycle in a small lake. Proc. Nat. Acad. Sci. USA 33: 148–153.

James, R. T., B. L. Jones, and V. H. Smith. 1995a. Historical trends in the Lake Okeechobee ecosystem. II. Nutrient budgets. Arch. Hydrobiol. 107: 25–47.

James, R. T., V. H. Smith, and B. L. Jones. 1995b. Historical trends in the Lake Okeechobee ecosystem. III. Water quality. Arch. Hydrobiol. 107: 49–65.

Jansson, M., H. Olsson, and K. Pettersson. Phosphatases: origin, characteristics and function in lakes. Hydrobiol. 170: 157–176.

Janus, L. L., D. M. Soballe, and B. L. Jones, 1990. Nutrient budget analyses and phosphorus loading goal for Lake Okeechobee, Florida. Verh. int. Verein. Limnol. 24: 538–546.

Jassby, A. D., C. R. Goldman, and J. E. Reuter. 1995. Long-term change in Lake Tahoe (California-Nevada, U.S.A.) and its relation to atmospheric deposition of algal nutrients. Arch. Hydrobiol. 135: 1–21.

Jassby, A. D., J. E. Reuter, R. P. Axler, C. R. Goldman, and S. H. Hackley. 1994. Atmospheric deposition of nitrogen and phosphorus in the annual nutrient load of Lake Tahoe (California-Nevada). Water Resources Research 30: 2207–2216.

Jones, J. R. and M. F. Knowlton. 1993. Limnology of Missouri reservoirs: an analysis of regional patterns. Lake Reserv. Manag. 8: 17–30.

Kenney, B. C. 1982. Beware of spurious self-correlations! Water Resources Research 18: 1041–1048.

Kiilsgaard, C. W., C. M. Rohm, S. M. Pierson, and J. M. Omernik. 1996. Total phosphorus regions for lakes in the northeastern United States. Lake and Reserv. Manag. NEED PAGES

Kratzer, C. R. and P. L. Brezonik. 1984. Application of nutrient loading models to the analysis of trophic conditions in Lake Okeechobee, Florida. Environ. Manag. 8: 109–120.

Larsen, D. P., D. W. Schultz, and K. W. Malueg. 1981 Summer internal phosphorus supplies in Shagawa Lake, Minnesota. Limnol. Oceanogr. 26: 740–753.

Lean, D. R. S. 1973. Phosphorus dynamics in lake water. Science 179: 678–680.

Lean, D. R. S. and E. White. 1983. Chemical and radiotracer measurements of phosphorus uptake by lake plankton. Canad. J. Fish. Aquat. Sci. 40: 147–155.

Lean, D. R. S., A. P. Abbot, M. N. Charlton, and S. S. Rao. 1983. Seasonal phosphate demand for Lake Erie plankton. J. Great Lakes Res. 9: 83–91.

Lerman, A. 1978. Chemical exchange across sediment-water interface. Ann. Rev. Earth Planet. Sci. 6: 281–303.

Lyche, A., T. Andersen, K. Christofferen, D. O. Hessen, P. H. Berger Hansen, and A. Klysner. 1996. Mesocosm tracer studies. 1. Zooplankton as sources and sinks in the pelagic phosphorus cycle of a mesotrophic lake. Limnol. Oceanogr. 41: 460–474.

McDowell, W. H., D. F. Gatz, J. M. Ondov, C. D. Pollman, and J. W. Winchester. 1997. Atmospheric deposition into south Florida, Advisory panel final report. Findings from the conference on: Atmospheric Deposition into South Florida: Measuring Net Atmospheric Inputs of Nutrients, October 20–22, 1997. South Florida Water Management District, West Palm Beach, FL.

Minear, R. A. 1972. Characterization of naturally occurring dissolved organophosphorus compounds. Environ. Sci. Technol. 6: 921–927.

Moore, P. A., Jr. and K. R. Reddy. 1994. Role of Eh and pH on phosphorus geochemistry in sediments of Lake Okeechobee, Florida. J. Environ. Qual. 23: 955–964.

Moore, P. A., Jr., K. R. Reddy, and D. A. Graetz. 1991. Phosphorus geochemistry in the sediment-water column of a hypereutrophic lake. J. Environ. Qual. 20: 869–875.

National Research Council. 1996. Freshwater Ecosystems: Revitalizing Educational Programs in Limnology. National academy Press, Washington, D.C., 364 p.

Newman, S. and K. R. Reddy. 1992. Sediment resuspension effects on alkaline phosphatase activity. Hydrobiol. 245: 75–86.

Newman, S. and K. R. Reddy. 1993. Alkaline phosphatase activity in the sediment-water column of a hypereutrophic lake. J. Env. Qual. 22: 832–838.

Newman, S., F. J. Aldridge, E. J. Phlips, and K. R. Reddy. 1994. Assessment of phosphorus availability for natural phytoplankton populations from a hypereutrophic lake. Arch. Hydrobiol. 130: 409–427.

Ogburn, R. W., III. 1984. Phosphorus dynamics in an acidic, soft water Florida lake. Ph.D. dissertation, Univ. of Florida, Gainesville, 152 p.

Ogburn, R. W., III, and P. L. Brezonik. 1986. Examination of the oligotrophication hypothesis: phosphorus cycling in an acidic Florida lake. Water Air Soil Poll. 30: 1001–1006.

Ogburn, R. W., III, P. L. Brezonik, and J. J. Delfino. 1987. Effect of pH on phosphorus release during macrophyte (Eleocharis sp.) decomposition. Water Resources Bull. 23: 829–831.

Olila, O. G. and K. R. Reddy. 1993. Phosphorus sorption characteristics of sediments of shallow eutrophic lakes of Florida. Arch. Hydrobiol. 129: 45–65.

Olila, O. G. and K. R. Reddy. 1995. Influence of pH on phosphorus retention in oxidized lake sediments. Soil Sci. Soc. Am. J. 59: 946–959.

Olila, O. G. and K. R. Reddy. 1997. Influence of redox potential on phosphate-uptake by sediments in two subtropical eutrophic lakes. Hydrobiol. 345: 45–57.

Olila, O. G., K. R. Reddy, and W. G. Harris. 1995. Forms and distribution of inorganic phosphorus in sediments of two shallow eutrophic lakes in Florida. Hydrobiol. 302: 47–161.

Omernik, J. M. 1977. Nonpoint-source stream level relationships: a nationwide study. EPA Ecol. Res. Ser., EPA-600/3-77-105, U. S. EPA, Environ. Res. Lab., Corvallis, OR, 151 p.

Omernik, J. M., D. P. Larsen, C. M. Rohm, and S. E. Clark. 1988. Summer total phosphorus in lakes: A map of Minnesota, Wisconsin, and Michigan. Environ. Manag. 12: 815–825.

Osgood, R. A. 1988. Lake mixis and internal phosphorus dynamics. Arch. Hydrobiol. 113: 629–638.

Perdue, E. M. and C. R. Lytle. 1983. Distribution model for binding of protons and metal ions by humic substances. Environ. Sci. Technol. 17: 654–660.

Pettersson, K. B. Boström, and O.-S. Jacobsen. 1988. Phosphorus in sediments—speciation and analysis. Hydrobiol. 170: 91–102.

Pollman, C. D. 1983. Internal loading in shallow lakes. Ph.D. dissertation, Univ. of Florida, Gainesville, 191 p.

Pollman, C. D. 1999. Hydrologic and geochemical controls on acid neutralizing capacity in two acidic seepage lakes in Florida. J. Hydrol. (submitted).

Pollman, C. D. and W. M. Landing. 1999. Atmospheric deposition of phosphorus in Florida: results from the FAMS program. (unpublished results).

Pollman, C. D., T. M. Lee, W. J. Anderson, L. A. Sacks, S. A. Gherini, and R. K. Munson. 1991. Preliminary analysis of the hydrologic and geochemical controls on acid-neutralizing capacity in two acidic seepage lakes in Florida. Water Resources Research 27: 2321–2335.

Prepas, E. E. 1983. Orthophosphate turnover time in shallow productive lakes. Canad. J. Fish. Aquat. Sci. 40: 1412–1418.

Preston, S. D. 1983. Numerical analyses of a eutrophic lake chain. M.S. thesis, Univ. of Florida, Gainesville, 194 p.

Reddy, K. R., M. M. Fisher, and D. Ivanoff. 1996. Resuspension and diffusive flux of nitrogen and phosphorus in a hypereutrophic lake. J. Environ. Qual. 25: 363–371.

Rigler, F. H. 1964. The phosphorus fractions and the turnover time of inorganic phosphorus in different types of lakes. Limnol. Oceanogr. 9: 511–518.

Rigler, F. H. 1968. Further observations inconsistent with the hypothesis that the molybdenum blue method measures orthophosphate in lake water. Limnol. Oceanogr. 13: 7–13.

Salonen, K., R. I. Jones, H. De Haan, and M. James. 1994. Radiotracer study of phosphorus uptake by plankton and redistribution in the water column of a small humic lake. Limnol. Oceanogr. 39: 69–83.

Sampson, C.E., P.L. Brezonik, and E. Weir. 1994. Effects of acidification on chemical composition and chemical cycles in a seepage lake: inferences from a whole-lake experiment, pp. 121–159 in Environmental Chemistry of Lakes and Reservoirs, L.A. Baker (Ed.), Am. Chem. Soc., Washington, D.C.

Sas, H. (Ed.) 1989. Lake restoration by reduction of nutrient loading: expectations, experiences, extrapolations. Academia Verlag Richardz, St. Augustin, Germany, 497 p.

Shannon, E. E. and P. L. Brezonik. 1972. Limnological characteristics of north and central Florida lakes. Limnol. Oceanogr. 17: 97–110.

Sheng, Y. P. 1998. The effects of hydrodynamic processes on phosphorus distribution in aquatic systems. Chapter 16, this book.

Vadstein, O., O. Brekke, T. Andersen, and Y. Olsen. 1995. Estimation of phosphorus release rates from natural zooplankton communities feeding on planktonic algae and bacteria. Limnol. Oceanogr. 40: 250– 262.

Van der Loeff, M. M. R., L. G. Anderson, P. O. J. Hall, A. Iverfeldt, A. B. Josefson, B. Sundby, and S. F. G. Westerlund. 1984. The asphyxiation technique: an approach to distinguishing between molecular diffusion and biologically mediated transport at the sediment-water interface. Limnol. Oceanogr. 29: 675–686.

Van Rees, K. C., K. R. Reddy, and P. S. C. Rao. 1996. Influence of benthic organisms on solute transport in lake sediments. Hydrobiol. 317: 31–40.

Vollenweider, R. A., 1969. Möglichkeiten und Grenzen elementarer Modelle der Stoffbilanz von Seen. Arch. Hydrobiol. 66: 1–36.

Vollenweider, R. A. 1975. Input-output models with special reference to the phosphorus loading concept in limnology. Schweiz. Z. Hydrol. 37: 53–84.

Welch, E. B. and G. D. Cooke. 1995. Internal phosphorus loading in shallow lakes: importance and control. Lake and Reserv. Manag. 11: 273–281.

Williams, J. D. H., J. K. Syers, S. S. Shukla, R. F. Harris, and D. E. Armstrong. 1971. Levels of inorganic and total phosphorus in lake sediments as related to other sediment parameters. Environ. Sci. Technol. 5: 1113–1120.

4 Phosphorus in Florida's Ecosystems: Analysis of Current Issues

K.R. Reddy, E. Lowe, and T. Fontaine

4.1 ABSTRACT

Phosphorus is often one of the major nutrients limiting the productivity of terrestrial, wetland, and aquatic ecosystems. In Florida, anthropogenic P loads from urban and agricultural activities in terrestrial ecosystems have increased the trophic state of associated aquatic systems to the eutrophic or hypereutrophic condition. Because many of these systems are hydrologically linked, the extent of eutrophication has increased through time as impacts have been transferred downstream. Although many of the fate processes within a specific ecosystem have been determined, the transfer of P across landscapes and between ecosystems is poorly understood. In this paper we present some key issues related to the sources, transformations, and transport of P within and across ecosystems, as related to surface water quality. Selected examples from various hydrologic units of Florida will be presented.

Phosphorus is imported into terrestrial ecosystems through fertilizers, organic wastes, wastewater, and animal feeds. Some of this P is exported out of the system

1-56670-331-X/99/$0.00+$.50

as part of the product produced, some is exported to adjacent aquatic systems, but the majority remains within the system. Long histories of P rich material application have built up soil P levels in many ecosystems, and the residual P may be sufficient to meet all the P requirements of the crops grown on these soils. Application of organic wastes based on crop nitrogen needs usually result in application of P in excess of crop needs. Soils contaminated with P as a result of intensive use or abandoned agricultural lands with long history of fertilization can continue to release P even after the farming activities on these lands are stopped.

Wetlands can function as sources or sinks for P, depending on the type of wetland and the loading rates. Most of the P added is retained within the system, resulting in accumulation of large reserves of P. Wetlands created on agricultural lands can function as sources of P for some period unless the bioavailable P is stabilized. Similarly, lakes can function as sinks for P and store large amounts of P in sediments, but their sediments can become a net source of P after a reduction in external P loading.

The major issues that need to be addressed are as follows:

1. What are the major sources of excess P?
2. How can the controllable sources of P be reduced?
3. What processes transport excess P from one ecosystem to another?
4. What are the transformations of excess P and how do they affect P bioavailability?
5. What are the thresholds of P concentration and P load at which unacceptable ecological changes occur?
6. What will be the response of P-enriched ecosystems to a reduction in the P load?

A brief discussion is provided on each of these issues using information available for some of Florida's major ecosystems.

4.2 INTRODUCTION

Phosphorus is one of the major nutrients limiting the productivity of terrestrial (upland), wetland, and aquatic ecosystems. Florida's ecosystems are sensitive to anthropogenic nutrient loads and many of its aquatic systems are now eutrophic or hypereutrophic. The quality of water leaving one ecosystem can significantly impact the water quality of another ecosystem (Fig. 4.1). Accordingly, urban, agricultural, and environmental management practices implemented in one system have the potential to impact adjacent systems. For example, drainage water resulting from agricultural practices in the Everglades Agricultural Area (EAA) have created eutrophic conditions in the Water Conservation Areas (WCAs) of the Everglades. Similarly, nutrient loads from the dairy industry have moved Lake Okeechobee toward hypereutrophic conditions. Holistic management practices should consider such indirect impacts when developing water quality improvement plans.

The objectives of this chapter are to (1) present an overview of key issues related to P management in Florida's ecosystems as related to surface water quality,

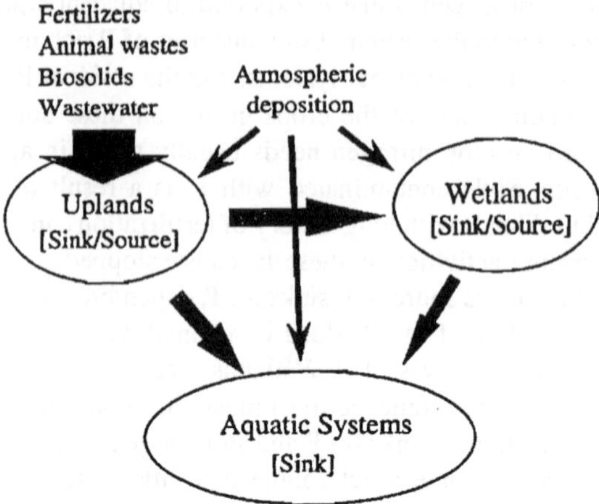

FIGURE 4.1 Schematic showing the P transfer among upland, wetland, and aquatic systems.

(2) provide examples of the experimental data developed to address these issues, and (3) identify key research and monitoring needs for integrated P management.

Wetlands and aquatic systems are important natural resources, because they provide habitat for diverse flora and fauna and water for many agricultural, domestic, and industrial activities. The water quality of these ecosystems is affected by land use, basin hydrology, geology, and water management practices. Although many point sources of P discharges to wetlands and aquatic systems have been controlled or reduced, nonpoint sources through surface and subsurface flow pose a greater danger in increasing loads to adjacent water bodies (Sharpley et al., 1994). The USEPA has identified agricultural operations as one of the major sources of P to lakes, rivers, and estuaries (Parry, 1998). To abate this problem, many state and federal agencies are in the process of developing watershed management strategies to reduce P loads into the water bodies. For example, the International Joint Commission between the U.S. and Canada effectively implemented several P management strategies to reduce loads to the Great Lakes (Rohlich and O'Connor, 1980). Similarly, the State of Florida has implemented several management strategies to reduce P loads to Lake Okeechobee (Bottcher et al., 1998). The USEPA is in the process of developing comprehensive national strategy to control nutrients from nonpoint sources (Parry, 1998).

Florida encompasses a surface area of approximately 149,900 km^2 with 36% occupied by forested land, 9% rangeland, 12% cropland (row crops, vegetables, citrus, sugarcane, and others), 15% pasture, 8% urban, and 27% wetlands and aquatic systems (Bottcher et al., 1998). Based on river basins, Florida can be divided into five regions: the Northwest region (covers area of the Panhandle of Florida), the Suwannee River Region, the Southwest region, the St. John's River Region, and the Kissimmee–Okeechobee–Everglades Region. Five Water Management Districts WMDs have been established by the State of Florida to manage the water quantity and quality of these regions (Fig. 4.2). Prominent examples of the State's water resources within these regions are the Kissimmee-Okeechobee-Everglades system,

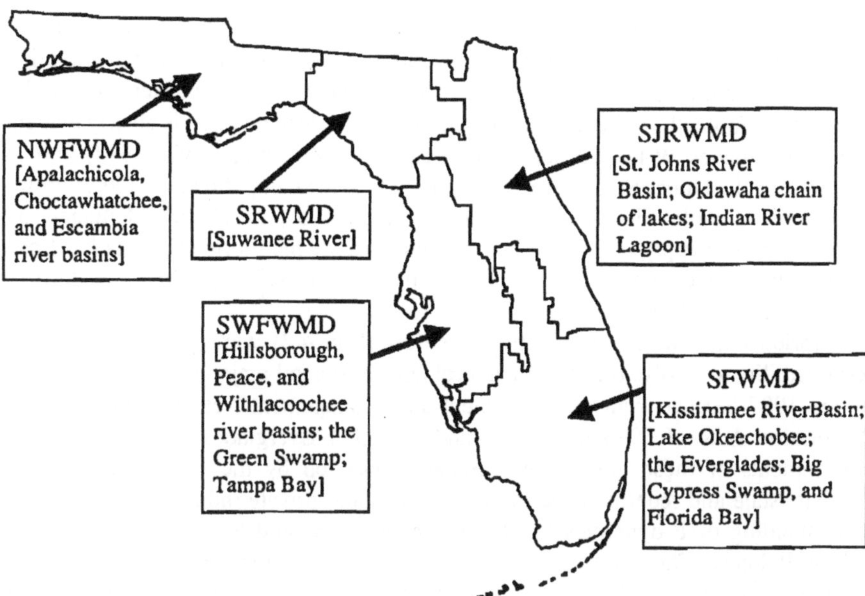

FIGURE 4.2 Florida map showing the boundaries of five water management districts and examples of key ecosystems within each region. NWFWMD = Northwest Florida Water Management District, SRWMD = Suwanee River Water Management District, SWFWMD = Southwest Florida Water Management District, SJRWMD = St. Johns River Water Management District, SFWMD = South Florida Water Management District.

the Florida Keys, the Indian River Lagoon, Wakulla Springs, the Green Swamp, the Apalachicola River and Bay, the Suwannee River, and the Floridian and Biscayne aquifers.

Florida's water resources are considerable. Because of its flat topography and high rainfall, 27% of its surface area is wetlands, lakes, and streams (Bottcher et al., 1998). It ranks third in the United States in precipitation with an annual average rainfall of 1350 mm/yr and has more available groundwater, proportionally, than any other state. Florida has 27 first-magnitude springs, more than any other state, more than 1,700 streams and rivers, and about 7,800 freshwater lakes (Fernald and Patton, 1984). With their associated uplands, these water resources support productive and economically significant ecosystems. For example, sport fishing alone is valued at $500 million annually in the Indian River Lagoon (Florida Sea Grant, 1993). Florida's biodiversity is nationally recognized. It ranks among the top three states in the continental U.S. in the total number of species of plants and animals (Florida Biodiversity Task Force, 1993). This biological wealth attracts many nature enthusiasts, and the combined economic value of consumptive and nonconsumptive water uses has been estimated at $5.2 billion statewide (Cox et al., 1994).

Cultural eutrophication is a major threat to Florida's valuable water resources. As a result of P enrichment, many of Florida's lakes are moving from oligotrophic to either mesotrphic or eutrophic conditions (Canfield, 1981). Eutrophication of natural wetlands is also a serious concern in Florida. Phosphorus enrichment also converted many oligotrophic wetlands into eutrophic wetlands, resulting in alteration of plant and microbial communities, increased productivity, and nutrient accumulation. For example, the effect of P loading on wetland eutrophication is clearly evident in the Everglades Water Conservation Areas (SWIM, 1992; Davis, 1994) and in wetlands of the Upper St. Johns River Basin (SJRWMD, unpublished). Other prominent and valuable Florida ecosystems threatened by, or already damaged by increases in the supply of P and other nutrients include the Florida Everglades, Lake Okeechobee, Lake Apopka and the Harris Chain of Lakes, Tampa Bay, Florida Bay, and Apalachicola Bay.

Florida's five water management districts (Fig. 4.2) were given the primary responsibility to resolve the ecological problems of several major aquatic ecosystems by the 1987 Surface Water Improvement and Management Act, the Lake Apopka Restoration Acts of 1985 and 1996, and the 1994 Everglades Forever Act. The requirements of these laws have stimulated research by the water management districts, state and federal agencies, universities, and private industry to develop an understanding of P dynamics in Florida's ecosystems and to develop methods to reduce P loads. These research projects have studied how physical, biological, geological and chemical factors interact to regulate the fate and transport of P in soils/sediments and water components of terrestrial, wetland and aquatic ecosystems (e.g., St. Johns River Basin, Oklawaha River Basin, Lake Apopka, Lake Okeechobee Basin, Lake Okeechobee, the Everglades, and Florida Bay).

The major issues confronting water managers are

1. What are the major sources of excess P?
2. How can the controllable sources of P be reduced?
3. What processes transport excess P from one ecosystem to another?
4. What are the transformations of excess P and how do they affect P bioavailability?
5. What are the thresholds of P concentration and P load at which unacceptable ecological changes occur?
6. What will be the response of P-enriched ecosystems to a reduction in the P load?

In this chapter, we will consider each of these issues using information available for some of Florida's major ecosystems.

4.3 WHAT ARE THE MAJOR SOURCES OF EXCESS P?

Most of the P that enters aquatic ecosystems stems from activities in the upland portion of the drainage basin. Phosphorus is added to uplands in fertilizers, organic solids (sewage sludge, animal wastes, composts, and crop residues), wastewater, and feeds (Table 4.1). Some of this P is exported out of the drainage basin as a part of the product produced, but the majority remains within the basin and can contribute to the eutrophication of streams, lakes, and estuaries (Boggess et al., 1995; Sharpley et al., 1996).

TABLE 4.1
Phosphorus Imports from Various Sources into State of Florida (see text for assumptions made to estimate P inputs)

Source	Annual import (Mg/P/yr)	Percent of total
Fertilizer	42,660	70
Biosolids	3,040	5
Wastewater	4,930	8
Compost	?	?
Animal manures	4,670	8
Atmospheric deposition	6,000	10
Natural weathering of minerals	?	?

Approximately 42,660 Mg of fertilizer P was used during 1996 in Florida. Fifty percent of this was applied within the boundaries of SFWMD, 20% in the SJRWMD, 17% in the SWFWMD, 8% in the NWFWMD, and 5% in the SRWMD (FLDACS, 1996). Fertilizer P is primarily in inorganic form, which is bioavailable and can be a major source of P for many ecosystems. For example, fertilizer P accounted for 51% of P imports to the Okeechobee Basin (Boggess et al., 1995). Unfortunately, similar estimates are not available for other drainage basins, despite the potential importance of this input.

Feed supplements are another significant source of P. Although statewide estimates of the magnitude of their use are not available, they can be a major contributor of P. For example, in the Lake Okeechobee drainage basin, 49% of the P load to the lake stemmed from feeds used by the dairy industry (Boggess et al., 1995).

Organic solids are in the form of biosolids (commonly known as "sewage sludge"), animal wastes, composts, and crop residues. Florida produces about 253,000 dry tons of biosolids. About two-thirds of the supply is applied to land, and the remainder is disposed in land fills and incinerators. The P content of biosolids is in the range of 0.4% to 5.3% (mean = 1.8% dry weight basis) (Obreza, 1997). In biosolids, only 10% to 30% of the total P is present as organic P (Wolf and Baker, 1985). Assuming that about two-thirds of biosolids produced are applied on land, the estimated P added from this source to Florida's ecosystem is approximately 3,040 Mg/P/yr.

In 1995, about 22 million Mg of MSW was produced in Florida; 55% of this waste can be composted (Hinkley and Goven, 1996). In 1992, about 0.5 million Mg of animal manure and 2.8 million Mg of yard trimmings were available for composting (Cooperative Extension Office, IFAS, University of Florida). Phosphorus available in animal manures obtained from various livestock operations in Florida was estimated at 4,700 Mg/yr (Landers et al., 1998). The MSW, animal manure, and yard waste can potentially produce approximately 5.4 million Mg of usable compost. Phosphorus in organic solids occurs as both inorganic and organic forms. The proportion of these forms in organic solids is important to P availability for plant uptake and for transport in leaching and surface runoff. Data obtained for different types of manures indicate that the ratio of soluble P to total P is in the range of 0.24 to 0.70, suggesting that a significant portion can be associated with manure particles (Reddy et al., 1978).

The volume of municipal wastewater discharges has been about 6.76 million m^3/day, with about 16% applied to land, and 17% discharged through deepwell injection. Assuming an effluent total P concentration of 2 mg/L, the estimated P load is about 13,520 kg P/day (4,930 Mg P/yr). Approximately 50% are discharged into bays, rivers, and wetlands, while the remainder goes to the ocean.

Phosphorus inputs through atmospheric deposition are estimated to contribute about 20 to 80 mg P/m^2 yr into Florida (Redfield, 1998). At this rate, annual load to Florida's ecosystem is estimated to be approximately 3,000 to 12,000 Mg P/yr. Rates below 30 mg P/m^2 yr are found near coastal and remote areas, 30 to 50 mg P/m^2 yr are found in mixed land uses, and >50 mg P/m^2 yr are associated with urban and agricultural areas (Redfield, 1998). Because of the uncertainties associated with these estimates, we used a more conservative value of 40 mg P/m^2 yr for P input through atmospheric deposition. Phosphorus added through this source is significant and is higher than P added through waste materials (Table 4.1). Under natural conditions, background levels of P can also be contributed by parent material through weathering and dissolution of rocks and minerals of variable solubilities. For example, high background levels of P in surface waters in many areas of the southwest region of Florida is because of dissolution of P from native phosphatic minerals. In order to assess the P impacts, we need to know the relative contribution of these natural sources as compared to P imports through fertilizers, organic solids, and feeds. This should be given serious consideration, when determining P budgets and in developing management regulation and strategies to restore ecosystems.

4.4 HOW CAN THE CONTROLLABLE SOURCES OF EXCESS P BE REDUCED?

The above analysis shows that most of the major sources of P are controllable. How, then, can P use be reduced while still accomplishing its intended purposes? For fertilizers, there is good evidence that application rates are often higher than crop requirements for optimum productivity. Long histories of fertilizer and other P-rich material application has built up soil P levels in many ecosystems, and the residual P may be sufficient to meet some or all the P requirements of the crops grown on these soils. In many areas of Florida, soil test P values exceed crop requirements. For example, organic soils used for vegetable production in the Lake Apopka Basin have sufficient available P, and crops grown on these soils do not respond to application of P fertilizers. Several best management practices (BMPs) are now recommended to improve the P use efficiency by crops and to reduce P in runoff and drainage (Bottcher et al., 1998; Izuno and Whalen, 1998). One BMP is to use soil tests to determine P fertilizer needs, rather than relying on historic application rates.

Agricultural scientists, and some farmers, use plant and soil analyses to determine the P status and requirements of crops. The most common approach is to use soil test procedures, which involve extraction of soils with selected chemicals. The amount of P extracted is related to crop yields to determine the P fertility of soils.

These relationships have been developed for various crops and soil types. For mineral soils (Entisols, Spodosols, and Ultisols), soils are extracted with Mehlich I (0.025 M H_2SO_4 + 0.05 M HCl) reagents for 5 min. (soil to solution ratio: 1:4), and filtered solutions are analyzed for soluble P. For organic soils (Histosols), soils are extracted with water (soil to solution ratio of 1:100), and filtered solutions are analyzed for soluble P. Such tests, however, are not directly based on the capacity of soils to retain or release P, and its ultimate effect on the surrounding environment.

Extensive soil testing on mineral soils has shown that many crops do not respond to additional P fertilization, if Mehlich I-P levels are >31 mg P/kg (Kidder et al., 1997). At Mehlich I-P levels of <10 mg P/kg P fertilization is recommended at a level needed to optimize crop yields, and at P levels between 10 to 30 mg P/kg, lower rates of fertilizer P are recommended (Kidder et al., 1997). For pastures, bahia grass grown on flatwood soils, decreasing fertilizer P application rates from 48 to 24 kg P/ha yr did not decrease forage yields (Fig. 4.3) (Rechcigl and Bottcher, 1995). Although these soils contained Mehlich I-P in the range of 9 to 13 mg P/kg soil, they did not respond to added fertilizer P, suggesting that further validation is needed between soil test values and crop response. Despite the poor responses of crops to additional P applications, and recommendations to reduce P fertilization, many farmers continue to apply P at rates beyond crop needs (Kidder et al., 1991). Over 50% of the fertilizer P purchased in 1996 is used in the South Florida Water Management District Region, where aquatic systems are sensitive to P loading.

Acceptable land application rates of organic wastes are usually calculated by consideration of the fate and transport of N in the given soil-water-plant system. In Florida, because of the P-sensitivity of many aquatic systems, serious consideration should be given to application of organic solids based on the fate and transport of P. In many cases, applications of organic solids based on crop N needs result in application of P in excess of agricultural requirements, resulting in adverse impacts on surface and groundwater (McCoy et al., 1986). For example, the average N content of biosolids is about 3% of the dry weight (range = 0.6 to 7.5%), and the

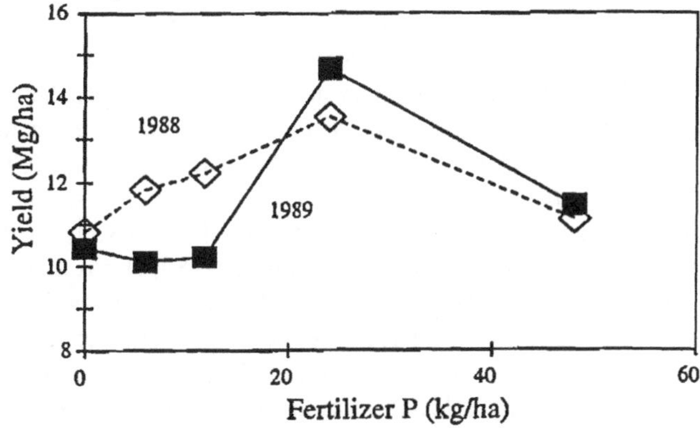

FIGURE 4.3 Bahia grass (Paspalum notatum) yield as influenced by phosphorus fertilizer application (Rechcigl and Bottcher, 1995).

average P content is about 1.8% (range = 0.4 to 5.3%). This low N/P ratio means that when the land application rates of biosolids are based on the N content, there is a potential for creating water quality problems in land areas with soils having poor P retention capacity. On the other hand, if the application rates are determined based on P needs, rates would be much lower and supplemental N may need to be applied as inorganic fertilizers to meet the N demand. Organic waste loading rates should be based on site soil characteristics, the bioavailability of N and P, and the hydrologic characteristics of the site.

Another critical issue in application of P-rich substances is to what extent the added P is retained by soils. Unlike C and N, much of the P added through fertilizers and organic solids is retained within the system. In many of Florida's upland ecosystems, the long-term use of the land by agricultural and animal industries has increased soil P levels. The soil P content is higher in surface soil than subsoil. Soil total P content can be highly variable. For mineral soils, (soil bulk density of about 1.5 g/cm^3) total P content can range from 20 to 2000 mg P/kg, and in organic soils, (soil bulk density of 0.1 to 0.3 g/cm^3) total P content can vary from 300 to 1500 mg P/kg. For example, in the Okeechobee Basin, the total P content of surface soil (A horizon) was 1900 mg/kg in intensive areas and holding areas of cattle (Graetz and Nair, 1995). In the same soil profile, subsoil total P concentrations were in the range of 150 to 180 mg P/kg. In natural areas and lands used for pasture, soil total P in surface horizons was in the range of 30 to 150 mg P/kg (Graetz and Nair, 1995).

The amount of P accumulation depends on the capacities of soils to retain P through adsorption and precipitation reactions. In mineral soils (Spodosols, Entisols, and Ultisols) inorganic P typically constitutes 50 to 90% of total P, whereas in organic soils (Histosols), inorganic P typically constitutes only 10 to 50% of total P. Inorganic P forms are associated with amorphous, and poorly crystalline Fe and Al in acid soils, and with Ca compounds in alkaline soils. Organic P forms are associated with sugar phosphates, nucleic acids, adenosine phosphates, phospholipids, inositols, and fulvic and humic acids.

Long-term application of P can decrease a soil's capacity to retain P. Soils have finite capacity to retain P through sorption and precipitation reactions (Berkheiser et al., 1980; Rhue and Harris, 1998). Phosphorus sorption processes have been extensively studied using various soil types and soil minerals. General conclusions from these studies are that P sorption in acid mineral soils can be correlated to amorphous and poorly crystalline forms of Fe and Al (oxalate extractable Fe and Al) and crystalline forms of Fe and Al [citrate-dithionate-bicarbonate (CDB) extractable Fe and Al]. In alkaline-calcareous soils, P retention is associated with Ca minerals. Several studies have been conducted on Florida soils to determine P retention capacities (Yuan and Lucas, 1982; Burgoa, 1989; Harris et al., 1996; Nair et al., 1998). Using various soils, Harris et al. (1996) developed an index of relative P adsorption (RPA) capacities. The RPA is defined as the ratio between the absolute amount of P sorbed to the maximum amount of P that can be retained. This ratio has been used to characterize selected Florida soils for their P retention capacities (Table 4.2). Surface horizons of Spodosols and Entisols (clean sands) have RPAs in the range of 0.05 to 0.26, as compared to Ultisols and Entisols (with coated sands), with RPAs of 0.74 and 0.48, respectively. Most of the Spodosols are in south and

TABLE 4.2
Relative Phosphorus Adsorption (RPA) of Soils (Harris et al., 1995)

Soil	A	E	Bh/C/Bt
Spodosols	0.26	0.08	0.96 (Bh)
Entisols			
Clean sands	0.05	—	0.01 (C)
Coated	0.48	—	0.47 (C)
Ultisols	0.74	0.60	0.96 (Bt)

RPA = absolute amount of P adsorbed/maximum amount of P retained

central Florida where soils are intensively used for agriculture, and many of these areas are considered to be potential sources of P contamination. For example, in the Okeechobee Basin, Spodosols in areas intensively used for dairies and beef pastures showed very little or no P retention. Low P sorption capacities were reported for A and E horizon of Spodosols and relatively high capacities in subsurface Bh horizon (Nair et al., 1998).

Although soils can accumulate P as a result of continual loading, their capacity to buffer dissolved P in the soil porewater may decrease with time, resulting in elevated P levels in the drainage water. This effect can be evaluated by determining the EPCo (equilibrium P concentration mg/L), for a soil (Fig. 4.4). A soil retains P only if the influent water P concentrations is >EPCo for that soil. If the water moving through the soil has a P concentration of <EPCo, the soil releases P. Earlier studies with soils loaded with manures showed that increased P loading decreased soil P retention capacity, and increased EPCo (Reddy et al., 1978). Soil test procedures (Olsen, Bray–1, and Mehlich I) provided reasonably good estimates of labile P, EPC_0, and algae-available P, for a diverse group of noncalcareous agricultural soils (Wolf et al., 1985). For Spodosols in the Okeechobee Basin, Nair et al. (1998) showed high EPCo for soils impacted by P loading (Table 4.3).

TABLE 4.3
Influence of Land Use on Total P, Relative Phosphorus Adsorption (RPA), and Equilibrium P Concentration (EPCo) of Soils (Nair et al, 1997)

	A	E	Bh	Total P	EPCo
Land use		RPA		mg/kg	mg/L
Intensive	N/A	0.03	0.38	2330	5.0
Holding	0.03	0.20	0.42	181	1.4
Pasture	0.05	0.08	0.47	31	0.1
Beef	0.07	N/A	0.37	31	0.1
Forage	0.07	0.02	0.34	23	0.2
Native	0.47	0.21	0.36	18	0.1

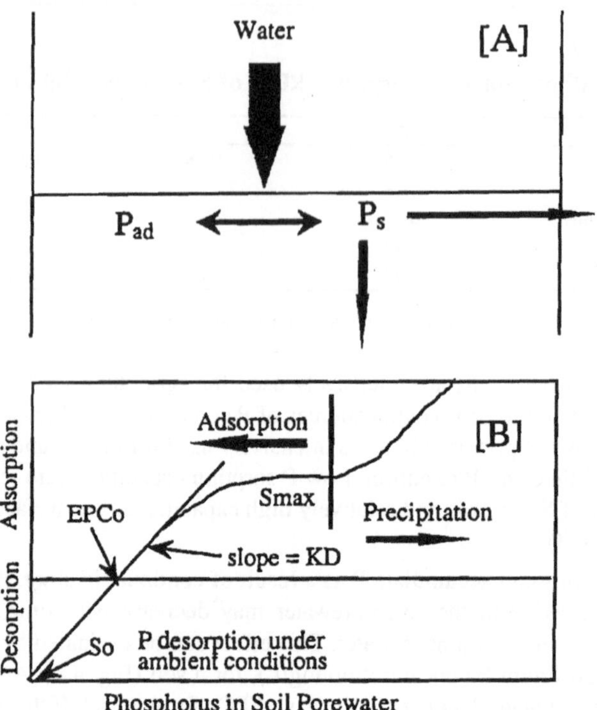

FIGURE 4.4 Schematic showing phosphorus sorption isotherm for soils. [A] = phosphorus transport processors; [B] = phosphorus sorption isotherm of upland soils. KD = phosphorus adsorption coefficient, Smax = phosphorus sorption maximum, EPCo = equilibrium P concentration at which point adsorption equals desorption. Pad = phosphorus adsorbed phase, Ps = phosphorus in solution.

Soils in upland ecosystems are not adequately characterized with respect to their long-term capacity for P retention or release. Results available at this time cannot be spatially extrapolated to watershed or basin scales unless relationships between P retention characteristics and easily measurable soil properties are developed. Data on easily measurable soil properties such as organic matter content, and extractable P, Fe, Al, and Ca, can be obtained in a cost-effective manner for a large number of sites within the ecosystem.

In addition to chemical processes, P retention by soils is also regulated by several biological processes including uptake and release by higher plants and microorganisms. In P limited soils, a significant portion of P can be tied up in organic pools, and the turnover of this pool through the activities of extracellular enzymes can play a significant role in regulating the bioavailable P. Forms of organic P, and their breakdown in upland soils are discussed in detail by Newman and Robinson (1998) and Wetzel (1998). Phosphorus retention in organic pools, especially as inositol phosphate and P bound to fulvic and humic acids, represents a stable, long-term P sink. Biological processes can, thus, be very important in regulating mobility of P especially in organic soils, and surface horizons of mineral soils.

Upland systems can be managed for maximum P retention especially when soils are used for land application of organic wastes. Some of these waste materials contain metallic cations that can increase the overall P retention capacity of the soil. Wastes low in metallic cations can be amended with P binding chemicals (such as alum, ferric chloride, or lime) before their application on the land (Moore and Miller, 1994). Additional research is needed to determine the utility of chemical amendments in reducing the bioavailability of pools of P in organic wastes. As discussed earlier, long term application of fertilizers and wastes increases the total and bioavailable P of soils. The stability of this stored P depends on soil physico¬chemical properties, management practices, and hydrology. Limited data available for selected sites is not sufficient to extrapolate to landscape level. Studies suggest that a large proportion of P stored in upland soils is in stable forms (Graetz and Nair, 1995). However, it is critical to determine the fraction of P that is mobile or released into water.

4.5 WHAT PROCESSES TRANSPORT EXCESS P FROM ONE ECOSYSTEM TO ANOTHER?

4.5.1 Uplands

The majority of Florida's P load is applied to upland ecosystems. Phosphorus from mineral upland soils can be transported via surface and subsurface flow. In poorly drained flatwood soils, both surface and subsurface runoff (lateral flows) can be important, especially during heavy rainfall events. Where the spodic horizon is shallow, there is greater potential for P transport through surface runoff (Campbell et al., 1995). Areas farmed on organic soils (Lake Apopka Basin, the Everglades Agricultural Area) are typically artificially drained through a network of drainage canals. Phosphorus transport in these soils primarily occurs through subsurface flow. Although much of the P added to an upland ecosystem is retained within the soil, the amount of P that leaves the system through surface and subsurface flow is sufficient to affect the water quality of adjacent wetlands and aquatic ecosystems. The amount of P leaving an upland system is affected by a number of factors including, soil type, hydrology, land use, and management practices. Measurement of P outputs is expensive, and often P discharges of only a few of the watersheds are adequately characterized. For example, P discharges from the EAA and vegetable farms adjacent to Lake Apopka have been quantified. Simulation models have been used to estimate P transport from the Okeechobee Basin (Campbell et al., 1995). Our understanding of P transport has improved through the development of database and simulation models, especially for upland ecosystems in the St. Johns River Region and Kissimmee-Okeechobee-Everglades Region. However, such a database is not available for other regions of Florida. Phosphorus transport can be quantified and predicted through the use of simulation models, but only after thorough testing and validation with experimental data.

Soils with high levels of P as a result of intensive use, such as abandoned agricultural lands with long histories of fertilization, can continue to release P even after the farming activities on these lands stop. For example, land use activities around dairies and beef ranches in the Lake Okeechobee Basin have resulted in P enrichment of surface soils, especially in soils of intensive areas and holding areas surrounding milking barns of dairies. Several of these dairies in the Lake Okeechobee Basin have been bought by the State of Florida, in order to reduce P inputs to the lake. Agricultural lands in central and south Florida have also been acquired by state agencies to protect adjacent aquatic systems. Some examples are vegetable farms in the Oklawaha River Basin, Lake Apopka Basin, Upper St. Johns River basin and the Everglades. Some of these acquired lands have been converted into wetlands for habitat restoration and water quality improvement. However, long term farming in these areas created P enriched soils that can release P into drainage water for long periods of time after farming has ceased.

4.5.2 Wetlands

At the landscape level, wetlands often form a critical interface between uplands and adjacent water bodies, as all of these ecosystems are hydrologically linked. Consequently, a thorough understanding of P dynamics in wetlands is critical to understanding the processes which transport P from uplands to lakes and streams. Water and associated contaminants (such as P) are transported from uplands by either subsurface or surface flow. Surface flow can include first-, second-, and third-order streams as well as the associated riparian floodplains, marshes and swamps. In low gradient systems, streams are largely composed of interconnected marshes and swamps. To improve drainage in agricultural areas, ditches are often cut to connect isolated wetlands. The resulting flow of water in the drainage basin follows a complex path through wetlands, ditches and streams. Thus, P loading to the receiving aquatic system depends on the retention capacity of several components of the basin.

Wetlands can function as sources or sinks for P, depending on the type of wetland and the loading rates. Most of the P added is retained within the system, resulting in accumulation of large reserves of P. Thus, retention is defined as the capacity of wetlands to remove water column P through physical, chemical and biological processes, and retain it in a form not readily released under normal conditions. Retention of P by wetlands decreases the load to downstream aquatic systems (Reddy et al., 1996). Wetlands not only store P but also transform P from biologically available forms into nonavailable forms and vice versa. Thus, it is important to include the contribution of wetlands in retaining P when developing best management practices for a drainage basin. Land use practices in uplands, along with processes occurring in wetlands, should all be considered in nutrient management options for a water body.

Wetlands, as low lying areas in the landscape, receive P inputs from all adjacent uplands (agricultural and urban activities). If the integrity of an upland area is compromised, it is likely that it will soon be reflected in the integrity of associated wetlands. However, if the integrity of the wetland is compromised the effect may not be immediately reflected in the condition of the upland, since materials transfer is largely toward the wetlands. The response of a wetland to P inputs varies and depends on the wetland type (e.g., forested, marsh with emergent macrophytes), as compared to lakes and streams which respond to P inputs more rapidly because of more mixing. For many natural wetlands, external P loads (especially nonpoint sources) have not been characterized. Phosphorus inputs from the EAA and other nonpoint sources into adjacent natural wetlands (Water Conservation Areas 1, 2a and 3a) have been estimated to be about 347 Mg P/yr, and rainfall inputs to these areas have been about 272 Mg P/yr (SWIM, 1992). On an areal basis, loading rates from the EAA and rainfall were: 0.23, 0.25 and 0.13 g P/m^2 yr for WCA-1, WCA-2a, and WCA-3a, respectively (SWIM, 1992). Phosphorus inputs to wetlands located in the lower Kissimmee River and Taylor Creek/Nubbin slough watersheds were estimated to be in the range of 2.8 to 4.0 g/m^2 yr (Reddy et al. 1996).

Natural and constructed wetlands intentionally used as buffers to retain P are usually managed to improve their overall performance, and to maintain wetland integrity. The extent of management required depends upon the P retention capacity of the wetlands and the desired effluent quality. Management scenarios can vary, depending on the type of wetland. For example, wetlands used for municipal wastewater treatment are usually small. They can be managed efficiently by altering the hydraulic loading rate or integrating them with conventional treatment systems. Large-scale systems can be managed by controlling P loads.

A wide range of P loads (0.1 to 1000 g P/m^2 yr) is used in treatment wetlands (Kadlec and Knight, 1996). Analyzing water quality data from several treatment wetlands, Kadlec and Knight (1996) observed that an increase in the P loading rate increased the effluent P concentration. Richardson et al. (1997) in their analysis of water quality data from treatment wetlands, showed that at P loading rates of approximately 1 g P/m^2 yr, the effluent P concentration was approximately 40 µg/L. However, the data presented by Richardson et al., (1997) shows a high degree of variability between loading rates 0.1 to 10 g P/m^2 yr, suggesting that it is too simplistic to establish *one* loading rate to determine the lower limit of wetlands to remove P. Clearly wetlands can not reduce P levels below background levels. Although simple input/output analysis found in the literature provides some general guidance relative to P loading rates and effluent P concentrations, they can not be directly applied to a particular wetland, without due consideration of site-specific conditions.

Phosphorus added to wetlands is retained through biotic processes such as assimilation by vegetation, periphyton, and microorganisms, and through abiotic processes such as sedimentation, sorption, and precipitation. Steady P loading increases the P content of all components of wetlands (vegetation, soil, and periphyton), and results in distinct horizontal and vertical gradients in soils and water columns (Koch and Reddy, 1992, DeBusk et al., 1994). Phosphorus loading also increases bioavailable P pools in soil, and decreases soil capacity to buffer porewater P at the original level. Addition of P also increases microbially mediated processes, resulting in short-term storage of P in microbial cells as polyphosphates. In P enriched wetlands, a significant amount of stored P is in labile pools, as low C/P ratios of detrital tissue favor rapid decomposition and release. The lability and stability of detrital tissue is an important factor regulating wetland P dynamics as cycling of detrital material can maintain eutrophic conditions in a wetland, even after external loads are curtailed. These conditions can increase the EPCw (equilibrium P concentration between soil and water column) of the soil, which determines the direction of P flux between soil and overlying water column (Fig. 4.5). If the inflow concentrations are decreased below the EPCw (as a result of reduction in P loads), wetland soils could function as a steady source of P to the water column until a new equilibrium is attained.

When agricultural lands are converted into wetlands, the residual fertilizer P stored in these soils can be rapidly released upon flooding. For example, vegetable farms in the Lake Apopka Basin converted into wetlands released P into the overlying water column, even 32 months after flooding (Table 4.4). In soils of the Lake Griffin flow-way, up to 30% of the total P was present in bioavailable pool that can be

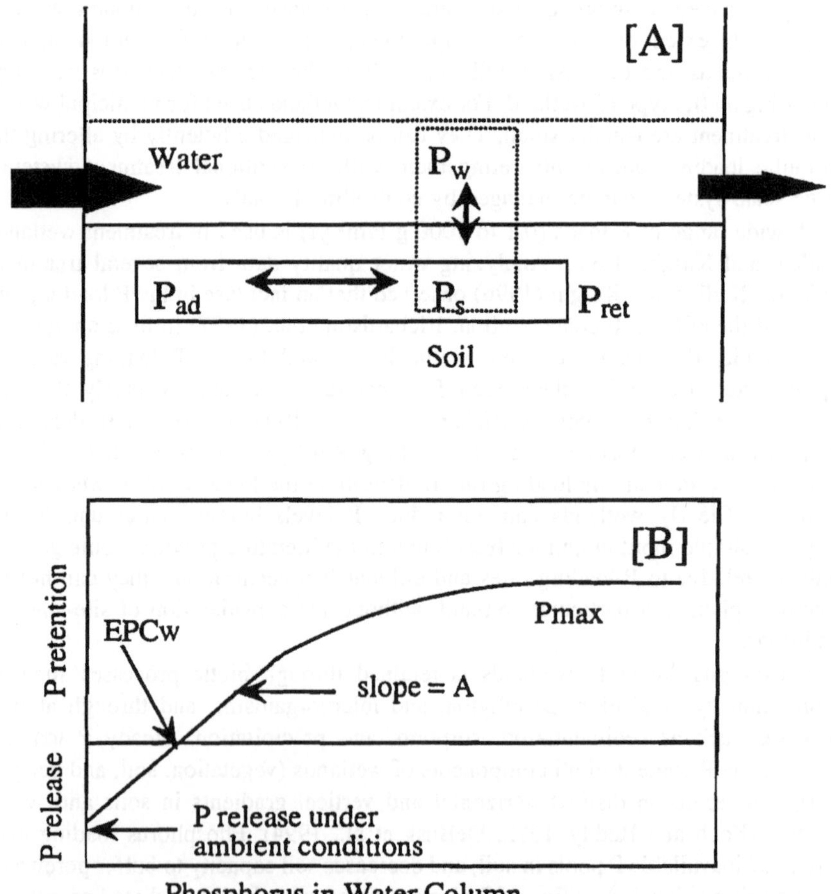

FIGURE 4.5 Schematic showing phosphorus retention isotherms for wetland soils and lake sediments. [A] = phosphorus exchange processes; [B] = phosphorus sorption isotherms for wetland soils and lake sediments expressed on areal basis. A = phosphorus retention coefficient, EPCw = equilibrium P concentration in the water column at which point P retention equals release, Pmax = phosphorus retention maximum (Reddy et al., 1995).

potentially released into the water column (Reddy et al, 1997). Even after 30 weeks after flooding, soils from the flow-way maintained an average EPCw value of 0.22 mg P/L, suggesting that these soils cannot reduce effluent P levels below 0.22 mg P/L (Reddy et al., 1997). However, filtration of particulate P can cause removal of total P below this value if the sedimentation rate exceeds the rate of P release (SJRWMD, unpublished).

TABLE 4.4
Diffusive Flux of Soluble P from Soil to Overlying Floodwater in Selected Wetlands Created on Agricultural Lands (D'Angelo and Reddy, 1998)

Site	Time after flooding (months)	Soluble P flux potential (mg/m^2 day)
Lake Apopka (50 m from inflow)	3	0.3
	8	0.3
	13	0.1
	32	0.6
Lake Apopka (3000 m from inflow)	3	5.5
	8	3.1
	13	2.4
	21	1.1
Sunnyhill Farm	48	2.0
	60	0.7
Emeralda	12	0.9–1.3
Knights (ENR)	10	0.3–9.2
	16	0.6–1.0

Bioavailable P in agricultural lands converted to wetlands can be stabilized through application of phosphate binding chemicals such as alum, ($Al_2(SO_4)_3$) ferric chloride ($FeCl_3$), and lime ($Ca(OH)_2$) (Ann, 1996). In addition, establishment of macrophyte communities may stabilize soil porewater P, and reduce overall P flux to overlying water column.

Phosphorus is often one of the major nutrients limiting the productivity of wetland ecosystems. Much of the P added to wetlands is retained within the system, resulting in accumulation of large reserves in detrital tissue and soil, which can serve as a P source for a long period of time, even after external loads are reduced. Similarly, wetlands created on agricultural lands can function as sources of P for some period until the bioavailable P is stabilized. When considering wetlands for storing P, several issues needed to addressed.

1. What was the natural P loading rate and resultant trophic state?
2. What is the relative bioavailability of various forms of P in wetland soils and lake sediments?
3. How stable is the stored P, and under what conditions will it be released back into the water column?

4. What is the long-term assimilative capacity of these systems?
5. How long can the stored P maintain eutrophic conditions once external loads are curtailed?

4.5.3 Aquatic Systems

Aquatic systems, such as lakes are the final recipients of P discharged from adjacent uplands and wetlands. As P moves through uplands and wetlands, it undergoes various biogeochemical transformations, and most of it remains within these systems. In terms of the P mass balances (of upland ecosystems), the amount of P transferred to lakes may not be significant. However, these relatively small losses may be enough to create eutrophic conditions in adjacent lakes. Addition of P to lakes increases algal productivity and decreases dissolved oxygen and biodiversity, and changes in these parameters are often used as indicators of eutrophication. Since algae can obtain C and N from the atmosphere, P regulates the growth of algal biomass, and ultimately the trophic state of the lake. Phosphorus and other nutrients stored in algal biomass are cycled within the water column during decomposition, and the P associated with recalcitrant dead algal biomass settles and becomes an integral part of the sediments. Although sediments usually function as a major storage reservoir of P, they can at times function as sources of P to the water column.

The P concentration in lakes depends on inputs from tributaries, groundwater and the atmosphere, interactions between the bottom sediments and the overlying water column, exchanges between the vegetated (littoral) and open (limnetic) zones, and internal biogeochemical processes in the sediment and water column. Management strategies for the restoration of lakes must address several issues.

1. What was the natural loading of P to the lake and the resultant trophic state?
2. What is the relative bioavailability of various forms of P in the sediment and water column?
3. What is the relative contribution of P loads from internal sources as compared to external sources?
4. What is the time span required for these systems to reach their background condition after external loads are curtailed?

4.6 WHAT ARE THE TRANSFORMATIONS OF P AND HOW DO THEY AFFECT P BIOAVAILABILITY?

Phosphorus is discharged into surface waters in both organic and inorganic forms. The relative proportion of each of these forms depends on the P source, the soils in the drainage basin, and the land uses. For example, drainage and runoff water from areas with organic soils can have a greater proportion of total P in organic form than inorganic forms. Conversely, runoff water and drainage from mineral soils contain P mostly in inorganic form, with large amounts associated with suspended sediments. Some of the issues related to P concentrations of water discharged from uplands are

(1) what is the ratio of inorganic P to organic P? (2) what proportion of total P is present in bioavailable form, and (3) what proportion of total P is stable and resistant to biological breakdown?

Continuous, long-term application of P also increases the proportion of P which is bioavailable. Soil P undergoes various transformations as it cycles through inorganic pools (Fe and Al, Ca and Mg associated minerals) and organic pools (through plants, animals, microbes, and soil organic matter). Predicting these transformations requires a detailed understanding of the soil characteristics, chemical equilibria, and rates of transformation of organic and inorganic pools, and how P availability is affected by various management strategies. With respect to water quality, the amount of inorganic and organic P present in a bioavailable pool is critical, since these pools of P can be readily mobilized. However, there is no single suitable method to quantify the fraction of total P that is bioavailable P. The bioavailable P is defined as a form of organic and inorganic P that can be readily used by biota including higher plants, algae, and microbes. Numerous chemical and biological methods have been used to determine bioavailable P in soils (Sharpley, 1991, 1993), but many of these methods yield results that are "operationally defined." Analysis of selected surface water samples from the Upper St. Johns River Basin indicated that about 70% of total P was bioavailable as estimated by the Fe-oxide strip method and the acid hydrolyzable method (Fig. 4.6, Reddy, K.R., unpublished results).

Floodplain areas, which have long been drained and farmed, are now being acquired by state and federal agencies with the intent to restore lost habitat, and to reduce the pollution of adjacent water bodies. In many cases, the management goal for these areas is to recreate wetlands or, if soil subsidence has been severe, to create lakes. Reflooding of these lands, however, can temporarily create water quality problems as the residual fertilizer P, and the P released from oxidation of organic material, is rapidly released into the water column. With time after flooding, hydrophytic vegetation becomes established. Uptake of P by plants, and microbial proc-

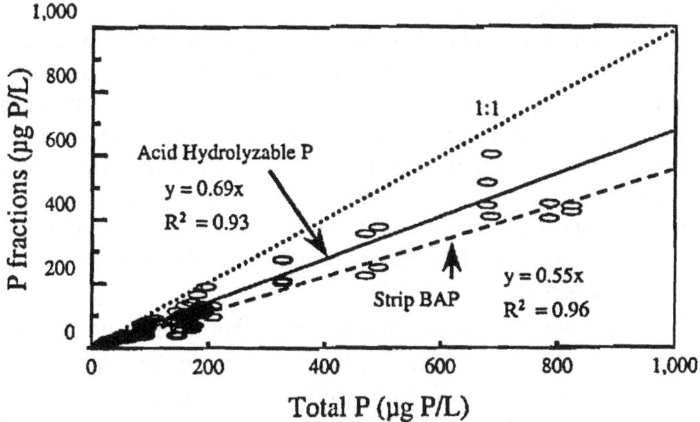

FIGURE 4.6 Relationship between total P and bioavailable P (measured by acid hydrolysis and Fe-oxide strip method) of surface water in the Upper St. Johns River Basin (K.R. Reddy, unpublished results).

esses gradually reduces the bioavailable P. In Florida, agricultural lands currently being considered for conversion into wetlands are dominated by organic soils. For example, farmlands proposed for conversion into treatment wetlands (15,000 ha) to buffer agricultural drainage water from the EAA have organic soils used for sugarcane production. Similarly, several thousand hectares of agricultural lands with organic and mineral soils are flooded in central Florida to create wetland habitats. The unique soil conditions created by flooding influences the transformation and availability of both soil organic and inorganic P. Earlier studies have shown that flooding and anaerobic decomposition processes increased the rate of soluble P production (Reddy, 1983; D'Angelo and Reddy, 1994a and b), and soluble N and C (Reddy, 1982), thus adversely impacting surface water. In contrast, macrophytes and periphyton can also convert labile inorganic P into organic P through uptake and storage, stabilize the soil porewater P, and reduce the P concentrations of surface water. However, long-term stable storage of P in organic pools depends on the quality of detrital material accreted on the soil surface.

4.7 WHAT ARE THE THRESHOLDS OF P CONCENTRATION AND P LOAD AT WHICH UNACCEPTABLE ECOLOGICAL CHANGES OCCUR?

In P-limited wetlands and lakes, the P concentration of surface water is very low, as added P is rapidly assimilated into biota and adsorbed onto suspended particles. The small fraction of bioavailable P undergoes rapid turnover, and is efficiently used by the organism present in the water column. As the system is loaded with P, the surface water P concentration increases only after the P uptake by organisms reaches saturation level. Threshold P concentrations that cause ecological changes in lakes are fairly well known (Correll, 1998), but very limited information is available for wetlands. The threshold P concentration for a lake or a wetland depends on the capacity of the system to assimilate P without causing significant ecological changes. However, many systems are weakly buffered and the biotic communities respond to P loading rapidly. In the Everglades, microbial and algal communities responded rapidly while responses of macrophytes followed later (McCormick and O'Dell, 1996). For the Everglades WCAs, substantial changes in periphyton communities have been noted at a water column total P concentration of 10 to 30 μg/L (Vymazal and Richardson, 1995; McCormick and O'Dell, 1996). Similarly, addition of P to an oligotrophic or mesotrophic lake can result in shift of plankton species, with later shifts by other organisms.

The phosphorus concentration of the water column is often used to assess the trophic status of a wetland or a lake. So, how reliable are the established concentration limits in determining threshold for wetlands and aquatic systems? The Vollenweider model, commonly used for lakes, takes mass loadings into consideration (Vollenweider, 1976). If P concentration is used, Correll (1998) points out that total P should be used as criteria rather than dissolved reactive P (DRP), because of its rapid turnover in the water column. In eutrophic lakes such as Lake Apopka, the

DRP levels are <5 μg/L, while total P concentrations range from 100 to 200 μg/L (Reddy and Graetz, 1990). The low DRP concentration in this lake reflects rapid turnover resulting in high planktonic activity in the water column.

Surface water total P concentration of 165 Florida lakes surveyed ranged from 3 to 834 μg/L (Canfield, 1981), with total P concentration of 75% of the lakes in excess of 10 μg/L. Consequently, the majority of these lakes are classified as either mesotrophic or eutrophic. There is evidence, however, that the trophic state of subtrophical lakes is lower than that of temperate lakes for any given level of P (Salas and Martino, 1991).

Most natural wetlands are not used for reducing P levels of agricultural or urban runoff water. However, in those cases where they have been, eutrophication has become a major issue, because of adverse impacts on wetland plants and animal communities.

Because most interest in wetlands with respect to P has been in its removal, little attention has been paid to the effects of P on wetland trophic state. As a result, the conceptual basis for management of wetlands lags well behind that for lakes. Lowe and Keenan (1997) point out that P effect in wetlands are patterned and localized unlike effects in lakes that are unpatterned and generalized. They propose that management of eutrophication of wetlands should, therefore, focus on managing the size of the zone effects rather than the P concentration. In many cases the extent of impacts resulting from P loading have been difficult to measure, with the notable exception of the Everglades. In developing P loading strategies, we need to address the following issues:

1. What are the impacts of P loading on natural wetlands?
2. Are some wetlands more tolerant (or have greater assimilatory capacities) than others and, if so, why?
3. What level or size of impact is acceptable?
4. Can we reverse eutrophication effects and restore wetlands to their original condition?
5. To what extent is it legally permissible to load P into wetlands?

At present, state and federal agencies are actively involved in restoration of the Everglades through various P control strategies (Izuno and Whalen, 1998; Moustafa et al., 1998). The Everglades restoration program should provide insights that can be applied to other wetland restoration programs in Florida and other subtropical regions of the world.

4.8 WHAT WILL BE THE RESPONSE OF P-ENRICHED ECOSYSTEMS TO A REDUCTION IN THE P LOAD?

Long-term P loading saturates the biotic (uptake by microbes, algae, and higher plants) and abiotic (sorption on particulate matter) components of wetland and lakes, resulting in increased P levels in the water column and underlying soils or sediments.

The response of ecosystems to reduction in external P loads depends on their internal reserve of P and its stability and bioavailability. Thus, characterization and identification of internal P reserves is essential to determine the time required for recovery.

Once external loads are curtailed, P enriched wetland soils are exposed to an overlying water column with low P concentration. This creates steep gradients in the soluble P concentration between soil and water column. Phosphorus enriched areas have a larger mass of bioavailable soil P than unimpacted areas. For example, when soil cores from impacted areas of the Everglades Water Conservation Area 2a were flooded with low P (<5 μg P/L) water, rapid P flux into overlying water was observed (Fig. 4.7). It is estimated that the impacted soil can sustain this rate of P flux for about five years. This estimate was based on the assumption that about 25% of the total P in the top 30 cm of soil is mobile and potentially can diffuse into the overlying water column. At present, there are major data gaps exist for determining the factors that regulate the recovery of P-impacted wetlands to their historical condition.

Many of Florida's lakes are eutrophic, and some have reached hypereutrophic status. Based on N/P (mass ratio), Canfield (1981) concluded that N may be limiting in hypereutrophic lakes. Thus, these lakes may not respond to small reductions in P loading. External P loads to many Florida lakes have not been quantified. Two shallow lakes, Lake Apopka (central Florida) and Lake Okeechobee (south Florida) have been extensively studied (see Stites and Reddy, 1998; Steinman et al., 1998). Lake Apopka (surface area = 125 km^2) receives P loads of 62 Mg P/yr from vegetable farms (84%), atmospheric deposition (8%), and other sources (8%) (Stites et al., 1997). Lake Okeechobee (surface area = 1,732 km^2) receives annual P loads of 518 Mg/yr (Steinman et al., 1998). The average annual total P concentration in Lake Okeechobee increased from near 50 μg/L in the 1970s to the current 90 μg/L, as a result of P loading from the upstream basin (Flaig and Havens, 1995). In Lake Apopka, the P concentration increased from <60 μg/L to more than 200 μg/L in

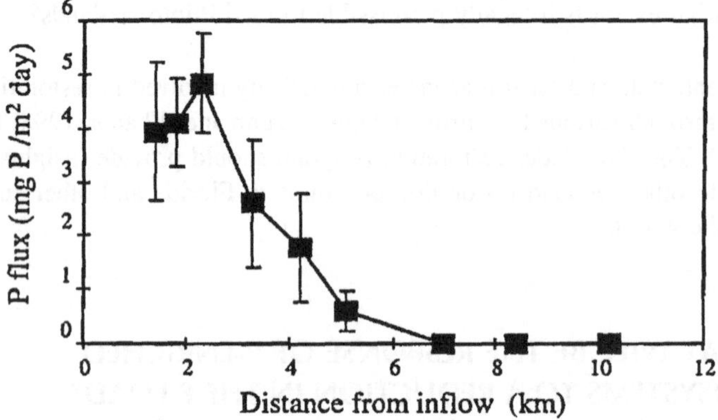

FIGURE 4.7 Phosphorus flux from soils impacted by external P loading in the Everglades (site = Water Conservation Area 2a). Flux was measured on soil cores obtained at several locations as a function of distance from inflow (Fisher and Reddy, 1999, unpublished results).

response to farm discharges, which began in the 1940s (Lowe et al., submitted). For both lakes, intensive efforts are being made to reduce P loads through establishment of loading criteria and implementation of BMPs.

Nonpoint sources of P often dominate loads that lead to the eutrophication of aquatic systems, including lakes. Thus, in many situations, alternative land use management practices have been implemented in an effort to reduce the overall P load to receiving water bodies. For example, historic agricultural management practices north of Lake Okeechobee significantly impacted the water quality of the lake. As a result, this shallow subtropical lake may be moving from a naturally mesotrophic state to a hypereutrophic state (Flaig and Havens, 1995; James et al., 1995). The time needed for lakes to recover after P loads are reduced varies (Sas, 1991). For example, lakes with high sediment P levels may not respond as rapidly to external P load reduction, as lakes with low background P levels (Welch and Cooke, 1995). Shallow lakes apparently respond nonlinearly to P-load reduction with rapid shifts in lake characteristics possible once threshold P concentration are reached and with alternative stable status at intermediate P levels (Moss et al., 1996; Klinge et al., 1994).

Both biotic and abiotic reactions regulate the dissolved P concentration of the water column. In eutrophic lakes, much of the dissolved reactive P added is rapidly assimilated by algae. Dead algal cells, along with the particulate inorganic and organic solids, accrete on bottom sediments. Sediment bound P accretion rate increases with P loading. For example, in Lake Okeechobee, P accretion rates have increased about four-fold since the 1900s (from about 0.25 g P/m^2 yr before 1910 to about 1 g P/m^2 in the 1980s) (Brezonik and Engstrom, 1997). The concentration of total P and its P fractions are generally higher in recent sediments, and decrease with depth, suggesting the influence of increased P loading. Although accretion of sediment bound P suggests that P flux is downward (i.e., from water column to sediments), the dissolved reactive P flux is upward (i.e., from sediments to water column) in response to concentration gradients established at the sediment-water interface.

In shallow lakes, P flux across the sediment-water interface occurs in different modes, depending on meteorological and hydrodynamic conditions. During calm days, when vertical turbulent mixing and bottom shear stress is insufficient to resuspend surface sediment, dissolved P moves via passive diffusion and advection. The processes affecting P exchange in this mode include the following:

1. Diffusion and advection resulting from wind-driven currents
2. Diffusion and advection resulting from flow and bioturbation
3. Processes within the water column (mineralization, sorption by particulate matter, and biotic uptake and release)
4. Diagenetic processes (mineralization, sorption and precipitation/dissolution) in bottom sediments
5. Redox conditions (O_2 content) at the sediment–water interface

During windy periods, resuspension of P from buried sediments may be an important mode of P transfer to the water column. Because sediment resuspension events are transitory, P flux resulting from this process may occur at short-time scales at a rapid rate, as compared to diffusive flux. Estimates of P flux that are due to diffusion only (based on concentration gradients) are shown in Table 4.5. The relative importance of P transfer resulting from diffusive flux and resuspension flux must be quantified to accurately estimate annual P flux from internal sources (see Sheng, 1998 for detailed discussion). This internal load (once the result of external load) can extend the time required for ecosystems to reach their historical condition (Sas, 1989; Chapra and Canale,1991) (Fig. 4.8). Such lag time for recovery should be considered in developing management strategies for an aquatic system.

TABLE 4.5
Diffusive Flux of Soluble P from Sediments to Overlying Water in Selected Aquatic Systems

Aquatic system	Diffusive flux mg P/m² day	Reference
Lake Apopka	1–5.3	Reddy et al., 1996
Lake Okeechobee		Moore et al., 1998
Mud zone	0.1–1.9	
Peat zone	0.2–2.2	
Sand zone	0.1–0.5	
Littoral zone	0.6–1.5	
Lake Barco	0.02–0.05	Reddy and Fisher (unpublished results)
Tampa Bay	1.9–6.3	
Indian River Lagoon	0.7–1.7	

Decisions regarding management and restoration of wetlands and aquatic systems are often difficult and controversial as they involve regulating P loads from external sources. Implementation of P reduction goals can have significant costs and economic impacts. Thus, we need to have a thorough understanding of the dynamics of physical, chemical, and biological processes that regulate water quality within

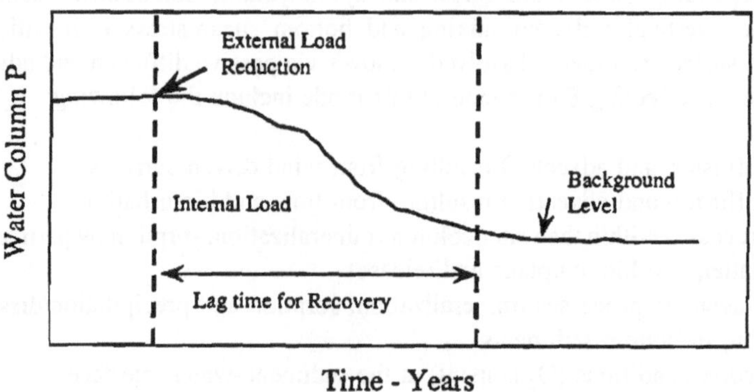

FIGURE 4.8 Schematic showing the influence of internal phosphorus load on water column P concentration and the lag-time for recovery after external loads are reduced.

these ecosystems. This scientific foundation is needed to develop management models, which can be used as decision-making tools, and to evaluate the responses of wetlands and aquatic systems to reduction in P loads. The key questions often asked are

1. Will the wetlands and aquatic systems respond to P load reduction?
2. If so, how long will it take for these systems to recover and reach their historical condition?
3. Are there any economically feasible management options to speedup the recovery process?

Although many similarities exist among different wetlands and aquatic ecosystems with respect to physical, chemical, and biological processes, the management strategies required for each ecosystem are site specific. Each wetland or lake system may respond differently to P load reductions, depending on their historical P loading record and the existing hydrology and geology of the site. Historical water quality data and paleoecological information can be used to determine the characteristics of an original water system quality and ecological conditions. This sets the limit to what can reasonably be achieved. Phosphorus enrichment of sediments and their physicochemical properties can also provide an indicator of the time required for recovery. For example, if Lake Okeechobee sediments are enriched with P to the degree observed in Lake Apopka and other hypereutrophic lakes, then the time required for recovery of Lake Okeechobee would be long. However, P enrichment alone is not an indication of potential recovery time; rather, the proportion of P in the bioavailable pool and the stability of stored P also influence recovery time and impact on water quality. In Lake Apopka, for example, despite many decades of high loading of bioavailable P, the P in sediment is largely (>80%) unavailable. Reduction in P loads through implementation of BMPs are apparently leading to improvements of total P in Lake Apopka and Lake Okeechobee (Flaig and Havens, 1995).

4.8.1 General Discussion

The Florida Department of Environmental Protection (DEP) has developed the following working definition of ecosystem management: "Ecosystem management is an integrated, flexible approach to management of Florida's biological and physical environments–conducted through the use of tools such as planning, land acquisition, environmental education, regulation, and pollution prevention–designed to maintain, protect, and improve the state's natural, managed, and human communities" (FLDEP, 1994). This definition is adapted from the dominant ecosystem management themes presented by Grumbine (1994). One aspect of ecosystem management involves water quality, more specifically, P management within the ecosystem, to reduce eutrophication of water resources. The effectiveness of P management within the ecosystem can be measured by knowing

1. The baseline P concentration of surface water, sediments, and biota
2. The mass P imports and export (budgets) within and between ecosystems

3. The status of biogeochemical indicators of water quality
4. The P transfer mechanisms within and between ecosystems
5. The extent to which P fate and input can be predicted by easily measured water quality parameters (water, soil and sediment, vegetation, and periphyton and plankton) and environmental variables
6. The criteria that will be used to evaluate the ecosystem recovery after restoration plans are implemented

Although state agencies are investing vast amounts of resources for water quality monitoring, many ecosystems remain inadequately monitored to evaluate the adverse impacts of P loading. Many of these monitoring efforts are limited in scope and cannot be readily extrapolated to the whole ecosystem. Very few monitoring programs capture spatial and temporal variations in P concentration in surface water, soils, and sediments. Since sampling analysis is often performed by several agencies, standardization in parameters to be measured, in methods, and in quality assurance is needed.

For effective P management, P budgets need to be developed for the ecosystem. For example, P budgets for Lake Apopka, Lake Okeechobee, the Lake Okeechobee Basin and the Everglades are available (see example of P budget for the Okeechobee Basin) (Fig. 4.9). However, similar budgets are not available for other ecosystems, within each of the five WMDs. This type of information is essential for large-scale understanding of the relative importance of sources and sinks at an ecosystem scale, and for developing P control strategies.

In addition to the P concentrations in water and soil; several biogeochemical characteristics of water, soils and sediments; vegetation and periphyton and plankton can be used as indicators of P impacts on an ecosystem. Selection of these indicators should be based on their ease of measurement, and their sensitivity to changes in P concentration of the water column or P loading.

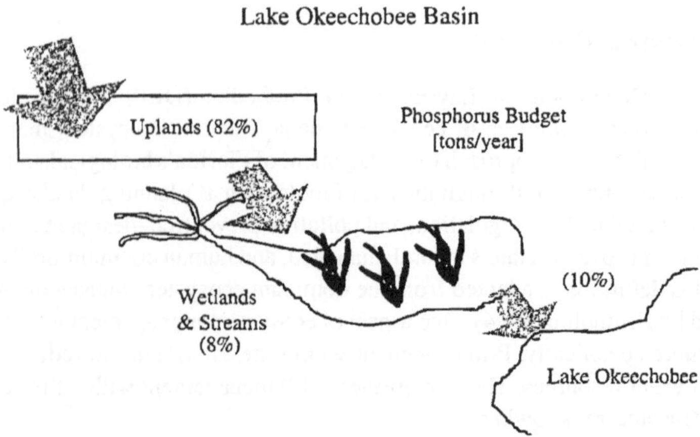

FIGURE 4.9 Schematic showing the P transfer and budgets in upland, wetland and lake ecosystems.

In large-scale ecosystem management and restoration, it is useful to determine what portions of the landscape best indicate the integrity of the entire ecosystem including uplands, wetlands and aquatic habitats. For example, certain patches of an ecosystem (e.g., wetlands or lakes) may contain the best record of overall ecosystem integrity, because of their position in the landscape. Once it is decided on the indicator ecosystem unit or patch, it is necessary to determine what biogeochemical characteristics of the selected area provide the most efficient indicator of pollutant impacts. For example, there is evidence that the rates of certain biogeochemical processes in wetlands reflect changes in materials budget long before such changes are reflected in population of higher organisms. Thus, indicators of biogeochemical processes, such as microbial respiration, organic matter turnover, organic N and P mineralization, phosphate adsorption and desorption, denitrification, and sulfate reduction are some examples, that can provide early indications of impending ecological changes (Reddy and D'Angelo, 1996).

Evaluation of P transfer within and between ecosystems requires knowledge of landscape ecology, hydrology and biogeochemistry. Although, in-situ processes of P and associated nutrients are well understood, the linkage of these processes in P transfer from one unit of an ecosystem to another (e.g. from uplands to wetland) requires further study, especially at the landscape level. At present, P is known to be the key nutrient affecting the productivity of Florida ecosystems. However, we should not ignore the interactive effects of P on other elemental cycles (such as C and N), that may have comparable effects on ecosystems. Phosphorus control strategies should be developed in the context of overall ecosystem response. Although biogeochemical processes may be sensitive and reliable indicators of P impacts on an ecosystem, their measurement can be time-consuming and expensive. The concentration of certain chemical substrates, intermediates, and end products of ecologically important biogeochemical processes may provide rapid and inexpensive indicators of the rates of these processes. If these simple measurements made on an indicator ecosystem unit are well correlated with the related processes, the resulting empirical equations can be used to transfer the process level information to landscape level (Fisher, 1997). To evaluate the P impacts and successes of restoration efforts, it is helpful to develop a fundamental understanding of the biogeochemical process regulating the ecosystem function. Risk assessment is only as good as the information/knowledge available at the time. Lack of understanding of the factors that affect the biogeochemical processes regulating the fate and transport of P and associated nutrients decreases the certainty of an assessment.

4.9 FUTURE RESEARCH NEEDS

Effective P control strategies can be strengthened greatly by improving our understanding of the processes controlling the fate and transport of P within and between ecosystems. Research conducted during the past 10 years on selected Florida's ecosystems provided the basis for developing P control strategies for those drainage basins. Only a few ecosystems have been studied in detail, and it is not known how well rates and processes in these systems can be transferred to other ecosystems. In Florida's ecosystems, P movement is rapid, because of unique soil types and their poor P retention capacity, and rapid movement of water through the soil profile. Many wetlands and aquatic systems are located adjacent to upland ecosystems with active urban and agricultural development, and become recipients of P loads.

The following are some of the key research needs to address P issues in Florida' ecosystems.

1. Quantify P budgets within each ecosystem.These budgets should quantify the amount of P imports and exports for each region in Florida. This should be done for each District's boundaries.
2. Determine historical background levels of P for each ecosystem. This requires extensive paleoecological research or monitoring, especially in unimpacted areas of an ecosystem to determine natural sources of P.
3. Quantify P inputs from wet and dry atmospheric sources in each region of Florida.
4. Establish reference sites, representing an upland, wetland, and aquatic ecosystem for each region. These sites can be used to conduct long-term research and monitoring, as well as assessment of impacts to other sites.
5. Standardize methods for selected parameters to be measured in all ecosystems. Methods followed by scientists and managers within each region should be comparable.
6. Develop calibrated soil test procedures of P availability in uplands used for agriculture, which include water quality as one criteria.
7. Land application of wastes (organic solids) should be tailored properly to soil and hydrologic characteristics of the site, and the composition of the waste. The amount of bioavailable P in the waste should be one of the criteria in determining P application rates.
8. Develop soil/sediment criteria for easily measurable biogeochemical properties. This can be used to determine P impacts and long-term implications.
9. Determine hydrologic pathways governing P movement within uplands, and transfers to adjacent wetland or aquatic ecosystems.
10. Quantify the influence of P loading on other nutrient cycles, and their feedback in regulating eutrophication of wetlands and aquatic systems.
11. Determine the influence of internal P load on recovery of an ecosystem. For uplands, determine how long P stored in these soils will continue to be released even after agricultural activities and P imports are minimized. For wetlands and aquatic ecosystems, determine how long it takes for these ecosystems to recover after all external loads are curtailed.
12. Develop criteria to use in wetlands and aquatic ecosystems as indicators to determine the success (or lack of success) of P management in adjacent uplands systems.
13. Determine the relationships between biogeochemical processes and easily measurable indicators for different ecosystems. Use the resulting relationships to extrapolate process level information to ecosystem scale.
14. Develop empirical (statistical) and mechanistic models to synthesize experimental data and to aid prediction of impacts and recovery.

15. Integrate monitoring data on biogeochemical indicators into predictive models to assess system behavior at the landscape (which includes multiple ecosystems such as uplands, wetlands, and lakes) level.

4.10 ACKNOWLEDGMENT

This is to acknowledge the contribution of Florida Agricultural Station Journal Series No. R-06623.

REFERENCES

Ann, Yoeng-Kwan. 1996. Phosphorus immobilization by chemical amendments in a constructed wetland. Ph.D. Dissertation. University of Florida.

Berkheiser, V.E. Street, J.J., Rao, P.S.C. and Yuan, T.L. 1980. Partitioning of inorganic orthophosphate in soil-water systems. In: CRC Critical Reviews in Environmental Control. CRC Press, Boca Raton, Fl., pp. 179–224.

Brezonik, P.L. and D.R. Engstron. 1997. Modern and historic accumulation rates of phosphorus in Lake Okeechobee, Florida. Journal of Paleolimnology (in press).

Boggess, C.F., E.G. Flaig and R.C. Fluck. 1995. Phosphorus budget-basin relationships for Lake Okeechobee tributary basins. Ecol. Eng. 5:143–162.

Bottcher, A.B., T.K. Tremwell, and K.L. Campbell. 1998. Phosphorus management in flatwood (Spodosols) soils. In. Phosphorus Biogeochemistry of Subtropical Ecosystems: Florida as a Case Example. K.R. Reddy, G.A. O'Connor and C.L. Shelske (eds.) Lewis Pub. (In press).

Burgoa, B. 1989. Phosphorus spatial distribution, sorption and transport in a Spodosol, Ph.D. Dissertation, University of Florida, Gainesville, Fl.

Campbell, K.L., J.C. Capece, and T.K. Tremwel. 1995. Surface/subsurface hydrology and phosphorus transport in the Kissimmee River Basin, Florida. Ecol. Eng. 5:301–330.

Canfield, D.E. 1981. Chemical and trophic state characteristics of Florida lakes in relation to regional geology. Florida Agr. Expt. J. Series. 3513. Univ. of Florida. Pp. 444.

Chapra, S. and R. Canale. 1991. Long-term phenomenological model of phosphorus and oxygen for stratified lakes. Water Res. 25:707–715.

Correll, D.L. 1998. The role of phosphorus in the eutrophication of receiving waters: A Review. J. Environ. Qual. 27:261–266.

Cox, J., R., Kautz, M. MacLaughlin, and T. Gilbert. 1994. Closing the gaps in Florida's wildlife habitat conservation systems. Office of Environmental Services, Florida Game and Fresh Water Fish Commission, Tallahassee, Florida 239 pp.

D'Angelo, E. M. and K. R. Reddy. 1994. Diagenesis of organic matter in a wetland receiving hypereutrophic lake water. I. Distribution of dissolved nutrients in the soil and water column. J. Environ. Qual. 23:937–943.

D'Angelo, E. M. and K. R. Reddy. 1994. Diagenesis of organic matter in a wetland receiving hypereutrophic lake water. II. Role of inorganic electron acceptors in nutrient release. J. Environ. Qual. 23:928–936.

DeBusk, W.F., K.R. Reddy, M.S. Koch, and Y. Wang. 1994. Spatial distribution of soil nutrients in a northern Everglades marsh: Water Conservation Area 2A. Soil Sci. Soc. Am. J. 58:543–552.

FLDACS. 1996. State of Florida Department of Agriculture and Consumer Services. Report on fertilizer materials and fertilizer mixtures consumed in Florida.

FLDEP. 1994. Beginning ecosystem management. Florida Department of Environmental Protection. Report. pp. 39.

Fernald, E.A. and D.J. Patton. 1984. Water resources atlas of Florida. Institute of Science an Public Affairs, Florida State Univ. Tallahassee, Fl. pp. 291.

Fisher, M.M. 1997. Estimating landscape scale flux of phosphorus using geographic information systems (GIS). M.S. Thesis, University of Florida.

Flaig, E.G. and K.E. Havens. 1995. Historical trends in the Lake Okeechobee ecosystem. 1. Landuse and nutrient loading. Arch. Hydrobiol. 107:1–24.

Florida Biodiversity Task Force. 1993. Conserving Florida's Biodiversity. Report to Governor Lawton Chiles. Office of the Governor, Tallahassee, Florida. 43pp.

Florida Sea Grant. 1993. Summary of the Indian River Lagoon economics related to recreational fishing. Florida Sea Grant Program, Gainesville.

Graetz, D.A. and V.D. Nair. 1995. Fate of phosphorus in Florida spodosols contaminated with cattle manure. Ecol. Eng. 5:163–181.

Grumbine, R.E. 1994. What is ecosystem management? Conservation Biology 8:27–38.

Harris, W.G., R.D. Rhue, R.B. Brown, and R. Littell. 1996. Phosphorus retention as related to morphology of sandy coastal plan materials. Soil Sci. Soc. Am. J. 60:1513–1521.

Hinkley, W., and P. Goven. 1996. Solid Waste Management in Florida. Bureau of Solid and Hazardous Waste, Division of Waste Management, Florida Department Environmental Protection, Tallahassee, FL pp. 124.

Izuno, F. and P.J. Whalen. 1998. Phosphorus management in organic (Histosols) soils. *In*. Phosphorus Biogeochemistry of subtropical Ecosystems: Florida as a Case Example. K.R. Reddy, G.A. O'Connor and C.L. Shelske (eds.) Lewis Pub. (In press).

James, R.T., V.H. Smith, and B.L. Jones. 1995. Historical trends in the Lake Okeechobee Ecosystems. II. Nutrient Budgets. Archiv fur Hydrobiologie/Supplement (Monographische Beitrage) 107:25–47.

Kadlec, R.H. and R.L. Knight. 1996. Treatment wetlands CRC press, Boca Raton, FL.

Kidder, G., E.A. Hanlon, and C.G. Chambliss. 1997. UF/IFAS standardized fertilization recommendations for agronomic crops. Fact Sheet SL-129. Florida Cooperative Extension Service, IFAS, University of Florida. pp.7

Kidder, G., G.J. Hochmuth, D.R. Hensel, E.A. Hanlon, W.A. Tilton, J.D. Dilbeck, and D.E. Schrader. 1991. Potatoes: Horticulturally and environmentally sound fertilization of Hastings Area producers. U.S. Dept. of Agriculture special project No. 91-EWQI-9292.

Klinge, M, M. Grimm, and S. Hosper. 1994. Eutrophication and ecological rehabilitation of Dutch lakes: Explanation and preduction by a new conceptual framework. In: Living with Water. Proceedings of the 1994 International Conference on Integrated Water Resources Management, RAI, Amsterdam, The Netherlands.

Koch, M.S., and K.R. Reddy. 1992. Distribution of soil and plant nutrients along a trophic gradient in the Florida Everglades. Soil Sci. Soc. Am. J. 56:1492–1499.

Landers, C.H., D. Moffitt, and K. Alt. 1998. Nutrients available from livestock manure relative to crop growth requirements. USDA-NRCS Resource Assessment and Strategic Planning working paper 98-1. http:/www.nhq.nrcs.usda.gov/land/pubs/nweb.html.

Lowe, E.F. and L.W. Keenan. 1997. Managing phosphorus-based, cultural eutrophication in wetlands: a conceptual approach. Ecol. Eng. 9:109–118.

McCormick, P.V. and M.B. O'Dell. 1996. Quantifying periphyton responses to phosphorus in the Florida Everglades: A synoptic experimental approach. J.N. Am. Benthol. 15:450–468.

McCoy, J.L., L.T. Sikora, and R.R. Weil. 1986. Plant availability of phosphorus in sewage sludge compost. J. Environ. Qual. 15:405–409.

Moore, P.A., and D.M. Miller. 1994. Decreasing phosphorus solubility in poultry litter with aluminum, calcium, and iron amendments. J. Environ. Qual. 23:325–330.

Moore, P.A., K.R. Reddy, and M.M. Fisher. 1998. Phosphorus flux between sediment and overlying water in Lake Okeechobee, Florida. J. Environ. Qual. (in review).

Moss, B., J. Madgwick, and G. Phillips. 1996. A guide to the restoration of nutrient-enriched shallow lakes. Broads Authority, Norfolk, U.K.

Moustafa, M.Z. and M.J. Chimney. 1998. Phosphorus retention by the Everglades nutrient removal project; An Everglades stormwater treatment area. *In*. Phosphorus Biogeochemistry of Subtropical Ecosystems: Florida as a Case Example. K.R. Reddy, G.A. O'Connor and C.L. Shelske (eds.) Lewis Pub. (In press).

Nair, V.D., D.A. Graetz, and K.R. Reddy. 1998. Influence of dairy manure on phosphorus retention capacity of spodosols. J. Environ. Qual. (in press)

Newman S. and J.S. Robinson. 1998. Forms of organic phosphorus in water, soils and sediments. *In*. Phosphorus Biogeochemistry of Subtropical Ecosystems: Florida as a Case Example. K.R. Reddy, G.A. O'Connor and C.L. Shelske (eds.) Lewis Pub. (In press).

Obreza, T.A. 1997. Nutrient management for residuals application sites. In. Biosolids Management in Florida. Florida Dept. Environmental Protection, Tallahassee, FL pp. 23–28.

Parry, R. 1998. Agricultural phosphorus and water quality: A. U.S. Environmental Protection Agency Perspective. J. Environ. Qual. 27:258–261.

Rechcigl, J.E. and A.B. Bottcher. 1995. Fate of phosphorus on bahiagrass (*Paspalum notat*um) pastures. Ecol. Eng. 5:247–259.

Reddy, K. R. 1979. Land areas receiving organic wastes: Transformations and transport in relation to nonpoint source pollution. Environmental Impact of Nonpoint Source Pollution (eds. M. R. Overcash and J. M. Davidson). Ann Arbor Sci., Ann Arbor, MI, p. 243274.

Reddy, K.R., and E.M. D'Angelo. 1996. Biogeochemical indicators to evaluate pollutant removal efficiency in constructed wetlands. Wat Sci. Tech. 35:1–10.

Reddy, K.R., E.G. Flaig, and D.A. Graetz. 1996. Phosphorus storage capacity of uplands, wetlands and streams of the Lake Okeechobee basin. Agriculture, Environment and Ecosystems 59:203–216.

Reddy, K.R., Fisher, M.M., Ivanoff, D. 1996. Resuspension and diffusive flux of nitrogen and phosphorus in a hypereutrophic lake. J.Environ. Qual. 25:363–371.

Reddy, K. R., M. R. Overcash, R. Khaleel, and P. W. Westerman. 1980. Phosphorus adsorption-desorption characteristics of two soils used for disposal of animal wastes. J. Environ. Qual. 9:8692.

Reddy, K.R. J.S. Robinson, and Y. Wang. 1997. Phosphorus release potential of constructed wetlands in the Emeralda Marsh Conservation Area. Final Report submitted to St. Johns River Water Management District, Palatka, FL p. 77.

Reddy, K. R. 1982. Mineralization of nitrogen in organic soils. Soil Sci. Am. J. 46:561566.

Reddy, K. R. 1983. Soluble phosphorus release from organic soils. Agriculture, Ecosystems, and Environment 9:373382.

Reddy, K.R. and D.A. Graetz. 1990. Internal nutrient budget for Lake Apopka. Final Rep. Spec. publ. SJ-91-SP6. St. Johns River Water Management District, Palatka, Fl.

Redfield, G.W. 1998. Quantifying atmospheric deposition of phosphorus: A conceptual model and literature review for environmental management. South Florida Water Management District. Techn. Publ. WRE #60 pp. 35 West Palm Beach, FL.

Richardson, C.J., S. Qian, C.B. Craft and R.G. Qualls. 1997. Predictive models for phosphorus retention in wetlands. Wetland ecology and Management. 4:159–175.

Rhue, D. and W.G. Harris, 1998. Phosphorus sorption/desorption reactions in soils and sediments. *In*. Phosphorus Biogeochemistry of Subropical Ecosystems: Florida as a Case Example. K.R. Reddy, G.A. O'Connor and C.L. Shelske (eds.) Lewis Pub. (In press).

Rochlich, G.A. and D.J. O' Connor. 1980. Phosphorus management for the Great Lakes. Final Rep. Phosphorus Management Strategies Task Force. PLUARG Tech. Rep. Int. Joint Commission, Windsor, ON, Canada.

Sas, H. 1989. Lake restoration by reduction of nutrient loading. Academia Verlag Richarz, St. Augustin.

Salas, H. and P. Martino. 1991. A simplified phosphorus trophic state model for warm-water tropical lakes. Water Res. 25:344–350.

Sharpley, A.M. 1991. Soil phosphorus extracted by iron-aluminum-oxide-impregnated filler paper. Soil Sci. Soc. Am. J. 55:1038–1041.

Sharpley, A.N. 1993. An innovative approach to estimate bioavailable phosphorus in agricultural runoff. J. Environ. Qual. 22:597–601.

Sharpley, A.N., Chapra, S.C., Wedepohl, R., Sims, J.T., Daniel, T.C. and K.R. Reddy. 1994. Managing agricultural phosphorus for protection of surface water: Issues and options. J. Environ. Qual. 23:437–451.

Sharpley, A.N., T.C. Daniel, J.T. Sims, and D.H. Pote. 1996. Determining environmentally sound soil phosphorus levels. J. Soil Water Conserv. 51:160–166.

Steinman, A.D., K.E. Havens, N.G. Aumen, R.T. James, K.R. Jin, and J. Zhang. 1998. Phosphorus in Lake Okeechobee: Sources, sinks, and strategies. In. Phosphorus Biogeochemistry of Subtropical Ecosystems: Florida as a Case Example. K.R. Reddy, G.A. O'Connor and C.L. Shelske (eds.) Lewis Pub. (In press).

Stites, D.L, M. Coveney, L. Battoe, and E. Lowe. 1997. An external phosphorus budget for Lake Apopka. Tech. Memorandum. St. Johns River Water Management District, Palatka, FL. pp. 86.

SWIM. 1992. Surface water improvement and management plan for the Everglades: Supporting information document. South Florida Water Management District (SFWMD). West Palm Beach.

Vollenweider, R.A. 1976. Advances in defining critical loading levels of phosphorus in lake eutrophication. Mem. 1st. Ital. Idrobiol. 33:53–83.

Vymazal, J.C.B. and C.J. Richardson. 1995. Species composition, biomass, and nutrient content of periphyton in the Florida Everglades. J. Phycol. 31:343–354.

Welch E., and G.D. Cook. 1995. Internal phosphorus loading in shallow lakes: Importance and control. Lake Reserv. Manage. 11(3):273–281.

Wetzel, R.G. 1998. Organic phosphorus mineralization in soils and sediments. *In*. Phosphorus Biogeochemistry of Subtropical Ecosystems: Florida as a Case Example. K.R. Reddy, G.A. O'Connor and C.L. Shelske (eds.) Lewis Pub. (In press).

Wolf, A.M. and D.E. Baker. 1985. Criteria for land spreading of sludges in the northeast: Phosphorus. Pages 39–41 in D.E. Baker, D.R. Bouldin, H.A. Elliott, and J.R. Miller (eds.) *Criteria and recommendations for land application of sludges in the northeast.* Penn. Agric. Exp. Stn. Res. Bull. No. 851. The Pennsylvania State University, University Park.

Wolf, A.W., D.E. Baker, H.B. Pionke, and H.M. Kunishi. 1985. Soil tests for estimating labile, soluble, and algae-available phosphorus in agricultural soils. J. Environ. Qual. 14:341–348.

Yuan, T.L. and Lucas, D.E. 1982. Retention of phosphorus by sandy soils as evaluated by adsorption isotherms. Soil Crop Sci. Soc. Fla. Proc., 41:195–201.

5 Introduction to Soils of Subtropical Florida

Willie Harris and Wade Hurt

5.1 ABSTRACT

Subtropical Florida contains a variety of ecological communities, including expansive fresh-water wetlands. Ecological balance within and among communities is sensitive to nutrient availability, and nutrient flux is greatly influenced by soils. Hence, the understanding and prediction of human ecological impacts in subtropical Florida require an assessment of soils.

Soil distribution in the region relates to parent material, topography, and vegetation. Marine-derived quartz sand and $CaCO_3$ are the main parent materials. These materials yield highly contrasting soils, but have in common the fact that neither alters to form other minerals. Thickness of sand decreases seaward of topographic inflections from the last Pleistocene shoreline, and sand is very thin or absent in the southern extremes of the region. Topography influences soil mainly through its effects on hydrology and vegetation. The broad trough of the Everglades, for example, has maintained hydrologic conditions favorable for saw grass marsh over a sufficient time to form a vast area of organic soils (Histosols).

Soil orders which occur in subtropical Florida, in order of areal extent, are Histosols, Spodosols, Entisols, Alfisols, Mollisols, and Inceptisols. They are distin-

1-56670-331-X/99/$0.00+$.50

guished from one another by organic matter content (Histosols) and presence or absence of diagnostic horizons. Most Histosols formed from saw grass roots and rhizomes. Spodosols, characterized by illuvial C and Al, are associated with fluctuating water tables and sandy textures common to flatwoods. Entisols tend to be either deep sands with no diagnostic horizon within 2 m depth, or featureless accumulations of periphyton-precipitated $CaCO_3$ (marl). Some marl Entisols may soon be reclassified as Histosols based on their biogenic origin. Alfisols, which have a finer-textured subsurface horizon, are common in sloughs and depressions, particularly in areas east of Lake Okeechobee. A thick, dark, high-base surface horizon is definitive of Mollisols, which occur on the Keys and along river systems and lake margins. Inceptisols commonly have thick organic surface horizons that are too thin to be Histosols. Most soils of the region are wet enough to classify in aquic suborders, with notable exceptions being sandy Entisols of the central ridge.

Management of agricultural nutrients in subtropical Florida is challenged by prevalent hydrologic and soil conditions that are unfavorable to nutrient retention. The pathway of nutrients from point of application to vulnerable water bodies is often short, and the retardation via soil interaction, minimal.

5.2 INTRODUCTION

Subtropical Florida is a varied region with respect to soils, landforms, and plant communities. It contains the Everglades, a vast system of marshes, sloughs, tree islands, and cypress forests that extends southward from Lake Okeechobee to Florida Bay (VanArman et al., 1984). The Everglades encompasses one of the largest systems of fresh water (saw grass) marsh in the world, along with huge expanses of organic soils associated with the saw grass ecosystem. Soils dominated by $CaCO_3$ (marl) also occur extensively in the Everglades and adjacent ecosystems. Such massive accumulations of marl and organic matter in a warm, humid climate require a restrictive set of hydrological conditions whereby very broad areas have water above to just below the soil surface much of the time. These conditions are in part attributable to the nearly level topography and very low elevation (above mean sea level) of the areas. The potentiometric and conductive characteristics of south Florida aquifers are also important factors. Not all of subtropical Florida is wet, however. The region also includes the rolling, deep, droughty sands of the sand pine scrub on the central ridge and coastal strand, and the nearly level multicolored sands of the flatwoods.

Plant communities have evolved in adaptation to soil and hydrologic conditions of subtropical Florida, such that there are natural associations among plants, soils, and hydrology. These associations have been disrupted, however, by human activities in many areas. Some effects of human intervention are obvious, such as in the cases of agricultural fields and subdivisions. Other effects are more subtle, and may be beyond the view of agricultural or urban activities. The ecological balance within the plant communities of subtropical Florida is particularly sensitive to nutrient availability. In effect, artificial enrichment of certain nutrients can eliminate adaptive advantages of species associated with certain soils and landforms, and thereby initiate ecological shifts (Federico er al., 1981, Allen et al., 1982, Koch and Reddy,

1992). Soils are the major repository of natural and agriculturally applied nutrients. Their chemical and hydraulic properties strongly control nutrient availability and flux. Hence, the understanding and prediction of human ecological influences in subtropical Florida require an assessment of soils.

The purpose of this chapter is to provide basic information about the soils of subtropical Florida that can serve as a reference and "point of departure" for the other chapters of this book. "Subtropical Florida," as used in this chapter, is defined as the region with coldest monthly mean temperature, corrected to sea level, of between 16° and 20° C for one or more months. This includes the area of Florida south of latitude 27.50 (Fig. 5.1).* Other definitions concerning the subtropics are available (Bailey, 1996).

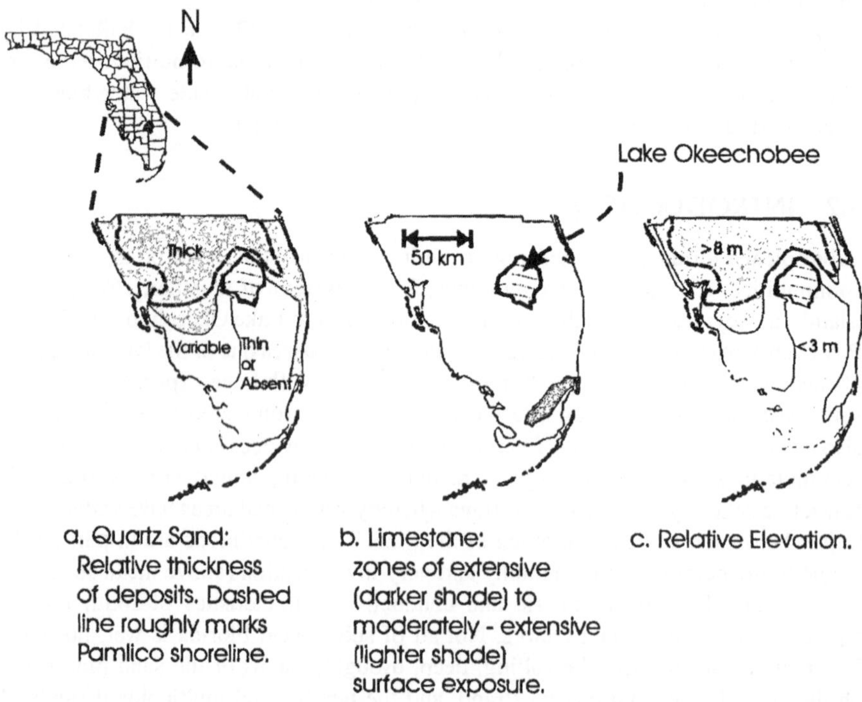

FIGURE 5.1 General maps of selected geographic characteristics that relate to soil and plant distributions in subtropical Florida. The light gray zone of 1c (<3 m) corresponds roughly to the Everglades saw grass marsh (compare with Fig. 5.2), in addition to mangrove communities along the coast.

* Note: We have, for all maps presented, used county boundaries that in our judgement most closely correspond to the approximate northern extent of the region we have defined in the text as subtropical Florida. These maps are intended only to enable comparison of overall trends in several relevant geographic aspects of the region. They were developed after studying a number of geographic information sources (Cooke, 1945; Jones, 1948; Healy, 1975; Scott, 1980; Florida Soil Conservation Service, 1989; Snyder et al., 1990; Gleason and Stone, 1994; Meindl, 1996). Fine detail has been eliminated by "smoothing" to fit the small scale and exclusively instructional purpose of the maps.

A great deal of the soil-related information conveyed in this chapter was gained through the efforts of soil survey professionals. Modern soil survey reports are (or will soon be) available for the following counties in subtropical Florida: Broward (Pendleton et al., 1984), Charlotte (Henderson et al., 1984), Collier (Liudahl et al., 1998), Dade (Noble et al., 1996), DeSoto (Cowherd et al., 1989), Glades (Carter et al., in press), Hardee (Robbins et al., 1984), Hendry (Belz et al., 1990), Highlands (Carter et al., 1989), Lee (Henderson et al., 1984), Manatee (Hyde and Huckle, 1983), Monroe (Hurt et al., 1995), Okeechobee (Carter et al., in press), Palm Beach (McCollum et al., 1978), Sarasota (Hyde et al., 1991) and St. Lucie (Watts and Stankey, 1980). For a comprehensive coverage of Florida soils on a statewide basis, we refer readers to the chapter entitled *Soils* (Brown et al., 1990), in *Ecosystems of Florida* (R.L. Myers and J.J. Ewel, editors).

The following section is an overview of the factors most relevant to soil formation and distribution in subtropical Florida. It is intended as a foundation for the subsequent discussions pertaining to the soils themselves.

5.3 FACTORS AFFECTING SOIL DISTRIBUTION IN SUBTROPICAL FLORIDA

The factors having the greatest control over the geographic differentiation of soils in subtropical Florida are parent material, topography, and vegetation. Two other classical factors, climate and time, have an important effect on soil weathering intensity (rate and duration), but are less definitive on a local scale within this region than are the other three factors. Soil forming factors are conceptually considered to be independent variables with respect to their influence on soils, however, they are not necessarily independent from each other, as will be evident from the following discussion.

5.3.1 Parent Material

Quartz sand and calcium carbonate (of limestone origin) are the two major soil parent materials of subtropical Florida. We consider mucks (organic materials), that are prevalent in the region, to be soils rather than parent materials. The soil-vs.-parent material distinction is not clear-cut for marl, which is a silt-sized $CaCO_3$-dominated material formed by microbially induced precipitation within Ca-rich shallow water environments (discussed later). We consider the initial *in-situ* accretion of marl, in the presence of emergent plants, as a soil-forming process. However, marl that is reworked by tides or storms would be sediment, and potential parent material for soil.

Both quartz sand and limestone are derived from marine sediments of relatively recent origin (mainly Quaternary and Holocene). Sandy materials are thick in the northern part of the region, and extending for some distance down the Atlantic coast (Fig. 5.1a). The region of greatest thickness corresponds roughly to the last Pleistocene shoreline (Pamlico) (Cooke, 1945; Healy, 1975), south of which the sand deposits becomes appreciably thinner. There are areas near the southern tip of the Florida peninsula that are so devoid of quartz sand that bare limestone is extensively exposed at the land surface (Fig. 5.1b). Marl-over-limestone is also a common sequence as the southern tip is approached.

Quartz grains in the region are mainly sand-sized, and marl particles, silt-sized. The terms "sand" and "quartz" are often used interchangeably in Florida, as are the terms "marl" and "calcitic muds." However, mineralogy/particle-size generalizations are sometimes not justified. For example, the sand in some areas of subtropical Florida contains appreciable quantities of carbonate shell fragments. Quartz, where it occurs in the southern reaches of the region, overlies the older limestone. In some areas it is well mixed with secondarily formed marl materials within the soil.

A scarcity of clay-sized aluminosilicate minerals is particularly characteristic of soils in sand-depleted areas. Secondary phyllosilicates such as smectite and sepiolite are present in detectable amounts in only a small minority of the soils. Clay, where it does occur, may be mixed in with the organic materials, or incorporated in solution channels and crevices of limestone bedrock outcropping at the land surface. Some clay, probably of detrital origin, is associated with most of the sandy materials.

Quartz and $CaCO_3$ (mainly calcite, but some aragonite is also present) are "strange bedfellows" in a geochemical sense. Quartz is a ubiquitous and stable mineral in soils, whereas $CaCO_3$ is normally rare in humid regions where acidic soil conditions commonly prevail. However, the two minerals have in common the fact that neither undergoes alteration to secondary clay-sized minerals. Thus, most soils in subtropical Florida start out clay-poor and stay that way.

Soils that form in quartz sand are obviously quite distinct from those associated with carbonate materials. Quartz imparts little chemical control to soils, whereas carbonate is a strong alkaline buffer. Quartz persists under all drainage conditions, but marl soils are restricted to wet environments where they can form and remain stable under the influence of underlying limestone. Carbonates are quickly lost from soils in leaching environments, even if limestone is subjacent. The formation of subsurface soil horizons enriched in metals and carbon is a widespread occurrence in quartz sands on flatwoods landforms throughout Florida, but these horizons do not form in carbonate parent material even when the latter occur under similar geomorphic and hydrologic circumstances.

5.3.2 Topography

Topography is closely related to soil parent material in subtropical Florida, but it has its own influence on soil formation through control of hydrology. There is an elevational break roughly corresponding to the Pamlico shoreline, above which occur the deeper sands (as alluded to above) of the Pamlico and older Pleistocene marine terraces (Healy, 1975) (Fig. 5.1a and c). Poorly drained flatwoods is the predominant landform corresponding to the terraces and areas of deep sands, but other landforms are interspersed within the flatwoods "matrix." The latter include better-drained sand hills, and more poorly drained sloughs (sometimes called "flats"). Flood plains of variable drainage are also prevalent along river systems. More recent sands associated with the contemporary Atlantic shoreline form strands of intermediate elevations. Lowest elevations commonly occur where sands are thinnest, and where marl or bare limestone rock are the dominant parent material. However, the southern tip of the Atlantic coastal ridge, a local topographic high, also is an area of exposed limestone (Fig. 5.1b and c).

The apparently flat landscape so prevalent in subtropical Florida belies a profound topographic control on the distribution of ecological communities, as well as on the soils associated with those communities (see next section). In no place is this more evident than in the Everglades and the adjacent region of cypress swamps to the west known as Big Cypress. These landscapes have no perceptible slope for as far as can be seen, but abrupt plant-community boundaries attest to subtle changes in the elevation of the underlying substrate relative to that of the water table. Plants effectively amplify subtle elevational changes by virtue of their sensitivity to water conditions, as discussed below.

5.3.3 Vegetation

Knowledge of how vegetative patterns relate to soils and landscapes is highly relevant to the ecological understanding of subtropical Florida. Plant distribution is the most readily observable landscape feature in a region where extremely low topographic relief can make the direct distinction between landforms virtually impossible. A few centimeters difference in elevation in subtropical Florida can make the difference between a flooded and nonflooded soil surface, and produce a not-so-subtle boundary between herbaceous and woody species. This boundary also usually corresponds to significant soil differences, in large part because of the influence of the vegetation on the soil.

There is a variety of natural plant communities in subtropical Florida (Fig. 5.2). The plants making up each community have a mutually adaptive advantage within the soil-topographic-hydrologic setting. Community boundaries can be abrupt and readily determinable, in which case they are commonly obvious even to an untrained observer. They can also be gradual and less precisely established. In the latter case, the same plants grow within both communities along the transitions. Also, some plants are sufficiently adaptable and competitive that they are commonly found in a number of communities. Plants are not just passive ecological opportunists, but themselves exert an appreciable influence on the local environment. For example, their evapotranspiration rate can be a significant factor in the local water budget. Also, accreting plant remains make up the bulk of the soil over extensive areas, particularly those areas dominated by saw grass marsh. In effect, plant communities and associated components (soils, micro-organisms, wildlife, etc.) comprise ecological subsystems within the overall ecology of the region. Additional information about Florida ecosystems is contained in *26 Ecological Communities of Florida* (Florida Soil Conservation Service. 1989) and in *Ecosystems of Florida* (R.L. Myers and J.J. Ewel, editors).

It is important to understand that some ecological associations are less consistent than others, and are due both to natural- and human-related confounding factors. There are areas of subtropical Florida, particularly to the north and along the east coast, where only small patches of climax vegetation have not been supplanted by agricultural and urban uses. Also, a variety of soils may occur within and among areas of a given plant community that are due mainly to parent material variations.

However, the hydrological variations that tend to have a strong influence on plant adaptation also affect soils. Thus, some soils have a much greater likelihood to occur under a given plant community than other soils.

The next few paragraphs of this section briefly describe the plant communities that are reasonably common in subtropical Florida. General soil trends are also presented, but soils and their distributions will be discussed in greater detail in the following section. A generalized distribution map (Fig. 5.2) shows areas of concentration for selected communities that are relatively extensive.

Saw Grass Marsh

This community is most extensive in the Everglades, but smaller areas occur throughout subtropical Florida. Areas appear as open expanses of saw grass *(Cladium jamaicense)* with little or no other vegetative species. Saw grass is favored by frequent shallow flooding, a prevalent condition in the region. The unusually large areas of saw grass in the Everglades owe their existence to the nature of the landscape: a very broad, very flat land surface at near sea-level elevation (compare Fig.1c and Fig. 5.2). The long-term (about 5,000-year) life span of the Everglades saw grass communities has resulted in the formation of organic soils (discussed below). Much of the original saw grass community has been converted to sugarcane production near Lake Okeechobee and to vegetable crop production in south Dade County. Invasive species such as cattail *(Typha)*, Brazilian pepper *(Schinus terebin-*

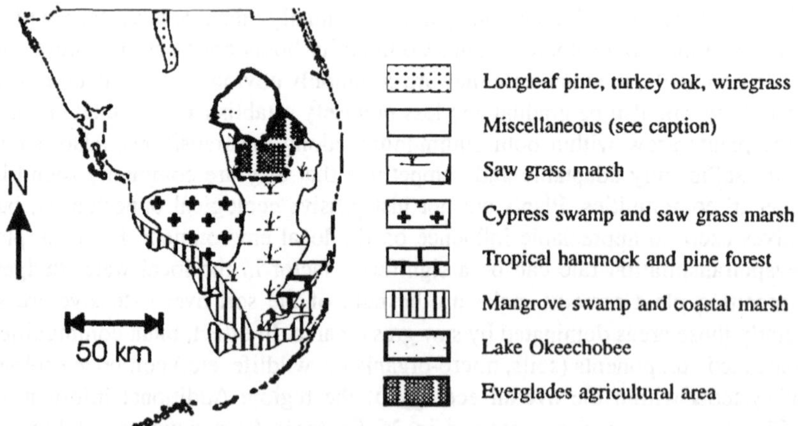

FIGURE 5.2 General map showing areas in subtropical Florida where important plant communities occur most extensively (not exclusively!). The saw grass marsh is the Everglades, and corresponds to a zone of very low elevations shown in Fig. 5.1c. The "miscellaneous" delineation shown in white is an area where agricultural and urban activities have extensively supplanted native plant communities. The dominant landform of this area is flatwoods, but many other associated landforms occur as well. The longleaf pine-turkey oak hills community is delineated despite its widespread conversion to citrus because it provides a useful visual reference to the central (Lake Wales) ridge of peninsular Florida.

thifolius), and melaleuca *(melaleuca quinquenervia)* have become established in some areas influenced by agriculture. Cattails are linked to nutrient enrichment, Brazilian pepper to rock plowing, and melaleuca to hydrologic modifications. However, these plants are tenacious opportunists in a general sense.

Coastal Strand Community

This community is so named because of its association with coastal landforms. The natural vegetation includes a variety of low-growing grasses, vines and herbaceous plants, with few trees or large shrubs. Trees (pine, palm, and oak) and shrubs often occur in stunted form because of the natural forces of wind, salt and blowing sand. Coastal strands occur along the Atlantic Ocean from Brevard County to Key West in Monroe County and along the Gulf of Mexico from Manatee County to Naples in Collier County. Isolated areas of this system also occur in Everglades National Park, most notably along Highlands Beach and on Cape Sable. These ecosystems are oriented parallel to coastal beaches, bays, and sounds and encompass the area affected by salt spray from the ocean, gulf, and salt water bays. They occur on nearly level to strongly sloping land. The soils associated with coastal strands are variable with respect to drainage and extent of development, though most (not surprisingly) have very sandy textures.

Sand Pine Scrub

Sand pine *(Pinus clausa)* are most common in the northern parts of subtropical Florida, inland from the coast and in the central portion of the area. Individual areas of sand pine scrub are generally small in extent (less than 100 hectares). These areas typically have smaller, wetter ecosystems interspersed throughout. Sand pine scrub communities occur on nearly level to strongly sloping land. They are easily identified by the even-aged stands of sand pine with thick scrubby oak understory. Ground cover under the trees and shrubs is often sparse, and large areas of barren soil are often noticeable. In other cases, the sand pine are scattered or absent and scrubby oaks are the dominant vegetation. The plants of this community are those that adapt well to low available water and periodic natural fire. Soils tend to be deep, sandy, droughty, and nutrient-poor.

Longleaf Pine-Turkey Oak Hills

This community occurs mostly in the northwestern part of the region. Plants have similar drought and fire tolerance to those of the sand pine scrub community. Individual areas vary widely in size, occupying nearly level to steep slopes. Logged areas tend to have larger and greater numbers of turkey oaks *(Quercus laevis)*. Soils, like those of the sand pine scrub, tend to be deep and sandy. Much of this ecosystem has been converted to citrus.

Flatwoods Community

This community name stems from its strong association with flatwoods landforms, and the fluctuating water table conditions associated with them. Flatwoods landforms, as the name implies, are nearly level. They cover more land area than any other landform in subtropical Florida, but the native plant community has been extensively converted to pasture, citrus, and vegetable crops. The community consists of an overstory of slash pine *(Pinus elliottii)* with an understory dominated by saw palmetto *(Serenoa repens)*. Soils underlying these communities in most areas are deep, sandy, nutrient-poor, poorly drained, and typically have subsurface accumulations of aluminum and carbon (as discussed in the following section). However, flatwoods near the southern extreme of the Florida peninsula and along the highest elevations of the Florida Keys, which are sand-poor areas (Fig. 5.1a), may have shallow soils and limestone outcrops. Community composition of the latter flatwoods varies somewhat from that of the more typical deep-sand flatwoods, however, the pine-and-palmetto dominance is evidence that hydrology is more influential than parent material in the case of these species.

Cabbage Palm/Oak Hammock

Areas of this community occur in Martin and Glades counties and in Highlands, Okeechobee, and surrounding counties as small isolated areas. They occupy slightly concave to slightly convex, nearly level land and are easily identified by the dense canopy of cabbage palm, live oak, and laurel oak. This community normally includes an overstory of cabbage palm *(Sabal palmetto)* and oaks, with a sparse park-like understory dominated by saw palmetto. A variety of poorly to somewhat poorly drained soils occur under this community. Some of areas have been converted to citrus, vegetable crops, and pasture.

Tropical Hammock

This community occurs on elevated areas in the Everglades and on ridges of the Florida Keys. Areas tend to be small and isolated. Tropical hammocks are commonly interspersed with flatwoods communities. They normally occupy nearly level land. Shallow depth (<50 cm) to bedrock is dominant. Areas are recognized by dense "jungle-like" clumps or strands of small- to medium-sized trees. Typifying plants include poison wood *(Metopium toxiferum)*, gumbo limbo *(Bursera simaruba)*, mastic *(Sideroxylon foetidissimum)*, tamarin *(Lysiloma latisiliqua)*, locust berry *(Byrsonima lucida)*, marlberry *(Ardisia* spp.), and wild coffee *(Psychotria nervosa)*. Soils are commonly organic because of the build-up of leaf litter, in the absence of sand, over limestone or coral. This community is unique in that it promotes the formation of relatively well-drained organic soils within a warm, humid climate. Most organic soils of the region and elsewhere, by contrast, form under very poorly drained conditions. Few areas of this ecosystem have been converted to urban or agricultural uses, owing mainly to their isolated or protected status.

Scrub Cypress

A large area west of the Everglades (Fig. 5.2), generally known as Big Cypress Swamp, is characterized by scrub cypress communities. They appear as broad areas of marsh with draft cypress (less than 7 m tall) scattered throughout. Scrub cypress are maintained by extreme seasonal changes in water saturation and low levels of plant nutrients. They occur on flat to slightly depressional land and are easily identified by a sparse canopy of even-sized cypress trees and marshy understory. The natural vegetation of this ecosystem is low growing grasses, vines and herbaceous plants with few cypress trees or large shrubs. Cypress trees and shrubs often occur in stunted form because of the wetness. Typifying plants include bald cypress *(Taxodium distichum)*, pond cypress *(Taxodium distichum* var. *nutans)*, waxmyrtle *(Myrica cerifera)*, cabbage palm, air plants *(Tillandsia setacea)*, maidencane *(Panicum hemitomon)*, and clubhead cutgrass *(Leersia hexandra)*. A variety of poorly and very poorly drained soils are associated with this ecosystem. Few areas have been converted to crops or pastures, but some areas have been converted to urban uses.

Cypress Swamp

Areas of this community occur interspersed with other communities such as Flatwoods and Sloughs. They form strand-like to oval-shaped clusters of trees within depressional landscapes. An individual community often appears dome-like, because of the tendency for taller trees to occur toward the center of the cluster. The predominant species is bald cypress, which is often the only plant found in large numbers. A variety of soils may occur with this ecosystem, but they have in common very poor drainage. Both organic and inorganic soils are found in this community.

Mangrove Swamp

Mangrove swamps occur extensively along saltwater shorelines throughout subtropical Florida. Shorelines that support this ecosystem are back bays and estuary fringes that have mild wave action. Mangrove communities are easily identified by the dominantly thick canopy of mangrove trees of about 4 to 10 m height and little other vegetation. Also, red mangrove *(Rhizophora mangle)*, with conspicuous prop roots, is the most seaward plant. Other plants that occur in this community include black mangrove *(Avicennia germinans)*, white mangrove *(Laguncularia racemosa)*, and button mangrove *(Conocarpus erectus)*, along with leather fern *(Acrpstocji, aureum)* and sea purslane *(Sesuvium portulacastrum)*. Mangrove swamps are probably second only to saw grass marshes in terms of areal occurrence of organic soils. Marl soils also support this community.

Salt Marsh

This community mostly occurs in small isolated areas along the Atlantic and Gulf Coasts, and along tidal rivers. Areas appear as open expanses of grasses, sedge, and rushes with a matrix of interconnected shallow tidal creeks and streams. They are easily identified by the lack of trees, few shrubs, and dominance of salt tolerant grasses and grasslike plants. Species include seashore saltgrass *(Distichlis spicata)*, seashore dropseed *(Sporobolus virginicus)*, seashore paspalum *(Paspalum vaginatum)*, smooth cordgrass *(Spartina alterniflora)*, gulf cordgrass *(Spartina spartinae)*, marshhay cordgrass *(Spartina patens)*, big cordgrass *(Spartina cynosuroi-*

des), and black neddlerush *(Juncus roemeiranus)*. Very poorly drained organic soils predominate.

Swamp Hardwood

This community occurs along the Peace and Caloosahatchee Rivers, and in Corkscrew and Fakahatchee Swamps. It is also found in other small isolated areas throughout subtropical Florida, except the southeast portion of the region. The vegetation is primarily deciduous trees. Annual flooding, thick canopies, and dense understories are characteristic of this community. Typifying plants include red maple *(Acer rubrum)*, American elm *(Ulmus americana)*, water tupelo *(Nyssa aquatica)*, swamp tupelo *(Nyssa sylvatica biflora)*, bald cypress, buttonbush *(Cephalanthus occidentalis)*, dahoon holly *(Ilex cassine)*, redbay *(Persea borbonia)*, cinnamon fern *(Osmunda cinnamomea)*, lizard's tail *(Saururus cernuus)*, and royal fern *(Osmunda regalis)*. Poorly to very poorly drained inorganic and organic soils underlie this community.

Slough Community

Sloughs, also known as "flats," occur throughout subtropical Florida as small areas in association with other landforms. The plant community is named for the landform because of the strong association between the two. Communities appear as open expanses of grasses, sedges, rushes, and various herbaceous species with scattered pine trees and shrubs. They occur on flat to slightly concave surfaces and are easily identified by the sparseness of trees and shrubs and dominance of grasses and grasslike vegetation. They occupy a part of the landscape that is subject to flowing water across the surface during periods of high rainfall. Typifying plants under native conditions include little blue maidencane *(Amphicarpum muhlenbergianum)*, chalky bluestem *(Andropogon capillipes)*, bottlebrush threeawn *(Aristida spiciformis)*, Florida threeawn *(Aristida rhizomophora)*, bluejoint panicum *(Panicum tenerum)*, toothachegrass *(Ctenium aromaticum)*, south Florida slash pine *(Pinus elliottii densa)*, slash pine *(Pinus elliottii)*, widgeon grass *(Ruppia maritima)*, milkwort *(Polygala)*, and waxmyrtle *(Myrica cerifera)*. The exotic tree melaleuca is the dominant vegetation of many areas, however, especially where the hydrology has been altered.

5.4 IMPORTANT SOILS OF SUBTROPICAL FLORIDA

5.4.1 General Aspects of Soil Classification

Seven of the eleven soil orders recognized at the highest categorical level in the U.S. soil classification system (Soil Survey Staff, 1996) are found in Florida. Soil differentiation closely follows the variations in parent material, topography, and vegetation discussed above. The influence of topography is manifested most profoundly through its control of hydrology, which is an important determinant of vegetation and soil genesis in subtropical Florida.

The soils of subtropical Florida have been mapped and characterized in accord with the USDA soil taxonomic system (Soil Survey Staff, 1996). In effect, most of the modern inventory of soil information is "catalogued" by the classes of that system. It is therefore expedient to refer to these classes in our groupings and discussions of soils in this chapter. Below we provide a rudimentary explanation of the system. A comprehensive presentation of its rationale, evolution, structure, and criteria has been published under the title *Soil Taxonomy, A Basic System of Soil Classification for Making and Interpreting Soils Surveys* (Soil Survey Staff, 1975). A series of abbreviated updates have been published subsequently, the most recent as of this writing being *Keys to Soil Taxonomy: Seventh Edition* (Soil Survey Staff, 1996).

The USDA soil classification system is hierarchical, with six categorical levels: order (highest), suborder, great group, subgroup, family, and series (lowest). The hierarchical nature of the systems means that classification must proceed from the most general category (order), to the more restrictive categories. That is, one must first know the order before determining the suborder, and the suborder before determining the great group, etc. This means that the classification process can be terminated at any category from the order downward, and that the term for the resultant class will reflect the criteria for all categories to the level classified.

The nomenclature of USDA soil taxonomy is structured such that the class term includes a "formative element" for any category for which classification has been performed, down to the family level. Thus the class name conveys information about the morphological, chemical, and physical characteristics of a soil meeting the criteria of that class for those familiar with the system. For example, if a soil has sufficient organic carbon content to a sufficient depth, it would be classified in the order of Histosols (as further discussed below). There are several suborders of Histosols, one of which, Saprists, is common in subtropical Florida. The suborder term retains the formative element "ist" to indicate the Histosol order, and includes in this case "sapr" to indicate a highly decomposed state of the organic matter in the soil.

Determination of soil order, the first step in classification, requires reference to the "key to soil orders" contained in the latest "Keys to Soil Taxonomy" (Soil Survey Staff, 1996). The procedure is to begin at the top of the key and to determine if the requirements are met for each successive order as presented in the key's listing of definitive criteria. The first "match" identifies the appropriate order. The real key to soil orders is quite detailed. However, we present an abbreviated key (Table 5.1) that is intended to illustrate the basic application of such a key and to introduce the differentiating criteria of the soil orders of subtropical Florida.

Classification to successively lower categories requires equally systematic reference to the *Keys to Soil Taxonomy*, involving greater and greater detail with regard to the properties of the soil. For example, a soil in fitting the Spodosol order might be in the Aquod or Orthod suborder, depending on wetness of the soil. It might fit the Alaquod or Epiaquod great group depending upon the amount of iron in the spodic horizon. It then might fit into one of many subgroups depending upon a wide variety of soil features and properties, and so on down to the series level. The series category contains the greatest number of classes, since "series" is the least general level within a hierarchical system. In effect, series are the most narrowly defined classes, and are intended to reflect locally important interpretive differentia. More than 200 series are used to describe the soils in subtropical Florida.

TABLE 5.1
Key to Soil Orders that Occur in Subtropical Florida (abbreviated)

A. Soils that consist of organic materials (muck or mucky peat) in at least two-thirds of the thickness above bedrock, with no mineral soil layer as thick as 10 cm, or are saturated for most of the year, and organic materials constitute half of the upper 80 cm of soil (extensive in subtropical Florida).	Histosols
B. Other soils that have a spodic horizon (an acidic subsurface horizon, usually black to dark reddish brown, in which organic material has accumulated in combination with aluminum and iron) 10 cm or more thick (extensive in subtropical Florida).	Spodosols
C. Other soils that have a dark (usually black to very dark gray) mineral surface horizon that is more than 25 cm thick (10 cm if directly underlain by bedrock) and that has a base saturation of 50% or more (or minor extent in subtropical Florida).	Mollisols
D. Other soils that have an argillic or kandic horizon and a base saturation of 35% or more. Argillic and kandic horizons are subsurface zones of clay accumulations from overlying horion(s). Base saturation is the ration of "base" cations (Ca^{2+}, Mg^{2+}, K^+, Na^+) to acid cations (H_3O^+, Al^{3+}) on the exchange complex, times 100 (extensive in subtropical Florida).	Alfisols
E. Other soils that have horizon development exemplified by color and/or structure, or that have umbric or histic epipedons (of minor extent in subtropical Florida).	Inceptisols
F. Other soils (extensive in subtropical Florida).	Entisols

We have largely confined discussion in this chapter to the two highest soil-taxonomic levels, order and suborder, because they constitute convenient soil groupings for the geographic scale addressed. However, as an example of classification nomenclature to the series level, consider the Myakka series. Soils of this series are in the family of sandy, siliceous, hyperthermic Aeric Alaquods, in the subgroup of Aeric Alaquods, in the great group of Alaquods, in the suborder of Aquods, and in the order of Spodosols. The series name (Myakka, Smyrna, Ona, etc.) is usually based on a geographic location near the site where the series was originally established, and hence does not (if used alone) convey any systematic information about the soil.

Six of the seven soil orders that occur in Florida are found in subtropical Florida (Table 5.1, Fig. 5.3). These orders are discussed below, along with their suborders that are of greatest importance in the region.

5.4.2 Histosols

Histosols comprise a technically defined soil class that roughly corresponds to the generic concept of "organic soils." They are the dominant soils of the Everglades (saw grass marsh communities) and coastal mangrove swamps of subtropical Florida (Figs. 5.2 and 5.3). Hence, they occupy vast areas. Taxonomically, they meet specific requirements for thickness of organic soil materials (Table 5.1). The latter are defined as having an organic carbon minimum that is scaled to the clay content, such that 12% organic carbon is the lower limit for soils with no clay, and 18% for soils with 60% clay or more. Histosols, therefore, are not required to be dominated by organic

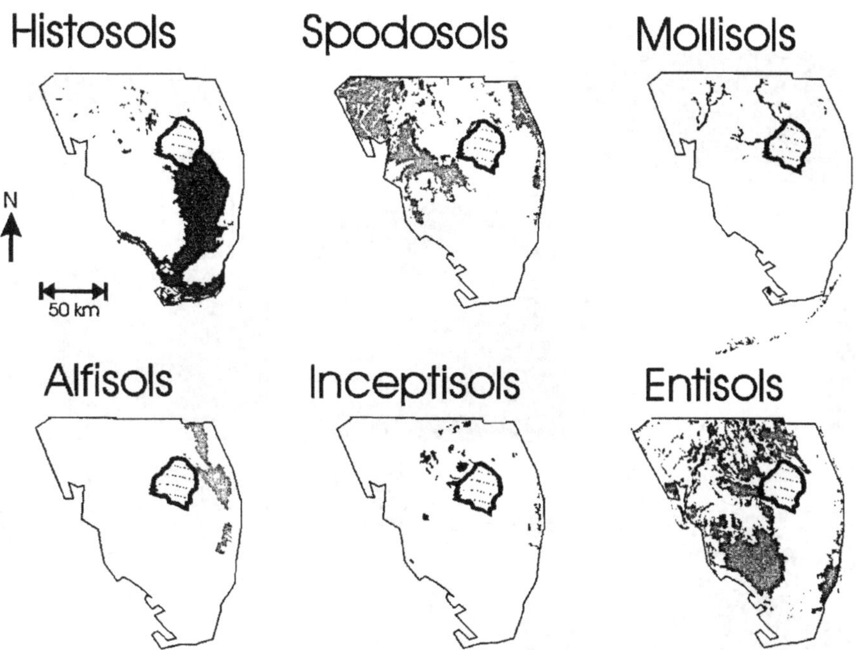

FIGURE 5.3 General maps showing areas for which the designated soil suborder is the most abundant among those that occur. Delineations are not necessarily dominated by the designated suborder. The lighter gray zone to the south on the Entisols map distinguishes Udorthents, associated with limestone outcropping and rock plowing. These maps are based on the State Soil Geographic Data Base (STATSCO) generated by the USDA Natural Resources Conservation Service, and were accessed with the help of Mr. C. G. Hoogeweg.

carbon on a mass basis. However, organic components are dominant on a volume basis and largely control the properties of the soil. Characteristic properties of Histosols include very low bulk density, high moisture retention, low bearing strength, high cation exchange capacity, and dark colors (Fig. 5.4). These properties are influenced by such factors as the dominant vegetative contributor to organic accumulations, extent of organic matter decomposition, and amount and type of inorganic components.

Histosols form via the accumulation of the decay products of vegetation. Significant organic matter accumulation requires that biomass production exceed biodegradation. The formation of the great majority of Histosols in subtropical Florida results from a high biomass production of emergent marsh vegetation in conjunction with low biodegradation rates in the constantly anaerobic soil environment (barring artificial drainage; see below). The predominant vegetation associated with Histosol formation in subtropical Florida is saw grass (really a sedge, *Cladium jamaicensis*), but more succulent aquatic plant species can be important in some localities (e.g., the eastern Everglades, Meindl, 1996).

The distribution of Histosols in subtropical Florida is ultimately linked to topography, through its control of hydrology, ergo, vegetation. The most extensive area

FIGURE 5.4 Some typical soils that occur in subtropical Florida. Top left: Saprist, showing dark organic horizon. Top middle: Aquod, showing a white E horizon overlying a black Bh (Spodic) horizon. Top right: Aqualf, showing dark surface horizon (Ap), light E horizon, and argillic horizon which is darkest in the upper part. Lighter colors in the argillic are attributable to reducing conditions and to the presence of secondary carbonates. Lower left: Aquoll, showing thick dark mollic epipedon. Lower right: Aquent, predominated by marl. These photographs were taken during the course of the Florida Cooperative Soils Survey Program, and were provided by Mr. Frank Sodek, Dr. V.W. Carlisle, and Dr. M.E. Collins.

of Histosol domination (Fig. 5.3) corresponds closely to the broad topographic trough underlying the Everglades (Fig. 5.1c), where hydrologic conditions are hospitable to saw grass (Fig. 5.2). The Everglades is a Holocene-aged system. By most estimates, organic accumulation has been occurring over roughly the last five thousand years (Meindl, 1996). The relatively stable conditions of quiescent, shallow water with cyclic seasonal drying have prevailed during this period to produce the Everglades and the Histosols that occur there. The convergence of these conditions over such a large area for thousands of years is a globally unique phenomenon. Saw grass has thrived in the phosphorus-limited marsh environment, aided by natural fires (ignited by lightning) during dry cycles that serve to eliminate encroaching woody species (Estevez et al., 1984). The primary organic accumulation in saw-grass-derived Histosols originates from the roots and rhizomes of the plants (Gleason and Stone, 1994).

The Histosols of subtropical Florida vary in thickness from a few centimeters to several meters. They generally overlie limestone, marl, or mineral soil material (mainly sand). Their thickness, in the absence of human intervention, relates to the nature of the hydrological cycle and its influence on the vegetation over time (Estevez et al., 1984. Gleason and Stone, 1994). There is a trend toward thinner soils progressing southward, with deepest soils occurring just south of Lake Okeechobee. A marked degree of subsidence, i.e., reduction in Histosol thickness, has resulted from human water control practices during the twentieth century (Light and Dineen, 1994. Meindl, 1996). Artificial drainage promotes subsidence by increasing the rate of biochemical oxidation of organic matter via aerobic organisms. Also, physical shrinkage can result from desiccation and the diminished force of groundwater buoyancy following drainage. Other artificial changes in the natural hydrological cycle, including impoundment, have ecological consequences as well (Estevez et al., 1984).

An important determinant of the properties of Histosols is the extent to which vegetative remains have decomposed within the soil. Three categories of organic soils materials are designated to reflect degree of decomposition: *fibric* (least decomposed), *hemic,* and *sapric* (most decomposed). Distinction between three of the Histosol suborders, *Fibrists, Hemists* and *Saprists,* is based on that type of organic material is dominant. Fibric materials have >3/4 fibers by volume, hemic materials, between 3/4 and 1/6, and Sapric materials, <1/6. Fiber content is determined after "rubbing," a technique used in the field assessment of Histosols. The soil textural class *mucky peat* corresponds to hemic materials, and *muck,* to Sapric materials. Saprists are the most abundant Histosols of subtropical Florida. Areas of Histosols are sometimes called "peatlands," but in the Everglades there is very little peat as technically defined by soil scientists. The term *peat* is restricted to fibric soil materials.

A fourth suborder of Histosols, *Folists,* also occurs in subtropical Florida, but only to a minor extent. Folists are the only Histosols that are never saturated with water in their upper profile for more than a few days. Folists tend to be thin (<1 m), forming under trees primarily from leaf litter, twigs, and branches (Brown et al., 1990) on the Florida Keys and on tree islands within the Everglades and southeastern Dade County.

5.4.3 Spodosols

Spodosols also occur extensively in subtropical Florida (Fig. 5.3). They have sandy texture and a Spodic (Bh) horizon (Table 5.1). A Spodic horizon is a dark subsurface zone of accumulated carbon and associated metals (mainly Al). A light-gray to white eluvial (E) horizon, a zone of metal and carbon depletion, generally overlies the Spodic. Spodosols of the region form in sandy parent materials on flatwoods landforms, and usually are poorly to somewhat poorly drained. Depths to the top of Bh-horizons vary from <25 cm to >2 m, though those > 2 m in depth are not considered in soil classification. Bh-horizon thicknesses vary from 10 cm to >1 m. Bedrock, sand, or an argillic horizon (a zone of clay accumulation) may occur at varying depths beneath the spodic horizon.

Spodosols of Florida are strongly associated with fluctuating water table, that is a characteristic of flatwoods hydrology. They do occur on portions of other landforms where water tables fluctuate within a sandy material, but flatwoods are by far the most extensive occurrence of this condition. Flatwoods Spodosols, though wet by "ordinary" standards, are often the driest soils to be found within the very wet landscapes of subtropical Florida. The role of the water table influence on Spodosol formation in Florida is still an open question, since in other regions Spodosols are commonly well-drained. However, the general mechanism of Spodosol genesis entails a dissolution of metals via complexation with organic acids in the upper eluvial zone (E horizon), followed by downward leaching, and culminating in accumulation of carbon and metals via precipitation deeper in the soil. This process usually leads to a distinctive "white-over-black" morphology, and only occurs in sandy materials. Hence, Spodosols are not common in the southern part of the region where sands are thin or absent (Fig. 5.4, Fig. 5.1a). However, they commonly underlie flatwoods plant communities (slash pine, palmetto, etc.) in most areas.

Aquods are Spodosols (*od* is the formative element) that are wet for extended periods of most years (as the *Aqu* formative element for the suborder conveys) unless they have been artificially drained. They are the most extensive soils in Florida flatwoods, except in the minor areas of sand-poor flatwoods to the extreme south. *Orthods* also occur in the subtropical Florida, but to a lesser extent. They are better drained than Aquods, but are usually not well drained. The Bh horizons of some Orthods are located deep (1 to 2 m) in the soil.

5.4.4 Entisols

Soils lacking, within 2-m depth, diagnostic horizons and other features that are specifically defined and required for other orders of the USDA taxonomic system default to Entisols. The pervasive concept of Entisols being young and/or unweathered soils is not applicable to most of the Entisols of Florida. The majority of Florida Entisols are Psamments (*ent* is the formative element), found to a large extent on old, highly leached landscapes. Most Florida Psamments contain minerals such as kaolinite and gibbsite, which occur as coating on sand grains, and which attest to the intensity of weathering that these soils have experienced. Psamments often occur contiguously on the same landscapes with Ultisols (mapped on the central ridge just north of our boundary for subtropical Florida), an order associated with intense weathering, such that the Psamments are stratigraphically continuous with the commonly thick (but <2 m) E horizons of Ultisols. In effect, Ultisols are not taxonomically permitted to have an E horizon of thickness > 2 m, and when that happens the soil defaults to an Entisol. Obviously, such arbitrary classification boundaries do not correspond to weathering boundaries.

The soils that classify as Entisols in subtropical Florida are extensive, but are not representative of Florida as a whole. Most Entisols of the region are *Aquents* rather than Psamments. These Aquents, in turn, comprise two major groups of highly contrasting soils, one dominated by sand (*Psammaquents*), and the other by $CaCO_3$ in the form of marl (*Fluvaquents*). The genesis and environmental setting of these two Great Groups of Aquents are highly contrasting as well.

Psammaquents are restricted to areas of relatively deep marine sand (Fig. 5.1a), which constitutes their parent material. They are common on low marine terraces between the Everglades and the Atlantic coastal ridge, particularly in tidal marshes, sloughs, and depressions (Noble et al., 1996). Some Psammaquents have Bh horizons which do not technically qualify as Spodic horizons by USDA criteria because of color (not dark enough) or pH (too high).

The marl soils of subtropical Florida, though currently classified as Fluvaquents, are primarily of biological origin. However, they have the potential to be redistributed as sediments via the action of tides and storms. Marl accumulates from microbially induced precipitation of $CaCO_3$ (Gleason and Spackman, 1974) within the limestone-influenced flood waters of the extensive saw grass marsh ecosystems. Communities of these microorganisms, which form floating or sunken mats called *Periphyton*, occur extensively in association with saw grass (Gleason and Stone, 1994). The $CaCO_3$ accumulates over time as detritus within the quiescent but frequently flooded environment. The underlying limestone maintains a high Ca activity in the flood water, which enables $CaCO_3$ precipitation in conjunction with photosynthesis-related CO_2 flux. This scenario explains the exceptional case of accumulation, rather than dissolution, of carbonates in a humid subtropical climate. Accumulations of up to 2 m have been reported (Gallatin et al., 1958), with a thickening trend toward the south. It is likely that marl soils will soon be reclassified to better reflect their biogenic origin. A proposed suborder of Histosols, Limnists, would include soil materials composed of marl, coprogenous earth, and diatomaceous earth. Mineral marl soils would become Limniaquents under this proposal.

Psamments, *Orthents*, and *Arents* also occur in subtropical Florida, though less extensively than Aquents. Quartzipsamments (the only Psamment Great Group which occurs in Florida) are prevalent on well to excessively drained landscapes generally associated with longleaf pine-turkey oak hills and sand pine scrub communities. They are restricted mainly to the northern part of the region and to Atlantic coastal areas. Some Psamments of the region have brownish colors because of the presence of metal-oxide-cemented clay-and-silt coatings on grains, while others are essentially white because of the absence of these coatings. Most of the white Psamments have underlying Bh horizons that are either too deep or too light-colored for these soils to be classified as Spodosols. Orthents are extensive in subtropical Florida, but owe their existence to the rock-plowing of limestone outcrops (Fig. 5.1b). They consist of shallow pulverized rock on well to somewhat poorly drained landscapes, typically vegetated by pine forests (flatwoods community) or tropical hammocks. Soils forming in materials recently disturbed by human activities are classified as Arents, which are most concentrated in heavily urbanized areas.

5.4.5 Alfisols

Alfisols are relatively extensive in subtropical Florida (Fig. 5.3). They have an argillic horizon, and a base saturation within prescribed depths (most commonly 1.25 m below the upper boundary of the argillic) of 35% or more. The base saturation criterion separates Alfisols from Ultisols; the two orders are morphologically indistinguishable. Depths to argillic horizons vary from <25 cm to 2 m and thicknesses vary from <20 cm to >2 m. Bedrock or other material may occur at varying depths beneath the argillic horizon.

The genesis of Alfisols involves translocation of clay from the upper to lower soil zones to ultimately form argillic horizons. Not all the clay in argillic horizons is transported; some was present in the parent material and some may have formed residually. However, the net result of the clay movement is that there is appreciably less clay in upper horizons than deeper ones. Verification that some of the clay in a horizon has been transported from above is sometimes difficult in the coarse-textured soils common to Florida. Careful examination sometimes reveals clay coatings in pore channels and clay bridging between sand grains, which are micromorphological features commonly considered as evidence for clay movement. The relatively high base saturation of Alfisols is an indicator that they have not been extensively leached, at least in the lower part. Some of the Alfisols in subtropical Florida have free (secondarily precipitated) carbonates within and below the argillic horizon. The presence of carbonates is a good indicator of wet conditions and minimal leaching.

Aqualfs are the most common Alfisols (*alf* is the formative element) of Florida. They are wet for extended periods during most years unless they have been artificially drained. The most extensive occurrence of Aqualfs is in areas east of Lake Okeechobee and the Everglades (Fig. 5.3). Aqualfs are common to wet landforms such as sloughs and depressions, and occur to some extent in the wetter areas of flatwoods. The distribution of Aqualfs relates more to parent material and hydrology than to vegetation.

Udalfs also occur in the region. They are drier than Aqualfs, but their ability to retain plant-available water varies widely, generally decreasing with increasing depth to the argillic horizon.

5.4.6 Mollisols

Mollisols have dark surface horizons, called *Mollic epipedons*, which meet specific thickness requirements (Table 5.1). These horizons, though dark, do not meet the organic carbon requirement for organic soil material (see Histosols discussion above). Generally, mollic epipedons must be 25 cm or more thick, but can be thinner if they directly overlie a root-restrictive layer such as bedrock (the latter being limestone in subtropical Florida). A high base saturation (>50%) is also a requirement for Mollisols. Limestone bedrock and argillic horizons are sometimes present.

Soils of the Mollisol suborder have a varied genesis, even within subtropical Florida. Mollisols are far more prevalent in semi-arid prairie regions such as the American midwest, where they occur on landscapes ranging from well to poorly drained. Prairie-grass vegetation produces thick, dark surface horizons because the decay of grass roots results in effective incorporation and preservation of organic matter. The formation of such horizons under a warm, humid climate such as that of subtropical Florida requires special circumstances to "tip the balance" toward a greater degree of organic matter accumulation and incorporation. Wetness is a soil condition that favors organic matter accumulation, as previously discussed. It is therefore not surprising that very poorly to poorly drained sites are the most likely to be occupied by Mollisols in Florida. The presence of basic nutrients also promotes organic matter accumulation by supporting high biomass production. Florida Mollisols occur on landscapes and parent materials in which basic cations, particularly Ca, have not been leached out of the soil zone. Both continuously wet conditions and the presence of limestone at a shallow depth would favor higher Ca activities in soils.

Aquolls are the Mollisols (*oll* is the formative element) that are wet for extended period of most years unless they have been artificially drained to reduce saturation. They are by far the most common form of Mollisols occurring in subtropical Florida. Aquolls are most concentrated in poorly drained areas along river systems (note dendritic patterns in Fig. 5.3) and lake margins. *Rendolls* are Mollisols that are underlain at a shallow depth (<20 cm) by limestone bedrock. They are better drained than Aquolls. Their occurrence is restricted to areas where there is little sand or marl present, including the Florida Keys (Fig. 5.3). Rendolls are commonly found under tropical hardwoods.

5.4.7 Inceptisols

Inceptisols are soils that have had some horizon development, but do not meet the criteria for the soil orders already discussed. Inceptisols include soils of a great diversity of morphologies and compositions, owing to their default-like second-from-the-botton status in the sequential keys to soil orders. The permissible expressions of soil formation for Inceptisols include color and structural changes, and clay translocation that is insufficient to meet criteria for Alfisols and Ultisols. Also, some soils with *Umbric* and *Histic epipedons* are classified as Inceptisols. The former epipedons are morphologically identical to Mollic epipedons except that base saturation is below 50%, and the latter are organic surface layers that are too thin to be classified as Histosols.

Genetically, Inceptisols are a "mixed bag." Some soils of the order are morphologically indistinguishable from soils of other orders (e.g., Mollisols), such that laboratory data are essential for the classification. On the other hand, some Inceptisols can be locally distinct, in that their development is appreciably less than soils of other orders with which they are associated on the landscape. For example, Inceptisols are commonly found on flood plains and young alluvial terraces, where their geographic segregation and relative youth make them quite distinguishable from soils of the surrounding older landforms. Most of the soils classified as Inceptisols in subtropical Florida are *Aquepts* (*ept* is the formative element) by virtue of being wet and having a Histic or Umbric epipedon. In many cases they could be considered as "almost Histosols" both in terms of taxonomy and landscape setting. They tend to occur at the margins of Histosol-dominated marshes and swamps.

5.5 SOIL PROPERTIES AND NUTRIENT RETENTION

Agricultural soils of subtropical Florida pose a nutrient management challenge because of their low nutrient retention capacities and hydrological settings. Neither the sandy nor organic soil materials retain anions such as phosphate as effectively as soils higher in metal oxides and other clay-sized minerals. The nearly level landscapes are characterized by rapid infiltration, significant lateral subsurface flow, and a high degree of ground- and surface-water interaction (Allen et al., 1982). Hence, the pathway of nutrients from point of application to vulnerable water bodies is often short, and the retardation via soil interaction, minimal.

Nitrate and phosphate are commonly the nutrients of greatest concern with respect to water quality. The gravity of nitrate contamination stems from the health risks it poses. There are important areas of excessively drained Quartzipsamments which are heavily used for citrus production, and consequently are at high risk of nitrate contamination of groundwater if not properly managed. Nitrate is less of a concern with regard to surface water quality because of denitrification in many watersheds and to the fact that nitrogen is seldom the limiting nutrient. The latter distinction applies to phosphate, which in turn does not pose the health concern of nitrate. Phosphate is a serious surface water quality concern in subtropical Florida because of the very low phosphorus retention capacity of most soils in conjunction with the extremely low native phosphorus concentrations. Lakes and wetlands, which are naturally adapted to low phosphorus activities, are therefore very sensitive to elevated levels of solution phosphorus.

The rest of this book is dedicated to phosphorus-related environmental issues in subtropical Florida and vicinity. We therefore will not address in this chapter the voluminous documentation of phosphorus impacts. Rather, we will briefly summarize a few general characteristics of the major agricultural soils that have an important bearing on their phosphorus retention. The concept of phosphorus retention encompasses both adsorption capacity (under specified solution concentration, temperature, etc.) and resistance to desorption (degree of hysteresis). Phosphorus retention capacity can be assessed in a number of ways and under a variety of discretionary conditions. Our comparisons of soils with respect to phosphorus retention capacity are intended only to convey relative tendencies.

There are soil materials occurring extensively in subtropical Florida that retain practically no phosphorus at all! These are materials composed of clean, white quartz sand. Some Psamments consist of white sand to depths of 2 m or more. Clean quartz sand grains also dominate the E horizons overlying the Spodic (Bh) horizons in Aquods. The presence of enough grain-coating material to impart a brownish hue to sandy soils significantly enhances phosphorus retention capacity (Harris et al., 1996), though this capacity is still relatively low in comparison to most other soils of the world.

Spodosols are soils of great internal contrast in phosphorus retention. The E horizons of Spodosols retain very little phosphorus, whereas the Spodic horizons (also sandy) have a relatively high phosphorus *adsorption* capacity (Nair et al., 1995. Nair and Graetz, 1995. Harris et al., 1996). The phosphorus retention behavior of Spodic horizons relates in part to the noncrystalline organo-metal components that have accumulated during Spodosol formation. These components have very high surface area and reactivity, such that they can have a large impact on chemical behavior even in relatively small amounts. However, they tend to retain a lower proportion of the phosphorus when exposed to solutions of reduced phosphorus concentrations than do other inorganic soil materials (Zhou et al., 1997). Apparently, the organically modified surfaces of these materials have a diminished tendency either for specific adsorption or other mechanisms proposed to explain the high degree of hysteresis for metal-oxide minerals. Spodosols pose a high-risk for phosphorus movement to lakes and streams because (1) the surface- and E-horizons are minimally retentive (2) the potential for lateral subsurface flow is high, (3) Bh horizons can be bypassed during periods of high water table, and (4) Bh horizons can serve as sources of phosphorus under some conditions (Burgoa, 1989. Zhou et al., 1997).

Phosphorus retention in Alfisols, and in some Aquolls and Aquods, is favored by the presence of argillic horizons, which are enriched in clay relative to other soils horizons of subtropical Florida. Some Aqualfs and Aquolls are alkaline and have free carbonates, conditions which at least theoretically could result in the precipitation of calcium phosphates. However, lateral subsurface flow during periods of high water table could (as in the case of Aquod Bh horizons) reduce the extent of phosphorus contact with the most reactive part of the soil (argillic horizon in the case of Aqualfs and Aquolls).

Histosols of the region have a high cation exchange capacity that helps to retain cationic nutrients. They also adsorb some phosphorus, particularly at high concentrations. However, adsorption of phosphorus on organic soil materials tends to be reversible. Hence, previously loaded Histosols can revert from a sink to a source, buffering against the lowering of phosphorus activity after external loading is reduced (Richardson and Vaithiyanathan, 1995). The phosphorus adsorption capacity of Florida Histosols correlates well with ash content (Porter and Sanchez, 1992), suggesting that inorganic components significantly affect phosphorus retention even for organic soils. Phosphorus mineralization, particularly as enhanced by accelerated organic matter decomposition associated with artificial drainage, is an important consideration for Histosols of subtropical Florida.

The fate of phosphorus in marl- (Aquents) and rock-plowed (Orthents) soils has not been thoroughly investigated. Some areas of rock-plowed soils have had a history (about 30 years in some cases) of phosphorus application, and would be interesting to study with regard to possible formation of thermodynamically stable forms of calcium phosphate. The latter are seldom detected in soils under normal circumstances, but high application rates and long-term field incubations of phosphate in essentially pure calcium carbonate constitute exceptional circumstances.

Risk assessment protocols developed in other areas can be inappropriate for subtropical Florida landscapes. For example, phosphorus retention indices that emphasize surface runnoff potential (e.g., Lemunyon and Gilbert, 1993) would not account for the risks from leaching and lateral subsurface flow, which are important pathways in subtropical Florida (Federico et al., 1981; Allen, 1987; Mansell et al., 1991).

REFERENCES

Allen, L.H. Jr. 1987. Dairy-siting criteria and other options for wastewater management on high water-table soils. Soil and Crop Sci. Soc. Fla. Proc. 47:108–127.

Allen, L. H. Jr., J.M. Ruddell, G. J. Ritter, F. E. Davis, and P. Yates. 1982. Land use effects on Taylor Creek water quality. Proc. Spec. Conf. on Environ. Sound Water and Soil Mgmt. Orlando, FL., July 20–23, 1982.

Bailey, R.G. 1996. Ecosystems geography. Springer-Verlag, NY, NY.

Belz, D.J., L.J. Carter, D.A. Dearstyne, and J.D. Overing. 1990. Soil Survey of Hendry County, Florida. USDA, SCS in cooperation with the University of Florida, Institute of Food and Agricultural Sciences, Agricultural Experiment Stations and Soil Science Department. U.S. GPO, Washington, DC.

Brown, R.B., E.L. Stone, and V.W. Carlisle. 1990. Soils. p. 35–69. *In* R.L. Myers and J.J. Ewel (eds.) Ecosystems of Florida. University of Central Florida Press, Orlando.

Burgoa, B. 1989. Phosphorus spatial distribution, sorption, and transport in a Spodosol. Ph.D. dissert., University of Florida, Gainesville, FL. 144 p.

Carter, L.J., D. Lewis, L. Crockett, and J. Vega. 1989. Soil Survey of Highlands County, Florida. USDA, SCS in cooperation with the University of Florida, Institute of Food and Agricultural Sciences, Agricultural Experiment Stations and Soil Science Department. U.S. GPO, Washington, DC.

Carter, L.J., D. Lewis, and J. Vega. In press. Soil Survey of Okeechobee County, Florida. USDA, NRCS in cooperation with the University of Florida, Institute of Food and Agricultural Sciences, Agricultural Experiment Stations and Soil and Water Science Department. U.S. GPO, Washington, DC.

Carter, L.J., D. Lewis, and J. Vega. In press. Soil Survey of Glades County, Florida. USDA, NRCS in cooperation with the University of Florida, Institute of Food and Agricultural Sciences, Agricultural Experiment Stations and Soil and Water Science Department. U.S. GPO, Washington, DC.

Cooke, C.W. 1945. Geology of Florida. Geological Bull. 29. Florida Geological Survey. Tallahassee, FL.

Cowherd, W.D., W.G. Henderson, E.J. Sheehan, and S.T. Ploetz. 1989. Soil Survey of DeSoto County, Florida. USDA, SCS in cooperation with the University of Florida, Institute of Food and Agricultural Sciences, Agricultural Experiment Stations and Soil Science Department. U.S. GPO, Washington, DC.

Estevez, E.D., B.J. Hartman, R. Kautz, and E. Purdum. 1984. Ecosystems of surface waters. p. 92–107. *In* E. Fernald and D. Patton (eds.) Water Resources Atlas of Florida. Florida State University.

Myers, R.L., and J.J. Ewel. 1990. Ecosystems of Florida. University of Central Florida Press. Orlando, FL.

Federico, A. C., F. E. Davis, K. G. Dickson, and C. R. Kratzer. 1981. Lake Okeechobee water quality studies and eutrophication assessment. South Florida Water Manage. Distric. Tech. Publ. 81–82. West Palm Beach, FL.

Florida Soil Conservation Service. 1989. Twenty six ecological communities of Florida. Florida Soil and Water Conservation Society. Gainesville, FL.

Gallatin, M.H., J.K. Ballard, C.B. Evans, H.S. Galberry, J.J. Hinton, D.P. Powell, E. Truett, W.L. Watts, G.C. Wilson Jr., and R.G. Leighty. 1958. Soil Survey of Dade County, Florida. USDA, SCS in cooperation with Florida Agricultural Experiment Station. USDA, SCS Washington, DC.

Gleason, P.J., and W. Spackman. 1974. Calcareous periphytin and water chemistry in the Everglades. p. 146–181. *In* (P.J. Gleason, ed.) Environments of south Florida, past and present. Memoir No. 2, Miami Geological Society, Miami.

Gleason, P.J., and P. Stone, 1994. Age, origin, and landscape evolution of the Everglades peatland. p. 149–198. *In* S.M. Davis and J.C. Ogden (eds.) Everglades: The ecosystem and its restoration. St. Lucie Press, Delray Beach, FL.

Harris, W.G., R.D. Rhue, G. Kidder, R.C. Littell, and R.B. Brown. 1996. P retention as related to morphology of sandy coastal plain soils. Soil Sci. Soc. Am J. 60:1513–1521.

Healy, H.G. 1975. Terraces and shorelines of Florida. U.S. Geological Survey and Florida Bureau of Geology. Tallahassee, FL.

Henderson, W.G. Jr., L.J. Carter, A.L. Moore, R.A. Stein, C.A. Wettstein, and H. Yamataki. 1984a. Soil Survey of Charlotte County, Florida. USDA, SCS in cooperation with the University of Florida, Institute of Food and Agricultural Sciences, Agricultural Experiment Stations and Soil Science Department. U.S. GPO, Washington, DC.

Henderson, W.G. Jr., L.J. Carter, A.L. Moore, R.A. Stein, C.A. Wettstein, and H. Yamataki. 1984b. Soil Survey of Lee County, Florida. USDA, SCS in cooperation with the University of Florida, Institute of Food and Agricultural Sciences, Agricultural Experiment Stations and Soil Science Department. U.S. GPO, Washington, DC.

Hurt, G.W., C.V. Noble, and R.W. Drew. 1995. Soil Survey of Monroe County, Keys Area, Florida. USDA, SCS in cooperation with the University of Florida, Institute of Food and Agricultural Sciences, Agricultural Experiment Stations and Soil Science Department. U.S. GPO, Washington, DC.

Hyde, A.G., and H.F. Huckle. 1983. Soil Survey of Manatee County, Florida. USDA, SCS in cooperation with the University of Florida, Institute of Food and Agricultural Sciences, Agricultural Experiment Stations and Soil Science Department. U.S. GPO, Washington, DC.

Hyde, A.G., G.W. Hurt, and C.A. Wettstein. 1991. Soil Survey of Sarasota County, Florida. USDA, SCS in cooperation with the University of Florida, Institute of Food and Agricultural Sciences, Agricultural Experiment Stations and Soil Science Department. U.S. GPO, Washington, DC.

Jones, L.A. 1948. Soils, geology, and water control in the Everglades region. University of Florida Agric. Experiment Station Bull 442. Gainesville, FL.

Koch, M.S., and K.R. Reddy. 1992. Distribution of soil and plant nutrients along a trophic gradient in the Florida Everglades. Soil Sci. Soc. Am. J. 56:1492–1499.

Lemunyon, J.I., and R.G. Gilbert. 1993. The concept and need for a P assessment tool. J. Prod. Agric. 6:483–486.

Light, S.S., and J.W. Dineen. 1994. Water control in the Everglades: A historical perspective. p. 47–84. *In* S.M. Davis and J.C. Ogden (eds.) Everglades: The ecosystem and its restoration. St. Lucie Press, Delray Beach, FL.

Liudahl, K., D.J. Belz, L. Carey, R.W. Drew, S. Fisher, R. Pate, and J.N. Schuster. 1998. Soil Survey of Collier County Area, Florida. USDA, NRCS in cooperation with the University of Florida, Institute of Food and Agricultural Sciences, Agricultural Experiment Stations and Soil and Water Science Department. U.S. GPO, Washington, DC.

Mansell, R.S., S.A. Bloom, and B. Burgoa. 1991. Phosphorus transport with water flow in an acid, sandy soils. *In* B. Jacob and M.Y. Corapcioglu (ed.) Transport processes in porous media. Kluwer Academic Publishers, Dorecht, The Netherlands.

McCollum, S.H., O.E. Cruz, L.T. Stem, W.H. Wittstruck, R.D. Ford, and F.C. Watts. 1978. Soil Survey of Palm Beach County Area, Florida. USDA, NRCS in cooperation with the University of Florida, Institute of Food and Agricultural Sciences, Agricultural Experiment Stations and Soil Science Department. U.S. GPO, Washington, DC.

Meindl, C.F. 1996. Environmental perception and the historical geography of the great American wetland: Florida's Everglades, 1895–1930. Doctoral dissertation, University of Florida.

Nair, V.D., and D.A. Graetz. 1995. Fate of phosphorus in Florida Spodosols contaminated with cattle manure. Ecol. Engin. 5:163–172.

Nair, V.D., D.A, Graetz, and K.M. Portier, 1995. Forms of phosphorus in soil profiles from dairies of south Florida. Soil Sci. Soc. Am. J. 59:1244–1249.

Noble, C.V., R.W.Drew, and J.D. Slabaugh. 1996. Soil Survey of Dade County Area, Florida. USDA, SCS in cooperation with the University of Florida, Institute of Food and Agricultural Sciences, Agricultural Experiment Stations and Soil Science Department. U.S. GPO, Washington, DC.

Pendleton, R.F., H.D. Dollar, L. Law, S.H. McCollum, and D.J. Belz.1984. Soil Survey of Broward County, Eastern Part, Florida. USDA, SCS in cooperation with the University of Florida, Institute of Food and Agricultural Sciences, Agricultural Experiment Stations and Soil Science Department. U.S. GPO, Washington, DC.

Porter, P.S., and C.A. Sanchez. 1992. The effect of soil properties on phosphorus sorption by Everglades Histosols. Soil Sci. 154:387–398.

Richardson, C.J., and P. Vaithiyanathan. 1995. Phosphorus sorption characteristics of Everglades soils along a eutrophication gradient. Soil Sci. Soc. Am. J. 59:1782–1788.

Robbins, J.M. Jr., R.D. Ford, J.T. Werner, and W.D. Cowherd. 1984. Soil Survey of Hardee County, Florida. USDA, SCS in cooperation with the University of Florida, Institute of Food and Agricultural Sciences, Agricultural Experiment Stations and Soil Science Department. U.S. GPO, Washington, DC.

Scott, T.M. 1980. The sand and gravel resources of Florida. Rep. of Investigations No. 90. Florida Bureau of Geology, Tallahassee, FL.

Soil Survey Staff, 1975. Soil taxonomy: a basic system of soil classification for making and interpreting soil surveys. U.S. Govt. Printing Office, Washington, DC.

Soil Survey Staff. 1996. Keys to soil taxonomy. 7th edition. USDA-Natural Resources Conservation Service.

Snyder, J.R., A. Herndon, and W.B. Robertson, Jr. 1990. South Florida rockland. p. 230–277. *In* R.L. Myers and J.J. Ewel (eds.) Ecosystems of Florida. University of Central Florida Press, Orlando.

Van Arman, J., D. Nealon, S. Burns, B. Jones, L. Smith, T. MacVicar, M. Yansura, A. Federico, J. Bucca, M. Knapp, and P. Gleason. 1984. South Florida Water Management District. p. 138–157. *In* E. Fernald and D. Patton (eds.) Water Resources Atlas of Florida. Florida State University.

Watts, F.C., and D.L. Stankey. 1980. Soil Survey of St. Lucie County Area, Florida. USDA, SCS in cooperation with the University of Florida, Institute of Food and Agricultural Sciences, Agricultural Experiment Stations and Soil Science Department. U.S. GPO, Washington, DC.

Zhou, M., R.D. Rhue, and W.G. Harris. 1997. Phosphorus sorption characteristics of Bh and Bt horizons from sandy coastal plain soils. Soil Sci. Soc. Am J. 61: 1364–1369.

Section III

Biogeochemical Transformations

6 Inorganic Forms of Phosphorus in Soils and Sediments

D.A. Graetz and V.D. Nair

6.1 ABSTRACT

Inorganic P exists in soils and sediments both as P containing minerals and in various amorphous forms. To understand the fate of P under various environmental conditions, it is necessary to characterize pools of P in soils and sediments. This is generally accomplished with sequential extraction schemes, which extract different phases or pools of P rather than specific minerals. Common inorganic P pools identified in sequential fractionation procedures include (1) loosely bound (also referred to as labile or exchangeable P), (2) fractions associated with Al, Fe, and Mn oxides and hydroxides, (3) the Ca- and Mg-bound fraction, and (4) minerals and organic material resistant to previous extractants. The procedures often incorporate estimates of organic as well as inorganic P.

Generally the objectives of P fractionation are to (1) provide insight into the fate and transformation of P added to soils as fertilizer or manure, (2) estimate the availability of P to plants for agronomic purposes, (3) estimate the potential for P movement from (erosion) and through (leaching) soils, and (4) provide information regarding the interaction between P in the sediments and the overlying water in the case of aquatic systems.

Examples of P fractionation data applied to Florida's subtropical soils, wetlands, and aquatic systems are presented to illustrate the fate and transport of P in these

1-56670-331-X/99/$0.00+$.50

ecosystems. It was found that a large amount of P (up to 80% of total P) has the potential to leave heavily manure-impacted upland regions although less than 10% of the total P was likely to leave a low manure-impacted pasture soil. Most of the P in the wetland soils and sediments were associated with inorganic Fe and Al, and are assumed to be stable with little possibility of being desorbed except under extended water-logged conditions. Concentrations of readily available P in sediments depended on sediment type. Littoral and peat sediments, rather than mud or sand/rock sediments, are more likely contributors to internal P cycling of lakes.

In addition to P fractionation schemes designed to extract somewhat specific forms of soil P, several extractants have been used routinely to extract a representation of "available P" for soil testing purposes. These extractants are generally designed to provide a measure of P (and in some cases, other elements) that can be correlated with crop response. Recently, traditional agronomic soil tests have been proposed for environmental purposes as well. From an environmental standpoint, results from these soil test procedures may be used to predict parameters such as equilibrium P concentration (EPC), labile P (resin-P), and algae-available P.

The concept of the degree of P saturation (DPS) has recently been introduced as an environmental measure of soil P available to be released to surface and subsurface runoff. The DPS relates extractable P to the P sorbing capacity of a soil. Although the methodology has been used successfully in the Netherlands and in parts of the United States, there is a need to identify methodologies which allow simple, inexpensive measurements of DPS under various environmental conditions.

6.2 INTRODUCTION

The fate and transport of phosphorus (P) in soils and sediments is, to a large degree, determined by the various forms of P present. Inorganic P is mostly found in combination with Al, Fe, Ca, and Mg. Phosphate minerals include apatite [$Ca_{10}(PO_4)_6F_2$], crandallite [$CaAl_3(PO_4)(OH)_2$], wavellite [$Al_3(PO_4)_2(OH)_3$], variscite [$Al_3(OH)_2H_2PO_4$] and strengite [$Fe(OH)_2H_2PO_4$]. Phosphorus may also be "fixed," through a variety of chemical reactions, by noncrystalline oxides of Fe and Al, which often occur as coatings on soil particles. An indirect involvement of organic C through complex formations with Fe and Al has been suggested by Reddy et al. (1999) and Nair et al. (1998) in cases where significant correlations with C, in addition to Fe and/or Al, were obtained for the P sorbing capacity of a soil. A knowledge of the distribution of P among chemical forms is useful for understanding P availability to upland and aquatic plants under varying soil/sediment environments, predicting the fate and transport of P in environmental studies, and evaluating the interaction of P between sediments and overlying waters.

6.3 PHOSPHORUS FRACTIONATION SCHEMES

Inorganic P in agricultural soils normally ranges from 50 to 75% of the total P (TP), but the percentage can vary from 10 to 90% (Sharpley, this volume). Difficulties in identification of specific P minerals have led to the development of chemical fractionation schemes that attempt to characterize the forms of P in soils and sediments. The schemes evolved from the assumption that certain chemical reagents preferentially extract discrete chemical forms (or at least pools) of P from soils and sediments. Common inorganic P pools identified in many sequential fractionation procedures include

1. Loosely bound (labile or exchangeable)
2. Fractions associated with Al, Fe, and Mn oxides and hydroxides
3. The Ca (and Mg) bound fraction
4. Minerals and organic material resistant to previous extractants

Because P in soils and sediments exists in such a diversity of amorphous and mineral forms, it is best to consider that the various reagents extract a pool of P, generally related to a given chemical group. Therefore, the chemical P forms are often considered operationally defined and are subject to broad interpretations.

Dean (1938), sought to develop an extractant for soil P that could be used for evaluating the plant availability of P. He was one of the first to suggest that fractionation of P could be a useful tool for identifying the forms of P in soil. His fractionation scheme consisted of two extractants, 0.25 *M* NaOH followed by 0.25 *M* H_2SO_4, which he suggested extracted Fe- and Al-P, and Ca-P forms, respectively. Later, Turner and Rice (1954) reported that neutral NH_4F would dissolve Al-P, but not Fe-P. Chang and Jackson (1957) developed a comprehensive fractionation procedure that has become one of the most widely used procedures for fractionating P in soils. Subsequently, a variety of fractionation schemes were developed to resolve various problems with the Chang and Jackson (1957) procedure or to accommodate other research needs. A selection of these procedures, along with some of the reasons for their development, are presented in Table 6.1. Many of the modifications were developed after Williams et al., (1971) showed that NH_4F did not reliably distinguish between Al- and Fe-P, especially in calcareous soils and sediments, and that some P solubilized by NH_4F was resorbed by CaF_2 formed during the extraction process.

The above schemes are often modified by others, and this adds to the problem of interpreting data obtained from different literature sources. Modifications include repeated extractions [0.1 *M* NH_4Cl extractions in the Hieltjes and Lijklema (1980) scheme; Pettersson and Istvanovics,1988; Nair et al., 1995] and changes in soil: solution ratio [1:10 ratio instead of 1:1000 soil/extractant ratio in 0.1 *M* NH_4Cl extractions in the Hieltjes and Lijklema (1980) scheme; Nair et al., 1995]. Substituting a single extractant [0.1 *M* KCl for 0.1 *M* NH_4Cl in the Hieltjes and Lijklema (1980) scheme; Koch and Reddy, 1992; Reddy et al. 1998] within a scheme is also common. Although all these modifications are justified within a particular research scheme, they pose problems when comparisons between the work of different authors are made. In spite of these drawbacks, P fractionation schemes offer a means of obtaining significant information on the P chemistry of soils and sediments (Nair et al., 1995).

Inorganic fractionation schemes generally incorporate some organic P estimates as well. However, Barbanti et al. (1994) contents that the organic pool is much more

TABLE 6.1
Examples of Phosphorus Fractionation Procedures Used To Identify P Forms or Pools in Soils and Sediments

Extraction Sequence	P Fraction	Comments
	Chang and Jackson (1957)	
1 *M* NH_4Cl	Water-soluble P and loosely-bound P; Exchangeable Ca	Previously no method was available for fractionation of soil inorganic P into each of the four principal chemical pools; Used data from Turner and Rice (1954) showing that neutral NH_4F can dissolve Al- P but not Fe-P.
0.5 *M* neutral NH_4F	Al-P	
0.1 *M* NaOH	Fe-P	
0.25 *M* H_2SO_4	Ca-P	
Citrate/dithionite	Occluded Fe-P	
0.5 *M* NH_4F	Occluded Al-P	
	Williams et al. (1971)	
0.1 *M* NaOH	Fe- and Al-bound P	Showed that NH_4F did not reliably distinguish between Al- and Fe-P and that some solubilized P may be resorbed by CaF_2 formed during the extraction with NH_4F.
Citrate-dithionite-bicarbonate	P resorbed from NaOH ext.	
0.5 *M* HCl	Ca-P	
	Hieltjies and Liklema (1980)	
1 *M* NH_4Cl	Removes carbonates and loosely-bound Ca	Found that removal of $CaCO_3$ by NH_4Cl is necessary to achieve satisfactory discrimination between Fe- and Al-P, and Ca-P in calcareous sediments.
0.1 *M* NaOH	Fe- and Al-bound P	
0.5 *M* HCl	Ca-P	
	Hedley et al. (1982)	
Anion exchange resin	Most biologically available P_i	Scheme used to identify the soil P fractions which are altered by cropping practices. Redistribution of P in soils can be inferred, i.e., P_i moving to immobilized P_i or P_o forms or mineralization of immobilized forms.
0.5 *M* $NaHCO_3$	Labile P_i and P_o	
0.1 *M* NaOH	P_i and P_o held by chemisorption to Fe and Al components	
Sonification in 0.1 *M* HaOH	P_i and P_o held within soil aggregates	
1 M HCl	Mainly Apatite-type minerals	
$H_2SO_4 + H_2O_2$	More chemically stable P_o and insoluble P_i forms	
	Ruttenberg (1992)	
1 *M* $MgCl_2$ (pH 8)	Exchangeable or loosely sorbed P	Developed to evaluate diagenesis of P in marine sediments; separates authigenic apatite from detrital apatite; Solved secondary adsorption on to residual surfaces by washing residue with $MgCl_2$ between extraction steps.
Citrate-dithionite-bicarbonate (pH 7.6)	Easily reducible or reactive ferric Fe-bound P	
Acetate buffer (pH 4)	Authigenic apatite	
1 *M* HCl	Detrital apatite	
Ashed at 550° C; 1 *M* HCl	Organic P	

complex, and less–well defined, than the inorganic pools. The organic fraction is usually considered cumulatively (Aspila et al., 1976; Golterman and Booman, 1988), or divided into operational groups (van Eck, 1982, Ingall et al., 1990; De Groot, 1990; Oluyedun et al., 1991). The NaOH fraction of Hieltjes and Lijklema (1980) yields TP after Kjeldahl digestion and the difference between TP and inorganic P is a measure of the organic P in the fraction (Nair et al., 1995; Olila et al., 1995). The P in the residual fraction, that is the fraction not previously extracted by any of the chemical reagents, is primarily organic P (Ruttenberg, 1992). Newman and Robinson (this volume) discuss the different techniques available to quantify organic P forms in soils and sediments in greater detail. Fractionation schemes have been used to evaluate the fate of P in a variety of climatic, soil and land-use situations. Generally, the objectives of these studies are to

1. Provide insight into the fate and transformation of P added to soils as fertilizer or manure
2. Estimate the availability of P to plants for agronomic purposes
3. Determine the potential for P movement from (erosion) and through (leaching) soils
4. Provide information regarding the interaction between P in the sediments and the overlying water at aquatic systems

Fractionation of P forms has been particularly useful in understanding the transformations of P added to soil, either in inorganic fertilizers or organic amendments such as manures (Table 6.2). Information about transformations between P pools in the soil aids our understanding, not only of the availability of P to plants, but the potential for P transport from and through the soil. The fate of P in manure-amended soils is of interest because of the tendency to apply large amounts of manure to the land. This often results in the accumulation of excessive amounts of P in the soil profile and to environmental concerns about the fate of this P with regard to water quality.

Leinweber et al. (1997) used sequential extractions to analyze both manure (chicken and pig) and soil from a densely populated livestock area in Germany. About 60% of the manure-P was extracted by the various extractants and represented, primarily, inorganic P. Labile P (resin- and $NaHCO_3$-P) accounted for 24 to 39% of the total P in the manure. In the soils, NaOH-P tended to dominate, indicating that the manure-P was interacting with pedogenic oxides and humic substances in the soils. Zhang and MacKenzie (1997) used P fractionation and path analysis to compare the behavior of fertilizer and manure-P in soils. They suggested that P behaves differently when added as manure, compared to inorganic fertilizer, which may affect the depth of P movement through the soil profile. Eghball et al. (1996) found that P from manure moved deeper in the soil than P from chemical fertilizer in long-term studies. Simard et al. (1995) found that a significant portion of the P moving downward in soils receiving substantial amounts of animal manure accumulated in labile forms, i.e., water-soluble, Mehlich-3, and $NaHCO_3$ extractable P forms.

Fractionation and path analysis has also been used by Beck and Sanchez (1994) to evaluate differences in P pools between unfertilized and fertilized cropping systems on P deficient soils in the Amazon basin (Table 6.2). They used a modified version of the Hedley et al. (1982) fractionation scheme. Inorganic P explained 96% of the variation in the levels of available P in the fertilized system, but in the unfertilized system, organic P was the primary source of available P. The NaOH

TABLE 6.2
Application of Selected P Fractionation Schemes in Recent Years

Fractions	Results	Reference
	Manure-Amended Soils	
Resin, $NaHCO_3$, NaOH, H_2SO_4, Residual	Most soils had larger proportions of NaOH-P and residual P, indicating reaction of manure-derived P with pedogenic oxides and humic substances.	Leinweber et al., 1997.
Resin, $NaHCO_3$ P_i and P_o, NaOH P_i and P_o, Sonic P_i and P_o, HCl P_i, Residual	In manure-amended soils, labile P (Resin, $NaHCO_3$, NaOH -P_o) increased in the A horizon while resilient P (NaOH-P_i and HCl-P) were the major sinks in the B and C horizons.	Simard et al., 1995.
$NaHCO_3$ P_i and P_o, 0.1 *M* NaOH P_i and P_o, 1 *M* HCl, H_2SO_4-H_2O_2, Residual	Changes in P fractions with long-term corn monoculture were related to both added manure and inorganic fertilizer. Organic P forms were less related to available P in this study.	Zhang and MacKenzie, 1997.
	Arable Soils	
Resin, $NaHCO_3$ P_i and P_o, NaOH P_i and P_o, Sonic P_i and P_o, HCl P_i, Residual	NaOH P_i acted as the main sink for P in the fertilized system; Organic P was the main source of available P in the nonfertilized system.	Beck and Sanchez, 1994.
NH_4Cl, NH_4F, NaOH, H_2SO_4	Fe-P was a major P sink, especially in soils with a higher organic matter content; Ca-P and Al-P were the major contributors of plant-available P.	Subramaniam and Singh, 1997.
Resin, $NaHCO_3$ P_i and P_o, NaOH P_i and P_o, Sonic P_i and P_o, HCl P_i, Residual	Plant uptake of P was significantly related to labile P_i ($NaHCO_3$ P_i) and moderately labile P_o (NaOH P_o) fractions.	Sattell and Morris, 1992.
	Natural Ecosystems	
Resin, $NaHCO_3$ P_i and P_o, NaOH P_i and P_o, Sonic P_i and P_o, HCl P_i, Residual	The Hedley fractionation provides a valuable index of the relative importance of biological process to soil P content across a soil weathering gradient.	Cross and Schlesinger, 1995.
	Sediments	
NH_4F, NaOH, HCl; Biologically available P (BAP) in 0.1 *M* NaOH	Sediment BAP in drainage ditches was comparable to values for topsoils in adjacent fields; Greater than 90% of sediment P was NH_4F-P or NaOH-P.	Sallade and Sims, 1997.

extractable P pool was the main sink for fertilizer P in the soils they studied. Similar results were found in tropical Alfisols (Sri Lanka), (Sattell and Morris, 1992). Forty-five percent of the total P was in the NaOH fraction and the greatest portion (72%) of this P was organic. Plant uptake of P was related ($R^2 = 0.71$) to labile inorganic P ($NaHCO_3$-P_i) and moderately labile organic P (NaOH-P_0). The NaOH extractable P pool, especially Fe-P, was also a major P sink in heavily fertilized soils in Norway, although Ca-P and Al-P were the major contributors to the plant available P fraction in these soils (Subramaniam and Singh, 1997). Although P fractionation schemes are most often used to evaluate the P status of agricultural soils, Cross and Schlesinger (1995) have used the Hedley et al. (1982) procedure to study P cycling by biological versus geochemical processes during soil development. They found that across a weathering gradient in natural ecosystems, NaOH- and sonicated NaOH-P increased with weathering as more of the P became geochemically fixed to Fe and Al oxides in the soil. However, the biological P fractions (organic P) became an increasing proportion of the available-P with increased weathering.

The forms of sediment P draining agricultural fields affect P retention in these ditches, as well as P lost in drainage water. Sallade and Sims (1997) used a modified form of the Chang and Jackson (1957) procedure as part of a study to identify agricultural drainage ditches with significant potential for nonpoint source pollution of surface waters in Delaware's Inland Bays' watershed. An average of 95% of the inorganic P in the surface (0-5 cm) sediments of the ditches was extracted by NH_4F (Al-P; 69%) and NaOH (Fe-P; 26%). Significant portions of these fractions can be considered at least moderately labile. Ditches were thus considered as significant contributors of P to overlying water.

6.4 SOIL TEST P: AGRONOMIC VERSUS ENVIRONMENTAL ISSUES

In addition to the fractionation schemes designed to extract specific forms of soil P, several extractants have been used routinely to extract "available P" for soil testing purposes. These extractants are generally designed to provide a measure of P (and in some cases, other elements) that can be correlated with crop response. Extractants used in soil tests are generally classified into three categories as noted in Table 6.3 (Kuo, 1996).

Recently, these traditional agronomic soil tests have been proposed for environmental purposes as well (Sims, 1997). From an environmental standpoint, soil test results could be used to predict parameters such as equilibrium P concentration (EPC), labile P (resin-P), and algae-available P (Wolf et al. 1985). Relationships between southeastern soils (all Ultisols) and the above parameters are shown in Table 6.4. Significant correlations between agronomic soil tests and dissolved reactive P (DRP) and bioavailable P (BAP) in runoff were also observed in an Ultisol (silt loam) in Arkansas (Pote et al. 1996). In this case, soil samples were taken from the 0–2 cm depth to represent the soil layer that would be most interactive with surface runoff. Advantages that these soil tests have over other potential methods of evaluating P pollution potential include (1) their routine use in many commercial and governmental laboratories,(2) their relatively low cost, and (3) their relevance

TABLE 6.3
Extractants Commonly Used in the U.S. to Evaluate P Availability for Agronomic Purposes

Category	Composition	Examples	Comments
Dilute concentrations of strong acids	0.05 *M* HCl + 0.0125 *M* H_2SO_4 0.015 *M* NH_4F + 0.2 *M* CH_3COOH 0.25 *M* NH_4NO_3 + 0.013 *M* HNO_3 0.03 *M* NH_4F + 0.025 *M* HCl	Mehlich-1 Mehlich-3 Bray P_1	Fluoride is included to complex Al and prevent adsorption by Ca.
Dilute concentrations of weak acids	0.54 *M* CH_3COOH + 0.7 *M* $NaC_2H_3O_2$ (pH 4.8)	Morgan	
Buffered alkaline solutions	0.5 M $NaHCO_3$ (pH 8.5) 1 *M* NH_4HCO_3 + 0.005 *M* DTPA	Olsen	DTPA facilities extraction of microelements

TABLE 6.4
Relationships between Labile P, EPC, and Algae-Available P and Mehlich-1 Soil Test Values for Southeastern U.S. Ultisols (Wolf et al., 1985)

P Parameter (y)	Equation	r^2	Comments
Labile P: coarse textured soils, mg/kg	$y = -1.20 + 0.80x$	0.90	Not recommended for use on soils with textures coarser than sandy loams or on soils that have received continuous additions of manure.
Labile P: medium textured soils, mg/kg	$y = -0.47 + 0.97x$	0.93	
EPC, μg/L	$y = -39.77 + 5.66x$	0.85	
Algae-available P, mg/kg	$y = 41.29 + 3.51x$	0.86	Relationship influenced by the soil's active Al (oxalate-extractable) content.

to farmers from a crop yield standpoint and, hopefully, from an environmental standpoint.

Extractable soil P can also be used to estimate the degree of P saturation of a soil. This concept is based on the fact that as the amount of P adsorbed by the soil increases, the EPC in the soil solution also increases (Sharpley, this volume). Studies in the Netherlands (Breeuwsma and Silva, 1992) have shown that P concentrations in the soil solution can approach a critical concentration well before the soil is completely saturated with P. They have developed a test referred to as the degree of P saturation (DPS) which relates the soil P sorption capacity (PSC) to an extractable P concentration as follows:

$$DPS = \frac{\text{Extractable soil P}}{\text{P sorption maximum}} \times 100$$

The extractable P (P_{ox}) is determined by extraction with 0.2 *M* ammonium oxalate buffered at pH 3.0. Phosphorus sorption capacity is determined by standard P adsorption isotherms, such as the Langmuir isotherm. Due to the time involved in determining adsorption isotherms, the PSC in the Dutch test is estimated from oxalate-extractable Al (Al_{ox}) and Fe (Fe_{ox}) values, and the DPS expressed as:

$$DPS = \frac{P_{ox}}{0.5(Fe_{ox} + Al_{ox})} \times 100$$

The DPS relates extractable P to the P sorption capacity of the soil (Breeuwsma and Silva, 1992; Moore et al., 1997), and is used as an indicator of soil P available to be released to runoff. Recent research has shown that the extractable P in soils, such as P_{ox}, influences the amount of P in both runoff and subsurface drainage (Pote et al., 1996, Sharpley et al.,1996, Nair et al., 1998).

From an environmental standpoint, erosion of P-laden soil particles in surface runoff from agricultural lands has received most of the attention with regard to P and water quality. However, the vertical and lateral transport of P through soil profiles, in which drainage water ultimately enters surface water, has recently been identified as a potential source of P to surface waters in certain types of landscapes, such as in the Atlantic Coastal Plain (Mozaffari and Sims, 1994), and in Florida Spodosols (Allen, 1987; Mansell et al., 1991; Campbell et al., 1995). Sims et al. (1998) reviewed ongoing research in the Atlantic Coastal Plain of the USA (Delaware), the midwestern USA (Indiana), and eastern Canada (Quebec) and emphasized the need for soil and water conservation practices to minimize P losses in subsurface runoff.

In the Dutch studies (Breeuwsma and Silva, 1992), soils with more than 25% of the sorption capacity saturated maintained a P concentration of 0.1 mg P L^{-1} in groundwater. Lookman et al. (1995) showed that the relative size of the quickly desorbing P pool, i.e., extractable P, increased with increasing DPS. Their study also suggested that nearly all P_{ox} was potentially desorbable and that DPS was directly pertinent to the estimation of P leaching losses from P-laden agricultural soils.

Sharpley (1995) suggested calculating P sorption saturation as the percentage of P sorption maximum (calculated using Langmuir isotherms) extractable as Mehlich-3 P. He showed that P sorption saturation of surface soils, represented by the percentage of P sorption maximum as Melhich-3 P for ten soils, ranged from 23.5 to 31.3%.

In the Southeastern and Mid-Atlantic States, Mehlich-1 (or double-acid extractable P) is used routinely as the soil test P and, therefore, it might be practical to utilize Mehlich-1 P to determine the DPS. Sallade and Sims (1997) used the DPS criteria (based on the ratio of Mehlich-1 P to Langmuir P sorption capacity) to evaluate sediment properties important to P release. They calculated a mean DPS value of 4.5%; however, when calculations were based on the ratio of bioavailable P (BAP) (0.1 *M* NaOH extractable) to P sorption capacity, the DPS values for the same 40 sediments gave a mean of 43%. This difference suggested a need to establish a relationship between DPS and P release to overlying water, and they concluded that the 25% value proposed by the Dutch researchers may not be valid under all circumstances.

6.5 APPLICATIONS OF FRACTIONATION SCHEMES IN FLORIDA'S SUBTROPICAL ENVIRONMENT

The Lake Okeechobee watershed in Florida, U.S.A., with high P imports, is one of the most widely researched areas in the subtropics. Land uses within this watershed are primarily dairy farming and beef ranching. Phosphorus from the upland regions moved through the wetlands into the lake, resulting in its current eutrophic state. Application of inorganic fertilizers for improved forage production, adds to the P load of this watershed (Boggess et al., 1995). Examples of P fractionation studies, using the Hieltjes and Lijklema (1980) fractionation procedure, in uplands and wetlands in the Lake Okeechobee watershed, and in the lake sediments, are discussed below.

6.5.1 THE UPLANDS

Nair et al. (1995) identified the P forms in the soil profiles of differentially manure-impacted soils in the Okeechobee watershed. All soils were Spodosols, and soils were collected by horizon, A, E, Bh, and Bw. Description of Spodosols and a typical Spodosol profile are provided by Campbell and Capece (this volume). Total P (TP) in the surface (A horizon) soils varied from 30 mg P kg^{-1} in non-manure impacted (native) soils to about 2930 mg P kg^{-1} in the heavily impacted (intensive component) manured soils. Studies were also conducted on abandoned dairies, once subjected to intense dairy farming. Other component soils in this study included pasture areas, used for grazing, and forage areas, used for forage production. Table 6.5 shows the percentage of each fraction within a horizon for the various components.

There is no statistical difference in the percentage of labile P (NH_4Cl-extractable P), the P that would most likely move from the A horizon of the various components. More P will be lost from the heavily manure-impacted intensive areas with high TP values, than from the less impacted pasture, forage and native areas (Table 6.5). It is also of interest to note that the P will continue to be lost from dairies that have been abandoned for considerable periods of time (Nair et al., 1995). The P that leaves the surface horizon might be lost through surface and subsurface drainage (Campbell et al., 1995), and the portion that reaches the spodic (Bh) horizon will be held as Al- and Fe-associated P (Table 6.5), either in the inorganic or in the organic fraction. The high percentages of HCl-extractable P (Ca- and Mg-associated P) in the A horizon of the intensive dairy components are also of potential concern. Such P could be continuously extracted by NH_4Cl (Nair et al., 1995) or by water (Graetz and Nair, 1995), suggesting that about 80% of the total soil P had the potential to move eventually with drainage water into Lake Okeechobee. Harris et al. (1994), despite high Ca-P (up to 13,000 mg P kg^{1} on a clay basis), found no crystalline P in the clay x-ray diffraction analysis. The clay was dominated by Si, phytoliths and their amorphous degradation products which have very low P sorption capacity. Vivanite and a calcium phosphate mineral resembling poorly crystalline apatite were found in a stream sediment in the watershed (Harris et al., 1994).

TABLE 6.5
Percentage of P Fractions within a Horizon for the Components, Intensive (Abandoned and Active), Pasture, Forage, and Native (adapted from Nair et al., 1995)

Component	Labile P, %	Al/Fe-P, %	Org.-P, %	Ca/Mg-P, %	Res. P, %	Total P, mg kg^{-1}
			A horizon			
Aband. int.	9.5 a	8.0 c	2.8 c	69.9 a	9.8 a	2933
Active int.	10.5 a	6.5 c	1.6 c	67.6 a	13.7 a	3028
Pasture	8.9 a	21.5 b	41.4 a	15.1 b	13.1 a	275
Forage	6.0 a	23.6 b	41.1 a	15.0 b	14.4 a	47
Native	11.1 a	47.1 a	18.9 b	10.6 b	12.2 a	31
			E horizon			
Aband. int.	39.9 a	15.7 a	11.4 b	24.6 a	8.5 a	105
Active int.	37.8 a	10.2 a	21.5 b	21.5 a	9.1 a	136
Pasture	12.0 b	13.6 a	55.5 a	9.5 b	9.4 a	28
Forage	5.5 bc	15.1 a	55.0 a	16.1 ab	8.2 a	17
Native	1.8 c	19.7 a	54.9 a	15.7 ab	8.0 a	15
			Bh horizon			
Aband. int.	3.4 b	74.4 ab	11.5 bc	8.5 a	2.2 b	535
Active int.	11.0 a	78.0 a	0.4 c	7.2 a	3.4 b	228
Pasture	0.6 b	71.3 ab	19.9 ab	4.3 a	3.9 b	113
Forage	1.0 b	56.7 bc	28.0 ab	6.6 a	7.8 a	50
Native	0.5 b	46.5 c	41.0 a	5.0 a	7.0 a	66
			Bw horizon			
Aband. int.	3.5 ab	63.4 a	21.3 ab	6.0 a	5.8 a	219
Active int.	13.9 a	49.1 a	8.3 b	9.1 a	19.6 a	159
Pasture	0.5 ab	46.9 a	35.4 a	4.9 a	12.4 a	45
Forage	0.4 b	44.3 a	33.6 a	7.2 a	14.6 a	56
Native	0.0 b	48.0 a	38.7 a	7.2 a	6.1 a	52

Mean values for a given fraction within a horizon followed by the same letter are not different ($p = 0.05$) using the Waller-Duncan k-ratio multiple comparison procedure.

6.5.2 The Wetlands

The P from the uplands, if diverted through wetlands, will be retained by the wetlands, at least partially. Reddy et al. (1995) characterized wetland soils and sediments from the Lake Okeechobee watershed using a modified Hieltjes and Lijklema (1980) fractionation scheme where the 1 *M* NH_4Cl extactant was replaced by 1 *M* KCl. The labile P pool in stream sediments ranged from 0.1% (TP = 357 mg P kg^{-1}) to 2.3% (TP = 32 mg P kg^{-1}) and in wetland soils 0.1% (TP = 347 mg P kg^{-1}) to 1.1% (TP = 190 mg P kg^{-1}). A drained wetland in their studies had a larger labile P pool, 9.5% of TP (351 mg P kg^{-1}). Most of the P in the wetland soils and sediments was associated with inorganic Fe and Al, with the amounts ranging from 20% (TP = 45 mg P kg^{-1}) to 70% of the TP (TP = 93 mg P kg^{-1}) for stream sediments and from 17% (TP = 987 mg P kg^{-1}) to 37% of the TP (TP = 294 mg P

kg^{-1}) in wetland soils. Phosphorus associated with crystalline Fe and Al oxides are likely to be stable, with the possibility of being desorbed only under extended waterlogged conditions (Reddy et al., 1995). Most of the wetland soils and stream sediments characterized in their study had low Ca- and Mg-associated P (0.5 *M* HCl extraction). The organic P pool (extracted by 0.1 *M* NaOH), is biologically reactive with the potential of being hydrolyzed to bioavailable forms. Values ranged from 6% (TP = 877 mg kg^{-1}) to 57% (TP = 139 mg P kg^{-1}) of the TP. Total P in the wetland soils and sediments was greater than in the native upland areas (30 mg P kg^{-1}). Whether wetlands and stream sediments act as sources or sinks for P depends on the physic-chemical characteristics of the soils and sediments as well as the water column above it (Richardson, 1985). Reddy et al. (1995) noted that wetland soils adjacent to streams had a higher capacity to retain P than stream sediments, and concluded that wetland/stream systems function as sinks for P under most conditions, except possibly under periods of high rainfall. Wetlands and streams offer long flow paths between uplands and the receiving water bodies. Thus, they should be efficient sinks for the P fractions that have the potential to leave the uplands (Reddy et al., 1996).

6.5.3 Interaction of P between Sediments and Overlying Waters in the Lake

If P eventually reaches Lake Okeechobee, what would be the potential for the lake sediments to release P into the overlying water? The sediments in the lake were classified as mud (44%), sand and rock (28%), littoral (19%) and peat (9%) (Reddy et al., 1991). Concentrations of readily available P (defined as the sum of porewater P and NH_4Cl-extractable P) in the mud sediment was 2% of the TP (1140 mg P kg^{-1}); 9.7% of the TP (1109 mg P kg^{-1}) was in the littoral sediment and 17.4% of the TP (449 mg P kg^{-1}) in the peat (Olila et al., 1995). Although the percentage of the readily available P in the mud sediment was lower than that of a hypereutrophic lake, Lake Apopka located in central Florida, the percentage of the readily available P in the littoral and peat sediments were comparable to those of Lake Apopka.

Olila et al. (1995) also compared their results to TP and inorganic P fractions reported in literature for sediments from selected lakes, based on the Hieltjes and Lijklema (1980) fractionation scheme. They concluded that the concentrations of readily available P (NH_4Cl-P) in the Florida lakes were similar to the P concentrations in Lake Balaton (Hungary), Lake Nieuwkoop (The Netherlands), and Lakes Vallentunasjon and Norrviken (Sweden). Lake Apopka, with its large sediment P reservoir, and the sediments of the littoral and peat areas of Lake Okeechobee with high concentrations of readily available P, may be important contributors to internal P cycling of the lakes (Olila et al., 1995).

6.6 SUMMARY AND CONCLUSIONS

The total P content of a soil or sediment provides little information regarding the behavior of P in the environment, both from agronomic and environmental standpoints. However, P minerals in soils and sediments are difficult to measure quantitatively. In addition, much of the P exists in amorphous solids or sorbed onto soil coatings not detected by standard instrumental techniques, such as x-ray analysis. Because of these problems, several sequential fractionation schemes, based on the assumption that certain reagents can preferentially extract chemical forms of P from soils and sediments, have been developed which provide much needed information regarding the P status of soils and sediments. Common inorganic P pools identified in sequential fractionation procedures include 1) loosely bound (also referred to as labile or exchangeable P), 2) fractions associated with Al, Fe, and Mn oxides and hydroxides, 3) a Ca- and Mg-bound fraction, and 4) minerals and organic material resistant to previous extractants (Chang and Jackson, 1957; Williams et al., 1971; Hieltjies and Liklema, 1980). Other procedures, such as Hedley et al. (1982), make provisions to evaluate both organic and inorganic P pools.

Much of the early use of these fractionation procedures was for agronomic purposes, i.e., evaluating the transformations of P in fertilizer and manure that affect plant availability of P. More recently, they have been used to determine the potential for P movement from soils via erosion and subsurface lateral movement and to provide information about the interaction of P between sediments and overlying water in aquatic systems.

Criticism of P fractionation procedures include that they are operationally defined and are subject to broad interpretations, that the classification of soil P as Ca-P, Fe-P, and Al-P is overly simplistic, and that they are not capable of distinguishing among various P reaction products in soils and sediments. However, considering the objectives of the studies in which P fractionation is generally used, P fractionation offers a means of obtaining significant information on the P chemistry of soils and sediments.

REFERENCES

Allen, L.H. 1987. Dairy-siting criteria and other options for wastewater management on high water-table soils. Soil Crop Sci. Soc. Fla. Proc. 47:108–127.

Aspila, K.I., H. Agemian, and A.S.Y. Chau. 1976. A semi-automated method for the determination of inorganic, organic, and total phosphorus in sediments. Analyst. 101:187–197.

Barbanti, A., M.C. Bergamini, F. Frascari, S. Miserocchi, and G. Rosso. 1994. Critical aspects of sedimentary phosphorus chemical fractionation. J. Environ. Qual. 23:1093–1102.

Beck, M.A., and P.A. Sanchez. 1994. Soil phosphorus fraction dynamics during 18 years of cultivation of a typic Paleudult. Soil Sci. 34:1424–1431.

Boggess, C.F., E.G. Flaig, and R.C. Fluck. 1995. Phosphorus budget-basin relationships for Lake Okeechobee tributary basins. Ecol. Eng. 5:143–162.

Breeuwsma, A., and S. Silva. 1992. Phosphorus fertilisation and environmental effects in the Netherlands and the Po region (Italy). Rep. 57. Agric. Res. Dep. The Winand Staring Centre for Integrated Land, Soil and Water Research. Wageningen, The Netherlands.

Campbell, K.L., J.C. Capece, and T.K. Tremwel. 1995. Surface/subsurface hydrology and phosphorus transport in the Kissimmee River Basin, Florida. Ecol. Eng. 5:301–330.

Campbell, K.L., and J.C. Capece. Hydrologic processes influencing phosphorus transformations and transport. Chapter 11, this book.

Chang, S.C., and M.L. Jackson. 1957. Fractionation of soil phosphorus. Soil Sci. 84:133–144.

Cross, A.F., and W.H. Schlesinger. 1995. A literature review and evaluation of the Hedley fractionation: Application to the biogeochemical cycle of soil phosphorus in natural systems. Geoderma. 64:197–214.

Dean, L.A. 1938. An attempted fractionation of the soil phosphorus. J. Agri. Sci. 28:234–246.

De Groot, C.J. 1990. Some remarks on the presence of organic phosphates in sediments. Hydrobiologia 207:303–309.

Eghball, B., G.D. Binford, and D.D. Baltensperger. 1996. Phosphorus movement and adsorption in a soil receiving long-term manure and fertilizer application. J. Environ. Qual. 25:1339–1343.

Golterman, H.L., and A. Booman. 1988. Sequential extraction of iron-phosphate and calcium phosphate from sediments by chelating agents. Verh. Int. Ver. Limonol. 23:904–909.

Graetz, D.A., and V.D. Nair. 1995. Fate of phosphorus in Florida Spodosols contaminated with cattle manure. Ecol. Eng. 5:163–181.

Harris, W.G., H.D. Wang, and K.R. Reddy. 1994. Dairy manure influence on soil and sediment composition: implications for phosphorus retention. J. Environ. Qual. 23:1071–1081.

Hedley, M.J., J.W.B. Stewart, and B.S. Chauhan. 1982. Changes in inorganic and organic soil phosphorus fractions induces by cultivation practices and by laboratory incubations. Soil Sci. Soc. Am. J. 46:970–976.

Hieltjes, A.H.M., and L. Lijklema. 1980. Fractionation of inorganic phosphates in calcareous sediments. J. Environ. Qual. 9:405–407.

Ingall, E.D., P.A. Schroeder, and R.A. Berner. 1990. The nature of organic phosphorus in marine sediments: New insights from P NMR. Geochim. Cosmochim. Acta. 54:2617–2620.

Koch, M.S., and K.R. Reddy 1992. Distribution of soil and plant nutrients along a trophic gradient in the Florida Everglades. Soil Sci. Soc. Am. J. 56:1492–1499.

Kuo, S. 1996. Phosphorus. *In* J.M. Bartels (ed) Methods of Soil Analysis. Part 3—Chemical Methods. Soil Science Society of America, Inc. American Society of Agronomy, Inc., Madison, Wisconsin, USA. pp. 5:869–919.

Leinweber, P., L. Haumaier and W. Zech. 1997. Sequential extractions and ^{31}P-NMR spectroscopy of phosphorus forms in animal manures, whole soils and particle-size separates from a densely populated livestock area in northwest Germany. Biol. Fertil. Soils. 25:89–94.

Lookman, R., D. Freese, R. Merckx, K. Vlassak, and W.H. van Reimsdijk. 1995. Long-term kinetics of phosphate release from soil. Environ. Sci. Technol. 29:1569–1575.

Mansell, R. S., S. A. Bloom, and B. Burgoa. 1991. Phosphorus transport with water flow in acid, sandy soils. *In* Jacob B. and M. Y. Corapcioglu (eds.). Transport Processes in Porous Media. Kluwer Academic Publishers, Dorecht, The Netherlands. pp. 271–314.

Moore, P.A. Jr., B.C. Joern, and T.L. Provin. 1997. Improvements needed in environmental soil testing for phosphorus. *In* Sims, J. Thomas (ed.) Soil Testing for Phosphorus: Environmental Uses and Implications. SERA-IEG 17 USDA-CREES Regional Committee: Minimizing Agricultural Phosphorus Losses for Protection of the Water Resource.

Mozaffari, M., and J.T. Sims. 1994. Phosphorus availability and sorption in an Atlantic Coastal Plain watershed dominated by animal-based agriculture. Soil Sci. 157:97–107.

Nair, V.D., D.A. Graetz, and K.M. Portier. 1995. Forms of phosphorus in soil profiles from dairies of south Florida. Soil Sci. Soc. Am. J. 59:1244–1249.

Nair, V.D., D.A. Graetz, and K.R. Reddy. 1998. Dairy manure influences on phosphorus retention capacity of Spodosols. J. Environ. Qual. 27:522–527.

Newman, S., and J.S. Robinson. Forms of organic phosphorus in water, soils, and sediments. (this volume).

Olila, O.G., K.R. Reddy, and W.G. Harris. 1995. Forms and distribution of inorganic phosphorus in sediments of two shallow eutrophic lakes in Florida. Hydrobiologia. 302:147–161.

Oluyedum, O.A., S.O. Ajayi, and G.W. Van Loon. 1991. Methods for fractionation of organic phosphorus in sediments. Sci. Total Environ. 106:243–252.

Pettersson K., and V. Istvanovics. 1988. Sediment phosphorus in Lake Balaton—Forms and mobility. p. 25–41. *In*: Psenner, R., and A. Gunatilaka (eds.). Proceedings of the First International Workshop on Sediment Phosphorus. Advances in Liminology, Heft 30, Stuttgart.

Pote, D.H., T.C. Daniel, A.N. Sharpley, P.A. Moore, Jr., D.R. Edwards, and D.J. Nichols. 1996. Relating extractable soil phosphorus to phosphorus losses in runoff. Soil Sci. Soc. Am. J. 60:855–859.

Reddy, K.R., M. Brenner, M.M. Fisher, and D.B. Ivanhoff. 1991. Lake Okeechobee Phosphorus Dynamics Study: Biogeochemical Processes in the Sediments. Vol. III. Final Report to the South Florida Water Management District, West Palm Beach, FL. Contract No. 531-m88-.445-A4. Soil Science Department, IFAS, University of Florida, Gainesville.

Reddy, K.R., O.A. Diaz, L.J. Scinto, and M. Agami. 1995. Phosphorus dynamics in selected wetlands and streams of the Lake Okeechobee Basin. Ecol. Eng. 5:183–208.

Reddy, K.R., E.G. Flaig, and D.A. Graetz. 1996. Phosphorus storage capacity of uplands, wetlands and streams of the Lake Okeechobee Watershed, Florida. Agri. Ecosyst. Environ. 59:203–216.

Reddy, K.R., R.H. Kadlec, E. Flaig, and P.M. Gale. 1999. Phosphorus retention in streams and wetlands: A review. Crit. Rev. Environ. Science and Technol. 29:1–64.

Richardson, C.J. 1985. Mechanisms controlling phosphorus capacity in freshwater wetlands. Science: 228:1424–1427.

Ruttenberg, K.C. 1992. Development of a sequential extraction method for different forms of phosphorus in marine sediments. Limnol. Oceanogr. 37:1460–1482.

Sallade, Y.E., and J.T. Sims. 1997. Phosphorus transformations in the sediments of Delaware's agricultural drainageways: I. Phosphorus forms and sorption. J. Environ. Qual. 26:1571–1579.

Sattell, R.R., and R.A. Morris. 1992. Phosphorus fractions and availability in Sri Lankan Alfisols. Soil Sci. Soc. Am. J. 56:1510–1515.

Sharpley, A.N. 1995. Dependence of runoff phosphorus on soil phosphorus. J. Environ. Qual. 24:920–926.

Sharpley, A.N. 1998. Global issues of phosphorus in terrestrial ecosystems. (this volume).

Sharpley, A.N., T.C. Daniel, J.T. Sims, and D.H. Pote. 1996. Determining environmentally sound soil phosphorus levels. J. Soil and Water Conserv. 51:160–166.

Simard, R.R., D. Cluis, G. Gangbazo, and S. Beauchemin. 1995. Phosphorus status of forest and agricultural soils from a watershed of high animal density. J. Environ. Qual. 24:1010–1017.

Sims, J.T. 1997. Soil Testing for Phosphorus: Environmental Uses and Implications. SERA-IEG 17 USDA-CREES Regional Committee: Minimizing Agricultural Phosphorus Losses for Protection of the Water Resource.

Sims, J.T., R.R. Simard, and B.C. Joern. 1998. Phosphorus loss in agricultural drainage: Historical perspective and current research. J. Environ. Qual. 27:277–293.

Subramaniam, V., and B.R. Singh. 1997. Phosphorus supplying capacity of heavily fertilized soils I. Phosphorus adsorption characteristics and phosphorus fractionation. Nutrient Cycling in Agroecosystems. 47:115–122.

Turner, R.C., and H.M. Rice. 1954. Role of fluoride ion in release of phosphate adsorbed by Al and Fe hydroxide. Soil Sci. 74:141–148.

van Eck, G.T.M. 1982. Forms of phosphorus in particulate matter from the Hollands Diep/Haringvliet, The Netherlands. Hydrobiologia 92:665–681.

Williams, J.D.H., J.K. Syers, R.F. Harris, and D.E. Armstrong. 1971. Fractionation of soil inorganic phosphate in calcareous lake sediments. Soil Sci. Soc. Am. Proc. 35:250–255.

Wolf, A.M., D.E. Baker, H.B. Pionke, and H.M. Kunishi. 1985. Soil tests for estimating labile, soluble, and algae-available phosphorus in agricultural soils. J. environ. Qual. 14:341–348.

Zhang, T.Q., and A.F. MacKenzie. 1997. Changes in soil phosphorus fractions under long-term corn monoculture. Soil Sci. Soc. Am. J. 61:485–493.

7 Phosphorus Sorption/Desorption Reactions in Soils and Sediments

R.D. Rhue and Willie G. Harris

7.1 ABSTRACT

The purpose of this chapter is to provide an overview of adsorption/precipitation reactions that control P solubility and mobility in soils and sediments. Sorption is generally characterized by a rapid initial uptake followed by long periods during which the rate of sorption decreases. The initial rapid-uptake phase probably results from ligand exchange. Several models ascribe the slow-uptake phase to diffusion of P into the interior of soil particles, the diffusion being driven by chemical and electrical potential gradients. Diffusion of phosphate along surfaces to sites of decreasing accessibility hs also been proposed to explain the slow-uptake phase. However, the identification of both amorphous and crystalline metal-phosphates following sorption suggests that precipitation can be an important mechanism under certain conditions. Sorption of phosphate is strongly influenced by redox potential. The reduction and dissolution of iron and its reprecipitation to form ferrous minerals is thought to be the dominant process controlling P solubility in anaerobic systems. Oxalate-extractable iron may be the most important measure of changes in P sorption capacity when soils are flooded. Strong evidence suggests that sorption and release of P in some lake sediments is strongly related to redox-dependent changes in microbial physiology.

1-56670-331-X/99/$0.00+$.50

7.2 INTRODUCTION

Phosphorus (P) has been implicated as a major factor in the eutrophication of lakes and rivers. P interaction with soils and sediments is, therefore, of primary interest to both soil and water scientists. The world's ever growing population will continue to place greater demands on the soil, both in terms of food production and waste disposal. Disposal of wastewater and animal manures in some areas of the world have already exceeded the capacity of soils to hold P, resulting in serious pollution threats to ground and surface water bodies. Downward movement through the vadose zone to groundwater, then laterally into streams and lakes is a major transport route for P in many sandy soils found in Florida. Understanding the chemical and physical interactions of P with soils and sediments is, thus, critical to minimizing the impact that land application of fertilizers and wastes has on the eutrophication of lakes and streams. The fact that soil bodies are dynamic chemical and biological systems in a constant state of flux and disequilibrium further complicates the task of predicting the solubility and mobility of P within them.

Although P is ubiquitous in nature, its concentration in pore waters and aquatic systems is usually quite low. This results from the fact that P adsorbs strongly to oxides and other mineral surfaces commonly found in soils and aquifers, as well as forming precipitates of low water solubility with commonly occurring cations such as Fe, Al, and Ca. The purpose of this paper is to provide an overview of sorption/desorption reactions that can affect P solubility and mobility in soils and sediments.

7.3 THE OPERATIONAL DEFINITION OF SORPTION

A sorption reaction involves the removal of P from solution by concentrating it in, or on, a solid phase. Experimentally, one mixes a soil or sediment sample with a P solution of known initial concentration, equilibrates the mixture for a period of time, and then measures the amount of P remaining in solution. The difference between initial and final P concentrations is taken as sorption, which can be either positive or negative in magnitude. Loss of P from solution can result from either adsorption or precipitation. Adsorption is defined as the concentration of a substance at an interface; in the case of phosphate, this interface is the solid-liquid interface. In contrast, precipitation is the removal of two or more components from solution by their mutual combination to form a new solid-phase. The major difference between adsorption and precipitation is one of geometry: adsorption being a two-dimensional process, precipitation a three-dimensional process (Corey, 1981). Distinguishing between adsorption and precipitation is difficult because precipitation is often preceded by adsorption that is then followed by the nucleation of a "surface precipitate." Clearly, however, precipitation will not occur if the solution is under-saturated with respect to the new solid phase. Many measurements of P adsorption maxima in soils should be regarded as suspect because the equilibrating solutions were supersaturated with respect to one or more phosphate solid phases (Corey, 1981). Because adsorption and precipitation represent different mechanisms of P retention, reporting "adsorption maxima" from such studies is inappropriate.

7.4 ADSORPTION ISOTHERMS

An adsorption isotherm describes the equilibrium relationship between the concentrations of adsorbed and dissolved species at a given temperature. The three most common descriptions of P sorption are the Langmuir, Freundlich, and Tempkin equations. All three were originally derived to describe gas adsorption on solid surfaces (Berkheiser et al., 1980). In their original forms, they were incapable of accounting for electrical interactions between surface charge and anion charge, speciation of the adsorbate, or competition among various adsorbing species that might be present. Although soil scientists have adapted and modified these equations and used them to describe anion adsorption from aqueous solution, they are still limited in their ability to describe P sorption. Not surprisingly, these equations have proven less than satisfactory in describing P sorption over a wide range of experimental conditions.

7.4.1 Langmuir Equation

The Langmuir equation assumes that adsorption occurs at specific sites on the surface and that once these sites are filled, no further adsorption occurs, i.e., monolayer adsorption. Furthermore, the heat of adsorption is assumed to be constant with no lateral interactions between adsorbate molecules as the monolayer is filled. A commonly used, linear form of the Langmuir equation is

$$\frac{C}{S} = \frac{1}{kS_{max}} + \frac{C}{S_{max}} \tag{1}$$

where C is solution P concentration, S is the amount of sorbed P, S_{max} is the adsorption maximum, and k is a constant related to the bonding energy between phosphate and the surface.

7.4.2 Freundlich Equation

The Freundlich equation was originally introduced as a strictly empirical expression to describe adsorption phenomena. More recently, Sposito (1980) showed that the Freundlich equation could be rigorously derived for trace adsorption of an ion participating in an exchange reaction. The derivation is based on the assumption that the surface is heterogeneous and that adsorption by each class of exchange sites can be described by a Langmuir equation. The Freundlich equation takes the form:

$$(S = kC^n) \tag{2}$$

where k and n are empirical constants, with $n \leq 1$. A modified form of the Freundlich equation replaces S with $S + S_o$, where S_o is a constant that corrects for the amount of previously adsorbed P (Bache and Williams, 1971).

7.4.3 Tempkin Equation

The Tempkin equation can be derived from the Langmuir equation by replacing the constant heat of adsorption, Q, with a distribution function such that

$Q = Q_o(1 - \alpha\theta)$ where Θ is the fractional surface coverage and Q_o and α are constants (Berkheiser et al., 1980). The Tempkin equation takes the form:

$$\frac{S}{S_{max}} = \frac{RT}{b}\ln(AC) \tag{3}$$

where R is the ideal gas constant, T is absolute temperature, and A and b are empirical constants.

Adsorption equations commonly used to describe phosphate adsorption are often special cases of the general expression:

$$S = \sum \frac{b_i K_i^{\beta_i} C^{\beta_i}}{1 + K_i^{\beta_i} C^{\beta_i}}, \qquad i = 1, 2, \ldots n \tag{4}$$

where b_i and β_i are constants (Goldberg and Sposito, 1984).

Substituting n = 1 or 2 results in the conventional or 2-site Langmuir equations, respectively. The Freundlich equation results if $n = 1$, $0 < \beta_i < 1$, and $K_i^{\beta_i} C^{\beta_i} << 1$. These authors cautioned that adsorption isotherms like these are probably best regarded as strictly empirical equations in which the constants are derived by curve-fitting procedures. Because of the empirical nature of these equations, "goodness of fit" should not be confused with "correctness" of a particular sorption model.

These models also suffer from the disadvantage that their parameters must be regarded as unknown functions of pH and ionic strength, both of which impact P sorption. Other experimental variables that can affect P sorption include temperature, the concentration and composition of the background electrolyte, aeration, the solution P concentration used, soil-to-solution ratio, competition from other anions, and the amount and timing of previous P additions. The effects of these variables on P sorption isotherms were discussed by Berkheiser et al. (1980).

7.5 P SORPTION MECHANISMS

P sorption is generally characterized by a high initial sorption rate that decreases as a function of time (Fig. 7.1). Slow uptake of P can continue for months or years. Rate expressions that have been used to describe uptake kinetics include zero, first, second, and third order equations, parabolic and other fractional order equations, and exponential expressions such as the Elovich equation. A plot of reciprocal rate, Z, versus time (Fig. 7.2) is generally sigmoidal, and exhibits features of all of the above rate expressions. For example, parabolic and other fractional rate expressions adequately describe the Z function over short reaction times (Region I), whereas first-order kinetics describe the behavior in region III of the curve. The relatively linear sorption in region II is described by the Elovich equation quite well. However, none of these expressions has proven useful over the entire time scale for sorption. Pavlatou and Polyzopoulos (1988) succeeded in describing P sorption kinetics with an expression that incorporated parabolic, Elovich, and exponential behavior, and concluded that diffusion was the rate-determining step. The sorption behavior depicted in Fig. 7.1 has been interpreted as a biphasic reaction, an initial, rapid

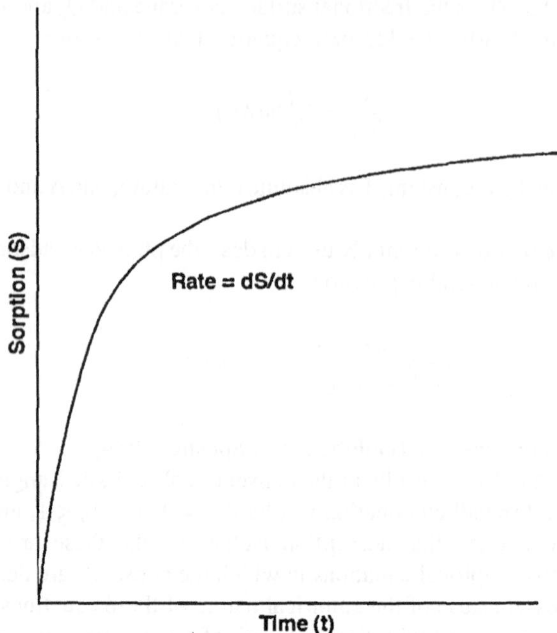

FIGURE 7.1 Phosphorus sorption vs. time for a typical soil.

sorption is thought to last on the order of minutes or hours, followed by a slow sorption reaction lasting weeks or months.

7.5.1 The Rapid Phase of P Sorption

The predominant mechanism during the initial, rapid reaction is thought to be adsorption, and mechanistic models have been developed that describe this uptake phase fairly well. Phosphorus adsorption occurs by both ion exchange and ligand exchange.

Ion Exchange

Ion-exchange results from the electrostatic attraction of phosphate anions to positively charged sites that exist on variable-charge surfaces below the zero point of charge. Ion exchange reactions are reversible, nonspecific, and extremely rapid. Ion exchange probably accounts for only a small fraction of the adsorbed P in soils and sediments; rather, most of the rapidly sorbed P is held by ligand exchange.

Ligand Exchange

Ligand exchange is a mechanism by which a phosphate anion replaces a surface hydroxyl that is coordinated with a metal cation in a solid phase. The most reactive surfaces for phosphate adsorption appear to be iron and aluminum oxyhydroxides,

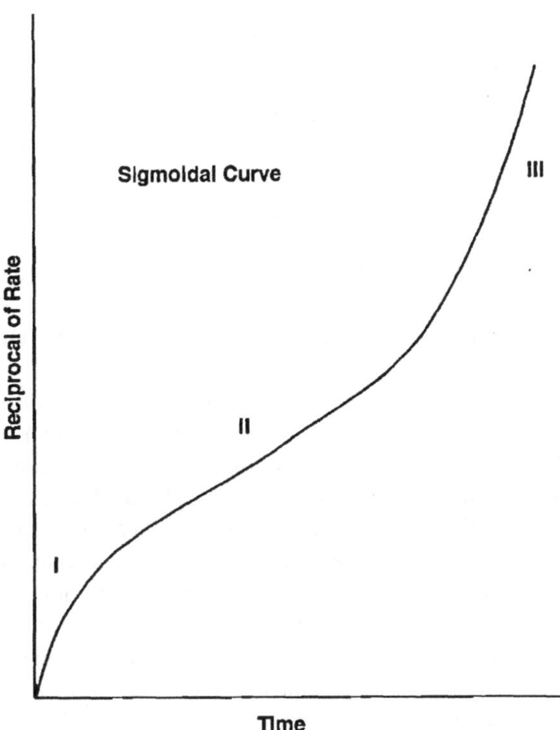

FIGURE 7.2 Typical behavior for the reciprocal of P sorption rate vs. time for soil. See text for discussion of Regions I, II, and III.

followed by edge sites on layer silicates. Research suggests that carbonate minerals in soil play a minor role in P sorption, even when coated with iron oxides (Hamad et al., 1992). Phosphate apparently does not react directly with organic matter, although at least one common anion, borate, has been shown to do so (see McBride, 1994). However, phosphate can react with iron and aluminum that is associated with soil and sediment organic matter, as evidenced by the close relationship between organically complexed Fe and Al and P sorption parameters (Yuan and Lavkulich, 1994, Zhou et al., 1997). McBride (1994) attributed this type of bonding to the formation of a "Type A" ternary complex where a metal cation, such as Al^{+3} or Fe^{+3}, bonds simultaneously to functional groups on organic matter and to an anion such as phosphate.

Ligand exchange is also referred to as "specific" adsorption and differs from ion-exchange in several important aspects. McBride (1994) summarized these differences as follows:

1. Adsorption by ligand exchange is accompanied by the release of OH^-.
2. Ligand exchange shows a high degree of specificity.
3. The adsorption step often occurs much more rapidly than the desorption step, leading to hysteresis in the isotherm.

4. Adsorption of anions by ligand exchange is accompanied by an increase in surface negative charge.

In the case of P sorption, hysteresis is often taken to imply irreversibility. However, this can be an over simplification because both P sorption and desorption can be very slow processes in soils, continuing for times much longer than those commonly used in laboratory studies (Barrow, 1983a).

Several types of hydroxyl ions occur on metal oxide surfaces with differing reactivity toward phosphate (Fig. 7.3). Reactivity of the hydroxyls shown in Fig. 7.3 toward phosphate increases in the order: ol > hydroxo > aquo. The aquo form dominates in acidic solutions, and is thought to exchange readily with anions such as phosphate.

Surface ligands in soils and sediments are not restricted to hydroxyls and water. Ligands such as sulfate, silicate, etc., can also exchange with P or influence P sorption. In soils dominated by amorphous iron and aluminum compounds, P effectively replaced sulfate and silicate as well as hydroxyl ions during the rapid sorption phase (Pardo and Guadalix, 1990). At low P concentrations, P sorption was accompanied mostly by release of adsorbed sulfate with some silicate. As more P was sorbed, more hydroxyls were displaced. However, molar ratios of total anions released to P sorbed were < 1, indicating the importance of aquo ligands in P adsorption. Miller et al. (1989) found that flow-generated partition coefficients were greater than corresponding batch-generated values and attributed this in part to removal of competitive antecedent species in the flow system. Their results indicated that curvature in Langmuir plots [Eq. (1)] may result from competition between P and displaced ligands. In one instance, where the competing anion was known to be arsenate, modification of the Langmuir model to include competitive sorption effects was shown to greatly improve its ability to describe P sorption.

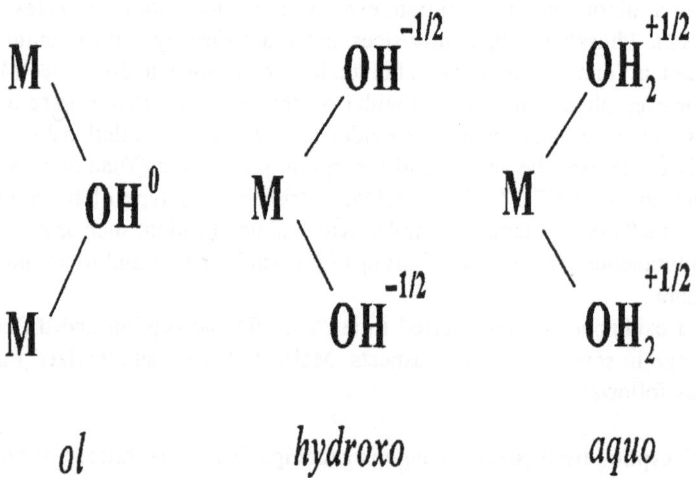

FIGURE 7.3 Three types of surface hydroxyls found on hydrous oxides of iron and aluminum and the surface charge associated with each type (see McBride, 1994).

Organic anions also compete strongly with phosphate for ligand exchange sites in soils (Afif et al., 1995, Kafkafi et al., 1988). Dissolved organic matter (DOM) derived from vetch and clover inhibited P sorption by acid soil, whereas DOM derived from animal manure did not (Ohno and Crannell, 1996). Inhibition at low DOM concentrations was attributed to competition at ligand exchange sites on the soil surface. At higher concentrations of DOM, Al was solubilized by complexation reactions between DOM and structural Al. The higher molecular weight of animal-derived DOM was thought to be a factor in its inability to compete with P for sorption.

Energetics

The adsorption of phosphate from solution by metal oxides has been referred to as *chemisorption.* Selwood (1975) defines chemisorption as adsorption that involves some kind of electronic interaction between adsorbent and adsorbate. Chemisorption exhibits several characteristics that distinguish it from physical adsorption. Among these are heats of adsorption at least one order of magnitude greater than normal heats of condensation, specific involvement of the adsorbent in the sorption reaction, restriction to monolayer coverage of the surface, and the existence of an activation energy. Not all chemisorption reactions exhibit all of these characteristics, including phosphate adsorption on metal oxides. The following represents a ligand exchange reaction involving soluble Fe^{+3}:

$$FeOH^{+2}(aq) + H_2PO_4^-(aq) = FeH_2PO_4{}^{+2}(aq) + OH^-(aq) \tag{5}$$

The free energy change, ΔG, for this reaction at pH 5 and a $p(H_2PO_4^-)$ of 3 is +2.18 kcal/mol. If ligand exchange involving structural iron on the surface of an oxide exhibits similar energetics, as has been suggested (see McBride, 1994), then the ΔG for phosphate adsorption should also be on the order of 2 kcal/mole. To put this in perspective, free energy changes for alkali and alkaline earth metal cation exchange on montmorillonite have been measured in the range of 1 to 2 kcal/mol (Gast, 1972; Laudelout et al., 1968; Udo, 1978), which is similar to those for ligand exchange.

Sorption of phosphate on two reference silicate clays and a titanium oxide at pH values below their respective zero points of charge was endothermic, emphasizing the significance of the entropy contribution to the free energy of adsorption (Malati et al., 1993). A ring closure mechanism involving two adjacent adsorption sites (i.e., a bidentate or binulcear bridging complex) was proposed to account for the increase in entropy needed to make the overall free energy of adsorption negative. Hundal (1988) also reported endothermic P adsorption but attributed sorption to precipitation. He interpreted the increase in enthalpy to energy required to form critical clusters of phosphate on the mineral surface, a heterogenous nucleation state required to initiate precipitation. In the cases of Malati et al. (1993) and Hundal (1988), P sorption increased as temperature increased, consistent with an endothermic reaction. However, Barrow (1983a) reported just the opposite effect of temperature for systems aged sufficiently that no net uptake of P was occurring. Froelich (1988) also stated that, for suspended sediments that have reached equilibrium with solution P concentrations, an increase in temperature will decrease the amount of P sorbed. These contradictory results suggest that the initial adsorption reaction may be endothermic, but that, with time, the sorption mechanism changes and becomes exothermic.

The precipitation of Al phosphate,

$$Al^{3+}(aq) + PO_4^{-3}(aq) = AlPO_4(s) \quad (6)$$

is also an endothermic reaction ($\Delta H° = +74.6$ kJ/mol), as is precipitation of β-tricalcium phosphate ($\Delta H° = +62.5$ kJ/mol). Thus, both the initial adsorption and the precipitation of P appear to be endothermic, a situation that is not usually associated with chemisorption. These data also indicate that heats of adsorption will not distinguish between adsorption and precipitation mechanisms.

Reversibility

Specific adsorption of some anions onto metal oxide surfaces appears to be completely reversible (see McBride, 1994). Sorption of other anions appear to be almost completely irreversible, requiring drastic changes in the chemical environment to induce desorption. The low energy associated with ligand exchange reactions like that shown in Eq. (5) would seem to be consistent with reversibility. Barrow (1983b), for example, has argued that if time of contact between a P solution and soil is properly taken into account, P adsorption can be shown to be reversible. Little or no hysteresis between P sorption and desorption has been reported for some soils (Mansell et al., 1977; Tiessen et al., 1993). The rapid equilibration of $^{32}PO_4$ with adsorbed P suggests that some fraction of the sorbed P remains exchangeable, and thus reversibly sorbed, for long periods of time. The apparent lack of reversibility in some cases could be the result of either diffusion of adsorbed phosphate into the solid phase or into pores within aggregates, or the formation of insoluble precipitates. Desorption of these latter fractions occurs very slowly and would complicate the study of adsorption reversibility.

Surface Complexation Models

As discussed earlier, the conventional Langmuir and Freundlich equations do not take into account pH and ionic strength effects on anion adsorption. These limitations have prompted the development of surface complexation models that have explicit dependence on these variables and on P concentration. While such models represent a conceptual advance over empirical equations of the Langmuir and Freundlich type, they do not as yet provide definitive theoretical treatment of P adsorption (Goldberg and Sposito, 1984). In addition to pH and ionic strength effects, the models consider both solution and surface speciation and adsorption effects on surface charge. These models all incorporate concepts of diffuse double layer theory and of the existence of a finite number of adsorption sites on the surface (resulting in a Langmuir-type adsorption maximum). The models differ primarily in the number of planes into which the various solution species can adsorb, resulting in either multicapacitance or single capacitance configurations. Bowden et al. (1977) were probably the first to apply such a model to sorption of anions onto metal oxide surfaces. Their model was an adaptation of the Stern model for the electric double layer, in which the charge resulting from specific adsorption of anions like phosphate lies in a plane some distance from the surface and whose potential varies with the amount of phosphate adsorbed, pH, and concentration of electrolyte. The Bowden model consists of a series of five equations in five unknowns (three values for planar charge and two for electric potential). Four parameters must be evaluated from simple acid-base titration curves of the metal oxide in the absence of a specifically adsorbing anion. Finally, a parameter defining the number of anion binding sites and a suite of binding constants, one for each surface anionic species, must be defined. However, after this was done, the model was able to describe quite well phosphate adsorption by goethite in the pH range from 2 to 11.

Bar-Yosef et al. (1988) interpreted P sorption on Ca- and K-clays using an empirical, pH-dependent, competitive adsorption model that took into account the presence of individual P species (i.e., $H_2PO_4^{-1}$, HPO_4^{-2}, PO_4^{-3}, etc.) and pH effects on surface charge. P adsorption was greater at higher (vs. lower) ionic strengths and in the presence of Ca-clay (compared with K-clay). Both effects were attributed to changes in the diffuse double layer and associated changes in accessibility of P to surface sites. Failure of the model to adequately describe P adsorption by montmorillonite clay at pH < 6.0 was attributed to flocculation of the clay by edge-to-face associations, blocking access to adsorption sites on clay edges. Flocculation effects were not included in the model.

Berkheiser et al. (1980) cited a number of studies in which the effects of ionic strength and cation species on P sorption were consistent with those reported by Bar-Yosef et al. (1988). However, in one report extrapolation of the sorption data to very long reaction times led to the conclusion that these variables had no effect on the final sorption equilibrium, although they did affect the rate at which equilibrium was attained.

The Slow Phase of P Sorption

Two mechanisms have been proposed to explain the slow phase of P sorption: diffusion (either into soil particles or to surface sites of limited accessibility) and precipitation (either by direct heterogenous nucleation or following the dissolution of the host solid, following an initial adsorption reaction).

Diffusion

Van Riemsdijk and De Haan (1981) believed that, at sufficiently high P concentrations, the slow uptake of P by metal oxides resulted from the diffusion of phosphate through metal-phosphate coatings that formed on the surface of the oxides during the initial, rapid uptake phase. Diffusion through this coating was assumed to be rate limiting, thus explaining the continuing uptake, or release, of P over long reaction times. These authors evaluated a number of rate laws for P sorption under conditions of constant supersaturation with respect to metal-phosphate solubility and showed that sorption did not follow first-order precipitation kinetics. The best description of sorption was obtained with an equation that had been successfully applied to the oxidation kinetics of metals, a process thought to resemble phosphate adsorption under conditions of constant supersaturation:

$$\frac{dS}{dt} = k_1 \exp(-k_2 S) \tag{7}$$

where t = time, S = the amount sorbed, and k_1 and k_2 are functions of solution P. The integrated form of the rate equation predicts that sorption is a linear function of ln(t) for t > 1h, which was shown to be true for P sorption by a sandy soil.

Van Riemsdijk et al. (1984) later presented a model of P sorption based on the diffusion process described above. The model predicted that, for several assumed geometries and at constant supersaturation, sorption was proportional to the product of P concentration and time, Ct. Furthermore, sorption data obtained at several P concentrations and soil-to-solution ratios yielded a single curve from which P sorption rates could be calculated for any combination of solution P concentration, sorbed P, and time.

Barrow (1983a) described a model in which an initial adsorption of P induces a diffusion gradient toward the interior of the particle, i.e., a solid-state diffusion process. A surface complexation model similar to that of Bowden et al. (1977) was used to describe the initial adsorption reaction between phosphate and soil. An increase in surface activity of phosphate initiates diffusion, with gradients in chemical and electrical potentials providing the driving forces. The model is able to describe the effects of phosphate concentration, pH, temperature, and time of contact on sorption. It is also able to describe the effects of period of prior contact, soil-to-solution ratio, temperature, and time on desorption. The model predicts that phosphate that has reacted with soil for a long period is not "fixed" but can be slowly recovered if a low enough surface activity can be induced.

Freese et al. (1992) evaluated two models for phosphate sorption that differed with respect to the forms of Fe and Al oxides, i.e., amorphous vs. crystalline, that were considered and whether the phosphate initially sorbed was taken into account. Both P sorption and desorption were related to soil contents of oxalate-extractable Fe and Al. Oxalate-extractable P was a major part of total soil P. Freese et al. (1995) later reported a significant correlation between the phosphate sorption maximum and the sum of amorphous iron and aluminum. Lookman et al. (1995) modeled long-term phosphate desorption from soils using two discrete P "pools," one available and one strongly fixed. Field soils were used that had received large amounts of P in the past as a result of excessive applications of manure. The release kinetics for each P pool were described using a first-order rate equation. The slow rate of P release from the second pool was attributed to either dissolution or to diffusion from interior sites within sesquioxide aggregates. The sum of the fast and slow pools was equal to the oxalate-extractable P content of the soil. The fact that 15 to 70% of the oxalate-extractable P was desorbed after 1600 h was taken as evidence that all oxalate-extractable P is potentially desorbable in sandy soils; i.e., no irreversibly fixed oxalate-extractable P exists. Yuan and Lavkulich (1994) also found a good correlation between oxalate-extractable iron and aluminum and P sorption.

Diffusion of P into soil particles is limited to amorphous and poorly crystalline solids, since penetration of well crystallized solids by the relatively large phosphate ion would be severally restricted. The effects of solid-phase crystallinity on P uptake during the slow phase of sorption was discussed by Parfitt (1989).

The slow uptake of P has also been attributed to diffusion along surfaces to sites of decreasing accessibility (Willet et al., 1988). Evidence for diffusion into mesopores also comes from the work of Cabrera et al. (1981), who reported that the reaction of P with lepidocrocite continued for longer times than with goethite. Their lepidocrocite samples consisted of small crystals that formed rather large aggregates containing mesopores. With crystalline goethite that had no mesopores or vacant sites within the crystal lattice, Parfitt (1989) reported that there was virtually no slow P sorption. Madrid and Arambarri (1985) attributed differences in P adsorption and desorption rates for lepidocrocite and hematite to differences in surface porosity. Lookman et al. (1994) worked with samples of amorphous Al oxide having varying degrees of aggregation. They observed a slow P uptake phase by well aggregated Al oxide and concluded that sorption was limited by diffusion through pores within aggregates. Their conclusion was based at least partly on the fact that "bulk" aluminum phosphate, i.e., tetrahedrally coordinated Al, was not observed in MAS NMR signals, thus ruling out precipitation as the predominant sorption mechanism.

Precipitation

Precipitation of a new solid phase is initiated when the critical concentration for nucleation of seed crystals is exceeded. The critical concentration is always greater than the equilibrium mineral solubility because energy is required to maintain the solid-liquid interface associated with the seed crystals. This interfacial energy must be balanced by increases in the overall free energies of the dissolved constituents that form the new solid phase. The resulting degree of supersaturation required to maintain a stable crystal decreases as the size of the crystal increases. This is shown in Equation [8], relating the degree of supersaturation to crystal radius (Corey, 1981):

$$\frac{K_{so}(r)}{K_{so}} = \frac{2\sigma V}{rRT} \tag{8}$$

where K_{so} = solubility product for very large crystals, $K_{so}(r)$ = solubility product for crystals of radius r, V = molar volume of the solid phase, R = the ideal gas constant, T = absolute temperature, and σ is the interfacial free energy density.

Homogeneous nucleation refers to the formation of seed crystals by direct combination of constituents in the solution phase, and heterogenous nucleation to nucleation on a preexisting surface. In the latter case, adsorption of one or more constituents occurs before nucleation and crystal growth can proceed. Heterogenous nucleation is probably the predominant mechanism occurring in soils and sediments, because of a preponderance of available surface area and the fact that the critical solution concentration for heterogeneous nucleation is generally less than that for homogeneous nucleation. Adsorption should control the aqueous solubility at P concentrations below the critical concentration needed for nucleation. On the other hand, precipitation should control aqueous solubility if the P concentration has exceeded the critical concentration. Unfortunately, it is difficult to determine the critical concentration for heterogeneous nucleation in a particular soil or sediment at any given time, because it depends on the concentrations of all constituents needed to form precipitates and on the properties of the host mineral.

There is laboratory evidence that phosphate precipitation is accompanied by the dissolution of the host mineral following an initial adsorption reaction. Dissolution of the host mineral provides the metal cations that can subsequently precipitate with P to form new solid phases. At low P concentrations, P was adsorbed onto amorphous Al hydroxide as an inner-sphere surface complex with a maximum of one phosphate per surface Al atom (Lookman et al., 1994). However, at higher P concentrations, there was a large turnover of Al hydroxide to Al phosphate. This was interpreted as a "weathering" process during which Al-O-Al bonds were broken and surface layers of the hydroxide literally dissolved, providing Al ions for precipitation with P. Martin et al. (1988) attributed P sorption on naturally occurring goethite crystals to partial dissolution of the mineral and precipitation of iron phosphate from solution. They isolated and identified crystallites of the mineral griphite, an Fe-Mn hydroxy phosphate. Adsorbed phosphate was not detected on the goethite surface above a concentration of 0.1 atomic percent, and no phosphate penetration into the goethite was observed. They concluded that phosphate retention in Fe oxide systems was be due primarily to precipitation rather than adsorption. Nanzyo (1984) concluded that a "bulk solid" similar in character to aluminum phosphate gel formed after P sorption onto an alumina gel, and later reported (Nanzyo, 1986) that phosphate sorbed on iron hydroxide gel at pH < 4.9 reacted further to form iron phosphates.

Direct evidence of contemporary phosphate-mineral precipitation in soils and sediments is rare, even in the case of soils heavily amended with soluble forms of P (Pierzynski et al., 1990a, b, and c; Harris et al., 1994). However, Harris et al. (1994) reported that the mineral vivianite, ($Fe_3(PO_4)_2 \cdot 8H_2O$) did precipitate in the sediment of a stream receiving runoff from a dairy barn. This mineral was detected visually because of its distinctive blue color as it appeared on the surface of sediment aggregates. It was identified by X-ray diffraction only after tedious collection using a dissecting microscope. Precipitation of the vivianite was the direct result of phosphate enrichment from the dairy, because the native soils of the area are sandy and have extremely low P retention capacities. Vivianite has been reported in other settings as well (Dell, 1973; Lindsay et al., 1989; Slanskey, 1986). Froelich (1988) cited several cases where positive identification of new phosphatic solid phases was obtained for sediments.

Some wastewater treatment approaches to phosphate precipitation mimic soil and sediment conditions and verify that P can be removed via precipitation in porous media. An example where evidence for a precipitation mechanism was convincing comes from a study of P removal from municipal waste water (Aulenbach and Meisheng, 1988). At Lake George, New York, sand beds had been removing essentially 100% of the P from municipal wastewater during continual use for 45 years. With a P sorption capacity of only 5.6 mg P/kg, adsorption sites in the sand beds should have become P saturated in less than 1 year. The mechanism of P removal by these sands was shown to be precipitation. Sand grains were coated with $CaCO_3$

and contained significant amounts of $CaHPO_4$ and $AlPO_4$ that were detectable by X-ray diffraction techniques.

Prospective precipitation of P phases in soils and sediments is often evaluated indirectly because of the impracticality of direct assessment. Indirect approaches include solution speciation modeling and laboratory experiments designed to simulate soil systems. Froelich (1988) stated that attempts to relate P concentrations in natural waters to solubility of some mineral have been unconvincing. The large number of possible phosphate minerals (>350, Nriagu, 1984) makes it very likely that one could be found whose solubility might control the P activity for any solution analyzed. Results of laboratory and greenhouse studies do not necessarily apply to soils, either. For example, early closed-system experiments investigating prospective fertilizer reaction products (e.g., Kittrick and Jackson, 1955) resulted in precipitation of microcrystalline variscite ($AlPO_4 \cdot 2H_2O$). This was in harmony with early conceptualizations on aluminum phosphate formation in soils (e.g., Cole and Jackson, 1950; Lindsay et al., 1959), and consequently variscite has become an icon in the soils literature to represent the aluminum phosphate solid phase. However, there are arguments that variscite precipitation is unlikely in soils (Nriagu, 1976; Hsu, 1982a and b; Hsu, 1993), and we know of no direct verification of its formation in soils as a reaction product of P amendments. Furthermore, the most common aluminum phosphate mineral found in weathered phosphatic soils is wavellite $[Al_3(OH)_3(PO_4)_2 \cdot 5H_2O]$ (Altschuler, 1973; Flicoteaux and Lucas, 1984; Wang et al, 1989).

Another indirect means of inferring precipitation is through the interpretation of sorption data. For example, when one plots sorption data according to Eq. (1), it is not uncommon to find inflection points at high solution P concentrations. Hundal (1988) observed three inflection points that corresponded closely with saturation with respect to β-tricalcium phosphate, octacalcium phosphate and, finally, dicalcium phosphate dihydrate. The differential isosteric heat of adsorption also revealed discontinuities that corresponded to these inflection points.

Coprecipitation of P has been proposed as an important self-cleaning mechanism for freshwater bodies. As much as 97% of the phosphate from the epilimnion of one lake was attributed to coprecipitation of inorganic P and calcite with incorporation of some of the surface P into the bulk structure as the calcite crystals grew (House, 1990). The phosphate coprecipitation rate was shown to be linearly related to the calcite precipitation rate. In a similar study, solubility data suggested that the phosphate concentration in a *high rate pond* was controlled by amorphous tricalcium phosphate having a solubility product equal to $10^{-25.2}$ (Moutin et al., 1992). Phosphate solubility diagrams and mineral equilibria calculations were used to support the conclusion that P sorption by Lake Apopka sediments at pH > 8.5 was due to P coprecipitation with $CaCO_3$ as nonapatitic Ca-P compounds (Olila and Reddy, 1995).

There are important distinctions between precipitation of phosphates of anthropogenic origin and precipitation that has occurred in soils developed from phosphatic parent materials. Phosphate minerals that occur abundantly in the weathered zone of some phosphatic soils are not found in heavily fertilized soils. Differences relate to the variety of phosphate forms that are applied to soils and to the shorter duration over which transformations have taken place. Phosphates in highly weathered phosphatic soils have obviously approached equilibrium "end-points" (e.g., dominance of the most stable phases under prevailing conditions) more closely than have those that have been in soils for only a few years or decades. Rates of formation are particularly important in controlling the reaction products of phosphate amendments. Heterogeneous and dynamic conditions of chemical potentials and physical states within soils, both at macro- and micro-scales, are critical considerations as well. Interestingly, the P-sorption capacity of phosphatic soils can be relatively high, because the residual phosphate is in the form of highly stable minerals and hence is noncompetitive for surface sorption reactions involving other soil components.

7.6 PHOSPHORUS CHEMISTRY IN FLOODED SOILS AND SEDIMENTS

Phosphorus chemistry in flooded soils and sediments differs markedly from that in oxidized upland soils. Under reducing conditions, transformations of iron play a dominant role in P chemistry. The reduction of Fe^{+3} to the more soluble Fe^{+2} and its reprecipitation to form ferrous minerals is thought to be the dominant process controlling P solubility in anaerobic systems (Holford and Patrick, 1979). In microbially active systems, flooding leads to depletion of oxygen and nitrate and eventually to reduction of metals, such as manganese and iron, contained in oxide minerals. Upon reduction of ferric oxide minerals, water soluble and exchangeable concentrations of Fe^{+2} increase markedly. The dissolution of iron minerals is accompanied by increases in concentrations of both adsorbed and water soluble P (Holford and Patrick, 1979). In acid systems, reduction is generally associated with an increase in pH. However, water-soluble P tends to decrease as pH increases toward neutrality in anaerobic systems (Holford and Patrick, 1979). This is attributed to sorption of P by ferrous iron that precipitates to form new ferrous oxide mineral phases. These new solid phases apparently have a higher surface area than the original ferric oxide minerals, as indicated by the fact that P sorption capacities and oxalate-extractable iron concentrations of soils and sediments are often greater after reduction (Patrick and Khalid, 1974; Khalid et al., 1977). Oxalate-extractable iron, which is a measure of amorphous and poorly crystalline forms of iron oxides, is considered by some to be the most important measure of changes in P sorption capacity when soils are flooded (Khalid et al., 1977). Oxalate treatment has been shown to nearly eliminate the ability of reduced soils and sediments to adsorb P. In addition to adsorption of P by ligand exchange reactions with newly formed ferrous oxide minerals, P may also precipitate directly with Fe^{+2} to form minerals such as vivianite, $Fe_3(PO_4)_2\ 8H_20$. With further reduction of soil and sediment systems, sulfide may form and can react with Fe^{+2} to form FeS. Because the solubility of FeS is lower than that of ferrous oxide minerals, precipitation of FeS will drive the dissolution of these latter minerals, resulting in the release of sorbed P once again to pore waters at very low redox potentials.

Moore and Reddy (1994) studied P dynamics in mud sediments taken from Lake Okeechobee, Florida. Because this lake is fairly shallow (mean depth 2.7 m), the sediment/water interface is generally oxic, with the possible exception of summer months when oxygen demand is high and winds are calm. Even when the interface is oxic, a sharp gradient in redox potential can develop over distances of only a few centimeters into the sediment. The authors found that water-soluble Fe and P concentrations increased with depth along this redox potential gradient in a manner consistent with iron and P transformations discussed above. Fe^{+2} and P then diffused along this gradient to the sediment/water interface. They concluded that oxidation and precipitation of iron at the interface controlled P solubility under oxic conditions and limited P flux into the overlying water. Deeper within the sediment, P solubility was believed to be controlled by the more soluble ferrous iron and/or calcium phosphate precipitation. Their data showed that large fluxes of P from sediment into overlying water could occur during periods of low dissolved oxygen levels.

While dissolution and reprecipitation of iron oxide minerals leads directly to changes in soluble P concentrations, the role of the microbial biomass and its transformations under changing redox conditions cannot be overlooked. As stated by Gachter et al. (1988), it has not been proven that the only source of P released to solution when sediments become reduced is inorganic iron oxide minerals. These authors present strong indirect evidence that sorption and release of P in some lake sediments is controlled at least partly by redox-dependent changes in microbial physiology. Their argument is supported by the fact that a large percentage of total P in lake sediments can be of microbial origin, sedimentary microorganisms are capable of storing large quantities of P as polyphosphates that can be released under anaerobic conditions, anoxic releases of iron and P from lake sediments is often uncoupled; i.e., release of P does not necessarily occur concomitantly with that of iron, uptake of P from solution by unsterilized lake sediments can be much greater than that by sterilized sediments, and sedimentary microorganisms are capable of depleting solution P concentrations from >100 μmoles/L initially to < 0.1 μmoles/L.

7.7 SUMMARY

Chemical and physical interactions of P with soils and sediments are critical to minimizing the impact that land application of fertilizers and wastes has on the eutrophication of lakes and streams. This paper provides an overview of reactions that affect P solubility and mobility in soils and sediments.

Sorption of phosphate entails its removal from solution and incorporation into the solid phase via adsorption or precipitation. Phosphate generally shows rapid initial sorption followed by a long period during which the rate steadily decreases. The initial rapid uptake is generally regarded as resulting from exchange of surface ligands (e.g., water, hydroxyl, sulfate, silicate, and so on) by phosphate ions. The slow uptake phase has been attributed to both diffusion and precipitation reactions. An increase in surface activity of phosphate arising from ligand exchange could induce solid-state diffusion of phosphate into the interior of particles or aggregates, the diffusion being driven by chemical and electrical potential gradients. Slow uptake could also result from diffusion of phosphate through metal-phosphate coatings that form initially on the surface of soil solids. Kinetic expressions based on diffusion through coatings under conditions of constant supersaturation described sorption on sandy soil material well. Diffusion of phosphate through porous structures of solid aggregates to sites of decreasing accessibility has also been proposed. Evidence suggests a strong relationship between reversibly sorbed P and amorphous Fe and Al, as measured by oxalate extraction. Models of P transport have been created to capture both the rapid and kinetically controlled components of phosphate sorption.

Phosphate adsorption commonly is hysteretic to some degree, as determined by laboratory adsorption/desorption isotherms. Hysteresis does not necessarily constitute truly irreversible adsorption but may be attributed instead to differences in rates of adsorption and desorption. Hysteresis could also indicate a change in sorption mechanism with time. There is some evidence that most, if not all, adsorbed phosphate could be desorbed, given enough time and sufficiently low phosphate solution activities. Adsorption of phosphate should be a low-energy reaction, based on comparable ligand exchange reactions in solution.

Precipitation can occur when solution phosphate activities exceed the solubility of discrete phosphate phases. Heterogeneous nucleation following initial adsorption can result in the formation of a new solid phase. The identification of both amorphous and crystalline phosphates following sorption experiments with well characterized metal oxides suggests that precipitation can be the dominant sorption mechanism under certain conditions. However, direct verification of phosphate precipitates in soils and sediments is usually precluded by insufficient mass for practical solid-state detection. Precipitation of thermodynamically stable phosphate minerals is often slow and may be inhibited by various soil solution constituents.

Sorption and desorption of phosphate associated with Fe is also strongly influenced by redox potential. The reduction and dissolution of iron and its reprecipitation to form ferrous minerals is thought to be the dominant process controlling P solubility in anaerobic systems. The dissolution of iron minerals is accompanied by increases in concentrations of both adsorbed and water soluble P, and by increases in P sorption capacity. Oxalate-extractable iron may be the most important measure of changes in P sorption capacity when soils are flooded. While dissolution and reprecipitation of iron oxide minerals exerts a major influence on changes in soluble P concentrations, the role of the microbial biomass and its transformations under changing redox conditions cannot be overlooked. Strong evidence suggests that sorption and release of P in some lake sediments is strongly related to redox-dependent changes in microbial physiology.

7.8 ACKNOWLEDGMENT

This is to acknowledge the contribution of Florida Agricultural Station Journal Series No. R-06686.

REFERENCES

Afif, E., V. Barron, and J. Torrent. 1995. Organic matter delays but does not prevent phosphate sorption by cerrado soils of Brazil. Soil Sci. 159:207–211.

Altschuler, Z.S. 1973. The weathering of phosphate minerals—geochemical and environmental aspects. p. 33–96. In E.J. Griffith et al. (ed.) Environmental Phosphorus Handbook. John Wiley and Sons, NY.

Aulenbach, D.B., and N. Meisheng. 1988. Studies on the mechanism of phosphorus removal from treated wastewater by sand. J. Water Pollution Control Fed. 60:2089–2094

Bache, B.W., and E.G. Williams. 1971. A phosphate sorption index for soils. J. Soil Sci. 22:289–301.

Barrow, N.J. 1983a. A mechanistic model for describing the sorption and desorption of phosphate by soil. J. Soil Sci. 34:733–750.

Barrow, N.J. 1983b. On the reversibility of phosphate sorption by soils. J. Soil Sci. 34:751–758.

Bar-Yosef, B., U. Kafkafi, R. Rosenberg, and G. Sposito. 1988. Phosphorus adsorption by kaolinite and montmorillonite: I. Effect of time, ionic strength, and pH. Soil Sci. Soc. Am. J. 52:1580–1585.

Berkheiser, V.E., J.J. Street, P.S.C. Rao, and T.L. Yuan. 1980. Partitioning of inorganic orthophosphate in soil-water systems. Critical Reviews in Chemistry. CRC Press, Inc., Boca Raton, FL.

Bowden, J.W., A.M. Posner, and J.P. Quirk. 1977. Ionic adsorption on variable charge mineral surfaces. Theoretical-charge development and titration curves. Aust. J. Soil Sci. 15:121–134.

Cabrera, F., P. De Arambarri, L. Madrid, and G.G. Toca. 1981. Desorption of phosphorus from iron oxide in relation to pH and porosity. Geoderma 26:203–216.

Cole, C.V., and M.L. Jackson. 1950. Solubility equilibrium constant of dihydroxy-aluminum dihydrogen phosphate relating to a mechanism of phosphate fixation in soils. Soil Sci. Soc. Am. Proc. 15:84–89.

Corey, R.B. 1981. Adsorption vs. precipitation. p. 161–182. In M.A. Anderson and A.J. Rubin (ed.), Adsorption of Inorganics at Solid-Liquid Interfaces. Ann Arbor Science, Inc., Ann Arbor, MI.

Dell, C.I. 1973. Vivianite: An authigenic phosphate mineral in Great Lakes sediments. p. 1027–1028. In Proc. 16th Conf. Great Lakes Res., Int. Assoc. Great Lakes Res., Ann Arbor, MI.

Flicoteaux, R., and J. Lucas. 1984. Weathering of phosphate minerals. p. 292–317. In J.O. Nriagu and P.B. Moore (ed.) Phosphate Minerals. Springer-Verlag, New York, NY.

Freese, D., W.H. van Riemsdijk, and S.E.A.T.M. van der Zee. 1995. Modelling phosphate-sorption kinetics in acid soils. Eur. J. Soil Sci. 46:239–245.

Freese, D., S.E.A.T.M. van der Zee, and W.H. van Riemsdijk. 1992. Comparison of different models for phosphate sorption as a function of the iron and aluminum oxides of soils. J. Soil Sci. 43:729–738.

Froelich, P.N. 1988. Kinetic control of dissolved phosphate in natural rivers and estuaries: A primer on the phosphate buffer mechanism. Limnol. Oceanography 33:649–668.

Gachter, Rene, J.S. Meyer, and A. Mares. 1988. Contribution of bacteria to release and fixation of phosphorus in lake sediments. Limnol. Oceanogr. 33:1542–1558.

Gast, R.G. 1972. Alkali metal cation exchange on Chambers montmorillonite. Soil Sci. Soc. Am. Proc. 36:14–19.

Goldberg, S., and G. Sposito. 1984. A chemical model of phosphate adsorption by soils: I. Reference oxide minerals. Soil Sci. Soc. Am. J. 48:772–778.

Hamad, M.E., D.L. Rimmer, and J.K. Syers. 1992. Effect of iron oxide on phosphate sorption by calcite and calcareous soils. J. Soil Sci. 43:273–281.

Harris, W.G., H.D. Wang, and K.R. Reddy. 1994. Dairy manure influence on soil and sediment composition: implications for P retention. J. Environ. Qual. 23:1071–1081.

Holford, I.C.R., and W.H. Patrick, Jr. 1979. Effects of reduction and pH changes on phosphorus sorption and mobility in an acid soil. Soil Sci. Soc. Am. J. 43:292–297.

House, W.A. 1990. The prediction of phosphate coprecipitation with calcite in freshwaters. Water Res. 24:1017–1023.

Hundal, H.S. 1988. A mechanism of phosphate adsorption on Narrabri medium clay loam soil. J. Agric. Sci. 111:155–158.

Hsu, P.H. 1982a. Crystallization of variscite at room temperature. Soil Sci. 133:305–313.

Hsu, P.H. 1982b. Crystallization of Fe (III) phosphate at room temperature. Soil Sci. Soc. Am. J. 46:928–932.

Hsu, P.H. 1993. Effects of aluminum and phosphate concentrations and acidity on the crystallization of variscite at 90 C. Soil Sci. 156:71–78.

Kafkafi, U., B. Bar-Yosef, R. Rosenberg, and G. Sposito. 1988. Phosphorus adsorption by kaolinite and montmorillonite: II. Organic anion competition. Soil Sci. Soc. Am. J. 52:1585–1589.

Khalid, R.A., W.H. Patrick, Jr., and R.D. DeLaune. 1977. Phosphorus sorption characteristics of flooded soils. Soil Sci. Soc. Am. J. 41:305–310.

Kittrick, J.A., and M.L. Jackson. 1955. Application of solubility product principles to the variscite/kaolinite system. Soil Sci. Soc. Am. Proc. 19:455–457.

Laudelout, H., R. Van Bladel, G.H. Bolt, and A.L. Page. 1968. Thermodynamics of heterovalent cation exchange reactions in a montmorillonite clay. Trans. Faraday Soc. 64:1477–1488.

Lindsay, W.L., P.L.G. Vlek, and S.H. Chien. 1989. Phosphate minerals. p. 1089–1130. In J.B. Dixon and S.B. Weed (ed.) Minerals in Soils Environments. Soil Sci. Soc. Am., Madison, WI.

Lindsay, W.L., M. Peech, and J.S. Clark. 1959. Solubility criteria for the existence of variscite in soils. Soil Sci. Soc. Am. Proc. 23:357–360.

Lookman, R., D. Freese, R. Merckx, K. Vlassak, and W. H. van Riemsdijk. 1995. Long-term kinetics of phosphate release from soil. Environ. Sci. Tech. 29:1569–1575.

Lookman, R., P. Grobet, R. Merckx, and K. Vlassak. 1994. Phosphate sorption by synthetic amorphous aluminum hydroxides: A 27Al and 31P solid-state MAS NMR spectroscopy study. Eur. J. Soil Sci. 45:37–44.

Madrid, L., and P. De Arambarri. 1985. Adsorption of phosphate by two iron oxides in relation to their porosity. J. Soil Sci. 36:523–530.

Malati, M.A., R.A. Fassam, and I.R. Henderson. 1993. Mechanism of phosphate interaction with two reference clays and an anatase pigment. J. Chem. Tech. Biotechnol. 58:387–389.

Mansell, R.S., H.M. Selim, P. Kanchanasut, J.M. Davidson, and J.G.A. Fiskell. 1977. Experimental and simulated transport of phosphorus through sandy soils. Water Resources Res. 13:189–194.

Martin, R.R., R. St.C. Smart, and K. Tazaki. 1988. Direct observation of phosphate precipitation in the goethite/phosphate system. Soil Sci. Soc. Am. J. 52:1492–1500.

McBride, M.B. 1994. Environmental Chemistry of Soils. Oxford University Press, New York, NY.

Miller, D.M., M.E. Summer, and W.P. Miller. 1989. A comparison of batch- and flow-generated anion adsorption isotherms. Soil Sci. Soc. Am. J. 53:373–380.

Moore, P.A., Jr., and K.R. Reddy. 1994. Role of Eh and pH on phosphorus geochemistry in sediments of Lake Okeechobee, Florida. J. Environ. Qual. 23:955–964.

Moutin, T., J.Y. Gal, H. El Halouani, B. Picot, and J. Bontoux. 1992. Decrease of phosphate concentration in a high rate pond by precipitation of calcium phosphate: Theoretical and experimental results. Water Res. 26:1445–1450.

Nanzyo, M. 1984. Diffuse reflectance infrared spectra of phosphate sorbed on alumina gel. J. Soil Sci. 35:63–69.

Nanzyo, M. 1986. Infrared spectra of phosphate sorbed on iron hydroxide gel and the sorption products. Soil Sci. Plant Nutrition 32:51–58.

Nriagu, J. 1976. Phosphate-clay mineral relations in soils and sediments. Can. J. Soil Sci. 13:717–736

Nriagu, J. 1984. Phosphate minerals: their properties and general mode of occurrence. p. 1–5. In J.O. Nriagu and P.B. Moore (eds.) Phosphate Minerals. Springer-Verlag, New York, NY.

Ohno, T., and B.S. Crannell. 1996. Green and animal manure derived dissolved organic matter effects on phosphorus sorption. J. Environ. Qual. 25:1137–1143.

Olila, O.G., and K.R. Reddy. 1995. Influence of pH on phosphorus retention in oxidized lake sediments. Soil Sci. Soc. Am. J. 59:946–959.

Pardo, M.T., and M.E. Guadalix. 1990. Phosphate sorption in allophanic soils and release of sulphate, silicate, and hydroxyl. J. Soil Sci. 41:607–612.

Parfitt, R.L. 1989. Phosphate reactions with natural allophane, ferrihydrate, and goethite. J. Soil Sci. 40:359–369.

Patrick, W.H., and R.A. Khalid. 1974. Phosphate release and sorption by soils and sediments: Effect of aerobic and anaerobic conditions. Science 186:53–55.

Pavlatou, A., and N.A. Polyzopoulos. 1988. The role of diffusion in the kinetics of phosphate desorption: The relevance of the Elovich equation. J. Soil Sci. 39:425–436.

Pierzynski, G.M., T.J. Logan, S.J. Traina, and J.M. Bigham. 1990a. Phosphorus chemistry and mineralogy in excessively fertilized soils: Quantitative analysis of phosphorus-rich particles. Soil Sci. Soc. Am. J. 54:1576–1583.

Pierzynski, G.M., T.J. Logan, S.J. Traina, and J.M. Bigham. 1990b. Phosphorus chemistry and mineralogy in excessively fertilized soils: Descriptions of phosphorus-rich particles. Soil Sci. Soc. Am. J. 54:1583–1589.

Pierzynski, G.M., T.J. Logan, and S.J. Traina. 1990c. Phosphorus chemistry and mineralogy in excessively fertilized soils: Solubility equilibria. Soil Sci. Soc. Am. J. 54:1589–1595.

van Riemsdijk, W.H., L.J.M. Boumans, and F.A.M. de Haan. 1984. Phosphate sorption by soils: I. A model for phosphate reaction with metal-oxides in soil. Soil Sci. Soc. Am. J. 48:537–541.

van Riemsdijk, W.H., and F.A.M. de Haan. 1981. Reaction of orthophosphate with a sandy soil at constant supersaturation. Soil Sci. Soc. Am. J. 45:261–266.

Selwood, P.W. 1975. Chemisorption and Magnitization. Academic Press, New York, NY.

Slanskey, M. 1986. Geology of sedimentary phosphates. Elsevier, New York, NY.

Sposito, G. 1980. Derivation of the Freundlich equation for ion exchange reactions in soils. Soil Sci. Soc. Am. J. 44:652–654.

Tiessen, H., M.K. Abekoe, I.H. Salcedo, and E. Owusu-Bennoah. 1993. Reversibilty of phosphorus sorption by ferrugenous nodules. Plant and Soil 153:113–124.

Udo, E.J. 1978. Thermodynamics of potassium-calcium and magnesium-calcium exchange reactions on a kaolinitic soil clay. Soil Sci. Soc. Am. J. 42:556–560.

Yuan, G., and L.M. Lavkulich. 1994. Phosphate sorption in relation to extractable iron and aluminum in Spodosols. Soil Sci. Soc. Am. J. 58:343–346.

Wang H. D., W. G. Harris, and T. L. Yuan. 1989. Phosphate minerals in some Florida phosphatic soils. Soil Crop Sci. Soc. Fla. Proc. 48:49–55.

Willet, I.R., C.J. Chartres, and T.T. Nguyen. 1988. Migration of phosphate into aggregated particles of ferrihydrite. J. Soil Sci. 39:275–282.

Zhou, M., R.D. Rhue, and W.G. Harris. 1997. Phosphorus sorption characteristics of Bh and Bt horizons from sandy coastal plain soils. Soil Sci. Soc. Am. J. 61:1364–1369.

8 Forms of Organic Phosphorus in Water, Soils, and Sediments

Susan Newman and J. Stephen Robinson

8.1 ABSTRACT

The mineralization of labile forms of organic P (OP) is a critical link for internal P cycling and, to a large degree, may determine the productivity of an ecosystem. However, OP is typically identified simply as the difference between total P (TP) and inorganic P (IP). This chapter provides an overview of techniques available to identify and quantify different forms of OP, and summarizes results from Florida ecosystems, emphasizing wetland and aquatic systems.

The ability of OP compounds to act as either a sink or a source for bioavailable P is dependent on their molecular structure. Therefore, the first task in examining the role of OP in internal cycling is to identify its different forms. Organic P identification can be separated into groups based on biological availability, chemical fractionation, and physicochemical separation. Bioavailable P has been estimated using both direct (uptake of radiotracers) and indirect measures such as resins, chemical extraction of P, and enzymatic hydrolysis. Conventional characterization of OP forms in soils and sediments is through chemical fractionation in which alkali media are used to extract different fractions of OP based on the mechanism by which the OP fraction is incorporated in the other soil or sediment components. These operationally defined characterizations can be used to monitor turnover rates of labile

1-56670-331-X/99/$0.00+$.50

and nonlabile forms of OP and to make broad comparisons among different regions. More direct quantification of OP forms can be obtained by considering physicochemical characteristics at the molecular level. For example, obtaining ^{31}P-nuclear magnetic resonance (NMR) spectra on alkaline extracts is the most commonly used technique for the quantification of broad classes of OP. The diagnostic property in ^{31}P-NMR spectroscopy is the chemical shift of the P nuclei. Other methods for direct identification of OP compounds include enzymatic hydrolysis and physical separation based on molecular weight using chromatography. The separation and identification of OP compounds is an expensive and time-consuming process and, to date, <50% of P forms are thought to have been identified. This paper discusses the advantages and disadvantages of the different techniques and the data they produce to highlight existing knowledge and identify future areas of research.

8.2 INTRODUCTION

The TP pool of aquatic and wetland ecosystems is often dominated by OP, which can comprise >50% of sediment TP (Reddy et al., 1998) and as much as 90% of water column TP (Rigler, 1964). The accumulation and subsequent mineralization of OP is, therefore, an important component of the P cycle (see Wetzel, this volume) and may be the factor controlling productivity in P-limited systems. To assess the role of OP in ecosystem productivity, it is necessary to distinguish between OP that is resistant to mineralization and that which is readily available to support biological growth. In soils and sediments, OP is commonly identified using sequential and nonsequential chemical fractionation procedures (Sommers et al., 1972; Hedley et al., 1982), where alkali and acid media are used to extract different forms of OP. The advantage of these fractionations is the quantitative definition of OP as dissolved, labile, moderately labile, and recalcitrant or highly resistant. However, the disadvantage of fractionation schemes is that P forms are operationally defined and do not typically identify specific OP compounds.

In some cases, particularly for predictive purposes, it may be necessary to obtain more detailed information about the compounds that comprise the different OP pools. There are several techniques used to isolate and identify specific OP constituents, including chromatographic fractionation (Minear, 1972; Weimer and Armstrong, 1979), ^{31}P NMR (Newman and Tate, 1980; Condron et al., 1985; Bedrock et al., 1994), and direct enzymatic hydrolysis (Herbes et al., 1975; Cotner and Heath, 1988). With the exception of enzymatic hydrolysis, techniques to identify OP employ the use of more sophisticated equipment than those used for operational descriptions.

This paper is divided into three sections: (1) indirect measurement of OP, (2) direct identification and measurement of OP, and (3) recommendations for techniques to identify and measure OP in subtropical ecosystems, such as Florida. In each of these sections, examples are provided to explain the individual procedures, identifying both the benefits and limitations of each technique. The distribution and mobility of OP are influenced by climatic conditions such as temperature and moisture (Harrison, 1987); thus, data collected from Florida ecosystems will be used, wherever available, to follow the theme of this book.

8.3 INDIRECT MEASUREMENT OF ORGANIC PHOSPHORUS

The identification of P in surface water is often based on its reactivity with molybdate, particle size, and ease of hydrolysis (Cembella et al., 1984). Using these operational definitions, inexpensive and useful information about the chemical properties and potential mobility of the separate OP fractions are obtained.

8.3.1 Chemical Reactivity

Of the few forms distinguished, soluble reactive P (SRP) is the most labile form of P and often the focus of P cycling research (Hutchinson and Bowen, 1950; Rigler, 1956). Although frequently cited as representing inorganic P (IP), SRP is operationally defined. During the commonly used molybdate blue method of SRP analysis the sample pH is reduced to <2, which may result in the hydrolysis of OP, such that hydrolyzed OP is measured in conjunction with IP. Soluble reactive P concentrations in the surface water of lakes and wetlands may be low, while dissolved and total OP are abundant. In hypereutrophic Lake Apopka, and in nutrient-unenriched sections of the Everglades, SRP is frequently close to the minimum detection level of 1 μg L^{-1} (APHA, 1995), whereas dissolved OP (DOP) can comprise >50% of the TP pool (Table 8.1) and >90% of total soluble P (Newman, 1991). Thus, in these types of systems, we need to measure OP in addition to SRP and TP. Organic P in surface waters is measured in both filtered and unfiltered water and is chemically defined as the difference between TP and acid-hydrolyzable and molybdate-reactive P (Fig. 8.1, APHA, 1995). In Lake Apopka surface water, this analytical scheme determined that OP dominated both the total and dissolved P fractions, followed by acid-hydrolyzable and then reactive P (Fig. 8.2).

TABLE 8.1
Organic P Distribution in Selected Florida Ecosystems

	Location		
	Everglades	L. Apopka	L. Okeechobee
Water column (DOP as % of TP)	6–67[a]	3–37[b]	24[c]
Soils/sediments (TOP as % of TP)	37–70[d]	50[e]	20–75[f]

[a] McCormick et al., 1996, South Florida Water Management District, unpublished data.
[b] Newman, 1991; Charrick et al., 1993.
[c] South Florida Water Management District, unpublished data.
[d] DeBusk et al., 1994; Reddy et al., 1998.
[e] Reddy and Graetz, 1990.
[f] Reddy, 1991.

8.3.2 Size Fractionation

In combination with chemical reactivity, size fractionation is frequently used in the indirect determination of organic matter where the organic components are charac-

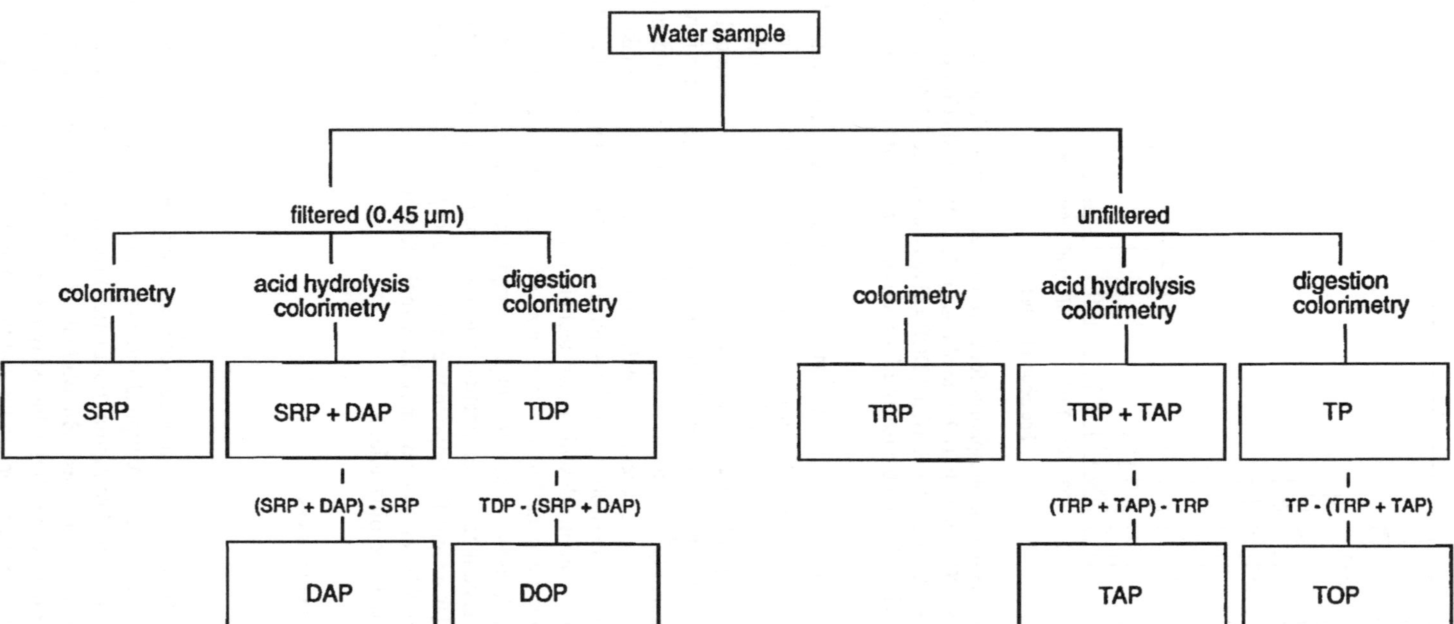

FIGURE 8.1 Analytical procedure used to determine different fractions of phosphorus in natural water (modified from APHA, 1995). TP = total P, TOP = total organic P, TAP = total acid hydrolyzable P, TRP = total reactive P, TDP = total dissolved P, DOP = dissolved organic P, DAP = dissolved acid hydrolyzable P, and SRP = soluble reactive P.

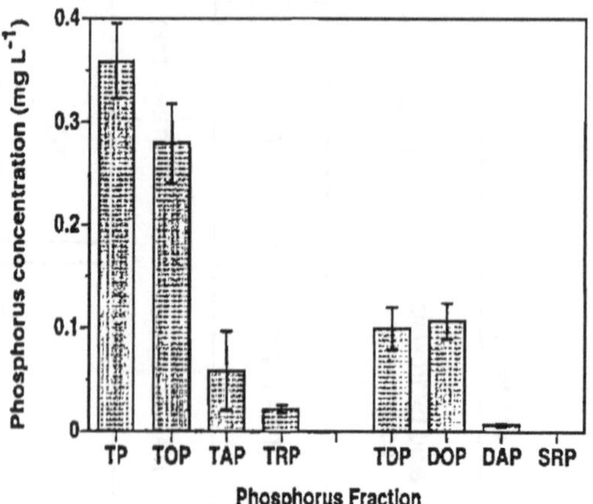

FIGURE 8.2 The distribution of phosphorus forms in natural water at eight sites in Lake Apopka in October 1989. TP = total P, TOP = total organic P, TAP = total acid hydrolyzable P, TRP = total reactive P, TDP = total dissolved P, DOP = dissolved organic P, DAP = dissolved acid hydrolyzable P, and SRP = soluble reactive P *(modified from Newman, 1991).*

terized based on their molecular weight. Minear (1972) eluted lake water DOP through gel columns to separate OP into low and high molecular weight fractions. His study showed that up to 20% of DOP had a molecular weight ≥50,000. Lean (1973) observed that filterable (0.45 μm) P compounds in lake water were colloidal with a high molecular weight. However, the identification of P compounds in water, soils, and sediments as high molecular weight based on their inability to penetrate the gel particles is defined regardless of which gel was used, which can make cross comparisons more difficult (Broberg and Persson, 1988). Molecular weight distinctions have also been combined with other techniques to provide a quantitative approach, which will be considered in the next section under chromatographic separation. In general, high molecular weight, complex structures, such as humic acids, are highly resistant to mineralization and tend to accumulate. Conversely, simple, low molecular weight P compounds are more susceptible to hydrolysis and thus more labile (see Wetzel, this volume). Low molecular weight compounds are, therefore, more likely to be bioavailable.

8.3.3 Chemical Extraction

Bioavailable P

Bioavailable P, both IP and OP, in surface water and sediments has been identified using bioassays and extractions. A typical bioassay is conducted by growing a P-starved algal species (cultured or natural population) for several days in the surface water or soil suspension of interest. Bioavailable P may be determined following radiotracer uptake, such as the Rigler bioassay (Rigler, 1966) or calculated as the difference between P concentrations in the solution measured before and after algal growth (USEPA, 1978; Sharpley et al., 1991; Krogstad and Løvstad, 1991). With the exception of radiotracer studies, the long duration of these incubations presents a serious limitation in the use of algal growth assays; as a result, researchers have developed chemical extractions to estimate algal available IP and OP. Sodium hydroxide- and bicarbonate-extractable P are the most common chemical indices of algal-available P. For example, Sharpley et al. (1991) and Young et al. (1985) determined that the growth of P-starved *Selenastrum capricornutum*, grown with agricultural runoff and lake tributary sediments as the sole sources of P, was significantly correlated with sodium hydroxide-extractable P ($r^2 = 0.61$ to 0.95). A review of the different extracts to measure bioavailability in freshwater sediments and water concluded that in soil samples and tributary water particulate matter sodium hydroxide-extractable P was found to be the best predictor of algal-extractable P (Boström et al., 1988). But in other studies, Boström et al. determined that only a minor portion of sodium hydroxide-extractable P was utilized by algae.

A recent study conducted in Lake Apopka suggested that water-soluble and nitrilotriacetic acid (NTA)-extractable P were good indicators of sedimentary P immediately available to support algal growth (Kenney, 1997). In that study, Kenney extracted sediment P using sodium hydroxide, water, hot water (100° C), iron oxide strips, hydrochloric acid, and NTA. Calibrating the algal growth on sediment P with that achieved with IP, he determined that, in sediment with > 40% organic matter, extraction using NTA provided the best indicator of algal-available P. Kenney also concluded that if these extractions were not calibrated with algal growth assays, water-extractable and NTA-extractable P could be used to represent minimum and maximum bioavailable P, respectively. Sodium hydroxide-extractable P was not able to successfully predict bioavailable P in this system.

Although the identification of specific OP compounds in the environment is the primary focus of this chapter, it is also important to assess organic forms of P stored intracellularly. The ability of algal cells to store surplus P is well documented and is typically identified using hot-water extraction. In this technique, the phytoplankton sample is boiled in water, and the filtrate is analyzed for molybdate-reactive P (Fitzgerald and Nelson, 1966). Aldridge (1994) determined that 50% of P available to support Lake Okeechobee phytoplankton growth was represented by hot-water-extractable P. The technique of Fitzgerald and Nelson (1966) was modified to incorporate the measurement of both acid-hydrolyzable along with reactive forms of hot-water-extractable P to examine surplus P cycling in *Peridinium* blooms in Lake Kinneret, Israel (Wynne and Berman, 1980). These researchers determined that hot-water-extractable P contributed approximately 40% of total cell P, and that the nonreactive portion of this P pool contained phosphate ester bonds, as evidenced by the release of IP following incubation with alkaline phosphatase enzymes. In Lake Apopka phytoplankton, 60% of total cell P was stored as hot-water-extractable P, and reactive P contributed approximately 15 to 26% of TP (Newman et al., 1994).

Sequential Fractionation

The conventional approach to P identification in soils and sediments is through extraction with different chemicals. Phosphorus groups are extracted using alkali and/or acid media and characterized based on the mechanism by which the OP fractions interact with other soil or sediment components (Sommers et al., 1972; Bowman and Cole, 1978; Hedley et al., 1982). Chemicals used as extraction solutions for P include sodium bicarbonate (labile/available P), sodium hydroxide (organic or Fe- and Al-bound P), and hydrochloric acid (Ca- and Mg-bound P, Fig. 8.3). Various versions of these procedures have been used to examine P distribution in Florida wetland and aquatic soils (Reddy and Graetz, 1990; Newman and Reddy, 1993; Qualls and Richardson, 1995; Ivanoff et al., 1998). Sequential fractionation procedures were used to determine that P enrichment in both wetlands and streams resulted in an increase in labile P (Koch and Reddy, 1992; Gale et al., 1994; Qualls and Richardson, 1995), with the highest concentrations located near inflow points (e.g., Fig. 8.4). This identification of labile fractions allows the documentation of shorter term ecosystem changes, compared to monitoring ecosystem responses based on TP changes.

In organic soils, such as those found in the Everglades, fractionation procedures have indicated that P is primarily stored as OP with additional storage as Ca-bound P (Fig. 8.4, Qualls and Richardson, 1995, Reddy et al., 1998). This information has a direct application, and has been used in restoration efforts, in which constructed wetlands have been used to remove P from agricultural runoff before it enters the Everglades. These constructed wetlands are managed to enhance the creation of natural storage products identified from P fractionation such as peat accretion (OP accumulation) and calcium phosphate precipitation.

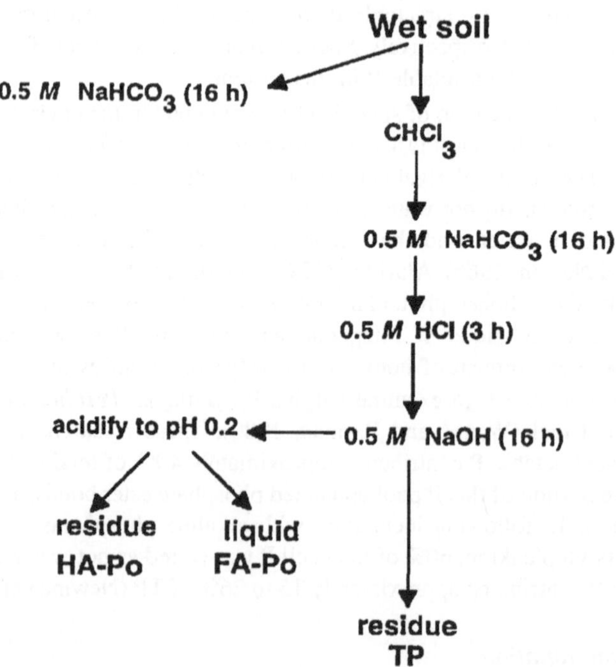

FIGURE 8.3 Phosphorus fractionation procedure used to identify organic P in wetland soils (modified from Ivanoff et al., 1998).

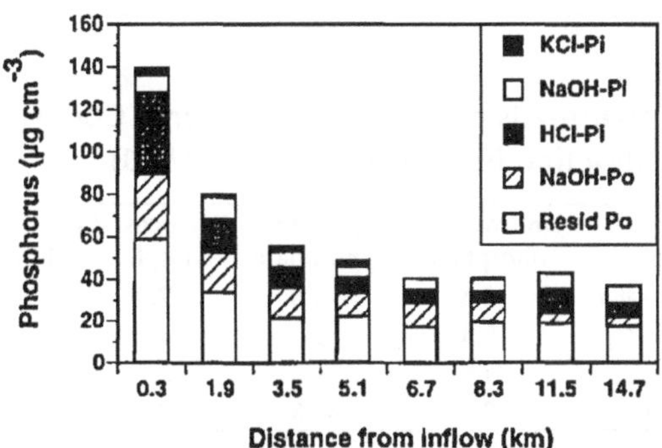

FIGURE 8.4 Phosphorus fractions determined along a nutrient gradient in soils in the northern Everglades (modified from Reddy et al., 1991). KCl = readily available (labile) P, NaOH–Pi = Fe-/Al-bound P, and HCl = Ca-bound P.

The origins of P fractionation techniques are typically equated to the work of Chang and Jackson (1957) who developed a scheme to identify different forms of IP. However, the use of fractionation procedures specifically focused on the OP components of soils and sediments has been studied less intensively. In a recent review, Tiessen et al. (1994) observed that the nature of OP estimated from fractionation is less well defined than those in inorganic fractionation and some care should be taken in interpreting data obtained through chemical extractions. Tiessen et al. also emphasized that it is unlikely that any extraction procedure can provide an exact separation of different P compounds. Extraction processes can be vigorous and may result in the hydrolysis of individual P compounds which would then be measured as SRP and not OP. In a recent study, approximately 40% of D-glucose-6-phosphate and *p*-nitrophenol phosphate were hydrolyzed when extracted in 0.5 *M* NaOH and 1 *M* HCl, respectively (Ivanoff et al., 1998, Table 8.2). The remaining 14 of the 16 OP compound and extractant solution combinations examined resulted in OP hydrolysis ranging from 0.2 to 5.8% (Ivanoff et al., 1998). An additional problem with the use of these extraction procedures is that P recovery during extraction is sensitive to the soil:solution ratio, extraction ratios of 50:1 and 100:1 appear to be the range used most frequently (Ivanoff et al., 1998). However, despite these limitations, chemical extractions remain the most popular methods in the investigation of OP in soils. The combination of these techniques with more direct methods shows a great deal of potential and will be discussed in the next section of this paper.

8.4 DIRECT MEASUREMENT OF ORGANIC PHOSPHORUS

Indirect methods provide us with a means of grouping OP into pools of different stability, and allow us to characterize P as extractable or hydrolyzable by a specific extractant. However, these methods do not define the molecular structure of the organic moieties, i.e., alcohol, ester, and so on. For further interpretation of OP, it

TABLE 8.2
Percent Hydrolysis of Organic Phosphorus Compounds during Extraction (from Ivanoff et al., 1998).

Extractant	p-nitrophenyl phosphate	glycero-phosphate	D-glucose-6-phosphate	Phytic acid
Deionized water	2.8	1.0	2.0	0.2
0.5 *M* $NaHCO_3$	0.3	2.7	5.8	3.2
1 *M* HCl	37.8	1.2	2.1	0.3
0.5 *M* NaOH	0.2	40.8	2.2	0.2

is necessary to improve the level of P identification and to use more direct methods of analysis. These methods are designed to isolate the different components of OP and define their molecular structure. Inositol phosphates, nucleic acids, and phospholipids generally comprise up to 50% of OP in soils, but the remainder of the OP constituents are unidentified (Anderson, 1980). Several methods have been proposed for direct OP identification and measurement, including direct enzyme hydrolysis (Herbes et al., 1975), liquid chromatography (Minear, 1972; Weimer and Armstrong, 1979), and ^{31}P-NMR (Newman and Tate, 1980; Condron et al., 1985; Bedrock et al., 1994).

8.4.1 Enzyme Hydrolysis

Enzyme hydrolysis can be considered a method of biological extraction, analogous to the chemical extraction discussed above, with the exception that enzymes bind and hydrolyze specific types of bonds. Hydrolysis provides a direct indication of the molecular structure of the compounds in whose reactions they act as catalysts. Herbes et al. (1975) characterized soluble OP in water based on its hydrolysis with enzymes. The water was incubated with alkaline phosphatase, phosphodiesterase, and phytase, and the release of SRP was measured. Up to 50% of the soluble OP was hydrolyzed by phytase, and this enzyme hydrolyzable fraction was composed of both low and high molecular weight compounds. In Florida, acid phosphatase was added to two Spodosol soil extracts, and only 20 to 30% of water-soluble OP was hydrolyzable by acid phosphatase (Fox and Comerford, 1992). Simple monoesters were therefore considered to be a minor contributor to DOP in these soils.

The enzyme addition technique has also been applied in several lake studies (Boavida and Heath, 1988; Cotner and Heath, 1988). These researchers examined phosphatase activity not to identify the forms of OP but to determine how much of phytoplankton P demand could be met by OP in the surface water, i.e., how much was bioavailable. In Lake Plußsee, the enzymatic release of P satisfied the P uptake of microplankton (Chrost and Overbeck, 1987). In contrast, in Lake Apopka, no P was released following the addition of alkaline phosphatase to lake water samples (Newman, 1991). Enzyme addition studies have suggested that the enzymatic release of SRP from OP may be seasonal (Cotner and Heath, 1988), and does not always indicate that OP was a significant source of P to support algal growth (Boavida and Heath, 1988).

8.4.2 Chromatography

Chromatography is a fairly popular technique to examine OP in freshwaters, because it is rapid, simple, reasonably inexpensive, and has wide applicability. Although several chromatographic techniques are available, including thin-layer, paper, and electrophoresis, liquid chromatography using Sephadex gel columns has been most widely used (Broberg and Persson, 1988). In this procedure, a liquid or gaseous phase is passed through a porous gel with a known pore size. Molecules exceeding the gel pore size are excluded and thus pass through the column more quickly than smaller molecules that penetrate the gel. This procedure provides an indirect measure of OP. However, researchers have also eluted known OP compounds through different columns and, by comparing elution profiles of a sample to the standards, it is possible to identify OP compounds. Alternatively, post-chromatographic analyses of eluted components can be conducted. Minear (1972) subjected the eluted high molecular weight compounds to DNA analysis and also exposed this fraction to a DNA specific enzyme. Using this technique, he determined that 40 to 60% of DOP from lake samples could be accounted for by DNA or its fragments. However, chromatographic separations are not as straightforward as they appear; there are numerous stationary phase and eluent combinations, making it difficult to compare and evaluate results across studies. Chromatographic separations are also sensitive to sample handling, and their results may vary significantly among different ecosystems (Broberg and Persson, 1988).

8.4.3 Nuclear magnetic resonance

Although chromatographic techniques have also been used to examine OP in soils (Cosgrove, 1962; Steward and Tate, 1971; Gerritse, 1978; Tate, 1979), ^{31}P-NMR analysis of OP in alkaline extracts is the most common method used to identify broad classes of OP compounds in soil. Nuclear magnetic resonance uses the principle that the absorption of radio-frequency radiation by a nucleus in a magnetic field can be correlated with the structure of the molecule in which it occurs. ^{31}P-NMR uses the chemical shift of the P nuclei as a means to identify and quantify the OP compounds. Chemical shift in NMR is similar to frequency, which is used in infrared spectra. The identification of the species is based on the comparison of the spectra of an unknown sample to that of known compounds, because nuclei of different P groups, e.g., ortho-P monoesters, have distinct chemical shift values.

^{31}P-NMR has typically been used to characterize OP in upland mineral soils (Newman and Tate, 1980; Hawkes et al., 1984; Condron et al., 1985) and has recently been used to study OP in organic soils (Bedrock et al., 1994; Dai et al., 1996). Dai et al. (1996) identified six forms of P using ^{31}P-NMR: phosphonates, inorganic orthophosphate, orthophosphate monoesters, orthophosphate diesters, pyrophosphates, and polyphosphates.

One limitation of solution-state ^{31}P-NMR has been the use of an extractant that will extract maximum concentrations of soil P without causing hydrolysis; typically, 0.5 *M* NaOH has been used (Newman and Tate, 1980). Recently, the technique was improved by a change in the extraction solution to 0.25 *M* NaOH-0.05 *M* EDTA

(Bowman and Moir, 1993; Cade-Menun and Preston, 1996). This is a mild extraction that had a good correlation with acid-base sequential extractions and dry combustion followed by acid extractions (Bowman and Moir, 1993). A primary advantage of the incorporation of EDTA is that the chelating effect of EDTA increases the extraction efficiency of sodium hydroxide by breaking down organo-metallic P complexes. In addition, the EDTA may chelate Fe in the soil extract and thus reduce the paramagnetic interference during the ^{31}P-NMR spectroscopy, minimizing line broadening and distortion of the NMR spectra. As with P fractionation procedures, one of the major limitations that have restricted the routine use of ^{31}P-NMR was the amount of time required to complete the analysis—typically 16 to 120 hours (Newman and Tate, 1980; Hawkes et al., 1984; Condron et al., 1985; Adams and Byrne, 1989; Bedrock et al., 1994). Incorporating these advances, Robinson et al. (submitted) used ^{31}P-NMR to examine the OP pools in three constructed Florida wetland soils. They determined that the combined extractant, NaOH-EDTA, produced signatures relatively rapidly (30 to 72 min) and extracted 32 to 39% of TP and 35 to 72% of OP. They also determined that the quality of the ^{31}P-NMR spectra was improved when the extraction of OP was preceded by the removal of labile IP (Fig. 8.5). The quality of the spectra was enhanced for all three major P pools identified: inorganic ortho-P, ortho-P monoesters, and ortho-P diesters.

In addition to the procedures mentioned above, there is also a new direct measurement of OP being developed for use in the Florida Everglades. The technique is designed to determine the molecular weight and principal functional groups of the organic matter to which P is bound. It will be used to examine P in agricultural runoff water entering the Everglades and in a constructed wetland designed to treat this agricultural runoff (Salters et al., 1997). This research will identify the organic

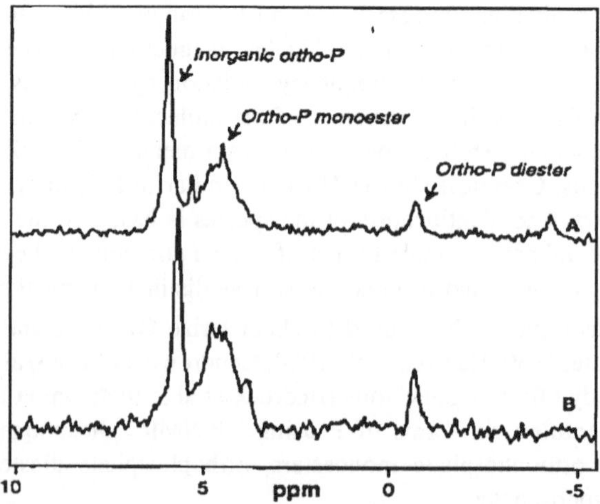

FIGURE 8.5 Stack plot showing P forms in Lake Apopka marsh soils identified using 31P-NMR: (a) distribution of inorganic ortho P and phosphomonoesters and diesters in 0.25 *M* NaOH–0.5 *M* EDTA extracts, and (b) improved spectra following pretreatment with 1 *M* KCl (modified from Robinson et al, unpublished ms.).

matter molecular weight using capillary electrophoresis with polymer additives (CEPA, Sullivan, 1994). The principal functional groups will be identified using *Fourier transform ion cyclotron mass spectrometry* (FT-ICR MS, Fievre et al., 1997b). Using CEPA, Fievre et al. (1997a), determined molecular weights of Suwannee River humic and fulvic acids, which were in good agreement with other measurements on the same material (Table 8.3). The potential strength of this technique is the ability of the sophisticated equipment to produce accurate separation and weight characterization of the P compounds. Currently, there are very few facilities available that have this type of equipment, so this procedure may have limited use over the next few years.

TABLE 8.3
Estimated Weights of Humic and Fulvic Acids (from Fievre et al., 1997a)

Sample	Migration time (min)	Molecular weight	Accepted molecular weight
Suwannee River fulvic acid	4.40	1700	1460
Suwannee River humic acid	4.73	4300	4390

8.5 RECOMMENDATIONS FOR FLORIDA AND OTHER SUBTROPICAL ECOSYSTEMS

The need to identify different forms of OP will vary depending on the goal of the study. However, to fully understand P cycling, the forms of OP need to be identified and their mobility and turnover rates evaluated. Therefore, more focus on the identification of P forms beyond SRP and TP is needed. Although the measurement of TDP is common, the ecological significance of this P pool has not been determined. This can be achieved by conducting more bioassays or by determining the IP and OP constituents of this pool. In addition, the surplus P pool inside the algal biomass should also be considered. Hot-water-extractable P appears to be a good indicator of phytoplankton surplus P (Newman et al., 1994; Aldridge 1994). To ensure that the complete pool of surplus P is identified, we recommend that both acid-hydrolyzable and reactive pools be measured to consider the organic forms of polyphosphates (Newman et al., 1994).

Several of the techniques discussed in this chapter rely on the use of chemical extractions to isolate the OP compounds. The selection of the extractant is the key to these procedures. If the goal is evaluate short-term algal-available P, a milder extraction such as water or NTA (Kenney, 1997) should be considered. However, if the sediments have <40% organic matter and the goal is long-term nutrient availability, stronger extracting solutions are appropriate. Ivanoff et al. (1998) demonstrated that extractions may result in the hydrolysis of the OP compounds in question. In particular, the commonly used 0.5 *M* sodium hydroxide appears too harsh. Leinweber et al. (1997) studied the solubility and forms of P in manures from chicken and pigs; whole soil samples; and clay-, silt-, and sand-size separates from arable and grassland soils. Solution ^{31}P-NMR spectra of 0.5 *M* sodium hydroxide extracts from manures and some soil samples showed greater signal intensities for orthophosphate and monoester P than 0.1 *M* sodium hydroxide extracts. This can be explained by alkaline hydrolysis of phosphate diesters at higher sodium hydroxide concentrations and/or by preferential extraction of diesters at lower concentrations. Therefore, we recommend against using 0.5 *M* sodium hydroxide in high-OP Florida soils.

Soil fractionation procedures are more powerful techniques when conducted in conjunction with an identification procedure such as ^{31}P-NMR. Sodium hydroxide (0.25 *M*) combined with EDTA appears to offer some advantages as an extraction technique used in combination with ^{31}P-NMR. Although Robinson et al. (unpublished ms.) extracted only 35 to 72% of OP in three constructed wetlands, the presence of clay minerals in these soils may have reduced the amount of P recovered due to the strong sorption characteristics of clay mineral-humate complexes. In addition, the possible presence of sparingly soluble P-containing compounds, such as calcium phytate and inorganic phosphates, may explain the variation in the efficiency of P recovery. Despite this variation, the NaOH-EDTA recovered a greater amount of P than did the more conventionally used 0.5 *M* sodium hydroxide. Therefore, we recommend that this method be evaluated on a wider range of Florida soils.

In addition, the quality of the ^{31}P-NMR spectra was improved when the extraction of OP was preceded by the removal of labile IP (Robinson et al., unpublished ms.). The increase in the proportion of ortho-P diester in two of the constructed wetland soils following sodium bicarbonate extraction, indicates that not all ortho-P diesters are removed in this extractant. Consequently, sodium bicarbonate extraction may underestimate the size of the labile OP pool in these soils. Guggenberger et al. (1996) showed a close relation between macroporous anion-exchange resin-OP and ^{31}P-NMR estimates for diester-P in two European soils. It was concluded that the resin used in this study isolates a structurally and functionally reasonably uniform pool of potentially labile soil OP. Therefore, in order to improve further the quality of ortho-P monoester spectra in NaOH-EDTA extracts, we suggest a prior extraction of the soil with macroporous anion-exchange resin.

Our ability to predict whether OP will act as a source or sink for P is currently limited by the extent to which the different forms are identified. Future research should continue to develop identification techniques and determine the biological role of OP identified compounds so their ecological significance can be evaluated. In addition, focus should be placed on refining techniques, such as ^{31}P-NMR, so the signal-to-noise ratio is increased.

8.6 ACKNOWLEDGMENTS

This manuscript was improved by comments from Tom Fontaine, Karl Havens, Delia Ivanoff, Paul McCormick, Al Steinman, Dave Struve, and two anonymous reviewers.

REFERENCES

Adams, M.A., and L.T. Byme. 1989. Phosphorus-31 NMR analysis of phosphorus compounds in extracts of surface soils from selected Karl (*Eucalyptus diversicolor* F. Muell.) forests. Soil Biol. Biochem. 2:523–528.

Aldridge, F.J. 1994. Application of nutrient enrichment bioassays to evaluate spatial and temporal limiting-nutrient patterns and to estimate surplus phosphorus concentration in Lake Okeechobee, FL. Ph.D. diss., Univ. of Florida, Gainesville, FL.

American Public Health Association (APHA). 1995. Standard methods for the examination of water and wastewater. 19th ed. APHA, Washington, DC.

Anderson, G. 1980. Assessing organic phosphorus in soils. In The role of phosphorus in agriculture, (F.E. Khasawneh, E.C. Sample and E.J. Kampreth, eds.) pp. 411–431. Am. Soc. Agr., Madison, WI.

Bedrock, C.N., M.V. Cheshire, J.A. Chudek, B.A. Goodman, and C.A. Shand. 1994. Use of ^{31}P-NMR to study the forms of phosphorus in peat soils. Sci. Total Environ. 152:1–8.

Boavida, M.J., and R.T. Heath. 1988. Is alkaline phosphatase always important in phosphate regeneration? Arch. Hydrobiol. 111:507–518.

Boström, B., G. Persson, and B. Broberg. 1988. Bioavailability of different phosphorus forms in freshwater systems. Hydrobiologia 170:133–155.

Bowman, R.A., and C.V. Cole. 1978. An exploratory method for fractionation of organic phosphorus from grassland soils. Soil Sci. 125:95–101.

Bowman, R.A., and J.O. Moir. 1993. Basic EDTA as an extractant for soil organic phosphorus. Soil Sci. Soc. Am. J. 57:1516–1518.

Broberg, O., and G. Persson. 1988. Particulate and dissolved phosphorus forms in freshwater: composition and analysis. Hydrobiologia 170:61–90.

Cade-Menun, B.J., and C.M. Preston. 1996. A comparison of soil extraction procedures for ^{31}P-NMR spectroscopy. Soil Sci. 161:770–785.

Carrick, H.J., C.L. Schelske, F.J., Aldridge, and M.F. Coveney. 1993. Assessment of phytoplankton nutrient limitation in productive waters: application of dilution bioassays. Can. J. Fish. Aquat. Sci. 50:2208–2221.

Cembella, A.D., N.J. Antia, and P.J. Harrison. 1984. The utilization of inorganic and organic phosphorus compounds as nutrients by eukaryotic microalgae: a multidisciplinary perspective: Part 1. CRC Crit. Rev. Microbiol. 10: 317–391.

Chang, S.C. and M.L. Jackson. 1957. Fractionation of soil phosphorus. Soil Sci. 84:133–144.

Chrost, R.J., and J. Overbeck. 1987. Kinetics of alkaline phosphatase activity and phosphorus availability for phytoplankton and bacterioplankton in Lake Plußsee (north German eutrophic lake). Microb. Ecol. 13:229–248.

Condron, L.M., K.M. Goh, and R.H. Newman. 1985. Nature and distribution of soil phosphorus as revealed by a sequential extraction method followed by ^{31}P nuclear magnetic resonance analysis. Soil Sci. 36:199–207.

Cosgrove, D.J. 1962. Forms of inositol hexaphosphate in soils. Nature 194:1265–1266.

Cotner, J.B., and R.T. Heath. 1988. Potential phosphate release from phosphomonoesters by acid phosphatase in a bog lake. Arch. Hydrobiol. 111:329–338.

Dai, K'o H, M.B. David, G.F. Vance, and A.J. Krzyszowska. 1996. Characterization of phosphorus in a spruce-fir Spodosol by phosphorus-31 nuclear magnetic resonance spectroscopy. Soil Sci. Soc. Am. J. 60:1943–1950.

DeBusk, W.F., K.R. Reddy, M.S. Koch, and Y. Wang. 1994. Spatial distribution of soil nutrients in a northern Everglades marsh: water conservation area 2A. Soil Sci. Soc. Am. J. 58:543–552.

Fievre, A., L. Bennett, W.T. Cooper. 1997a. Characterization of aquatic humic substances by capillary electrophoresis. Poster presentation at the 27th International Symposium of Environmental Analytical Chemistry. Jekyll Island, GA, June 1997.

Fievre, A., T. Solouki, A.G. Marshall, and W.T. Cooper. 1997b. High-resolution Fourier transform ion cyclotron resonance mass spectrometry of humic and fulvic acids by laser desorption/ionization and electrospray ionization. Energy Fuels 11:554–560.

Fitzgerald, G.P., and T.C. Nelson. 1966. Extractive and enzymatic analyses for limiting or surplus phosphorus in algae. J. Phycol. 2:32–37.

Fox, T.R., and N.B. Comerford. 1992. Rhizosphere phosphatase activity and phosphatase hydrolyzable organic phosphorus in two forested Spodosols. Soil Biol. Biochem. 24:579–583.

Gale, P.M., K.R. Reddy, and D.A. Graetz. 1994. Phosphorus retention by wetland soils used for wastewater disposal. J. Environ. Qual. 23:370–377.

Gerritse, R.G. 1978. Assessment of a procedure for fractionating organic phosphates in soil and organic materials using gel filtration and H.P.L.C. J.Sci. Food. Agric. 29:577–586.

Guggenberger, G., B.T. Christensen, G. Rubaek, and W. Zech. 1996. Land-use and fertilization effects on P forms in two European soils: Resin extraction and P-31-NMR analysis. Eur. J. Soil Sci. 47: 605–614.

Harrison, A.F. 1987. Soil organic phosphorus; a review of world literature. C.A.B. International. Wallingford, Oxon, UK.

Hawkes, M.J., D.S. Powlson, E.W. Randall, and K.R. Tate. 1984. A ^{31}P nuclear magnetic resonance study of the phosphorus species in alkali extracts of soils from long-term field experiments. J. Soil Sci. 35:35–45.

Hedley, M.J., J.W.B. Stewart, and B.S. Chauhan. 1982. Changes in inorganic and organic soil phosphorus fractions induced by cultivation practices and by laboratory incubations. Soil Sci. Soc. Am. J. 46:970–975.

Herbes, S.E., H.E. Allen, and K.H. Mancy. 1975. Enzymatic characterization of soluble organic phosphorus in lake water. Science 187:432–434.

Hutchinson, G.E., and V.T. Bowen. 1950. Limnological studies in Connecticut. IX. A quantitative radiochemical study of the phosphorus cycle in Linsley Pond. Ecol. 31:194–203.

Ivanoff, D.B., K.R. Reddy, and S. Robinson. 1998. Chemical fractionation of organic phosphorus in selected Histosols. Soil Sci. 163:36–45.

Kenney, W.F. 1997. A comparison of chemical assays for the estimation of bioavailable phosphorus in Lake Apopka sediments. M.S. thesis, Univ. of Florida, Gainesville, FL.

Koch, M.S., and K.R. Reddy. 1992. Distribution of soil and plant nutrients along a trophic gradient in the Florida Everglades. Soil Sci. Soc. Am. J. 56:1492–1499.

Krogstad, T., and O. Løvstad. 1991. Available soil phosphorus for planktonic blue-green algae in eutrophic lake water samples. Arch. Hydrobiol. 122:117–128.

Lean, D. R. 1973. Movements of phosphorus between its biologically important forms in lakewater. J. Fish. Res. Bd. Can. 30:1525–1536.

Leinweber, P., L. Haumaier, and W. Zech. 1997. Sequential extractions and P-31-NMR spectroscopy of phosphorus forms in animal manures, whole soils and particle-size separates from a densely populated livestock area in northwest Germany. Biol. Fert. Soils 25: 89–94.

McCormick, P.V., P.S. Rawlik, K. Lurding, E.P. Smith, and F.H. Sklar. 1996. Periphyton-water quality relationships along a nutrient gradient in the northern Everglades. J. N. Am. Benthol. Soc. 15:433–449.

Minear, R.A. 1972. Characterization of naturally occurring dissolved organophosphorus compounds. Environ. Sci. Technol. 6:431–437.

Newman, R.H., and K.R. Tate. 1980. Soil phosphorus characterization by ^{31}P-nuclear magnetic resonance. Commun. Soil Sci. Plant Anal. 11:835–842.

Newman, S. 1991. Bioavailability of organic phosphorus in a shallow hypereutrophic lake. Ph.D. diss., Univ. of Florida, Gainesville, FL.

Newman, S., and K.R. Reddy. 1993. Alkaline phosphatase activity in the sediment-water column of a hypereutrophic lake. J. Environ. Qual. 22:832–838.

Newman, S., F.J. Aldridge, E.J. Phlips, and K.R. Reddy. 1994. Assessment of phosphorus availability for natural phytoplankton populations from a hypereutrophic lake. Arch. Hydrobiol. 130:409–427.

Qualls, R.G., and C.J. Richardson. 1995. Forms of soil phosphorus along a nutrient enrichment gradient in the northern Everglades. Soil Sci. 160:183–198.

Reddy, K.R. 1991. Lake Okeechobee phosphorus dynamics study: biogeochemical processes in the sediments. Report submitted to South Florida Water Management District, West Palm Beach, FL. Contract # 531-M88-0445.

Reddy, K.R., and D.A. Graetz. 1990. Internal nutrient budget for Lake Apopka. Submitted to St. Johns River Water Management District, Palatka, FL. Contract # 15-150-01-SWIM.

Reddy, K.R., W.F. DeBusk, Y. Wang, R.D. DeLaune, and M. Koch. 1991. Physico-chemical properties of soils in the water conservation area 2 of the Everglades. Report submitted to South Florida Water Management District, West Palm Beach, FL.

Reddy, K.R., R.D. Delaune, W.F. DeBusk, and M.S. Koch. 1993. Long-term nutrient accumulation rates in the Everglades. Soil Sci. Soc. Am. J. 57:1147–1155.

Reddy, K.R., Y. Wang, W.F. DeBusk, M.M. Fisher, and S. Newman. 1998. Forms of soil phosphorus in selected hydrologic units of Florida Everglades ecosystems. Soil Sci. Soc. Am. J. in press.

Rigler, F.H. 1956. A tracer study of the phosphate cycle in lake water. Ecol. 37:550–562.

Rigler, F.H. 1964. The phosphorus fractions and turnover time of inorganic phosphorus in different types of lakes. Limnol. Oceanogr. 6:165–174.

Rigler, F.H. 1966. Radiobiological analysis of inorganic phosphorus in lake water. Verh. Int. Ver. Limnol. 16:465–470.

Robinson, J.S., C.T. Johnston, and K.R. Reddy. Combined chemical and ^{31}P-NMR spectroscopic analysis of phosphorus in wetland organic soils. Soil Sci. submitted.

Salters, V.J.M., W.T. Cooper, Y.P Hsieh, W.M. Landing, A.G. Marshall, L. Proctor, and Y. Wang. 1997. The speciation and sources of dissolved phosphorus in the Everglades. Poster Presentation at the Phosphorus Biogeochemistry in Florida Ecosystems Symposium, July 1997, Clearwater Beach, Fl.

Sharpley, A.N., W.W. Troeger, and S.J. Smith. 1991. The measurement of bioavailable phosphorus in agricultural runoff. J. Environ. Qual. 20:235–238.

Sommers, L.E., R.F. Harris, J.D.H. Williams, D.E. Armstrong, and J.K. Syers. 1972. Fractionation of organic phosphorus in lake sediments. Soil Sci. Soc. Am. Proc. 36:51–54.

Steward, J.H., and M.E. Tate. 1971. Gel chromatography of soil organic phosphorus. J. Chrom. 60:75–82.

Sullivan, L.R. 1994. Novel applications of capillary electrophoresis in biogeochemical analyses. M.S. thesis, Florida State Univ., Tallahassee, FL.

Tate, K.R. 1979. Fractionation of soil organic phosphorus in two New Zealand soils by use of sodium borate. N. Z. J. Sci. 22:137–142.

Tiessen, H., J.W.B. Stewart, and A. Oberson. 1994. Innovative soil phosphorus availability indices: assessing organic phosphorus. In Soil testing: prospects for improving nutrient recommendations. Soil Science Society of America Special Publication 40.

U.S. Environmental Protection Agency (USEPA). 1978. The *Selanastrum capricornutum* Printz. algal assay bottle test. EPA-600/9-78-018. Office of Research and Development, USEPA, Corvallis, OR.

Weimer, W.C., and D.E. Armstrong. 1979. Naturally occurring organic phosphorus compounds in aquatic plants. Environ. Sci. Technol. 13:826–829.

Wetzel, R.G. Organic phosphorus mineralization in soils and sediments. Chapter 9, this book.

Wynne, D., and T. Berman. 1980. Hot water extractable phosphorus-an indicator of nutritional status of *Peridinium cinctum* (Dinophyceae) from Lake Kinneret (Israel)?. J. Phycol. 16:40–46.

Young, T.C., J.V. DePinto, S.C. Martin, and J.S. Bonner. 1985. Algal-available particulate phosphorus in the Great Lakes basin. J. Great Lakes Res. 11:434–446.

9 Organic Phosphorus Mineralization in Soils and Sediments

Robert G. Wetzel

9.1 ABSTRACT

Organic phosphorus constitutes about half of the total P of the biosphere. Much of the organic P resides in living and senescent plant organic matter or is associated in organic compounds immobilized with soil particles. Mineralization rates are masked by this large pool, from which a small but critical quantity of organic P is rapidly recycled among the soil, plants, and microbes. The organic P transformations are controlled by a combination of P concentrations in solution and biological activities, the most important of which are microbial alterations of redox reactions and bonding to particles. As the concentrations or biological availability of the inorganic pool decreases, much of the P resides in organic compounds and usually results in an increase in the rates of recycling between the inorganic and organic P pools.

In aquatic ecosystems, P availability depends to a great extent on importation from the drainage basin. Both physical and biotic processes in lakes, streams, and wetlands tend toward concentration of P in inorganic and organic pools of the sediments. Exchange rates between various deposits of P and the interstitial water of the sediments depend on local adsorption and diffusion coefficients and their alterations by enzyme-mediated reactions of the microbiota. In organic-rich sediments not illuminated by sunlight, high sediment demand for oxygen results from microbial respiratory metabolism, slow rates of diffusion from overlying water, and

1-56670-331-X/99/$0.00+$.50

inorganic elements such as Fe(II) that accumulate in reduced form. On a molar ratio of phosphate liberated per unit of organic carbon mineralized by bacteria, the availability of alternate electron acceptor compounds such as sulfate largely control mineralization rates and release of sorbed and bonded P from sediments. Microbial conversion of reactive sediment Fe compounds to iron sulfides by sulfate-reducing bacteria can lead to more efficient release of Fe-associated phosphate than does direct microbial Fe(III) oxide reduction.

Epipelic algal photosynthesis, as well as oxygen release from the roots of emergent aquatic plants into the rhizosphere, markedly alter redox conditions of the sediments on a distinct diurnal basis. These rapid (minutes) shifts in the depth of the oxidized microzone (by as much as a factor of 100) and redox conditions in the sediments, in turn, in a highly dynamic sediment system, rapidly alter both mineralization rates of organic P pools and inactivation of inorganic P by oxidative reactions. Of the several fractions of organic P, a low molecular weight fraction cycles rapidly (hours) with soluble forms. Dissolved and colloidal organic P fractions released from microbiota are cycled much more slowly, particularly if sorbed to inorganic and detrital particles.

9.2 INTRODUCTION

The phosphorus (P) cycle is outwardly simple biologically. P is assimilated as phosphate and is released directly as phosphate during metabolism by excretion or during mineralization of organic P-containing compounds. The oxidation state of P is constant (+V) regardless of chemical transformations. Despite the apparent simplicity of the P cycle, P is highly reactive and much of the biospheric P resides in inorganic pools that are unavailable to biota by most biological processes.

Biological metabolism regulates P cycling both directly and indirectly. Mineralization of organic P results in the release of phosphate. Bacterial metabolism indirectly alters the availability and accessibility of both inorganic and organic P by changes to the redox conditions of the environment. Because terrestrial soils are largely oxidative, and nearly all aquatic sediments are reducing, major differences emerge between these environments on the rates and controls of P cycling.

About half of the total P in the biosphere is in organic form. This quantity varies, however, in relation to both the abundance and availability of phosphate. As the inorganic pool that is accessible to biota decreases, as is commonly the case in aquatic ecosystems, much of the P is in organic compounds. As a result, an increase in the rates of recycling between inorganic and organic P pools is frequently found in aquatic ecosystems.

9.3 ORGANIC PHOSPHORUS IN ORGANISMS

Phosphorus is a major constituent of macromolecules, particularly in nucleic acids (DNA and RNA), in phospholipids of membranes, and as monoesters of a variety of compounds, particularly those involved in biochemical pathways. In growing microorganisms, more than half of the organic phosphates are in nucleic acids, whereas phospholipids and monoesters constitute the remainder in varying proportions (Magid et al., 1996). As we have seen (Newman and Robinson, this volume), many organisms can store orthophosphate or polyphosphates. In plants, inositol hexaphosphate (IHP) can form a major storage compound for P, particularly in seeds.

9.4 ORGANIC PHOSPHORUS IN SOILS

Organic P extracted from soils differs considerably from that in living organisms with domination by monoesters (Table 9.1). Presumably, as organic P is released by secretion or lysis from organisms into the soil, and if not utilized metabolically immediately, it can be stabilized in the soil matrix. The bonding mechanism can occur through the carbon moiety or through reactions for the P moiety, particularly in association with low molecular weight phosphomonoesters.

TABLE 9.1
Organic P Fractions (% of Total Organic P) in Growing Organisms and Soils

	E. coli	Fungi	Spirodella	Nicotiana	Soils
Nucleic Acids	65	58	60	52	2
Phospolipids	15	20	30	23	5
Monoesters	20	22	10	25	>50

Source: from Magid et al., 1996, and after numerous sources
Note: Approx. half of the organic P of soils remains chemically unidentified.

Up to 50% of total extractable organic P can consist of various inositol phosphates. For example, inositol hexaphosphate (IHP) adsorbs more strongly to iron oxide of soils than does orthophosphate (Anderson et al., 1974). Other organic P compounds, such as phosphodiesters, adsorb to the soil matrix less strongly and therefore are more likely to be vulnerable to attack by phosphatases and to cycle biologically (Magid et al., 1996). This differential stabilization results in inositol phosphates and several other monoesters dominating (80 to 90%) organic P compounds of soils. Organic P compounds are associated with positively charged sites on organic matter, clay particles, or cations in soil solutions. The soil particles, however, lose much of their P absorption capacity, and sorption onto organic matter is controlled via an organic matter-metal-P complex (Zhou et al., 1997). The high adsorption capacity of Spodosol Bh horizons, for example, appears to result from association with aluminum-organic matter complexes. These organic matter-metal-P complexes tend to release P readily at low solution P concentrations (Zhou et al., 1997). Only small amounts of phospholipids, nucleic acids, and traces of diesters, triesters, and phosphatidylcholine are found in the organic P pools (Kowalenko, 1978; Condron et al., 1985). As much as one-third of the inositol P can be complexed with humic and fulvic acids.

As a result of large quantities of organic P being immobilized in soils, and only a small portion of the total organic P being biologically active, mineralization rates are masked. The result is rapid recycling of P among the soil, plants, and microbes and microfauna. The dynamic organic P cycle of soils involves organic P transformations that are controlled by P concentrations in solution and biological activities.

In contrast to the aquatic ecosystems discussed below, in which most P is imported and recycled, in soil systems, phosphate ions are slowly dissolved from P-containing minerals. A portion of this solution inorganic P pool will be precipitated (Fig. 9.1), much of which will be unavailable in occluded forms within the soils (Stewart and Tiessen, 1987; Magid et al., 1996). In addition, much of the organic P is stabilized in relatively recalcitrant organic compounds that are relatively inaccessible to enzymatic hydrolysis.

Organic P is released directly by secretion, such as from plant roots along with other organic compounds, or cell lysis (Fig. 9.1). Such compounds can be assimilated directly or following enzymatic hydrolysis as phosphate (see discussion below). In addition, under P limitation lateral roots exude organic acids (citrate, malate, succinate) that increases P solubilization (Johnson et al., 1996).

Active cycling of organic P in soils is largely mediated by microbial metabolism. To understand cycling of organic P, it is essential to evaluate the organically bound P within compounds of the microbial biomass and the organismal dynamics, particularly the bacteria, as they influence organic P dynamics. Microbial metabolism can influence the rate of mineralization of organic P in two primary ways: (a) direct biochemical mobilization by extracellular or periplasmic hydrolases, particularly from inositol phosphates, and (b) bacterial and other microbial metabolism within the soil alters ionic composition, redox, pH, and other conditions that, in turn, can alter the efficacy of P binding to particulate organic matter and clays and other soil particles, as well as solubility of organic P and its potential enzymatic hydrolysis. Because of the high reactivity of solubilized P, the distance between sites of enzymatic hydrolysis of organic P and sites of uptake must be short, often in immediate

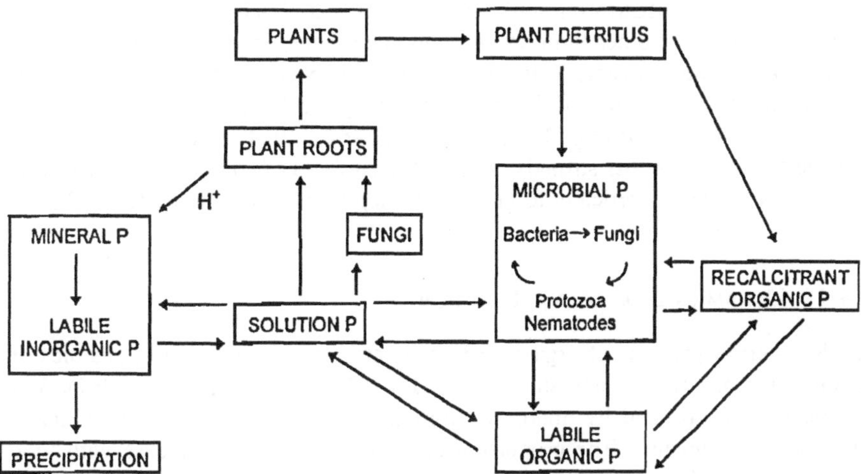

FIGURE 9.1 Primary couplings among phosphorus pools and their fluxes within the phosphorus cycle of soils. (Modified from Stewart and Tiessen, 1987.)

juxtaposition. In studies of the distribution of phosphatase activity of soil in proximity to plant roots, increases in both acid and alkaline phosphatase activities were found at the soil-root interfaces (Tarafdar and Jungk, 1987). The phosphatase activity extended from the root surface for distances of 1 to 3 mm. Inactivation by reaction with the soil matrix increases with time after release and with available surface areas associated with decreasing size of soil aggregates (Linquist et al., 1997a).

The P content of the biota within soil is overwhelmingly concentrated in the fungi (approx. 69%) and bacteria (approx. 30%) (Magid et al., 1996). Although the total amount of P contained in the microbial organisms is a small percentage (3 to 14%) of the total soil organic P, much of it is available to plants inhabiting the soils. The rates of mineralization of organic P, however, are primarily mediated by the bacteria and protozoa (Table 9.2) and coupled directly to rates of organic carbon and total nitrogen mineralization (Linquist et al., 1997b). Even though the higher trophic levels contribute little to the total biomass or organic P, or to the direct rates of mineralization, their activities can alter the population dynamics of the microbes and thus indirectly alter the composite rates of net mineralization of organic P.

All workers in cycling of organic P in soils emphasize the importance of methodology in biasing interpretations. Fungi and especially bacteria are relatively immobile, and less than 3% of the pore-space of soils is colonized by microorganisms (Elliott et al., 1980; van Veen and Kuikman, 1990). Many of these organisms—usually many more than half—are dormant. Analytical techniques extract forms of P from many soil particle surfaces that are not utilized or readily available to soil microorganisms. Movement of water and P solutes is largely through macropores (>100 μm), but most movement from the macropores to mesopores (0.5 to 100 μm), where nearly all microbial processes occur, depends on diffusion. Hence, proximity of organisms potentially utilizing P regenerated by mineralization of organic P is essential to minimize losses by physical and chemical sorptive processes to soil particles.

Processes that enhance microbial metabolism (e.g., loading of labile dissolved organic substrates or other limiting nutrients) commonly increase microbial P and decrease leachable inorganic P. Because of the reactivity and adsorptivity of inorganic P, much (often > 95%) of the P released by leaching with drainage of soil solutions is as organic P (e.g., Hanappel et al., 1964a, 1964b; Magid et al., 1992). Organic P tends to be less strongly adsorbed to surfaces of soil particles; therefore, it is likely to migrate vertically to greater soil depths as well as to be leached from soils.

9.4.1 Dynamics of Organic P in Soils

The long-term dynamics of organic P in soils are closely coupled to dynamics of organic carbon, nitrogen, and sulfur (Tate, 1984; Stewart and Tiessen, 1987). Soil organic P is determined by major soil-forming factors (parent minerals, climate, topography, organisms). The proportion of organic P tends to increase with time until in later weathering stages when iron and aluminum complexation of inorganic P with low P solubility dominates. Both inorganic and particularly organic P are then lost by leaching and erosion. Although losses of P are minimal within many

TABLE 9.2
Net Mineralization of P (kg ha^{-1} y^{-1}) by Different Organismal Groups in Two Depth Layers in Soils under Conventional Management

	0–10 cm	10–25 cm
Microbes		
Bacteria	3.14	5.05
Fungi	0.22	0.36
Protozoa		
Amoebae	0.84	1.01
Flagellates	0.05	0.04
Nematodes		
Herbivores	0.0	0.0
Bacterivores	0.29	0.56
Fungivores	0.01	0.02
Predators	0.07	0.06
Microarthropods		
Cryptostigmatic mites	0.0003	0.0003
Noncryptostigmatic mites	0.0006	0.0007
Bacterivorous mites	0.0003	0.003
Predatory mites	0.0007	0.001
Nematophagous mites	0.0	0.0
Predatory collembola	0.0	0.0
Fungivores	0.014	0.022
Annelids		
Enchytraeids	0.01	0.075
Earthworms	0.0	0.0
Total mineralization	4.65	7.19

Source: from Magid et al., 1996, after data of Ryszkowski et al., 1989.

natural ecosystems, losses of soil organic P and its mineralization often increase markedly upon cultivation concomitant with changes in vegetation monocultures, reduced biodiversity, suppressed nutrient cycling, and increased loading of mineral fertilizers.

The two dominating biological P pools are total soil organic P and the microbial biomass P (Di et al., 1997). The total soil organic P consists of several different P pools and is insensitive to rapid short-term P dynamics. Improved insights into the mineralization rates of organic P were obtained by radioisotopic labeling of inorganic P pools in soil and monitoring changes in specific activities of the pool over time (e.g., Walbridge and Vitousek, 1987). Inorganic P released from microbial mineralization was assumed to dilute the labile P pool in the soil in proportion to the rates of organic P mineralization. However, some of the inorganic P can be inactivated by adsorption to soil particles or can be assimilated microbially and transformed to organic P. Obviously, many pathways of P transformation (Fig. 9.1) are occurring simultaneously.

Clearly, root activities of plants in soils can be enhanced by increased availability of P. However, spatial variations in soil fertility and temporal seasonal variations in P uptake are large and confound simple generalities (Di et al., 1997). The roles of mycorrhizae in uptake of P by plants are unclear. Although few differences in the uptake of inorganic P have been found between plants with and without mycorrhizae, Jayachandran et al. (1992) have shown that the presence of both endo- and ecto-mycorrhizal fungi can stimulate net mineralization of organic P in soil, likely via the production and release of extracellular phosphatase enzymes.

Soil organic P that is stabilized by sorption or reactive processes to soil particles can reenter by two means.

1. *Direct utilization of the organic compounds containing the organic P.* These reactions result in carbon losses via respired CO_2.
2. *Biochemical mineralization by extracellular enzymatic phosphatase hydrolysis of P esters.*

Free enzymes are readily inactivated, however, by binding to organic and inorganic substrates. Under overall P limitations, however, the generation and release rates of phosphatases increase, and rates of recycling of organic P increase. Much of the organic P is associated with microbial cells.

As soils mature and the readily available P decreases, the rates of recycling of organic P by microbiota accelerate and become increasingly important for the availability of soil P to plants. Because so much of the organic P of soils is bound within relative unreactive components, the total organic P of soils is relatively stable in comparison to content of C and N. Thus, with long-term weathering and increased Fe and Al activities, organic P and soil organic matter are reduced, as much of the P is tightly occluded and losses of organic P increase by leaching.

9.5 ENZYMATIC UTILIZATION OF ORGANIC P IN SOILS AND SEDIMENTS

Bacteria and microalgae growing on soil particles and within interstitial water of sediments of aquatic ecosystems are able to use exogenous organic P compounds through enzymatic hydrolysis of terminal phosphate groups. A broad spectrum of dissolved organic P compounds can sustain growth of these organisms, including glycerophosphate, D-glucose 6-phosphate (G-6-P), adenosine 5'-monophosphate (AMP), cytidine 5'-monophosphate (CMP), guanosine 5'-monophosphate (GMP), adenylic acids, and phosphonate compounds (Cembella et al., 1983, 1984).

The availability of phosphate monoesters requires enzymatic cleavage of the ester linkage joining the inorganic P group to the organic moiety. Such hydrolysis is achieved by phosphomonoesterases (phosphatases) bound to or within the cell membrane or by dissolved enzymes released extracellularly within the interstitial solution adjacent to the cells. There is a general lack of specificity for the organic moiety that include, beyond phosphate monoesters, substrates such as diesters and phosphoanhydrides such as pyrophosphates, ADP, and ATP. The phosphatases of bacteria, algae, fungi, and higher plants are constitutive inorganic P-irrepressible enzymes, which tend to be intracellular and function in intermediary metabolism or are phosphate ester-induced inorganic P-repressible enzymes, which are usually membrane-bound and function in extracellular P cleavage (Cembella et al., 1983). Both types can occur simultaneously in the same organism. Although free, dissolved phosphatase is short lived (Reichardt et al., 1967; Pettersson, 1980), substantial quantities of phosphatase, often exceeding 50% of the total, can occur in dissolved phase free of cell membranes (Wetzel, 1981; Stewart and Wetzel, 1982a; Cotner and Heath, 1988; Rai and Jacobson, 1993).

The enzyme 5'-nucleotidase catalyzes the hydrolysis of phosphoryl groups from 5'-nucleotides, such as ATP, from the dissolved organic P pool in natural waters (Ammerman and Azam, 1985, 1991a, 1991b; Tamminen, 1989; Cotner and Wetzel, 1991a, 1991b; Bentzen and Taylor, 1991). Although the synthesis of the external enzyme alkaline phosphatase, discussed above, is often repressed in high phosphate environments, 5'-nucleotidase is not. Greater 5'-nucleotidase activity has been associated with bacterial-size small microbiota (<1 μm), particularly in environments low in dissolved organic phosphorus concentrations. In high-phosphate environments, most of the phosphate regenerated by 5'-nucleotidase is not assimilated immediately. In comparisons of different enzyme activities and rates of hydrolysis of different organic compounds, 5'-nucleotidase activity was found to have greater relative importance to P cycling in an oligotrophic lake than in a moderately eutrophic lake (Cotner and Wetzel, 1991b). These data suggested that nucleotidase activity may be more important to P regeneration in oligotrophic habitats than phosphatase activity.

Certain complex humic macromolecules of both low and high molecular weights contain P, such as phosphate esters and inositol hexaphosphate (e.g., Anderson and Hance, 1963; Koenings and Hooper, 1976). Humic compounds can be partially altered by photochemical oxidation with low-intensity ultraviolet light (both UV-B and UV-A portions of the spectrum) such as occurs in sunlight (Francko and Heath, 1979; Stewart and Wetzel, 1981; Wetzel et al., 1995). Phosphate adsorbed to ferric-humic compounds can be released by UV-induced photoreduction of ferric iron (Francko and Heath, 1982; Jones et al., 1988).

Limitation by P invariably induces high exoenzymatic phosphatase activity among many algae and bacteria. Dissolved organic compounds, particularly phenolic compounds that originate from lignin and cellulosic structural components of higher plants, result in relatively recalcitrant dissolved humic substances. These humic compounds complex chemically with many enzymes, particularly phosphatases that are both competitively and noncompetitively inhibited (Wetzel, 1993; Kim and Wetzel, 1993). The humic-enzyme complexes can be stored in an inactivated state for some time, relocated within the aquatic ecosystem, and subsequently be partially reactivated by exposure to natural ultraviolet light (Wetzel, 1993; Boavida and Wetzel, 1998).

Production of phosphatase by bacteria and algae was markedly stimulated by low concentrations of dissolved humic compounds of low molecular weight (Stewart and Wetzel, 1982b). Evidence indicates that these humic substances can act as sequestering agents for phosphate ions, organophosphorus compounds, and iron, thereby reducing P and possibly iron availability to microorganisms (Münster, 1994; Shaw, 1994; cf. review of Wetzel, 1983, 1998). In acidified environments where aluminum [Al(III)] and iron [Fe(III)] concentrations are often markedly increased, production of acid phosphatases is enhanced (Jansson, 1981). Although, at low pH, the Al(III) and Fe(III) do not affect the enzymes directly, the metal ions combine with the phosphate group on the substrate and inhibit enzymatic hydrolysis. Organisms may compensate, at considerable energetic costs, for these losses by increased production of phosphatases.

9.6 MINERALIZATION OF ORGANIC PHOSPHORUS IN AQUATIC SEDIMENTS

Discussion of mineralization and fluxes of organic P separate from the organisms in which much of it resides, and separate from large pools of inorganic and organic P associated with sediment particles, is difficult and confusing because fluxes among these compartments are so rapid. Much focus on organic P of aquatic systems, especially lakes and reservoirs, has been directed to exchange of P between sediments and the overlying water. An appreciable amount of the total loading of P to sediments is in organic forms. Commonly, an apparent net movement of P into sediments occurs in most wetlands and lakes. The effectiveness of the net P sink to the sediments and the rapidity of processes regenerating the P by mineralization in both inorganic and organic forms depends on an array of physical, chemical, and biological factors. The P content of the sediments is often several orders of magnitude greater than that of the overlying water.

Phosphorus inputs into lake and wetland sediments occur by several mechanisms (Williams and Mayer, 1972; Boström et al., 1988a; Wetzel, 1990).

1. Sedimentation of P minerals imported from the drainage basin
2. Adsorption or precipitation of P with inorganic compounds by coprecipitation with Fe and Mn, adsorption to clays, oxyhydroxides, and similar materials, and associations with carbonates
3. Sedimentation of P with organic matter imported from the drainage basin or produced within the lake or wetland

About one-third of the P of sediments is as organic compounds (Williams and Mayer, 1972; Boström et al., 1982). Abundance of organic P is greatest in the surficial sediments and declines with depth as a greater percentage is in apatite and nonapatite P materials. Of the organic P fraction, much is likely associated with humic-fulvic acid complexes. Less than 10% of the organic complexes are composed of inositol- and hexaphosphates. Within lake basins, however, the total organic P content of the sediments can vary greatly in relation to basin morphology, hydraulic loadings, water hydrodynamics, trophic state, and other factors.

Once the P is within the sediment, exchanges across the sediment-water interface are regulated by mechanisms associated with mineral-water equilibria, sorption processes (particularly ion exchange), oxygen and other electron acceptor-dependent redox interactions, and the physiological and behavioral activities of many biota from bacteria to fish (Fig. 9.2). Exchange rates between various deposits of P and the interstitial water of the sediments depend on local adsorption and diffusion coefficients and their alteration by enzyme-mediated reactions of the microbiota.

Phosphate is bound to particles by physical adsorption as well as chemical binding of different strengths (complex, covalent, and ionic bonds). Physical and chemical mobilization include desorption, dissolution of P-containing compounds, particularly assisted by microbially mediated acidity, and ligand exchange mechanisms between P and hydroxide ions or organic chelating compounds (Boström et al. 1982; Stumm and Morgan 1996). Microbial biochemical mobilization processes include mineralization by hydrolysis of phosphate-ester bonds, release of P from living cells as a result of changing environmental conditions, particularly redox, and autolysis of cells.

9.6.1 *Movement of P Across the Sediment-Water Interface*

Movement of P from the sediments to the overlying water depends on hydrodynamic and biotic mechanisms that transport dissolved inorganic and organic P from the sediments. Because of steep concentration gradients of P between interstitial water and the overlying water, molecular diffusion is a primary transport mechanism (Fig. 9.2). Currents from wind-induced water turbulence can disrupt gradients and resuspend sediment particles. Similarly disturbance of sediments by benthic invertebrates living on or in the sediments and by bottom-feeding fishes can, when occurring in large densities, cause appreciable bioturbation of sediments. Microbial generation of gases, particularly CO_2, CH_4, and N_2, can accumulate, form bubbles, and rise to the surface. Such ebullition can disrupt gradients and accelerate diffusion of P compounds upward. The metabolism and growth of plants living on and within the sediments can both suppress and enhance the transport of P across the sediment-water interface. Each of these processes is discussed in detail by Wetzel (1983, 1990, 1998).

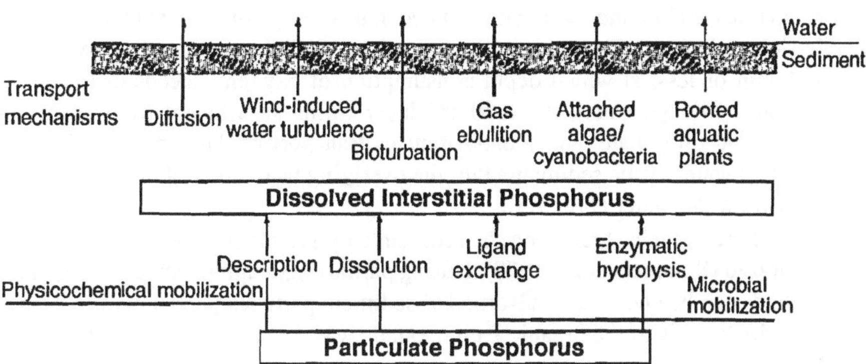

FIGURE 9.2 Processes involved in the mobilization of P from particulate stores into dissolved states of interstitial water of the sediments, and transport across the sediment-water interface into overlying water. (Modified extensively from Boström et al., 1982, 1988, and Wetzel, 1990.)

9.6.2 *Microbial Controls*

Most, but not all, of the mineralization processes of organic P compounds in sediments are enzymatically hydrolyzed by bacteria. Both the types of microbial communities and their rates of metabolism are strongly regulated by the dynamics of the markedly stratified redox conditions within sediments. This subject has been reviewed extensively; the ensuing summary is directed toward mineralization of organic P compounds.

Very few data exist on the occurrence and distribution of phosphatases or other organic P hydrolyzing enzymes in sediments or associated with bacteria of sediments. Expectedly, a positive correlation has been found between bacterial biomass and alkaline phosphatase activity in surficial lake sediments (Reichardt, 1978). At several centimeters below the sediment-water interface, correlations were less strong between enzyme activities and bacterial biomass. A higher affinity of phosphatases was found in sediments where a low mobility of sediment P occurred (Pettersson, unpublished; cited in Boström et al., 1982). The lowest substrate affinity with highest half-saturation constants was found in organic sediments with low chemical adsorption capacities that release P under anaerobic conditions. Low availability of P in sediments induced production of phosphatase activity with high affinity for substrates.

The oxygen content of the microzone (several millimeters) at the sediment-water interface is influenced predominately by bacterial metabolism but can be altered appreciably by the metabolism of algae, fungi, planktonic invertebrates that migrate to and live within the interface, and by sessile benthic invertebrates. Microbial degradation of dead particulate organic matter that settles onto the sediments is the primary consumer of oxygen. Much of the more labile organic fractions, largely of plant origin, is released from the senescing tissues before and while settling to the sediments. Hence, the sediments often receive organic P residues that are relatively resistant to rapid decomposition.

High sediment demand for oxygen results from respiratory metabolism, slow rates of diffusion, and inorganic elements [e.g., Fe(II)] that accumulate in reduced form when released into the sediment from decomposition of organic matter. In most organic-rich sediments that are not illuminated by sunlight, sediment is oxidized only to 1 mm or less, at which depth consumption of oxygen exceeds diffusional replacement. Although small, the oxidized layer forms an efficient trap for iron, manganese, and phosphate, and greatly reduces transport of these materials from the interstitial water of the sediments into the overlying water as well as scavenging phosphate from the water (Fig. 9.3).

Much of the organic P reaching the sediments by sedimentation is decomposed and hydrolyzed (Sommers et al., 1970). During partial degradation, however, organic P compounds can be chemically adsorbed to sediment particles (Rodel et al., 1977). Enzymatic hydrolysis of the organic P compounds was appreciably reduced by such complexation.

Much of the sediment P is inorganic (e.g., apatite), derived from the drainage basin and phosphate adsorbed to clays and Al and ferric hydroxides (Frevert, 1979a, 1979b; Detenbeck and Brezonik, 1991; Andersen and Jensen, 1992; Danen-Louwerse et al., 1993), and particularly covalently bonded to hydrated Fe oxides (Golter-

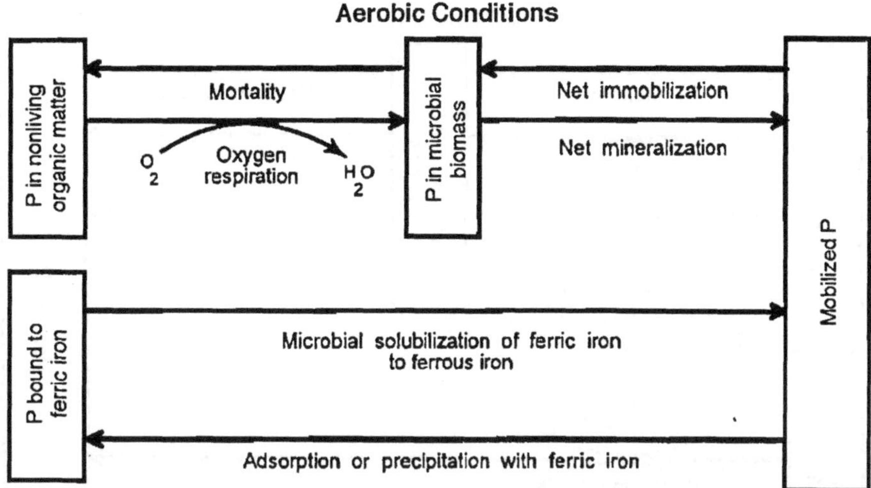

FIGURE 9.3 Potential transfer processes among phosphorus pools in lake sediments under aerobic conditions. (Modified from Boström et al., 1988a.)

man, 1995). Additionally, phosphate coprecipitates with Fe, Mn, and carbonates (Otsuki and Wetzel, 1972; Boström et al., 1988a). Within the sediments, the reduction of ferric hydroxides and complexes results in ferrous Fe and adsorbed phosphate to be mobilized into the interstitial water (Fig. 9.4). Under very reducing conditions where sulfate is reduced to hydrogen sulfide, some FeS is precipitated. Sufficient Fe can be removed by FeS precipitation to enhance the release of phosphate into the interstitial waters (Wetzel, 1983, 1998).

When mobilization of sediment-associated phosphate is based on a molar ratio of PO_4^{3-} liberated per unit of organic carbon mineralized by bacterial metabolism, the availability of alternate electron acceptor compounds, such as sulfate, becomes more important within the interstitial waters of the sediments than the oxidative conditions of the overlying water (Hasler and Einsele, 1948; Sugawara et al., 1957; Caraco et al., 1990; Roden and Edmonds, 1997). The relative PO_4^{3-} release from sediments can be significantly higher as sulfate concentrations increase from natural or anthropogenic sources, particularly in oligotrophic, softwater lakes.

Iron sulfide formation coupled to sulfate reduction can reduce the abundance of Fe compounds that can complex PO_4^{3-} into sediment porewater. When sulfate content is low or absent in anaerobic sediments, microbial reduction generates Fe(II) compounds from microbial reduction of Fe(III) oxide. PO_4^{3-} can be retained effectively with Fe(II) compounds, but Fe-associated PO_4^{3-} is quantitatively released when amorphous Fe(III) oxide of sediment is converted to iron sulfides (Roden and Edmonds, 1997). Conversion of reactive sediment Fe compounds to iron sulfides by sulfate-reducing bacteria lead to more efficient release of Fe-associated PO_4^{3-} than does direct microbial Fe(III) oxide reduction (Fig. 9.5).

Iron also interacts with dissolved humic substances to bind P at acidic to near-neutral pH values, irrespective of redox conditions, in the surficial sediments of

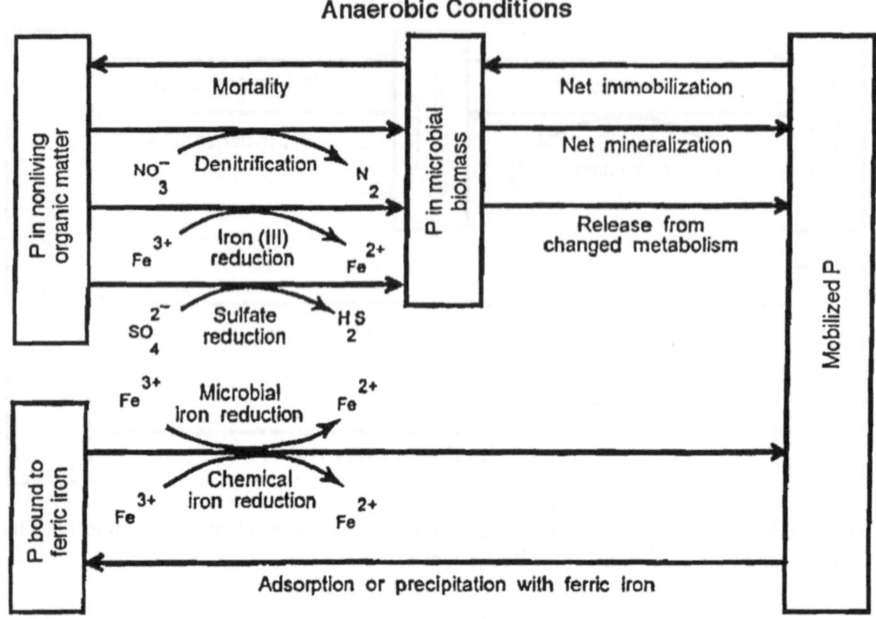

FIGURE 9.4 Potential transfer processes among phosphorus pools in lake sediments under anaerobic conditions. (Modified from Boström et al., 1988a.)

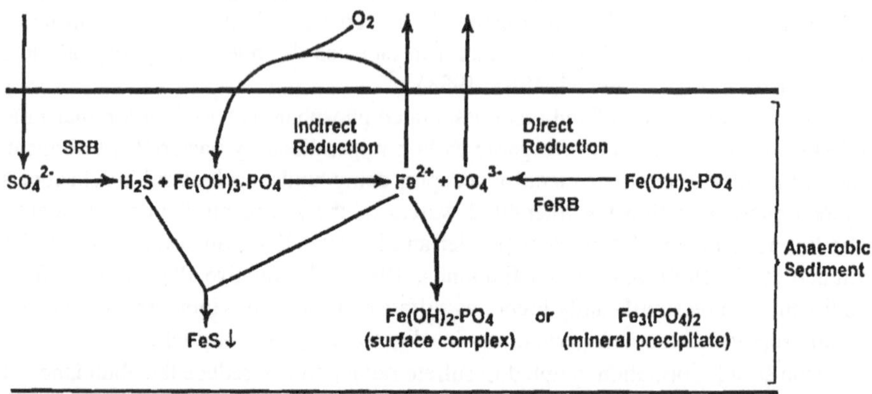

FIGURE 9.5 Interactions of sulfate on the reduction of Fe(III) and release of Fe(II) and phosphate. *SRB* = sulfate-reducing bacteria; *FeRB* = iron-reducing bacteria. (Modified from Roden and Edmonds, 1997.)

certain soft waters (Jackson and Schindler, 1975; Francko, 1986). This process should enhance the retention of P by sediments in waters in which high loadings of Fe and humic substances occur.

Other processes can contribute to the composite release of Fe-associated PO_4^{3-} from sediments. For example, uptake of excessive amounts of P by aerobic bacteria (e.g., Fleischer, 1986) and storage as polyphosphates could be rapidly degraded and released with the onset of anaerobic conditions (DeMontigny and Prairie, 1993; Gächter and Meyer, 1993). Because organic sediments are aerobic for only one or very few millimeters, this contribution is likely small in comparison to other mechanisms.

Additionally, solubilization of phosphate precipitates by aerobic bacteria can occur by carbohydrate metabolism and release of organic acids (Harrison et al., 1972). The organic acids function as both to make precipitates soluble and to act as chelating agents for cations and thus allow mobilization of phosphates. Of the phosphate precipitates analyzed, solubilization of $CaHPO_4 > Ca_3\ (PO_4)_2 > FePO_4 > Mg_3\ (PO_4)_2 > Al_3\ (PO_4)_2$.

The sorption capacities of Fe(III) oxide decrease as pH levels increase above 6.5 (Stumm and Morgan, 1996). Large increases in pH values can occur within the sediments for several millimeters as a result of diurnal fluctuations in photosynthesis by epipelic algae (Fig. 9.6) and by submersed aquatic macrophytes and associated epiphytic algae. Simultaneous increases in oxygen concentrations (Fig. 9.6) and adsorption to or with photosynthetically-induced precipitation of $CaCO_3$ (e.g., Mickle and Wetzel, 1978) can counteract the availability of desorbed PO_4^{3-}.

If the sediments receive light, even at very low intensities (<30 μmol quanta m^{-2} s^{-1}), photosynthesis of epipelic algal communities growing on the sediments can quickly (minutes) produce high, often markedly supersaturated (200 to 300% saturation), concentrations of oxygen within the community usually less than 2 mm in thickness (Fig. 9.6) (Revsbech et al., 1983; Carlton and Wetzel, 1987). This oxygen can diffuse several millimeters into the interstitial water of the supporting sediments at rates greater than are consumed by bacterial respiration and chemical oxidations. By this mechanism, diurnal changes occur in the oxidized microzone of the sediments from fully oxidized with a five- to tenfold increase in oxidized depth during

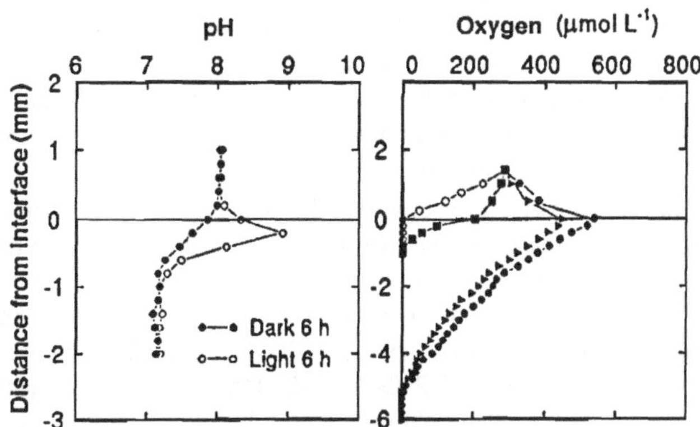

FIGURE 9.6 Distribution of pH (left, illumination 30 μmol quanta m^{-2} s^{-1}) and dissolved oxygen (right) immediately above and within sediments colonized by epipelic microalgae and bacterial communities. Oxygen microprofiles (right) during darkness (o——o), and after 1 h (■——■), 8 h (△——△), and 10 h (●——●) illumination with 10 μmol quanta m^{-2} s^{-1}. (Extracted from data of Carlton and Wetzel, 1987, 1988.)

the daylight hours to fully reducing at nighttime. It was shown experimentally that the rate of P efflux from sediments to the overlying water is inversely related to the extent of sediment oxygenation and the magnitude of epipelic algal photosynthesis (Carlton and Wetzel, 1988). Although much of the reduced P efflux is caused by direct chemical redox changes, the microbial community also assimilates and complexes nutrients in organic compounds (cf. Kelderman et al., 1988; Hansson, 1989).

In bacterially regulated (nonilluminated, nonphotosynthetic) organic sediments, rates of bacterial metabolism and oxygen consumption increased with increasing temperatures, particularly above 17° C (Kamp-Nielsen, 1975). Above this temperature, release of P to the overlying water occurred as the oxidized microzone deteriorated even though dissolved oxygen occurred in the overlying water. Presumably, diffusion of oxygen from the overlying water was insufficient to compensate for the microbial consumption. In contrast, Carlton and Wetzel (1988) found that when sediments were weakly illuminated (even less than 10 μmol quanta m^{-2} s^{-1}) and supported a modest epipelic algal community, oxygen production at the surface of the sediments and *within* the sediments quickly (minutes) supersaturated, oxidized the sediments as well as increased pH markedly, and suppressed phosphate release into the overlying water. This process was reversed within minutes following darkness. Importantly, the effectiveness of the diurnal shifts between highly aerobic to totally anaerobic sediments was optimal at 17° C and above and became weakly evident at 11° C or less (Carlton and Wetzel, 1988).

9.6.3 Microbial Storage and Release of P

Rapid uptake of P by bacteria occurs under aerobic conditions and results in storage of P in polyphosphate granules (Marais et al., 1983; Florentz et al., 1984). Under anaerobic conditions P of polyphosphates are likely hydrolyzed for ATP and energy storage products (e.g., poly-β-hydroxybutyrate) synthesis (Wentzel et al., 1986). An excess of PO_4^{3-} in the cytoplasm can induce increased osmotic pressures and diffusion out of the cells. Similarly Fleischer (1983, 1986) demonstrated that facultative anaerobic bacteria are able to rapidly assimilate P from solubilization of Fe(III) and release excess P during shifts between aerobic and anaerobic conditions. Phosphorus of microbial biomass in surficial sediments can be substantial, in the range of 10 to 15% of the total phosphorus of sediments (Boström, 1988).

Clearly, particulate organic P contains the bulk of the total organic P of sediments. In addition to releases as reactive inorganic soluble phosphate (= orthophosphate), which has an extremely short turnover time, two soluble organic P fractions are common (Lean, 1973; Boström, et al. 1988b). The soluble organic fractions include a low molecular weight (approx. 250 D) organic P compound (XP) and soluble macromolecular colloidal P (> 5 × 10^6 D). An exchange mechanism predominated between the inorganic phosphate and the particulate fractions, but some P is excreted by microorganism as the XP compounds (Fig. 9.7). Polycondensation of the XP compound can produce the high molecular weight colloidal compound.

Exchange mechanisms and rate constants for fluxes, illustrated in Figure 9.7, are taken largely from studies in water rather than sediments. Very few kinetic analyses exist for fluxes of organic P within sediments. Extrapolation of rates from

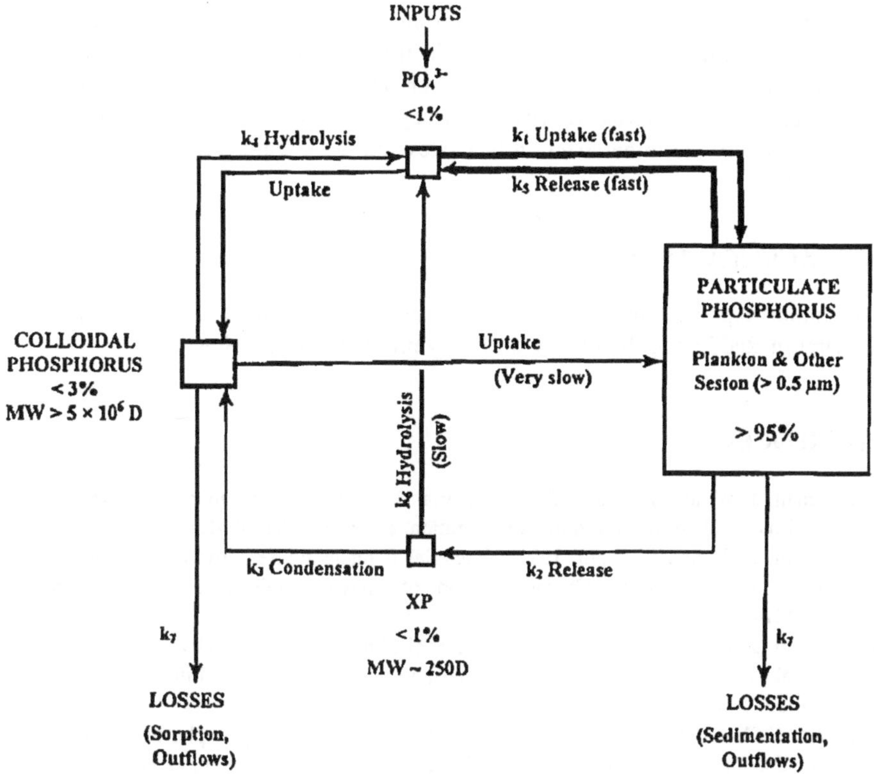

FIGURE 9.7 Exchange mechanisms among phosphate and the particulate fractions. (Modified and expanded from Lean 1973.)

the water to those within sediments may be done cautiously to yield some insights on possible rates. As in water, most of the organic P in sediments is in particulate particles, either in microbial mass or adsorbed to inorganic or detrital organic particles (cf. Rijkeboer et al., 1991). The release of extracellular organic P compounds (XP) is common, but their condensation to high molecular weight colloidal P compounds is very rapid (minutes). Both XP and colloidal-P fractions are not readily utilized as direct P sources by microorganisms, and they resist rapid attack from enzymes capable of hydrolyzing the most common organic P compounds (Olsson and Jansson, 1984). Release of inorganic phosphate by particulate organic matter of microbiota is generally much greater than excretion of soluble organic P (XP). Mass balance calculations suggest k_5 to be about 70 times greater than k_2 (Fig. 9.7). The direct hydrolysis of XP to phosphate was insignificant in comparison to release via the colloidal form.

Phosphorus mobilization within sediments by mineralization processes can result in large quantities of inorganic and organic P in soluble and stabilized forms. Adsorption processes and fixed pools in biota and organic detritus can persist for long periods of time. Such accumulation can fluctuate suddenly on a seasonal basis as chemical environmental conditions of the overlying water alter retention properties of the sediments. In addition, however, much more rapid flux dynamics occur in many sediments as mediated by a combination of photosynthetic and mineralization processes that are intricately coupled on a diurnal basis. The introduction of natural or artificial alternative electron acceptors, such as nitrate or sulfate, can radically modify the mineralization flux rates and retention capacities of organic P in sediments.

9.7 ACKNOWLEDGMENT

The author expresses his appreciation for the encouragement and enthusiasm of K. Ramesh Reddy and the review of this synthesis by Susan Newman.

REFERENCES

Ammerman, J. W. and F. Azam. 1985. Bacterial 5'-nucleotidase in aquatic ecosystems: A novel mechanism of phosphorus regeneration. Science 227:1338–1340.

Ammerman, J. W. and F. Azam. 1991a. Bacterial 5'-nucleotidase activity in estuarine and coastal marine waters: Characterization of enzyme activity. Limnol. Oceanogr. 36:1427–1436.

Ammerman, J. W. and F. Azam. 1991b. Bacterial 5'-nucleotidase activity in estuarine and coastal marine waters: Role in phosphorus regeneration. Limnol. Oceanogr. 36:1437–1447.

Andersen, F. Ø. and J. S. Jensen. 1992. Regeneration of inorganic phosphorus and nitrogen from decomposition of seston in a freshwater sediment. Hydrobiologia 228:71–81.

Anderson, G. and R. J. Hance. 1963. Investigation of an organic phosphorus component of fulvic acid. Plant Soil 19:296–303.

Anderson, G., E. G. Williams, and J. O. Moir. 1974. A comparison of the sorption of inorganic orthophosphate and inositol hexaphosphate by six acid soils. J. Soil Sci. 25:51–62.

Bentzen, E. and W. D. Taylor. 1991. Estimating organic P utilization by freshwater plankton using [^{32}P]ATP. J. Plankton Res. 13:1223–1238.

Boavida, M. J. and R. G. Wetzel. 1997. Efficacy of phosphatase activity in freshwater ecosystems: Inhibition by dissolved humic substances and UV reactivation. Freshwat. Biol. (Submitted)

Boström, B. 1988. Relations between chemistry, microbial biomass and activity in sediments of a polluted vs. a nonpolluted eutrophic lake. Verh. Internat. Verein. Limnol. 23:451–459.

Boström, B., M. Jansson, and C. Forsberg. 1982. Phosphorus release from lake sediments. Arch. Hydrobiol. Beih. Ergebn. Limnol. 18:5–59.

Boström, B., J. M. Andersen, S. Fleischer, and J. Jansson. 1988a. Exchange of phosphorus across the sediment-water interface. Hydrobiologia 170:229–244.

Boström, B., G. Persson, and B. Broberg. 1988b. Bioavailability of different phosphorus forms in freshwater systems. Hydrobiologia 170:133–155.

Caraco, N., J. Cole, and G. E. Likens. 1990. A comparison of phosphorus immobilization in sediments of freshwater and coastal marine systems. Biogeochemistry 9:277–290.

Carlton, R. G. and R. G. Wetzel. 1987. Distributions and fates of oxygen in periphyton communities. Can. J. Bot. 65:1031–1037.

Carlton, R. G. and R. G. Wetzel. 1988. Phosphorus flux from lake sediments: Effects of epipelic algal oxygen production. Limnol. Oceanogr. 33:562–570.

Cembella, A. D., N. J. Antia, and P. J. Harrison. 1983. The utilization of inorganic and organic phosphorous compounds as nutrients by eukaryotic microalgae: A multidisciplinary perspective: Part 1. CRC Crit. Rev. Microbiol. 10:317–391.

Cembella, A. D., N. J. Antia, and P. J. Harrison. 1984. The utilization of inorganic and organic phosphorous compounds as nutrients by eukaryotic microalgae: A multidisciplinary perspective: Part 2. CRC Crit. Rev. Microbiol. 11:13–81.

Condron, L. M., K. M. Goh, and R. H. Newman. 1985. Nature and distribution of soil phosphorus as revealed by a sequential extraction method followed by ^{31}P nuclear magnetic resonance analysis. J. Soil. Sci. 36:199–207.

Cotner, J. B. and R. T. Heath. 1988. Potential phosphate release from phosphomonoesters by acid phosphatase in a bog lake. Arch. Hydrobiol. 111:329–338.

Cotner, J. B. and R. G. Wetzel. 1991a. Bacterial phosphatase from different habitats in a small, hardwater lake. In: R. J. Chróst, Editor. Microbial Enzymes in Aquatic Environments. Springer-Verlag, New York. pp. 187–205.

Cotner, J. B. and R. G. Wetzel. 1991b. 5'-nucleotidase activity in a eutrophic lake and an oligotrophic lake. Appl. Environ. Microbiol. 57:1306–1312.

Danen-Louwerse, H., L. Lijklema, and M. Coenraats. 1993. Iron content of sediment and phosphate adsorption properties. Hydrobiologia 253:311–317.

DeMontigny, C. and Y. T. Prairie. 1993. The relative importance of biological and chemical processes in the release of phosphorus from a highly organic sediment. Hydrobiologia 253:141–150.

Detenbeck, N. E. and P. L. Brezonik. 1991. Phosphorus sorption by sediments from a soft-water seepage lake. 2. Effects of pH and sediment composition. Environ. Sci. Technol. 25:403–409.

Di, H. J., L. M. Condron, and E. Frossard. 1997. Isotope techniques to study phosphorus cycling in agricultural and forest soils. Biol. Fertil. Soils 24:1–12.

Elliot, E. T., R. V. Anderson, D. C. Coleman, and C. V. Cole. 1980. Habitable pore space and microbial trophic interactions. Oikos 35:327–335.

Fleischer, S. 1983. Microbial phosphorus release during enhanced glycolysis. Naturwissenschaften 70:415.

Fleischer, S. 1986. Aerobic uptake of Fe(III)-precipitated phosphorus by microorganisms. Arch. Hydrobiol. 107:269–277.

Florentz, M., P. Granger, and P. Hartemann. 1984. Use of ^{31}P nuclear magnetic resonance and electron microscopy to study phosphorus metabolism of microorganisms from wastewater. Appl. Environ. Microbiol. 47:519–525.

Francko, D. A. 1986. Epilimnetic phosphorus cycling: Influence of humic materials and iron on coexisting major mechanisms. Can. J. Fish. Aquat. Sci. 43:302–310.

Francko, D. A. and R. T. Heath. 1979. Functionally distinct classes of complex phosphorus compounds in lake water. Limnol. Oceanogr. 24:463–473.

Francko, D. A. and R. T. Heath. 1982. UV-sensitive complex phosphorus: Association with dissolved humic material and iron in a bog lake. Limnol. Oceanogr. 27:564–569.

Frevert, T. 1979a. Phosphorus and iron concentrations in the interstitial water and dry substance of sediments of Lake Constance (Obersee). I. General discussion. Arch. Hydrobiol./Suppl. 55:298–323.

Frevert, T. 1979b. The p_e redox concept in natural sediment-water systems; its role in controlling phosphorus release from lake sediments. Arch. Hydrobiol./Suppl. 55:278–297.

Gächter, R. and J. S. Meyer. 1993. The role of microorganisms in mobilization and fixation of phosphorous in sediments. Hydrobiologia 253:103–121.

Golterman, H. L. 1995. The role of the iron hydroxide-phosphate-sulfide system in the phosphate exchange between sediments and overlying water. Hydrobiologia 297:43–54.

Hannapel, R. J., W. H. Fuller, S. Bosma, and J. S. Bullock. 1964a. Phosphorus movement in a calcareous soil. I. Predominance of organic forms of phosphorus in phosphorus movement. Soil Sci. 97:350–357.

Hannapel, R. J., W. H. Fuller, and R. H. Fox. 1964b. Phosphorus movement in a calcareous soil. II. Soil microbial activity and organic phosphorus movement. Soil Sci. 97:421–427.

Hansson, L.-A. 1989. The influence of a periphytic biolayer on phosphorus exchange between substrate and water. Arch. Hydrobiol. 115:21–26.

Harrison, M. J., R. W. Pacha, and R. Y. Morita. 1972. Solubilization of inorganic phosphate by bacteria isolated form Upper Klemath lake sediment. Limnol. Oceanogr. 17:50–57.

Hasler, A. D. and W. G. Einsele. 1948. Fertilization for increasing productivity of natural inland waters. Trans. N. Amer. Wildl. Conf. 13:527–555.

Jackson, T. A. and D. W. Schindler. 1975. The bio-geochemistry of phosphorus in an experimental lake environment: Evidence for the formation of humic-metal-phosphate complexes. Verh. Internat. Verein. Limnol. 19:211–221.

Jansson, M. 1981. Induction of high phosphatase activity by aluminum in acid lakes. Arch. Hydrobiol. 93:32–44.

Jayachandran, K., A. P. Schwab, and B. A. D. Hetrick. 1992. Mineralization of organic phosphorus by vesicular-arbuscular mycorrhizal fungi. Soil Biol. Biochem. 24:897–903.

Johnson, J. F., D. L. Allan, C. P. Vance, and G. Weiblen. 1996. Root carbon dioxide fixation by phosphorus-deficient *Lupinus albus*: Contribution to organic acid exudation by proteoid roots. Plant Physiol. 112:19–30.

Jones, R. I., K. Salonen, and H. de Haan. 1988. Phosphorus transformations in the epilimnion of humic lakes: Abiotic interactions between dissolved humic materials and phosphate. Freshwat. Biol. 19:357–369.

Kamp-Nielsen, L. 1975. A kinetic approach to the aerobic sediment-water exchange of phosphorus in Lake Esrom. Ecol. Modelling 1:153–160.

Kelderman, P., H. J. Lindeboom, and J. Klein. 1988. Light dependent sediment-water exchange of dissolved reactive phosphorus and silicon in a producing microflora mat. Hydrobiologia 159:137–147.

Kim, B. and R. G. Wetzel. 1993. The effect of dissolved humic substances on the alkaline phosphatase and the growth of microalgae. Verh. Internat. Verein. Limnol. 25:129–132.

Koenings, J. P. and F. F. Hooper. 1976. The influence of colloidal organic matter on iron and iron-phosphorus cycling in an acid bog lake. Limnol. Oceanogr. 21:684–696.

Kowalenko, C. G. 1978. Organic nitrogen, phosphorus and sulfur in soils. In: M. Schnitzer and S. U. Khan, Editors. Soil Organic Matter. Elsevier Publishers, New York. pp. 95–136.

Lean, D. R. S. 1973. Movements of phosphorus between its biologically important forms in lake water. J. Fish. Res. Bd. Can. 30:1525–1536.

Linquist, B. A., P. W. Singleton, R. S. Yost, and K. G. Cassman. 1997a. Aggregate size effects on the sorption and release of phosphorus in an Ultisol. Soil Sci. Soc. Am. J. 61:160–166.

Linquist, B. A., P. W. Singleton, and K. G. Cassman. 1997b. Inorganic and organic phosphorus dynamics during a build-up and decline of available phosphorus in an Ultisol. Soil Sci. 162:254–264.

Magid, J., N. Christensen, and H. Nielsen. 1992. Measuring phosphorus fluxes through the root zone of a layered sandy soil: Comparisons between lysimeter and suction cell solution. J. Soil. Sci. 43:739–747.

Magid, J, H. Tiessen, and L. M. Condron. 1996. Dynamics of organic phosphorus in soils under natural and agricultural ecosystems. In: A. Piccolo, Editor. Humic Substances in Terrestrial Ecosystems. Elsevier Science B.V., Amsterdam. pp. 429–466.

Marais, G. V. R., R. E. Loewenthal, and I. P. Siebritz. 1983. Observations supporting phosphate removal by biological excess uptake—a review. Wat. Sci. Technol. 15:15–41.

Mickle, A. M. and R. G. Wetzel. 1978. Effectiveness of submersed angiosperm-epiphyte complexes on exchange of nutrients and organic carbon in littoral systems. I. Inorganic nutrients. Aquat. Bot. 4:303–316.

Münster, U. 1994. Studies on phosphatase activities in humic lakes. Environ. Int. 20:49–59.

Olsson, H. and M. Jansson. 1984. Stability of dissolved ^{32}P-labelled phosphorus compounds in lake water and algal cultures—resistance to enzymatic treatment and algal uptake. Verh. Internat. Verein. Limnol. 22:200–204.

Otsuki, A. and R. G. Wetzel. 1972. Coprecipitation of phosphate with carbonates in a marl lake. Limnol. Oceanogr. 17:763–767.

Pettersson, K. 1980. Alkaline phosphatase activity and algal surplus phosphorus as phosphorus-deficiency indicators in Lake Erken. Arch. Hydrobiol. 89:54–87.

Rai, H. and T. R. Jacobsen. 1993. Dissolved alkaline phosphatase activity (APA) and the contribution of APA by size fractionated plankton in Lake Schöhsee. Verh. Internat. Verein. Limnol. 25:85–89.

Reichardt, W. 1978. Responses of phosphorus remobilizing *Cytophaga* species to nutritional and thermal stress. Verh. Internat. Verein. Limnol. 20:2227–2232.

Reichardt, W., J. Overbeck, and L. Steubing. 1967. Free dissolved enzymes in lake waters. Nature 216:1345–1347.

Revsbech, N. P., B. B. Jørgensen, T. H. Blackburn, and Y. Cohen. 1983. Microelectrode studies of the photosynthesis and O_2, H_2S, and pH profiles of a microbial mat. Limnol. Oceanogr. 28:1062–1074.

Rijkeboer, M., F. de Bles, and H. J. Gons. 1991. Role of sestonic detritus as a P-buffer. Mem. Ist. Ital. Idrobiol. 48:251–260.

Rodel, M. G., D. E. Armstrong, and R. F. Harris. 1977. Sorption and hydrolysis of added organic phosphorus compounds in lake sediments. Limnol. Oceanogr. 22:415–422.

Roden, E. E. and J. W. Edmonds. 1997. Phosphate mobilization in anaerobic sediments: Microbial Fe(III) oxide reduction versus iron-sulfide formation. Biogeochemistry (In press)

Ryszkowski, L., J. Karg, B. Szpakowska, and I. Zyczynska-Baloniak. 1989. Distribution of phosphorus in meadow and cultivated field ecosystems. In: H. Tiessen, Editor. Phosphorus Cycles in Terrestrial and Aquatic Ecosystems. Saskatchewan Institute of Pedology, Saskatoon. pp. 178–192. (Cited in Magid, et al. 1996.)

Shaw, P. J. 1994. The effect of pH, dissolved humic substances, and ionic composition on the transfer of iron and phosphate to particulate size fractions in epilimnetic lake water. Limnol. Oceanogr. 39:1734–1743.

Sommers, L. E., R. F. Harris, J. D. H. Williams, D. E. Armstrong, and J. K. Syers. 1970. Determination of total organic phosphorus in lake sediments. Limnol. Oceanogr. 15:301–304.

Stewart, A. J. and R. G. Wetzel. 1981. Dissolved humic materials: Photodegradation, sediment effects, and reactivity with phosphate and calcium carbonate precipitation. Arch. Hydrobiol. 92:265–286.

Stewart, A. J. and R. G. Wetzel. 1982a. Phytoplankton contribution to alkaline phosphatase activities. Arch. Hydrobiol. 93:265–271.

Stewart, A. J. and R. G. Wetzel. 1982b. Influence of dissolved humic materials on carbon assimilation and alkaline phosphatase activity in natural algal-bacterial assemblages. Freshwat. Biol. 12:369–380.

Stewart, J. W. B. and H. Tiessen. 1987. Dynamics of soil organic phosphorus. Biogeochemistry 4:41–60.

Stumm, W. and J. J. Morgan. 1996. Aquatic Chemistry: Chemical Equilibria and Rates in Natural Waters. 3rd Edition. Wiley Interscience, New York. 1022 pp.

Sugawara, K., T. Koyama, and E. Kamata. 1957. Recovery of precipitated phosphate from lake muds related to sulfate reduction. Chem. Inst. Fac. Sci. Nagoya Univ. 5:60–67.

Tamminen, T. 1989. Dissolved organic phosphorus regeneration by bacterioplankton: 5'-nucleotidase activity and subsequent phosphate uptake in a mesocosm enrichment experiment. Mar. Ecol. Progr. Ser. 58:89–100.

Tarafdar, J. C. and A. Jungk. 1987. Phosphatase activity in the rhizosphere and its relation to the depletion of soil organic phosphorus. Biol. Fertil. Soils 3:199–204.

Tate, K. R. 1984. The biological transformations of P in soil. Plant Soil 76:245–256.

van Veen, J. A. and P. J. Kuikman. 1990. Soil structural aspects of decomposition of organic matter by micro-organisms. Biogeochemistry 11:213–234.

Walbridge, M. R. and P. M. Vitousek. 1987. Phosphorus mineralization potentials in acid organic soils: Processes affecting $^{32}PO_4^{3-}$ isotope dilution measurements. Soil Biol. Biochem. 19:709–717.

Wentzel, M. C., L. H. Lötter, R. E. Loewenthal, and G. v. R. Marais. 1986. Metabolic behaviour of *Acinetobacter* spp. in enhanced biological phosphorus removal—a biochemical model. Water SA 12:209–224.

Wetzel, R. G. 1981. Long-term dissolved and particulate alkaline phosphatase activity in a hardwater lake in relation to lake stability and phosphorus enrichments. Verh. Internat. Verein. Limnol. 21:337–349.

Wetzel, R. G. 1983. Limnology. 2nd Edition. Saunders College Publishing, Philadelphia. 860 pp.

Wetzel, R. G. 1990. Land-water interfaces: Metabolic and limnological regulators. Verh. Internat. Verein. Limnol. 24:6–24.

Wetzel, R. G. 1993. Humic compounds from wetlands: Complexation, inactivation, and reactivation of surface-bound and extracellular enzymes. Verh. Internat. Verein. Limnol. 25:122–128.

Wetzel, R. G. 1998. Limnology: Lake and River Ecosystems. 3rd Edition. Academic Press, New York. (In press)

Wetzel, R. G., P. G. Hatcher, and T. S. Bianchi. 1995. Natural photolysis by ultraviolet irradiance of recalcitrant dissolved organic matter to simple substrates for rapid bacterial metabolism. Limnol. Oceanogr. 40:1369–1380.

Williams, J. D. H. and T. Mayer. 1972. Effects of sediment diagenesis and regeneration of phosphorus with special reference to lakes Erie and Ontario. In: H. E. Allen and J. R. Kramer, Editors. Nutrients in Natural Waters. J. Wiley & Sons, New York. pp. 281–315.

Zhou, M., R. D. Rhue, and W. G. Harris. 1997. Phosphorus sorption characteristics of Bh and Bt horizons from sandy coastal plain soils. Soil Sci. Soc. Am. J. 61:1364–1369.

Section IV

Biogeochemical Responses

10 Influence of Phosphorus Loading on Microbial Processes in the Soil and Water Column of Wetlands

K.R. Reddy, J.R. White, A. Wright, and T. Chua

10.1 ABSTRACT

Phosphorus is often the limiting nutrient in oligotrophic wetlands. Increased P loading to these systems results in a zone with nutrient nonlimiting conditions near the source, and nutrient limiting conditions further from the source. Between these two extremes, there exists a gradient in quality and quantity of organic matter, nutrient accumulation, microbial communities, and biogeochemical cycles, resulting in diversity of microbial consortia and associated processes. These gradients are observed in the detrital layer and in the soil and water columns of many wetlands, most notably in the Everglades.

Microbial processes and associated gradients can be good candidates for indicators of ecological integrity, because they are potentially very sensitive. They are likely to be highly reliable indicators of ecosystem change, as changes at such fundamental levels can affect all higher organisms utilizing the ecosystem. Changes at higher levels, such as a decline in populations of a suite of higher organisms, may be due to factors that affect only a small portion of the biota. Changes in microbially

1-56670-331-X/99/$0.00+$.50

mediated processes can provide early warning signals or even be used to predict comprehensive alteration of biota.

Several biogeochemical parameters and associated processes are affected by P loading to P limited areas of the Everglades. Total and bicarbonate extractable P were higher in the detrital layer and soils of the impacted site than the unimpacted site. The C:P ratio of detritus and soils decreased by over 50% as a consequence of P inputs into the system. Clearly, P loading has enriched the soil in forms of P. Microbial biomass (MB) C, N, and P were also higher in the detrital layers and surface soils (0 to 10 cm depth) in P enriched areas. The P limitation to the microbial biomass has been lowered in the impacted area as the ratio of MBP/P_{total} in the detrital layer has decreased from 27 to 16%. A similar 50% decrease in MBP/P_{total} was observed in the 0 to 10 cm soil interval. Increased MB has, in turn, led to higher rates of microbial mediated processes that regulate the biogeochemical cycling of C, N, and P. Breakdown of organic matter has increased, evidenced by greater microbial respiration rates and higher activity of some extracellular enzymes. Net mineralization rates or releases of inorganic N and P were higher in soils and detrital layers at the elevated P site, increasing nutrient availability to higher plants. Nitrogen fixation, nitrification, and denitrification rates were also higher in the impacted area. Overall, P loading has increased the size of the microbial pool and organic matter turnover rates, which had led to a greater release of inorganic N and P and is further driving eutrophication.

10.2 INTRODUCTION

Wetlands host complex microbial communities, including bacteria, fungi, protozoa, and viruses. The size and the diversity of microbial communities are directly related to the quality and the quantity of resources available in the system. Soil microbial communities generally respond rapidly to perturbations such as external nutrient loading, hydrologic alterations, and fire. Anthropogenic nutrient loading from point or nonpoint sources to a P-limited wetland system such as the Everglades can alter physical, chemical, and biological properties and processes in the soil and water, which in turn can influence ecosystem function and productivity.

Many freshwater wetlands are open systems receiving inputs of C and nutrients from upstream portions of the watershed including, agricultural, and urban areas. Prolonged nutrient (such as N and P) loading to wetlands can result in distinct gradients in floodwater and soil. The degree of nutrient enrichment depends on mass loading and hydraulic retention time. This enrichment effect can be seen in many subtropical freshwater wetlands, most notably in the Everglades (Davis, 1991; Reddy et al., 1993; Craft and Richardson, 1993a, b; DeBusk et al., 1994). The Florida Everglades wetlands are historically lownutrient systems but are currently affected by nutrient loading from the adjacent Everglades Agricultural Area (EAA). A large body of information is available on productivity and aboveground nutrient cycling in freshwater wetlands (e.g., Davis and van der Valk, 1983; Howard-Williams, 1985). However, little is known of the impact of exogenous nutrients on the diversity and the ecology of microbial decomposers and periphyton communities, associated biogeochemical cycling, and long-term storage and bioavailability of nutrients. Continual nutrient loading to an oligotrophic wetland results in a zone of high nutrient availability or nonlimiting nutrient conditions near the input, and low nutrient availability or nutrient limiting conditions furthest from the input point. Between these two extremes, there exists a gradient in quality and quantity of organic matter, nutrient accumulation, microbial communities, composition, and biogeochemical cycles.

Low-nutrient systems are characterized by low external loading of nutrients and relatively closed, efficient elemental cycling (Odum, 1969, 1985). Microbial activity and plant productivity are therefore nutrient limited. In response, vascular plants, periphyton, and microbial communities are extremely efficient in utilizing and conserving nutrients through reallocation and uptake of nutrients at very low nutrient concentrations. Plant detritus in these systems generally has high C: N: P mass ratios (e.g., *Cladium* leaves, C:N:P ratio = 2520:63:1, Koch and Reddy, 1992), and decomposition of this material results in conditions where microbial and periphytic communities outcompete vascular plants for nutrients. Nutrients are held in tight, closed cycles whose efficiency enables maintenance of energy flow. The overall turnover rate of high C:N:P ratio organic matter is usually slow, and long-term decomposition may be both C and nutrient limited (Davis, 1991; DeBusk and Reddy, 1998). Environmental factors such as watertable fluctuations and fire can result in pulsed release of nutrients, which may provide a significant source of plant available nutrients in low nutrient systems (Lodge et al., 1994).

High-nutrient wetland systems are characterized by rapid turnover of C and nutrients, and by open elemental cycling, where nutrient inputs often exceed demand. These systems are low stress and contain varying degrees of internal cycling. In response, vascular plants and microbial/periphyton communities are less efficient in nutrient utilization. Plant detritus in affected areas generally has low C:N:P mass ratios (e.g., *Cladium* leaves, C:N:P ratio 360:9:1, Koch and Reddy, 1992), and high net mineralization or release of nutrients during decomposition results in decreased importance of internal cycling by microbes and plants, as compared with nutrient loading from external sources.

Although loading of anthropogenic nutrients stimulates the growth of aquatic vegetation in wetlands, a significant portion of the nutrient requirements may be met through remineralization during decomposition of organic matter. The rate of organic matter turnover and nutrient regeneration is influenced by hydroperiod (Happell and Chanton, 1993), characteristics of the organic substrates (Webster and Benfield, 1986; Enriquez et al., 1993; DeBusk and Reddy, 1998), supply of electron acceptors (D'Angelo and Reddy, 1994a, b) and addition of growth limiting nutrients (Button, 1985; McKinley and Vestal; 1992, Amador and Jones, 1995) (see Fig. 10.1.)

The Everglades is one of the unique subtropical wetland ecosystems of the world, and it evolved biologically from organic matter accumulation in a low-nutrient environment, within a limestone depression (Davis, 1943; Gleason, 1974). Historically, the major source of nutrients to the Everglades has been from atmospheric deposition, with minimum secondary nutrient inputs through infrequent sheet flooding in the northern Everglades from Lake Okeechobee. Nutrient limitation, hydrology, and fire are some of the key factors in establishment of the endemic Everglades flora, which has adapted to a low-nutrient environment (Davis, 1991). Nutrient

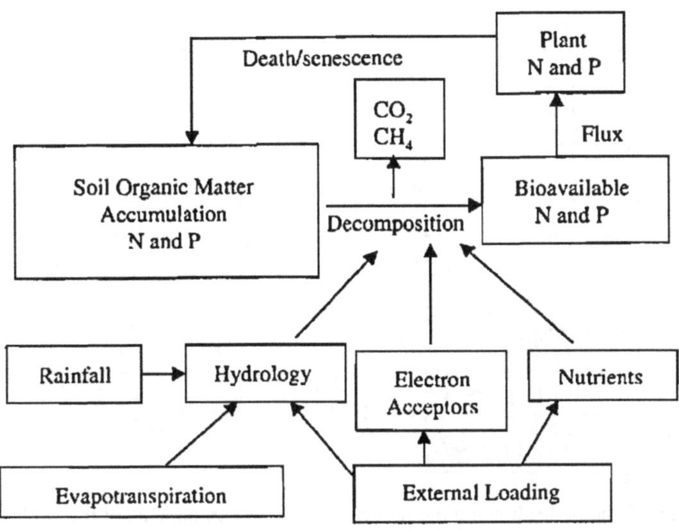

FIGURE 10.1 Schematic showing the regulators of organic matter decomposition and nutrient release in wetlands.

loading to Water Conservation Areas (WCAs) of the northern Everglades has not only altered the vegetational communities but also increased nutrient accumulation (Davis, 1991; DeBusk et al., 1994; Newman et al., 1997). Although the effects of nutrient loading and hydrology on changes in plant communities are clearly evident, very limited information is available on the influence of these factors on biogeochemical processes regulating nutrient availability and cycling in impacted and unimpacted areas. Nutrient accumulation rates of 0.11 to 1.14 g P m^{-2} yr^{-1} and 5.4 to 24.3 g N m^{-2} yr^{-1} have been reported for the Everglades (Craft and Richardson, 1993a, b; Reddy et al., 1993). The highest accumulation rates were noted in areas closer to the source of nutrient inputs, and the lowest accumulation rates occurred in areas farthest from the input points.

To evaluate nutrient impacts in wetlands, we must identify the portion of the wetland that responds rapidly, that accurately represents the impact of external loading, and that provides warning signals of ecosystem health. Changes in plant community structures are often slow and, by the time visual changes are observed, the system is severely damaged. Water column nutrient concentrations change rapidly and can be highly variable. Water column nutrients are in direct contact with the microbial communities associated with periphyton and plant detritus in the water column, and changes in composition and activities of these communities and materials may provide an indication of recent (<3 years) impact from added nutrients (Fig. 10.2). Since plant detritus and soil components function as the major storages and supplies of essential nutrients to biota, it is important to identify and quantify the key biological and chemical processes affected by nutrient loading. Carbon, N, and P cycles regulate many biogeochemical processes in the soil and water column of an ecosystem, and the majority of these biogeochemical processes are mediated by microbial activities. The biogeochemical processes that respond rapidly to nutri-

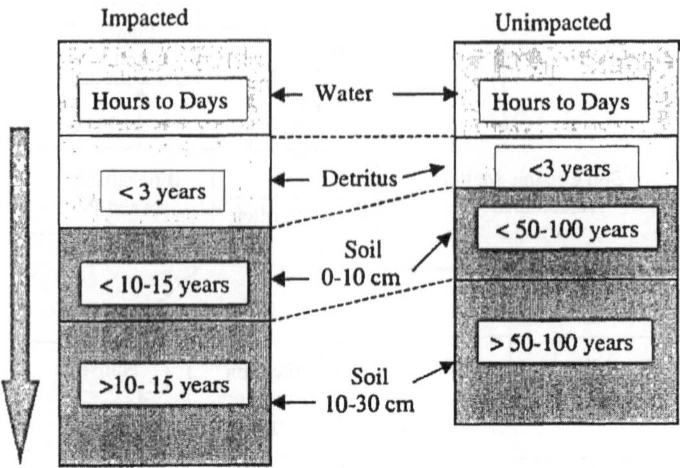

FIGURE 10.2 Schematic showing the relative age of the components of the wetland profile as affected by nutrient loading.

ent loading and related physicochemical properties of soil, detritus, and water column can be used as indicators to determine nutrient impacts. Similar approaches have been used in upland ecosystems to characterize soil quality (Torstensson, 1997).

The objectives of this chapter are to (1) review current research on the influence of P loading on selected microbial processes regulating C, N, and P cycles in the Florida Everglades, with significance to other subtropical wetlands, and (2) develop a simple impact index to determine the relative sensitivity of various biogeochemical properties to nutrient loading. This chapter summarizes the results of various studies conducted along a nutrient enrichment gradient in WCA-2a of the Everglades. Detrital layer at the impacted site consists of partially decomposed dead cattail plant tissue and floc and, at the reference site, detrital layer essentially contained partially decomposed saw grass tissue and attached benthic periphyton. Discussion is restricted to sites referred to as *impacted* (<2.3 km from source of nutrient loading) and reference (8 to 10 km from source of nutrient loading) (Fig. 10.3).

10.3 AEROBIC AND ANAEROBIC INTERFACES

Wetlands offer aerobic-anaerobic interfaces in the water column (e.g., surfaces of detrital plant tissue and benthic periphyton mats), at the soilwater interface, and in the root zone of aquatic macrophytes. The juxtaposition of aerobic and anaerobic zones in wetlands supports a wide range of microbial populations and associated metabolic activities, with oxygen reduction occurring in the aerobic interface of the substrate, and reduction of alternate electron acceptors in the anaerobic zone (Reddy and D'Angelo, 1994). Under continuously saturated soil conditions, vertical layering of different metabolic activities can be present, with oxygen reduction occurring at and just below the soilfloodwater interface. Oxygen profiles measured in WCA-2a

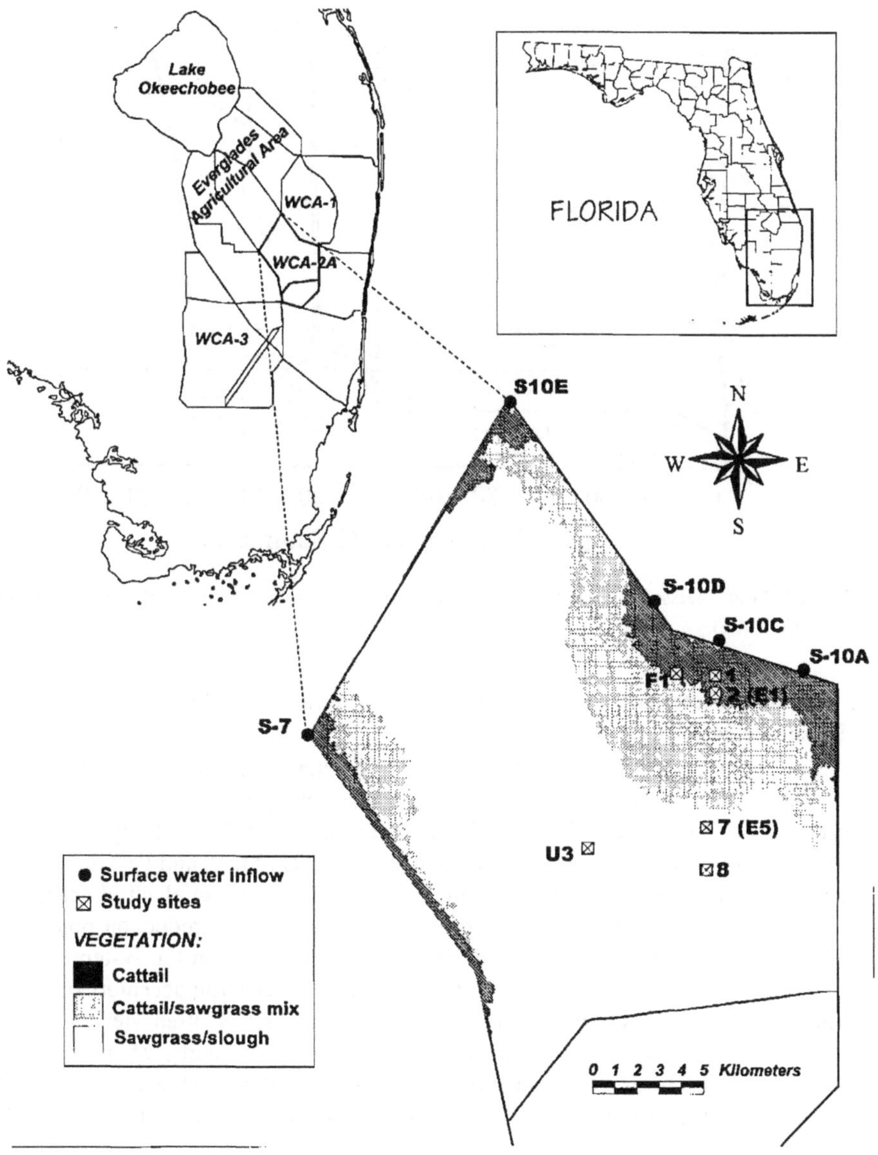

FIGURE 10.3 Map showing Water Conservation Area 2a (WCA-2a) of the Everglades. The relative position of the impacted sites F-1 and E-1 (2.3 km from the nutrient loading point) and reference sites U-3 and E-5 (10.1 km from the nutrient loading point) are shown.

soil cores (Fig. 10.4) provided evidence that algal photosynthesis maintains aerobic conditions in the litter layer during daytime, whereas oxygen depletion results from respiration at night, inducing anaerobic conditions (DeBusk, 1996). Much of the aerobic decomposition of plant detritus occurs in the water column; however, the supply of oxygen may be insufficient to meet demands and drive certain microbial

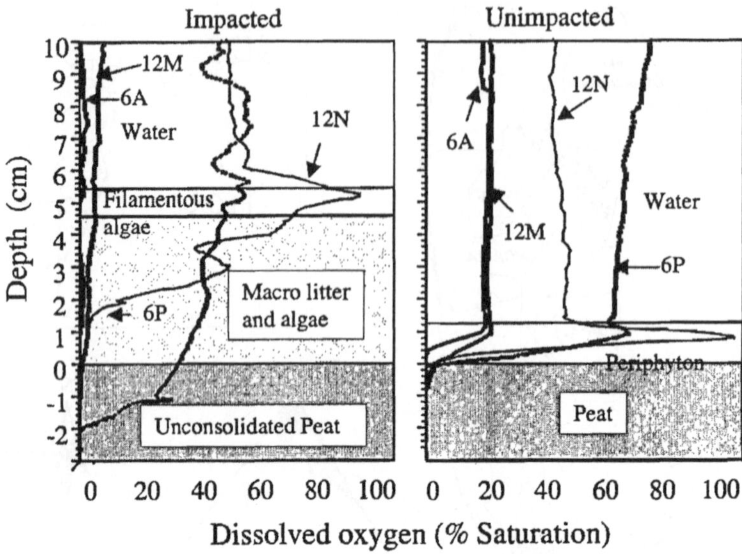

FIGURE 10.4 Dissolved oxygen profiles in soil-detrital-water column of the Everglades (DeBusk, 1996).

groups to use alternate electron acceptors, e.g., nitrate, oxidized forms of Fe and Mn, sulfate, and HCO_3.

Soil drainage adds oxygen to the soil, while other inorganic electron acceptors may be added through hydraulic loading to the system. Draining of a wetland soil accelerates organic matter decomposition due to the introduction of oxygen deeper into the profile. In many wetlands, the influence of NO_3^-, Mn^{4+}, and Fe^{3+} on organic matter decomposition is minimal, as the concentration of these electron acceptors is usually low. The demand for electron acceptors of greater reduction potentials (NO_3^-, Fe, and Mn) is high, and they are depleted rapidly from the system. Long-term sustainable microbial activity is then supported by electron acceptors of lower reduction potentials (sulfate and HCO_3). Methanogenesis is often viewed as the terminal step in anaerobic decomposition in freshwater wetlands, whereas sulfate reduction is viewed as dominant process in coastal wetlands. However, both processes can function simultaneously in the same ecosystem and compete for available substrates (D'Angelo and Reddy, 1998).

10.4 CHEMICAL COMPOSITION OF DETRITAL AND SOIL LAYERS

Agricultural drainage water loading has a minimal impact on concentrations of organic C and N in detritus and soils (Table 10.1). However, net C and N accumulation can increase with nutrient loading as a result of increased primary productivity and N assimilation (Reddy et al., 1993). Unlike C and N, P added to a wetland is usually retained within the system in organic and inorganic pools. Phosphorus

TABLE 10.1
Influence of Phosphorus Loading on Selected Parameters in the Detrital and Surface Soil Layer of the Everglades Water Conservation Area 2a

Parameter	Units	Sampling stations IS	RS	Impact index (log [IS/RS])
Detrital layer				
Ash content	g kg^{-1}	9.0	15.5	–0.24
Total C	g kg^{-1}	441	411	0.0 3
Total N	g kg^{-1}	25.8	23.9	0.03
Total P	mg kg^{-1}	1608	486	0.52
C/N		174	174	0
C/P		270	850	–0.51
N/P		16	49	–0.49
Extractable NH_4–N	mg kg^{-1}	448	272	0.22
Bicarbonate extractable P	mg kg^{-1}	95	8	1.07
0 to 10 cm soil layer				
Bulk Density	g cm^{-3}	0.06	0.06	0
Ash content	%	10.6	13.2	–0.09
Total C	g kg^{-1}	420	443	–0.02
Total N	g kg^{-1}	27.6	28.9	–0.02
Total P	mg kg^{-1}	1461	484	0.48
C/N		15	15	0
C/P		290	915	–0.51
N/P		19	60	–0.50
Extractable NH_4–N	mg kg^{-1}	114	123	–0.03
Bicarbonate extractable P	mg kg^{-1}	23	7	0.52

Values are averages of the sampling periods of Feb. 1996, Aug. 1996, and Mar. 1997. IS = sampling stations at 1.4 and 2.3 km from inflow; RS = sampling stations at 8 and 10.1 km from inflow.

loading can dramatically increase total P content of detritus and surface soil. For example, P loading resulted in accumulation of P enriched detrital material on the soil surface (Fig. 10.5). Total P concentrations increased approximately, three- to fourfold in impacted sites of WCA-2a (Table 10.1). Both extractable NH_4-N and bicarbonate extractable P increased in the detrital layer of the impacted site, reflecting a high rate of organic matter decomposition (Table 10.1).

10.5 EXTRACELLULAR ENZYME ACTIVITY

Decomposition of organic matter involves a stepwise conversion of complex organic molecules to simple organic and inorganic constituents as a result of processes including

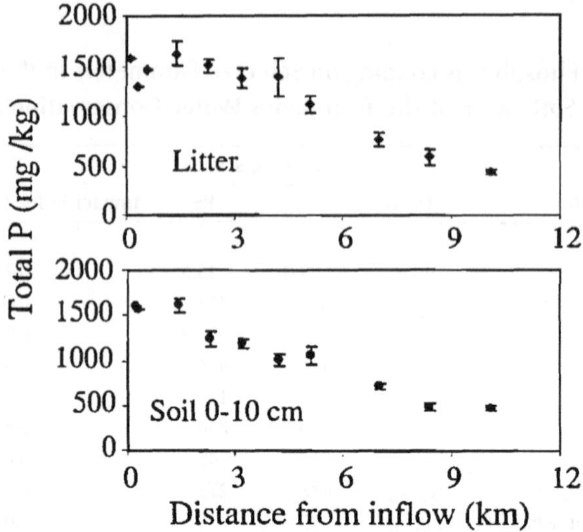

FIGURE 10.5 Total P content of detrital layer and surface soil along a nutrient enrichment gradient of the Everglades.

1. Abiotic leaching (Benner et al., 1985) and fragmentation (Boulton and Boon, 1991)
2. Extracellular enzyme hydrolysis (Cunningham and Wetzel, 1989; Sinsabaugh, 1994)
3. Aerobic and anaerobic catabolic activity of heterotrophs (Fig. 10.6) (Oremland, 1988; Kerner, 1993)

Steps 1 and 2 are generally considered to be limiting in the overall decomposition of plant detritus.

During the initial stages of microbial decomposition, complex polymers of plant detritus and soil organic matter are hydrolyzed through the activity of extracellular enzymes into simple organic molecules. These low molecular weight compounds are directly transferred into microbial cells, oxidized, and used as an energy source (Chrost, 1991). Nutrient enrichment can significantly influence the production of extracellular enzymes as a result of increased activity of microbes and plant productivity. Most models for plant detritus decomposition link degradation rates to variations in climate or composition of detrital materials rather than microbial activity (Sinsabaugh, 1994). Measurement of enzyme activity can be linked to detritus characteristics and decomposition rates. Thus, changes in extracellular enzyme activity may prove to be a sensitive indicator of wetland eutrophication.

Measurement of enzymes in detritus and soil samples obtained from WCA-2a showed that alkaline phosphatase (APA) decreased with P enrichment, whereas B-D glucosidase increased with P enrichment (Table 10.2) (Wright and Reddy, 1996). High B-D glucosidase activity reflects the accumulation of organic substrates derived

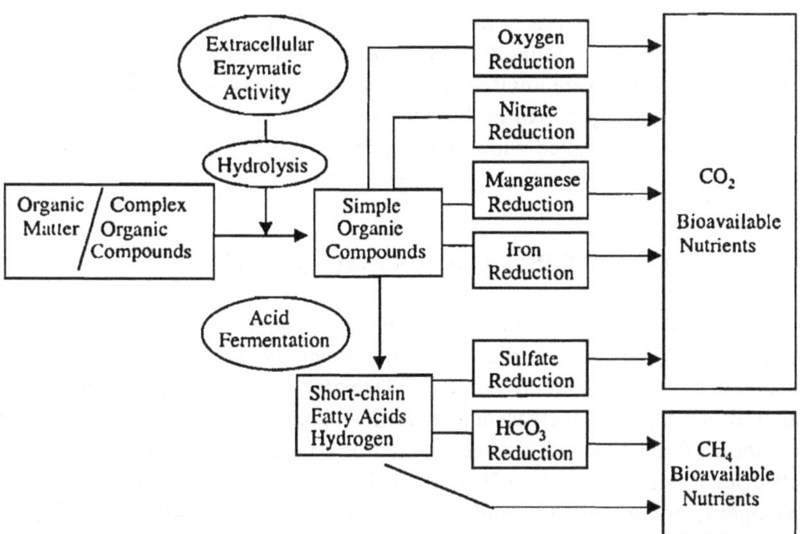

FIGURE 10.6 Schematic showing the pathways involved in decomposition of organic matter in wetland soils.

TABLE 10.2
Extracellular Enzyme Activity in Detrital and Surface Soil Layers of the Everglades Water Conservation Area 2a

Extracellular enzyme activity	Units	Sampling stations IS	Sampling stations RS	Impact index (log [IS/RS])
Detrital layer				
B-D glucosidase	mg p-nitrophenol kg^{-1} h^{-1}	2055	562	0.56
Protease	mg tyrosine kg^{-1} h^{-1}	9317	6447	0.16
Alkaline phosphatase	mg p-nitrophenol kg^{-1} h^{-1}	1965	9856	–0.70
Arylsulfatase	mg p-nitrophenol kg^{-1} h^{-1}	2324	2904	–0.10
Phenol oxidase	mmoles dicq kg^{-1} h^{-1}	48	31	0.19
0 to 10 cm soil layer				
B-D blucosidase	mg p-nitrophenol kg^{-1} h^{-1}	1056	448	0.40
Protease	mg tyrosine kg^{-1} h^{-1}	5192	4441	0.07
Alkaline phosphatase	mg p-nitrophenol kg^{-1} h^{-1}	1277	3016	–0.37
Arylsulfatase	mg p-nitrophenol kg^{-1} h^{-1}	1701	1286	0.12
Phenol oxidase	mmoles dicq kg^{-1} h^{-1}	13	9	0.16

Values are averages of the sampling periods of Feb. 1996, May 1996, Aug. 1996, and Mar. 1997. IS = sampling stations at 1.4 and 2.3 km from inflow; RS = sampling stations at 8 and 10.1 km from inflow.

from P-enriched plant detritus at the impacted site. High levels of inorganic P at the impacted site apparently inhibited APA production. Arylsulfatase and phenol oxidase and protease activities were unaffected by P loading. Activities of all enzymes were higher in the detrital layer than in the 0 to 10 cm soil layer. Phosphatase activity in periphyton mats was affected within two to three weeks after initiation of P loading to unimpacted Everglades soil (Newman et al., 1998).

10.6 MICROBIAL POPULATIONS AND BIOMASS

Microbial communities respond to environmental impact by changes in size and composition of populations and a decrease in diversity (Drake et al., 1996). Taxonomically distinguishable microbes can replace each other as biogeochemical cycles are altered as a result of increased/decreased availability of substrates for catabolic and anabolic functions and changes in redox conditions. Surface soils from P-enriched areas yielded 10^3 to 10^4 times higher numbers of anaerobes—including methanogens, sulfate reducers, and H_2 and acetate consumers—than did soils from unimpacted sites (Table 10.3) (Drake et al., 1996). Conversely, similar numbers of aerobes were present at impacted and reference sites. High numbers of anaerobes at the impacted site suggest oxygen limitation and intense reducing conditions, as reflected by low redox potentials at impacted site (Drake et al., 1996).

TABLE 10.3
Selected Microbial Groups in Impacted and Reference Sites of the Everglades Water Conservation Areas

Microbial population	Units	Sampling stations		Impact index (log [IS/RS])
		IS	RS	
General Anaerobes	total count	4×10^7	4×10^7	−0.1
Anaerobes				
Acetate producers	total count	4×10^{11}	5×10^7	3.9
H_2 consumers	total count	1×10^8	4×10^7	3.4
CO_2 consumers	total count	2×10^5	2×10^7	2.0
Methanogens	total count	3×10^{11}	2×10^7	6.2
Sulfate reducers	total count	4×10^{11}	5×10^7	3.9

Station F-1 is approx. 2.3 km from the inflow point in WCA-2a. Station 3A is from the interior of WCA-3a (Drake et al., 1996).

The soil microbial biomass is the key component of an ecosystem responsible for organic matter decomposition, nutrient cycling, and energy flow (Wardle, 1992). The pool size of microbial biomass reflects the microbial population in soils. The ratio of microbial biomass C to total organic C (MBC/TOC) has been related to the soil C availability and the tendency for a soil to accumulate organic C (Sparling, 1992). Phosphorus loading increased the microbial biomass pool of detritus and surface soils of WCA-2a (DeBusk and Reddy, 1998). Microbial biomass C was greater in recently accreted detrital plant tissue than the surface 0 to 10 cm soil (Table 10.3). Microbial biomass C in the detrital layer was approximately twofold greater in the P-enriched area than the reference site, suggesting that increased P loading increased the microbial population size (Table 10.4). Indices based on microbial activity and organic C have been proposed to provide an operationally defined response of soil microbial populations to substrate quality and environmental conditions (Anderson and Domsch, 1989).

Microbial communities assimilate and recycle nutrients within the biomass pool as needed. Soils that maintain high microbial biomass are not only capable of storing more nutrients but also greater cycling within the system (Torstensson et al., 1998). Microbial biomass N (MBN) was higher in detrital and surface soil layers of the impacted site than the reference site (Table 10.4). The N assimilation coefficient, (the ratio of MBN to total Kjeldahl N, MBN/TKN), was higher for detritus than 0 to 10 cm soil for both the impacted and reference sites. High N assimilation coefficient at impacted sites suggests high N demand by microbes, and possible N limitation. Low detrital or soil N/P ratio at the impacted sites suggests high N demand in relation to P availability (Table 10.1).

Nutrient limitations affect the overall nutrient status of the ecosystem. The most limiting nutrient will be held in the closed microbial compartment and will be unavailable to higher plants. Phosphorus has historically been the limiting nutrient to the microbial biomass in the Everglades and, hence, has controlled the size and activity of the microbial pool. In turn, P has been limiting for macrophytes, and the Everglades ecosystem evolved and sustained itself under these historically low-nutrient conditions. The P assimilation coefficient (MBP/TP) for the impacted site ranged from 0.08 to 0.2, as compared with 0.15 to 0.3 for the reference site (Table 10.4). This comparison suggests that P limitation to the microbial communities is less at the inflow points and consequently more P is released into soil porewater. The C/P ratio of detritus and surface soils was 270 and 291, respectively, for the impacted site, as compared with 850 and 915 for the reference site (Table 10.1). The MBP/TP ratio can serve as an indicator of the P assimilation efficiency of microbes. A high ratio implies P limitation and rapid assimilation, and storage of bioavailable P in microbial biomass, as was observed in unimpacted sites of the Everglades (Chua and Reddy, 1997). A highly significant correlation was observed between alkaline phosphatase and the P assimilation coefficient, suggesting P limitation in the system (Chua and Reddy, 1997).

10.7 MICROBIAL RESPIRATION

Microbial respiration reflects the activity of microorganisms under ambient substrate and nutrient concentrations. Addition of low levels of P to a P-limited system can increase microbial activity (Amador and Jones, 1995). Thus, measurement of microbial activity can provide a sensitive indicator of nutrient loading to such an ecosystem. Both aerobic and anaerobic microbial respiration rates were greater in the detrital layer than in the 0 to 10 cm soil layer (DeBusk and Reddy, 1998; Wright and Reddy, 1998). Respiration rates measured within two hours on freshly collected

TABLE 10.4
Influence of Phosphorus Loading on Selected Parameters Related to Microbial Biomass and Microbial Respiration Carbon Cycling Processes in Detrital and Surface Soil Layers of the Everglades Water Conservation Area 2a

		Sampling stations		Impact index
Indicator/process	Units	IS	RS	(log [IS/RS])
Detrital layer				
MBC	g kg^{-1}	16.8	7.9	0.33
MBC/TOC		0.038	0.019	0.30
MBN	mg kg^{-1}	1323	778	0.23
MBN/TKN		0.051	0.033	0.19
MBP	mg kg^{-1}	319	150	0.33
MBP/TP		0.20	0.30	–0.18
Microbial respiration	mg C kg^{-1} h^{-1}			
Aerobic		171	105	0.21
Anaerobic		163	64	0.41
Metabolic quotient	h^{-1}			
Aerobic		0.010	0.013	–0.11
Anaerobic		0.009	0.009	0.05
0–10 cm soil layer				
MBC	g kg^{-1}	5.7	4.4	0.11
MBC/TOC		0.014	0.010	0.15
MBN	mg kg^{-1}	455	344	0.12
MBN/TKN		113	0.012	0.12
MBP	mg kg^{-1}	0.08	76	0.17
MBP/TP			0.15	–0.27
Microbial respiration	mg C kg^{-1} h^{-1}	136		
Aerobic		109	80	0.23
Anaerobic			57	0.28
Metabolic quotient	h^{-1}	0.024		
Aerobic		0.019	0.018	0.12
Anaerobic			0.013	0.16

Values are averages of the sampling periods of Feb. 1996, Aug. 1996, and Mar. 1997. IS = sampling stations at 1.4 and 2.3 km from inflow; RS = sampling stations at 8 and 10.1 km from inflow.

detritus and surface soil samples were greater under drained conditions than flooded conditions (Wright and Reddy, 1998). Approximately a twofold increase in respiration rates was observed in the detritus layer of the impacted site, as compared to the reference site (Table 10.4). Anaerobic respiration represented approximately one-third of aerobic respiration in detritus and soil samples of the Everglades (DeBusk and Reddy, 1998). Addition of P accelerated the rate of microbial respiration, apparently by an increased supply of electron donors from labile detrital plant tissue. Total P and lignocellulose content of the detrital tissue and soil organic matter were key variables regulating organic C mineralization in WCA-2a soils (DeBusk and Reddy, 1998).

The metabolic quotient, or specific respiration rate (qCO_2), is the ratio of the basal respiration rate (as CO_2C) per unit microbial biomass C (MBC), and provides an indication of efficiency of microbial in utilization of available substrates (Anderson and Domsch, 1993, Wardle, 1993, Ohtonen, 1994). The qCO_2 has been used as a response variable reflecting the effects of temperature, soil management, ecosystem succession and heavy metal stress, and is apparently a considerably more sensitive parameter than MBC/TOC (Anderson and Domsch, 1993). Low qCO_2 in detritus of the impacted site (Table 10.3) suggests poor efficiency of microbes in utilization of available substrates. Minimal differences in qCO_2 were observed for 0 to 10 cm soil of impacted and reference sites.

10.8 ORGANIC NITROGEN AND PHOSPHORUS MINERALIZATION

Organic N mineralization occurs through (1) hydrolytic deamination of amino acids and peptides, (2) degradation of nucleotides, and (3) metabolism of methylamines by methanogenic bacteria (King et al., 1983). Both potentially mineralizable N (PMN) and substrate induced N mineralization (SINM) in detritus and soil layers were higher at the impacted site than the reference site (Table 10.5) (White and Reddy, 1997). Substrate induced N mineralization indicates the activity of microbes and associated enzymes involved in organic N mineralization. This concept is similar to substrate induced respiration (SIR), typically used to estimate the size of microbial populations (Anderson and Domsch, 1989). Low C/N ratio and P nonlimiting conditions at the impacted site resulted in higher rates of PMN. Phosphorus loading increased rates of organic N mineralization. As the microbial P needs are satisfied, a significant portion of N needs of macrophytes and microbial communities are met through the mineralization of organic N. Increased rates of organic N mineralization result in elevated levels of NH_4–N in soil porewater (Fig. 10.7). Since P requirements of macrophytes are met in the impacted area, growth of macrophytes may be regulated to some extent by the supply of inorganic N, and thus the rate of organic N mineralization. The (NH_4–N/SRP) ratio in the soil porewater of the impacted site is in the range of 2 to 4, as compared to 200 to 400 for the unimpacted, reference site. The high NH_4–N/SRP ratio at the reference site reflects P limitation in the system.

Phosphorus added to a wetland or released during decomposition of soil organic matter is usually retained in the system through sorption and precipitation reactions. Phosphorus stored in vegetation during the active growth period is potentially released during winter months as a result of leaching and decomposition. Substrate induced P mineralization (SIPM) can indicate activity of microbes and associated enzymes involved in organic P mineralization. Hydrolysis of added substrates such as glucose-6-phosphate was not influenced by P enrichment in the litter layer. In the 0 to 10 cm soil layer, substrate hydrolysis decreased by approximately 67% at the

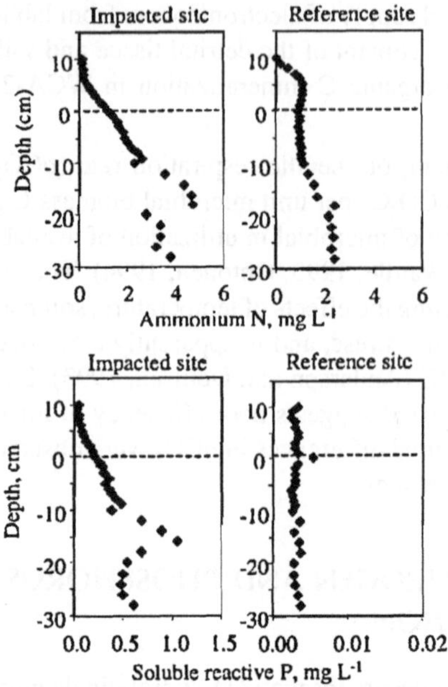

FIGURE 10.7 Ammonium N and soluble reactive P concentration of the soil porewater in the impacted and reference sites of WCA-2a of the Everglades (White and Reddy, unpublished results; Fisher, 1997).

impacted site. The mineralization coefficient (SIMP/MBP) also decreased by 54 and 74% in the litter and 0 to 10 cm soils, respectively, at the impacted site. These decreases suggest that P is not limiting to microbes. High substrate P hydrolysis at the reference site was attributed to limitation of labile inorganic P and high APA. Potentially mineralizable organic P (PMP) was also higher at the impacted site for both litter and surface 0 to 10 cm soil layer, reflecting high porewater P concentrations (Fig. 10.7).

10.9 NITRIFICATION, DENITRIFICATION, AND NITROGEN FIXATION

Nitrification is the biological oxidation of reduced N forms (NH_4–N) to a more oxidized state (NO_3–N). Nitrification is a key process in the N budget of wetland systems, since the NO_3^- formed is assimilated by many species. More importantly, nitrification provides the NO_3^- for denitrification, a major N removal mechanism in wetlands. In wetlands, nitrification can occur in (1) the water column, (2) the surface-oxidized soils, and (3) the oxidized rhizosphere of plants. Nitrification rates were higher at the impacted site than the reference site for both detrital and surface soil layers (White and Reddy, 1998, unpublished results). Since NH_4–N and alkalinity were not limiting at either site, low rates of nitrification at the reference site were probably due to P limitation of the microbial pool.

TABLE 10.5
Influence of Phosphorus Loading on Selected Parameters Related to Organic N and P Mineralization in Detritus and Surface Soil Layer of the Everglades Water Conservation Area 2a

Indicator/process	Units	Sampling stations IS	RS	Impact index
Detritus layer				
PMN (anaerobic)	mg kg^{-1} d^{-1}	162	65	0.40
SINM (D-alanine)	mg kg^{-1} h^{-1}	59	14	0.62
SINM/MBN	h^{-1}	0.044	0.018	0.39
PMP (anaerobic)	mg kg^{-1} d^{-1}	15.4	2.2	0.85
SIPM (glucose 6-phosphate)	mg kg^{-1} h^{-1}	29.5	24.4	0.08
SIPM/MBP	h^{-1}	0.1	0.19	–0.28
0–10 cm soil layer				
PMN (anaerobic)	mg kg^{-1} d^{-1}	46	26	0.25
SINM (D-alanine)	mg kg^{-1} h^{-1}	34	12	0.45
SINM/MBN	h^{-1}	0.075	0.35	0.33
PMP (anaerobic)	mg kg^{-1} d^{-1}	2.2	2.4	–0.03
SIPM (glucose 6-phosphate)	mg kg^{-1} h^{-1}	7.1	18.3	–0.41
SIPM/MBP	h^{-1}	0.09	0.27	–0.48

Values are averages of the sampling periods of Feb. 1996, Aug. 1996, and Mar. 1997. IS = sampling stations at 1.4 and 2.3 km from inflow; RS = sampling stations at 8 and 10.1 km from inflow.

Denitrification is a respiratory process where facultative anaerobic bacteria use NO_3^- (or NO_2^-) in the absence of O_2 as the terminal electron acceptor during the oxidation of organic C, resulting in the production of gaseous end products such as N_2O and N_2. The denitrification enzyme activity (DEA) is used to quantify the activity of denitrifying populations and is an indicator of the denitrification potential of soils (Tiedje, 1994). Actual denitrification rates under field conditions are usually regulated by NO_3^- concentration, since available C is not limiting in most wetlands. The DEA of detrital and soil layers was approximately threefold higher at the impacted site than at the reference site (Table 10.6). Much higher DEA values were observed at impacted sites <1 km from the inflow (White and Reddy, 1997). The denitrifying population increases with NO_3^- loading, and high DEA values at the impacted site reflect NO_3^- loading from drainage water. It is likely that the DEA in the Everglades soils is limited by NO_3^- concentration rather than P concentration.

Phosphorus loading to wetlands increases P concentration of periphyton mats and changes its species composition and nutrient content (McCormick and O'Dell, 1996). Decrease in N/P ratio of periphyton mats may promote N_2 fixing species. Thus, biological N_2 fixation in periphyton mats be a sensitive indicator of evaluate P loading impacts on wetlands. The acetylene (C_2H_2) reduction technique has been used by researchers to indirectly estimate the rate of N_2 fixation. The stoichiometry of the reduction of acetylene (C_2H_2) to ethylene (C_2H_2) implies that 3 moles of C_2H_2 is reduced for every mole of N_2 fixed. Biological N_2 fixation was higher in periphyton mats obtained from the impacted site (F-1 station, 2 km from inflow) than the reference site (U-3, 11 km from inflow) (Table 10.6).

10.10 IMPACT INDEX

The usefulness of these indicators in evaluating impacts of nutrient loading on ecosystem health depends on an indicator's natural variability within the system. Uncertainty in evaluation is not only due to spatial variability, but also due to the dynamic nature of many of these processes that complicate the extrapolation of laboratory results to field conditions. Temporal variation of these indicators can introduce additional uncertainty in the interpretation of the results. In an ecosystem with distinct gradients, impacts can be described by using the relationship developed between biogeochemical processes and associated easily measurable parameters. However, these relationships should be based on data collected on sites with a wide range of physical, chemical, and biological properties, and loading impacts. Results presented in this chapter are based on four sites (two sites each for impacted and reference sites) and parameters measured in two seasons (wet and dry seasons) of a wetland ecosystem impacted by P loading, thereby limiting extrapolation of results. However, the concepts presented can be useful in evaluating nutrient impacts on wetlands.

For comparison purposes, a simple impact index was calculated for each process or parameter measured:

$$\text{Impact index} = \log [\text{IS/RS}]$$

where [IS] is the rate or concentration of a parameter measured at an impacted site, and [RS] is the rate or concentration of a parameter measured at a reference site. The log [IS/RS] provides an index value of 0, which indicates no change, a negative value indicates a decrease and a positive value indicates an increase. For example, a value of "1," represents a tenfold change in concentration of a parameter or rate of a process, relative to the reference site (Fig. 10.8*a*). This approach allows ranking of the parameters or processes most affected by nutrient loading/disturbance. Impact indices should be viewed in the context of spatial variability within impacted and unimpacted sites. Field replicate variability in relatively small areas (< m^2) is process or parameter dependent and can be highly variable. For example, enzyme activities in the detrital layer ranged from 27 to 100%, whereas variability in microbial respiration rates was <20% (Wright and Reddy, 1996, 1998). Experimental technique variability in measurement of indicator parameters measured at WCA-2a sites is <10% for both in situ and laboratory conditions.

The calculated impact index values (Tables 10.2 through 10.6) aid in normalizing rates of various biogeochemical processes and concentrations of various parameters into a common format. Data presented in these tables are not adequate to perform multivariate statistical analysis, including principal component analysis. However, data presented in this chapter aids in developing a qualitative ranking of processes

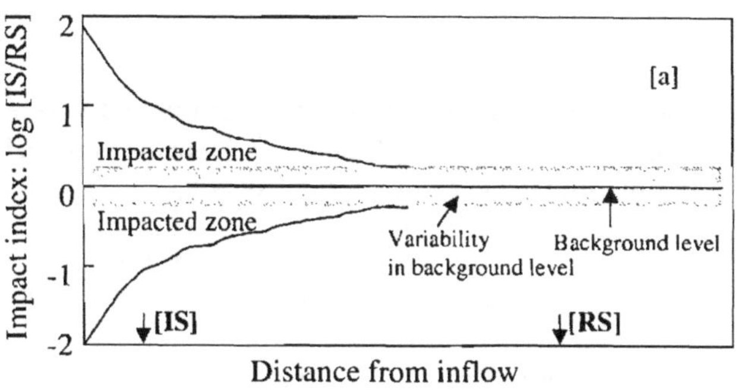

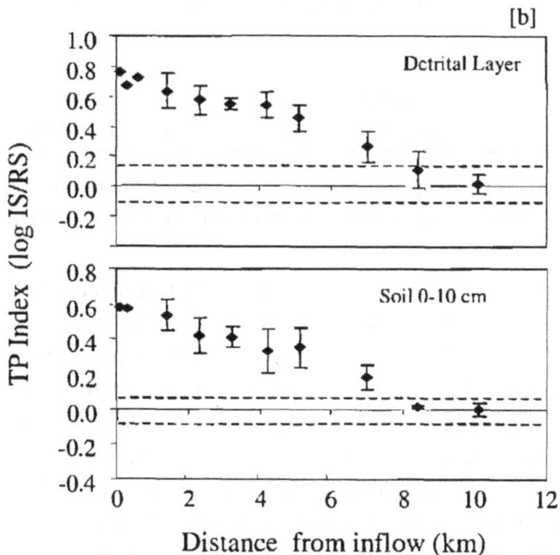

FIGURE 10.8 (a) Schematic showing the concept of impact index as a function of distance from inflow source and (b) calculated impact index for total P content of detrital and soil layers. Error bars represent standard deviations. Dashed lines represent variability at the reference site.

and parameters based on their sensitivity to P loading. Because of variability of uncertainty, the index values were grouped into four broad groups of positive and negative impacts. We assume that impact index values in the range of –0.1 to 0.1 are within the experimental variability of many of the processes and parameters measured. For example, the variability of the impact index for total P was <0.2 for the detrital layer and <0.1 for the surface soil of the reference site, respectively (Fig. 8*b*). However, for some parameters such as microbial numbers (usually reported on log-scale), the variability could be much higher. Biogeochemical processes and parameters measured on the detrital layer were more sensitive to nutrient loading than those measured on surface soil (Table 10.7). The detrital component probably represents most recent (nonsteady state) impacts, whereas the surface soil may represent long-term impacts (Fig. 10.2). Microbial communities associated with

TABLE 10.6
Influence of Phosphorus Loading on Selected Parameters Related to Nitrification–Denitrification and N_2 Fixation in Detritus and Surface Soil Layer of the Everglades Water Conservation Area 2a

Indicator/process	Units	Sampling stations IS	RS	Impact index
Detrital layer				
Nitrification	mg kg^{-1} d^{-1}	148	101	0.17
DEA	mg kg^{-1} d^{-1}	2.17	1.06	0.31
N_2 fixation	µmoles C_2H_4 kg^{-1} h^{-1}	510	140	0.56
0–10 cm soil layer				
Nitrification	mg kg^{-1} d^{-1}	122	76	0.20
DEA	mg kg^{-1} d^{-1}	1.01	0.49	0.31

Values are averages of the sampling periods of Feb. 1996, Aug. 1996, and Mar. 1997. IS = sampling stations at 1.4 and 2.3 km from inflow; RS = sampling stations at 8 and 10.1 km from inflow.

detrital layer are in direct contact with water column nutrients and should respond rapidly to changes in water chemistry. However, impact index values obtained on soils may be more reliable, because they represent long-term steady-state conditions.

Results presented in this chapter provide a simple strategy for integrated evaluation of P impacts on wetlands, using soil biogeochemical processes and parameters as potential indicators. Although biogeochemical processes may be sensitive and reliable indicators of wetland integrity, their measurement can be time consuming and expensive. We have found, however, that concentrations of certain chemical substrates, intermediates, and end products can function as surrogates for biogeochemical processes (Reddy and D'Angelo, 1996; Reddy et al., 1998; D'Angelo and Reddy, 1998; DeBusk and Reddy, 1998). Furthermore, relationships between indicators and processes may provide reliable estimates of ecosystem health. Simple strategies presented in this chapter provide a tool to normalize the process level information into a common format for possible integration into predictive models. To evaluate impacts, a reliable database is needed for several reference sites (background level of impacts) in various geographical regions as classified by wetland vegetation, soils, and hydrology. The strategies presented could be used in different types of wetlands to assess restoration and remediation efforts, and potentially could be used as a screening tool to choose how best to use limited restoration resources.

10.11 CONCLUSIONS

Organic matter accumulation in wetlands is a result of greater net primary productivity and slower rates of decomposition than terrestrial ecosystems. Consequently, organic matter accretion provides a long-term storage compartment for nutrients including C, N, and P. The Everglades ecosystem has adapted to low external nutrient

TABLE 10.7
Impact Indices and Relative Sensitivity of Various Biogeochemical Processes/Indicators Measured in Detrital and Soil Layers Impacted and Reference Sites

Biochemical indicator/process	Detrital/soil layer	Impact index, log [IS/RS]	Relative sensitivity to phosphorus loading
APA	D	−1.0 to −0.50	High
C/P ratio	D, S		
SIPM	S	−0.50 to −0.25	Medium
(SIPM/MBP)	D, S		
(MBP/TP)	S		
APA	S		
(MBP/TP)	D	−0.25 to −0.10	Low
Ash content	D		
Metabolic quotient	D		
Arylsulfatase	O		
Ash content	S	−0.10 to 0.10	Negligible
Total C and N	D, S		
Metabolic quotient	S		
C/N ratio	D, S		
Extractable NH_4-N	S		
SIPM	D		
PMP	S		
General aerobes	S		
Protease	S		
Protease	D	0.10 to 0.25	Low
Aerobic microbial respiration	D, S		
MBC	S		
MBN	D, S		
Extractable NH_4-N	D		
MBN/N total	D, S		
Nitrification	D		
Arylsulfatase	S		
Phenol oxidase	D, S		
Total P	S	0.25 to 0.50	Medium
B-D glucosidase	S		
MBC	D		
Anaerobic microbial respiration	D, S		
PMN, SINM	D, S		
Nitrification	D		

TABLE 10.7
Impact Indices and Relative Sensitivity of Various Biogeochemical Processes/Indicators Measured in Detrital and Soil Layers Impacted and Reference Sites (continued)

Biochemical indicator/process	Detrital/soil layer	Impact index, log [IS/RS]	Relative sensitivity to phosphorus loading
N_2 fixation	D		
DEA	D		
SINM/MBN	D, S		
MBP	D		
Total P	D	0.50 to 1.0	High
B-D glucosidase	D		
N_2-fixation	D		
SINM	D		
Bicarbonate extractable P	D, S		
PMP	D		
Anaerobes[a]		>1.0	Very High
-Acetate producers	S		
-H_2 consumers	S		
CO_2 consumers	S		
Methanogens	S		
Sulfate reducers	S		

D = detrital layer, S = 0 to 10 cm soil, APA = alkaline phosphatase activity, C/P = carbon to phosphorus mass ratio, SIPM = substrate induced organic P mineralization, MBP = microbial biomass P, PMP = potentially mineralizable P, MBC = microbial biomass C, MBN = microbial biomass N, PMN = potentially mineralizable N, SINM = substrate induced organic N mineralization, and DEA = denitrification enzyme activity.

[a]Anaerobes were not measured in detrital layer.

loads, relying primarily on low P-limited production of plant biomass and slow recycling of organic substrates by a small, nutrient-limited microbial pool. The once nutrient-limited microbial pool has increased in size due to the availability of nutrients, particularly P, in the inflow water and currently mediate the large release of nutrients from deposited organic matter.

Several microbial processes were affected by P enrichment of the soil and litter. Recently accreted plant detritus and floc showed a high degree of positive response to P loading on various microbial processes. Increases in P loading in the impacted areas decreased C/P ratio of litter and soil, resulting in increased rate of organic N and P mineralization. Although P may no longer be limiting to biota in the impacted zone, a high degree of P enrichment may induce N limitation. The spread and growth of *Typha* in the impacted zone may be directly linked to the increased supply of bioavailable nutrients liberated through enhanced decomposition of P-enriched litter and organic matter.

Future work should be directed to determine the extent of microbial diversity in litter and surface soils and their functional role in regulating the productivity and water quality of wetlands. More specifically, we must identify and isolate individual species that are sensitive to anthropogenic impacts and can be subsequently used as indicator species. For example, molecular techniques are now available to identify various species of sulfate reducing bacteria that may serve as sensitive indicators of community change as result of nutrient loading. Although impact indices developed are useful in identifying the processes most affected by loading, additional research must be conducted to determine spatial and temporal effects on these indices. The indices values can be integrated into models based on Geographic Information Systems (GISs) to evaluate impacts at landscape level.

10.12 ACKNOWLEDGMENTS

The Florida Agricultural Experiment Station Journal Series No. R-06581 is acknowledged. This chapter summarizes the research contributions of several graduate students and research staff of the Wetland Biogeochemistry Laboratory, University of Florida. We thank E.M. D'Angelo and W.F. DeBusk for participating in discussions during development of the ideas presented in this chapter. Research in part is supported by a contract from the South Florida Water Management District (SFWMD). We thank Drs. Paul McCormick and Sue Newman of SFWMD for their assistance and support during the course of field experiments. Data on biological nitrogen fixation are provided by P. Inglett and E.M. D'Angelo. We also thank Dr. A. Ogram, Soil and Water Science Department, for his critical review of this manuscript.

REFERENCES

Amador, J.A., and R.D. Jones. 1993. Nutrient limitations on microbial respiration in peat soils with different total phosphorus content. Soil Biol. Biochem. 25:793–801.

Amador, J.A. and R.D. Jones. 1995. Carbon mineralization in pristine and phosphorus-enriched peat soils of the Florida Everglades. Soil Science 159:129–141.

Anderson, T.H., and K.H. Domsch. 1993. The metabolic quotient for CO_2/qCO_2 as a specific activity parameter to assess the effects of environmental conditions, such as pH, on the microbial biomass of forest soils. Soil Biol. Biochem. 25:393–395.

Anderson, T.H. and K.H. Domsch. 1989. Ratios of microbial biomass carbon to total organic carbon in arable soils. Soil Biol. Biochem. 21:471–479.

Benner, R.M., M.A. Moran, and R.E. Hodson. 1985. Effects of pH and plant source on lignocellulose biodegradation rate in two wetland ecosystems, the Okeefenokee Swamp and a Georgia salt marsh. Limnol. Oceanogr. 30:489–499.

Boulton, A.J. and P.I. Boon. 1991. A review of methodology used to measure leaf litter decomposition lotic environments: Time to turn over an old leaf. Aust. J. Mar. Freshwater Res. 42:1–43.

Button, D.K. 1985. Kinetics of nutrient-limited transport and microbial growth. Microbiological Reviews. 49:270–297.

Chróst, R.J. 1991. Environmental control of the synthesis and activity of aquatic microbial ectoenzymes, Chap. 3. p. 2959. *In* R. J. Chróst (ed.) Microbial Enzymes in Aquatic Environments. Springer Verlag, New York, NY.

Chua T.A. and K.R. Reddy. 1997. Short-term mineralization of organic phosphorus along a nutrient gradient. Agronomy Abstracts. 89th Annual meeting of American Society of Agronomy, Anaheim, CA Oct. 26–30, 1997. Ameri. Soc. Agr. Madison, WI.

Craft, C.B. and C.J. Richardson. 1993a. Peat accretion and phosphorus accumulation along a Eutrophication gradient in the northern Everglades. Biogeochemistry 22:133–156.

Craft, C.B. and C.J. Richardson. 1993b. Peat accretion and N, P and organic C accumulation in nutrient-enriched and unenriched Everglades peatlands. Ecological Applications 3:446–458.

Cunningham, H.W., and R.G. Wetzel. 1989. Kinetic analysis of protein degradation by a freshwater wetland sediment community. Appl. Environ. Microbiol. 56:1963–1976.

D'Angelo, E.M. and K.R. Reddy. 1994a. Diagenesis of organic matter in a wetland receiving hypereutrophic lake water. I. Distribution of dissolved nutrients in the soil and water column. J. Environ. Qual. 23:937–943.

D'Angelo, E.M. and K.R. Reddy. 1994b. Diagenesis of organic matter in a wetland receiving hypereutrophic lake water. II. Role of inorganic electron acceptors in nutrient release. J. Environ. Qual. 23:928–936.

D'Angelo, E.M. and K.R. Reddy. 1998. Influence of electron acceptors on biogeochemical properties of wetland soils: Microbial activity and organic carbon mineralization. Soil Biol. Biochem. (in review).

Davis, J.H., Jr. 1943. The natural features of southern Florida, especially the vegetation, and the Everglades Fl. Geol. Surv. Bull. 25. Florida Geol. Surv., Tallahassee.

Davis, C.B. and A.G. van der Valk. 1983. Uptake and release of nutrients by living and decomposing *Typha glauca* Godr. tissues at Eagle Lake, Iowa. Aquat. Bot. 16:75–87.

Davis, S. M. 1991. Growth, decomposition, and nutrient retention of *Cladium jamaicense* Crantz and *Typha domingensis* Pers. in the Florida Everglades. Aquat. Bot. 40:203224.

DeBusk, W.F. 1996. Organic matter turnover along a nutrient gradient in the Everglades. Ph.D. Dissertation, University of Florida.

DeBusk W.F. and K.R. Reddy. 1998. Turnover of detrital organic carbon in a nutrient-impacted Everglades marsh. Soil Sci. Soc. Am. J. (in press).

DeBusk, W.F., K.R. Reddy, M.S. Koch and Y. Wang. 1994. Spatial distribution of soil nutrients in the northern Everglades. Soil Sci. Soc. Am. J. 58:543–552.

Drake, H.L., N.G. Aumen, C. Kuhner, C. Wagner, A. Grießhammer, and M. Schmittroth. 1996. Anaerobic microflora of Everglades sediments: Effects of nutrients on population profiles and activities. Appl. Environ. Microbiol. 62:486–493.

Enriquez, S., C.M. Duarte, K. Sand-Jensen. 1993. Patterns in decomposition rates among photosynthetic organisms: the importance of detritus C:N:P content. Oecologia 94:457–471.

Fisher, M.M. 1997. Estimating landscape scale flux of phosphorus using geographic information systems (GIS). M.S. Thesis. University of Florida.

Gleason, P.J., A.D. Cohen, H. K. Brooks, P. Stone, R. Goodrick, W. G. Smith, and W. Spackman, Jr. 1974. The environmental significance of holocene sediments from the Everglades and saline tidal plain. pp. 287–341. In P. J. Gleason (ed) Environments of South Florida: Present and Past. Miami Geol. Soc., Miami, Fl.

Happell, J.D. and J.P. Chanton. 1993. Carbon remineralization in a north Florida swamp forest: effects of water level on the pathways and rates of soil organic matter decomposition. Global Biogeochemical Cycles 7:475–490.

Howard-Williams, C. 1985. Cycling and retention of nitrogen and phosphorus in wetlands: a theoretical and applied perspective. Freshwater Biol. 15:391–431.

Kerner, M. 1993. Coupling of microbial fermentation and respiration processes in an intertidal mudflat of the Elbe estuary. Limnology and Oceanography. 38:314–330.

King, G.M., M.J. Klug and D.R. Lovley. 1983. Metabolism of acetate, methanol, and methylated amines in intertidal sediments of Lowes Cove, Maine. Appl. Environ. Microbiol. 45:1848-1853.

Koch, M., and K.R. Reddy. 1992. Distribution of soil and plant nutrients along a trophic gradient in the Florida Everglades. Soil Sci. Soc. Am. J. 56:1492–1499.

Lodge, D.J., W.H. McDowell, and C. P. McSwiney. 1994. The importance of nutrient pulses in tropical forests. Tree 9:384–387.

McCormick P.V. and M.B. O'Dell. 1996. Quantifying periphyton responses to phosphorus in the Florida Everglades: A Synoptic-experimental approach. J.N. Am. Benthol. Soc. 15:450–468.

McKinley, V.L. and J.R. Vestal. 1992. Mineralization of glucose and lignocellulose by four arctic freshwater sediments in response to nutrient enrichment. Applied and Environmental Microbiology 58:1554–1563.

Newman S., P.V. McCormick, and J.G. Backus. 1998. Phosphatase activity as an early warning indicator of wetland eutrophication: Problems and prospects In: I. Hernandez and B.A. Whitton (eds). Phosphatases in Environment, Kluwer Academic Pub. (in press).

Newman, S., K.R. Reddy, W.F. DeBusk and Y. Wang. 1997. Spatial distribution of soil nutrient in a northern Everglades marsh: Water Conservation Area 1. 61:1275–1283.

Odum, E. P. 1969. The strategy of ecosystem development. Science 164:262–270.

Odum, E.P. 1985. Trends expected in stressed ecosystem. Bioscience 35:419–422.

Ohtonen, R. 1994. Accumulation of organic matter along a pollution gradient: application of Odum's theory of ecosystem energetics. Microb. Ecol. 27:43–55.

Oremland, R.S. 1988. Biogeochemistry of methanogenic bacteria. 641707. In: Biology of Anaerobic Microorganisms. Ed. A.J.B. Zehnder. John Wiley & Sons, New York.

Reddy, K.R., R.D. Delaune, W. F. DeBusk, and M. Koch. 1993. Longterm nutrient accumulation rates in the Everglades wetlands. Soil Sci. Soc. Am. 57:1145–1147.

Reddy, K.R. and E.M. D'Angelo. 1994. Soil processes regulating water quality in wetlands. *In* Global Wetlands—Old World and New (W. Mitsch ed.) p. 309–324. Elsevier Publ. New York.

Reddy, K.R. and E.M. D'Angelo. 1996. Biogeochemical indicators to evaluate pollutant removal efficiency in constructed wetlands. Wat. Sci. Tech. 35:1–10.

Reddy, K.R., Y. Wang, W.F. DeBusk, M.M. Fisher and S. Newman. 1998. Forms of soils phosphorus in selected hydrologic units of Florida Everglades ecosystems. Soil Sci. Soc. Am. J. (in press).

Sinsabaugh, R.L 1994. Enzymatic analysis of microbial pattern and processes. Biol. Fertil. Soils.17:69–74.

Sparling, G.P. 1992. Ratio of microbial biomass carbon to soil organic carbon as a sensitive indicator of changes in soil organic matter. Aust. J. Soil Res. 30:195–207.

Tiedje, J.M. 1994. Denitrifiers. *In* Methods of soils analysis: Part 2. Microbiological and biochemical properties. SSSA Book Series No. 5 pp. 245–267. Soil Sci. Soc. Am. Madison, WI.

Torstensson L. 1997. Microbial assays in soils: In Soil Ecotoxicology. J. Tarradellas, G. Bitton, and D. Rossel (eds) CRC Lewis Publ., Boca Raton, pp. 207–234.

Torstensson, L., M. Pell, and B Stenberg. 1998. Need of a strategy for evaluation of arable soil quality. Ambio. 27:4–8.

Wardle, D.A. 1992. A comparative assessment of factors which influence microbial biomass carbon and nitrogen in soil. Biological Reviews 67:321–358.

Wardle, D.A. 1993. Changes in the microbial biomass and metabolic quotient during leaf litter succession in some New Zealand forest and scrubland ecosystems. Functional Ecology 7:346–355.

Webster J.R. and E.F. Benfield. 1986. Vascular plant breakdown in freshwater ecosystems. Ann. Rev. Ecol. Syst. 17:567–594.

White J.R. and K.R. Reddy. 1997. Nitrogen transformations along a eutrophic gradient in the Northern Everglades. Agron. Abstracts. 89th Annual Meetings of Amer. Soc. Agron. Oct. 26–30, 1997.

White, J.R. and K.R. Reddy. 1998. Nitrification and denitrification potentials in Everglades wetland soils. (unpublished results).

Wright, A.L. and K.R. Reddy. 1996. Influence of nutrient loading on extracellular enzyme activity in soils of the Florida Everglades. Agron. Abstracts. pp. 331. Annual Meetings of Amer. Soc. Agron. Nov. 3–8, 1996. Indianapolis, IN.

Wright, A.L. and K.R. Reddy. 1998. Microbial respiration in soils and litter under flooded and drained conditions in the Everglades. (Unpublished results).

11 Effects of Phosphorus Enrichment on Structure and Function of Sawgrass and Cattail Communities in the Everglades

S.L. Miao and W.F. DeBusk

11.1 ABSTRACT

Anthropogenic phosphorus (P) loading has resulted in increased soil P concentrations over a significant portion of the northern Everglades. Changes in macrophyte community structure and function associated with soil P enrichment are examined by reviewing recent field and laboratory studies in the Everglades conducted by the South Florida Water Management District and University of Florida. Of primary interest is the widespread replacement of the native sawgrass marsh by cattails in

1-56670-331-X/99/$0.00+$.50

Water Conservation Area 2A (WCA-2A). The spatial distribution of cattails (*Typha domingensis*) in WCA-2A correlates strongly with soil P concentration. Comparative studies in P-enriched and reference areas of WCA-2A demonstrated increased macrophyte production and photosynthesis in the P-enriched area. Field P gradient studies indicate that macrophyte communities in the reference area of WCA-2A are P limited. Plant tissue nutrient concentration, storage, and distribution in macrophyte communities changed significantly along the P gradient. In addition, the rate of detrital organic carbon turnover increased with P enrichment. Sawgrass (*Cladium jamaicense* Crantz) and cattail exhibit contrasting resource allocation patterns that are characteristic of plants adapted to low- and high-nutrient environments, respectively. Soil P enhanced cattail rhizome expansion and leaf growth after disturbances. The results suggest that a shift from a P-limited to a P- enriched system in WCA-2A has contributed significantly to the recent spread and dominance of cattail, a previously restricted native species, in P-enriched areas of the marsh.

11.2 INTRODUCTION

Phosphorus (P) is of great interest in the study of wetland ecosystems because of its many transformations and its ability to be retained in the wetland (Kadlec, 1987). P is often in limited supply for plant growth (Schlesinger, 1991), particularly in oligotrophic wetlands (Heilman, 1968). Excessive P loading has important effects on physiological and ecological processes of macrophytes at the level of the individual, population, and community, and eventually leads to changes at the level of the ecosystem. Furthermore, plants develop different growth and resource allocation patterns that allow them to adapt to P availability in their immediate environments. Plants growing in low-P habitats generally have low photosynthetic and growth rates but high nutrient-use efficiency (Chapin et al., 1982). In contrast, plants growing in high-P environments, in general, have high photosynthetic and growth rates but low nutrient-use efficiency. The effects of P availability on the ecophysiology and life history of plants can ultimately alter competitive interactions and, hence, species composition and dominance of plant communities (Tilman, 1982; Craft et al., 1995; Newman et al., 1996).

The historic Florida Everglades was a P-limited system (Steward and Ornes, 1975, 1983; Koch and Reddy, 1992; Davis and Ogden, 1994). Nutrient loading to the Everglades occurred primarily through rainfall, supplemented by occasional pulses of nutrients from Lake Okeechobee overflow (Davis, 1943; Parker, 1974). Major vegetation communities included sawgrass (*Cladium jamaicense* Crantz) marsh, wet prairies, sloughs and tree islands (Harshberger, 1914; Harper, 1927; Small, 1932; Davis, 1943; Hofstetter and Parsons, 1979). The sawgrass marsh was the dominant plant community in terms of total area, accounting for nearly two-thirds of the vegetative cover in the Everglades (Davis, 1943; Loveless, 1959; Rutchey and Vilchek, 1994, 1998). The historical predominance of sawgrass is evidence of its low nutrient requirements.

Recent human development in the vicinity of the Everglades and surrounding watershed has created changes in nutrient loading and hydrology (Davis and Ogden, 1994). A large area of the northern Everglades presently known as the Everglades

Agricultural Area (EAA), was drained and converted to agricultural production during the first half of this century. As a result, nutrients from organic soil mineralization, along with additional nutrients from fertilizers, have been transported via drainage canals into the Water Conservation Area (WCAs) for over 30 years. Loading of agricultural drainage into the WCAs has been accompanied by nutrient enrichment of water, soil, and vegetation (Davis and Ogden, 1994). During this time, major shifts in plant community structure have occurred in the Everglades, most notably the encroachment of cattail into native sawgrass and slough communities (Davis and Ogden, 1994; Rutchey and Vilchek, 1994, 1998; Jensen et al., 1995; Bartow et al., 1996; Newman et al., 1998). Although the causes for the replacement of sawgrass by cattail (*Typha domingensis* Pers.) are complex, a number of investigations have linked increased P loading from the EAA and altered hydrology to the expansion of cattails in the Everglades (Urban et al., 1993; Davis 1989, 1991, 1994; Koch and Reddy, 1992; Reddy et al., 1993; DeBusk et al., 1994; Craft and Richardson 1993, 1997; Bartow et al., 1996; Newman et al., 1996; Doren et al., 1997; Newman et al., 1998; Miao and Sklar, 1998).

It is hypothesized that cattail may take advantage of increased nutrients to successfully invade areas that were dominated by sawgrass and sloughs. Several recent studies conducted in WCA-2A by the South Florida Water Management District and University of Florida explored how an existing P gradient affects (1) physiology, growth, and resource allocation of individual plants, (2) nutrient storage and distribution in macrophyte communities, and (3) turnover of detrital material in the ecosystem. Controlled experiments in both field and greenhouse were conducted to determine the responses of sawgrass and cattail to nutrient gradients. Field studies at sites with high, intermediate and low soil P concentrations (Table 11.1) established patterns between altered environments and cattail abundance. Controlled experiments were used to (1) isolate the individual effects of altered hydroperiod and increased nutrient loading, (2) identify the synergistic effects of both of these factors, and (3) determine the underlying mechanisms controlling the observed responses. The experimental approach was twofold: to examine responses in the structure and function of plant communities as well as the response of individual plants.

TABLE 11.1
Field Sites in WCA-2A Used for Studying the Effect of Soil P Concentration on Growth and Nutrient Storage in Cattail and Sawgrass Stands (see Fig. 11.1 for locations)

Field site	Distance from S-10C inflow (m)	Soil total P (mg kg^{-1})
F1	2,200	1,367
E1	2,200	1,313
F4	6,800	806
E4	6,800	660
U3	10,900	518
U1	14,450	475

11.3 CHANGES IN THE STRUCTURE AND FUNCTION OF PLANT COMMUNITIES ALONG A P GRADIENT

Field study sites were located in WCA-2A, a 447 km^2 region of the northern Everglades (Fig. 11.1). Most of the hydraulic loading to WCA-2A occurs through

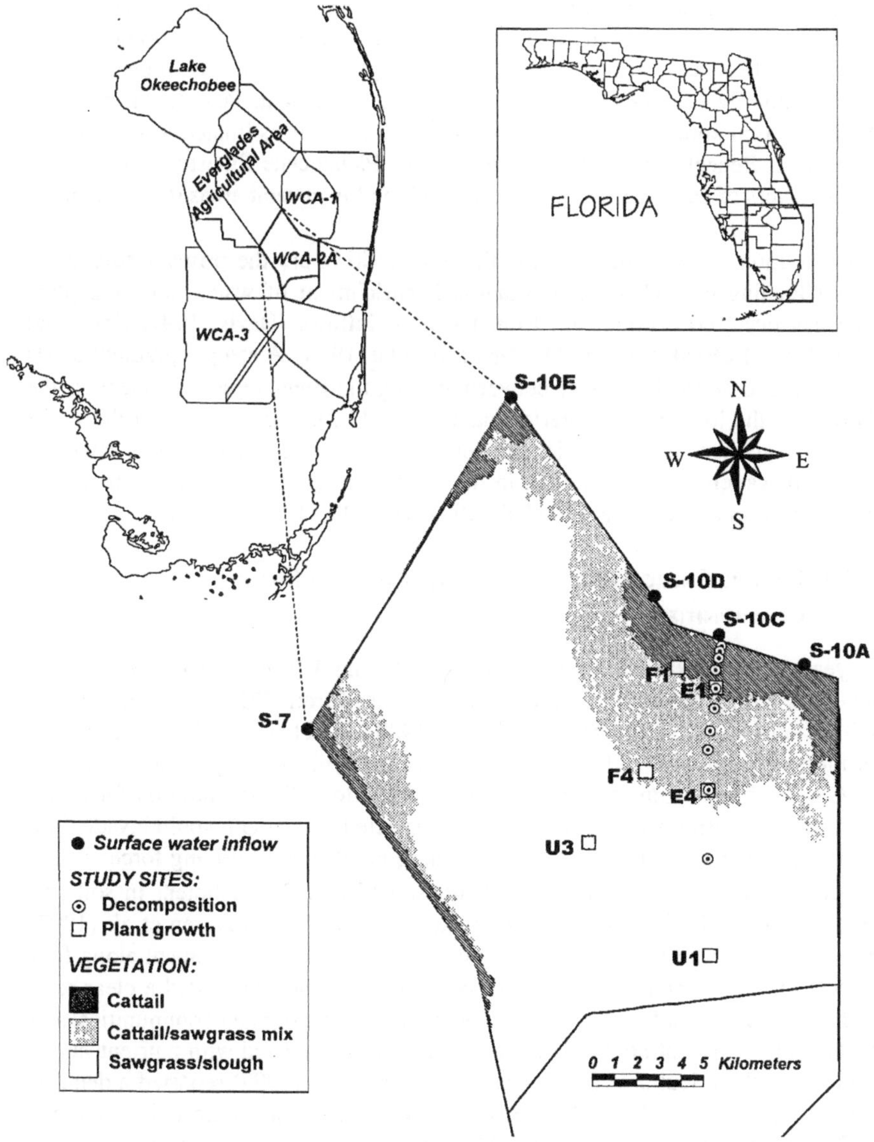

FIGURE 11.1 Field sites in Everglades Water Conservation Area 2A (WCA-2A) for macrophyte growth and decomposition studies. Spatial distribution of cattail/ sawgrass mix and sawgrass communities based on Rutchey and Vilchek, 1994.

the S-10 water control structures along the Hillsboro Canal into the northern portion of the study area. The general direction of flow is from north to south. Water depth is usually less than 1 m and varies considerably, both seasonally and annually, with occasional dry periods (SFWMD, unpublished data). The bulk of the surface outflow is through three control structures at the south end of WCA-2A, leading into WCA-3.

Soil in WCA-2A is currently not mapped by soil survey but is generally characterized as Everglades and Loxahatchee peats (Gleason et al., 1974). Everglades peat, the most common soil in the Everglades, is associated with the sawgrass marsh community. It is dark brown, finely fibrous to granular, with circumneutral pH, relatively high N content, and low SiO_2, Fe, and Al content. Peat depth in WCA-2A ranges from about 1 to 2 m, and age of basal peats is estimated to be 2,000 to 4,800 years. Beneath the peat lies a bedrock of Pleistocene limestone, with intermediate layers of calcitic mud, sandy clay, and sand in several areas (Gleason et al., 1974).

In addition to hydraulic loading, the S-10 inflows are the primary sources of nutrient loading to WCA-2A. Substantial P enrichment of water, soil, and plant tissue has occurred downstream from the S-10 inflows (Davis, 1991; Koch and Reddy, 1992; DeBusk et al., 1994; Miao and Sklar, 1998). A steep P gradient exists in surface and subsurface peat, between the high-nutrient region adjacent to the inflows and the low-nutrient interior marsh (Fig. 11.2). Coincident with the water and soil P gradient is a transition of dominant marsh vegetation, from sawgrass marsh and scattered aquatic slough in the interior, to cattail and mixed emergents near the inflows (Rutchey and Vilchek, 1994) (Fig. 11.1).

11.3.1 Cattail Encroachment into Sawgrass and Slough Communities

The yearly cattail expansion rate has increased from 1% in 1973 to 4% by 1987 (Wu et al., 1997). Monotypic cattail stands increased from 422 hectares in 1991 to 1647 hectares in 1995 (Rutchey and Vilchek, 1998). Wu et al. (1997) used a Markov transition probability model to quantify the dynamics of the rapid cattail expansion in WCA-2A. They estimated that the threshold soil total P concentration for accelerated cattail invasion was > 650 mg kg and expected that cattail would invade 50% of WCA-2A in another 6 to 10 years, assuming no change in driving forces.

Several recent field studies have demonstrated that P enrichment in soil and water is a major driving force of cattail expansion in WCAs. Doren et al. (1997) studied the pattern of marsh community composition and soil P content along four transects in WCAs, and in Everglades National Park. They suggested a clear relationship between cattail expansion, decline of natural marsh plant communities, and elevated soil P concentrations. These trends are also correlated with nutrient inputs associated with agriculture runoff. Craft and Richardson (1997) reported a detailed investigation into the relationship between soil nutrients, including P, base cations (Ca, Mg, K, Na), and metals (Al, Fe, Mn), and cattail encroachment into sawgrass communities in WCA-2A. The study confirmed that a soil P gradient originating from agriculture drainage pumped into northern WCA-2A existed in this area. The frequency of cattail and other species was positively correlated with the soil P

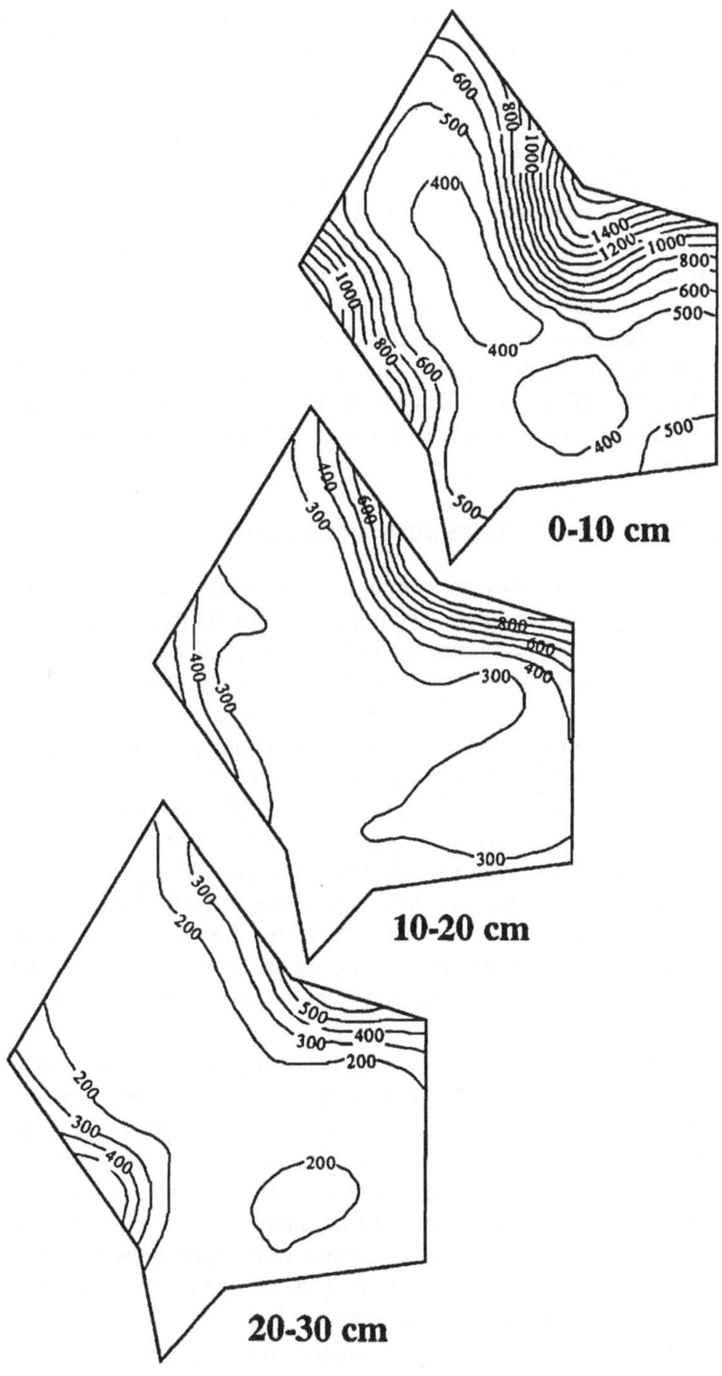

FIGURE 11.2 Spatial distribution of total P in WCA-2A peat at three depth intervals (DeBusk et al., 1994).

gradient, whereas sawgrass was inversely correlated with soil P. Craft and Richardson's study (1997) suggested that P enrichment of the peat favored cattail encroachment into sawgrass communities. In the areas with elevated soil P storage such as the Holey Land and Rotenberger Wildlife Management Areas, however, the major causal factor for cattail expansion was different from that of WCA-2A. Cattail expansion was largely controlled by hydrology in Holey Land and by muck fires in Rotenberger (Newman et al., 1998). It should be emphasized that the presence of high soil storage of P in these two areas is *a priori* condition for cattail expansion. Urban et al. (1993) found that sawgrass and cattail densities in WCA-2A were affected by soil and water nutrients, hydrology, and fire. Cattail increased at both nutrient-enriched and unenriched sites following fire, largely because of the nutrient release that was stimulated by fire. Again, this study showed that P loading into WCA-2A best explained cattail density fluctuations at the sites closest to the levee, whereas hydrology best explained sawgrass density fluctuations at the site distant from the levee.

11.3.2 Changes in Photosynthesis and Biomass

Photosynthesis is a sensitive functional ecosystem-level process and one of two biological routes (photosynthesis and microbial immobilization) by which soluble phosphate is converted into organic biomass. Thus, knowledge of photosynthesis can aid in the development of a model of P flux through Everglades communities and organisms. To examine the relationship between photosynthesis and soil P, leaf photosynthetic rates of sawgrass and cattail mature plants along the P gradient were measured in October, 1996 (Miao, unpublished data). The net photosynthetic rate (Ps) of both species had a nonlinear relationship with soil P concentrations (Fig. 11.3). The Ps increased with increasing soil P between 450 and 1000 mg kg^{-1}, then decreased in response to soil P greater than 1000 mg kg^{-1}. The maximum rate for both species occurred where soil P was around 1000 mg kg^{-1}. Both sawgrass and cattail showed a similar Ps in the reference areas where soil P concentration was approximately 450 mg kg. However, the Ps for cattail was approximately 47% greater than for sawgrass where soil P concentrations were > 500 mg kg^{-1} ($P < 0.05$).

There was no clear trend for stomatal conductance as a function of soil P for either species (Miao, unpublished data) (Fig. 11.3). On average, cattail stomatal conductance was approximately 78% greater than for sawgrass. Koch and Rawlik (1993) obtained similar results in WCA-2A. Leaf stomatal conductance is an important factor controlling transpiration that greatly affects evapotranspiration of entire plant communities. Greater stomatal conductance in cattail may result in greater water loss in the areas where sawgrass stands were replaced by cattail. Thus, the replacement of sawgrass by cattail may lead to alterations of the Everglades water budget, which has a fundamental impact on the hydrology of the system.

Biomass (g m^{-2}) of macrophyte communities is commonly used as an important criterion for plant community structure. In wetlands, dead and decomposed biomass plays an important role in nutrient cycling and peat formation, acting as a nutrient sink. Miao and Sklar (1998) studied both above- and below-ground biomass of

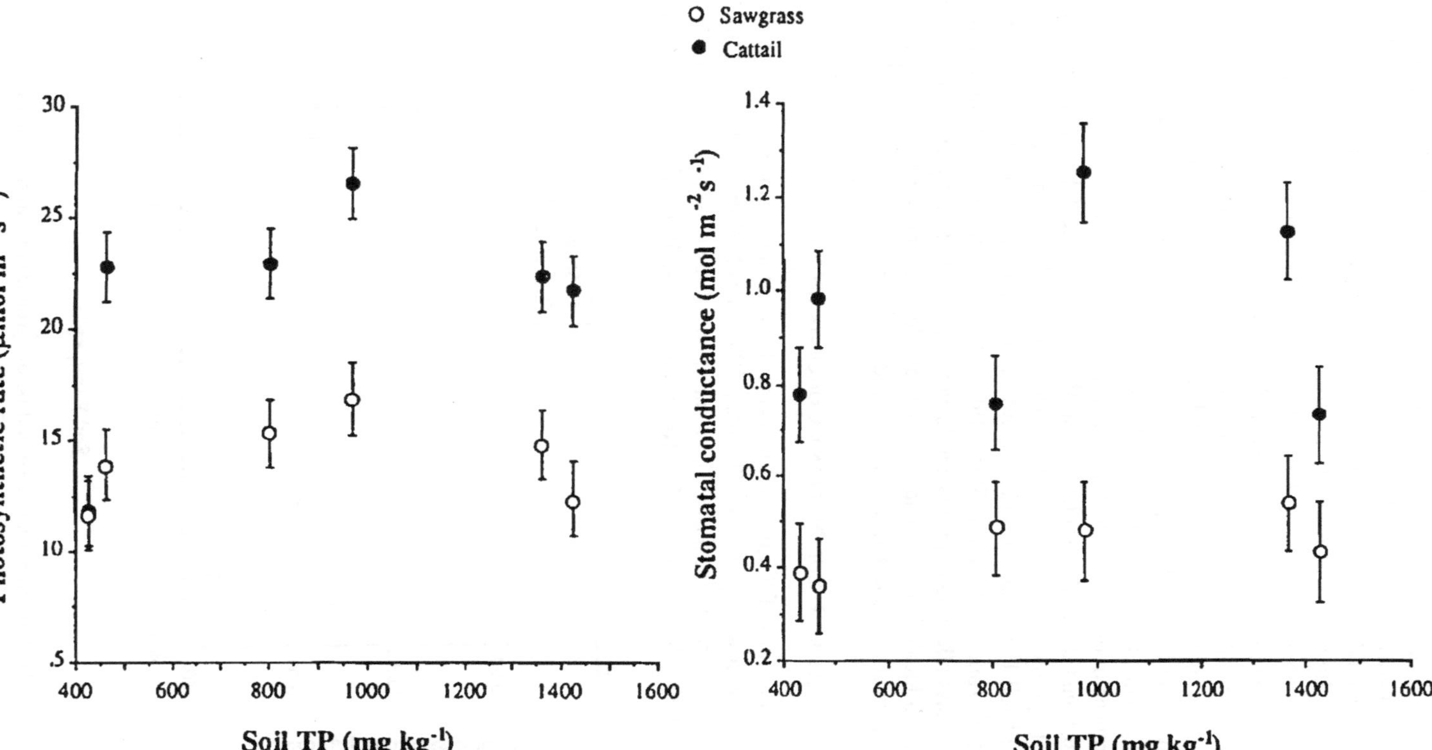

FIGURE 11.3 Net photosynthetic rates and stomatal conductance of sawgrass and cattail plants growing along a P gradient in Everglades WCA-2A.

sawgrass and cattail along the nutrient gradient in WCA-2A. Plant community biomass in sawgrass and cattail stands did not show a positive relationship with soil P (Fig. 11.4). Cattail biomass was similar ($P > 0.05$) across the P gradient, although slightly higher biomass was found in the area with low soil P. However, sawgrass biomass in the area with 518 mg kg^{-1} soil P was significantly lower relative to all other areas. Sawgrass biomass was generally greater than for cattail communities, although sawgrass plants exhibited lower Ps relative to cattail plants. Possible reasons for lower standing biomass in cattail stands are that (1) cattail plants have a higher leaf turnover rate (Davis, 1991), and (2) cattail leaves have greater inner air spaces and water content (Miao, unpublished data).

Differences in biomass allocation between both species in response to P enrichment were found along the gradient as well as in experimental studies (Miao and Newman, in preparation). Overall, sawgrass biomass allocation was similar across the range of soil P with about 60% to leaves and 40% to below-ground roots, shoot bases, and rhizomes (Miao and Sklar, 1998). In contrast, cattail plants exhibited flexible biomass allocation to roots and leaves in response to different soil P. In low soil P, cattail allocated more biomass to roots and less to leaves, whereas in high soil P, cattail allocated more to biomass leaves and less to roots.

11.3.3 Changes in Tissue Nutrient Concentrations

Tissue concentrations of limiting nutrients are determinants of the structure of plant communities in low-nutrient soil (Tilman, 1982, 1990; White, 1973). Although several studies examined the tissue P concentrations of sawgrass and cattail in WCA-2A (Toth, 1987, 1988; Davis, 1991; Koch and Reddy, 1992; Craft et al., 1995),

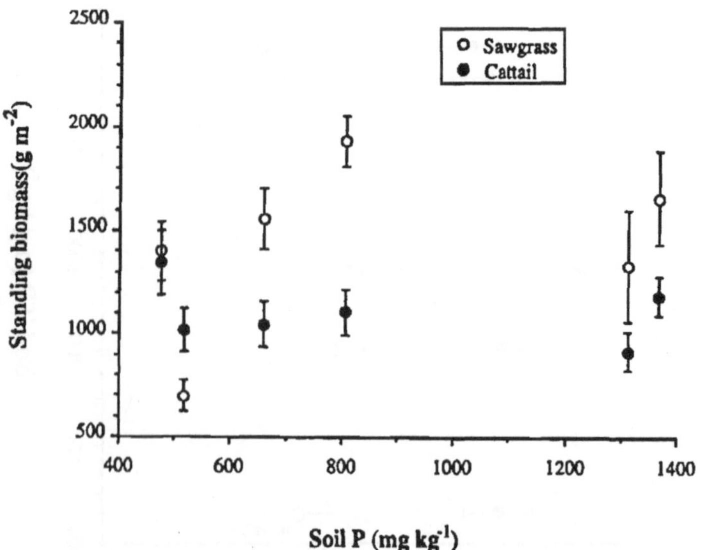

FIGURE 11.4 Standing biomass (above and below ground) of sawgrass and cattail communities along a P gradient in Everglades WCA-2A.

measurements were not conducted over a wide range of soil P and did not include tissues with storage function. The nutrient concentration of components with storage function, such as shoot bases, is an important consideration for understanding nutrient allocation at the whole-plant level. Species from high- and low-nutrient habitats may develop different growth and resource allocation patterns, consistent with their corresponding resource availability.

Nutrient concentrations (C, N, and P) of both sawgrass and cattail in tissues with growth, uptake, and storage function were examined for the first time along a P gradient by Miao and Sklar (1998). They found that P concentrations in macrophyte tissues were sensitive to soil P rather than soil N in the Everglades. This is not surprising, because soil N concentration is generally high for the Everglades peat soil (Koch and Reddy, 1992). Overall, P concentration of leaves, roots, rhizomes, and shoot bases exhibited a positive correlation with soil P ($P < 0.0001$) (Fig. 11.5). Positive correlations with surface water P were also found. Numerous studies have correlated the nutrient concentration of submerged macrophytes with that of the surrounding water (Gerloff and Krumbholz, 1966; Gossett and Norris, 1971; Boyd and Walley, 1972). Poor or nonexistent correlations were found in some studies when attempts were made to compare concentration of nutrients in soil with concentration in emergent macrophytes (Stake, 1967; Boyd and Hess, 1970). However,

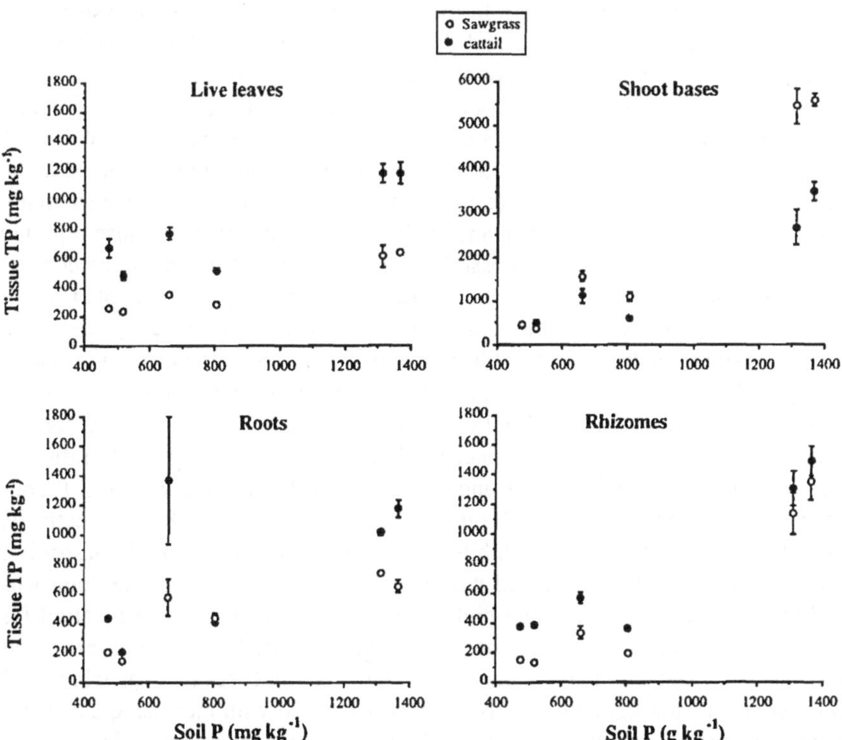

FIGURE 11.5 Tissue P concentration in sawgrass and cattail plants growing along a P gradient in Everglades WCA-2A.

a significant correlation of plant tissue nutrient concentration with soil P was found in Theresa Marsh (Klopatek, 1975). The significance of the correlation of macrophyte tissue P with soil and surface water P in WCA-2A is that tissue P appears to be a suitable indicator of soil and surface water P availability. Among the four plant components studied, the shoot base showed the greatest increase in P concentrations with increasing soil and water P. Thus, P concentrations in shoot bases may be a better indicator than leaves of soil and surface water P availability in the Everglades.

Differences in tissue nutrient concentration between species or within species may vary depending on the specific tissue being compared. In theory, plants from low-nutrient environments exhibit proportionally lower biomass and nutrient allocation to growth but greater allocation to storage (Tilman, 1982). This allocation pattern may increase plant survivorship if the excess nutrients are stored until needed. In contrast, plants from high-nutrient habitats may exhibit a greater nutrient concentration in tissues such as leaves with growth functions to maximize the growth rate. This hypothesis was not rejected when comparing sawgrass and cattail in WCA-2A. Compared to sawgrass, cattail consistently had greater P concentrations in leaves, roots, and rhizomes, whereas sawgrass had greater P storage in shoot bases except at the lowest soil P site. For example, in the area with higher soil P (518 mg kg^{-1}), tissue P concentrations of sawgrass vs. cattail were 259 vs. 676 mg kg^{-1} for leaves, 211 vs. 440 mg kg^{-1} for roots, 152 vs. 373 mg kg^{-1} for rhizomes, and 452 vs. 379 mg kg^{-1} for shoot bases (Fig. 11.5). In the historic Everglades, a low-P system with frequent disturbances such as fire and drought (Wade et al., 1980), it would have been advantageous for plants to have a greater capacity for resource storage.

Sawgrass and cattail leaf and root P concentrations along the soil P gradient (Miao and Sklar, 1998) are generally consistent with other studies in the WCA-2A (Table 11.2). However, there is considerable variability among studies. For example, root P concentrations of both species in Heilman et al. (in press) were substantially greater than those reported in Koch and Reddy (1992) but similar to those reported in Miao and Sklar (1998). The possible reasons for the differences include (1) different sampling time (seasonal variation), (2) different ages of sampled plants (age variation), (3) different number of sampling events, (4) differences in sampling techniques such as inclusion of dead leaves, and (5) microtopographical differences.

Tissue N/P ratio can be used as an index of nitrogen and phosphorus balance in plants. The N/P, C/P, and C/N ratios varied with soil P, species, and tissues (Table 11.3). In areas with soil P concentrations over 1000 mg kg^{-1}, N/P ratios of sawgrass and cattail averaged 10:1 and 8:1, respectively. In areas with soil P concentrations below 518 mg kg^{-1}, N/P ratios increased to 44:1 and 18:1, respectively. Among the tissues studied, roots exhibited the greatest values of N/P for both species, while N/P and C/P ratios decreased with increasing soil P, and little variation was found for C/N. Sawgrass exhibited higher N/P, C/P, and C/N ratios than cattail for all plant tissues except the shoot base. Interestingly, cattail shoot bases exhibited higher N/P, C/P, and C/N ratios relative to sawgrass shoot bases in areas where soil P was above 518 mg kg. Although N/P ratios decreased with increasing soil P for both species, sawgrass N/P ratios exhibited much greater variation than cattail. This was also found by Koch and Reddy (1992) in a study of macrophyte nutrient concentration in relation to soil P in the Everglades.

TABLE 11.2
Tissue P Concentrations of Sawgrass and Cattail from Existing Studies Conducted in WCA-2A of the Everglades

Plant type		P concentration: Unenriched site, mg g^{-1}	P concentration: Enriched site, mg g^{-1}	Reference
Sawgrass	Whole	0.22	1.94	Steward and Ornes, 1983
		0.20	0.70	Davis, 1991
	Shoot	0.2–0.3	1.0–1.2	Richardson, 1991
		0.22	0.75	Koch and Reddy, 1992
		0.33–0.41	0.65–1.41	Heilman et al., in press
		0.24	0.64	Miao and Sklar, 1998
	Root	0.2–0.3	0.8–2.3	Richardson, 1991
		0.12	0.75	Koch and Reddy, 1992
		0.29–0.89	2.32–5.25	Heilman et al., in press
		0.15	0.65	Miao and Sklar, 1998
Cattail	Whole	0.5	1.5	Davis, 1991
	Shoot	0.5–0.8	1.3–3.2	Richardson, 1991
		0.75	2.3	Koch and Reddy, 1992
		0.69–0.72	1.52–1.70	Heilman et al., in press
		0.49	1.18	Miao and Sklar, 1998
	Root	0.3–0.4	1.3–3.2	Richardson, 1991
		0.4	0.8	Koch and Reddy, 1992
		0.47–0.73	2.22–3.46	Heilman et al., in press
		0.72	1.18	Miao and Sklar, 1998

11.3.4 Changes in Nutrient Storage and Distribution in Plant Communities

Quantifying nutrient storage in various components of vegetation is important to understanding nutrient cycling processes at the ecosystem level. While nutrients in wetlands may be lost from the system by various biogeochemical processes, some nutrients are accumulated within the system through sedimentation or organic matter accumulation. In peat-dominated wetlands, a major portion of the nutrients is stored in live and detrital plant tissue, microbial biomass, and soil organic matter (Reddy et al., 1993).

Nutrient storage in macrophyte communities along the soil P gradient in WCA-2A, calculated as mass per unit area (g kg^{-1}), was reported by Miao and Sklar (1998) (Fig. 11.6). In general, P storage in both communities increased significantly with increasing soil P, but C storage decreased. Total N storage decreased with increasing soil P for cattail communities, but no significant change occurred for sawgrass communities with the exception of one site. Total N and C storage in sawgrass communities was greater than in cattail communities, which probably resulted from slower decomposition rates of sawgrass plants. For sawgrass communities, approximate P storage ranged from 2 to 80 kg ha^{-1}, N from 50 to 800 kg ha^{-1}, and C from

TABLE 11.3
N/P, C/P, and C/N Ratios in Plant Tissues of Sawgrass (S) and Cattail (C) Populations Growing along a Phosphorus Gradient in WCA-2A

	Leaves		Roots		Rhizome		Shoot base	
Soil TP, mg kg^{-1}	S	C	S	C	S	C	S	C
			N/P					
1367	24.5	17.8	54.7	30.4	11.3	12.2	4.0	6.1
1313	25.8	15.6	40.2	39.8	13.7	14.8	5.2	9.5
806	48.9	30.8	71.1	72.6	58.5	32.9	19.0	28.2
660	45.2	22.9	54.9	50.4	36.9	27.4	12.8	19.8
518	65.1	29.7	144.4	53.8	89.3	33.6	72.2	32.6
475	56.3	25.3	132.3	95.0	81.5	33.2	73.7	33.0
			C/P					
1367	1878	1037	1957	968	1004	761	215	328
1313	2162	921	1626	1146	1124	856	228	472
806	4303	2238	2858	2937	6219	3072	1190	1992
660	3532	1497	2937	1674	4167	1940	825	1192
518	5060	2447	8182	1775	9439	2869	3600	2687
475	4747	1824	6153	2825	8111	3026	2966	3178
			C/N					
1367	76.8	58.1	36.3	31.7	90.2	61.8	52.4	54.0
1313	82.0	59.6	40.3	29.0	82.8	58.0	43.0	49.9
806	87.5	74.1	40.2	41.3	107.8	94.52	62.8	70.8
660	78.5	64.9	51.5	33.1	112.5	70.7	64.3	60.3
518	77.6	83.1	56.6	32.3	106.8	87.0	51.3	80.8
475	84.2	71.6	46.1	31.6	99.8	92.1	40.1	96.1

2,500 to 40,000 kg/ha (Fig. 11.6). For cattail communities, approximate P storage ranged from 5 to 55 kg ha^{-1}, N from 25 to 800 kg ha^{-1}, and C from 2,500 to 40,000 kg ha^{-1}.

P, N, and C storages were positively correlated ($P < 0.001$, Fig. 11.7). For the regressions of P vs. N storage and P vs. C storage, three distinct groups characterized by different slopes were found for sawgrass communities. The slope of the P vs. N regressions decreased with decreasing soil P. This suggests that P storage for sawgrass communities was largely limited by soil P, for the same N storage. For example, when N storage was 40 g kg^{-1}, P storage of sawgrass communities in the high soil P area (1,313 mg kg^{-1}) was approximately four times greater than for the low soil P area (475 mg kg^{-1}). Similar relationships were found for cattail communities, except that only two different slopes were found. However, the regressions of N vs. C storage relationship were similar along the soil P gradient for both species (Fig. 11.7).

The distribution of P, N, and C storage in live, dead, and partially decomposed dead biomass did not vary significantly between species and among soil P. Approximately 50% of P and about 30 to 40% of N and C were stored in live biomass in all areas. However, there was a difference in P storage in live tissues. P storage in

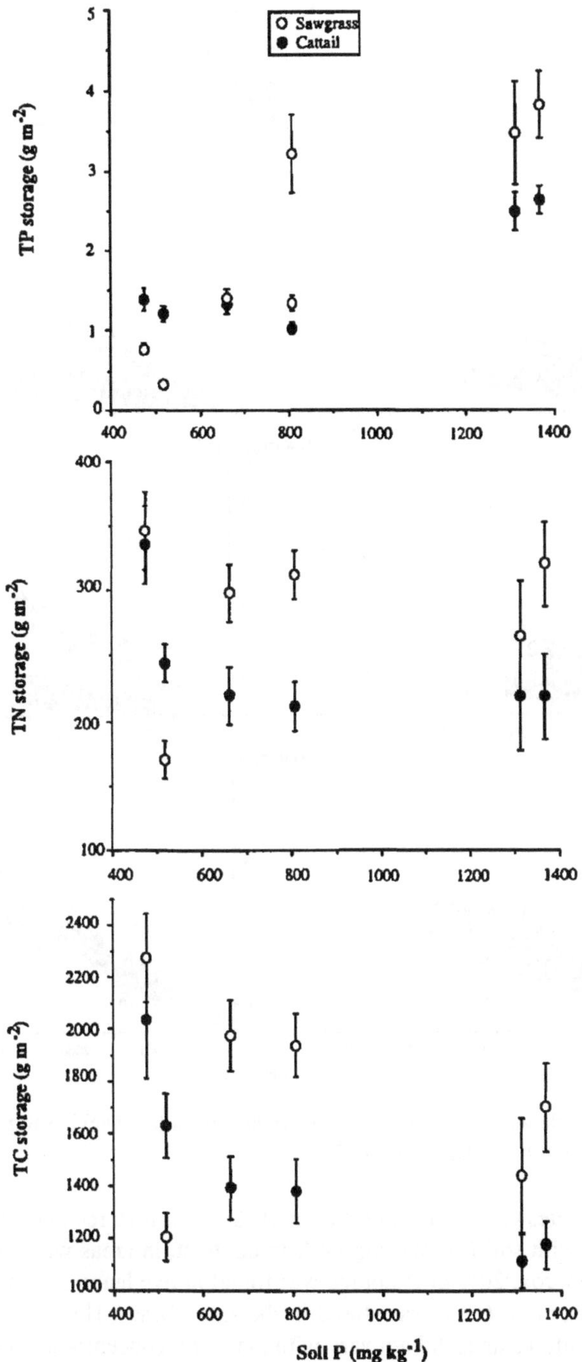

FIGURE 11.6 Mean (±1 SE) areal tissue nutrient content (storage) in sawgrass and cattail communities along a P gradient in WCA-2A.

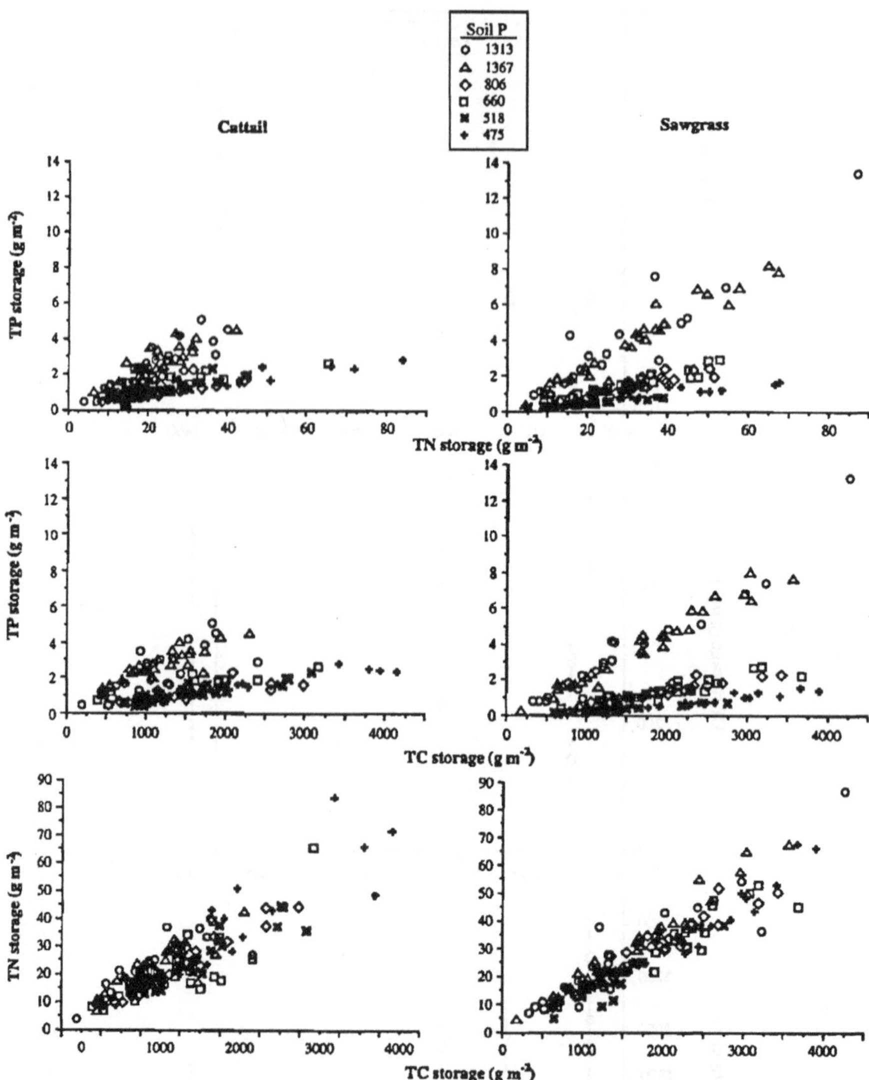

FIGURE 11.7 Relationships between areal nutrient content (storage) in sawgrass and cattail communities along a P gradient in WEC-2A.

live sawgrass tissues varied with soil P. Over 60% of P was stored in the live sawgrass leaves in areas with soil P < 500 mg kg^{-1}. In contrast, in areas with soil P > 1,300 mg kg^{-1}, only approx. 20% of P storage was found in live leaves. In these areas, the majority of P (approx. 60%) was stored in the shoot bases. However, P storage in live tissues of cattail stands did not vary with the soil P concentration approximately 50 to 60% of P in leaves for cattails along the soil P gradient. N and C storage distributions among the live tissues were similar along the gradient for sawgrass and cattail.

11.3.5 Changes in Turnover of Detrital Material

Turnover of plant detritus is an important component of nutrient and energy cycling in the marsh ecosystem. Organic C in decomposing plant tissue is a major source of energy to the heterotrophic microbial community, and subsequently, to the higher trophic levels of the marsh ecosystem. Detrital turnover rate to a great extent regulates nutrient availability to macrophytes and is a key factor in determining short-term retention and long-term storage of nutrients in vegetation and soil. A laboratory study was conducted using plant detritus collected along the soil P gradient in WCA-2A to determine the effects of detrital P concentration and lignocellulose content on the rate of organic C turnover in the soil litter layer and dead plant biomass (DeBusk, 1996).

Ten field sampling sites were established along a 10-km transect extending from the S-10C inflow into the interior marsh of WCA-2A (Fig. 11.1). Standing dead (attached to the plant) leaves from either cattails, sawgrass, or both species, based on occurrence at the sampling sites, and composite samples of the soil litter layer were analyzed for total P, lignin, and cellulose content. Organic C mineralization in the standing dead and litter material was determined by measuring CO_2 evolution associated with aerobic microbial respiration in the detritus samples (DeBusk, 1996). Mineralization rates measured in the laboratory thus represented the maximum, or potential, rates of detrital C turnover expected at the corresponding field sites.

Study results indicated that P concentration in standing dead leaves of cattail and sawgrass plants and in the soil litter layer significantly increased (linear regression analysis $P < 0.05$) between the low soil P interior marsh and the high soil P area near the inflows (Fig. 11.8). Elevated P concentration in the soil litter layer near the inflow was primarily the result of P immobilization by microbial decomposers utilizing the large pool of labile C (DeBusk, 1996). Mean P concentration was significantly lower in standing dead material (both cattail and sawgrass) than in the litter layer. Low P concentration in standing dead plant tissue results from rapid loss of soluble and readily mineralizable constituents. In contrast, P concentration in plant detritus was substantially higher in the soil litter layer as a result of microbial immobilization of P in surface and soil water, especially near the S-10 inflows (DeBusk, 1996).

There was no significant spatial trend in either lignin or cellulose content of plant detritus along the soil P gradient. However, mean cellulose content of standing dead material was significantly higher ($P < 0.05$) in cattail than in sawgrass, which was in turn significantly greater than cellulose content of the soil litter layer (Fig. 11.9). Similarly, mean lignin content of the soil litter layer was significantly higher than the lignin content of standing dead material, but there was no significant difference between lignin content of cattail and sawgrass standing dead material.

Mineralization of organic C in cattail and sawgrass standing dead tissue was significantly correlated with P concentration in the substrate (Fig. 11.10). The C mineralization rate in cattail standing dead material was more greatly affected by changes in substrate P concentration than in sawgrass, even though mean C mineralization rates were not significantly different between cattail and sawgrass. This

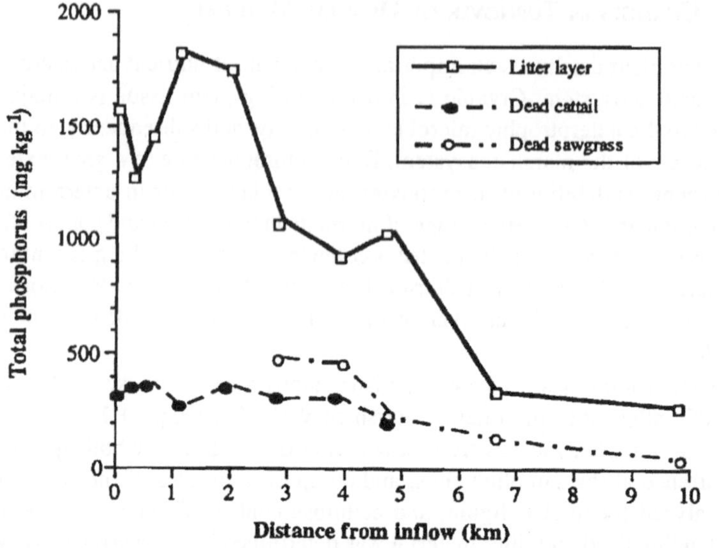

FIGURE 11.8 Total phosphorus concentration of standing dead leaves of cattail and sawgrass and the soil litter layer as a function of distance down gradient from the S-10C inflow in WCA-2A.

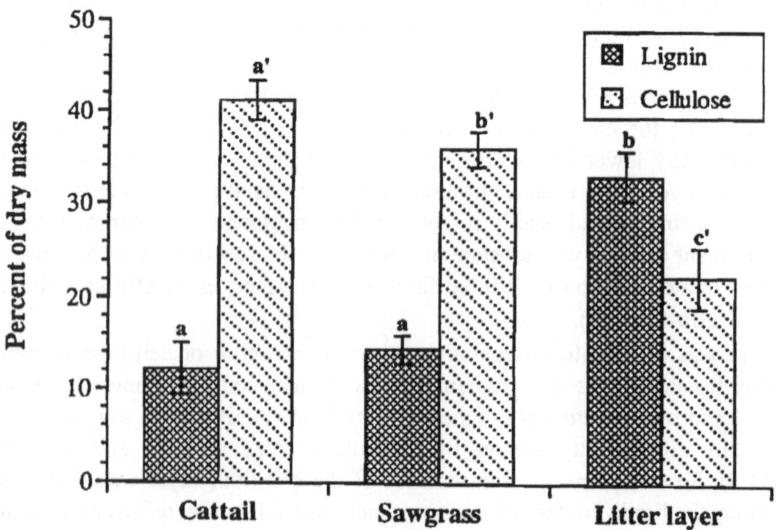

FIGURE 11.9 Lignin and cellulose content of standing deal material (cattail and sawgrass) and the soil litter layer along the WCA-2A soil P gradient. Each data point (bar) represents the mean of representative samples along the transect (see text); standard deviation is denoted by error bars. Values with the same letter designation were not significantly different ($P < 0.05$).

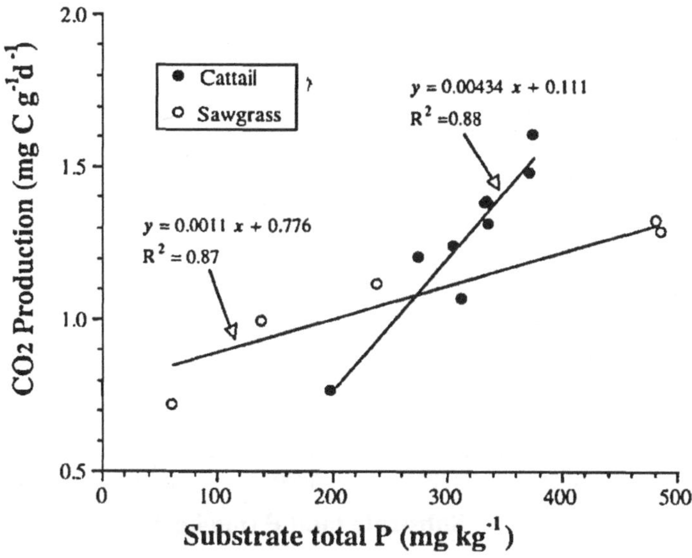

FIGURE 11.10 Microbial CO_2 production (C mineralization) as a function of substrate total P concentration in laboratory incubations of cattail and sawgrass standing dead leaves.

difference in response to P content may be because of the comparatively higher content of cellulose in the cattail standing dead tissue. Nevertheless, P concentration in the substrate accounted for about 88% of the variability in CO_2 production for both cattail and sawgrass.

Increased detrital P concentration also corresponded to an increase in C mineralization (CO_2 production) in the litter layer (Fig. 11.11). Total P concentration accounted for 85% of the variability in the CO_2 production rate, for all sampling sites. Rate of CO_2 production in the litter layer ranged from 0.4 to 1.9 mg C g^{-1} d^{-1} at the 10 sampling sites, compared to a range of 0.7 to 1.6 mg C g^{-1} d^{-1} for cattail and sawgrass standing dead material at the same sites. Despite the significantly higher lignin content of the litter layer compared with standing dead material, mean CO_2 production rates for litter and standing dead at the 10 sites were not significantly different. Mineralization of C in standing dead material was evidently limited by its low tissue P content. On the other hand, comparison of CO_2 production rates in standing dead and litter samples with similar total P concentrations (Figs. 11.10 and 11.11) indicates that C mineralization was more rapid in standing dead material, presumably because of its lower lignin and higher cellulose content. It should be noted that C mineralization in standing dead material may also be moisture limited under field conditions, in contrast to the conditions imposed during laboratory incubations.

The high C quality and low-nutrient (N and P) content of the standing dead material suggest that newly fallen litter would act as a short-term nutrient sink, through uptake and assimilation into soil microbial biomass. In fact, substantial immobilization of N and P by dead cattail and sawgrass leaf tissue placed in the

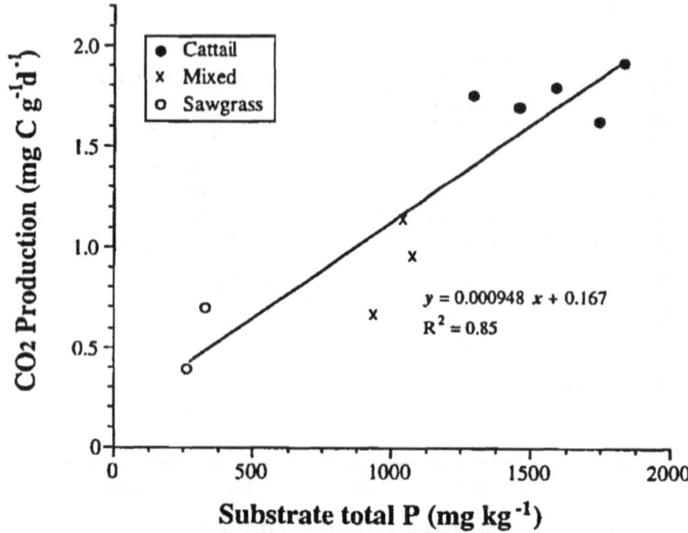

FIGURE 11.11 Microbial CO_2 production (C mineralization) as a function of substrate total P concentration in laboratory incubations of soil litter. Data for cattail-dominated, sawgrass-dominated, and mixed stands are denoted by separate symbols.

soil litter layer was observed during a short-term (six-month) decomposition study along the WCA-2A P gradient (DeBusk, 1996).

Although the rate of organic C turnover in detrital pools increased in response to P enrichment, it may not be assumed that organic C storage has decreased in high soil P areas of WCA-2A. In fact, there is evidence, based on Cs dating of intact soil cores, that the rate of organic matter accumulation has increased in the high soil P areas (Craft and Richardson; 1993 Reddy et al., 1993). The increased accumulation rate can be accounted for by the increased net primary production in these areas. Increased inputs of organic matter into the marsh detrital pool may offset accelerated turnover of this pool, thereby increasing the size of the pools. It is also possible that turnover rate may be attenuated by low O_2 availability in some areas with high soil P. A study by Belanger (1989) provided evidence that nutrient-enriched areas of WCA-2A, which were previously dominated by aerobic conditions in the water column and litter layer, have become primarily anaerobic because of high sediment oxygen demand.

Nutrient enrichment of previously oligotrophic areas of WCA-2A has resulted in increased C mineralization potential in soil and plant detrital pools, due to enhancement of overall substrate quality. Mineralization of organic C was significantly affected by C quality (lignin and cellulose content) and P content of the substrate. Carbon quality in individual detrital pools did not vary significantly along the P gradient; therefore, within-pool variability in the potential C mineralization rate was a function of substrate P enrichment.

Plant standing dead material and the soil litter layer are potentially very active (rapid turnover) C pools in the Everglades marsh because of their relatively high C availability. However, decomposition of standing dead tissue of cattails and sawgrass is most likely nutrient- and moisture-limited. The litter layer is also potentially important in short-term nutrient cycling. With a relatively high C and nutrient availability to support a large population of microbial decomposers, the litter layer may serve as either a source or sink for nutrients.

11.4 EXPERIMENTAL STUDIES: EFFECTS OF SOIL AND SURFACE WATER PHOSPHORUS CONCENTRATION ON MACROPHYTE GROWTH

Transect studies and modeling efforts have documented cattail expansion in the Everglades that is associated with both P-enriched and disturbed environments. Experimental studies support a close relationship between cattail growth and P enrichment and indicate that cattail is a competitively superior species under enriched conditions (Newman et al., 1996).

11.4.1 Growth of Transplanted Cattail Plants at Enriched and Unenriched Site

A field growth experiment was conducted at high and low P sites in WCA-2A (Miao and Newman, in preparation). Cattail seedlings germinated from seeds collected in WCA-2A, approximately 50 cm in height, were transplanted to two slough sites and monitored for approximately six months. These two sites had different soil P (1571 and 401 mg kg^{-1}, respectively) but similar water depths. At the end of the experiment, plants grown at the high P site exhibited approximately 170% increase in growth rate and more than tenfold increase in biomass compared with plants at the low P site (Fig. 11.12). On average, each of the original plants produced approximately

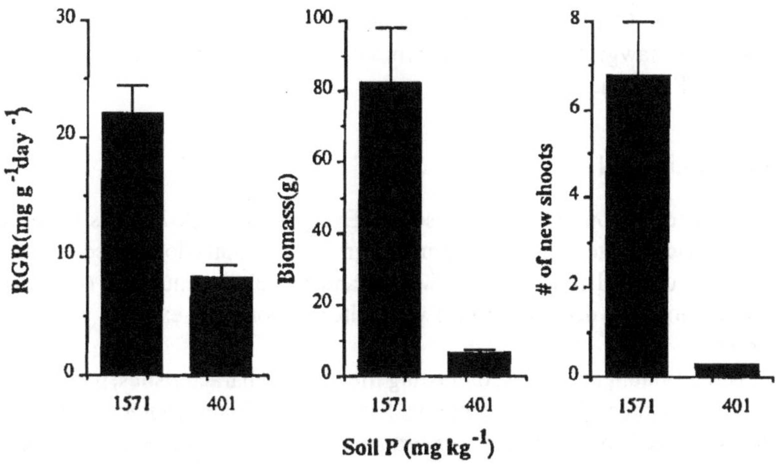

FIGURE 11.12 Relative growth rate (RGR), biomass, and shoot production for cattail plants grown at unenriched and enriched sites.

seven shoots at the high P site, whereas no shoot production was observed at the low P site. Slow growth is characteristic of plants adapted to low-nutrient habitats (Chapin et al., 1987). Higher growth rate and biomass production of plants grown at the high P site corresponded with a greater Ps. The Ps of plants at the high P site was significantly greater than at the low P site. In addition, plants grown at the high P site exhibited much higher tissue P concentration than those grown at the low P site. The Ps and tissue P concentrations in the transplanted seedlings corresponded to the trends observed in established cattail stands along the soil P gradient in WCA-2A.

11.4.2 Sawgrass and Cattail Regrowth after Leaf Removal

Differences between sawgrass and cattail in patterns of biomass allocation may result in differences in regrowth capability after disturbances. A study was set up in WCA-2A to test the following hypotheses:

1. Leaf regrowth after leaf removal will largely result from internal resource storage in sawgrass, but more likely will result from external soil P availability for cattail.
2. For low soil P, regrowth of both species will be similar, whereas for high soil P, cattail leaf growth will be greater than for sawgrass.

Sawgrass and cattail seedlings germinated from seeds collected in WCA-2A, approximately 50 cm in height, were transplanted to high and low P soil from WCA-2A. Six weeks after transplanting, apparent differences in plant height between plants in high and low soils were observed. At this time, leaves were removed just above the leaf meristem. Leaf regrowth rates were monitored for six to eight weeks by measuring the number and length of leaves. Overall, when plants were grown in the high-P soil, cattail exhibited significantly greater (about 75%) regrowth than sawgrass, whereas regrowth rates were similar when the two species were grown in the low-P soil (Fig. 11.13). This may demonstrate the ability of cattail to recover more quickly than sawgrass from certain types of disturbance (e.g., surface fires) in the presence of P enriched soils.

11.5 CONCLUSION

Increased P concentration in water and soil in Everglades WCA-2A was linked to the change in macrophyte vegetation from sawgrass- to cattail- dominated communities. The structure and function of sawgrass and cattail communities (including primary production and decomposition) were altered along the existing water and soil P gradient.

Sawgrass and cattail exhibited contrasting life history characteristics, particularly with respect to resource allocation patterns. Sawgrass exhibited characteristics typical of plants adapted to low-nutrient environments, whereas cattail was similar to plants adapted to high-nutrient environments. In particular, differences in seedling growth, biomass, and nutrient allocation were generally consistent with theories of terrestrial plant responses to contrasting nutrient habitats.

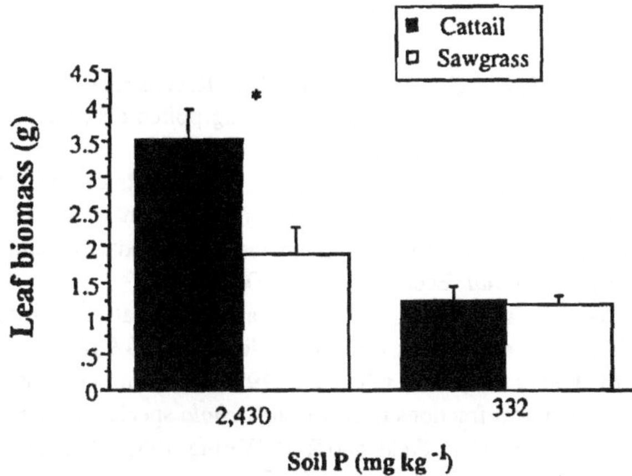

FIGURE 11.13 Leaf regrowth after leaf removal as a function of soil P for sawgrass and cattail plants.

Slow rhizome growth in a low-P environment suggested a limited cattail expansion. While regrowth of cattail in the high-P soil was greater relative to sawgrass following disturbance, a similar regrowth rate was observed for both species in a low-P environment. This suggests that cattail is not competitively superior in non-nutrient impacted areas of the Everglades marsh but gains a competitive advantage over sawgrass, in terms of photosynthesis and clonal growth, in nutrient-enriched areas. Results indicated that the shift from a P-limited to P-enriched system contributed significantly to the spread and dominance of cattails in WCA-2A.

Turnover of detrital material in the sawgrass and cattail marsh was significantly affected by P enrichment. Potential C mineralization rates in standing dead leaves were enhanced by high C quality (high cellulose content relative to lignin content) of the substrate, but limited by low nutrient availability, especially P. Potential C mineralization rates in the soil litter layer were attenuated, relative to the standing dead component, by somewhat lower C quality, but were significantly enhanced by P enrichment near the inflows.

Several factors in the historical Everglades (including extreme P limitation, fires, occasional periods of prolonged droughts, winter frosts, and herbivory) may limit cattail expansion. Anthropogenic pressures on the northern Everglades have significantly altered both nutrient (P) availability and hydroperiod during the last few decades. However, based on research results reported in this paper, P enrichment appears to be the principal factor responsible for the increased abundance of cattail in WCA-2A.

11.6 ACKNOWLEDGMENTS

We thank S. Newman for sharing her unpublished soil nutrient data, and P. V. McCormick, T. D. Fontaine, S. Newman, C. Fitz, K. Rutchey, Y. Wu, S. Gray, and two anonymous reviewers for their critical review of the manuscript.

REFERENCES

Bartow, S. M., C. B. Craft, and C. J. Richardson. 1996. Reconstructing historical changes in Everglades plant community composition using pollen distribution in peat. Journal of Lake and Reservoir Management 12:313–322.

Belanger, T. V., D. J. Scheidt, and J. R. Platko, II. 1989. Effects of nutrient enrichment on the Florida Everglades. Lake and Reservoir Management 5:101111.

Boyd, C. E. and L. W. Hess. 1970. Factors influencing shoot production and mineral nutrient levels in *Typha latifolia*. Ecology 51:296–300.

Boyd, C. E. and W. W.Walley. 1972. Production and chemical composition of *Saururus cernuus* L. at sites of different fertility. Ecology 53:927–932.

Chapin, F. S., J. M. Follett, and K. F. O'Connor. 1982. Growth, phosphate absorption, and phosphorus chemical fractions in two *Chionochola* species. J. Ecology 70:305–321.

Chapin, F. S., 2I, A. J. Bloom, C. B. Field, and R. H. Waring. 1987. Plant responses to multiple environmental factors. BioScience 37:49–57.

Craft, C. B. and C. J. Richardson. 1993. Peat accretion and phosphorus accumulation along a eutrophication gradient in the northern Everglades. Biogeochemistry 22:133–156.

Craft, C. B., J. Vymazal, and C. J. Richardson. 1995. Response of Everglades plant communities to nitrogen and phosphorus additions. Wetlands 15:258–271.

Craft, C. B. and C. J. Richardson. 1997. Relationships between soil nutrients and plant species composition in Everglades peatlands. J. Environ. Qual. 26:224–232.

Davis, J. H., Jr. 1943. The Natural Features of Southern Florida, Especially the Vegetation, and the Everglades. Florida Geological Survey, Tallahassee, Florida Bulletin 25.

Davis, S. M. 1989. Sawgrass and cattail production in relation to nutrient supply in the Everglades. Pages 325–341 *in* R. R. Sharitz and J. W. Gibbons, Editors. Fresh Water Wetlands and Wildlife. Office of Scientific and Technical Information, U. S. Department of Energy, Oak Ridge, Tennessee.

Davis, S. M. 1991. Growth, decomposition, and nutrient retention of *Cladium jamaicense* Crantz and *Typha domingensis* Pers. in the Florida Everglades. Aquatic Botany 40:203–224.

Davis, S. M. 1994. Phosphorus inputs and vegetation sensitivity in the Everglades. Pages 357–378 *in* S. M. Davis and J. C. Ogden, Editors. Everglades: The Ecosystem and Its Restoration. St. Lucie Press, Delray Beach, Florida, USA.

Davis, S. M. and J. C. Ogden. 1994. Everglades: The Ecosystem and Its Restoration. St. Lucie Press, Delray Beach, FL, USA.

DeBusk, W. F., K. R. Reddy, M. S. Koch, and Y. Wang. 1994. Spatial distribution of soil nutrients in a northern Everglades marsh: water conservation area 2A. Soil Sci. Soc. Am. J. 58:543–552.

DeBusk, W. F. 1996. Organic matter turnover along a nutrient gradient in the Everglades. Ph.D. diss. University of Florida, Gainesville.

Doren, R. F., T. V. Armentano, L. D. Whiteaker, and R. D. Jones. 1997. Marsh vegetation patterns and soil phosphorus gradients in the Everglades ecosystem. Aquatic Botany 30:1–19.

Gerloff, G. C. and P. H. Krumbholz. 1966. Tissue analysis as a measure of nutrient availability for the growth of angiosperm aquatic plants. Limnol. Oceanogr. 11:529–539.

Gleason, P. J., A. D. Cohen, P. Stone, W. G. Smith, H. K. Brooks, R. Goodrick, and W. Spackman, Jr. 1974. The environmental significance of holocene sediments from the Everglades and saline tidal plains. Pages 297–351 *in* P. J. Gleason, Editor. Environments of South Florida, present and past. Miami Geol. Soc., Coral Gables, FL.

Gossett, D. R. And W. E. Norris, Jr. 1971. Relationship between nutrient availability and content of nitrogen and phosphorus in tissue of the aquatic macrophytes *Eichornia crassipes* (Mart) Solms. Hydrobiologia 38:15–28.

Harper, R. M. 1927. Natural Resources of Southern Florida. Florida Geological Survey, Tallahassee, Florida 18th Annual Report.

Harshberger, J. W. 1914. The Vegetation of South Florida, South of 27 degrees 30' North, Exclusive of the Florida Keys. Trans. Wagner Free Inst. Sci. 7:49–189.

Heilman, P. E. 1968. Relationship of availability of P and cations to forest succession and bog formation in interior Alaska. Ecology. 49:331–336.

Heilman, E. R., K. R. Reddy, and N. B. Comerford. 1999. Phosphorus uptake kinetics of cattail (*Typha* spp.) and sawgrass (*Cladium jamaicense*). Aquatic botany (in review).

Hofstetter, R. H. and F. Parsons. 1979. The ecology of sawgrass in the Everglades of southern Florida. Pages 165–170 *in* R. M. Linn, Editor. Proceedings of 1st Conference on Scientific Research in the National Parks, US Department of Interior, National Park Service Transaction and Proceedings Series. Department of the Interior, National Park Service, Washington, DC.

Jensen, J. R., K. Rutchey, M. S. Koch, and S. Narumalani. 1995. Inland wetland change detection in the Everglades Water Conservation Area 2A using a time series of normalized remotely sensed data. Photogrammetric Engineering and Remote Sensing 61:199–209.

Kadlec, J. A. 1987. Nutrient dynamics and wetlands. Pages 393–419 *in* K. R. Reddy and W. H. Smith, Aquatic Plants for Water Treatment and Resource Recovery. Magnolia Publishing Inc., Orlando, FL.

Klopatek, J. M. 1975. The role of emergent macrophytes in mineral cycling in a freshwater marsh. Pages 367–393 *in* G. F. Howell, J. B Gentry, and M. H. Smith, Editors. Mineral Cycling in Southeastern Ecosystems. National Technical Information Science, Springfield, Virginia.

Koch, M. S. and K. R. Reddy. 1992. Distribution of soil and plant nutrients along a trophic gradient in the Florida Everglades. Soil Sci. Soc. Am. J. 56:1492–1499.

Koch, M. S. and P. S. Rawlik. 1993. Transpiration and stomatal conductance of two wetland macrophytes (*Cladium jamaicense)* and *(Typha domingensis*) in the subtropical Everglades. American Journal of Botany 80:1146–1154.

Loveless, C. M. 1959. A Study of the Vegetation of the Florida Everglades. Ecology 40:1–9.

Miao, S. L. and F. H. Sklar. 1998. Biomass and nutrient allocation of sawgrass and cattail along a nutrient gradient in Florida Everglades. Wetland Ecosystem and Management (in press).

Newman, S., J. B. Grace, and J. W. Koebel. 1996. The effects of nutrients and hydroperiod on mixtures of *Typha domingensis*, *Cladium jamaicense*, and *Eleocharis interstincta*: Implications for Everglades restoration. Ecological Application 6:774–783.

Newman, S., J. Schuette, J. B. Grace, K. Rutchey, T. Fontaine, K. R. Reddy, and M. Pietrucha. 1998. Factors influencing cattail abundance in the Northern Everglades. Aquatic Botany 60:265–280.

Parker, G. G. 1974. Hydrology of the predrainage system of the Everglades in southern Florida. Pages 18–27 *in* P. J. Gleason, Editors. Environments of South Florida: Present and Past II. Memoir 2. Miami Geological Society, Coral Gables, Florida.

Reddy, K. R., R. D. DeLaune, W. F. DeBusk, and M. S. Koch. 1993. Long-term nutrient accumulation in the Everglades. Soil. Sci. Soc. Am. J. 57:1147–1155.

Richardson, C. J. 1991. Effects of nutrients loadings and hydroperiod alterations on control of cattail expansion, community structure and nutrient retention in the water conservation areas of south Florida. Unpublished report submitted to Agricultural Area Environmental Protection District.

Rutchey, K. and L. Vilchek. 1994. Development of an Everglades vegetation map using a SPOT image and the Global Positioning System. Photogrammetric Engineering and Remote Sensing 60:767–775.

Rutchey, K. and L. Vilchek. 1998. Air photo-interpretation and satellite imagery analysis techniques for mapping cattail coverage in a northern Everglades impoundment. Journal of Photogrammetric Engineering and Remote Sensing. (in press).

Schlesinger W. H. 1991. Biogeochemistry, An Analysis of Global Change. Academic Press.

Small, J. K. 1932. Manual of the Southeastern Flora. University of NC Press, Chapel Hill, NC.

South Florida Water Management District. 1992. Surface water improvement and management for the Everglades: Supporting information document. SFWMD, West Palm Beach.

Stake, E. 1967. Higher vegetation and nitrogen in a rivulet in Central Sweden Schweiz. Z. Hydrol. 29:107–125.

Steward, K. K. and W. H. Ornes. 1975. The autecology of sawgrass in the Florida Everglades. Ecology 56:162–171.

Steward, K. K. and W. H. Ornes. 1983. Mineral nutrition of sawgrass (*Cladium jamaicense* Crantz) in relation to nutrient supply. Aquatic Botany 16:349–359.

Tilman, D. 1982. Resource Competition and Community Structure. Princeton University Press, Princeton, New Jersey.

Tilman, D. 1990. Mechanisms of plant competition for nutrients: The elements of predictive theory of competition. Pages 117–141 *in* J. Grace and D. Tilman, Editors. Perspective on Plant Competition. Academic Press, New York.

Toth, L. A. 1987. Effects of hydrologic regimes on lifetime productivity and nutrient dynamics of sawgrass. South Florida Water Management District, West Palm Beach, FL, USA Technical Publication 87-6.

Toth, L. A. 1988. Effects of hydrologic regimes on lifetime production and nutrient dynamics of cattail. South Florida Water Management District Technical Publication 88-6.

Urban, N. H., S. M. Davis, and N. G. Aumen. 1993. Fluctuations in sawgrass and cattail in Everglades Water Conservation Area 2A under varying nutrient, hydrologic and fire regimes. Aquatic Botany 46:203–223.

Wade, D. D., J. J. Ewel, and R. Hofstetter. 1980. Fire in south Florida ecosystems. U.S. Department of Agriculture, Forest Service. General Technical Report SE-17.

Wu, Y., F. H. Sklar, K. R. Rutchey. 1997. Analysis and simulations of fragmentation patterns in the Everglades. Ecological Application 7:268–276.

White, R. E. 1973. Studies on mineral ion absorption by plants. II. The interaction between metabolic activity and the rate of phosphorus uptake. Plant and Soil 38:509–523.

12 Influence of Phosphorus Loading on Wetlands Periphyton Assemblages: A Case Study from the Everglades

Paul V. McCormick and Leonard J. Scinto

12.1 ABSTRACT

Periphyton (attached algae and bacteria) is an important ecological component of shallow-water ecosystems. Although studies of periphyton in wetlands are lacking relative to other freshwater habitats, there is increasing evidence that periphyton can contribute significantly to key ecosystem processes (e.g., energy fixation, nutrient cycling and storage) in wetland habitats where sufficient light penetrates through the water column. Periphyton responds rapidly and predictably to increased phosphorus (P) loading and, thus, has been used widely as an indicator of changes in the trophic status in freshwater ecosystems.

Periphyton responses to increased P loading are illustrated by research being conducted in the Florida Everglades. Periphyton is abundant in oligotrophic areas of the Everglades and is responsible for much of the primary production and P storage in open-water habitats. Periphyton is strongly P-limited and accumulates excess P rapidly from the water column. Increased P loading stimulates periphyton productivity and growth rates and causes shifts in taxonomic composition from oligotrophic species to those capable of faster growth under P-enriched conditions.

1-56670-331-X/99/$0.00+$.50

Over longer time scales, periphyton abundance and productivity is reduced in enriched areas of the marsh as a result of increased growth of emergent macrophytes and the loss of open-water habitats.

Changes in the Everglades periphyton assemblage in response to increased P loading affect ecosystem processes by altering the food base available to aquatic consumers, reducing rates of aquatic primary productivity and oxygen production, and reducing rates of periphyton P removal and storage. Periphyton responses to P enrichment precede those of other biota (e.g., emergent macrophytes) and, thus, provide a valuable early indicator of eutrophication in the Everglades and other wetlands.

12.2 INTRODUCTION

Phosphorus (P) is a key element controlling global primary productivity, and the widespread application of this nutrient to increase soil fertility is responsible for the eutrophication of many freshwater ecosystems (Tiessen, 1995). The biological effects of increased P loading on lakes and rivers have been well documented and include excessive primary productivity, increased biological oxygen demand, and reduced biodiversity (National Academy of Sciences, 1969; Likens, 1972; Havens and Steinman, 1995). Most wetland studies have focused on the ability of these systems to remove nutrients such as P rather than the ecological impacts associated with these inputs (e.g., Howard-Williams, 1985; Moshiri, 1993; Olson, 1993). Water quality standards for wetlands have been proposed (USEPA, 1990) in an effort to afford these systems the same level of protection currently provided to other water bodies. However, the relationship between P loading and wetland changes has not been extensively studied.

Algae respond predictably to changes in water quality and have proven useful for monitoring changes in the trophic status of lakes and rivers (Patrick et al., 1954; Palmer, 1969; Lange-Bertalot, 1979; Stoermer, 1984; Schindler, 1987; Agbeti, 1992; Anderson et al., 1993; Christie and Smol, 1993). Algae have short life cycles and, thus, respond rapidly to increases in nutrient availability, often before other organisms are affected. Changes in algal growth rates and species composition affect biogeochemical cycles and energy flow to higher trophic levels and, thus, provide an indication of overall ecosystem function (Rosen, 1995; Lowe and Pan, 1996; McCormick and Cairns, 1997). Whereas algae occur primarily as floating cells or colonies (phytoplankton) in pelagic systems such as deep lakes, attached or floating mats of algae and bacteria (periphyton) are more common in shallow waters. Periphyton can be abundant even in macrophyte-dominated systems such as wetlands, although their use as indicators has been considerably more restricted (e.g., Swift and Nicholas, 1987; McCormick et al., 1996; Pan and Stevenson, 1996).

In this chapter we discuss (1) the ecological importance of periphyton in wetlands, (2) the response of periphyton to increased P loading in wetlands using the Florida Everglades as an example, and (3) how periphyton changes affect other wetland populations and processes and serve as an early indicator of eutrophication in these ecosystems. The Everglades is one of the few freshwater wetlands where periphyton responses to increased P loading have been the subject of intense scientific study and regulatory scrutiny. We believe that general patterns caused by P loading in the Everglades should be applicable to other wetlands, although we recognize that differences in abiotic and biotic conditions may limit the extent to which some responses can be extrapolated among ecosystems.

12.3 PERIPHYTON PROCESSES IN WETLANDS

Ecological characterizations of wetlands are typically based on hydrology, soils, and macrophyte productivity and species composition (e.g., Hammer, 1992; Mitsch and Gosselink, 1993). Periphyton is rarely included in wetland assessments despite evidence that it can be abundant and ecologically important in littoral habitats where sufficient light penetrates through the water column (Goldsborough and Robinson, 1996). Periphyton occurs in wetlands as (1) floating or subsurface mats (metaphyton), which may become entangled with floating and submersed vegetation, (2) thin films or more extensive growths attached to submersed and emergent macrophytes (epiphyton), and (3) contiguous mats, loose material, or motile cells associated with flooded or saturated soils (epipelon). Although there is relatively little quantitative information on periphyton processes in wetlands, considerable observational and experimental evidence from other freshwater ecosystems suggests that periphyton can contribute substantially to primary productivity, food web support, and nutrient cycling in wetland habitats (e.g., Stevenson et al., 1996).

Wetlands are characterized by extremely high rates of emergent macrophyte productivity (Likens, 1975; Richardson, 1979). High rates of periphyton productivity have also been measured in a few wetlands (Murkin, 1989; Goldsborough and Robinson, 1996; McCormick et al., 1998), but there are few instances where periphyton values can be compared directly with those for macrophytes (Robinson et al., 1997a, b). Dense stands of emergent or floating macrophytes restrict light penetration to the water surface to levels below the compensation point for algal growth (Goldsborough, 1993; Grimshaw et al., 1997). Therefore, periphyton growth is greatest in sparsely vegetated areas where it can account for the bulk of the primary production (e.g., Zedler, 1980; Browder, et al. 1982; Cronk and Mitsch, 1994). Consequently, the contribution of periphyton to wetland primary production is determined at least partly by the amount of available open-water habitat.

Using primary productivity rates alone to assess the relative contribution of macrophytes and periphyton to wetland food webs can be misleading and ignores issues of food quality and the feeding preferences of aquatic consumers. Only a small fraction of live macrophyte biomass is consumed by aquatic herbivores (Newman, 1991); most macrophyte production is converted to detritus, which may be consumed, accreted, or exported. Exceptions to this argument have come mainly from wetlands where dominant herbivores are either waterfowl or mammals (e.g., muskrats) (Murkin, 1989; Lodge, 1991). Periphyton typically represents a higher quality food than either living macrophyte tissue or detritus and is a preferred food source for various aquatic herbivores (Hann, 1991; Neill and Cornwell, 1992; Campeau et al., 1994; Lamberti, 1996). Therefore, a relatively large fraction of periphyton production may be consumed by herbivores, as shown for lotic ecosystems (McIntire, sediment transport model that simulates the resuspension and transport of fine-grained sediments (Sheng and Chen, 1991; Sheng, 1993; Chen and Sheng, 1994, 1995), and a phosphorus transport model that simulates the transport and transformation of various phosphorus species (Chen and Sheng, 1994, 1995). These models have been calibrated and verified with field data collected by University of Florida during fall 1988 and spring 1989 conditions, as well as storm conditions (Sheng et al., 1998). Model-simulated and measured SRP concentrations in Lake Okeechobee on June 16, 1989 are shown in Figs. 16.10 and 16.11. These results correlate well with the simulated and measured suspended sediment distributions in the lake, confirming the significant influence of suspended sediment dynamics on phosphorus transport. These model simulations were conducted with a 1-km numerical grid and a 15-minute time step, which are much finer than those used in the box model that uses 4 boxes for the entire lake and a 1-day time step (James and Bierman, 1995). Hence, the model is able to capture the effects of hydrodynamic and sediment processes on phosphorus dynamics without resorting to ad hoc parameter tuning. The model was also used to examine the effect of phosphorus load reduction on the water quality of Lake Okeechobee (Chen and Sheng, 1994). The results indicate that, even if the external loadings were completely eliminated, internal loading will increase water-column phosphorus concentration in Lake Okeechobee for several years.

16.7.5 Which Model Should Be Used?

The selection of a proper model for a nutrient-related study in an estuary or lake depends on several factors: objectives of the study, dominant processes controlling nutrients distribution in the water body, available data, available budget, and available time. Cost for a management-driven modeling study should be at least a small fraction (e.g., 0.1 to 1%) of the total budget. If the economic value of the ecosystem resources to be managed is on the order of $1 billion, then it is prudent to spend $1 million to $10 million on modeling studies.

16.8 CONCLUSIONS AND RECOMMENDATIONS

- Nutrient distribution in aquatic ecosystems is significantly influenced by the time-dependent circulation as well as residual circulation. Consequently, maximum nutrient concentration does not necessarily occur in the vicinity of the source of nutrient loading.
- The vertical density stratification cycle in deeper temperate aquatic ecosystems plays a major role in affecting the dissolved oxygen distribution in the lower layer of the water column that, in turn, affects the release of dissolved nutrients from the sediments.
- Transport of particulate nutrients is significantly influenced by sediment transport processes.
- Currents and waves in shallow subtropical and tropical estuaries often cause significant resuspension and vertical mixing of sediments, which enhances the resuspension of particulate nutrients from sediments, subsequent desorption of dissolved nutrients.

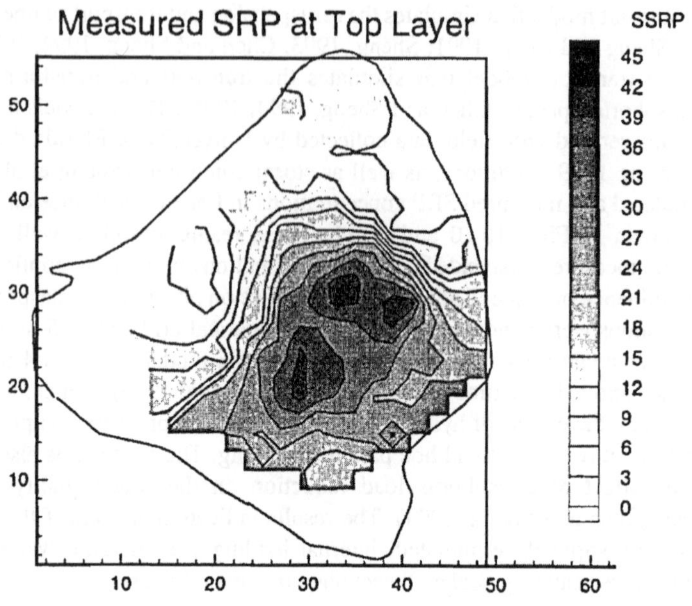

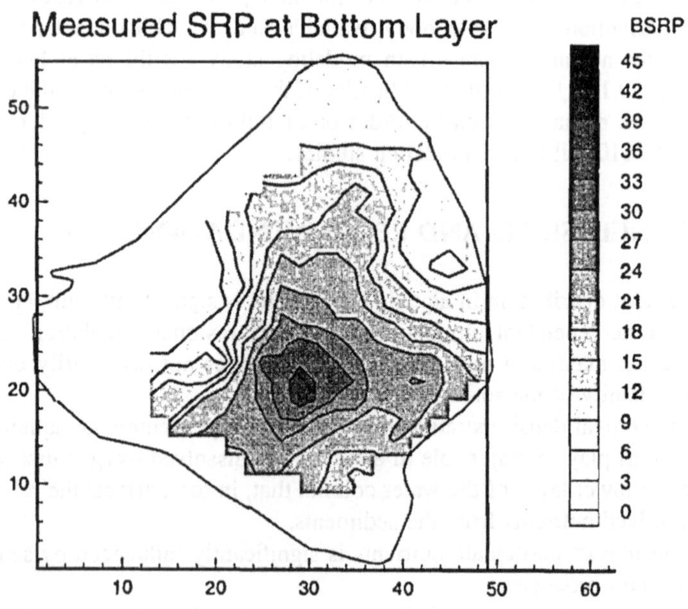

FIGURE 16.10 Contours of measured SRP concentration at near-surface and near-bottom layers in Lake Okeechobee at noon, June 16, 1989.

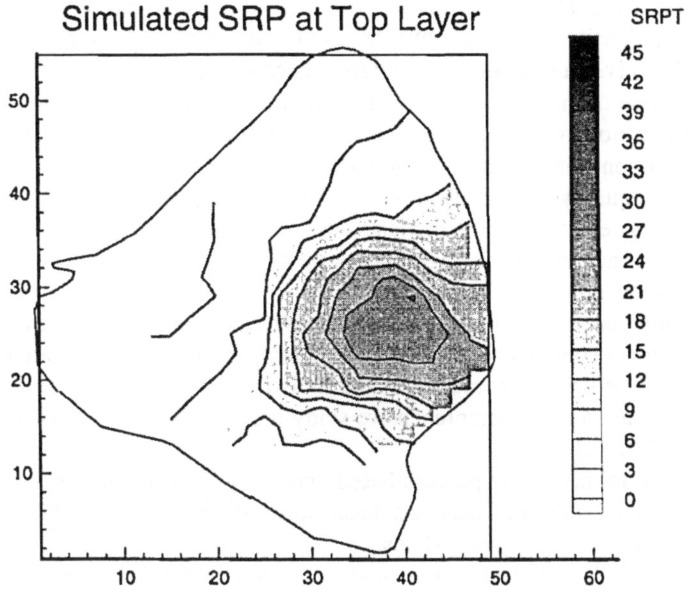

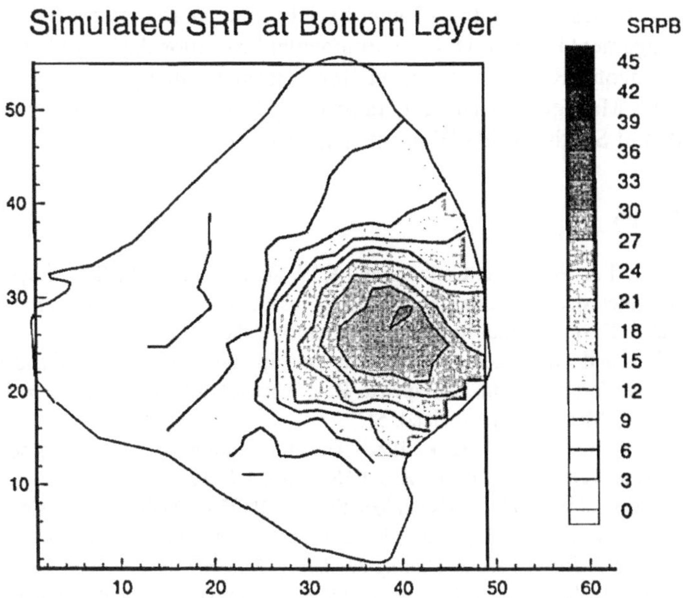

FIGURE 16.11 Contours of simulated SRP concentration at near-surface and near-bottom layers in Lake Okeechobee at noon, June 16, 1989.

- Fine sediments in estuaries and lakes can form lutocline (a thin layer of sharp density gradient) above the bottom that inhibits the vertical mixing of dissolved and particulate matters between water and sediment columns.
- Because of the significant effects of hydrodynamic processes on nutrient distribution in aquatic ecosystems, it is essential that management models of nutrient loading incorporate the hydrodynamic processes.
- Water quality models that, in a broad sense, include nutrient process models, eutrophication models, and pollutant load reduction models, should incorporate the significant effects of hydrodynamics on nutrient transport processes.
- Simple regression models do not address cause-effect relations. Box models do not contain adequate spatial and temporal resolutions. Both models should be used with caution as the tools for preliminary nutrient studies and to aid a more comprehensive study using multidimensional process-based models.
- A multidimensional process-based modeling system for assessing the impact of nutrient loads has been successfully developed for several subtropical aquatic ecosystems.
- Traditional nutrient loading models do not incorporate the effects of hydrodynamic processes, should be used with caution.

16.9 ACKNOWLEDGMENT

The various studies cited in this chapter have been supported by the St. Johns River Water Management District, U.S. Environmental Protection Agency National Center for Environmental Research, Sarasota Bay National Estuary Program, Southwest Florida Water Management District, Tampa Bay National Estuary Program, Florida Sea Grant, and South Florida Water management District.

REFERENCES

Ambrose, R.B., T.A. Wool, J.P. Connoly and R.W. Schanz. 1991. WASP4, a hydrodynamic and water quality model-model theory, user's manual, and programmers' guide. USEPA Environmental Research Laboratory, Athens, GA.

Boynton, W.R., and W.M. Kemp. 1993. Relationships between river flow and ecosystem processes/properties in Chesapeake Bay. Abstract, Estuarine Research Foundation Biannual Meeting, Hilton Head, SC.

Burban, P.-Y., W. Lick, and J. Lick. 1989. The flocculation of fine-grained sediments in estuarine waters. J. Geophys. Res. 94(C6):8323–8330.

Cerco, C.F., and T. Cole, 1994. User's guide to the CE-QUAL-ICM three-dimensional eutrophication model. Tech. Rept. EL-95-15, Waterways Experiment Station.

Chapra, S.C., 1997. *Surface Water-Quality Modeling.* McGraw Hill.

Chen, X., and Y.P. Sheng. 1994. Recent studies on hydrodynamics and sediment dynamics on nutrient transport processes in shallow lakes and estuaries. UF/COEL Tech. Rept., Coastal and Oceano. Eng. Dept., Univ. of Fla., Gainesville, FL.

Chen, X., and Y.P. Sheng. 1995. Application of a coupled 3-D hydrodynamics-sediment-water quality model. in Estuarine and Coastal Modeling, IV, Am. Soc. of Engrs. 325–339.

Coastal, Inc. 1996. Estimating critical external nitrogen loads for the Tampa Bay estuary: an empirically based approach to setting management targets. Technical Publication #06-96, Tampa Bay National Estuary Program, St. Petersburg, FL.

Fong, P., and M.A. Harwell. 1994. Modeling seagrass communities in tropical and subtropical bays and estuaries. Bull. Mar. Sci. 54(3):757–781.

Galegos, C.L. 1993. Development of optical models for protection of seagrass habitat. in Proceedings and Conclusions of Workshops on Submerged Vegetation Initiative and Photosynthetically Active Radiation, L.J. Morris and D. Tomasko, Eds., St. Johns River Water Management District, Palatka, FL. 77–90.

James, R.T., and V.J. Bierman. 1995. A preliminary modeling analysis of phosphorus and phytoplankton dynamics in Lake Okeechobee, Florida, I: model calibration. Water Research.

Lee, H., and Y.P. Sheng. 1993. Wind driven circulation in Lake Okeechobee—the effects of vegetation and thermal stratification. Tech. Rept., Coastal and Oceano. Eng. Dept., Univ. of Fla., Gainesville.

McPherson, B.F., and R.L. Miller. 1993. Causes of light attenuation in estuarine waters of southwest Florida. in Proceedings and Conclusions of Workshops on Submerged Aquatic Vegetation Initiative and Photosynthetically Active Radiation, L.J. Morris and D.A. Tomasko, Eds., Melbourne, FL. 227–236.

Martin, J., P.F. Wang, T. Wool, and G. Morrison. 1996. A mechanistic management-oriented water quality model for Tampa Bay," AscI Corporation and Southwest Florida Water Management District.

Reddy, K.R., Y.P. Sheng, and B. Jones. 1995. Lake Okeechobee phosphorus study: summary. *Volume I, Final Report for Contract No. C91-2554 to South Florida Water Management District,* Univ. of Fla., Gainesville.

Sheng, Y.P. 1982. Hydraulic applications of a turbulent transport model. Proc. 1982 ASCE Hyd. Div. Spec. Conf. on Applying Res. to Hyd. Practice, ASCE, Jackson, MS, pp. 106–119.

Sheng, Y.P. 1986. Modeling bottom boundary layer and cohesive sediment dynamics in estuaries and coastal waters," in Estuarine Cohesive Sediment Dynamics, A.J. Mehta, Ed., *Lecture Notes on Coastal & Estuarine Studies,* Am. Geophys. Union. 360–400.

Sheng, Y.P. 1987. On Modeling three-dimensional estuarine and marine hydrodynamics. in Three-Dimensional Models of Marine and Estuarine Dynamics, Elsevier Oceanographic Series, Elsevier. 35–54.

Sheng, Y.P. 1989. Evolution of a 3-D curvilinear-grid hydrodynamic model: CH3D, in Estuarine and Coastal Modeling, I, ASCE. 40–49.

Sheng, Y.P. 1993. Hydrodynamics, sediment dynamics, and their effects on phosphorus dynamics in Lake Okeechobee. in Nearshore, Estuarine, and Coastal Sediment Transport, Am. Geophys. Union, Coastal and Estuarine Studies Series. 558–571.

Sheng, Y.P. 1994. Modeling hydrodynamics and water quality dynamics in shallow waters. Keynote Paper, Proc.1st Intl. Symp. on Ecology and Engineering (ISEE), Univ. of Western Australia and Tech. Univ. of Malaysia.

Sheng, Y.P. 1996. A preliminary hydrodynamics and water quality model of Indian River Lagoon. UF/COEL Tech. Rept., Coastal and Oceano. Eng. Dept., Univ. of Fla., Gainesville, FL.

Sheng, Y.P. 1997. Pollutant load reduction models for estuaries. in Estuarine and Coastal Modeling, V, Am. Soc. Civil Engrs.

Sheng, Y.P. 1998a. Circulation in Charlotte Harbor estuarine system. in Proc. Charlotte Harbor Scientific Symp., Charlotte Harbor National Estuary Program.

Sheng, Y.P., and C. Villaret. 1989. Modeling the effect of suspended sediment stratification on bottom exchange processes. J. Geophys. Res. 94(C10):14429–14444.

Sheng, Y.P. and X. Chen. 1991. A three-dimensional numerical model of hydrodynamics, sediment transport, and phosphorus dynamics(LOHSP3D): theory, model development and documentation. Volume XII, Final Report for Contract No. C91-2393 to South Florida Water Management District, Coastal and Oceano. Eng. Dept., Univ. of Fla., Gainesville, FL.

Sheng, Y.P., and H.K. Lee. 1991. The effect of aquatic vegetation on wind-driven circulation in Lake Okeechobee. Rept. No. UFL/COEL-91-022, Coastal and Oceano. Eng. Dept., Univ. of Fla., Gainesville, FL.

Sheng, Y.P., and E.A. Yassuda. 1995. Application of three-dimensional circulation model to WASP modeling in Tampa Bay water quality. Final Report to Tampa Bay National Estuary Program.

Sheng, Y.P. and J. Davis. 1997. Circulation and transport in hypersaline Florida Bay. in Proc.25th Intl. Conf. on Coastal Eng., ASCE. 4242–4252.

Sheng, Y.P., K. Ahn, and J. Choi. 1991a. Wind-wave hindcasting and estimation of bottom shear stress in Lake Okeechobee. Rept. No. UFL/COEL-91-021, Coastal and Oceano. Eng. Dept., Univ. of Fla., Gainesville, FL.

Sheng, Y.P., D.E. Eliason, J. Choi, and H. Lee. 1991b. Numerical simulation of three-dimensional wind-driven circulation and sediment transport in Lake Okeechobee during Spring 1989. Rept. No. UFL/COEL-91-017, Coastal and Oceano. Eng. Dept., Univ. of Fla., Gainesville, FL.

Sheng, Y.P., X. Chen, and E.A. Yassuda. 1994. Wave-induced sediment resuspension in shallow waters. Proc.24th Intl. Conf. on Coastal Eng., Kobe, Japan. 3281–3294.

Sheng, Y.P., E.A. Yassuda, and C. Yang. 1995. Modeling the effect of nutrient load reduction on water quality. in Estuarine and Coastal Modeling, IV, Am. Soc.of Civil Engrs. 644–658.

Sheng, Y.P., S. Peene, and E.A. Yassuda. 1996. Circulation and transport in Sarasota Bay: the effect of tidal inlets in estuarine circulation and flushing quality. in Mixing in Estuaries and Coastal Seas, American Geophysical Union Coastal and Estuarine Studies Series. 184–210.

Sheng, Y.P., X. Chen, and E.A. Yassuda. 1997. On hydrodynamics and water quality dynamics in Tampa Bay. in Proceedings of TAMPA BASIS-3, Bay Area Scientific Symposium, Tampa Bay National Estuary Program and Tampa Bay Regional Planning Council. 295–314.

Sheng, Y.P., X. Chen, and S. Schofield. 1998. On sediment and phosphorus dynamics during episodic events. in Turbulence and Mixing in Lakes and Estuaries, J. Imberger. Ed., Am. Geophys. Union, In press.

Virnstein, R. 1993. Seagrass response to water quality. in Proceedings of An IRL Lagoon-Wide Modeling Workshop, Sheng et al., Eds., Univ. of Fla., Gainesville, FL. 137–147.

Vollenweider, R.A. 1968. The scientific basis of lake and stream eutrophication with particular reference to phosphorus and nitrogen as eutrophication factors. Tech. Rept. DAS/DSI/68.27, Organization for Economic Cooperation and Development, Paris.

Vollenweider, R.A. 1975. Input-output models with special reference to the phosphorus loading concept in limnology. Schweiz. Z. Hydrologie. 37:53–84.

Yassuda, E.A., and Y.P. Sheng. 1996. Integrated modeling of Tampa Bay estuarine system. Tech. Rept., Coastal and Oceano. Eng. Dept., Univ. of Fla., Gainesville, FL.

Yassuda, E.A. and Y.P. Sheng. 1997. Modeling dissolved oxygen dynamics in Tampa Bay during the Summer of 1991. in Estuarine and Coastal Modeling, V, ASCE.

Section VI

Phosphorus Management

17 Phosphorus Management in Flatwood (Spodosols) Soils

A.B. (Del) Bottcher, Terry K. Tremwel, and Kenneth L. Campbell

17.1 ABSTRACT

Flatwood soils dominate the Florida landscape (Fig. 17.1). These soils are typically Spodosols, which are nearly level, poorly drained, sandy soils that have high seasonal water tables and an organic Bh horizon. When properly drained, however, they are well suited for agricultural production. The high runoff from flatwood soils can cause high losses of phosphorus (P) compared to better-drained soils. Researchers in Florida have conducted numerous studies to discover the mechanisms of nutrient losses from agriculture on these soils. Reduction of P transport to watercourses has been the primary focus of these studies, because P has been found to be the nutrient limiting eutrophication in many Florida aquatic systems.

The flatwood soils support extensive beef, dairy, citrus, and vegetable operations. As a result, most of the research activities have focused on agricultural transport processes and control practices (known as *best management practices* or *BMPs*) for

1-56670-331-X/99/$0.00+$.50

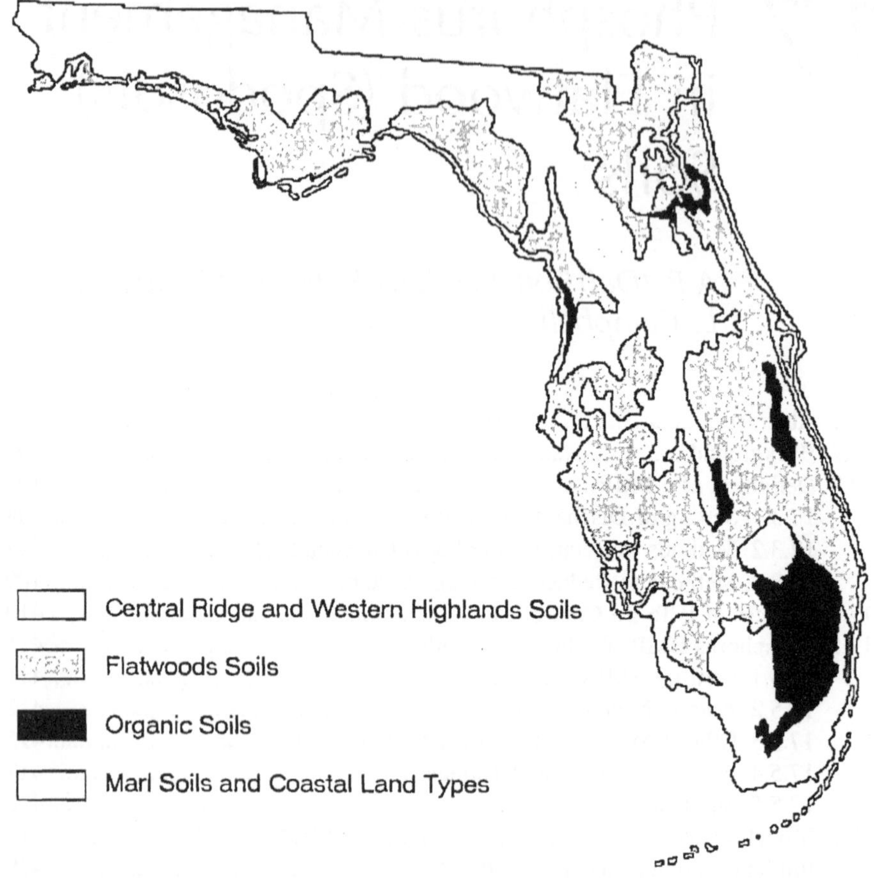

FIGURE 17.1 Major soil types of Florida.

P discharge. This chapter focuses on the BMPs that have been developed for agricultural water quality control in Florida, which is typical of many subtropical areas around the world. BMPs are discussed conceptually using field evaluation data and practical in-field design constraints. An itemized listing of the proposed BMPs is presented, as well as a discussion of implementation strategies.

17.2 BACKGROUND

Complaints of excessive aquatic weeds and algae blooms in Florida lakes and concern with wetland ecosystem modifications during the late 1960s and early 1970s prompted numerous studies. Studies on Lake Okeechobee using the Vollenweider model (Brezonik and Federico, 1975; Davis and Marshall, 1975) found the lake to be eutrophic with the potential to become hypereutrophic because of excessive phosphorus loading. Other studies within the Everglades marsh system found large areas of altered ecosystems resulting from anthropogenic influences (Everglades

SWIM Plan, 1992; Okeechobee SWIM Plan, 1993). These findings prompted extensive monitoring and research efforts throughout the state to better understand the nutrient cycling within the aquatic systems and land-source areas. Several of these studies have focused specifically on evaluating BMPs for reducing P losses (Everglades SWIM Plan, 1992; Okeechobee SWIM Plan, 1993; SWET, 1996).

In general, these studies concluded that the primary source of P to the natural water systems from rural watersheds is agricultural drainage. Because much of Florida's agriculture is on flatwoods, and these soils typically surround many of the water systems of concern, P control from flatwoods is of critical importance.

Flatwoods soils are typical of subtropical coastal plains and are characterized by the following properties:

- Nearly flat sandy soils (slopes $< 2\%$)
- Somewhat poorly to poorly drained soils
- High water tables, seasonally rising to the ground surface
- Spodosols dominating, which have an organic Bh horizon at 30 to 150 cm
- Very sandy low organic content ($<2\%$) in surface A horizon
- Very sandy E horizon that features loss of silicate clay, iron, and aluminum
- Low P adsorption capacity in upper layers
- High P adsorption and desorption in Bh layer
- Productive cropland when drained

These soil characteristics produce the unusual situation that P is held very weakly in the upper soil profile and therefore can be leached by either vertical or lateral flow. This feature makes flatwoods more vulnerable to P losses than most other soils.

Flatwood soils comprise about 80 million km^2 or about 51% of Florida (see Fig. 17.1). Other soils that are also nearly level and poorly drained are the organic (Histosols) and marl (Entisols) soils that are found primarily in southern Florida. These latter soils cover about 14% of Florida and would benefit from similar BMPs as the flatwoods. The better-drained soils of the central ridge and western highlands (Entisols, Ultisols, and Alfisols) make up about 27% of the state; the remaining 8% of the state is open water.

Forestry, beef, citrus, dairy, and vegetable operations dominate Florida flatwoods (see Table 17.1). The vegetable, row crop, dairy farms, and urban land have the highest P source loadings per unit area but cover a relatively small portion of the landscape. Although citrus and beef operations have relatively low P loss rates, the extensive areas of these land uses make them extremely important sources of P. The animal P source is primarily manure while for citrus, row crops, and vegetables; the source of P is from fertilizer. Residential and urban P losses are primarily from sewage disposal.

Many land-use-specific BMPs have been developed by UF/IFAS (Institute of Food and Agricultural Sciences, University of Florida) and the water management districts in Florida. Though some of these BMPs are not fully tested, they represent the current state of knowledge for reducing P losses from flatwood soils. The recommended BMPs from this accumulated work are presented in this chapter.

TABLE 17.1
Land Use Characteristics of Florida for 1992 (U.S. Dept. of Commerce, 1994)

Category	State of Florida (km^2)	Flatwoods soils (km^2)	Marl and organic soils (km^2)	Central ridge and western highlands soils (km^2)
Urban/residential	11,900	6,200	1,100	4,600
Forest and open	53,900	30,900	1,900	21,100
Wetlands	30,400	14,600	14,700	1,100
Water	12,100	—	—	—
Agriculture	43,400	25,000	4,200	14,200
Pasture	23,400	15,600	500	7,300
Vegetables	2,200	1,100	700	400
Citrus	6,600	2,900	450	3,250
Row crops	4,900	2,700	300	1,900
Poultry	300	100	10	190
Dairy	1,000	700	40	260
Sugarcane	2,000	100	1,900	0
Other	3,000	1,800	300	900
Total all land use	151,700	76,700	21,900	41,000

17.3 DEFINITION OF A BMP

Best management practices (BMPs) are those on-farm activities designed to reduce nutrient or sediment losses in drainage waters to an environmentally acceptable level, while simultaneously maintaining an economically viable farming operation for the grower. While some BMPs can positively impact the financial profitability of a farm, many will not. Where the cost of implementing certain BMPs puts an excessive financial burden on the farmer, the practices should be considered to be a BMP only if external funds are available to create an acceptable level of profit for the farm. In this chapter, BMPs for P control will be the focus of discussion, because they are currently considered to be the most critical for environmental protection of subtropical landscapes characterized by flatwood soils.

17.3.1 Setting of Discharge Standards

The level of P control needed is not well understood; therefore, specific P discharge standards have only been set for a few basins, mostly in south Florida. The problem with setting standards includes limited understanding of the extent of environmental impacts from P losses and limited data on the cost effectiveness of BMPs to reduce P losses. The difficulty and cost of adequately monitoring BMP compliance also limits the effectiveness of any abatement program.

In spite of these limitations, BMPs have been designed to try to bring the discharge levels of a contaminant into compliance with downstream water quality standards, which are based on concentration, total load, or both. In the case of Lake Okeechobee, the discharge standard for P is concentration-based, with the limit based on land use, whereas the standard for the Everglades Agricultural Area (EAA) is load-based. The EAA standard is an example of a standard based on the needs of the receiving water body. In contrast, the Lake Okeechobee standard considers the feasibility of obtaining P reductions based on the relative effectiveness of agricultural BMPs.

Phosphorus load reduction regulations are difficult to implement, because they require both water flow and P concentrations to be monitored. Such dual monitoring requirements can result in high measurement errors and added cost compared to concentration monitoring alone, but they are necessary if load estimates are needed. Concentration measurements alone typically provide poor load estimates, even if total flow is predictable, because of the natural water quality variability in samples collected. Therefore, some flow measurements or estimates may be needed to supplement concentration-based standards.

The use of a load- or concentration-based standard is also influenced by a BMPs impact on discharge versus concentration reductions. BMPs can reduce P load by either reducing P concentrations, water discharges, or both. For example, in nonirrigated sandy soils, water reductions are difficult and costly, so P standards in these areas are typically concentration based. In areas like the EAA, however, significant on-farm storage is feasible, and most farms have pumped discharge, allowing more effective use of a load-based standard.

17.3.2 Methods to Reduce Phosphorus Concentrations

Conceptually, there are only three ways to reduce nutrient concentrations in the runoff water from agricultural lands. They are:

- Reduce the amount of nutrients on the farm by minimizing nutrient inputs (fertilizers, feed, etc.) to the farm and maximizing nonrunoff nutrient outputs (optimize plant uptake and harvest) from the farm.
- Reduce the mobility of a nutrient by limiting water contact and/or reducing the solubility or erodibility of the source materials.
- Employ edge-of-field or farm predischarge treatment using biouptake, adsorption, deposition, or precipitation technologies, such as wetlands and/or chemical additives.

17.3.3 Methods to Reduce Discharge Volume

Reducing discharge volume is straightforward conceptually, but can be difficult to achieve in practice. Discharge volume reduction can only be obtained by two methods. They are:

- Increase the evapotranspiration (ET) from the farm, which can be achieved by maintaining optimal soil moisture through irrigation, growing water tolerate crops using higher water table levels, and increasing any open water surfaces, such as ponds or reservoirs.
- Decrease off-farm or groundwater irrigation water inputs to the farm by improved irrigation efficiency or by using stored runoff for irrigation supply.

Many of the flatwood soils are not irrigated, have very few water control structures, exhibit seasonal soil wetness (high water tables), and have limited water storage areas, all of which make discharge reductions difficult. In more developed areas, such as vegetable and citrus production sites, the irrigation inputs and pumped or gated outlet structures are ideal for optimizing on-farm water use, thereby reducing discharges.

Getting the farmers to understand the principles behind BMPs will make the design, acceptance, and implementation of farm-level BMP programs much easier. The pollutant source for potential runoff is the starting place for understanding the process; therefore, it will be presented in general terms before the specific BMPs are described. Presentation of the biogeochemical processes affecting nutrient transport and the impact of BMPs on P retention is beyond of the scope of this chapter. These topics are well covered by Campbell, et al. (1992) and Graetz and Nair (1994) in two major project reports for the South Florida Water Management District (SFWMD).

17.4 SOURCES OF P TO BE CONTROLLED

Phosphorus added to drainage water by agricultural operations originates from one of five sources: fertilizers, animal manures, mineralization of native organic and inorganic materials, irrigation supply water, or atmospheric deposition. These sources of P must be available if agriculture is to exist in Florida, and the relative importance of each source depends on the soil-plant system. Atmospheric deposition (1 to 2 kg P/ha/yr) is important only in native areas. The other four sources can vary significantly in amount but can be controlled by proper management or BMPs. As in most situations, problems occur with excess use, and nutrient management is no exception.

Control of P sources is particularly critical in flatwood soils compared to other locations, because of the low P sorption properties of the native soils, despite having a Spodic horizon. These sandy soils typically have very low clay, iron, and aluminum contents in the surface horizons, which limits P sorption. In many of the flatwood soils, P can be considered a mobile nutrient in the surface horizons only. If plant available P in a soil-plant system exceeds what the plants can use, then the excess P can be moved by water through surface and subsurface drainage within flatwood soils. Conceptually, the management approach is to keep the P sources in balance within the soil-plant environment.

The Spodosols, which are the predominant soil type in flatwoods, have a 0.15 to 0.5 m thick Bh horizon (spodic layer) at a depth of 0.3 to 2.0 m. Low hydraulic conductivity and relatively high organic matter, iron, and aluminum contents characterize this layer. Burgora (1989) and others have characteristic the P transport in Spodosols. They found the spodic layer can have a high capacity for adsorption of P, but the hydraulics of these soils greatly limits the contact of runoff water with this layer. The spodic horizon can create a shallow perched water table, causing most drainage to leave as surface runoff before contacting the spodic. Under native conditions (undrained), the surficial water table is normally around the spodic layer, further limiting the potential for drainage water contact. When these soils are drained, however, a much higher percentage of drainage can pass through the layer, thereby adsorbing additional P. The advantages of the increased adsorption are partially offset by the increased mineralization of the soil, since the spodic layer has a high P content naturally. When considering water management BMPs, a balance between adsorption and mineralization has to be considered.

17.5 PHOSPHORUS BMPS FOR FLATWOOD SOILS

The BMPs presented in this chapter are the result of numerous studies in Florida as well as generally accepted practices from other parts of the country. Where possible, the specific study or basis for the BMP is referenced. The BMPs are presented by P source and land use.

17.5.1 Fertility BMPs

The following fertility BMPs are applicable to all crops grown on flatwood soils, with a few noted exceptions. Specific fertilizer recommendations are crop dependent and are available from the local Florida Cooperative Extension Service. Fertility BMPs will be most effective on row crops and other high-value crops but should be considered for all fertilized land.

Calibrated Soil Testing

Calibrated soil testing (CST) is the procedure by which the actual crop response to fertilizer is determined as a function of the P level measured in the soil prior to fertilization. Laboratories providing CST can then provide fertilization recommendations. Refinement of CST has been a high research priority. For example, the study by Rechcigl et al. (1992) resulted in a 50% reduction in P recommendations for pasture grasses, with no expected yield reductions.

Banding of Fertilizer

The banding of P, limiting application to the rooted area of a crop, can significantly reduce fertility needs. Sanchez et al. (1990 and 1991) found that banding in some vegetable crops could reduce fertilizer application rates by 50%. Izuno and Bottcher (1991) further found that banding reduced P losses to water by as much as 20 to 30%. However, they also noted that banding is effective only for those crops with a limited root area, such as vegetables, citrus, and other row crops.

Prevention of Misplaced Fertilizer

Prevention of spills and the direct spreading of fertilizer into open drainage ditches can reduce P losses significantly. Because it takes as little as 0.4 kg to P/ha in drainage waters to exceed Everglades P standards, it is critical to prevent any concentrated spills that can be washed into nearby streams. Areas of intensive ditching, as found in the most intensive agricultural lands on flatwoods, are most vulnerable to direct fertilizer applications, because the ditches are in close proximity to the crops. Once P is in surface waters, methods of removal are less effective than in the soil-plant system.

Use of side-throw fertilizer spreaders along drainage ditches or appropriate spacing of drive lanes to prevent spread fertilizer from reaching ditches is a necessary practice. Turning off the spreader at the end of the field prevents overlapping applications of fertilizer during turns. Aerial fertilizer applications are more difficult to control but can be improved by proper flagging and pilot training.

Split Applications

Split application (multiple applications of P during the growing season) of P fertilizers and the use of slow-release forms of P fertilizers have limited application for field crops. Under special conditions, such as intensive vegetable production or sod production on sandy soils, split applications of phosphorus could be beneficial. Even for these special conditions, however, only two applications would be needed. Slow-release forms of phosphorus, such as rock phosphate, are normally inefficient for providing plant needs. For some special cases, phosphorus losses could be reduced from 0 to 5% by use of slow-release fertilizer or split applications of fertilizer. These management techniques are much more applicable to nitrogen fertilization on mineral soils than to P. For a general discussion of split application and other fertility topics, the reader is referred to UF/IFAS Circular 816 (Bottcher and Rhue, 1983), Circular 1171 (Bottcher et al., 1997), and Publication SL-129 (Kidder, et al., 1997).

17.5.2 Animal Manure BMPs

Animal manure BMPs are based on the concept of applying the manure on crops at an "agronomic" rate in accordance with the nutrient requirements determined by the CST procedure. Therefore, good manure management requires a farm to have sufficient cropland to handle the manure it generates. In cases where the animals are confined, a collection and distribution system is needed to deliver the manure evenly to cropland. If overgrazing, uneven grazing, or a poorly designed collection or distribution system exists, "hot spots" having excessive manure loadings will occur. For example, 10 m^2 of a high-intensity area (HIA) (animal density sufficient to cause bare-ground conditions) can produce a P load equivalent to that generated from 1 ha of improved pasture; i.e., about 1000 times greater per unit area. Rechcigl et al. (1992) found that a well managed, improved beef cattle pasture will meet the current P discharge standard of 0.35 mg/L set for the Okeechobee basin.

To effectively manage any manure utilization system, knowledge of the actual nutrient content of the manure is essential. This allows balancing of nutrients in the receiving soil-plant system. Data on manure characteristics are available (ASAE, 1995); however, collection and storage techniques can significantly change nutrient contents. Therefore, a good manure-testing program is needed. Such testing programs in North Florida and Lancaster, Pennsylvania have proven effective in balancing nutrient applications to crops.

The following specific manure BMPs apply primarily to dairy and, to a lesser extent, to beef cattle operations. The dairy related BMPs are currently required under

Florida Department of Environmental Protection "Dairy Rule" (Chapter 670 of DEP Administrative Code; FDEP, 1992) for the Okeechobee basin only. A statewide rule, which has been under development for the past few years, was recently been put on indefinite hold because of the current "reduce rules" policy of the Florida legislature.

Animal High-Intensity Area (HIA) Drainage Control

HIAs are nonvegetated areas with high animal densities near barns, shade structures, and feed areas that characteristically produce high runoff and P losses. However, in flatwood soils, the high water table and low percolation losses allow the surface and ground water from an HIA to be captured by constructing a perimeter ditch that has a pumped discharge. The collected water can then be pumped to a storage pond for later use in crop irrigation. This is essentially a recycling system if the crop is fed back to the animals. SFWMD has reported that this BMP has brought the majority of the dairies in the Okeechobee basin within P compliance. The dairies where this BMP is ineffective have historical accumulations of manures deposited outside the perimeter ditch before the installation of the BMP.

Collection and Distribution of Barn Manure

Manure deposited on impervious surfaces, such as concrete areas in and around barns, must be collected, stored, and delivered to cropland in a controlled fashion. Deposited manure can be either flushed to storage ponds or anaerobic lagoons or scraped onto concrete storage pads. The stored manure must be spread on crop land by either a mechanical spreader (scraped manure) or an irrigation system (flushed manure). The appropriate rate of P application to the crop is best determined by the CST procedure. However, because of the need to continuously apply manure to land application areas, a P balance method was developed by the USDA Natural Resources and Conservation Service in association with the University of Florida (NRCS, 1989). The manure P loading rates were based on estimated long-term grass uptake of P. They recommended loading rates of 50 and 67 kg-P/ha/yr for pastures and manure sprayfields (hayland), respectively.

Watering, Feed, and Shade Facilities Placement

These facilities normally have small HIAs develop around them. Such areas can be controlled by the following procedures, listed in order of greatest control:

- Move all watering, feed, and shade facilities into an area that has a manure collection and distribution system, such as barn or dairy HIAs with perimeter ditches.
- Move these facilities frequently enough to prevent the formation of a HIA.
- Place these facilities in upland locations so that long overland flow distances can significantly reduce P in runoff (see a later section titled *Maximizing Flow Distances for P Control*).

Feed Ration Control

Minimize the P in the animal feed. Work by Morse et al. (1992) has shown that P concentrations in feed rations may be reduced by as much as 30% while maintaining good milk production. Reduction in P ingestion results in a similar reduction in P excretion.

Fencing Animals from Ditches and Streams

Soil and plant assimilation of P is obviously limited when manure is deposited directly into a flowing stream or areas that are periodically flushed. Appropriate fencing should eliminate this possibility. Also, manure deposited near or on stream banks has a high potential for transport to the stream, especially when the animals disturb bank vegetation and stability. Therefore, fences should be placed at least 5 m from the top of the bank, because banks are often ill defined, and storm flows occasionally spread into the flood plain. Also, in some cases, the setback may need to be increased to ensure sufficient space for harvesting and maintenance equipment to operate along the bank. Harvesting of vegetation within the buffer will improve P removal.

Grazing Management

A critical factor in reducing P in runoff is keeping animal densities on pastures at levels that maintain appropriate nutrient balances. Table 17.2 shows the relative effects of grazing densities on runoff P concentrations. Values are estimated by the authors based on the review of water quality data provided by the South Florida Water Management District from tributaries of known land use and unpublished modeling work using Chemicals, Runoff, and Erosion from Agricultural Management Systems, CREAMS (Knisel, 1980). Also, grazing patterns or field rotation can be used to maximize P uptake by the animals and minimize HIA formation. Rotational grazing has been shown to be effective for improving forage quality and uptake for beef cattle in south Florida (Adjei, et al., 1987; Mislevy, et al., 1991).

TABLE 17.2
Estimated Animal Density and Runoff P Concentrations by Pasture Type

Land area	Animal density (cow/ha)	Concentration (mg/L)
Native	0	0.04–0.20
Pastures		
Native range	0.025–0.50	0.08–0.25
Semi-improved	0.50–2.5	0.10–0.30
Improved	1.2–2.5	0.25–0.50
Intensive	2.5–5	0.35–1.20
Holding areas	5–25	0.50–30
HIAs	25–250	30–900

Select High P Uptake Crops for Manure Application Areas

Crop uptake of P varies substantially, which means higher P uptake crops will require less acreage for spreading the manure (SCS, 1992). A balance between usability of the crop and cost saving associated with a smaller land application area will be required.

Composting

Composting is an alternative manure management approach that converts the manure to a marketable soil amendment (SCS, 1992). If markets are available, composting is an excellent choice for operations with limited land for spreading the manure on-site.

17.5.3 Other BMPs for Phosphorus Management

Crop Management

Profitability and nutrient runoff control both depend on a healthy and productive cropping system to assure maximum P uptake. Therefore, proper crop management programs, including cultural, water, nutrient, and pest (insects and weeds) management, are needed on every farm.

Irrigation and Drainage Management

Irrigation and drainage design and management serve to control the water status of the crop root zone as well as the rate and quantity of water leaving a field. Proper irrigation provides optimal P uptake conditions, while excessive irrigation significantly increases leaching, runoff, and related P losses. Good irrigation management is particularly critical for flatwood soils because of their high P leaching potential. Proper drainage is needed to ensure optimal moisture conditions during wet periods, while excessive drainage can cause drought conditions and increase aerobic organic matter decomposition, with its associated P release. In flatwood soils, drainage and irrigation are typically achieved by controlling the shallow water table using ditches and canals to move water to and from the field. Therefore, appropriate water management is achieved by a well designed and managed system of water conveyance and control structures, such as ditches, canals, weirs, culverts, and pump stations.

Observation float wells have been shown to be a valuable tool for managing water table elevations in flatwood soils (SWET, 1996). Soil and Water Engineering Technology, Inc., SWET, found that providing the farmer with visual knowledge of the water table position and a specific management procedure has provided significant reductions in irrigation water use and runoff.

Maximize Flow Distances for P Control

Table 17.3 shows estimates of P removal efficiency for various flow conditions, going from overland to channeled flow. The removal efficiencies are rough estimates made by the authors based on related overland flow studies for municipal waste treatment (Schanze, 1984) and observed stream-reach data. Therefore, Table 17.3 is included for illustrative purposes only, to show the relative advantages of different types of overland flow treatment. Phosphorus removal efficiencies are affected both by the P concentration in the runoff and the characteristics of the land it passes through. The type of vegetative cover is the most important land characteristic for P removal as seen in Table 17.3. Table 17.3 also illustrates that high P sources typically afford a higher percent reduction. This illustrates the benefit of buffer strips between high P source areas and stream systems. The P reduction rate is further influenced by the intensity of land use management. Areas having higher in-situ P conditions will offer less treatment for water passing through them. In summary, the further a P source is from the receiving stream and the more grass it has to pass through, the greater P reduction will be.

TABLE 17.3
Estimates of P Removal Efficiency in Flow System

	Percent removal/150 m	
	High P source (100 mg/L)	Low P source (1 mg/L)
Overland		
Woods	40	5
Grass		
native	80	10
improved pasture	80	3
Field ditches		
Bare	5	1
Grassed		
native	10	2
improved pasture	10	0.5
Streams/canals	0.5	0.1

Flow Way Buffer Strips

The effect of a high-P source is intensified by its proximity to a flow way. A vegetated buffer area between the P source and the flow way can partially mitigate these effects. Although P assimilation will occur in the buffer as water passes through it, the primary benefit of the buffer strip is to ensure that no source areas occur near the flow way, i.e., must fence animals out and have no fertilizer spread in area. The effectiveness of removing upslope P is limited by the width of the strip as seen in Table 17.3. For continuous treatment of upslope sources, buffer strips must be maintained and harvested to be fully effective.

Limit Drainage of Organic and/or Wetland Soils

As indicated earlier, drainage can increase mineralization of the organic matter through increased aeration of the soil. Because wetlands typically have high organic content, their drainage can lead to excessive P release during aerobic mineralization. Organic soils can mineralize between 20 to 80 kg-P/ha/yr. The management goal for cultivated organic soils is to maintain the highest possible water table and soil moisture while still maintaining a productive crop. Bottcher et al. (1997) have developed a BMP guidebook specifically for organic soils.

Alternative Land Use

In some situations, it may be necessary to change the land use to meet regulatory constraints. This could involve a minor or complete modification of the farming operation—for example, converting a dairy to a cattle ranch or changing vegetable production to sugarcane. If there is an economic advantage, then changing land use should be considered. This was the case for many dairies in the Okeechobee basin because of a government buy-out program.

17.5.4 Edge-of-Field/Farm Treatment

Edge-of-field or farm treatments have a strong appeal because farming operations are less constrained; however, they can be more expensive. Trade-offs between flexibility and cost should be evaluated before considering these systems. The principal edge-of-field treatment technologies are presented below.

Runoff Retention/Detention System

The storage of runoff water in a retention/detention area can reduce runoff by either increased evaporation rates or by recycling the stored water to the cropland, thereby reducing other irrigation water requirements and increasing crop ET. The storage ponds can also assimilate P by plant uptake, sediment deposition, and adsorption. However, long-term P assimilation is limited in small retention/detention ponds (Taylor, 1987; Burleson, 1988). If the ponds are allowed to drain through soil seepage, then much higher P adsorption to soils can be expected.

Use of Wetlands

Compared to retention/detention ponds, wetlands are much shallower and wider, thus increasing effective plant-to-water ratios and increasing P assimilation. Wetlands that are managed for high vegetative growth and organic debris accumulation can remove significant amounts of P if the following conditions are met (Hammer, 1989):

- Retention times and flow rates are managed to allow sufficient time for plant uptake, and no storm flushing occurs.
- Keep the wetland wet, since dry periods can release accumulated P. Wetlands require large amounts of land compared to other off-site treatments.

Chemical Treatment

Another choice is the addition of chemicals that precipitate and/or flocculate P compounds so they can be easily settled or filtered from the water. Chemical treatment would include chemical injection and deposition (pond, wetlands, etc.) components. Chemical treatment can be very effective but expensive.

17.5.5 Summary of BMPs

Table 17.4 is a summary table of the BMPs discussed. The table has additional information relating to the applicable crops and the reduction potential of the BMPs. However, in using these reduction ranges, one must understand the limitations. First, only a few of the listed BMPs have been field tested, and even those were tested for only a limited set of conditions. Therefore, most of the stated phosphorus reduction ranges are based on corollary data and the authors' knowledge of the physical and chemical processes of the region. The reduction ranges also reflect the variability of existing conditions among farms. For example, farms implementing a BMP for the first time can expect to experience the full benefit of that BMP, whereas those farms already practicing a specific BMP should expect no additional phosphorus reduction with continued implementation of that BMP. These ranges should be considered only as a guide. Additional research will be needed to further narrow these very broad reduction ranges.

17.6 MODELING TOOLS FOR DESIGN AND ASSESSMENT OF BMPS

Several computer models have been developed to assist in the assessment of BMP effectiveness; however, only one has been specifically designed for P transport in flatwood soils. The Everglades Agricultural Area Model (EAAMOD), developed by SWET (Bottcher et al., 1998), is specifically designed to simulate the various cropping systems and soil profile configurations found in flatwood and organic soils. The model has a user-friendly interface, called EAAWIN, to allow farm specific practices to be defined, including planting/harvest dates, P fertilization, and water management control practices (SWET, 1997a). EAAMOD simulates P and water movement in two dimensions and accounts for all important processes including: soil layering, ditch spacing, plant uptake, P adsorption/desorption, P mineralization, redox potential influences, etc. The model estimates P reduction potentials for most management scenarios for flatwood soils and can provide individual farmers and water resource managers with the information they need to make informed decisions.

Other larger-scale models have also been developed for assessing regional economic and operational impact of various BMP scenarios. The Watershed Assessment Model (WAM) developed by SWET (1997b) is a GIS-based watershed assessment model for Florida conditions. WAM simulates an entire watershed on a grid-cell basis (less than a hectare in size). The individual grid cells are simulated using detailed field models such as Groundwater Loading Effects of Agricultural Management Systems (GLEAMS) (Knisel, 1986) or EAAMOD. The simulated grid cell outputs are then routed through the stream system to the watershed outlet on time

TABLE 17.4
Summary of Flatwood BMPs

Name	Estimated P reduction range,[a] %	Crop[b]
Fertility BMPs		
Calibrated soil testing	0–50	SC, V, C, P, F, RC
Banding of fertilizer	0–40	V, C, RC
Misplaced fertilizer	0–15	All
Split application of fertilizer	0–10	All
Manure BMPs		
Dairy HIA drainage control	60–90	D, F
Barn manure handling	10–90	D, F
Watering, feed, and shade facilities placement	5–80	D, P
Feed ration control	0–20	D, P
Fencing ditches and streams	0–40	P
Grazing management	0–20	P
Select high P uptake crops	0–30	D, P
Composting	0–100	D
General BMPs		
Crop management	0–40	All
Irrigation and drainage management	0–30	All
Buffer strips	0–40	D P, RC, F, V
Limit drainage of wetlands	0–20	W
Use of wetlands	0–50	All
Chemical treatment	50–90	D, V
Alternative land use	0–60	D, V, RC, P

[a]Ranges are for individual farms after considering uncertainty and the variability of farm management unless otherwise noted.

[b]SC = sugar cane, V = vegetable, C = citrus, P = pasture, D = dairy, RC = row crop, S = sod, W = wetland, and F = forage.

scales as small as one hour. Because of WAMs ability to use the soil and land use information from the GIS, spatial and cumulative impacts of nutrient management programs can easily be assessed. The Lake Okeechobee Agricultural Decision Support System (LOADSS) and Interactive Dairy Model (Negahban et al., 1993) are similar models but are specifically designed for the Okeechobee basin. The LOADSS model has the additional advantage of providing estimates of economic impact from BMP implementation programs but does not provide the dynamic simulation of discharge and loads like WAM.

17.7 IMPLEMENTATION STRATEGIES FOR BMPS

The development of BMPs for P control does not mean BMPs will be implemented. Implementation requires that farmers clearly see the benefits of BMPs. Farmers must accept the BMPs as a necessary part of their operation. Acceptance is not always easy to obtain, but there are three general ways to achieve this in the agricultural community. They are:

1. Provide the farmers with sufficient, supportable evidence that the BMP will improve their farming operation (voluntary).
2. Provide the farmers with funds to entice them to implement the BMPs (incentive).
3. Make it law that they must implement BMPs (regulatory).

These strategies are listed in order of palatability to farmers. Each approach may achieve the desired water quality goals, but at varying rates of reduction and cost-effectiveness. These strategies has been tried in various parts of the country and in Florida with varying degrees of success. Typically, the voluntary approach has not worked as well as the incentive or regulatory approaches, but this was probably a result of poor education programs to promote acceptance of the BMPs.

Choosing the appropriate BMP implementation approach requires an understanding of the cost effectiveness of individual BMPs and the on-farm operational effects, so that the economic and operational impacts on the farmer can be assessed. If this assessment shows that certain BMPs are economically beneficial and operationally compatible with current conditions, the voluntary approach is all that is needed. However, if the assessment shows the BMPs are operationally compatible but potentially cost prohibitive, an incentive approach is needed. Finally, if the BMPs are found to be economical but not operationally compatible, the regulatory approach is best. In all cases, both the farmers and government agency staffs must be fully educated about BMP responses before acceptance of any approach can be achieved.

17.8 SUMMARY AND CONCLUSIONS

A variety of BMPs that reduce P losses from flatwood soils have been presented. The BMPs were presented from the perspective of the source of P they address and the principles behind controlling the source. Models for assisting in the assessment of BMPs were also briefly presented.

Generally, our current knowledge of BMPs is sufficient to develop effective basin-wide P control programs that will likely meet the water quality standards set by the State of Florida. The success of the Okeechobee Dairy Rule in bringing the majority of the dairies into compliance is evidence that appropriate BMPs will work. Additional research, however, is needed to better refine the costs and removal efficiencies that can realistically be achieved by BMPs. The cost of effective BMP programs will vary greatly across farms, leaving some farmers short of resources to fully implement the BMPs. In these cases, an incentive or cost-share program will be necessary. Acceptance of the benefits of a BMP program by all involved parties (farmers, government staffs, and environmentalists) is critical to the program's success.

Phosphorus control in Florida is technologically feasible, if the social and political obstacles are overcome. Good scientific data and education is the best way to overcome these obstacles.

REFERENCES

Adjei, M.B., P. Mislevy, K.H. Quesenberry, and W.R. Ocumpaugh. 1987. Grazing-frequency effects on forage production, quality, persistence and crown total nonstructural carbohydrate reserves of Limpograsses. Soil and Crop Science Society of Florida, Proceedings 47:233–236.

Allen, L.H., Jr., J.M. Ruddell, G.H. Ritter, F.E. Davis, and P. Yates. 1982. Land use effects on Taylor Creek water quality. In Environmentally sound water and soil management. Amer. Soc. of Civil Eng., New York, NY.

Bottcher, A.B., F.T. Izuno, and E.A. Hanlon. 1997. Procedural guide for the development of farm-level best management practices plans for phosphorus control in the Everglades Agricultural Area. UF/IFAS Circular 1177. University of Florida. Gainesville, FL.

Bottcher, A.B., and F.T. Izuno, (ed.). 1994. Everglades Agricultural Area (EAA): Water, soil, crop, and environmental management. University of Florida Press. Gainesville, FL.

Bottcher, A.B., N.B. Pickering, A.B. Cooper, B.M. Jacobson, B.L. Roy, and J. Slocum. 1998. EAAMOD—Everglades Agricultural Area model for high water table conditions. Proceedings of the 7th International Drainage Symposium. ASAE. St. Joseph, MI.

Brezonik, P.L., and A.C. Federico. 1975. Effects of backpumping from agricultural drainage canals on water quality in Lake Okeechobee. Technical Report Series 1(1). Florida Department of Environmental Regulation. Tallahassee, FL.

Burgora, B. A. 1989. Phosphorus spatial distribution, sorption, and transport in a Spodosol. Ph.D. Dissertation. University of Florida.

Burleson, R.W. 1988. Development of a continuous stormwater management model for agriculture in Florida's flatwoods resource area. Masters Thesis. Department of Agricultural Engineering, University of Florida, Gainesville, FL.

Campbell, K.L, T.K. Tremwel, J.C. Capece, and T.B. Cera. 1992. Biogeochemical behavior and transport of phosphorus in the Lake Okeechobee Basin: Area II final summary report. Deliverable 2.4.7. South Florida Water Management District. West Palm Beach, FL.

Campbell, K.L, T.K. Tremwel, A.B. Bottcher, and D.A. Graetz. 1993. Performance of selected BMPs for phosphorus reduction in high phosphorus source areas. Final Report to South Florida Water Management District, West Palm Beach, FL

Davis, F.E., and M.L. Marshall. 1975. Chemical and biological investigations of Lake Okeechobee, January 1973 June 1974. Interim report. Central and Southern Florida Flood Control District, Technical Publication No. 751. West Palm Beach, FL.

FDEP, 1992. Feedlot and dairy waste water treatment and management requirement. Florida Department of Environmental Regulation Administrative Code Chapter 62-670.500-540.

Federico, A.C., F.E. Davis, K.G. Dickson, and C.R. Kratzer. 1981. Lake Okeechobee water quality studies and eutrophication assessment. South Florida Water Management District, Technical Publication 812. West Palm Beach, FL.

Graetz, D.A., and V.D. Nair. 1994. Fate of P in Florida Spodosols contaminated with cattle manure. Ecological Engineering 5:163–181.

Gunsalus, B., E.G. Flaig, and G.H. Ritter. 1992. Effectiveness of agricultural best management practices implemented in the Taylor Creek/Nubbin Slough Watershed and the Lower Kissimmee River Basin. Proceedings of National RCWP Symposium—10 years of controlling nonpoint source pollution: The RCWP experience. EPA/625/R-92/006.

Hammer, D.A. (ed.). 1989. Constructed wetlands for wastewater treatment: municipal, industrial and agricultural. Lewis Publishers. Chelsea, MI.

Izuno, F. T., and A.B. Bottcher. May 1991. The effects of on-farm agricultural practices in the organic soils of the EAA on phosphorus and nitrogen transport—Screening BMPs for phosphorus loadings and concentration reductions. Final Report to South Florida Water Management District, West Palm Beach, FL.

Kidder, G., E.A. Hanlon, and C.G. Chambliss. 1997. UF/IFAS standardized fertilization recommendations for agronomic crops. UF/IFAS Publication SL-129. University of Florida, Gainesville, FL.

Knisel, W. G. 1980. CREAMS: A field scale model for Chemicals, Runoff, and Erosion from Agricultural Management Systems. U. S. Department of Agriculture, Conservation Report Number 26, 640 pp.

Knisel, W. G. 1993. GLEAMS: Groundwater Loading Effects of Agricultural Management Systems. UGA-CPES-BAED Publication no. 5.

Mislevy, P., G.W. Burton, and P. Busey. 1991. Bahiagrass response to grazing frequency. Soil and Crop Science Society of Florida, Proceedings 50:58–64.

Morse, D., H.H. Head, C.J. Wilcox, H.H. Van Horn, C.D. Hissem, and B. Harris. 1992. Effects of concentration of dietary phosphorus on amount and route of excretion. Journal of Dairy Science. 75:3039–3049.

Negahban, B., C.M. Fonyo, W.G. Boggess, J. Jones, K. Campbell, G.A. Kiker, E. Hamouda, E.G. Flaig, and H. Lal. 1993. LOADSS: A GIS-based decision support system for regional environmental planning. Conference proceedings of Application of Advanced Information Technologies: Effective management of natural resources. American Society of Agricultural Engineers. St. Joseph, MI.

NRCS. 1989. Internal technical memorandum. U. S. Department of Agriculture, Natural Resources and Conservation Service. Gainesville, FL.

Rechcigl, J.E, G.G. Payne, A.B. Bottcher, and P.S. Porter. 1992. Reduced phosphorus application on Bahiagrass and water quality. Agronomy Journal 84:463–468.

Sanchez, C. A., S. Swanson, and P. S. Porter. 1990. Banding to improve fertilizer use efficiency of lettuce. Journal of the American Society of Horticultural Science 115:581–584.

Sanchez, C. A., P. S. Porter, and M. F. Ulloa. 1991. Relative efficiency of broadcast and banded phosphorous for sweet corn produced on histosols. Soil Science Society of America Journal, 55:871–875.

Shanze, T. 1984. Overland flow treatment of municipal waste water in Florida. Master Thesis. University of Florida, Gainesville, FL.

SCS. 1992. Agricultural waste management field handbook. USDA Soil Conservation Service. Washington, DC.

SFWMD. 1989. Interim Surface Water Improvement and Management (SWIM) plan for Lake Okeechobee. Part I: Water quality & Part VII: Public information. South Florida Water Management District. West Palm Beach, FL.

SFWMD. 1992. Everglades SWIM plan. South Florida Water Management District. West Palm Beach, FL.

SFWMD. 1993. Update of Okeechobee SWIM plan. South Florida Water Management District. West Palm Beach, FL.

SWET (Soil and Water Engineering Technology, Inc.). 1996. Tri-county agricultural best management practices study. Final report – Phase II. Submitted to St. Johns River Water Management District. Soil and Water Engineering Technology, Inc. Gainesville FL.

SWET (Soil and Water Engineering Technology, Inc.). 1997a. EAAWIN Version 1.5 computer simulation model. Vol. 2 of the IFAS Phase V Annual Report to the Everglades Agricultural Area Environmental Protection District by Forrest Izuno and Ronald Rice. Everglades Research and Education Center, Belle Glade, FL.

SWET (Soil and Water Engineering Technology, Inc.). 1997b. Development of a GRID GIS based simulation model. Final report submitted to the St. Johns River Water Management District. Palatka, FL.

Taylor, Robert B., Jr. 1987. The use of retention/detention for controlling nitrogen and phosphorus in surface waters. Masters Thesis. Department of Agricultural Engineering, University of Florida, Gainesville, FL.

U.S. Department of Commerce. 1994. 1992 Census of agriculture—Volume 1 geographic area series–Part 9 Florida state and county data. Economics and Statistics Administration, Bureau of the Census, Washington, DC.

18 Phosphorus Management in Organic (Histosols) Soils

Forrest T. Izuno and Paul J. Whalen

18.1 ABSTRACT

By legislative mandate, the Everglades Agricultural Area (EAA) must achieve and maintain a minimum 25% reduction in total-P (TP) loading to receiving areas relative to the baseline years of 1978 to 1988 using "best management practices" (BMPs). Furthermore, all agricultural entities discharging to the South Florida Water Management District (SFWMD) water conveyance system are required to monitor drainage flows and TP concentrations. In addition, the agricultural community must support research to develop BMPs for achieving farm-level P load reductions.

Intensive research on agricultural P sources, usage, and fate began in 1986 on thirty-six 1.4 ha field plots located on three farms in the EAA. These plots were used to assess the feasibility and quantify the effects of altering drainage rates, modifying fertilizer application methods and rates, and using flooded rice production to reduce P concentrations and loads in the EAA. The studies showed that P load reductions in the 20 to 60% range could be expected on farm and basin levels if BMPs associated with these practices could be developed and implemented. The BMP menu developed is broken down into three main categories.

1. Fertilizer management
2. Water management
3. Particulate transport reduction

Included practices are banding fertilizers, following soil test fertilizer recommendations, avoiding inadvertent fertilizer misapplications and spills, reducing drainage

1-56670-331-X/99/$0.00+$.50

pumping, hydraulically blocking forms and installing booster pumps, redirecting drainage flow paths, and minimizing the generation and transport of particulate matter in the drainage stream.

From 1991 through the present, BMPs were developed for, and implemented at, ten farms across the EAA. The farms were selected as being representative of the sizes, soils, geology, crop systems, water management philosophies, and geographic locations prevalent in the EAA. Discharge volumes and P concentrations have been monitored for both pre- and post-BMP implementation phases. Comparisons have been made between pre- and post-BMP P loading data in a variety of ways including unit area load calculations, drainage pumping to rainfall volume ratios, P load to rainfall volume ratios, a computer simulation model, and the SFWMD's hydrologic adjustment model. All methods of comparison show that water management BMPs are currently the most effective means of reducing P loads. Farms are showing 20 to 70% reductions in P loads, depending on BMPs implemented and the method of determining reductions. Concurrently, basin-level P load reductions measured independently by the SFWMD are showing substantial reductions in P loading (ranging from 50 to 70%) at the primary EAA discharge points.

18.2 INTRODUCTION

South Florida has emerged as one of the most visible arenas in the world for water conflicts relative to the sustainability of agriculture, environmental preservation, and satisfying water quantity and quality demands by a rapidly growing population. Several factors make the artificially drained south Florida a natural flashpoint for myriad water and land issues, as well as an ideal region for the development, testing, and implementation of procedures for arriving at equitable solutions. These include: (1) the population explosion (requiring land for housing and recreation, and food and water for sustenance), (2) a unique oligotrophic ecological system, (3) a lucrative tourist industry, (4) a favorable climate and living conditions, and (5) an extremely productive and diverse agricultural industry.

Equitable resolution of the contentious issues has been further complicated in south Florida by the philosophical and moral dilemmas regarding the relative value of human life and comfort versus the preservation of the environment; the involvement of local, state, and federal legal documents and agents; the high levels of emotionalism that are often associated with issues of this nature; the paucity of timely and relevant scientific data; and the large amounts of money required. Ironically, these same apparent hindrances to the process are also largely responsible for the progress made in developing solutions to the Everglades Agricultural Area (EAA) specific total phosphorus loading problem, the worldwide consciousness developed with respect to the need for anticipating environmental/agricultural/urban/suburban land and water conflicts, and the manner in which these types of conflicts should be addressed. The unique hydrologic, geographic, geologic, agricultural, environmental, political, and socioeconomic characteristics of south Florida, along with the intensity of the conflicts, have made it difficult to rely on past case studies for assistance.

Water managers and the agricultural industry have taken the brunt of the blame for the undesirable ecological changes attributed to "unnatural" hydroperiods and elevated total phosphorus (TP) concentrations and loads in areas south of Lake Okeechobee. Hence, they have also accepted a major portion of the responsibility for mitigating the severity of the current and potential negative impacts associated with water quality and quantity management relative to agricultural production and human demand.

18.3 BACKGROUND

The EAA (Fig. 18.1) is a small portion (approximately 280,000 ha) of the original Everglades Region (approximately 1.94 million ha). Of the total land area, approximately 200,000 ha are cultivated. Sugarcane (*Saccharum* spp.) is the primary crop grown, with the remaining arable land dedicated, for the most part, to the production of winter vegetables, sod, and rice (*Oryza sativa* L.) (Coale and Glaz, 1992). The soils of the EAA are predominantly Histosols, underlain by marl and a bowl-shaped limestone bedrock formation. Water tables throughout the region are shallow, generally between 30 to 100 cm below the soil surface. By virtue of its subtropical climate, the EAA usually receives between 100 and 140 cm of rain annually, with the majority of the rainfall coming during a relatively small number of events (three to 10) despite well defined rainy and dry seasons. Although there is ample rainfall to supply south Florida's demand during a year, the flat topography and underlying geologic structure limits the ability to store water efficiently and effectively, with Lake Okeechobee being the only major reservoir for supply during periods of low rainfall.

The EAA exists solely by virtue of an extensive drainage network including canals, levees, dikes, and pumps. The canalization, diking, and resultant drainage of south Florida makes the region ideally suited for both habitation and agriculture. Unfortunately, that same beneficial drainage of the region has led to a disruption of natural hydroperiods and an influx of nutrients to areas south of the EAA.

Lake Okeechobee, the EAA canal system, and the Water Conservation Areas (WCAs) south of the EAA have been identified as being phosphorus (P) limited (SFWMD, 1992b). The drained Histosols of the EAA mineralize approximately 25 to 112 kg P/ha/yr (Daughtrey et al., 1973). Suburban and agricultural interests also apply P amendments for lawn and crop growth. The introduction of P from these "unnatural" sources to the WCAs is apparently negatively affecting the natural environment in the WCAs and is a potential threat to the Everglades National Park (SFWMD, 1992b). For a more detailed discussion of nutrient dynamics, crops, soils, the environment, and water quality and quantity characteristics in the EAA, the reader is referred to Bottcher and Izuno (1994) and SFWMD (1992b).

The Florida state government passed legislation to reduce P loading of the Everglades water system (Marjorie Stoneman Douglas Everglades Protection Act, 1991; amended in 1994 as the Everglades Forever Act). As a result of the Act, The Everglades BMP Regulatory Program (Chapter 40E-63) was developed (SFWMD, 1992a). The program requires that agricultural "best management practices" (BMPs) for achieving farm-level TP load reductions be implemented. Active research to develop effective and cost-efficient BMPs is also an integral part of the successful

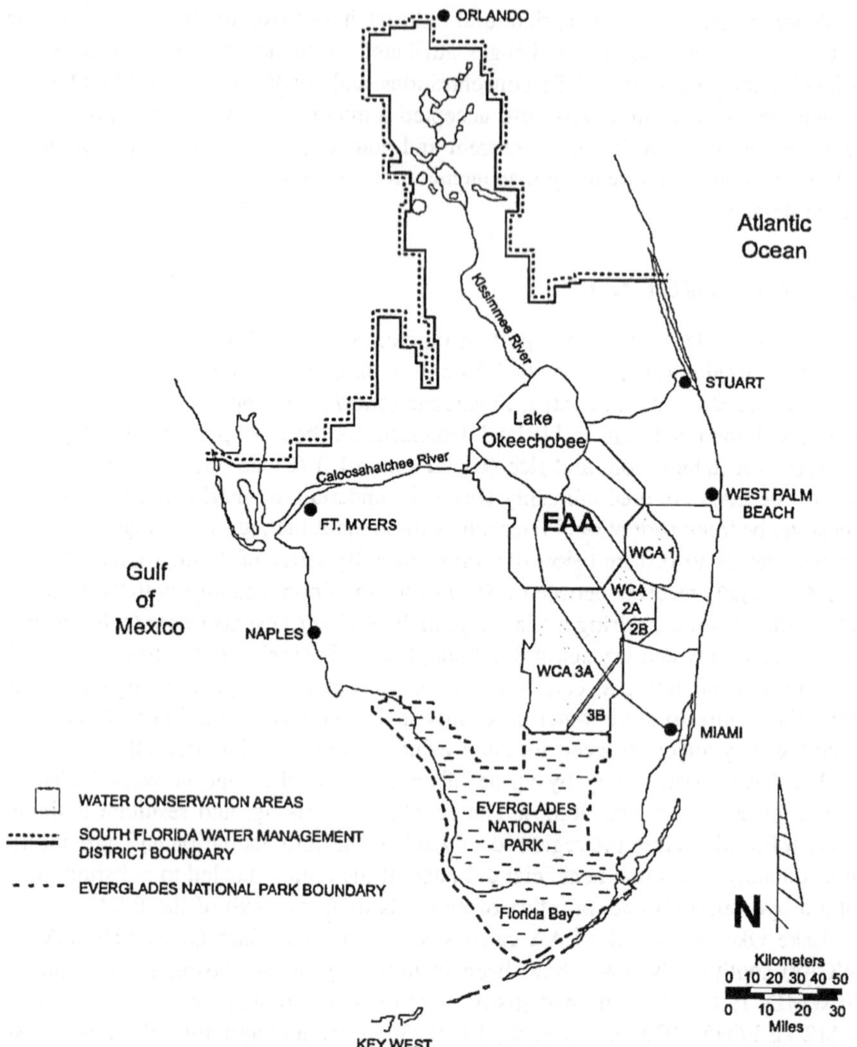

FIGURE 18.1 Location of the EAA with respect to the various geographic features and environmentally sensitive areas of south Florida.

implementation of the regulatory program. Additionally, all structures discharging drainage water to South Florida Water Management District (SFWMD) canals must be monitored for TP concentrations and flow volumes. Finally, Stormwater Treatment Areas (STAs) are to be developed and implemented for basin-level "polishing" (achieving further reduction in TP concentrations and loads) of agricultural drainage water prior to discharge into the WCAs. Essentially, the EAA must achieve and maintain a 75% TP load reduction relative to the basin-level baseline years of 1978 to 1988. Best management practices are to be used to achieve approximately one-third of the reduction, while the STAs are responsible for the remaining two-thirds.

18.4 REGULATORY ACTIONS TO CONTROL PHOSPHORUS LOADING IN THE EAA

During the period 1991 through 1992, the SFWMD developed the Everglades BMP Regulatory Program for the EAA. The BMP program was developed through a series of public workshops and roundtable discussions. The two-year effort resulted in Chapter 40E-63, Florida Administrative Code, which describes the intent, requirements, and compliance components of the Everglades BMP Regulatory Program (SFWMD, 1992a).

The BMP Program for the EAA is unique in that its goal is to achieve a 25% reduction in TP load for the entire EAA basin as a whole, rather than for individual farms or drainage subbasins. The primary water quality compliance monitoring to determine the annual EAA basin percent TP load reduction is conducted by the SFWMD (flow and TP concentration measurements) at SFWMD pump stations and water control complexes (Fig. 18.2).

During the rule development process, much discussion was focused on the determination of which landowners/drainage subbasins would be responsible for implementing additional BMPs if the mandated 25% TP load reductions for EAA basin compliance was not met. The SFWMD initially proposed that all drainage basins be treated equally. That meant that all dischargers would equally share the credit for reductions and equally share the responsibility for not meeting the targeted 25% TP load reduction for the EAA basin. Several landowners preferred to be able to demonstrate TP load reductions from individual properties. They felt that basins that were contributing the highest TP loads to the Everglades should be identified, and thus targeted, to implement additional BMPs should the EAA basin fall out of compliance. This is in contrast to requiring all areas to comply with the additional requirements. As a result, the Everglades BMP Rule requires water quality monitoring at each farm, and in the event that the 25% TP load reduction target is not achieved, the farms with the highest measured unit area P load discharged would be identified and targeted for implementation of additional BMPs. Once again, however, the farm and subbasin water quality monitoring program results would be used only for individual permit compliance if the EAA basin 25% TP load reduction target is not achieved. This phased approach would continue until the EAA basin again meets the annual 25% TP load reduction target level.

The major milestones established by the Everglades BMP Rule included:

- 1991 to 1992 .... Rule Development
- July 1993 .... All BMP permits to be issued
- January 1, 1995 .... Farm BMPs to be fully implemented
- May 1, 1995 to April 30, 1996 ...... First annual 25% basin TP load reduction compliance check

Political and legal mandates required immediate development and implementation of a BMP program prior to the completion of farm-level BMP implementation and efficacy verification research. Thus, the SFWMD was faced with the unenviable tasks of (1) establishing a base level of BMPs for each permit and (2) ensuring

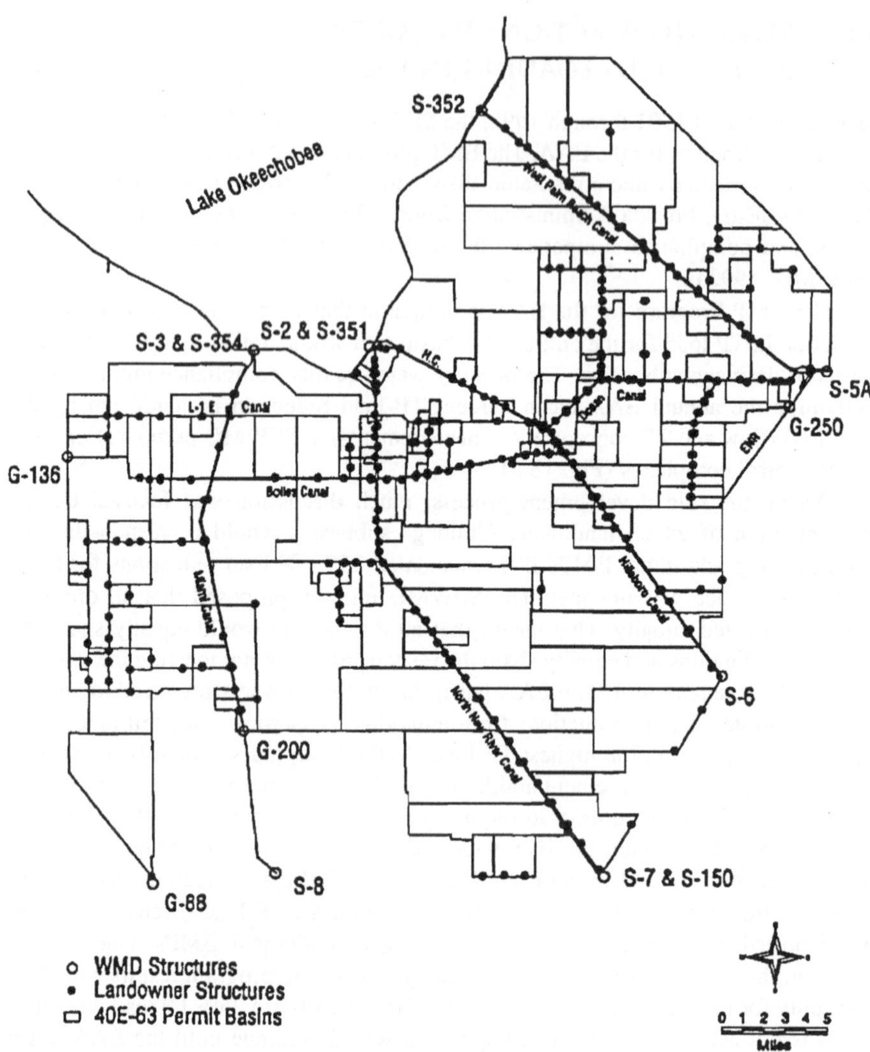

FIGURE 18.2 Water control structures within the EAA basin.

consistency between BMP plans for different landowners prior to receiving a complete scientific verification of the effectiveness of individual BMPs. To accomplish the above tasks, the Everglades BMP Rule required permit applications outlining farm-level BMP implementation plans to be filed by all landowners within the EAA, and a system of BMP "equivalents" was developed. The intent was to assign "points" to BMPs within three basic categories: (1) fertilizer management, (2) water management, and (3) sediment control, that when totaled would equal 25 "points," corresponding to the 25% reduction in TP load required by rule. The BMP list and points assigned to each BMP were based upon preliminary data and best professional judgment (Table 18.1).

TABLE 18.1
Best Management Practices Summary and BMP "Equivalent" Points

BMP	Points	Description
Water detention		Increased detention in canals, field ditches, soil profile, fallow fields, aquatic cover crop fields, prolonged crop flood: measured on an annual avg. basis—rainfall vs. runoff
0.5 inch detained	5	
1 inch detained	10	
Fertilizer application control	2.5	Uniform and controlled fertilizer application (e.g., direct application to plant roots by banding or side dressing; pneumatic, controlled-edge application such as AIRMAX)
Fertilizer content controls		Any combination
Fertilizer spill prevention	2.5	Formal spill prevention protocols (handling and transfer) Side-throw broadcast spreading near ditch banks
Soil testing	5	Avoid excess application by determining P levels needed
Plant tissue analysis	2.5	Avoid excess application by determining P levels needed
Split P application	5	Apply small P portions at various times during the growing season vs. entire application at beginning to prevent excess P from washing into canals (rarely used on cane in EAA)
Slow-release P fertilizer	5	Avoid flushing excess P from soil by using specially treated fertilizer that breaks down slowly, thus releasing P to the plant over time (rarely used in EAA)
Sediment controls		Each sediment control must be consistently implemented over the entire acreage
Any two	2.5	Leveling fields, ditch bank berm, sediment sump in canal, strong canal cleaning program, drainage sump in field ditches, slow field ditch drainage near pumps, sump upstream of drainage pump intake, cover crops, raised culvert bottoms, vegetation, other proposed by permittee
Any four	5	
Any six	10	
Other		
Pasture management	5	Reduce cattle waste nutrients in surface water runoff by "hot spot" fencing, providing watering holes, low cattle density, providing shade, pasture rotation, feed and supplement rotation, etc.
Improved infrastructure	5	Uniform drainage by increased on-farm control structures
Urban xeriscape	5	Lower runoff and P by using plants that require less of each
Detention pond littoral zone	5	Vegetative filtering area for property stormwater runoff
Other BMP proposed	TBD	Proposed by permittee and accepted by SFWMD

Twenty-five BMP "equivalents" or "points" was set as the minimum target BMP level. Utilizing the BMP equivalents approach allowed flexibility for each landowner to develop a BMP plan best suited for the site specific geographic and crop conditions. Table 18.2 presents an example that compares BMP equivalent plans for four different landowner permits. Although each basin had different land uses, soil types, and drainage capacities, BMP equivalent plans were developed and accepted.

After the permits are approved, post-permit verification of the approved BMP plans occurs on two levels: (1) BMP implementation reports, and (2) BMP field verification audits. Best management practice implementation reports are submitted

TABLE 18.2
Example of BMP "Equivalent" Plans for Four Different Landowner Basins

Basin A Sugar cane, deep soils	
BMP	**Points**
Water detention: 1.5 in	15
Fertilizer: soil testing	2.5
Fertilizer: spill and misapplication prevention program	2.5
Fertilizer: banding	5
Total	**25**

Basin B Sugar cane and vegetables, medium soils	
BMP	**Points**
Water detention, 1 in	10
Fertilizer, soil testing	2.5
Fertilizer: spill and misapplication prevention program	2.5
Fertilizer: pneumatic	5
Sediment controls, any four	5
Total	**25**

Basin C Sod, medium soils	
BMP	**Points**
Water detention: 1 in	10
Fertilizer: soil testing	2.5
Fertilizer: spill and misapplication prevention program	2.5
Sediment controls, any six	10
Total	**25**

Basin D Citrus, shallow soils	
BMP	**Points**
Water detention: 1 in	5
Fertilizer: soil testing	2.5
Fertilizer: spill and misapplication prevention program	2.5
Sediment controls, any four	5
Other: improved infrastructure	5
Other: low-volume drip irrigation	5
Total	**25**

annually to the SFWMD. These reports summarize not only the initial implementation of BMPs but ongoing BMP maintenance and documentation as well. During the office review portion of a BMP implementation audit, the SFWMD staff focuses on records that document (1) soil test results, (2) fertilizer recommendations and applications, (3) BMP training of farm personnel, (4) pump logs, and (5) other material that supports BMP implementation. In the field BMP implementation audit, SFWMD staff verify that the selected BMPs have been implemented. The observations range from (1) spoil on canal banks indicating canal cleaning was performed, (2) possessions of, or access to, fertilizer banding equipment, (3) possession of, or access to, land leveling equipment, (4) maintenance of vegetation on ditch banks (to reduce sedimentation), and (5) other observable evidence that supports BMP implementation. Site verifications allow SFWMD staff to work with the permittees by discussing BMP strategies and communicating areas of concern (if any exist). The BMP site verifications conducted thus far indicate that the permittees have implemented their respective BMP plans and that they are taking a proactive approach to reviewing and improving their plans where possible.

The proper calculation of any year's TP load reduction is more complex than a simple comparison of the average annual TP load from the base period to a current year's value. Because rainfall and surface water discharge are subject to large spatial and temporal changes in south Florida, an adjustment must be made to account for these variations when assessing BMP efficacy. These hydrologic variabilities could be large enough to obscure the measurement of effectiveness of BMPs in reducing TP loadings. The hydrologic adjustment attempts to factor out the impact of rainfall variability so that direct comparisons can be made between any year's TP load with that of the base period.

The approach used is to predict (or adjust) what the average annual TP load for the EAA would have been during the 1978 to 1991 pre-BMP base period (water years WY79–WY91) if the annual rainfall amount and distribution measured for a current year had occurred during the pre-BMP base period. The current year's TP load is then compared directly to the predicted base period average annual TP load to calculate the percent TP load reduction.

The resulting methodology is as accurate a measurement as possible of TP loads leaving the EAA basin that have resulted from either rainfall runoff or excess irrigation water discharged. The equations developed were able to effectively predict over 90% of the variability of TP loads during the base period ($r^2 = 0.91$). However, as is true with any prediction tools that deal with natural systems, they are not exact. The annual TP load reduction calculation, for example, has a degree of statistical variability. The standard error identified during the model calibration process was 0.183. The Everglades BMP Rule recognizes this degree of uncertainty and thus requires that the 25% TP load reduction target must be met (or exceeded) a minimum of only one year in three. Therefore, although annual percent TP load reduction values are calculated, examination of the TP load reduction for the EAA Basin is best determined by examining trends over time (a minimum three years is recommended).

The first year of the 25% TP load reduction compliance measurement program for the EAA Basin program did not occur until WY96. However, TP load reduction measurements have been conducted and reported over the past several years and provide an ongoing reporting system during the initial BMP start-up period. Table 18.3 summarizes the results of the EAA Basin TP load calculations according to procedures specified in the Everglades BMP Rule and the Everglades Forever Act. The EAA basin three-year trend has shown a flow-weighted average annual reduction of 51% TP loading.

The load trends presented in Fig. 18.3 appear to show reductions associated with key BMP implementation milestones. Total phosphorus concentrations (Fig. 18.4) also appear to follow those trends observed for TP loads.

Since the WY96 and WY97 measurements represent the first two full years of required BMP implementation throughout the EAA, the degree of confidence in being able to conclusively identify the reduction of TP loading across the EAA as being attributable to farm-level BMP implementation increases as the number of annual periods used in calculations approaches the 10-year length of the base period.

18.5 ON-FARM BMPS FOR ACHIEVING PHOSPHORUS LOAD REDUCTIONS

Best management practices for achieving farm-level TP load reductions in the EAA have been under development, testing, and implementation for the past 10 years.

TABLE 18.3
Summary of EAA Basin Total Phosphorus Calculations Conducted in Accordance with Procedures Specified in Rule 40E-63, F.A.C., and the Everglades Forever Act

	Time →					
	WY79 → WY88 (Base period)	WY89 → WY93 (Interim)	WY94	WY95	WY96	WY97
Three-year trend, P load reduction (flow weighted) (%)	**n/a**	**n/a**	**39**	**36**	**47**	**51**
Acres implemented with BMPs per Rule 40E-63 (%)	0	0[a]	15	63	100	100
Three-year trend, avg. annual P concentration (ppb)	n/a	n/a	121	128	109	104
Avg. annual P concentration (ppb)	173	166	112	116	98	98
Annual calculated P load reduction (%)	n/a	n/a	17	31	68	50

[a]Note: ~1992–1993 initiation of deep-well injection of domestic wastewater from Belle Glade, South Ban, and Pahokee 40E-61 Program BMP

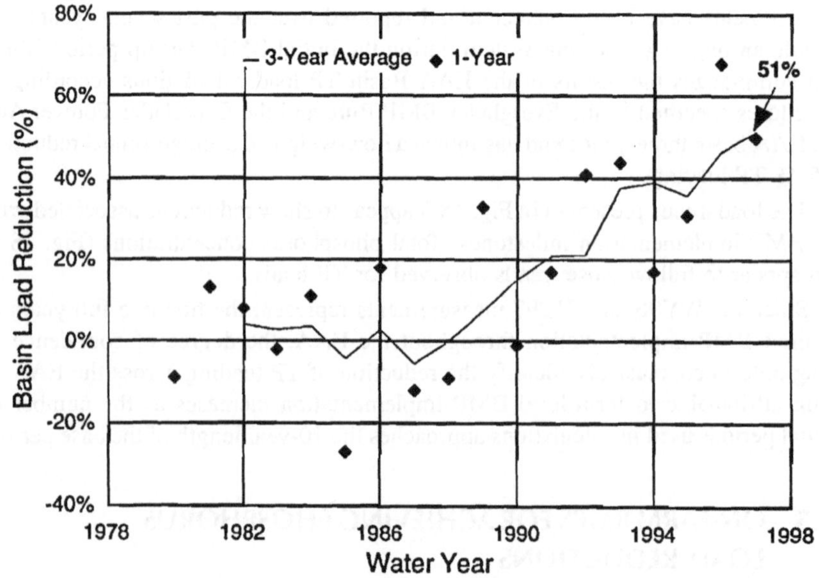

FIGURE 18.3 EAA basin total phosphorus load reduction calculations conducted in accordance with procedures specified in Rule 40E-63, F.A.C. and the Everglades Forever Act.

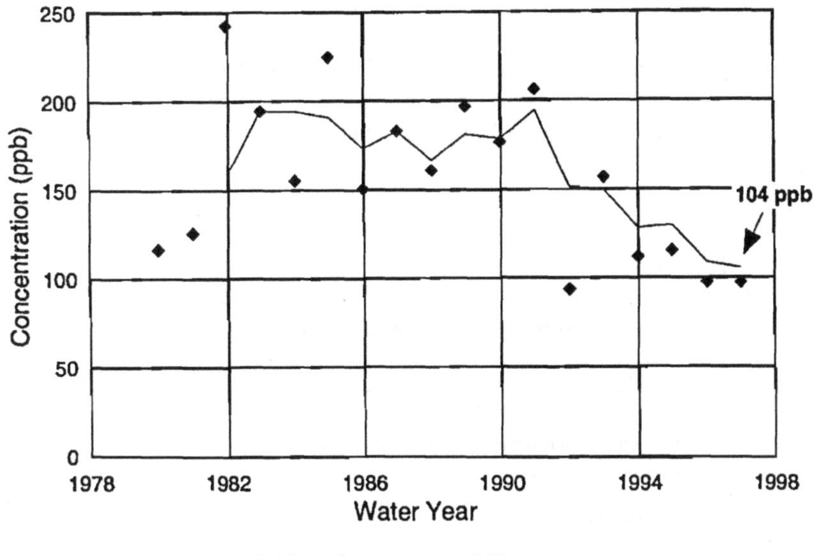

FIGURE 18.4 EAA basin total phosphorus concentration trend calculations conducted in accordance with procedures specified in Rule 40E-63, F.A.C. and the Everglades Forever Act.

During BMP development and testing, the scientific research process has been used to the greatest extent possible. At times, nonscientific external factors have required that the implementation process be expedited, relying on best scientific judgement and assumptions for legitimacy. Furthermore, the compressed time frame allotted to the research and development phases, and the haste with which positive results were demanded, led to less than ideal baseline TP loading data and the need to implement BMPs and assess their efficacy while under a rigorous regulatory schedule. As of 1997, first-generation TP load reduction BMPs have been implemented throughout the EAA and are being monitored for their efficacy at the farm and basin levels.

18.5.1 Research and Development

Organized efforts to develop and implement practical and effective BMPs for TP loading reduction from agricultural operations in the EAA began in 1985–1986 (Izuno and Bottcher, 1991). Four experiment sites, consisting of 8 or 12 large (0.70 ha) hydraulically similar plots, were established on typical organic soils on farmlands subject to typical agricultural production practices. Project principals identified four crop production systems that were hypothesized to be primary contributors to the apparently elevated TP concentration and load levels leaving farms in the EAA. Alternative management practices that appeared to be promising relative to TP concentration and load reductions for these agricultural operations were identified. A list of issues regarding quantification of the effects of agricultural production on TP concentrations and loads that required further investigation was also developed. The four primary conditions to be studied were (1) sugarcane vs. fallow fields, (2) "fast" vs. "slow" drainage rates, (3) amounts and application method of fertilizer for vegetable production, and (4) the use of rice as an aquatic crop following vegetable production to "cleanse" the soil of excess P. Total phosphorus and total dissolved phosphorus (TDP) load reductions determined in the study are summarized in Table 18.4.

TABLE 18.4
Total and Total Dissolved Phosphorus Load Reductions Resulting from Initial Large Plot Studies

Practices compared	TP load (%)	TDP load (%)
Fallow drained vs. sugarcane production	–12	–15
Broadcast vs. banded fertilizer for vegetables	–15	–21
Flooded fallow vs. flooded rice	–38	– 6
Fast vs. slow drainage[a]	–31	–31

[a]But concentrations reacted inversely (+40% for TP and +44% for TDP).

There were no statistically significant differences between TP, TDP, and particulate-P concentrations in drainage water emanating from drained fallow fields compared to sugarcane production fields (Coale et al., 1994a; Izuno and Bottcher, 1991). In fact, TP budgets indicated that the sugarcane production fields were potential net assimilators of TP. Results suggested that sugarcane production on otherwise fallow lands could enhance the basin-wide TP budget by physically removing P from the soil-water system.

Slower drainage rates resulted in higher TP concentrations in drainage water than faster rates (Coale et al., 1994b; Izuno and Bottcher, 1991). However, under higher drainage rate regimes, TP loading increased because of the higher volumes of water removed. Both drainage rates and volumes had major effects on TP concentrations and loads and, hence, offered avenues for achieving major TP reductions. However, the effects of individually altering either the drainage rate or volume to lower TP concentrations or loads had an inverse effect on the other. Hence, drainage rates and volumes must be conjunctively optimized to achieve maximum TP reductions. Factors to consider are the thickness and spatial uniformity of the aerated soil profile, the rate of field drainage, and the duration of the pumping event.

Because of the relatively small portion of the EAA planted to vegetable crops as compared to sugarcane, the majority of the negative publicity relating to agricultural nutrient amendments has centered on the sugarcane industry. Yet, P fertilizer usage for vegetable production greatly eclipses the amount of P amendment used for sugarcane production on a per hectare basis. Study results showed that when the comparably heavy applications (186 kg P/ha) of P were used, immediate and well defined spikes in drainage water TP and TDP concentrations occurred, especially when high-volume, high-intensity rainfall events directly following fertilizer applications triggered the pumping event. While vegetable fertilization practices increased TP concentrations and loads, the study also showed that the effects could be lessened by reducing fertilization rates and changing from traditional broadcast to banded applications (Izuno et al., 1991). Essentially, the amount of TP leaving the fields in drainage water was approximately equal to the amount of TP entering the fields in irrigation and rainwater. When fertilizer inputs and biomass exports were considered in the nutrient balance, P accumulation in the fields ranged from 71.5 to 161.7 kg P/ha/yr (Izuno and Bottcher, 1991; Izuno and Capone, 1995; Izuno et al., 1995).

While addressing the immediate TP concentration and load reduction requirements is vitally important, it is also necessary to consider achieving and maintaining long-term reductions. Much of the agricultural land in the EAA has been drained and cropped for years, leading to high background P levels in the soils resulting from fertilization and the oxidation of the drained organic soils. The ultimate objectives of any program such as this one should be to reduce the current high TP concentration pulse discharges of TP during drainage events as well as to achieve an area-wide TP_{in}/TP_{out} ratio of one or below.

The growth of rice, a harvestable aquatic crop, was studied relative to the hypothesis that rice grown with no fertilization following vegetables would (1) deplete residual nutrients in the soil-water system, (2) result in the export of P from the EAA in harvested grain, and (3) reduce soil oxidation and the subsequent subsidence of the soil and release of P through the mineralization process. Rice proved to be an effective option as long as the flood and draindown waters were properly handled (Izuno and Bottcher, 1991; Izuno and Capone, 1995; Izuno et al., 1995). Proper management of water for rice production requires (1) maintaining the flood, (2) using the rice fields as temporary storage and evapotranspiration (ET) sites for stormwater, and (3) ensuring that draindown water does not directly enter the off-farm drainage stream. In fact, the reuse of nutrient laden draindown waters as surface water "fertigation" for other crops being planted as rice is being harvested has the potential of being a substantiative combined fertilizer and water management BMP. However, mismanagement of the draindown waters can lead to major spikes in TP concentrations and loads during a drainage event.

This preliminary research led to the conclusion that the BMPs needed to focus on reducing loads attributable to both the dissolved and particulate P fractions, since either fraction can constitute up to about 70% of TP during any given event (Izuno and Bottcher, 1991). Furthermore, while fertilizer BMPs are effective, water management BMPs have the greatest potential for achieving the most substantial TP load reductions in the shortest amount of time. The most important goals of the water management BMP package should be to (1) achieve spatial drainage uniformity, (2) reduce farm discharge, and (3) ensure that only the highest quality water leaves the confines of the farm. By implementing all suggested BMPs, it was hypothesized that TP load reductions ranging from 20 to 60% could be realized (Izuno et al., 1995; Izuno and Capone, 1995; Izuno and Bottcher, 1991).

Reducing P-bearing bedload sediment and suspended particulate matter loads needs to be addressed in two parts. Bedload sediment transport reduction BMPs are currently being applied throughout the EAA. These practices include, but are not limited to (1) widening and deepening farm canals and ditches to reduce flow velocities, (2) the use of traditional sediment traps, (3) extensive ditch cleaning programs, (4) raising culvert inlets, and (5) stabilizing ditch and canal banks. Reducing the suspended particulate matter load is slightly more complicated, since most of the fine particulate matter originates internal to the ditches and canals. Floating plant detritus, algae, and plankton can contribute greatly to TP load (Stuck, 1997; Izuno and Rice, 1997). Practices to reduce TP loads attributed to these particulates are currently being developed. The practices will necessarily address (1) allowing the growth of aquatic life to reduce dissolved P, (2) controlling the plant growth rate and confining high growth areas to portions of the ditches and canals away from the major pump stations, and (3) the physical removal of the aquatic plants.

18.5.2 Implementation and Efficacy Verification

Ten farms ranging in size from 106 to 1865 ha, having from 1 to 5 pumps at the 1 or 2 main pump stations serving each farm, are being used for BMP implementation and efficacy quantification. The farms are representative of the EAA with respect to cropping systems, soil characteristics, water management practices, and geographic locations.

Total Phosphorus Concentrations

Total phosphorus concentrations observed during the baseline monitoring period from October 1992 through December 1994 are mapped in Fig. 18.5 (sampling period varies between sites because of the instrumentation installation schedule and staggered BMP implementation, with January 1, 1995 being the date by which all farms had to be in the BMP operational mode). Differences in concentrations that can be attributed to geographic location are distinguishable. Cropping systems also contribute greatly to the TP concentration variability. Essentially, TP concentrations for sugarcane monocultures range from 0.20 mg/L in the northeast (along the West Palm Beach Canal) to 0.07 mg/L in the southwest. All other sugarcane monoculture systems had average baseline TP concentrations of 0.07 or 0.08 mg/L. Farms with mixed cropping systems during the baseline period had average baseline TP concentrations ranging from 0.15 to 0.56 mg/L (the numerical average concentration for these sites was 0.29 mg/L). The single farm that is a vegetable monoculture had a baseline average TP concentration of 0.63 mg/L.

Total BMP monitoring period TP concentrations (1994 to current, period variable by site) for all sites are mapped in Fig. 18.6. The sugarcane monoculture sites, and those sites in the central and southwest portions of the EAA, showed little to no change in TP concentrations. The average TP concentration for one site increased from 0.18 to 0.31 mg/L because of the introduction of rice into the crop rotation without the necessary facilities to manage the drain down water. At one of the mixed crop sites, the average TP concentration also increased, probably resulting from the planting of the vegetable/rice rotation crops in field blocks immediately adjacent to the main farm pump stations. The average TP concentration at the vegetable monoculture site remained stable, but it decreased at the second sugarcane/vegetable rotation site.

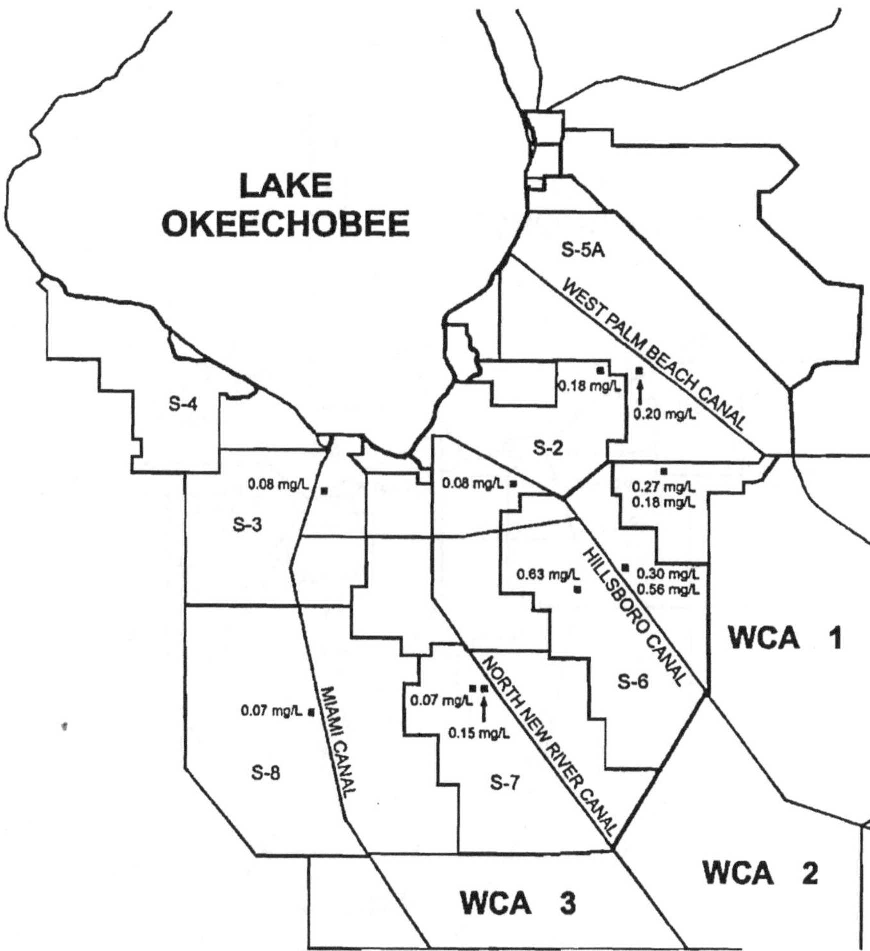

FIGURE 18.5 Averages of the P concentrations of the flow proportional samples collected at the sites. Values are representative of the total baseline monitoring periods for each site.

Comparisons of TP concentrations are for informational purposes and do not reflect the effects of hydrologic variability between sites and between years. Additionally, since monitoring for TDP was not done, the fractioning of TP into dissolved and particulate species cannot be accomplished. Hence, the within-site comparison of TP concentration numerical values merely reflects the fact that TP concentrations have not changed dramatically between the baseline and BMP periods. With the forthcoming addition of TDP analyses, both TDP and particulate-P fractions will be determined. The quantification of these TP fractions will serve as a diagnostic tool for determining the effectiveness of the BMPs and will identify in which fraction reductions are occurring. The fractioning into TDP and particulate-P will also aid in the development of additional BMPs, or refinements of existing BMPs, that can focus on either fraction to further reduce TP loads.

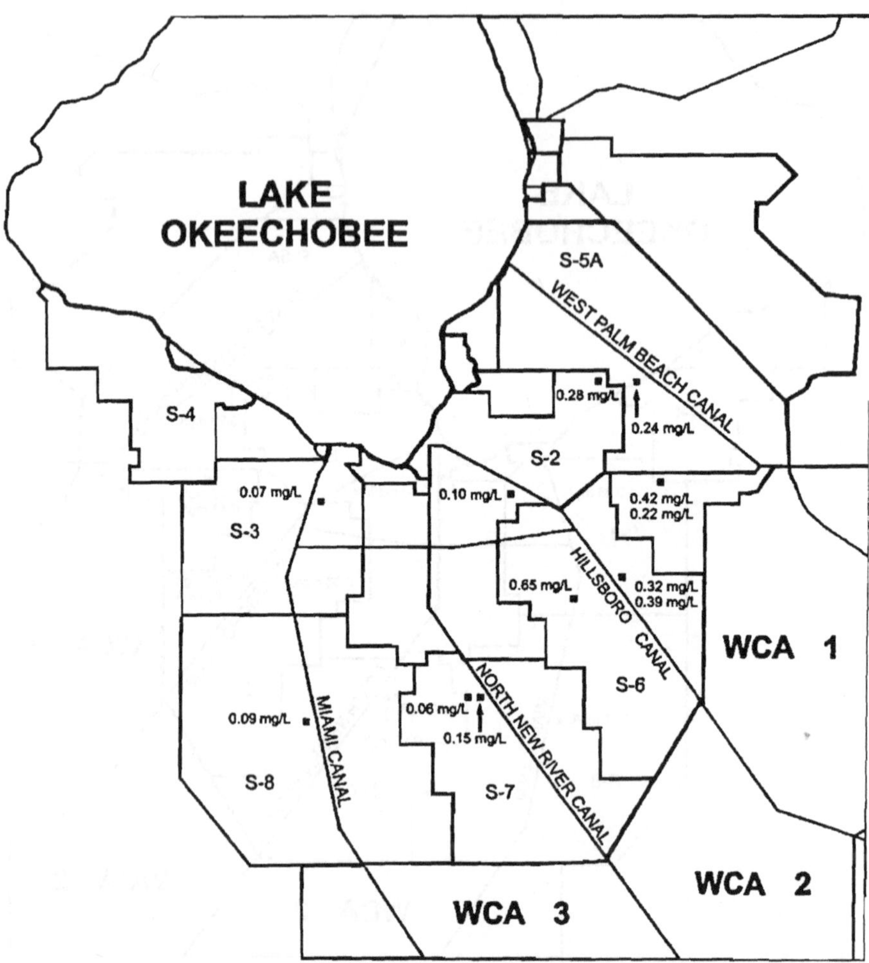

FIGURE 18.6 Averages of the P concentrations of the flow proportional samples collected at the sites. Values are representative of the total BMP monitoring periods through September 30, 1996 for each site, including only data collected after BMPs were installed. The default BMP implementation date is January 1, 1995 according to regulatory requirements.

Total Phosphorus Loads

The EAA Regulatory Program (Chapter 40E-63) set a TP load reduction criteria that must be met at the basin level. However, the rule bases incentives and disincentives on the performance of load reduction practices implemented upstream from the EAA basin main pumps at individual discharge points. Hence, it becomes extremely important to develop the ability to quantify TP load reductions achieved at the farm level and to be able to associate those reductions with individual or combinations of BMPs.

For comparison purposes, all farm load data must be adjusted for farm size and spatial and temporal hydrologic variability. The adjustment for farm size is accomplished by calculating unit area loads (UALs) for each farm (simply the total P load for the 12-month period divided by the farm size). Adjusting for hydrologic variability is more complex and requires the use of the SFWMD model (SFWMD, 1992a; Rice and Izuno, 1996). The model essentially establishes the 13-year period (1978 to 1991) as the baseline period and uses data for TP loads and rainfall at the main basin discharge points to establish baseline TP UAL vs. rainfall relationships. The baseline period TP load is adjusted for hydrologic variability (rainfall) that occurs during any given future year, allowing a direct comparison of annual TP loads with the pre-BMP baseline period. This hydrologic adjustment thereby extracts the variable annual rainfall factor out of the data sets being compared. While these relationships were developed from data specific to the main basin pump stations along the southern perimeter of the EAA, they are assumed to be appropriate for discharge points upstream in the same subbasins.

Table 18.5 shows the changes in adjusted UALs (AUALs) for all sites between consecutive water years (WY94 vs. WY95 and WY95 vs. WY96) and between the project pre- and post-BMP implementation periods (WY94 vs. WY96). A water year (WY), as defined by the South Florida Water Management District (SFWMD, 1997), is from May 1 to April 30 (e.g., WY94 spans the 12-month period from May 1, 1993 through April 30, 1994), corresponding to the wet-dry season cycle in south Florida. The WY94 vs. WY95 and WY95 vs. WY96 comparisons are transition periods during which growers were implementing and learning to use the BMPs. The WY94 vs. WY96 comparison is the truest comparison available of pre- and post-BMP implementation period TP loads since WY94 represents the period from May 1, 1993 through April 30, 1994 when few growers had actively implemented BMPs, and WY96 represents the period from May 1, 1995 through April 30, 1996 which falls entirely beyond the mandatory BMP implementation date of January 1, 1995.

The four sugarcane farms showed TP AUAL reductions ranging from 38.8% to 78.2% (numerical average of 66.5%), with only one site achieving less than a 70% reduction. That single site achieved a 38.8% TP AUAL reduction, still well above the legislatively mandated 25% reduction. This reduction is impressive considering that (1) the grower's options are limited, since baseline TP concentrations were low (0.08 mg/L), (2) the farm area is small (130 ha), and (3) drainage options have always been limited by a typically high stage in the receiving canal. To achieve the 38.8% reduction, the grower simply divided the farm into two water management blocks, where drainage adequacy for the back one-third of the farm is no longer designed to be achieved by the main farm pump station. Instead, a canal block and internal booster pump are being used to drain the back one-third of the farm, which the grower traditionally had trouble dewatering resulting from seepage from surrounding areas. Now, the grower drains the back of the farm into the front two-thirds of his land and operates the main pump only when absolutely necessary during large volume rainfall events.

The vegetable monoculture site achieved an 81.5% TP AUAL reduction by simply reducing drainage activity. During pre-BMP operations, the entire farm was

TABLE 18.5
Changes in Total Phosphorus AUAL for the 10 Experimental Farm Sites for WY94 through WY96 Using the SFWMD (1991) Hydrologic Adjustment Model

Experiment farm	Subbasin	WY94–WY95 (%)	WY95–WYWY96 (%)	WY94–WY96 (%)
1	S-5A	– 50	–56	–78
2	S-6	– 51	–63	–82
3	S-7	– 47	+15	–39
4	S-7	– 51	–53	–77
5	S-6	+ 5	+ 7	+12
6	S-8	+108	–42	+20
7	S-5A	– 58	–28	–70
8	S-6	– 20	–52	–62
9	S-6	– 71	– 7	–73
10	S-8	– 30	–66	–76

drained and water tables were held extremely low for days while planting operations were conducted. This active and excessive drainage included the direct discharge of the nutrient laden water from flooded fallow fields. Under BMP management, the flood waters are allowed to subside via lateral movement through the soil profile to farm ditches and canals and through ET. Once field water levels have subsided below the soil surface, active drainage begins. Additionally, the farm has been blocked into two equal parts allowing the temporary storage of draindown water from half the farm while the other half is being planted. After half of the farm is planted, water is routed back into the previously well-drained half, and cultivation activities commence on the remaining farm half. This system has allowed the grower to achieve a major reduction in discharge to off-farm canals.

The mixed crop sites showed much greater variability in TP AUAL reductions achieved. One of the sites has always been a sugarcane/vegetable rotation system. This large site (1012 ha) has vegetable fields scattered throughout, depending on the sugarcane harvest/replant schedule. By consolidating some of the vegetable fields into contiguous areas, reducing off-farm drainage pumping, and rerouting drainage water from the front of the farm around the outer boundaries of the farm such that it reenters the drainage stream after a long flow path, over-drainage of the fields near the pump station has been minimized and a 61.7% TP AUAL reduction has been achieved.

One of the sites was designated as a "reverse BMP" site during the first comparison period (WY94 to WY95). The grower unilaterally decided to go from basically a sugarcane monoculture (with a small pasture area) to a sugarcane/vegetable rotation. Upon doing so, primarily by virtue of the change in the volume of water discharged from the farm to maintain adequately low water tables for vegetable production, a 107.9% increase in TP AUAL resulted. Since that period, the grower has been phasing out vegetable production and TP AUALs should stabilize at pre-"reverse BMP" levels. The magnitude of the increase in TP AUAL vividly illustrates the effects of growing vegetables in rotation with sugarcane in a non-BMP situation where the entire farm is continuously drained to meet the water table requirements of a few vegetable fields. Had the grower implemented appropriate hydraulic BMPs, including blocking of the vegetable fields into a contiguous subfarm for water management purposes, vegetable production, at least to some extent (bounded by a reasonable ratio of vegetable to sugarcane area), would have been possible without the astounding TP AUAL increase.

The two sugarcane/rice rotation sites showed a 77.1% reduction and 11.9% increase in TP AUAL, respectively. The grower at the site showing a TP load reduction made many hydraulic capacity improvements, hydraulically blocked the rice fields, and planted a limited percentage of the total land area to rice. Hence, the grower realized a large reduction in TP AUAL and made a positive step toward bringing his TP_{in}/TP_{out} ratio closer to one. The grower at the other site, however, made few adjustments to discharge pump management since that part of his preexisting water management scheme had been minimized for years. The grower planted half of the farm area to rice and made no special provisions for handling the flood water draindown, thus greatly increasing the irrigation and subsequent drainage volumes necessary. Hence, rice production apparently had a negative effect on TP AUALs at this location. The site illustrates the importance of planning for the proper handling of draindown water when using rice as a vehicle for TP export from the EAA.

The final site has been a mixed cropping system for years prior to the start of this study, including the cultivation of sugarcane, sod, and a vegetable/rice rotation. A reduction in TP AUAL between the pre- and post-BMP periods (WY94 vs. WY96) of 69.7% has been achieved, primarily through (1) improving farm hydraulic capacities, (2) redefining the threshold criteria for initiating off-farm drainage pumping, and (3) splitting the farm into six potentially hydraulically isolated blocks for water management purposes. The grower can now pump water between any of the six hydraulic units by opening and/or closing the appropriate culverts and operating permanent and/or temporary booster pumps placed at strategic locations. In this way, the grower ensures that no vegetable or rice drainage water directly enters the drainage stream. Excess water is temporarily stored in sugarcane fields, while not elevating water tables to harmful levels for long periods of time. Assessments of water table levels required in each hydraulic unit are made and tended to prior to any off-farm discharge. Although demands on his farm managers have increased tremendously, the grower has managed to achieve a significant reduction in TP AUAL while retaining his traditional crop mix.

Further evidence of the efficacy of farm-level agricultural TP reduction BMPs can be seen in simple ratios of TP load to rainfall volume ratios. Based on a one- to two-year baseline data set (1992 to 1994), the ratios have been reduced by up to 35% (Izuno and Rice, 1997). Ratios of the volume of drainage water pumped to rainfall volume have decreased by up to 40% compared to the baseline period. Finally, average water table depths at the representative experiment farms monitored appear to have risen by 7.5 to 10 cm (3 to 4 in).

18.6 SUMMARY

The legislatively mandated 25% TP load reduction for the EAA basin is being exceeded according to SFWMD figures (51% TP load reduction for a three-year average spanning WY95 through WY97) (SFWMD, 1997). Total phosphorus concentrations have dropped to 98 g/L for WY96 and WY97, averaging 104 g/L for the three-year period WY95 to WY97 compared to 173 g/L for the base period of WY79 to WY91. At the farm level, the 10 representative study sites are showing a numerical average TP AUAL reduction of 69.2% between pre- and post-BMP implementation periods. Clearly, the implementation of BMPs is emerging as an effective method for achieving TP load reductions. Opportunities for even greater reductions are immense when one considers that particulate-P control BMPs have not yet been adequately developed and implemented. Additionally, the computer model (EAAMOD) that will allow the optimization of BMP selection and management, is still under development as part of this study. Finally, the long-term effects of the fertilizer BMPs and the efforts to reduce the basin-wide TP_{in}/TP_{out} loading ratio have not had adequate time to be fully evident in TP load reduction figures.

All current data and analyses demonstrate that TP load reduction BMPs are at least as effective as originally predicted by Izuno et al. (1995) and Izuno and Bottcher (1991). Future work toward quantifying reductions more accurately (Rice and Izuno, 1996), developing particulate-P control BMPs, and optimizing the implementation and operation of existing BMPs will add greatly to the ability to achieve a positive coexistence between agriculture and environmental concerns in south Florida.

REFERENCES

Bottcher, A.B. and F.T. Izuno, eds. 1994. *Everglades Agricultural Area: Water, Soil, Crop, and Environmental Management*. University Press of Florida. Gainesville, FL. 318 p.

Coale, F.J., F.T. Izuno, and A.B. Bottcher. 1994a. Sugarcane production impact on nitrogen and phosphorus in drainage water from an Everglades Histosol. J. of Environ. Qual. 23(1):116–120.

Coale, F.J., F.T. Izuno, and A.B. Bottcher. 1994b. Phosphorus in drainage water from sugarcane in the Everglades Agricultural Area as affected by drainage rate. J. of Environ. Qual. 23(1):121–126.

Coale, F.J. and B. Glaz. 1992. Sugar cane variety census: Florida 1991. Sugary Azucar. 87(11):27–33.

Daughtrey, Z.M., J.M. Gilliam, and E.J. Kamprath. 1973. Phosphorus supply characteristics of acid organic soils as measured by desorption and mineralization. Soil Science 115:18–24. Everglades Forever Act. 1994. Florida Statute Section 373.4592. Amendment of the 1991 Marjory Stoneman Douglas Everglades Protection Act. Tallahassee, FL.

Izuno, F.T. and A.B. Bottcher. 1991. The effects of on-farm agricultural practices in the organic soils of the EAA on nitrogen and phosphorus transport. Final Report submitted to the South Florida Water Management District. West Palm Beach, FL.

Izuno, F.T and R.W. Rice. 1997. Implementation and verification of BMPs for reducing P loading in the EAA. Phase V Annual Report submitted to the EAA-Environmental Protection District and the Florida Department of Environmental Protection. January.

Izuno, F.T., A.B. Bottcher, F.J. Coale, C.A. Sanchez, and D.B. Jones. 1995. Agricultural BMPs for phosphorus reduction in south Florida. Transactions of the Am. Soc. of Ag. Engineers 38(3):735–744.

Izuno, F.T. and L.T. Capone. 1995. Strategies for protecting Florida's Everglades: The best management practice approach. Water Science Tech. Vol. 31(8):123–131.

Izuno, F.T., C.A. Sanchez, F.J. Coale, A.B. Bottcher, and D.B. Jones. 1991. Phosphorus concentrations in drainage water in the Everglades Agricultural Area. J. of Environ. Qual. 20(3):608–619.

Marjory Stoneman Douglas Everglades Protection Act. 1991. Florida Statute Section 373.4592. Tallahassee, FL.

Rice, R.W. and F.T. Izuno. 1991. Techniques for assessing BMP effectiveness in the absence of historical baseline data. ASAE Paper No. 962127. Presented at the July 1996 ASAE Annual International Meeting. Phoenix, AZ.

SFWMD. 1992a. Everglades regulatory program: Everglades Agricultural Area. Part 1 of Chapter 40E-63 of the Florida Administrative Code. South Florida Water Management District. West Palm Beach, FL.

SFWMD. 1992b. Everglades surface water improvement and management plan. South Florida Water Management District. West Palm Beach, FL.

SFWMD. 1997. Everglades best management practices program annual report: Water year 1996 and water year 1997. P. Whalen, ed. South Florida Water Management District. West Palm Beach, FL. September.

Stuck, J.D. 1996. Particulate phosphorus transport in the water conveyance systems of the Everglades Agricultural Area. Ph.D. dissertation, University of Florida.

19 Long-Term Water Quality Trends in the Everglades

William W. Walker, Jr.

19.1 ABSTRACT

Long-term water quality and hydrologic monitoring data have provided important bases for defining the Everglades nutrient-enrichment problem, developing interim water quality standards and regulations, designing control measures, and evaluating the effectiveness of control measures. Specific monitoring and data-reduction procedures for determining compliance with interim and long-term objectives are built into the Settlement Agreement (USA et al., 1991), EAA Regulatory Rule (SFWMD, 1992b), and Everglades Forever Act (State of Florida, 1994). These procedures provide measures of performance for the phosphorus control program that are important from ecological, management, and legal perspectives.

Interpretation of monitoring data with respect to long-term or anthropogenic impacts is facilitated by application of a model, which attempts to differentiate long-term, hydrologic, and random variance components. The model has been used to develop tracking procedures for several Everglades locations.

Variations in flow, phosphorus concentration, and phosphorus loads at major structures in the EAA and WCAs over the 1978 to 1996 period are summarized. The structure, calibration, and application of a model for tracking ENP Shark River Slough inflow P concentrations are described. Interpretations and limitations of tracking results are described.

19.2 INTRODUCTION

Eutrophication induced by anthropogenic phosphorus loads poses a long-term threat to Everglades ecosystems. Impaired water quality and substantial shifts in microbial

1-56670-331-X/99/$0.00+$.50

and macrophyte communities have been observed in regions located downstream of agricultural discharges (Belanger et al., 1989; Nearhoof, 1992; Amador and Jones, 1993; Davis, 1994; Doren et al., 1997). This problem developed over a period of three decades following construction of the Central and Southern Florida Flood Control Project and drainage of wetland areas south of Lake Okeechobee to support intensive agriculture. As shown in Fig. 19.1, the Everglades Agricultural Area (EAA) is located between Lake Okeechobee and the Everglades Water Conservation Areas (WCAs).

In 1988, a lawsuit was filed by the federal government against the local regulatory agencies [Florida Department of Environmental Regulation and South Florida Water Management District (SFWMD)] for not enforcing water quality standards in Loxahatchee National Wildlife Refuge (LNWR) and Everglades National Park (ENP). The lawsuit ended in an out-of-court Settlement Agreement (SA) (USA et al., 1991) and federal consent decree in 1992.

The SA establishes interim and long-term requirements for water quality, control technology, and research. Generally, interim standards and controls are designed based on existing data and known technologies. The interim control program includes implementation of agricultural Best Management Practices (BMPs) and construction of wetland Stormwater Treatment Areas (STAs) to reduce phosphorus loads from the Everglades Agricultural Area (EAA) by approximately 80%, relative to a 1979 to 1988 baseline. Subsequently, SFWMD adopted the EAA Regulatory Rule (SFWMD,

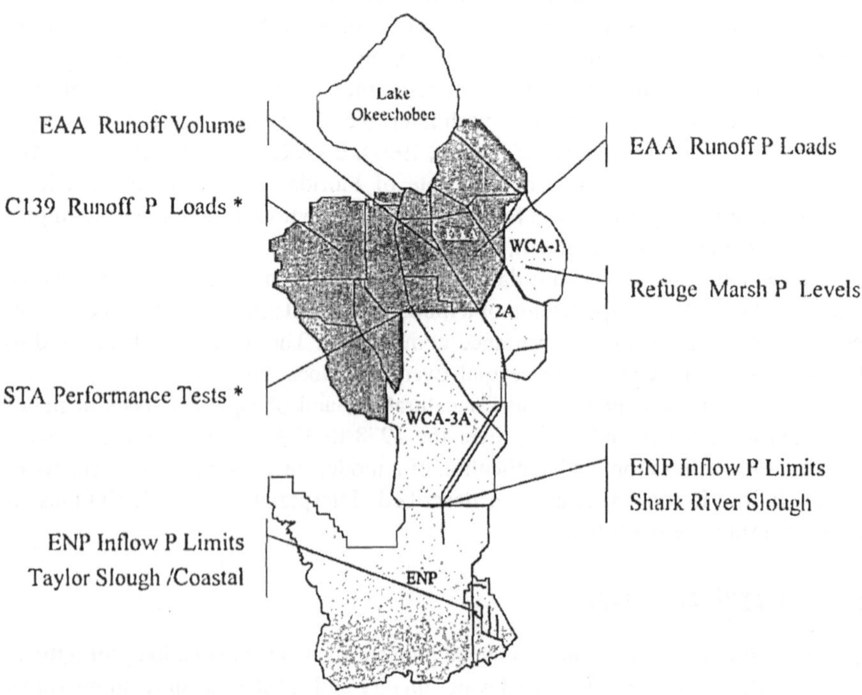

FIGURE 19.1 Regional map.

1992b; Whalen and Whalen, 1994), which requires implementation of BMPs in the EAA to achieve an annual-average phosphorus load reduction of at least 25%. The State of Florida (1994) passed the Everglades Forever Act, which defines a construction project and funding mechanism for STAs. Interim phosphorus standards will apply after interim control technologies are in place (1999 to 2006 for LNWR and 2003 to 2006 for ENP Shark Slough inflows). Long-term standards (>2006) and control technologies will be developed over a period of several years and require a substantial research effort to develop supporting data (Lean et al., 1992).

Long-term water quality and hydrologic monitoring data have provided important bases for defining the Everglades nutrient-enrichment problem, developing interim water quality standards and regulations, designing control measures, and evaluating the effectiveness of control measures. Specific monitoring and data-reduction procedures for determining compliance with interim and long-term objectives are built into the Settlement Agreement, EAA Regulatory Rule, and Everglades Forever Act. The procedures provide measures of performance for the control program that are important from ecological, management, and legal perspectives.

Interpretation of monitoring data with respect to long-term or anthropogenic impacts is facilitated by application of a model that attempts to differentiate long-term, hydrologic, and random variance components. The model has been used to develop tracking procedures for several Everglades locations (Fig. 19.1):

1. ENP Inflow P Limits (2 Basins) (USA et al., 1991, SFWMD, 1992a)
2. LNWR Marsh P Levels (USA et al., 1991, SFWMD, 1992a)
3. EAA Basin P Load Reductions (SFWMD, 1992b)
4. EAA Basin Runoff/BMP Replacement-Water Calculation (SFWMD, 1994)
5. C139 Basin Runoff and P Load (Walker, 1995a)
6. STA Performance Tests (Walker, 1996, FDEP, 1997)

Each procedure was developed within the constraints of historical data to accomplish a specific objective. They share a model structure that is generally applicable in situations where historical monitoring data are to be used as a frame of reference for interpreting current and/or future monitoring data. This would be the case when the management goal is to restore the system to its historical condition, to prevent degradation beyond its current condition, or to require improvement relative to its historical or current condition.

This chapter summarizes long-term variations in flow, total phosphorus concentration, and total phosphorus loads at major structures surrounding in the EAA and WCAs over the 1978 to 1996 period. The structure, calibration, and application of the model for tracking ENP Shark River Slough inflow P concentrations are described. Interpretations and limitations of model results are discussed.

19.3 DATA SOURCES

Water quality data summarized below have been collected by South Florida Water Management District (Germain, 1994) between 1978 and 1996. Hydrologic data collected by SFWMD, Corps of Engineers, U.S. Geologic Survey, and Everglades National Park have been extracted from SFWMD's DBHYDRO data base (SFWMD, 1996). Total phosphorus loads have been calculated using an algorithm that is similar to that described in the EAA Regulatory Rule (SFWMD, 1992b).

19.4 DATA SUMMARIES

Current and future control efforts target total phosphorus loads entering the WCAs from the EAA. Figures 19.2 and 19.3 show annual variations in discharge volume and flow-weighted mean phosphorus concentration, respectively, at major structures between Lake Okeechobee and ENP for Water Years 1978 through 1996. Variations in EAA rainfall and WCA water levels over the same period are shown in Fig. 19.4. An October to September Water Year is used for consistency with the tracking procedure for ENP inflow P limits (see below). Values are summarized for six locations. The term "EAA -> WCAs" primarily represents EAA runoff but also includes lake releases passing through the EAA and runoff from a portion of the C-139 basin. Although the S10s are outflow structures from WCA-1, discharges through these structures are heavily influenced by EAA runoff that is pumped into WCA-1 at S6 and flows along the southern perimeter of WCA-1 to the S10s.

Although minor WCA inflows and outflows are not represented, Figs. 19.2 and 19.3 provide general pictures of temporal and spatial gradients over the 19-year period. Temporal variations in flow and concentration at each location reflect variations in management, climate, and sampling/analytical error. Sorting out these factors is difficult; it is useful, however, to summarize predominant patterns in the data and describe potential causal mechanisms.

Patterns in the flow time series (Fig. 19.2) include the following:

1. Lower flows from EAA to the lake after 1985. These reflect implementation of the Interim Action Plan (IAP), which was designed to reduce P loads to the lake. This was accomplished by reducing backpumping from the EAA to the lake and increasing pumping from the EAA to the WCAs. Although IAP was adopted in 1979, full implementation did not occur until 1986.
2. Higher flows south of the EAA after 1991. These reflect (a) diversion of EAA runoff away from the lake [IAP (1)], (b) increased regulatory releases from Lake Okeechobee (especially in 1993), and (c) high rainfall in recent years (Fig. 19.4).

In each year between 1992 and 1996, discharges from the EAA to the WCAs exceeded the range experienced in 1978 to 1991. The 1995 discharge into ENP Shark Slough (S12s + S333) exceeded the 1978 to 1991 maximum by a factor of 2.5. Record high water levels were experienced in WCA-3A (Fig. 19.4) and ENP during this period. As demonstrated below, these conditions complicate the interpretation of recent water quality data.

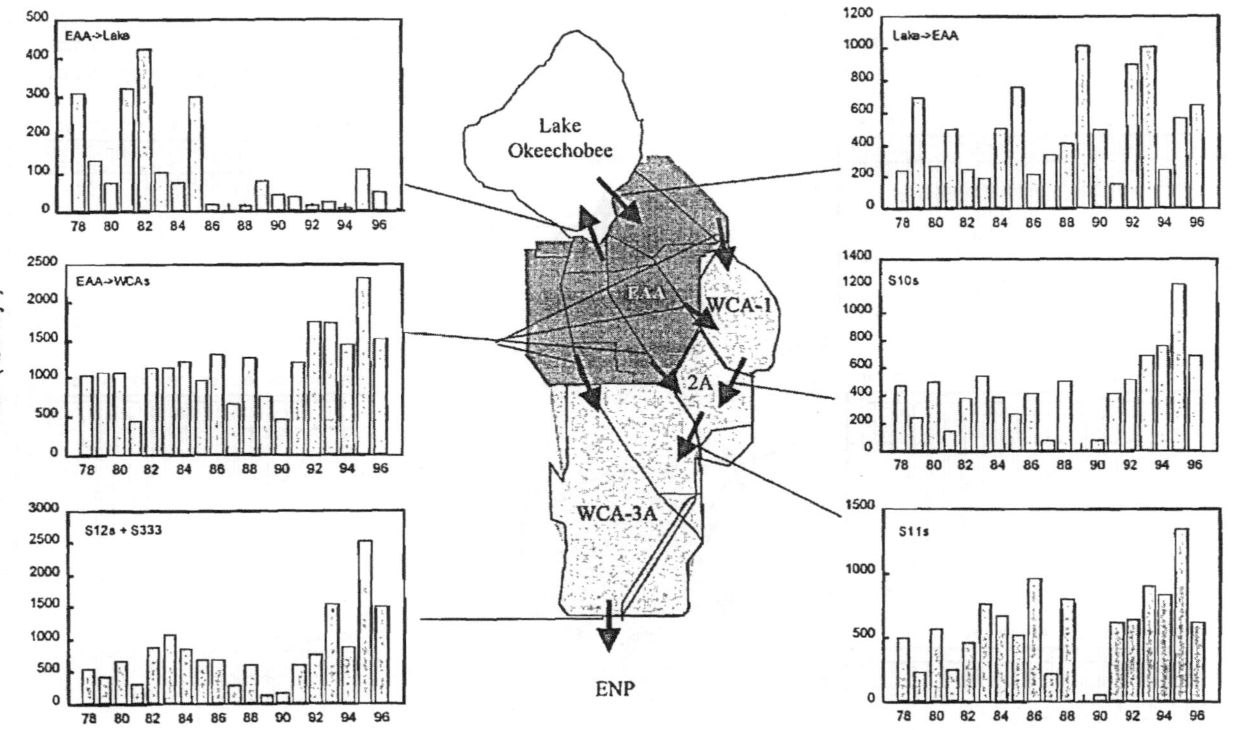

FIGURE 19.2 Flows expressed in 1000 acre-ft per year (kac-ft/yr), October–September water years. Lake –> EAA = discharge from lake to the EAA (structures S354, S351, and HGS5). EAA –> Lake = discharge from EAA to the lake (S2, S3, and S352). EAA –> WCAs = discharge from EAA to the WCAs (S5A, S6, S7, S150, S8, S200, and G250). S10s = discharge WCA-1 to WCA-2A (S10A, B, C, D, and E). S11s = discharge from WCA-2A to WCA-3A (S11A, B, and C). S12s + S333 = discharge from WCA-3A to ENP Shark Slough (S12A, B, C, D, and S333).

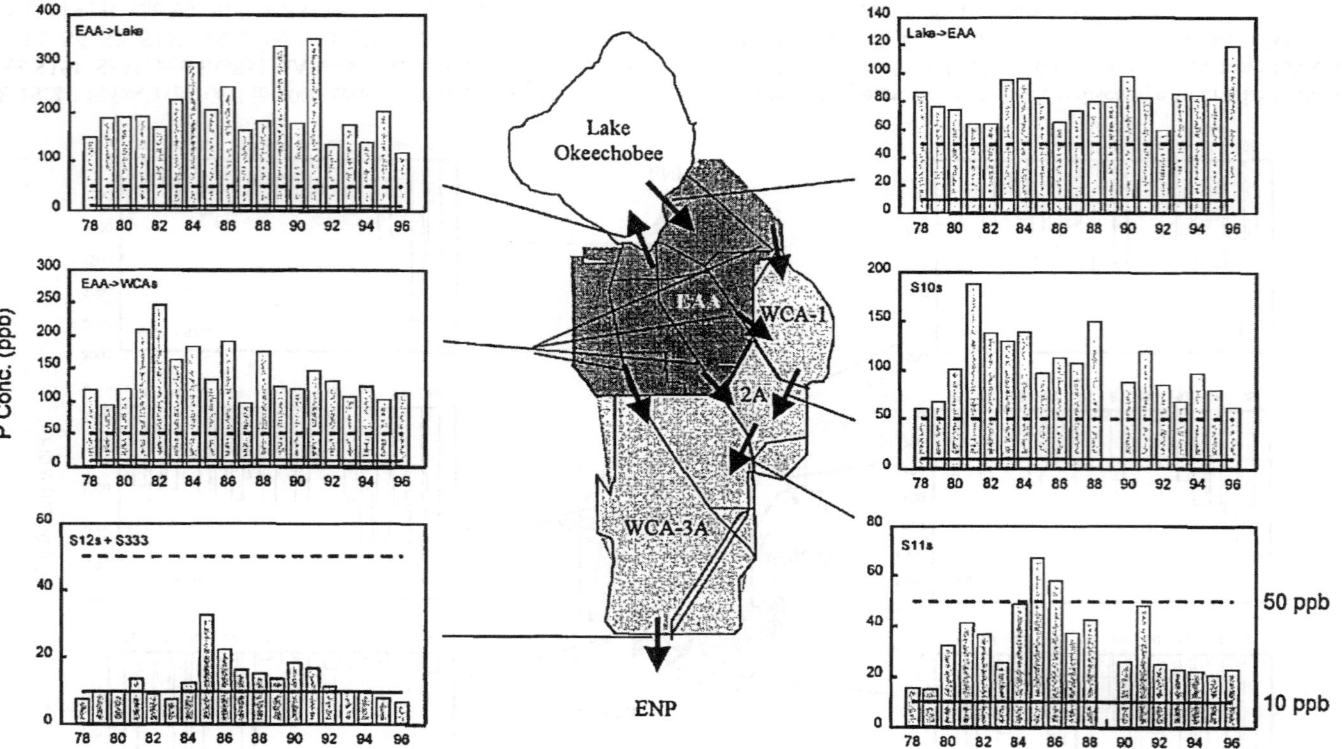

FIGURE 19.3 Annual flow-weighted mean phosphorus concentrations at major structures, 1978–1996. Concentrations expressed in parts per billion (ppb), October–September water years, horizontal dashed lines show 50-ppb design target for interim phosphorus control measures. Structure flows are defined in Fig. 19.2.

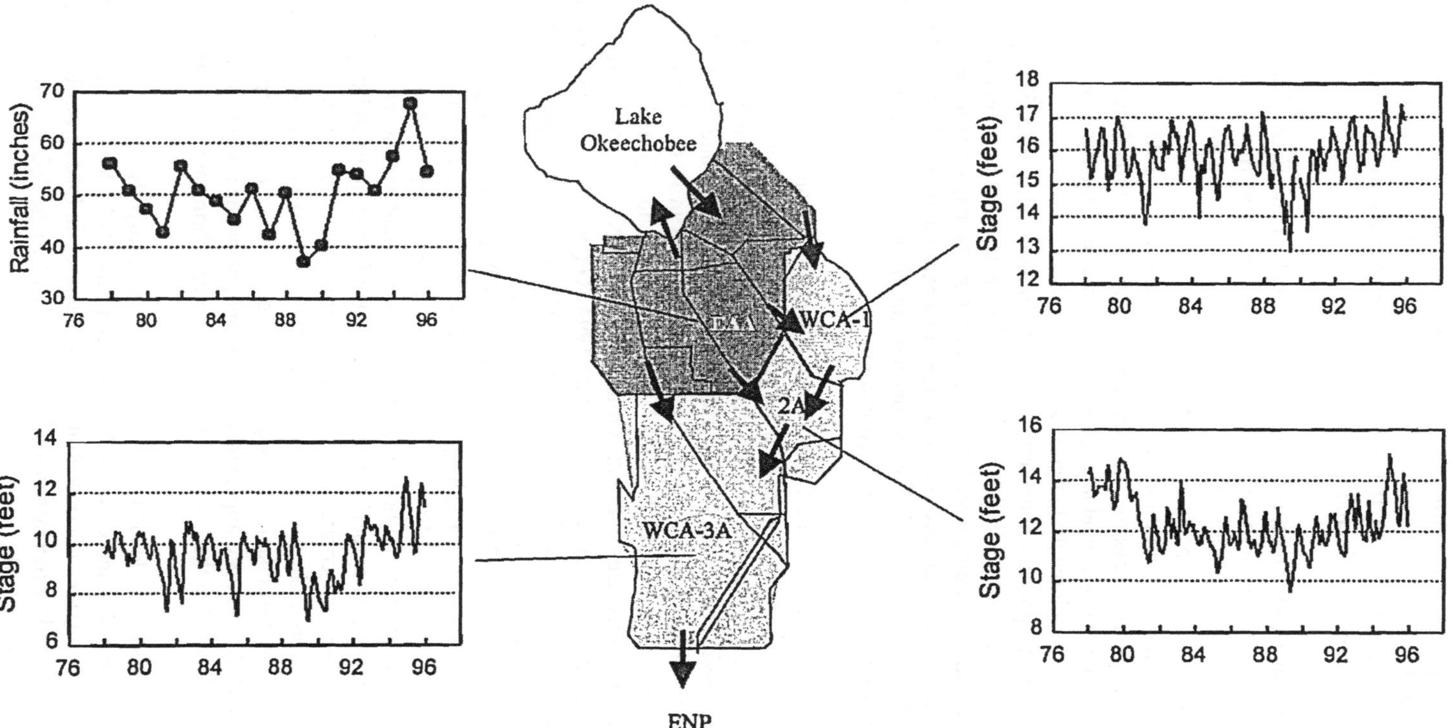

FIGURE 19.4 EAA Rainfall and WCA Stages, 1978–1996. EAA annual rainfall in inches, WCA monthly stages in feet (NGVD).

Total phosphorus concentrations (Fig. 19.3) are shown in relation to the 50-ppb interim control target established in the SA and EFA and the 10-ppb default long-term standard established in the EFA. Predominant patterns in the phosphorus concentration time series include:

1. Increasing concentrations in early years south of the EAA. These may reflect IAP, growth in agricultural land use, variations in water management (WCA regulation schedules, flow distribution), and long-term nutrient enrichment impacts on the WCAs.
2. Decreasing concentrations in later years south of the EAA. These may reflect control measures implemented after 1991, shifts in agricultural crops (away from vegetables), changes in water management, and control measures implemented after 1991.
3. Higher phosphorus concentrations at ENP inflows (S12s + S333) in Water Years 1985 and 1986. These reflected unusual operating conditions in which the structures were left open and WCA-3A stage was lowered. By decreasing contact between canal flows and adjacent marsh, this condition facilitated transport of phosphorus-rich runoff and lake releases through WCA canals to ENP inflow structures.

Of particular interest is the extent to which benefits of control measures implemented after 1991 are reflected in the concentration data. One such measure is adoption of the EAA Regulatory Rule (requiring a 25% reduction in EAA phosphorus load) in 1992. For the five-year period between May 1992 and April 1997, the EAA basin tracking procedure (SFWMD, 1992b) indicates an average load reduction of 46% relative to the May 1979–April 1988 base period and adjusted for variations in rainfall. The Everglades Nutrient Removal Project (ENR), a pilot-scale wetland treatment system in operation since August 1994 (Guardo et al., 1995; SFMWD, 1997), removed an additional 9% of the (post-BMP) EAA runoff phosphorus load that occurred between August 1994 and April 1997.

Precontrol (1978 to 1991) and postcontrol (1992 to 1996) averages for flow, phosphorus concentration, and phosphorus load are summarized in Fig. 19.5. With the exception of discharges from the Lake to the EAA, phosphorus concentration decreased at each location. Although this pattern is consistent with beneficial impacts of control measures, other mechanisms (in particular, higher flows and stages in the WCAs) may have also contributed to the apparent water quality improvements. Despite the apparent reductions relative to the 1978 to 1991 period, and regardless of the precise causal mechanisms, phosphorus concentrations in discharges from the EAA to the WCAs averaged approximately 100 ppb, twice the interim control target of 50 ppb established in the SA and used in designing Stormwater Treatment Areas (Walker, 1995).

Average phosphorus loads south of the EAA were slightly higher in the 1992 to 1996 period, because increases in flow (driven primarily by higher rainfall) more than offset the reductions in concentration. The decrease in loads from EAA to the Lake and corresponding increase in loads from the EAA to the WCAs reflect changes in water management (IAP).

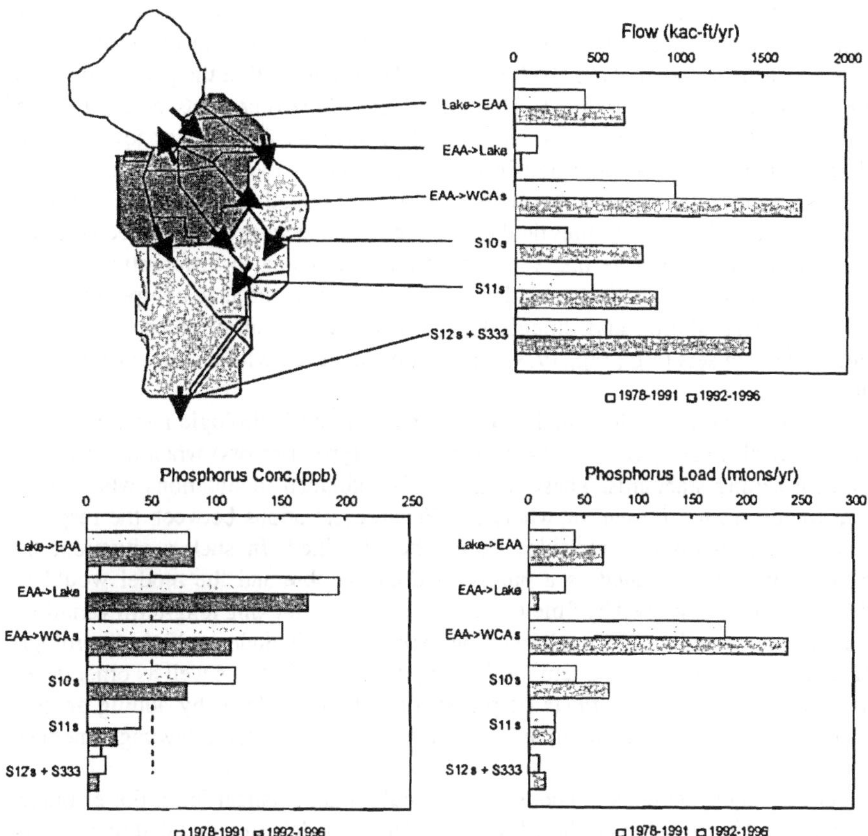

FIGURE 19.5 Comparison of 1979–1991 with 1992–1996 conditions at major structures.

19.5 TRACKING MODEL

Detection of trends or shifts in the long-term mean resulting from management activities or other anthropogenic factors is of primary concern to water-quality management. Interpretation of long-term data sets and establishment of interim standards at various Everglades locations (Fig. 19.1) have been facilitated by application of a general model that attempts to differentiate long-term, hydrologic, and random variance components. Explicit consideration of variability is the key to formulating a valid tracking procedure. The model has the following general form:

$$\text{Response} = \text{R Temporal Effect} + \text{Hydrologic Effect} + \text{Random Effect} \quad (1)$$

The response is the measurement to be tracked (e.g., concentration or load, averaged over appropriate spatial and temporal scales, linear or log-transformed). The *temporal effect* represents a long-term trend or step-change in the historical data (if present); this may reflect anthropogenic influences (e.g., land development, new point-source discharges, control measures, etc.). Theoretically, the temporal term could also reflect effects of long-term hydrologic or natural variations occurring over time scales that are too long to be identifiable within the period of record for the Response measurement. The *hydrologic effect* represents correlations of the Response with other measured variables, such as flow, water level, and/or rainfall (if present), as identified in the time frame of the analysis. The *random effect* is essentially an error term that represents all other sources of variance, including sampling error, analytical error, and variance sources not reflected in the temporal or hydrologic terms. While no attempt is made to differentiate precise causal mechanisms, explicit consideration of the above variance components provides better resolution and, hopefully, leads to more accurate interpretations of long-term data sets (e.g., Fig. 19.3) than can be achieved with a simple time-series analysis.

As demonstrated below, inclusion of temporal and hydrologic terms increases the statistical power (reduces risk of Type I and Type II errors) when the model is used for setting standards. These terms can be excluded in situations where long-term trends are not present or where significant correlations between the response variable and hydrologic variables cannot be identified. In such a situation, the response would be treated as a purely random variable and the model would be identical to that described by Smeltzer et al., (1989) for tracking long-term variations in lake water quality. The model can be expanded to include multiple hydrologic effects, interactions between temporal and hydrologic effects, as well as other deterministic terms. *Seasonal effects* (if present) can be considered by adding another term or eliminated by defining the response as an annual statistic (average, median, etc.).

The model is not constrained to any particular mathematical form. For example, hydrologic effects can be predicted by a simulation model, provided that uncertainty associated with such predictions *(random effects)* can be quantified. Everglades applications invoke relatively simple, multiple-regression models that provide direct estimates of parameter uncertainty. The hydrologic term provides a basis for adjusting historical and future monitoring data back to an average hydrologic condition, so that changes in the long-term mean (typically reflecting anthropogenic influences) can be tracked and not confused with random climatologic variability (e.g., wet-year vs. dry-year differences).

19.6 CALIBRATION TO SHARK RIVER SLOUGH

SA interim standards for ENP Shark River Slough were designed to provide a long-term-average, flow-weighted mean concentration equivalent to that present between March 1, 1978 and March 1, 1979, the legally-established base period consistent with ENPs designation as an Outstanding Florida Water (OFW). Analysis of monitoring data collected between December 1977 and September 1989 at five inflow structures (S12A, B, C, D, and S333) revealed significant increasing trends in phosphorus concentrations (Walker, 1991). To reduce possible influences of season and shifts in the flow distribution across the five inflow structures, the annual-average, flow-weighted mean concentration across all five structures was selected as a response variable and basis for the interim standard. Annual values for Water Years 1978 to 1990 (October-September) were used to calibrate a regression model of the following form:

$$C - C_m = b_1(T - T_m) + b_2(Q - Q_m) + E \quad (2)$$

where C = observed annual, flow-weighted-mean concentration total phosphorus (ppb), T = water year (October–September), Q = basin total flow (1000 acre-ft/yr), E = random error term, and $_m$ = subscript denoting average value of C, T, or Q in the calibration period.

Alternative Water Year definitions were investigated. The October to September definition was selected, because it provided the best data fit. Prior to calibration, biweekly concentration data used to calculate annual flow-weighted means were screened for outliers from a log-normal distribution while accounting for correlations between concentration and flow (Snedocor and Cochran, 1989); a single sample was rejected on this basis. Data from Water Years 1985 and 1986 were excluded from the calibration because of unusual operating conditions, which promoted discharge of high-phosphorus canal flows (vs. marsh sheet flows) through the inflow structures. The flow-weighted mean concentrations were 33 and 21 ppb, respectively, as compared with a range of 7 to 18 in other Water Years. These unusual operating conditions are not expected to be repeated.

When data from individual sampling dates are analyzed, total P concentrations in ENP Shark River Slough inflows are negatively correlated with water level in upstream WCA-3A (Walker, 1991). Similar negative correlations with water level are found at other structure and marsh sampling stations in the WCAs (Walker, 1995). At enriched marsh sites, P concentration tends to increase at low stage; this is thought to reflect (a) peat oxidation and subsequent P release from the soils into the water column during and following droughts, and (b) practical difficulties associated with obtaining representative marsh water samples at low stage. Higher water levels are thought to promote phosphorus uptake in the WCAs by increasing wetted area, increasing water residence time, and increasing the hydraulic exchange between canals and adjacent marsh areas. Lower stages tend to promote phosphorus transport through the WCAs by increasing the relative proportion of canal flow vs. marsh sheet flow.

When the data are analyzed on an annual basis, negative correlations with both WCA-3A stage ($r = -0.63$) and basin total flow ($r = -0.68$) are observed. The latter is used for hydrologic adjustment in the tracking procedure, because it provides a slightly better fit of the data. It is likely that flow is a partial surrogate for water-level effects. Consideration of both flow and stage does not improve the fit.

Table 19.1 lists calibration data and results. The model explains 80% of the variance in the historical data set with a residual standard error of 1.87 ppb. The fit is illustrated in Fig. 19.6. Figure 19.6a plots observed and predicted concentrations against time. The 80% prediction interval (10th, 50th, and 90th percentiles) is shown in relation to the observed data. Both regression slopes are significant at $p < 0.05$. The partial regression concept (Snedocor and Cochran, 1989) can be applied to elucidate temporal, hydrologic, and random variance components.

TABLE 19.1
Derivation of Interim Phosphorus Standards for ENP Shark River Slough Inflows

		Flow-Weighted-Mean Total P Concentration					
Water Year	Basin flow, kac-ft/yr	Observed, ppb	Predicted, ppb	Flow-adjusted, ppb	Detrended, ppb	50% target, ppb	90% limit, ppb
78	522.8	6.7	8.4	6.7	7.0	8.4	11.7
79	407.0	9.8	9.6	9.2	9.5	9.0	12.3
80	649.2	10.6	9.0	11.2	9.7	9.6	11.1
81	291.7	12.4	11.3	11.4	11.0	10.2	12.9
82	861.3	8.4	9.2	10.0	6.3	10.8	10.1
83	1061.3	7.0	8.9	9.5	4.4	11.4	9.4
84	842.8	12.0	10.5	13.4	8.7	12.0	10.2
87	276.6	15.9	14.9	14.8	10.9	13.8	13.0
88	585.5	15.6	14.1	15.9	10.0	14.4	11.4
89	116.9	13.5	16.9	11.6	7.3	15.0	14.0
90	148.2	18.1	17.3	16.3	11.2	15.6	13.8
Mean	523.9	11.8	11.8	11.8	8.7	8.7	11.8

Variables

C = Observed TP (ppb)

Q = Observed Flow (kac-ft/yr)

T = Water Year (October-September)

E = Random Error (ppb)

b_1, b_2 = Regression Slopes

SE = Regression Standard Error of Estimate (ppb)

m = Subscript Denoting Mean Value

Regression model

$$C = C_m + b_1 (T - T_m) + b_2 (Q - Q_m) + E = 11.8 + 0.5932 (T - 83.7) - 0.00465 (Q - 523.9) + E$$

Regression results

$R^2 = 0.80$ SE = 1.873 ppb $C_m = 11.8$ ppb $T_m = 83.7$

$Q_m = 523.9$ kac-ft/yr $b_1 = 0.5932$ $Var(b_1) = 0.02366$ $b_2 = -0.00465$

$Var(b_2) = -0.0046$ $Cov(b_1,b_2) = 0.00013$ $t_{\alpha,dof} = 1.397$ $n = 11$

C_Q = Flow-adjusted TP = $C + b_2 (Q_m - Q) = C - 0.00465 (523.9 - Q)$

C_T = Detrended TP (adjusted to $T_o = 78.5$) = $C + b_1 (T_o - T) = C + 0.5932 (78.5 - T)$

Target = $C_m + b_1 (78.5 - T_m) + b_2 (Q - Q_m) = 11.16 - 0.00465\,Q$

Limit = Target + S $t_{\alpha,dof}$ = $11.16 - 0.00465\,Q + 1.397\,S$

$$S = [SE^2 (1 + 1/n) + Var(b_1) (T_o - T_m)^2 + Var(b_2) (Q_c - Q_m)^2 + 2\,Cov(b_1,b_2) (T_o - T_m)(Q_c - Q_m)]^{0.5}$$
$$= [6.377 - 0.00591\,Q + 0.00000436\,Q^2]^{0.5}$$

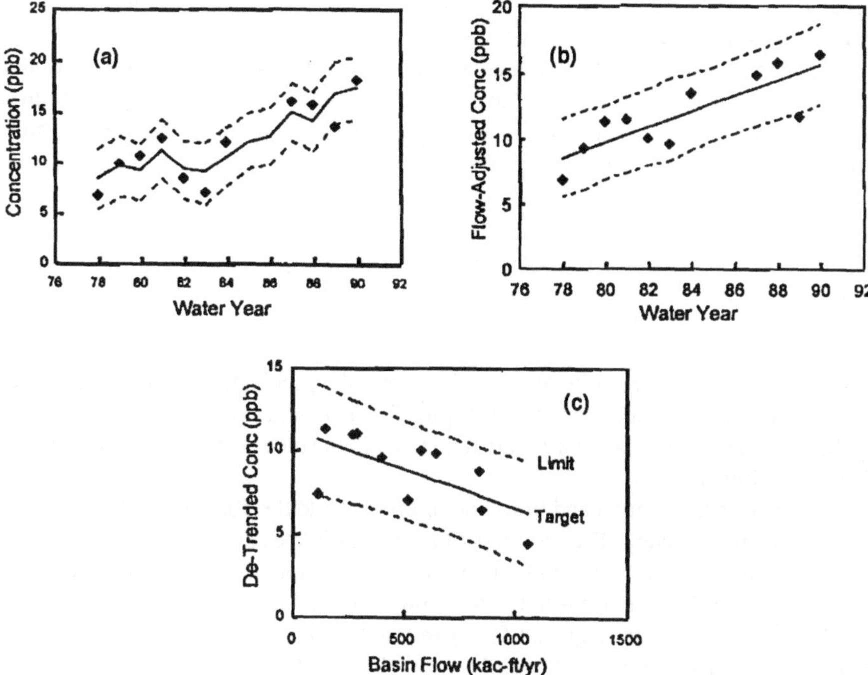

FIGURE 19.6 Model calibration to ENP Shark River Slough inflows, (a) observed, (b) adjusted to mean flow, and (c) adjusted to 1978–1979 conditions. October–September Water Years. Lines = 80% prediction intervals, squared = observed flow-weighted means.

The concentration measured in any year (C) can be adjusted back to an average flow condition (Q_m) using the following equation for flow-adjusted concentration (C_Q):

$$C_Q = C + b_2(Q_m - Q) \tag{3}$$

Figure 19.6b plots observed and predicted flow-adjusted concentrations against time. The long-term trend is more readily apparent in this display because effects of flow variations have been filtered out.

Similarly, the concentration measured in any year can be adjusted back to any base period (T_o) using the following equation for a time-adjusted or detrended concentration (C_T):

$$C_T = C + b_1(T_0 - T) \tag{4}$$

A base period (T_o) of 78.5 represents the 1978 to 1979 OFW time frame, as defined above. Figure 19.6c plots observed and predicted time-adjusted concentrations against flow. With the long-term trend removed in above manner, the negative correlation between concentration and flow is apparent. Figure 19.6c shows the predicted relationship between concentration and flow if the long-term mean were equivalent to that present in 1978 to 1979.

The model can be used to evaluate the likelihood that current monitoring results (C_c, Q_c) are equivalent to the 1978 to 1979 base period while accounting for hydrologic and random variations. This is accomplished using the following terms, which characterize the prediction interval for a 1978 to 1979 time frame under a given flow condition:

$$\text{Target} = (C_m + b_1(T_o - T_m) + b_2)(Q^c - Q_m) \tag{5}$$

$$\text{Limit} = \text{Target} + St_{\alpha dof} \tag{6}$$

where Target = 50th percentile of prediction interval = predicted mean (ppb), Limit = 90th percentile of prediction interval (ppb), S = standard error of predicted value (ppb), t = one-tailed student's t statistic, α = significance level = 0.10, and dof = degrees of freedom.

In Fig. 19.6c, the target and limit lines correspond to the 50th and 90th percentile predictions, respectively. The required parameter estimates and variance/covariance terms are derived from a standard multiple regression analysis (Snedocor and Cochran, 1989). If the current long-term flow-weighted mean is less than the 1978 to 1979 long-term mean (adjusted for hydrologic effects), there would be less than a 50% chance that the observed yearly mean (C_c) would exceed the target and less than a 10% chance that C_c would exceed the limit. The difference between the target and limit reflects the magnitude of the random effects term and uncertainty in model parameter estimates (b_1, b_2, C_m).

19.7 TYPE I AND TYPE II ERRORS

Under the terms of the settlement agreement, an exceedence of the limit in any year would trigger further scientific investigations which, in turn, may lead to implementation of additional phosphorus control measures. The significance level for the compliance test (0.10) represents the maximum Type-I error rate (probability of exceeding the limit if the future and 1978 to 1979 long-term means are exactly equal). Unless a model can be constructed to explain all of the variance in the data, there is no way to design a compliance test without explicitly adopting a maximum Type-I error. In this case, the 0.10 value was arrived at by negotiation and with the understanding that results of the test would be interpreted by a scientific panel in light of the inherent risk of Type I error.

Type II error (failure to detect an exceedence or excursion from the standard) is another unavoidable feature of compliance tests. In this case, a Type II error would occur when the actual long-term mean exceeds the 1978 to 1979 flow-adjusted mean but the measured annual value is still below the limit. Risk of Type II error depends on the specified maximum Type I error (10%), model error variance (random effects term), and the degree of excursion from the objective.

Figure 19.7 illustrates Type I and Type II error concepts. The probability that the measured annual mean exceeds the limit is plotted against the difference

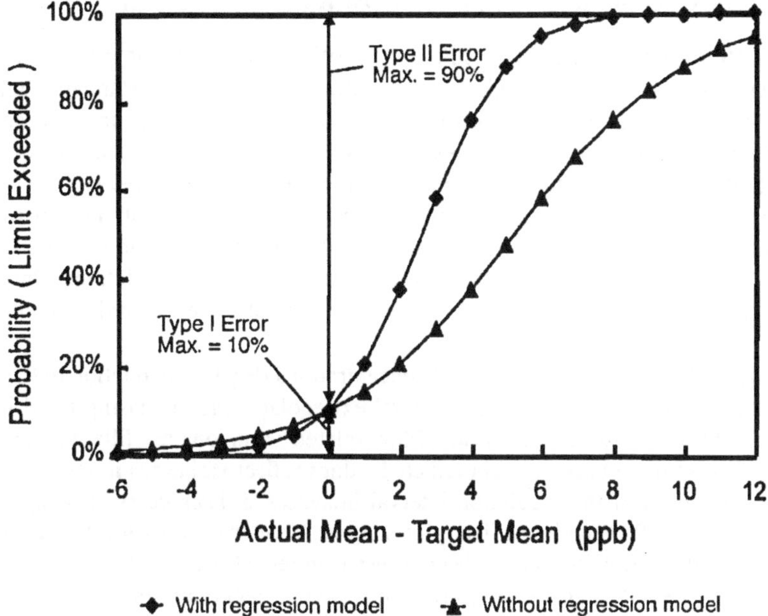

FIGURE 19.7 Type I and Type II error.

between the actual long-term mean and the objective (1978 to 1979 long-term mean). Probabilities are calculated using standard statistical procedures (Snedecor and Cochran, 1989; Walker, 1989). Type I errors (false exceedences) may occur when the actual long-term mean is below the objective. The risk of Type I error equals the probability shown on the left-hand side in Fig. 19.7 and has maximum value of 10% (by design). Type II errors (failure to detect exceedences) may occur when the actual mean exceeds the target. The risk of Type II error equals 100% minus the probability shown on the right-hand side of Fig. 19.7 and has a maximum value of 90%. As deviation from the target increases, risks of Type I and Type II errors decrease.

Probability curves are shown for two values of residual standard error in Fig. 19.7. Without applying the regression model, the random effects term in the model would have a standard deviation of 3.73 ppb (= standard deviation of annual flow-weighted means in the calibration period). With the regression model, the standard deviation is reduced to 1.87 ppb. Removing variance associated with trend and flow increases the probability of exceeding the limit when the long-term mean exceeds the objective. For example, if the true long-term mean were 5 ppb above the objective, the probability of an exceedence (measured annual value above limit) would be ~90% with the regression model, but only ~50% without the regression model. Risk of Type I error when the actual mean is below the objective is also lower with the regression model. The regression approach thus enables a more powerful compliance test than would result from treating the calibration data set as a random time series.

19.8 TRACKING RESULTS FOR SHARK RIVER SLOUGH

Figure 19.8 shows monitoring results for the Water Years 1991 to 1996 (six years following the 1978 to 1990 calibration period). Although interim standards will not be enforced until 2003, the procedure is useful for tracking responses to control measures implemented over the 1991 to 2002 period. As discussed above, these include the EAA Regulatory Program (1992) and ENR project (1994).

Figure 19.8a shows observed values before and after the calibration period in relation to the 80% prediction interval derived from the regression model (Table 19.1). Values in Fig. 19.8a reflect both long-term trend and flow variations. Observed values in 1992 to 1996 fall near the lower boundary of the 80% prediction interval (10th percentile).

Figure 19.8b shows flow-adjusted concentrations [Eq. (3)] in relation to the 80% prediction interval. The prediction interval extrapolates the increasing trend in the 1978 to 1990 data to the later years. Flow-related variations are filtered from this time series, so that observed and predicted values reflect variations in the long-term mean. The width of the prediction interval increases in later years, primarily as a result of higher flow regimes. The plot suggests that the increasing trend present during the calibration period has been arrested in recent years.

Figure 19.8c plots concentrations against flow in relation to the 80% prediction interval for 1978 to 1979 conditions. Observed values during the 1978 to 1991 calibration period have been adjusted to the 1978 to 1979 time frame [Eq. (4)]. The middle and upper values in the prediction interval correspond to the target and limit values at any flow. Compliance with the interim standards (when they are in effect) will require that the observed (unadjusted) flow-weighted means fall below the limit line in every year.

Under provisions of the settlement agreement, the maximum flow during the calibration period (1061 kac-ft/yr) is used to calculate the limit in years when the observed flow exceeds that value. This essentially prevents extrapolation of the regression beyond the calibration range. The dashed line in Fig. 19.8c shows the limit calculated according to this procedure. It is debatable whether this procedure provides a better estimate of the 90th percentile at high flows than the extrapolated (solid) line. The distribution of observed values after 1991 is such that the determination of "compliance" (if the standard were in effect) would be influenced only in the case of the extreme high-flow year (1995). In the remaining years, the system would have been in compliance in two out of five years (1994 and 1996), regardless of which limit line is used.

19.9 DISCUSSION

Extremely wet conditions experienced in recent years relative to the calibration period impose significant limitations on tracking results. As shown in Figs. 19.2 and 19.8, basin flows exceeded the maximum value experienced in the calibration period (1061 kac-ft/yr) in three out of six Water Years after 1990 (1993, 1995, and 1996). In these years, the model is being extrapolated beyond the range of the calibration data set. The extrapolation is particularly large in Water Year 1995, when the average

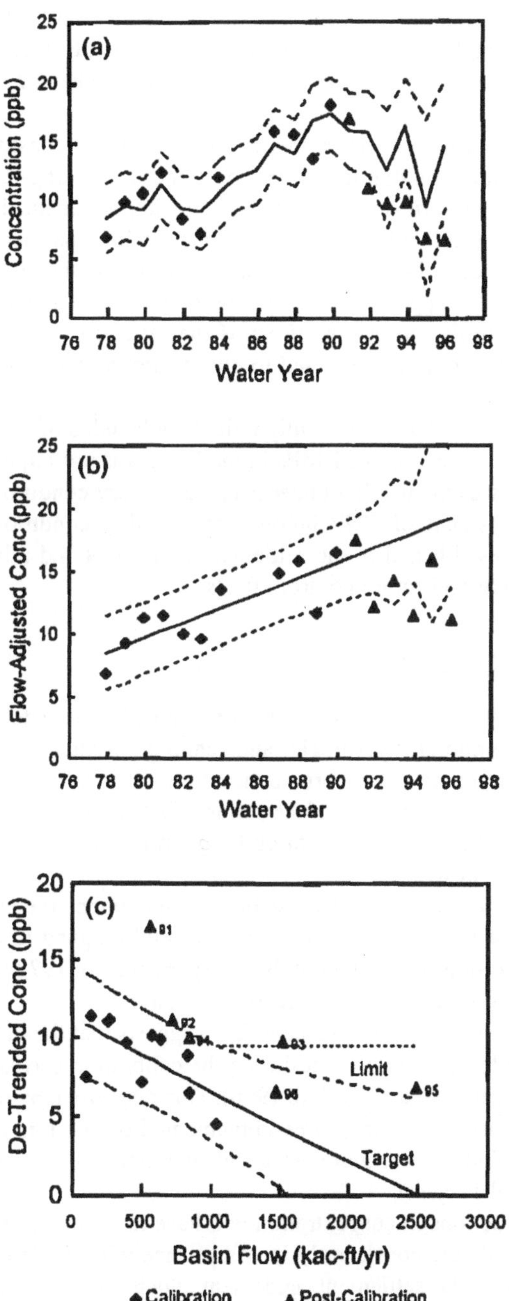

FIGURE 19.8 Model application to ENP Shark River Slough inflows, (a) observed, (b) adjusted to mean flow, and (c) adjusted to 1978–1979 conditions (calibration period), observed (1992–1996). October–September Water Years. Diamonds = observed flow-weighted means (1991–1996), lines = 80% prediction intervals.

flow exceeded the calibration maximum by approximately 2.5-fold. When the model is applied to these high-flow regimes, prediction uncertainty increases, as reflected by the wider prediction intervals in later years (Figs. 19.8a, 19.8b) and/or higher flows (Fig. 19.8c).

When recent data within the calibrated flow range are considered (WY 1992, 1994), Fig. 19.8b suggests that the increasing trend in the long-term mean present prior to 1991 has been arrested. Despite signs of improvement, it is unlikely that the interim control objective for ENP Shark Slough inflows has been achieved, since the flow-adjusted means in recent years are consistently above the 1978 to 1979 flow-adjusted mean (~ 8 ppb, Fig. 19.8b). Observed concentrations in 1992 to 1996 cluster around the limit line in Fig. 19.8c. If the interim objective were achieved, the observed values would be expected to cluster around the target line (center of 1978 to 1979 distribution).

Results for 1992 to 1996 do not fully reflect the benefits of existing controls. In particular, full implementation of BMPs in the EAA was not required until January 1996. Several years of monitoring under average and dry conditions will provide a more reliable assessment of ENP inflow water quality conditions in relation to interim objectives established in the settlement agreement and a basis for tracking responses to existing and future control efforts.

19.10 CONCLUSIONS

1. Interpretation of long-term data sets is facilitated by a model that explicitly considers temporal, hydrologic, and random variance components. The model for tracking phosphorus concentrations at inflows to ENP Shark River Slough explains 80% of the variance in the 1978 to 1990 calibration data using with terms representing long-term trend and correlation with basin annual flow.
2. Under terms of the state/federal settlement agreement, the model provides a basis for establishing interim control limits designed to achieve water quality conditions equivalent to those experienced in 1978 to 1979, while accounting for hydrologic and random variations.
3. Application of the model to recent data suggests that the increasing trend in the long-term mean present during the calibration period was arrested in postcalibration years (1992 to 1996). This response may reflect benefits of phosphorus-control measures implemented during this period (BMPs in the Everglades Agricultural Area and operation of the Everglades Nutrient Removal Project).
4. Further reductions in concentration will have to occur before ENP inflow concentrations are consistently in compliance with the interim objective established in the settlement agreement. Such reductions may occur as the system responds to full implementation of existing controls and to future controls.
5. Results for some recent years are limited by extremely wet conditions requiring extrapolation of the tracking model beyond the calibrated flow range.

6. Future monitoring under average and dry conditions will provide a more reliable assessment of ENP inflow water quality conditions in relation to interim objectives and a basis for tracking responses to existing and future controls.

REFERENCES

Amador, J.A., and R.D. Jones, 1993. Nutrient limitations on microbial respiration in peat soils with different total phosphorus content. *Soil Biol. Biochem.* 25: 792–801.

Jones, R.D. and J.A. Amador, 1992. Removal of total phosphorus and phosphate by peat soils of the Florida Everglades. *Can. J. Fish. Aq. Sci*_49:577–583.

Belanger, T.V., D. J. Scheidt, and J.R. Platko II, 1989. Effects of Nutrient Enrichment on the Florida Everglades. *Lake and Reservoir Management* 5(1): 101–111.

Davis, S.M., 1994. Phosphorus inputs and vegetation sensitivity in the Everglades. in Everglades—The Ecosystem and Its Restoration. S.M. Davis and J.C. Ogden (Editors), St. Lucie Press, Florida, 357–378.

Doren, R.F., T.V. Armentano, L.D.Whiteaker and R.D. Jones, 1997. Marsh vegetation patterns and soil phosphorus gradients in the Everglades ecosystem, Aquatic Botany 56(199):145–163.

Florida Department of Environmental Protection, 1997. Discharge Permit for Stormwater Treatment Area No. 6, Section 1. Issued to South Florida Water Management District.

Germain, G. J., 1994. Surface water quality monitoring network, South Florida Water Management District. South Florida Water Management District, Technical Memorandum DRE 317.

Guardo, M., L. Fink, T.D. Fontaine, S. Newman, M. Chimney, R. Bearzotti, G. Goforth, 1995. Large-scale constructed wetlands for nutrient removal from stormwater runoff: an Everglades restoration project. *Environmental Management*_19(6):879–889.

Lean, D., K. Reckhow, W. Walker, and R. Wetzel, 1992. Everglades nutrient threshold research plan. Research and Monitoring Subcommittees, Everglades Technical Oversight Committee, West Palm Beach, Florida.

Nearhoof, F.L., 1992. Nutrient-induced impacts and water quality violations in the Florida Everglades, Florida Department of Environmental Protection, Water Quality Technical Series, 3(4).

Smeltzer, E., W.W. Walker, and V. Garrison, 1989. Eleven years of lake eutrophication monitoring in Vermont: a critical evaluation. in Enhancing states' lake management programs, U.S. Environmental Protection Agency and North American Lake Management Society, 52–62.

Snedocor, G.W. and W.G. Cochran, 1989. *Statistical Methods.*_Iowa State University Press, Ames, Iowa, Eighth Edition.

South Florida Water Management District, 1992a. Surface water improvement and management plan for the Everglades, Appendix E, Derivation of phosphorus limits for Everglades National Park and phosphorus levels for Loxahatchee National Wildlife Refuge. West Palm Beach, Florida.

South Florida Water Management District, 1992b. Works of the District, Chapter 40-E63, Everglades Agricultural Area regulatory program, basin compliance. West Palm Beach, Florida.

South Florida Water Management District, 1994. Works of the District, Chapter 40-E63, Everglades Agricultural Area regulatory program, BMP replacement water rule. West Palm Beach, Florida.

South Florida Water Management District, 1997. Everglades nutrient removal project, 1996 monitoring report. submitted to Florida Department of Environmental Protection.

South Florida Water Management District, 1996. User's guide to accessing the SFWMD hydrometerological data base (HYDRO-PREP), DBHYDRO. Department of Water Resources Evaluation, West Palm Beach, Florida.

State of Florida, 1994. Everglades Forever Act. Tallahassee, Florida.

United States of America, South Florida Water Management District, Florida Department of Environmental Regulation, Settlement Agreement. United States District Court, Southern District of Florida, Case No. 88-1886-CIV-HOEVELER, July 1991.

Walker, W.W., 1989. LRSD.WK1—lake/reservoir sampling design worksheet., Software Package No. 3, North American Lake Management Society, Denver, Colorado.

Walker, W.W., 1991. Water quality trends at inflows to Everglades National Park. *Water Res. Bul.* 27(1): 59–72.

Walker, W.W., 1995a. Models for tracking C139 runoff and phosphorus load. prepared for South Florida Water Management District, West Palm Beach, Florida.

Walker, W.W., 1995b. Design basis for Everglades stormwater treatment areas. *Water Res. Bul.*, 31(4):671–685.

Walker, W.W., 1996. Test for evaluating performance of stormwater treatment areas. prepared for U.S. Department of the Interior, Everglades National Park.

Whalen, B.M. and P.J. Whalen, 1994. Nonpoint source regulatory program for the Everglades Agricultural Area," American Society of Agricultural Engineers, Paper No. FL94–101.

20 Techniques for Optimizing Phosphorus Removal In Treatment Wetlands

Thomas A. DeBusk and Forrest E. Dierberg

20.1 ABSTRACT

Increased interest in inexpensive phosphorus (P) control technologies for wastewaters and surface runoff streams has led to investigations on enhancing P removal capabilities of treatment wetlands. Wetland area requirements for P removal are high because long-term P sinks (e.g., sediment P accretion) occur at slow rates relative

1-56670-331-X/99/$0.00+$.50

to other contaminant removal processes. Two approaches can be used to enhance P removal effectiveness of treatment wetlands: vegetation harvest, in which P in either macrophyte or algal tissues is periodically removed from the system, or by using chemical amendments or substrates with a high affinity for P.

Small-scale laboratory and greenhouse studies have clearly demonstrated the promise of vegetation and chemical management to enhance P removal in treatment wetlands. For example, numerous small-scale harvested macrophyte and algae systems have exhibited mass P removal rates 30- to 50-fold higher than those achieved in nonharvested treatment wetlands. Additionally, a spectrum of chemicals, applied either to wetland water column, soils, or as a substrate in a subsurface flow wetland (SSF), have been found to markedly enhance wetland P removal. Despite favorable small-scale findings, vegetation and chemical management approaches have not achieved widespread use, because both run counter to the concept of wetlands as passive, natural treatment systems. As a consequence, full-scale wetland infrastructure and operational elements for incorporating vegetation and chemical management have not yet been refined.

Key issues for vegetation management include plant component selected for harvest (algae or floating, emergent or submerged macrophytes), infrastructure for plant cultivation and harvest, and biomass utilization approaches. Key issues for chemical P management include appropriate site for chemical deployment (wetland influent vs. effluent region, water vs. soils, SSF adsorption substrate), longevity of amendment, and fate and impact of chemical residuals. Additionally, both vegetation and chemical management approaches must be tailored around target wetland mass P removal rates and effluent P concentrations.

20.2 INTRODUCTION

Treatment wetlands have proven effective for removing a spectrum of pollutants from wastewaters and surface runoff streams (Reddy and Smith, 1987; Cooper and Findlater, 1990). Properly designed treatment wetlands can accommodate high influent concentrations of carbonaceous and nitrogenous contaminants and produce residual effluent levels as low as 8 mg L^{-1} biochemical oxygen demand (BOD) and 2 mg L^{-1} nitrogen (N) per liter (Kadlec and Knight, 1996; Kadlec et al., 1997). While phosphorus (P) is an element that can be reduced in treatment wetlands to concentrations two-orders of magnitude lower than N or carbon (C), cost-effective P removal has proven to be a challenge for treatment wetland technologies. Indeed, the area requirement for effective P removal is so high that it often limits the deployment of treatment wetlands for nutrient removal purposes.

A recent survey of treatment wetlands in Florida, most of which are surface flow (SF) wetlands, revealed at least ten systems that are greater than 100 ha in size (DeBusk and Krottje 1996). Because of the economic costs associated with high land requirements, sponsors of some of the largest treatment wetlands, such as the Apopka Marsh Flow-Way (Lowe et al. 1992), and the Everglades Nutrient Removal (ENR) Project (Guardo et al. 1995), are investigating the use of alternative management approaches (e.g., P-binding soil amendments) to enhance P removal.

Whether wetlands are used for treating domestic wastewater effluents, eutrophic lake waters, or agricultural runoff, there is a strong need to increase their P removal effectiveness: effluent P discharge limits are becoming more stringent in many regions, P is often the limiting nutrient in natural aquatic systems, and there exist no cost-effective "conventional" technologies for reducing P to extremely low effluent concentrations.

In this chapter, we review the opportunities of using two management techniques, vegetation harvest and chemical amendments, to facilitate wetland P removal. In this review, we first describe P removal mechanisms in treatment wetlands. Second, we evaluate aquatic plant harvesting as a management approach for enhancing P removal performance. Finally, we discuss the efficacy of chemical amendments to facilitate wetland P removal.

20.3 PHOSPHORUS REMOVAL PROCESSES IN TREATMENT WETLANDS

20.3.1 Biological Processes

Plant assimilation and microbial decomposition of organic matter are the key biological processes that influence P cycling in wetlands. There are no microbiologically mediated redox active P species that result in gaseous losses as there are in the N cycle. Phosphorus removal within wetlands therefore is confined to plant and soil/sediment storage. Microbial, algal, and macrophyte assimilation of P can be quite rapid, although the capacity of each of these system compartments for P storage differs markedly. These different system components also exhibit varying rates of P turnover through decomposition. In general, the system compartments with highest P uptake rates exhibit the quickest elemental turnover and lowest storage capacity (Richardson and Craft, 1993).

Phosphorus not released from wetland plant tissues during decomposition ultimately is stored in organic sediments. The magnitude and longevity of this P sink is related to chemical and physical processes in the sediment and at the sediment-water interface. On a global basis, wetland peats accumulate at an average rate of 1.0 to 2.0 mm yr^{-1}. Investigators in the Florida Everglades reported P accumulation in peats at a mean rate of 0.4 to 1.2 g P $m^{-2}yr^{-1}$ (Richardson and Craft, 1993; Reddy et al., 1993).

20.3.2 Chemical Processes

Some of the P that enters treatment wetlands, rather than being immediately assimilated by biota, is sequestered chemically. Phosphorus adsorption in sediments and soils depends on the pH and oxidation potential, the types and amounts of iron (Fe), aluminum (Al) and calcium (Ca) minerals, and the native soil P concentrations. Soil pH has a considerable influence on the chemistry of P retention in wetlands. In acid soils, P is fixed as Al and Fe phosphates, if the activities of these cations are high; P fixation is controlled by Ca activities in alkaline soils. Phosphorus retention is lowest in soils with slightly acidic to neutral pH (Reddy and D'Angelo, 1994).

Limited P removal (and even net P release) can occur because of shifts in chemical equilibria. The equilibria that regulate the degree of substrate P-binding include acid-base, reduction-oxidation, adsorption via ligand exchange, and precipitation (Faulkner and Richardson, 1989). The ability of wetland soils to adsorb P is therefore quite variable (Richardson, 1985; Geller et al., 1990; Reddy and D'Angelo, 1994; Kadlec and Knight, 1996).

An additional complexity in predicting P adsorption in wetland soils lies in evaluating the actual amount of contact between waterborne P and the soil adsorption sites. In SF wetlands, the inflow water generally contacts only the soil surface, where the adsorptive surface sites can readily be covered by autochthonous organic detritus. The soil adsorptive capacity of SF wetlands therefore may be quickly impeded or exhausted.

Subsurface-flow (SSF) wetlands, in which the wastewater is fed through a below-ground gravel or rock bed, provide a greater opportunity for continued exposure of P-laden water to substrate sorption sites. High P removal rates have been observed in short-term studies of SSF wetlands, even in nonvegetated systems. Phosphorus removal in nonvegetated gravel microcosms batch-loaded for 58 days with 9.4 cm d^{-1} of municipal wastewater effluent averaged 220 mg P $m^{-2}d^{-1}$ (Burgoon et al., 1991). In companion SSF microcosms containing a plastic substrate, presumably more inert to P adsorption, the P removal rate was much lower (1 mg P $m^{-2}d^{-1}$).

When subsurface flow occurs in natural wetlands, P adsorption by the substrate can be a prominent P sink. In cypress domes, for example, surface water at times permeates downward in the soil column. Dierberg and Brezonik (1985) reported mass P removals > 98.8% for wastewater-leached cypress dome soil columns that received from 12.3 to 19.6 g P per m^2 over a 21-month period. Regardless of the soil matrix and hydraulic movement, however, active P sorption sites within wetland soils eventually become saturated, resulting in a decrease in the P removal capacity over time (Faulkner and Richardson, 1989).

20.3.3 Short-Term Versus Long-Term Phosphorus Removal Processes

Short-term studies of SF and SSF wetlands, whether based on a simple inflow and outflow "black box" analysis or based on internal P removal processes, often provide system P removal estimates in excess of 100 mg P $m^{-2}d^{-1}$ (37 g P $m^{-2}yr^{-1}$) (Finlayson et al., 1986; Tanner et al., 1995; Tanner, 1996). By contrast, long-term processes that control permanent P storage in wetlands usually exhibit lower rates of P removal than the initial rapid uptake processes (adsorption and microbial and plant uptake) that are most often characterized in short-term studies (Richardson and Craft, 1993). Long-term P storage processes include peat production from undecomposed plant matter, chemical precipitation, and adsorption from the water column and subsequent deposition (and eventual burial) in the long-term peat storage compartment (Richardson and Craft, 1993; Kadlec and Knight, 1996).

Estimates of the long-term P mass removal rate in wetlands ranges from 0.40 g P $m^{-2}yr^{-1}$ in "oligotrophic" wetlands, up to an order of magnitude higher in eutrophic wetlands (Mitsch, 1992; Faulkner and Richardson, 1989). Richardson and Craft

(1993) reported a maximum long-term removal of 0.63 g P $m^{-2} yr^{-1}$, and an average of 0.40 g P $m^{-2} yr^{-1}$, for the peat-based Everglades.

The type of vegetation in a wetland influences long-term P storage, with forested wetlands thought to provide a consistently lower rate of P removal than emergent marshes (Kadlec and Knight, 1996). During its 12-year operational period, Reedy Creek Improvement District's Wetland Treatment System #1 (WTS1), a forested wetland, received an average P loading from treated municipal wastewater effluent of 40 mg P $m^{-2}d^{-1}$ (15 g P $m^{-2} yr^{-1}$) at mean annual influent concentrations ranging from 0.4 to 2.6 mg P L^{-1}. With the exception of two of its first four years of operation, this wetland exhibited a net export of P in the effluent during the remaining eight years of operation (DeBusk and Merrick, unpublished data).

Given the poor long-term P removal performance of treatment wetlands, how can the optimistic performance of short-term studies be explained? In one respect, this reveals the limitations to projecting long-term wetland P removal using short-term studies: ephemeral P sinks (adsorption) coupled with the fairly complex hydraulics of treatment wetlands can confound projections of long-term P removal performance. On the other hand, these studies point out the opportunity for deploying substrates with a strong affinity for P adsorption to enhance overall system P removal.

20.4 OPPORTUNITIES FOR ENHANCING PHOSPHORUS REMOVAL IN TREATMENT WETLANDS

There are several opportunities to manipulate the "natural" P cycle to enhance wetland contaminant removal performance. The following discussion relates to two practices: vegetation management, in which either macrophytes or algae (with sequestered P) are routinely harvested, and the use of chemical amendments to enhance P adsorption. Neither of these strategies has received widespread attention, largely because both run counter to the concept of wetlands as passive, natural water treatment systems (Kadlec and Knight, 1996). As noted previously, however, with the large land area requirements limiting the implementation of treatment wetlands for nutrient control, these management practices need to be scrutinized.

20.4.1 Can Vegetation Management Enhance Wetland Phosphorus Removal?

Numerous studies have documented the high productivity and rapid P uptake capability of aquatic vegetation, both macrophytes and algae (Reddy and DeBusk, 1985; Boyd, 1970; Davis et al., 1990). There exists, however, considerable controversy over whether or not biomass harvest can positively impact wetland mass P removal rates. There are a host of studies that demonstrate that macrophyte uptake constitutes only a minor portion of the P removal in the wastewater. Tanner (1996) observed dramatic differences in P uptake rates among six emergent macrophytes cultured on dairy wastewater but noted that species type did not influence overall P system removal (based on water inflow and outflow measurements). Total P removal in his

SSF mesocosms over a four-month period ranged from 370 to 430 mg TP $m^{-2}d^{-1}$, yet the maximum observed plant uptake was 150 mg TP $m^{-2}d^{-1}$, or only 35% of the observed P removal rate.

Upon close investigation, data from many studies reveal that the importance of plant uptake as a nutrient sink is related both to seasonal factors and to the wetland nutrient loading regime. Data from mesocosm studies in Florida demonstrate that in the summer, P uptake by large-leaved and small-leaved floating macrophytes composes a greater percentage of overall system P removal rate than during the winter (Table 20.1). Tanner et al. (1995) observed a similar seasonal trend in studies with the emergent macrophyte *Schoenoplectus validus*. Therefore, harvesting of macrophytes in the summer, when temperature and light conditions are favorable for rapid plant growth, may remove a substantial fraction of the P being sequestered in the wetland.

TABLE 20.1
Effects of Season on Mass P Removal by Selected Floating Macrophytes Cultured in Outdoor Microcosms in Florida, and Percentage of Observed Removal Due to Plant Uptake

	Summer		Winter	
	Mass removal, mg P $m^{-2}d^{-1}$	Percent due to plants	Mass removal, mg P $m^{-2}d^{-1}$	Percent due to plants
E. crassipes[a]	371	90	252	19
E. crassipes[b]	166	100	100	59
L. minor[a]	234	37	205	9
S. rotundifolia[a]	217	48	203	16
P. stratiotes[a]	297	73	205	35

[a]Reddy and DeBusk, 1985

[b]DeBusk et al., 1996

The nutrient loading rate to a wetland also appears to influence the importance of plant uptake to overall system P removal. Relative to "high nutrient" wetlands, systems with low nutrient availability (e.g., the Everglades) often exhibit slower rates for many P cycling processes (e.g., plant productivity and adsorption) as well as reduced P storage in various system compartments (e.g., standing crop of P in macrophytes). Results from mesocosm studies suggest that one net outcome of reduced nutrient conditions in wetlands can be increased importance of plant uptake as a short-term P sink. Reddy et al. (1987) cultured the submerged macrophyte egeria (*Elodea densa*) at several nutrient loading rates in outdoor mesocosms maintained at a 1.5 day hydraulic retention time (HRT). At a high (4 mg N L^{-1} and 0.8 mg P L^{-1}) loading rate, the system P removal rate was 131 mg P $m^{-2}d^{-1}$, with plant uptake accounting for 22% of the P removed. At a low (1 mg N L^{-1} and 0.2 mg P L^{-1}) loading rate, the mass P removal rate was lower (50 mg P $m^{-2}d^{-1}$), but plant uptake accounted for a greater percentage (64%) of the P removed from the water. Regardless of the nutrient status of the wetland, the key to vegetation management for P removal is to harvest plant species (macrophyte or algal) that exhibit moderate to high productivity and tissue P content.

20.4.2 Biomass Harvesting for Phosphorus Removal: Background

Aquatic biomass harvesting has never been embraced as a practical nutrient management tool for treatment wetlands. There is, however, an experience base on which to build. Observations with harvesting aquatic macrophytes and algae, and possible future opportunities, are described below.

Floating Macrophytes

Biomass harvesting of both large-leaved (e.g. *Eichhornia, Pistia, Hydrocotyle*) and small-leaved (*Lemna, Spirodela, Salvinia, and Azolla*) floating macrophytes was studied intensively in the 1980s because of interest in the use of aquatic biomass as a potential feedstock for methane production (Reddy and Smith, 1987). Many floating species are very productive and, unlike emergent macrophytes, the entire plant (roots + shoots) can be harvested readily.

During the mid 1980s, several operational-scale water hyacinth systems (0.6 to 12.2 ha in size) were established in Florida to remove nutrients from municipal wastewater treatment effluents (Stewart et al., 1987). A performance review documented a range of P removal rates from 50 to 280 mg P $m^{-2}d^{-1}$ for six water hyacinth-based treatment systems in Florida (DeBusk and Reddy, 1989). Even though these P removal rates are quite respectable, the concept of floating macrophyte harvest for nutrient removal fell out of favor in the late 1980s. Reasons for this include the requirement for specialized harvesting machinery to remove plants from culture ponds, lack of suitable use for the harvested biomass; monocultures of aquatic plants being subject to strong herbivore pressure from arthropod pests, and, finally, the recognition that many nutrient removal requirements (particularly N) could be accomplished effectively in wetland and aquatic systems without vegetation management.

Emergent Macrophytes

Like floating plants, emergent macrophytes are capable of high productivity and rapid P uptake (Boyd, 1970). Several studies have shown that P assimilation by emergent macrophytes is a function of the nutrient availability in the surrounding water and soil environment. The annual average P assimilation rate of *Typha domingensis* leaves in the Everglades reportedly ranges from 1.8 to 11.5 mg P $m^{-2}d^{-1}$ (Davis, 1991). Shoot P assimilation by *Typha* species in higher nutrient conditions is substantially greater: in dairy wastewaters, shoot P uptake rates of 47 mg P $m^{-2}d^{-1}$ were observed (DeBusk et al., 1996).

Emergent macrophytes present some difficulties with respect to biomass harvest. First, a substantial fraction of emergent macrophyte P storage can occur in below-ground roots and rhizomes (Tanner, 1996; DeBusk et al., 1995). Harvesting devices therefore must be able to collect the belowground tissues, or emergent species must be cultivated that can withstand repeated shoot harvest.

One means of making biomass harvesting more economically attractive is to grow commercially important species. DeBusk et al. (1995) described a P removal system in which seedlings of commercially attractive emergent macrophytes are cultured in a hydroponic nursery configuration. Phosphorus removal by *Pontederia cordata* in such a system used for dairy wastewater treatment was projected to be 26 mg P $m^{-2}d^{-1}$, sustainable over a nine-month period in south Florida. *P. cordata* presently has value as an aquascape plant, which provides an economic incentive to use this species as a harvestable P accumulator.

Adler et al. (1996) evaluated the ability of a mixture of moisture-tolerant grasses (reed canary grass, redtop, and rough bluegrass) to remove P to extremely low concentrations. The plants were cultured in a microcosm raceway containing an inert quartz sand and were fed in a subsurface fashion a nutrient medium containing 0.7 mg P L^{-1}. This system consistently achieved effluent P concentrations below 0.04 mg P L^{-1}. The harvested leaf clippings contained tissue P concentrations from 0.10 to 0.68%, with the harvested biomass accounting for 80% of the P removed from the nutrient medium. While this system more closely resembles a hydroponic farm than a wetland, it elucidates design features that facilitate efficient export of P in biomass, such as good contact between the nutrient medium and the roots, the use of a growth medium where P is the limiting nutrient in solution, and periodic harvest of the P stored in biomass.

Submerged Macrophytes and Periphyton

There exist few data that pertain to the harvesting of submerged macrophytes for P removal. Generally, relative to emergent and floating macrophytes, both the P uptake rate and P standing crop of submerged species is low (Getsinger and Dillon, 1984; Reddy et al., 1987). Submerged macrophytes also are more difficult than floating plants to harvest.

The use of attached algal communities (periphyton) for P control is a concept that has only recently been tested in outdoor systems. Davis et al. (1990) found that periphyton cultured in microcosms fed treated municipal wastewater effluent provided a mean P uptake rate of 89 mg P $m^{-2}d^{-1}$ over a one year period. Adey et al. (1993) reported comparable P removal performance (101 mg P $m^{-2}d^{-1}$) for a six month period in S. Florida using a mesocosm periphyton raceway that received a low nutrient (0.05 mg P L^{-1}) agricultural runoff. In the largest-scale study to date, a 0.1 ha periphyton raceway was used to treat municipal wastewater effluent in central California for one year. On average, this periphyton system reduced total P levels from 3.1 to 1.7 mg P L^{-1} and provided a mean annual mass P removal rate of 730 mg P $m^{-2}d^{-1}$ (Craggs et al., 1996).

Periphyton exhibit many of the benefits of floating macrophytes (high mass P removal rates, ease of harvest, direct assimilation of P from the water column) and, based on Adey's (1993) findings, may actually be superior to macrophytes for sequestering P under low nutrient conditions. Periphyton also develop a polyculture community on culture surfaces, which may increase their resilience to environmental perturbations. The economic value of periphyton biomass, however, remains to be developed.

20.4.3 Utilization of Phosphorus-Adsorbing Amendments in Wetlands

There exist several strategies for chemically enhancing P removal from wetlands. Chemicals can be added in either solid or liquid form to the wetland inflow, with the influent region of the wetland serving as a sedimentation basin for the resulting floc. Prior to the flooding of a SF wetland, chemicals also can be applied to the soils to either improve adsorption of P from the water or to reduce export of native soil P. Finally, SSF wetlands can be constructed with a below-ground substrate that has a strong affinity for P. Most research on amendments for enhancing wetland P removal has focused on this latter approach.

Both natural sediments/soils and artificial amendments have been evaluated as to their ability to enhance the P removal capacity of wetlands. Most substrates have been tested only in laboratory "bench-top" column adsorption and breakthrough experiments (Table 20.2). Reddy et al. (1995) used lab soil columns to assess the suitability of several compounds to increase P retention of SF wetland soils. Compared to a control, $FeCl_3$ (ferric chloride), $Al_2(SO_4)_3$ (alum), $Ca(OH)_2$ (lime), and $CaCO_3$ (calcium carbonate) amendments were effective at reducing P concentrations in the overlying floodwater and in the soil porewater. Respective compound application rates of 1.0, 4.0, 6.3 and 8.6 tons ha^{-1} were required to inhibit soil P release.

Alum and ferric chloride are the compounds that have received the most scrutiny as coagulation agents to be added to the wetland water column or influent wastewater stream (Brown and Caldwell, 1993). Effectiveness of coagulation agents for P removal is strongly dependent on target P levels, as well as concentrations of other constituents (e.g., particulate matter, dissolved organic compounds) in the wastewater stream.

Table 20.2 lists several P-adsorbing substrates that have been tested in greenhouse and field-scale treatment wetlands. It is important to point out that all of these investigations have been short term, so the longevity of the substrate for P removal is not yet well defined. Although the majority of investigations have examined single-substrate amendments for increased P removal efficiencies in treatment wetlands, some research is being targeted on combinations of P-adsorbing substrates, either as mixtures (James et al., 1992; Johansson, 1997) or as sequential pure beds (Jenssen et al., 1996; Johansson, 1997; Lemon et al., 1996; Wood and Hensman, 1989). In general, short-term studies demonstrate that substrate P removal usually exceeds plant P uptake when specialized P-adsorbing substrates are incorporated in treatment wetlands (Table 20.3).

In total, more than 20 substrates have been tested for their P removal capabilities at laboratory, greenhouse and field scales (Tables 20.2 and 20.3). Ferruginous sand, limestone, quick lime, hematite, and fly ash have yielded the highest P removals (>75%) in short-term (<12 mos.) pilot and field-scale studies. Another promising substrate for P removal in SSF wetlands is LECA® (Light Expanding Clay Aggregates), although the P adsorption capacity of this material reportedly varies widely

TABLE 20.2
Amendments Previously Tested for Enhancing P Adsorption in SF and SSF Wetlands

Laboratory tested	Greenhouse tested	Field tested
Dolomite	Pelletized fly ash	Gravel
Activated alumina	Fly ash	Peat
Activated red mud[a]	Shale	Limestone
Wollastonite tailings[c]		Iron ore tailings[b]
Sandstone (17% clay)		Fly ash
Burned dolomite		Quick lime
Limestone		Ferruginous sand
Clay aggregates (LECA®)		Clay aggregates (LECA®)
Fly ash		Alum
Blast furnace slag		Alum + lime
Quick lime		Polymerized alum sludge
Ferric chloride		
Alum		
Peat		
Peat + steel wool		
Peat + iron oxides		
Sand + steel wool		
Sand + iron oxides		

[a]Waste product in the Bayer process for extraction of alumina from bauxite

[b]Hematite

[c]Calcium metasilicate + ferrous metasilicate

with the formulation (Zhu et al., 1997). Jenssen et al. (1996) reported sustained TP removal (>97%) over 48 months in an SSF system containing LECA that was fed septic tank sewage effluent (Table 20.4). Johansson (1997), in contrast, reported that LECA was a poor substrate for P removal and concluded that the composition of the LECA particle explained the between-study differences in results. Jensen et al. (1996) used Norwegian LECA grains that had received a coating of lime to avoid aggregation of particles while Johansson (1997) used uncoated LECA particles. Phosphorus is thought to react with the Ca-based compounds on the LECA surface, either as an adsorbate or a precipitate (Jenssen et al., 1996).

20.4.4 Predicting Performance of Phosphorus-Adsorbing Substrates

Whether a chemical compound is added to water or deployed internally in the system (soil amendment in SF wetland, artificial substrate in SSF wetland), it will have a finite ability to immobilize P. In addition to the cost of the substance, critical considerations are the dose (if applied to water or soil) and the longevity (if used as a SSF wetland substrate). In SSF wetlands, for example, the purchase, transport and installation of the substrate typically constitutes greater than 50% of the total

TABLE 20.3
Compartmentalization of Phosphorus Removal between Substrate and Nonharvested Aquatic Plants in Natural and Constructed Wetlands Receiving Wastewaters

Substrate	Wetland type	Removal of added P, %		Reference
		Substrate	Plant	
Gravel	SSF	0–21	66–86	Rogers et al. 1990
Sand	SSF	11	52	Busnardo et al. 1992
Limestone; hematite	SSF	70–97	2–5	Axler et al. 1996
Ferruginous sand	SSF	>95	<5	Bucksteeg 1990
Shale	SSF	98	1	Drizo et al. 1997
Treated fly ash	SSF	71	<1	Dierberg and DeBusk 1994
Peat	Cypress dome	91	1	Dierberg and Brezonik 1983
Peat + litter	Mires[a]	96	3	Verhoeven 1986
Alum	SSF	32–60	<1	Davies and Cottingham 1993
Soil	SF		2.5	Herskowitz 1986 (cited in Kadlec and Knight [1996])
Sediment	Floating macrophyte	32	13	Fisher and Reddy 1987
Sediment	Floating macrophyte	30	25[b]	Fisher and Reddy 1987
None	Submerged macrophyte		16–21	Reddy et al. 1987
Sand	Submerged macrophyte		9	Dierberg and DeBusk 1995

[a]Natural mires not receiving wastewater discharge

[b]Plants were harvested

system cost (Reed et al., 1995). A P-adsorbing substrate that is used in an SSF wetland therefore must provide treatment for many years to be cost effective. Laboratory and mesocosm studies can be useful predictive tools for estimating P removal prior to the construction of the full-scale system.

Laboratory Studies

Chemical fractionations, adsorption isotherms, and breakthrough leaching columns in the laboratory provide relatively straightforward "first-order" estimates of the useful life of a proposed wetland substrate for P removal.

Chemical Fractionation Procedures

Richardson (1985) found that the P sorption capacity of wetland soils was best predicted by ammonium oxalate-extractable Al and Fe concentrations, with the Al component being a better predictor than Fe. Ammonium oxalate measures the P reactive forms of Fe and Al, which are amorphous and poorly crystallized. Reddy and D'Angelo (1997) derived an empirical relationship for a number of wetland soils and stream sediments in Florida that indicated about four moles of Fe + Al are required to bind one mole of P:

TABLE 20.4
Relevant Data for Substrates Used in Enhancing Phosphorus Removal in Greenhouse or Field-Scale SSF Wetlands

Substrate	Effluent treatment	Oper. time (mo)	T_w (d)	Concentration, mg P L^{-1}		Loading, mg P $m^{-2}d^{-1}$	P removal, %	Reference
				Inflow	Outflow			
River gravel	2° sewage	24	3–8	5–18	—	380 avg.	<10	Mann, 1990
Ferruginous sand	Septic tank effluent	—	—	13.7–20.7	0.3–0.8	60–250	98–99	Netter and Bischofsberger, 1990
Ferruginous sand	Septic tank effluent	60	14	11	—	220	97	Jenssen et al., 1996
LECA®	Septic tank effluent	48	14	10.4	—	190	97	Jenssen et al., 1996
Peat	Peat mining drainage	3	0.05–0.2	—	—	480–2880	−152–11	Ihme et al., 1991
Limestone	Aquaculture wastewater	2–4 per yr	3–7	1.04–13.3	0.12–1.75	20–440	75–95	Axler et al., 1996
Quick lime to root zone system	Dairy waste	3	—	70 avg.	6.0 avg.	700 avg.	89	Willadsen et al., 1990
Hematite	Aquaculture wastewater	2–4 per yr	3–7	1.04–13.3	0.04–0.07	20–440	96–97	Axler et al., 1996
Fly ash	Sewage	12	—	7.6	0.6–1.7	120–340	78–92	Wood and Hensman, 1989
Alum	1° sewage	—	2.5	8.0–11.1	4.1–6.8	770–1060	19–55	Davies and Cottingham, 1993
Alum + Lime	1° sewage	—	2.5	5.4–7.1	1.0–2.1	520–680	70–82	Davies and Cottingham, 1993
Shale gravel + dolomite sand	2° sewage	31	4	2.7	0.53	160	80	Lemon et al., 1995
Shale	Synthetic sewage	10	5	10–40	—	200–800	99	Drizo et al., 1997

Note: Operational time is provided in months, system hydraulic retention time (T_w) in days.

$$S_{max} = 0.24[\text{oxalate Fe + Al}] \qquad R^2 = 0.87 \ (n = 60)$$

where S_{max} is the maximum P retention capacity of soils, and oxalate Fe + Al is the ammonium oxalate extractable Fe and Al.

Adsorption Isotherms

The adsorption maxima from Langmuir or Freundlich isotherms provide an estimate of the life of a designed wetland having a characteristic P loading rate. An estimate of the upper limit of a substrate's P removal lifetime can be obtained by dividing the adsorption maximum (in units of $\mu g \ P \ g^{-1}$) by the P loading rate expressed as $\mu g \ P \ g^{-1} d^{-1}$ (James et al., 1992). Although this simplistic approach provides only an approximation of actual field adsorption maxima (Jenssen and Krogstad, 1988; Richardson and Marshall, 1986), it is still a useful measure for establishing upper boundaries and for comparing various soils and/or amended substrates.

Axler et al. (1996) used adsorption isotherms to predict the longevity of the hematite-based substrate that provided high rates of P removal (>96%) in SSF mesocosms in Minnesota (Table 20.4). They predicted that the hematite substrate could reduce the influent P concentration of 10 mg P L^{-1} to around 1 mg P L^{-1} for about seven years, based on a 2 cm d^{-1} hydraulic loading rate.

Breakthrough Leaching Columns

Sequential leachings of the substrate with P solutions will eventually end in a reduced capacity of the substrate to remove P. By characterizing the P loadings, column size, and the P removal performance after successive leachings, the useful life of a designed field-scale version can be estimated (James et al., 1992).

Microcosm/Mesocosm Experiments

With lab studies, it is difficult to characterize all of the factors (i.e., mass transport, microbial slimes, alternating oxidizing and reducing conditions, reaction kinetics, cumulative loading, and chemical precipitation) that influence long-term P removal rates in the field. Microcosm and mesocosm experiments are useful in determining effects of such factors, as well as in understanding relationships between system hydraulic characteristics and P removal performance. Several field investigations of SF wetlands have documented P removal *fronts,* which occur as adsorption sites of soils closest to the influent region become P saturated (Kadlec and Knight, 1996). Hydraulically, wetlands more closely approximate a plug flow reactor or a series of continuously stirred tank reactors (CSTR) than a single completely mixed reactor (Kadlec et al., 1993). Properly designed wetland mesocosms offer the promise, by incorporating water and soil sampling along gradients internal to the system, of providing better information on substrate longevity than can be obtained by analysis of inflow and outflow water quality data.

Microcosm and mesocosm studies also have been used to further our understanding of the interactions among plants, microorganisms, soils (Richardson and Marshall, 1986) and hydroperiod (Busnardo et al., 1992) in P cycling and sequestration within wetlands. For example, some studies have shown that the presence of plants can have a contravening effect on an added substrate. Davies and Cottingham

(1993) found more P removal was achieved with alum injection in nonvegetated gravel beds than in those containing macrophytes. Although no direct cause for the discrepancy was given by the authors, they noted that dosing alum directly into the macrophyte beds created mixing problems.

20.4.5 Design and Modeling Considerations for Wetlands Managed for Phosphorus Removal

Environmental Effects on Phosphorus-Removing Substrates

To design for efficient P removal in treatment wetlands, the internal P removal mechanisms need to be well understood so that P removal and storage processes can be managed and optimized (Richardson and Craft, 1993). Matching the P-adsorbing substrate to chemical characteristics of the waste stream is a critical step for optimizing P removal. Phosphate chemistry dictates that long-term P storage forms in soils are organic matter (50 to 70%) and calcium phosphate in alkaline wetlands, with Fe and Al binding P in acid wetlands (Richardson and Craft, 1993). Therefore, an initial step in identifying an appropriate P-binding substrate would be to define the chemical composition and pH of the waste stream to be treated. Similarly, P-removing amendments used internally in a wetland, such as a soil amendment or a SSF wetland substrate, must be compatible with a the range of environmental conditions likely to occur in the system. Reddy et al. (1995) recommended caution in using iron-based soil amendments in SF wetlands because of the possibility of anoxic soil conditions causing a release of iron-bound P. Management techniques for wetlands in which various P-adsorbing substrates are deployed have been proposed by several investigators (Table 20.5).

Hydraulic Conductivity

To be useful as a P removing substrate in SSF wetlands, a compound must have adequate hydraulic conductivity. Insufficient permeability of the substrate in relation to hydraulic loading is the factor responsible for most operational problems in SSF wetlands (Crites, 1994). Most studies of SSF systems recommend nominal substrate particle diameters of at least 1 cm (Reed et al., 1995). To overcome hydraulic constraints, many P-adsorbing amendments (e.g., LECA, fly ash) have been formulated into pellets.

Depending on the influent solids level, an initial coarse rock filter (or another suitable solids removal method) or an increased infiltration area system configuration may be advisable. Bavor and Schulz (1993) recommended that inlet solids loading to SSF wetlands should not exceed ≈40 g $m^{-2}d^{-1}$.

20.4.6 Mass Balance Modeling

Phosphorus retention modeling for wetlands is still in an early stage of development, although it is already a commonly used tool for sizing wetlands and estimating system hydraulic and contaminant loading rates. Mitsch et al. (1995) used a Vollenweider-type model in riparian constructed wetlands in the Midwest:

TABLE 20.5
Removal/Storage Processes and Management Techniques for Various P-Removing Amendments

Substrate	Removal/storage processes	Management techniques	References
Native soils	Fe and Al adsorption and precipitation in acid or neutral soils, Ca adsorption and precipitation in alkaline soils, and organic P production followed by sediment burial	Alternating oxidation and reduction by hydroperiod control	Faulkner and Richardson, 1989
		P loading rates < 5 g P m^{-2} yr^{-1} (< 1 g P m^{-2} yr^{-1} for Everglades)	Richardson and Craft, 1993
		Clay content < 5–10% to avoid hydraulic problems (only where subsurface flow occurs)	Geller et al., 1990
LECA®	Adsorption to Ca compounds or precipitated as Ca-phosphates	Maintain high pH	Jenssen et al., 1996
		Position LECA as the terminal process step to avoid biofouling	Johansson, 1997
		Careful selection of LECA since not all formulations provide equal P removal capacities	Zhu et al., 1997
Fe-enriched sand or peat	Adsorption/precipitation onto iron oxide mineral (poorly crystalline goethite, mixture of goethite and ferrihydrite, and paracrystalline form of FeOOH)	Maintain aerobic conditions, pH control probably also necessary	James et al., 1992
Alum	Adsorption to $Al(OH)_3$ floc or formation of $AlPO_4$	Potential sludge disposal requirements Dose into wetland inflow	Davies and Cottingham, 1993
Quick lime	Precipitation as calcium phosphate compounds	Maintain pH > 9.0	Willadsen et al., 1990
Pelletized fly ash	Precipitation as calcium phosphate compounds	Pelletizing required to promote P immobilization and maintain hydraulic conductivity in SSF systems	Dierberg and DeBusk, 1994

$$d[P]/dt = P_{in} - P_{out} - P_{ret}$$

where [P] = the total P concentration, P_{in} = the inflow of P concentration per unit time, P_{out} = the outflow of P concentration per unit time, and P_{ret} = the retention of P concentration per unit time.

Typically, P_{ret} is determined through the model calibration process, and can be described by a simple equation:

$$P_{ret} = k[P]$$

where k is the retention coefficient for P in reciprocal time units.

Mass balance models with a first-order areal uptake have proven to be predictive for P removal in SF treatment wetlands (Kadlec and Knight, 1996):

$$C_o = C_i \exp(-k/q)$$

where C_i = the inlet P concentration, C_o = the outlet P concentration, q = the hydraulic loading rate, and k = the long-term average first-order areal P removal rate constant.

The model equation applies to average performance over a long term, after adaptation trends (one to five years) are over (Kadlec and Knight, 1996). Corrections to the model need to be made if nonideal flow and significant hydrologic effects (e.g., ET, rainfall, and infiltration) are present.

Derived from an analysis of North American treatment systems, Kadlec and Knight (1996) reported k values of 13.1 ± 8.5, 3.1 ± 5.2, and 11.7 ± 4.2 m/yr for emergent marshes, forested wetlands, and SSF wetlands, respectively. New k values would need to be developed for emergent marshes that receive either chemical dosing at the inflow or the incorporation of P-adsorbing soil amendments. Similarly, SSF wetlands equipped with a substrate with a strong P affinity would exhibit a sharply higher k value. Routine vegetation harvest in either SF or SSF systems also would influence the k value.

20.5 MANAGING TREATMENT WETLANDS TO MAXIMIZE PHOSPHORUS REMOVAL

Results from laboratory, greenhouse, and field studies demonstrate that harvesting of vegetation and use of chemical amendments for enhancing wetland P removal is technically feasible. Prior to widespread, large-scale implementation, however, the economic feasibility and environmental impacts of these practices must be addressed. A synopsis of treatment opportunities, deployment strategies, and unresolved issues for these P management approaches follows.

20.5.1 VEGETATION HARVEST

Among all vegetation types, floating macrophytes and periphyton offer the greatest promise as harvestable P accumulators, since these plants are capable of rapid P uptake, they remove P directly from the water, and their biomass is easy to harvest. Alternatively, productive emergent grasses that can withstand repeated foliage cuttings may prove useful for P removal by biomass harvest. For either algae or macrophyte biomass management, the treatment wetland will need to be configured to accommodate specialized harvesting machinery.

It is highly likely that, over a long-term operational period, vegetation harvest can provide a dramatic increase in wetland mass P removal rates over that achievable by natural P sinks. However, both the P loading rate to the wetland and seasonal factors will influence the apparent effectiveness of vegetation harvest on mass P removal rates. Under the appropriate operational conditions, vegetation harvest should decrease wetland area requirements.

At present, the are no economically attractive uses for most types of aquatic biomass. This may not be a major economic constraint for wetlands for which area requirements can be dramatically reduced by plant harvest.

20.5.2 Chemical Amendments: Surface Flow Wetlands

Chemical amendments added to either the soils or inflow water of a SF wetland offer the possibility of enhancing P removal performance. In its simplest configuration, a chemical inflow treatment system would consist of an injector pump feeding a continuous flow of a coagulant (e.g., alum, iron salts) into the wetland inflow pipe. Mixing would occur in the inflow manifold, and flocculation would occur in the quiescent wetland influent region. Chemical soil amendments would be applied to soils prior to wetland flooding or during drawdown periods. Determination of the correct chemical doses will require a good understanding of the wetland water and soil chemistry.

Both water and soil amendments should increase mass P removal performance of the wetland, and may also provide lower effluent P concentrations than normally would be achieved in the system. Little is known of the stability or environmental effects of chemical flocs in the dynamic pH and redox environments of wetlands.

20.5.3 Chemical Amendments: Subsurface Flow Wetlands

Chemical substrates can be used to dramatically enhance the P removal performance of SSF wetlands. For wastewaters containing high levels of labile carbon (BOD), P-adsorbing substrates should be used only in the middle to outflow region of the beds to limit microbial biofilms from interfering with surface adsorption processes. SSF wetlands should be configured so that the substrate can be readily replaced or so that a new P-adsorbing bed can be constructed when the substrate's P adsorption capacity is depleted. Both P effluent quality and mass P removal rates can be improved with the use of P-adsorbing substrates. With the exception of infrequent substrate replacement, SSF wetlands equipped with P-adsorbing substrates should require no additional maintenance beyond that required for a conventional SSF system.

20.5.4 Compartmentalization of the Phosphorus Removal Unit Process

Treatment wetlands designs are evolving toward sequencing of unit processes to enhance N removal, and there are many circumstances where compartmentalization of the P removal component may be advantageous. For example, vegetation harvest may be practical only in the inflow region of a wetland, where plant productivity and tissue P concentrations are highest. For chemically enhanced P removal, it may be most effective to add chemicals such as alum to the wetland inflow and use a compartmentalized wetland influent region as a settling basin to trap the floc. In contrast, in some wetlands the influent chemical feed location may not guarantee effluent quality because of a contribution of P from downstream autochthonous sources (e.g., organic soils and detritus). In this circumstance, a chemical feed system situated closer to the wetland outflow will be better able to control effluent quality.

20.6 CONCLUSIONS

The high treatment wetland area requirements for P removal, relative to that for other wastewater constituents, have created a need to develop alternative P management techniques. Small-scale, short-term studies demonstrate that vegetation harvesting and chemical amendments can enhance the P removal effectiveness of wetlands, thereby reducing the area of the treatment system and facilitating its ability to meet stringent effluent P criteria. Whether these P management approaches can be accomplished cost-effectively in operational systems remains to be demonstrated.

REFERENCES

Adey, W., C. Luckett and K. Jensen. 1993. Phosphorus removal from natural waters using controlled algal production. Restoration Ecology 1: 29–39.

Adler, P.R., S.T. Summerfelt, D.M. Glenn and F. Takeda. 1996. Evaluation of a wetland system designed to meet stringent phosphorus discharge requirements. Wat. Env. Res., 68:836–840.

Axler, R.P., J. Henneck, S. Bridgham, C. Tikkanen, D. Nordman, A. Bamford, and M. McDonald. 1996. Constructed wetlands in northern Minnesota for treatment of aquaculture wastes. In: Constructed Wetlands in Cold Climates Symposium, Niagara-on-the-Lake, Ontario, CAN.

Bavor, H.J., and T.J. Schulz. 1993. Sustainable suspended solids and nutrient removal in large-scale, solid matrix, constructed wetland systems, pp. 219–225. Chapter 22. In: G.A. Moshiri (ed.), Constructed Wetlands for Water Quality Improvement, CRC Press, Inc., Boca Raton, FL.

Boyd, C.E. 1970. Production, mineral accumulation and pigment concentrations in *Typha latifolia* and *Scirpus americanus*. Ecology 51:285–290.

Brown and Caldwell Consultants. 1993. Analysis and Development of Chemical Treatment Processes. Technical Rept. to the South Florida Water Management District, West Palm Beach, FL.

Bucksteeg, K. 1990. Treatment of domestic sewage in emergent helophyte beds—German experiences and ATV-guidelines H 262, pp. 505–515. In: P.F. Cooper and B.C. Findlater (eds.), Constructed Wetlands in Water Pollution Control. Pergamon Press, New York.

Burgoon, P.S., K.R. Reddy, T.A. DeBusk and B. Koopman. 1991. Vegetated submerged beds with artificial substrates. II: N and P removal. Jour. Env. Eng. 117: 408–424.

Busnardo, M.J., R.M. Gersberg, R. Langis, T.L. Sinicrope, and J.B. Zedler. 1992. Nitrogen and phosphorus removal by wetland microcosms subjected to different hydroperiods. Ecol. Engineering 1:287–307.

Cooper, P.F. and B.C. Findlater, Eds. 1990. Constructed Wetlands in Water Pollution Control. 605 pp. Pergamon Press, Oxford, England.

Craggs, R.J., W.H. Adey, K.R. Jensen, M.S. St. John, F.B. Green and W.J. Oswald. 1996. Phosphorus removal from wastewater using an algal turf scrubber. Wat. Sci. Tech. 33:191–198.

Crites, R.W. 1994. Design criteria and practice for constructed wetlands. Wat. Sci. Tech. 29:1–6.

Davis, L.S., J.P Hoffman and P.W Cook. 1990. Production and nutrient accumulation by periphyton in a wastewater treatment facility. J. Phycol. 26:617–623.

Davis, S.M. 1991. Growth, decomposition, and nutrient retention of *Cladium jamaicense* Crantz and *Typha domingensis* Pers. in the Florida Everglades. Aquat. Bot. 40:203–224.

Davies, T.H., and P.D. Cottingham. 1993. Phosphorus removal from wastewater in a constructed wetland, pp. 315–320. In: G.A. Moshiri (ed.), Constructed Wetlands for Water Quality Improvement. CRC Press, Inc., Boca Raton, FL.

DeBusk, T.A. and K.R. Reddy. 1989. Wastewater nutrient removal in Florida using aquatic macrophytes. In: Proceedings, Biological Nitrogen and Phosphorus Removal: The Florida Experience II. University of Florida TREEO Center, Gainesville, FL, USA.

DeBusk, T.A., J.E. Peterson and K.R. Reddy. 1996. Use of aquatic and terrestrial plants for removing phosphorus from dairy wastewaters. Ecol. Eng. 5: 371–390.

DeBusk, T.A., J.E. Peterson and K.R. Jensen. 1995. Phosphorus removal from agricultural runoff: An assessment of macrophyte and periphyton-based treatment systems. In: K.L. Campbell, Ed., Proceedings, Versatility of Wetlands in the Agricultural Landscape Conference, pp. 619–626, Am. Society Agricultural Engineers.

DeBusk, T.A. and P. Krottje. 1996. The use of wetlands for wastewater treatment: A Florida overview, pp.189–194. In: Proc., 71st Florida Water Resources Conf., Florida Water Environment Assoc., Gainesville, FL.

Dierberg, F.E., and P.L. Brezonik. 1983. Nitrogen and phosphorus mass balances in natural and sewage-enriched cypress domes. J. Appl. Ecol. 20:323–337.

Dierberg, F.E., and P.L. Brezonik. 1985. Nitrogen and phosphorus removal by cypress swamp sediments. Water, Air, and Soil Poll. 24:207–213.

Dierberg, F.E., and T.A. DeBusk. 1994. Evaluation of two pelletized coal fly ashes as substrates for subsurface flow wetland wastewater treatment systems. 63 pp. Report submitted to the U.S. Dept. of Energy Technology Center, Morgantown, West Virginia.

Dierberg, F.E., and T.A. DeBusk. 1995. Phosphorus removal from agricultural runoff using an integrated wetland/chemical system. 84 pp. Final Report submitted to the U.S. Dept. of Agriculture under a Phase I SBIR contract. Office of Grants and Program Systems, Small Business and Innovation Research, Ag Box 2243, Washington, D.C. 20250–2243.

Drizo, A., C.A. Frost, K.A. Smith, and J. Grace. 1997. Phosphate and ammonium removal by constructed wetlands with horizontal subsurface flow, using shale as a substrate. Water Sci. Tech. 35:95–102.

Faulkner, S.P., and C.J. Richardson. 1989. Physical and chemical characteristics of freshwater wetland soils, pp. 41–72. In: D.A. Hammer (ed.), Constructed Wetlands for Wastewater Treatment: Municipal, Industrial, and Agricultural. Lewis Publishers, Inc., Michigan.

Finlayson, M., A. Chick, I. Von Oertzen, and D. Mitchell. 1987. Treatment of piggery effluent by an aquatic plant filter. Biol. Wastes 19:179–196.

Fisher, M.M., and K.R. Reddy. 1987. Water hyacinth (*Eichhornia crassipes* [Mart] Solms) for improving eutrophic lake water: water quality and mass balance, pp. 969–976. In: K.R. Reddy and W.H. Smith (eds.), Aquatic Plants for Water Treatment and Resource Recovery. Magnolia Pub. Inc., Orlando, FL.

Geller, G., and K. Kleyn, and A. Lenz. 1990. "Planted soil filters" for wastewater treatment: the complex system "planted solid filter", its components and their development, pp. 161–170. In: P.F. Cooper and B.C. Findlater (eds.), Constructed Wetlands in Water Pollution Control. Pergamon Press.

Getsinger, K.D. and C.R. Dillon. 1984. Quiescence, growth and senescence of *Egeria densa* in Lake Marion. Aquat. Bot. 20:329–338.

Guardo, M., L. Fink, T.D. Fontaine, S. Newman, M.J. Chimney, R. Bearzotti and G. Goforth. 1995. Large-scale constructed wetlands for nutrient removal from stormwater runoff: An Everglades restoration project. Environ. Mgmt. 19:879–889.

Herskowitz, J. 1986. Listowel Artificial Marsh Project Report. Ontario Ministry of the Environment. Water Resources Branch, Toronto. 253 pp.

Ihme, R. K. Heikkinen, and E. Lakso. 1991. Peat filtration, field ditches and sedimentation basins for the purification of runoff water from peat mining areas, pp. 25–48. In: Publications of the Water and Environment Res. Inst., No. 9. Natl. Board of Waters and Environment, Helsinki, Finland.

James, B.R., M.C. Rabenhorst, G.A. Frigon. 1992. Phosphorus sorption by peat and sand amended with iron oxides or steel wool. Water Environ. Res. 64:699–705.

Jenssen, P.D., and T. Krogstad. 1988. Particles found in clogging layers of wastewater infiltration systems may cause reduction in infiltration rate and enhance phosphorus adsorption. Wat. Sci. Tech. 20:251–253.

Jenssen, P.D., T. MÊhlum, and T. Zhu. 1996. Design and performance of subsurface flow constructed wetlands in Norway. In: Constructed Wetlands in Cold Climates Symposium, Niagara-on-the-Lake, Ontario, CAN.

Johansson, L. 1997. The use of Leca (Light Expanded Clay Aggregates) for the removal of phosphorus from wastewater. Wat. Sci. Tech. 35:87–93.

Kadlec, R.H., W. Bastiaens and D.T. Urban. 1993. Hydrological design of free water surface treatment wetlands. Pp. 77–86. In G.A. Moshiri, (Ed.), Constructed Wetlands for Water Quality Improvement. Lewis Publishers, Boca Raton, FL.

Kadlec, R.H., and R.L. Knight. 1996. Treatment Wetlands. Chapt. 14. CRC Press, Boca Raton, FL.

Kadlec, R.H., P.S. Burgoon and M.E. Henderson. 1997. Integrated natural systems for treating potato processing wastewater. Wat. Sci. Tech. 35:263–270.

Lemon, E., G. Bis, L. Rozema, and I. Smith. 1996. SWAMP pilot scale wetlands—Design and performance Niagara-On-The-Lake, Ontario. In: Constructed Wetlands in Cold Climates Symposium, Niagara-on-the-Lake, Ontario, CAN.

Lowe, E.F., L.E. Battoe, D.L. Stites, and M.F. Coveney. 1992. Particulate phosphorus removal via wetland filtration: an examination of potential for hypertrophic lake restoration. Environ. Manage. 16:67–74.

Mann, R.A. 1990. Phosphorus removal by constructed wetlands: substratum adsorption, pp. 97–105. In: P.F. Cooper and B.C. Findlater (eds.), Constructed Wetlands in Water Pollution Control. Pergamon Press, New York.

Mitsch, W.J., J.K. Cronk, X. Wu, R.W. Nairn, and D.L. Hey. 1995. Phosphorus retention in constructed freshwater riparian marshes. Ecol. Applications 5:830–845.

Mitsch, W.J. 1992. Landscape design and the role of created, restored and natural riparian wetlands in controlling nonpoint source pollution. Ecol. Engineering 1:27–47.

Netter, R., and W. Bischofsberger. 1990. Sewage treatment by planted soil filters, pp. 525–528. In: P.F. Cooper and B.C. Findlater (eds.), Constructed Wetlands in Water Pollution Control. Pergamon Press, New York.

Reddy, K.R. and W.F. DeBusk. 1985. Nutrient removal potential of selected aquatic macrophytes. J. Environ. Qual. 14:459–462.

Reddy, K.R., J.C. Tucker, and W.F. DeBusk. 1987. The role of egeria in removing nitrogen and phosphorus from nutrient-enriched waters. J. Aquat. Plant Manage. 25:14–19.

Reddy, K.R. and W.H. Smith (eds.) 1987. Aquatic Plants for Water Treatment and Resource Recovery. 1030 p. Magnolia Publishing, Inc., Orlando, FL.

Reddy, K.R., R.D. DeLaune, W.F. DeBusk and M. Koch. 1993. Long term nutrient accumulation rates in everglades wetlands. Soil Sci. Soc. Am. J. 57:1147–1155.

Reddy, K.R., and E.M. D'Angelo. 1994. Soil processes regulating water quality in wetlands, pp. 309–324. In: W.J. Mitsch (ed.), Global Wetlands: Old World and New. Elsevier Sciences.

Reddy, K.R., and E.M. D'Angelo. 1997. Biogeochemical indicators to evaluate pollutant removal efficiency in constructed wetlands. Wat. Sci. Tech. 35:1–10.

Reddy, K.R., E.M. D'Angelo, M.M. Fisher, Y. Ann, O.G. Olila, L.A.Schipper, D.L. Stites and M. Coveney. 1995. Nutrient Storage and Movement in the Lake Apopka Marsh: Phase I and I. St. Johns Water Management District Technical Report, Palatka, FL.

Reed, S.C., R.W. Crites and E.J. Middlebrooks. 1995. Natural Systems for Waste Management and Treatment, Second Edition. McGraw Hill, New York, NY.

Richardson, C.J. 1985. Mechanisms controlling phosphorus retention capacity in freshwater wetlands. Science 228:1424–1427.

Richardson, C.J., and P.E. Marshall. Processes controlling movement, storage, and export of phosphorus in a fen peatland. Ecol. Monogr. 56:279–302.

Richardson, C.J., and C.B. Craft. 1993. Effective phosphorus retention in wetlands: fact or fiction? Chapter 28. In: G.A. Moshiri (ed.), Constructed Wetlands for Water Quality Improvement. CRC Press, Inc., Boca Raton, FL.

Rogers, K.H., P.F. Breen, and A.J. Chick. 1990. Hydraulics, root distribution and phosphorus removal in experimental wetland systems, pp. 587–590. In: P.F. Cooper and B.C. Findlater (eds.), Constructed Wetlands in Water Pollution Control. Pergamon Press, New York.

Stewart, E.A., D.L. Haselow and N.M. Wyse. 1987. Review of operations and performance data on five water hyacinth-based treatment systems in Florida. pp. 279–288. In: K.R. Reddy and W.H. Smith (eds.), Aquatic Plants for Water Treatment and Resource Recovery. Magnolia Publishing, Orlando, FL.

Tanner, C.C., J.S. Clayton and M.P. Upsdell. 1995. Effect of loading rate and planting on treatment of dairy farm wastewaters in constructed wetlands—II. Removal of nitrogen and phosphorus. Wat. Res. 29:27–34.

Tanner, C.C. 1996. Plants for constructed wetland treatment systems—A comparison of the growth and nutrient uptake of eight emergent species. Ecol. Eng. 7: 59–83.

Willadsen, C.T., O. Riger-Kusk, and B. Qvist. 1990. Removal of nutritive salts from two Danish root zone systems, pp. 115–126. In: P.F. Cooper and B.C. Findlater (eds.), Constructed Wetlands in Water Pollution Control. Pergamon Press, New York.

Verhoeven, J.T.A. 1986. Nutrient dynamics in minerotrophic peat mires. Aquat. Bot. 25:117–137.

Wood, A., and L.C. Hensman. 1989. Research to develop engineering guidelines for implementation of constructed wetlands for wastewater treatment in southern Africa, pp. 581–589. In: D.A. Hammer (ed.), Constructed Wetlands for Wastewater Treatment: Municipal, Industrial, and Agricultural. Lewis Publishers, Inc. Chelsea, Michigan.

Zhu, T., P.D. Jenssen, T. MÊhlum, and T. Krogstad. 1997. Phosphorus sorption and chemical characteristics of lightweight aggregates (LWA)—potential filter media in treatment wetlands. Wat. Sci. Tech. 35:103–108.

21 Phosphorus Retention by the Everglades Nutrient Removal: An Everglades Stormwater Treatment Area

M.Z. Moustafa, S. Newman, T.D. Fontaine, M.J. Chimney, and T.C. Kosier

21.1 ABSTRACT

The Everglades Nutrient Removal (ENR) Project was constructed to reduce total phosphorus (TP) levels in stormwater runoff water from the Everglades Agricultural Area (EAA). The ENR Project operates as a once-through treatment system and has the capacity to process about one-third of the annual runoff that would otherwise be pumped directly into Water Conservation Area 1 (WCA-1). Water is first pumped from the Inflow Pump Station into a holding cell (buffer cell) and then distributed via gravity flow to two independent, parallel treatment trains separated by a transverse levee (Treatment Cells 1 → 3 and 2 → 4). Treatment Cells 1 and 2 are intended to remove the bulk of the nutrient load that enters the ENR Project, while Treatment Cells 3 and 4 are intended to accomplish the final polishing of the water to lower

1-56670-331-X/99/$0.00+$.50

nutrient concentrations. Besides providing the South Florida Water Management District (SFWMD) with practical experience in the construction, operation, and maintenance of a large-scale treatment wetland, the primary performance objective of the ENR Project is to reduce the influent TP mass by up to 75%. A secondary performance objective is to achieve a long-term annual flow-weighted mean TP concentration of no greater than 50 μg P L^{-1} at the outlet. Monitoring was conducted to document TP levels in the ENR Project for one year during construction phase (February 1993 to February 1994) and during the first 29 months of operation (from August 1994 to December 1996). The nutrient removal efficiency during operation was calculated in terms of decreases in both TP loads and concentrations. Monthly TP load removal ranged from 65 to 91% and averaged 80% for the entire period of record. Total P concentrations of influent water ranged from 66 to 201 μg P L^{-1}, while outlet concentrations ranged from 10 to 39 μg P L^{-1}. Analysis of TP mass and concentration reductions indicated that the ENR Project had exceeded its performance objectives during the first 29 months of operation.

21.2 INTRODUCTION

Wetlands are receiving increasing attention as effective systems for removing excessive nutrients from stormwater runoff before these waters enter downstream lakes, streams, and other aquatic habitats. Wetlands remove pollutants from water through a combination of processes.

1. Assimilation or adsorption by wetland sediments and biota
2. Mineralization
3. Export to groundwater (Howard-Williams, 1985; Baker, 1992)

Both physical and chemical pollutant removal mechanisms occur in wetlands including sedimentation, adsorption, precipitation and dissolution, filtration, biochemical interactions, volatilization and aerosol formation, and infiltration (Howard-Williams, 1985). Because of the many interactions between physical, chemical, and biological processes in wetlands, these mechanisms are not independent, and the dominant removal mechanism varies from wetland to wetland (Howard-Williams, 1985).

The Everglades Protection Area (EPA) includes Water Conservation Areas 1 (WCA-1, which is part of the Arthur R. Marshall Loxahatchee Wildlife Refuge), 2A, 2B, 3A, and 3B, and Everglades National Park, and encompasses what remains of a once larger Everglades ecosystem (Fig. 21.1a) (Light and Dineen, 1994; Lodge, 1994). There is growing concern in the regulatory, scientific, and environmental communities that the biotic integrity of the remaining Everglades is endangered. This concern stems from changes observed in plant and animal community composition in portions of the Everglades over the last several decades (e.g., Lodge, 1994; Ogden, 1994; Rutchey and Vilcheck, 1994). These changes have been attributed to disruption of the ecosystem's natural hydroperiod and eutrophication resulting from nutrient-rich stormwater runoff entering the EPA from the Everglades Agricultural Area (EAA). The EAA is a 240,000 ha (593,000 ac), highly productive irrigation drainage basin with a major production of sugarcane. Phosphorus (P) has been

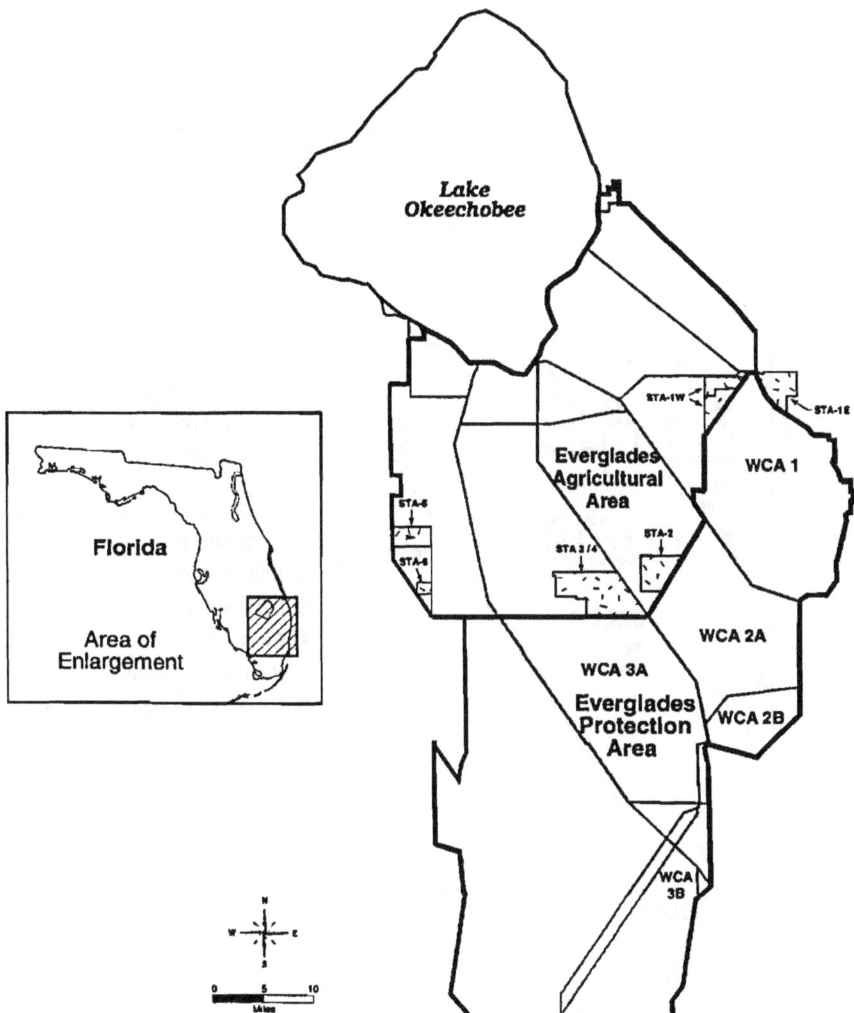

FIGURE 21.1a The Everglades Nutrient Removal (ENR) Project study area: location of the ENR Project and future Stormwater Treatment Areas (STAs) *(continues)*.

identified as the nutrient most responsible for algal and plant community composition changes in the EPA (McCormick and O'Dell, 1996; Koch and Reddy, 1992).

The Everglades Forever Act (EFA; Section 373.4592, Florida Statutes) was enacted by the Florida Legislature in 1994 and authorized a series of measures including construction of Stormwater Treatment Areas (STAs; Fig. 21.1a). It encourages the use of Best Management Practices (BMPs) within the EAA and setting of a threshold discharge limit for total phosphorus (TP), all of which are intended to ensure the protection and restoration of the remaining Everglades. A key component of South Florida Water Management District's (SFWMD) proposed Everglades

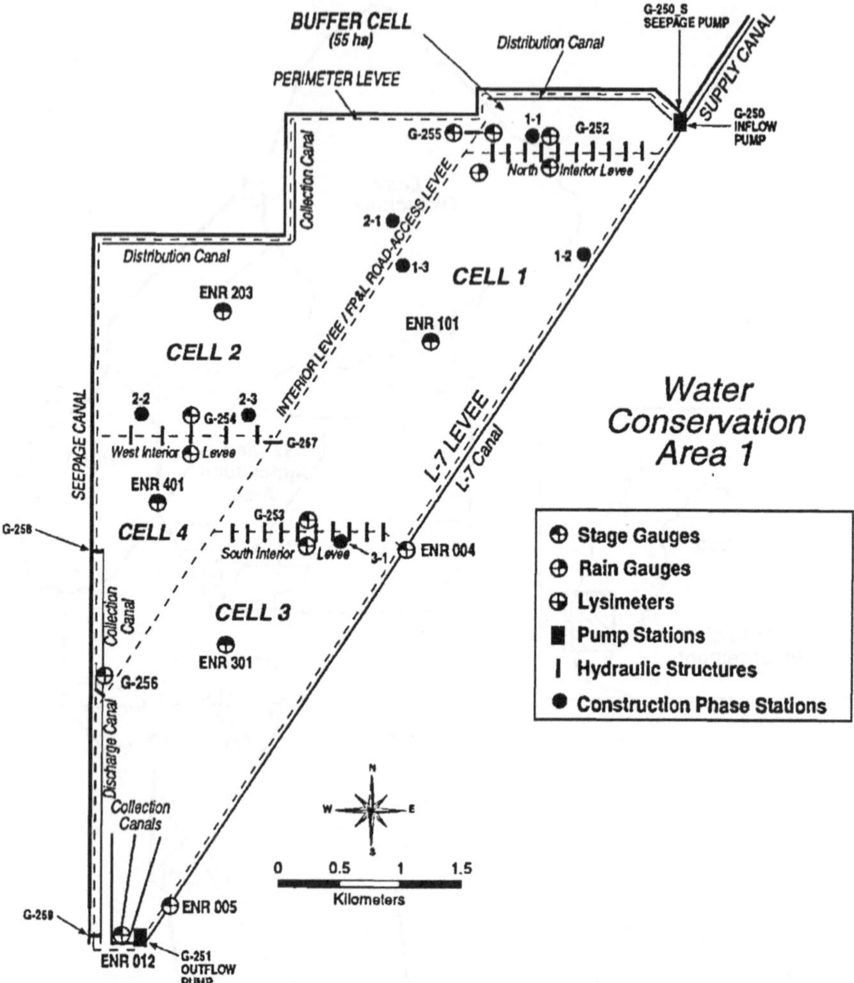

FIGURE 21.1b The Everglades Nutrient Removal (ENR) Project study area: location of inflow and outflow pump stations and other quality monitoring sites at the ENR Project.

Restoration Plan (Burns and McDonnell, 1994) is the construction of 16,000 ha of wetlands to serve as STAs. P-enriched EAA runoff will be discharged into these STAs, which will reduce phosphorus concentrations to acceptable levels before the water is released southward into the WCAs (Walker, 1995). The long-term nutrient removal mechanism that will operate in the STAs involves the initial incorporation of P into macrophyte tissue and the subsequent burial of this biomass in the bottom sediments as peat. Additional details on the theory and design basis for the STAs are provided in Kadlec and Newman (1992) and Walker (1995).

This chapter describes the treatment efficiency of the Everglades Nutrient Removal (ENR) Project in reducing total P (TP) levels in agricultural runoff within the marsh for one year during its construction (February 1993 to February 1994) and during the first 29 months of operation (August 1994 to December 1996). The effectiveness of the ENR Project in improving surface water quality was quantified by calculating the mass of TP retained within the wetland based on all inputs and outputs (i.e., a mass balance approach). The objectives of this research were to

1. Determine TP removal efficiency for the ENR Project and compare its observed performance to nutrient removal goals set for the project.
2. Evaluate TP retention in response to seasonally varying hydroperiod and nutrient loading.
3. Compare TP removal of the ENR Project to measures of nutrient removal efficiency reported for other freshwater wetlands.

21.3 STUDY SITE

The ENR Project is a 1,545 ha (3,818 acres) constructed freshwater wetland built by the SFWMD on land previously farmed in sugar cane, corn, and rice. The ENR Project is located approximately 25 km west of the city of West Palm Beach in south Florida (26° 38′ N and 80° 25′ W) on a site bordering the northwest corner of WCA-1 (Fig. 21.1b). Construction of the containment levees, pump stations, and other structural elements associated with the project was initiated in August 1991 and completed by September 1993. Portions of the wetland were flooded both before and during construction. Because of delays associated with obtaining regulatory discharge permits, flow-through operations did not begin until August 1994.

The ENR Project is operated as a once-through treatment system and has the capacity to process about one-third of the annual runoff that would otherwise be pumped directly into WCA-1. The primary source of water to the ENR Project is the S-5A drainage basin, which drains the northeastern portion of the EAA (Fig. 21.1a). Water is first pumped from the Inflow Pump Station (G250) into the Buffer Cell (54 ha) and then distributed via gravity flow to two independent, parallel treatment trains (i.e., Eastern Flow-way [Treatment Cells 1 and 3] and Western Flow-way [Treatment Cells 2 and 4]) that are separated by a transverse levee (direction of flow is Treatment Cells 1 → 3 and 2 → 4) (Fig. 21.1b). The Western Flow-way is approximately 60% the size of the Eastern Flow-way. Treatment Cells 1 and 3 have an aspect ratio of about 3:1, while the aspect ratio of Treatment Cells 2 and 4 and is about 2:1. The Buffer Cell provides hydraulic dampening of pumped runoff water and allows for independent water delivery into each treatment train. Approximately 67% of the water leaving the Buffer Cell was designed to enter the Eastern Flow-way via the G252 structures, while the remaining 33% of the flow would enter the Western Flow-way via the G255 structure. The division of the outflow from the Buffer Cell was based on the number and size of culverts at G252 (10) and G255 (5). A distribution canal was built along the north side of the Buffer Cell to assist in conveying water from the Inflow Pump Station to G255. Treatment Cells 1 and 2 were intended to remove the bulk of the nutrient load that enters the ENR Project, while Treatment Cells 3 and 4 would accomplish the final polishing of the water to further reduce nutrient concentrations. Water is discharged from the ENR Project at the Outflow Pump Station (G251) by pumping it over the L-7 levee into WCA-1. The Seepage Canal collects groundwater seepage from along the western and northern perimeter of the ENR Project and returns it to the Inflow Pump Station where it is pumped back into the project via the Seepage Return Pumps (G250_S). More complete descriptions of flow-through operations and the water quality, flow, and rainfall quantity monitoring networks within the ENR Project are provided in Abtew (1996), Abtew et al. (1995a, 1995b), and Guardo et al. (1995).

Treatment Cells 1 (525 ha) and 2 (414 ha) have been allowed to revegetate naturally; the dominant emergent macrophyte is cattail (*Typha domingensis* and *T. latifolia*). Treatment Cell 3 (404 ha) is a mixture of naturally recruited cattail and areas (131 ha) that were planted with species common to south Florida, i.e., arrowhead (*Sagittaria latifolia* and *S. lancifolia*), spikerush (*Eleocharis interstincta*), maidencane (*Panicum hemitomon*), pickerelweed (*Pontederia cordata)*, and sawgrass (*Cladium jamaicense*) (SFWMD, 1996). Treatment Cell 4 (146 ha) has been actively maintained through the selective use of herbicides as a periphyton/submersed macrophyte community that is dominated by coontail (*Ceratophylum demersum*) and southern naiad (*Najas quadalupensis*). Areas in Treatment Cells 1, 2, and 3 that were not initially colonized by emergent species during project construction also supported dense stands of *C. demersum* and *N. quadalupensis* by the end of this study. Water hyacinth (*Eichhornia crassipes*) and water lettuce (*Pistia stratiotes*) first appeared in northern areas of the project during construction and are becoming proportionally greater component of the plant community throughout the ENR Project (SFWMD, 1996, 1997).

21.4 METHODS

21.4.1 Construction-Phase Water Quality Monitoring

Surface grab samples were collected biweekly during construction (February 3, 1993 through February 23, 1994) at seven sites distributed throughout the ENR Project (Fig. 21.1b, Table 21.1). Sample sites were selected using two criteria: (a) continued access throughout construction, and (b) suitability of the location for documenting surface-water nutrient chemistry in established habitat types. All samples were analyzed for TP (EPA method 365.3) and soluble reactive P (SRP; SM method 4500 PF). Water depth was recorded at all sites on each sampling trip.

21.4.2 Operation-Phase Water Quality Monitoring

Flow-proportioned composite samples were collected using autosamplers at the Inflow, Outflow, and Seepage Return Pumps (Fig. 21.1b). These autosamplers were serviced, and samples were collected on a weekly basis. Surface grab samples were collected biweekly at all other ENR Project surface water quality stations. Composite rainfall samples were collected at a single wet-deposition sampler located along the G252 levee. This collector also was serviced and samples retrieved on a weekly basis. Seepage entering the ENR Project from WCA-1 was sampled on a quarterly basis at three shallow wells (~ 3 to 6 m deep) located along the L-7 levee. All water

TABLE 21.1
Location and Description of Water Quality Sites Sampled During Construction of the ENR Project (February 3, 1993 through February 23, 1994)

Station	Site Description	Latitude	Longitude
ENR1-1	Open-water area; approximately 300 m from Buffer Cell levee	26° 39′ 25.9″	80° 24′ 37.1″
ENR1-2	Open-water site	26° 38′ 39.9″	80° 25′ 21.6″
ENR1-3	Established cattail stand	26° 38′ 39.9″	80° 25′ 21.6″
ENR2-1	Cattail stand at the edge of an open-water area	26° 38′ 46.7″	80° 25′ 20.2″
ENR2-2	Open-water site	26° 38′ 06.8″	80° 26′ 25.2″
ENR2-3	Open-water site; recently flooded	26° 38′ 09.1″	80° 26′ 07.3″
ENR3-1	Established *Eleocharis* stand	26° 37′ 33.7″	80° 25′ 44.1″

quality samples were analyzed for TP (EPA method 365.3). The operation-phase data presented in this report represent the period from August 18, 1994 through December 31, 1996.

21.4.3 Water Budget Calculations

The water budget for the ENR Project was calculated based on (a) daily flow measurements at the Inflow, Seepage Return, and Outflow Pumps, (b) total daily rainfall collected at a network of automated tipping-bucket gauges located throughout the project, (c) daily estimates of surficial and subsurface seepage entering the ENR Project from WCA-1 along the L-7 levee, and (d) continuous evapotranspiration (ET) measurements at three automated lysimeters.

Flow was computed from pump run time and a flow/pump RPM rating curve developed for each set of pumps and expressed as a daily delivery rate ($m^3 \, sec^{-1}$). Rainfall was spatially averaged over the entire ENR Project utilizing Thiessen weighting coefficients developed for each rain gauge station in the network (SFWMD, 1996, 1997) and expressed as a total daily amount ($m^3 \, day^{-1}$).

$$V_{rain} = R \times A \tag{1}$$

where V_{rain} = daily total volume of rainfall over the entire ENR Project ($m^3 \, day^{-1}$), R = daily spatially-averaged rainfall over the entire ENR Project (m), and A = surface area of the ENR Project (m^2).

Seepage emerged along the toe of the L-7 levee (i.e., surficial seepage) and entered the ENR Project through 21 culverts as surface flow. Biweekly discharge measurements were made at each culvert from December 1994 through November 1995 (SFWMD, 1997). A regression relationship ($R^2 = 0.93$) was developed between the total volume of flow passing through these culverts, the stage in WCA-1, and the difference in stage between WCA-1 and the Eastern Flow-way of the ENR Project (Guardo, 1997). Daily surficial seepage were calculated using this model and expressed as a daily delivery rate ($m^3 \, day^{-1}$).

$$Qseep_{surface} = 0.217 \times [(H_{WCA-1}) - 4.57]^{1.311} \times \Delta H^{2.025} \times \text{Constant} \quad (2)$$

where $Qseep_{surface}$ = daily surficial seepage (m^3 day^{-1}), H_{WCA-1} = head elevation in WCA-1 (m), ΔH = head difference between the ENR Project and WCA-1 (m), and Constant = 86400 seconds day^{-1}.

The volume of subsurface seepage that entered the ENR Project was calculated using a regression ($R^2 = 0.96$) of simulated flow passing under the L-7 levee predicted from a multilayered groundwater model against corresponding stages in the Eastern Flow-way of the ENR Project and WCA-1 (Prymas, 1997) and expressed as a daily delivery rate (m^3 day^{-1}).

$$(Qseep_{subsurface} = 0.42) \times \frac{(H_{WCA-1})^{3.06}}{(H_{ENR})^{3.57}} \times \text{Constant} \quad (3)$$

where $Qseep_{subsurface}$ = daily subsurface seepage (m^3 day^{-1}) and H_{ENR} = head elevation in the ENR Project (m).

The sum of results from Eqs. (2) and (3) was the total seepage contribution from WCA-1 to the ENR Project water budget. Continuous ET data were collected at three automated lysimeters located in cattail, mixed marsh, and open-water areas of the project. Details on measurement and computation methodology of ET values are provided in Abtew (1996).

Total inflow and outflow volumes for the water budget were computed on a monthly basis by summing daily inflow (Inflow Pumps, Seepage Return Pumps, rainfall, and total seepage from WCA-1) and outflow (Outflow Pumps and ET). The residual difference between inflow and outflow volumes represented the cumulative measurement error associated with all terms in the water budget and any seepage not associated with WCA-1 (e.g., deep seepage loss from the ENR Project). This residual water volume was used to balance the water budget for each month.

21.4.4 Nutrient Budget Calculations

The effectiveness of a wetland in treating nutrient enriched water can be assessed using a simple input-output model in which the nutrient mass leaving the wetland is compared to the mass entering (i.e., a mass balance approach). The difference between nutrient input and output is the nutrient mass retained by the system (i.e., nutrient retention). A TP mass balance budget was developed for the ENR Project using a modified version of the program FLUX (Walker, 1989). We assumed that the entire surface area of the marsh was involved in the net retention of TP.

Daily TP loads were calculated by multiplying daily water volumes by the corresponding TP concentration in autosampler or rainfall samples collected during that time period; daily TP loads were then summed for each month. The average TP concentration (48 mg P m^{-3}) of all quarterly water quality samples collected from wells along the L-7 levee was used in all daily TP load calculations for seepage coming from WCA-1. Monthly TP loading rates (L_{in}) were calculated on an annualized basis as follows.

$$L_{in} = \frac{\sum_{n=1}^{m}\left\{\sum_{i=1}^{i=3}(C_{in} \times V_{in}) + D\right\}}{A} \times (365.25/m) \tag{4}$$

where L_{in} = annualized nutrient loading rate (g P m^{-2} yr^{-1}), m = number of days in the month, C_{in} = daily nutrient concentration for the Inflow Pumps (i = 1), Seepage Return Pumps (i = 2), and seepage from WCA-1 (i = 3) (g P m^{-3}), V_{in} = daily volume of influent water to the ENR Project at the Inflow Pumps (i = 1) Seepage Return Pumps (i = 2), and seepage from WCA-1 (i = 3) (m^3), D = daily nutrient mass in rainfall (g P m^{-3}), and, A = ENR Project surface area (m^2).

Nutrient output rates (L_{out}) also were calculated on an annualized basis for each month.

$$L_{out} = \frac{\sum_{n=1}^{m}(C_{out} \times V_{out})}{A} \times (365.25/m) \tag{5}$$

where L_{out} = annualized nutrient output rate (g P m^{-2} yr^{-1}), C_{out} = daily nutrient concentration at the Outflow Pumps (g P m^{-3}), and V_{out} = daily volume of water discharged from the ENR Project via the Outflow Pumps (m^3).

Monthly nutrient retention rates ($L_{retention}$) were calculated as the difference between all nutrient inputs (L_{in}) and outputs (L_{out}).

$$L_{retention} = L_{in} - L_{out} \tag{6}$$

Monthly nutrient removal efficiencies (TP%) were calculated by dividing nutrient retention rates by nutrient loading rates.

$$TP\% = \frac{L_{retention}}{L_{in}} \times 100 \tag{7}$$

The mass of nutrients removed by a wetland can be modeled using a first-order, area-specific equation of the general form.

$$Mass_{in} - Mass_{out} = k \times C \times A \times T \tag{8}$$

where $Mass_{in}$ = nutrient mass entering the marsh (g) during a time interval, $Mass_{out}$ = nutrient mass leaving the marsh (g) during a time interval, k = settling rate coefficient (m year^{-1}), C = average surface water nutrient concentration (g m^{-3}), and T = time interval (month or year).

Settling rate coefficients, k, were calculated using total monthly water mass and average flow-weighted nutrient concentrations for each month and expressed on an annualized basis as follows:

$$k = \left[\ln\left(\frac{C_{inflow}}{C_{outflow}}\right) \times \frac{(Q_{inflow} + Q_{outflow})/2}{A}\right] \times (365.25/m) \tag{9}$$

where C_{inflow} = monthly flow-weighted influent nutrient concentration to the ENR Project (g m^{-3}), $C_{outflow}$ = monthly flow-weighted effluent nutrient concentration from the ENR Project outflow (g m^{-3}), Q_{inflow} = monthly water mass entering the ENR Project (m^3), and $Q_{outflow}$ = monthly water mass exiting the ENR Project (m^3).

21.5 RESULTS AND DISCUSSION

21.5.1 Construction-Phase Water Chemistry

The ENR Project is composed of a surface layer of peat 1 to 2 m deep, underlain by several meters of carbonate rock, under which there is interbedded sand and limestone (Jammal and Associates, 1991). The predominant form of P in the soils is organic P (Reddy and Graetz, 1991), with some additional inorganic P applied as fertilizer during the 20 years the area was farmed. When drained, the organic material is mineralized, and subsequent flooding of drained soils results in an initial flush of nutrients from the soil into the overlying water (Reddy, 1983; Reddy and Rao, 1983). It was anticipated that the periodic flooding and draining necessary during the construction of the ENR Project would result in elevated P leaching from the soils to the overlying water column, compared to soils maintained under constant dry or wet conditions.

Mean TP and SRP concentrations generally exhibited similar patterns of temporal variation during construction (Fig. 21.2a). Four to five sample sites were dry from March to June 1993 and could not be sampled. The resulting small number of data points may account for the large standard errors observed for both parameters during this period. SRP concentrations were highly correlated with TP levels ($r = 0.93$) and, on average, accounted for 31% of TP (range 14 to 70%). Concentrations of both TP and SRP were greatest in March 1993, declined rapidly to relatively low levels by mid-April 1993 (which remained constant through August 1993), and then exhibited a second peak in September 1993. The second concentration peak coincided with the completion of construction and an increase in water levels throughout the project (Fig. 21.2b). TP and SRP concentrations declined after the September peak during which time TP concentrations fell below 50 μg P L^{-1} (Fig. 21.2a), the concentration that has been set as the STA interim discharge level.

The maximum TP (406 μg P L^{-1}) and SRP (207 μg P L^{-1}) concentrations measured in this study were similar to initial peak P concentrations observed after flooding of two other sections of the ENR Project. Before construction was started in 1991, Koch (1991) monitored the release of nutrients from soils to the overlying floodwater in fields at the north end of the ENR Project (now part of the Buffer Cell). A followup study conducted from September 1993 to February 1994 documented the release of nutrients in Treatment Cell 4 (Newman et al., 1997). However, unlike the results from this study, Koch (1991) and Newman et al. (1997) found that elevated TP levels in surface water were reduced to concentrations below 10 μg P L^{-1} within 34 months. The constant disturbance associated with construction at the sites sampled in this study appeared to have extended the timeframe by a number of months before equilibrium P concentrations were reached between the soils and the overlying water.

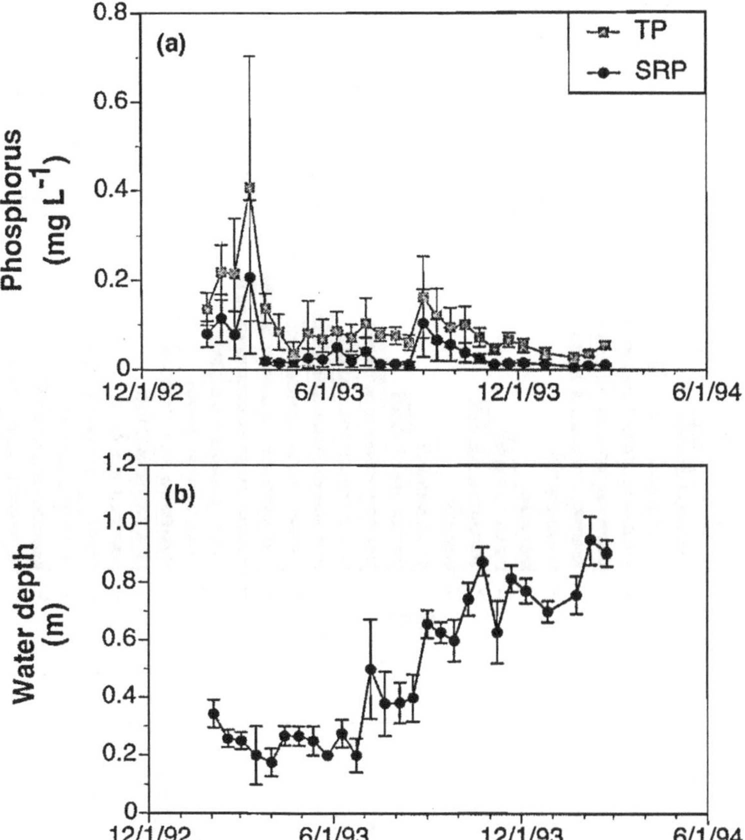

FIGURE 21.2 Mean total phosphorus (TP) and soluble reactive phosphorus (SRP) concentrations and water depths at stations sampled in the ENR Project during construction (February 1993 through February 1994). Bars represent ±1 standard error of the mean.

P concentrations monitored during the construction phase of the ENR Project provide important background information on the range of conditions to expect within future STAs. Data from this study, together with nutrient data from studies by Koch (1991) and Newman (et al., 1997) suggest the timeframe within which TP concentrations of 50 μg P L^{-1} may be achieved within the STAs if flooding occurs during construction.

21.5.2 Operation-Phase Water Budget

Flow at the Inflow and Outflow Pump Stations each month was always greater than the corresponding volumes of water associated with the other components of the water budget (Fig. 21.3a). Pumped inflow (Inflow Pumps + Seepage Return Pumps) accounted for 82% of flow into the ENR Project for the entire study, while rainfall and seepage from WCA-1 each made up 9%. Pumped outflow accounted for 69%

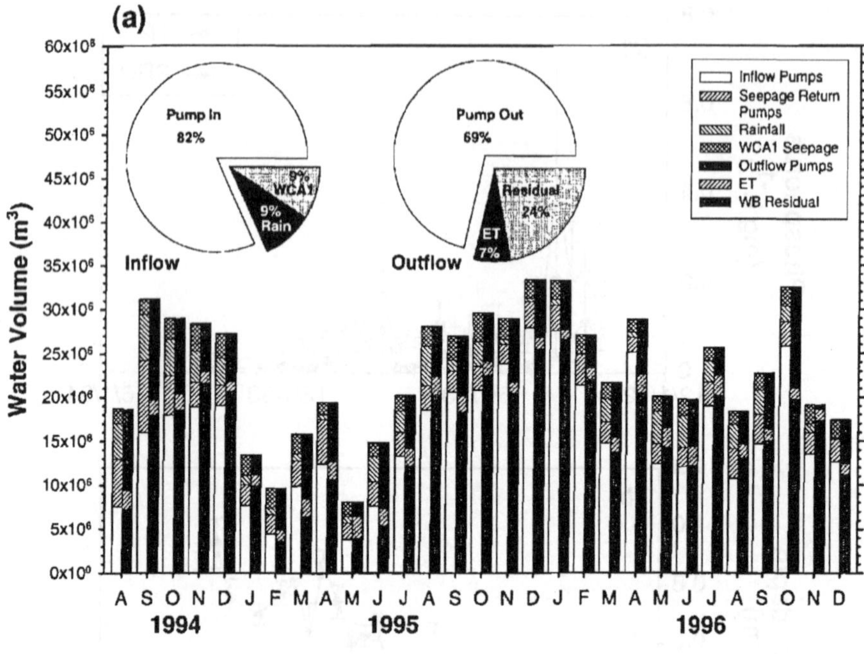

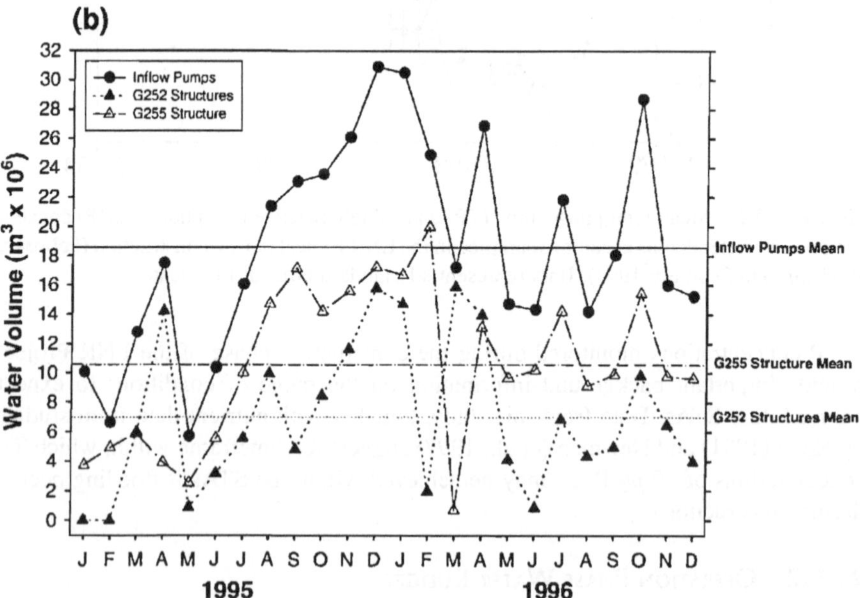

FIGURE 21.3 Water budget data for the ENR Project from August 1994 through December 1996: (a) inflow/outflow components for monthly water budgets (stacked bars) and inflow/outflow components for the entire 29 month water budget (pie charts), and (b) monthly and overall mean flow at the Inflow Pump Station and the G252 and G255 structures from January 1995 through December 1996.

of the total water loss from the marsh, while ET and the water budget residual accounted for 7 and 24%, respectively, of the remaining flow (Fig. 21.3a). Groundwater flows are currently being measured at a number of locations in the Eastern and Western Flow-ways to accurately determine the contribution of non-WCA-1 seepage to the water budget residual.

Flow through the distribution culverts at G252 and G255 was monitored with ultrasonic velocity meters (UVMs) beginning in January 1995. With the exception of three data points, monthly water volumes passing through G255 to Cell 2 were higher than the corresponding water volumes passing through G252 (Fig. 21.3b) to Cell 1. The mean water flow volume at G255 (10.6×10^6 m^3 $month^{-1}$) was greater than mean flow at G252 (7.0×10^6 m^3 $month^{-1}$). This represented a 60% and 40% division of the outflow from the Buffer Cell, which differed considerably from the intended outflow proportions of 33% through G255 and 67% through G252. The interior of the Buffer Cell became densely vegetated with emergent, submersed, and floating macrophytes during construction of the project, while the distribution canal along the north side of the Buffer Cell remained open to flow (see vegetation map in SFWMD [1997] report appendices). We speculate that this vegetation offered considerable resistance to flow within the Buffer Cell and forced a greater proportion of water down the distribution canal and through G255. However, as will be discussed later, the observed deviation in the flow distribution between G252 and G255 did not negatively affect the ENR Project's predicted TP removal performance.

21.5.3 Operation-Phase Nutrient Concentrations

Monthly flow-weighted mean TP concentrations at the Inflow Pump Station ranged from 66 to 201 μg P L^{-1} (Fig. 21.4a) and averaged 113 μg P L^{-1} (±5.1 SE) for the period of record. Monthly flow-weighted mean TP concentrations at the Outflow Pump Station were much lower and less variable, ranging from 10 to 39 μg P L^{-1} and averaged only 22 μg P L^{-1} (±1.6 SE). The range of monthly mean flow-weighted TP concentrations at the Seepage Return Pumps (12 to 51 μg P L^{-1}) were similar in magnitude to concentrations at the outflow and averaged 25 μg P L^{-1} (±1.6 SE). Volume-weighted monthly mean TP concentrations in rainwater fluctuated between 8 and 136 μg TP L^{-1}, averaging 32 μg TP L^{-1} (±5.2 SE).

One of the performance objectives for the ENR Project is to achieve a long-term flow-weighted mean concentration no greater than 50 μg P L^{-1} at the outlet. To date, the ENR Project has met this goal on both an overall basis and a monthly basis; note that, in addition to the low flow-weighted mean concentration (22 μg P L^{-1}), all monthly mean TP concentrations at the Outflow Pump Station were below 50 μg P L^{-1} throughout this study (Fig. 21.4a). Total P concentrations at the sampling stations located along the Eastern Flow-way further illustrate how effective the marsh was in reducing TP concentrations (Fig. 21.4b). In general, median and mean TP levels decreased between the Inflow and Outflow Pump Stations. Note that the range of TP concentrations at the outlet was substantially lower than inlet concentrations for the period of record (Fig. 21.4b).

A general assumption in the design of treatment wetlands is that, under ideal conditions, water will move through the system in sheet-flow fashion, which, in turn,

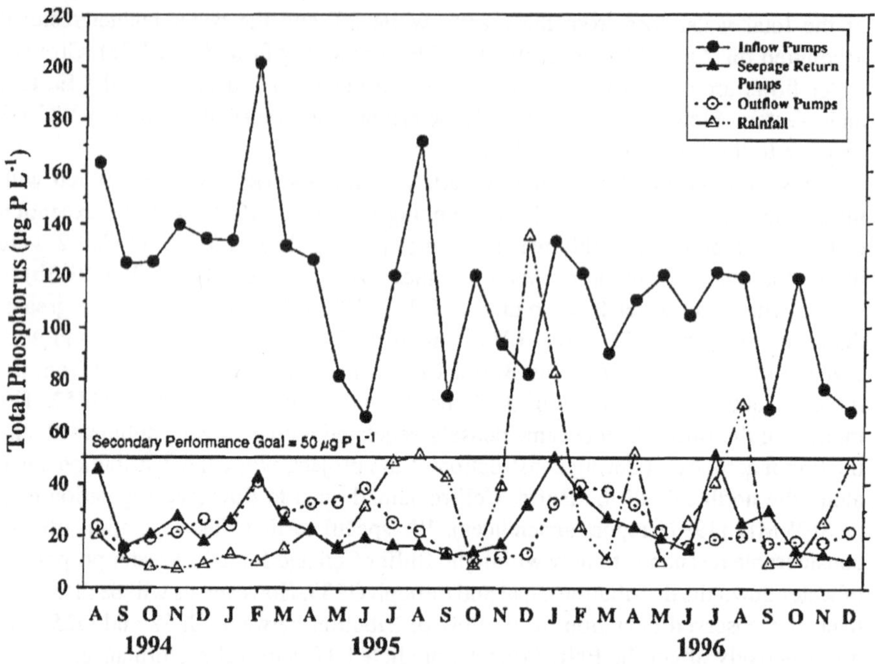

FIGURE 21.4a Temporal and spatial variation in total phosphorus concentrations at the ENR Project from August 1994 through December 1996. Part (a): time series plot of monthly flow volume-weighted mean TP concentrations at the Inflow Pumps, Seepage Return Pumps, and Outflow Pumps, and volume-weighted TP in rainfall *(continues).*

should promote uniform nutrient removal across the width of the marsh. Provided that sheet-flow was established, one might expect similar nutrient concentrations to occur at stations located relatively the same distance down the flow-path in the ENR Project (e.g., stations G252c and G252g on the levee between the Buffer Cell and Treatment Cell 1, stations G253c and G253g on the levee between Treatment Cells 1 and 3, and stations ENR305 and ENR306 at the bottom of Treatment Cell 3) (Fig. 21.1b). Inspection of TP data from these stations revealed that this idealized situation might not have been achieved; i.e., the ENR Project experienced some degree of hydrologic short-circuiting. Differences in mean TP concentrations occurred between all groups of stations (85 versus 76 μg P L^{-1} for the G252 stations, 50 versus 39 μg P L^{-1} for the G253 stations, and 48 versus 37 μg P L^{-1} for ENR305 and ENR306) (Table 21.2); differences between the G252 and G253 stations were statistically significant based on Student's t-Tests at a significance level (α) = 0.05 (Table 21.2). In addition, mean TP concentrations at stations in the interior of the cells (i.e., stations ENR102 and ENR103 and stations ENR302 and ENR303) (Fig. 21.1b) were higher relative to stations located immediately upstream (Fig. 21.4b). All of these results suggest that nonuniform flow, and by inference, nonuniform nutrient removal, existed within Treatment Cells 1 and 3. Hydrologic short circuiting was attributed to small differences in topography across each cell, variation in the

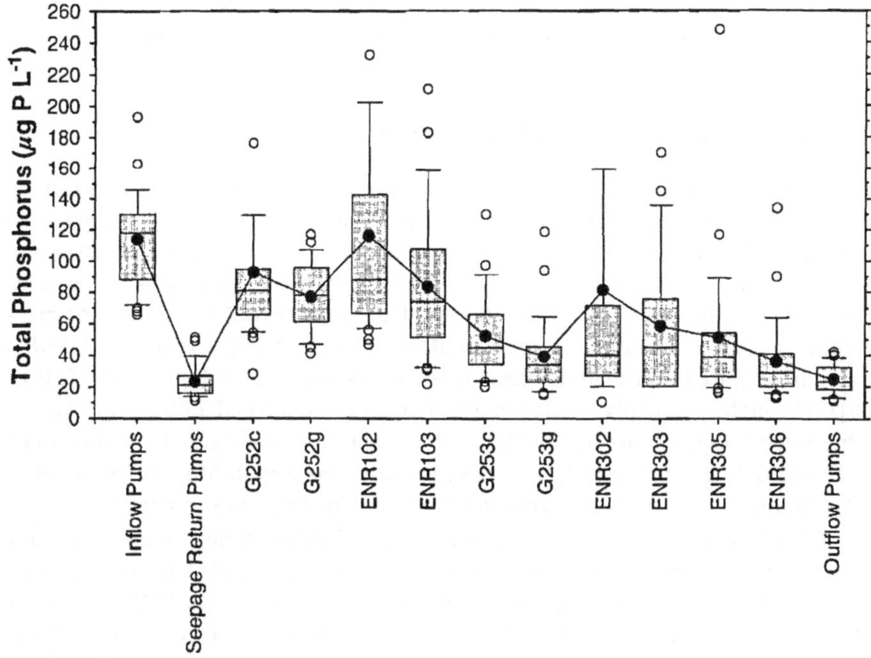

FIGURE 21.4b Temporal and spatial variation in total phosphorus concentrations at the ENR Project from August 1994 through December 1996. Part (b): Box plots of monthly mean total phosphorus concentrations at sampling stations located along the Eastern Flow-way, top and bottom of box = 75th and 25th percentiles, horizontal line inside box = median, vertical lines = 10th and 90th percentiles, solid circle = station mean, open circle = individual data values outside of the 10th and 90th percentiles.

TABLE 21.2
Summary Statistics for Student's t-Tests of Differences in Total Phosphorus Concentrations between Water Quality Stations on the G252 and G253 Levees and between Stations ENR305 and ENR306 in the ENR Project from August 1994 to December 1996

Station	N size	Mean µg P L⁻¹	Standard deviation µg P L⁻¹	Assumption of variance	t value	Degrees of freedom	Probability of a greater t value
G252c	173	85	51	Unequal	2.084	279.0	0.0380
G252g	170	76	30	Equal	2.076	341.0	0.0387
G253c	161	50	31	Unequal	3.392	321.8	0.0008
G253g	175	39	28	Equal	3.408	334.0	0.0007
ENR305	176	48	81	Unequal	1.570	311.2	0.1175
ENR305	176	37	56	Equal	1.570	350.0	0.1174

spatial distribution of aquatic vegetation, and the influence of the old agricultural drainage canals that remained largely unaltered when the ENR Project was constructed.

21.5.4 Operation-Phase Nutrient Budgets

The large differences between the monthly rates of TP mass entering and leaving the ENR Project (Fig. 21.5a) indicated that the marsh was consistently effective in removing TP from stormwater runoff and that most of the inflow TP mass was retained within the project. The month-to-month variability of the mass inflow rates during this study was considerably higher than the observed variability for mass outflow rates. The overall mean mass outflow rate (0.27 g P m^{-2} $year^{-1}$ [±0.03 SE]) was substantially lower than the mean mass inflow rate (1.55 g P m^{-2} $year^{-1}$ [±0.12 SE]). The highest monthly loading and retention rates (3.05 and 2.40 g P m^{-2} $year^{-1}$, respectively) occurred in January 1996, while the lowest loading and retention rates (0.31 and 0.21 g P m^{-2} $year^{-1}$, respectively) were observed in May 1995. Monthly TP retention rates were highly correlated with TP loading rates ($r = 0.97$).

All inflow sources to the ENR Project (i.e., Inflow Pumps, Seepage Return Pumps, WCA-1 seepage, and rainfall) carried TP into the project. The Inflow Pumps were by far the most important mass contributor and accounted for 91% of the total TP load (Fig. 21.5a). The other inflow sources accounted for approximately equal portions of the remaining TP mass: rainfall = 2.4%, WCA-1 seepage = 3%, and the Seepage Return Pumps = 3.6%. The proportions of TP mass associated with the various inflow sources were similar to their corresponding contributions to the total water budget (Fig. 21.3a).

A second performance objective for the ENR Project is to achieve a long-term average load reduction (i.e., removal efficiency) of 75% based on comparing inflow versus outflow TP mass. Monthly TP removal efficiencies ranged between 66% (March 1996) and 91% (October 1995) with an overall mean efficiency of 80% (±1.3% SE) (Fig. 21.5b). Inspection of the cumulative frequency distribution of these data indicated that ~75% of all monthly values were ≥75% removal efficiency (Fig. 21.5c). The ENR Project was a positive sink for TP throughout the entire period of operation and, to date, has exceeded its long-term TP load reduction goal.

Monthly TP removal efficiencies in the ENR Project varied dramatically several times during the course of this study (i.e., March to August 1995, December 1995 to June 1996, and October to November 1996) (Fig. 21.5b). This variability was associated to a limited degree with corresponding variation in nutrient loading to the marsh (Fig. 21.5a). The correspondence between low TP removal efficiency and low TP loading was most evident in May to June 1995; these months experienced a marked decrease in inflow TP concentrations (Fig. 21.4a), inflow TP loading rates (Fig. 21.5a), and TP removal efficiencies (Fig. 21.5b). A study of the performance of another treatment wetland located in south Florida (Boney Marsh) concluded that temporal variation in nutrient removal efficiency in this system corresponded closely with variation in hydraulic and/or nutrient loading (Moustafa et al., 1996). The lowest monthly nutrient retention in Boney Marsh occurred consistently in May or June over the ten-year study period, which coincided with the lowest monthly hydraulic

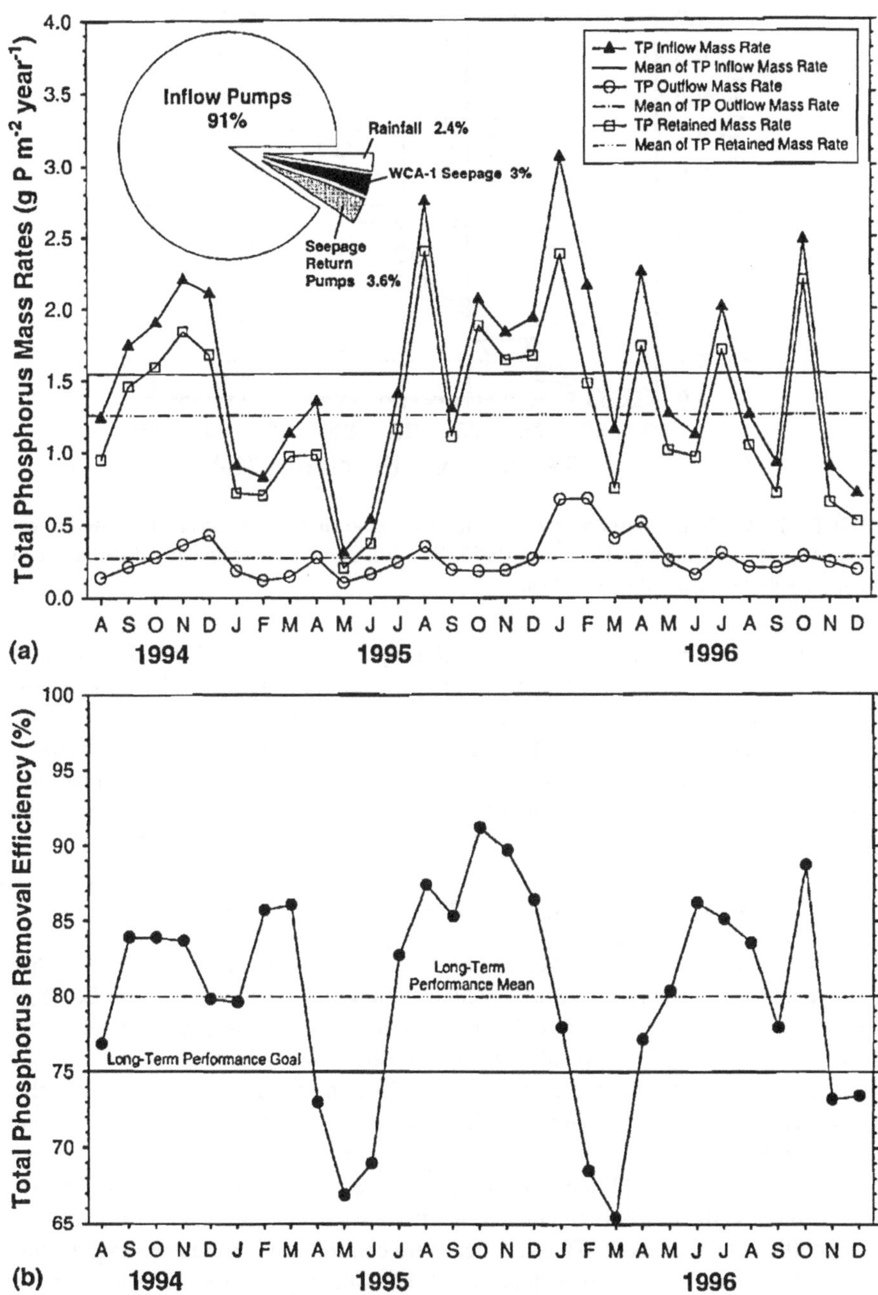

FIGURE 21.5 Total phosphorus mass rates and removal efficiencies at the ENR Project from August 1994 through December 1996. Part (a): monthly values and overall means for rates of phosphorus mass in, phosphorus mass out, and phosphorus mass retained by the marsh. Part (b): monthly values and overall mean for total phosphorus removal efficiencies relative to total phosphorus mass entering the marsh *(continues)*.

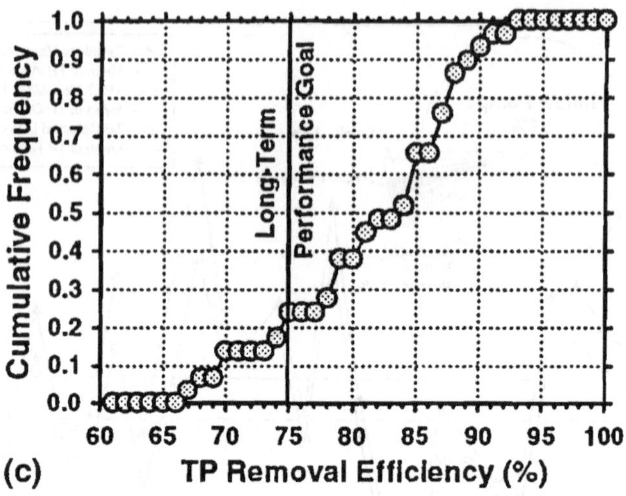

FIGURE 21.5 Total phosphorus mass rates and removal efficiencies at the ENR Project from August 1994 through December 1996. Part (c): cumulative frequency distribution of total phosphorus removal efficiencies.

and nutrient loading rates. Both the ENR Project and Boney Marsh data suggest that low nutrient retention resulted from low nutrient loading to these wetlands. However, the low nutrient removal efficiencies observed in the ENR Project during February to March 1996 (Fig. 21.5b) did not correspond with equally low nutrient loadings (Fig. 21.5a) or concentrations (Fig. 21.4a), which would suggest that over short time intervals (i.e., several months) other environmental factors may influence treatment performance.

21.5.5 Operation-Phase Settling Rate Coefficients (k)

Removal of nutrients by wetlands can be modeled as a first-order rate process that is governed by nutrient concentration and a settling rate coefficient [k; see Eq. (8)]; the settling rate in this case is a linear measure of nutrient retention. Kadlec and Newman (1992) present typical midrange k values for a number of natural and constructed wetlands that varied from 5.3 to 13.0 m yr^{-1}. They also analyzed the U.S. EPA's North American Wetlands Database and reported a mean k value of 7.9 m $year^{-1}$ for the entire database and 10.3 m $year^{-1}$ for wetlands in the southeast United States. Using models calibrated to historical peat accretion and water column phosphorus data, Walker (1995) calculated that the 90% confidence interval for settling rate coefficients in areas of WCA-2A that were continuously inundated ranged from 8.9 to 11.6 m $year^{-1}$ for the peat data and 11.3 to 14.8 m $year^{-1}$ for the water column data.

Annualized settling rate coefficients for the ENR Project ranged from 4.5 m $year^{-1}$ (May 1995) to 47.4 m $year^{-1}$ (December 1996) with mean and median values of ~24 m $year^{-1}$ (Fig. 21.6). In general, k values in the ENR Project were substantially higher than corresponding coefficients for Boney Marsh, the WCAs, and other

Florida wetlands (compare means, medians, and 25th and 75th percentiles in Fig. 21.6) or k values reported for the other wetlands discussed above. The high coefficients for the ENR Project were attributed to the fact that its vegetation community was still developing during this study. Although aerial photographic surveys indicated that the spatial coverage of aboveground vegetation was stabilizing (SFWMD, 1966, 1997), field observations revealed that beds of submersed macrophytes (and attached periphyton) were rapidly growing and expanding into new open-water areas throughout the marsh. We feel that the metabolic requirements of creating this new plant biomass were largely responsible for the enhanced TP removal rates observed during the first 29 months of project operation. These settling rates would become more stable when plant biomass stabilizes.

21.6 CONCLUSIONS

The ENR Project has provided the SFWMD with much practical experience in the construction, operation, and maintenance of a large-scale treatment wetland. Data from this and other studies conducted in the ENR Project during construction demonstrated a wide timeframe before surface water TP concentrations declined to low levels after inundation, which may have implications for nutrient stabilization in the STAs. Inspection of long-term averages for TP concentrations at the project outflow and inflow load reduction indicated that the ENR Project exceeded its performance objectives during the first 29 months of operation. The success of the ENR Project strongly suggests that the STAs will be effective in removing excess P from stormwater runoff from the EAA and can be expected to meet their design performance criteria. In this regard, the ENR Project has validated the incorporation of the STA s into SFWMD's Everglades restoration strategy.

21.7 ACKNOWLEDGMENTS

This investigation was conducted by the Ecosystem Restoration Department, SFWMD, West Palm Beach, FL. We thank the many personnel of the Water Resources Evaluation Department, SFWMD who participated in water quality sample and field data collection, sample analyses, and database management for the ENR Project.

REFERENCES

Abtew, W., M.J. Chimney, T. Kosier, M Guardo, S. Newman, and J. Obeysekera. 1995a. The Everglades Removal Project: a constructed wetland designed to treat agricultural runoff/drainage. Pages 45–56. In: K.L. Campbell (ed.), Versatility of wetlands in the agricultural landscape. American Society of Agricultural engineers, Tampa, FL.

Abtew, W., J. Obeysekera, and G. Shih. 1995b. Spatial variation of daily rainfall and network design. Transactions of the American Society of Agricultural Engineers 38: 843–845.

Abtew, W. 1996. Evapotranspiration measurements and modeling for three wetland systems in South Florida. Journal of the American Water Resources Association 32: 465–473.

Baker, L.P. 1992. Introduction to nonpoint source pollution in the United States and prospects for wetland use. Ecological Engineering 1:1–26.

Burns and McDonnell. 1994. Everglades Protection Project, Conceptual Design. Report for Contract C-3021, Amendment No. 2A prepared by Burns and McDonnell for the South Florida Water Management District, West Palm Beach, FL.

Guardo, M. 1997. Hydrologic balance for subtropical treatment wetland constructed for nutrient removal. Unpublished manuscript, South Florida Water Management District, West Palm Beach, FL.

Guardo, M., L. Fink, T.D. Fontaine, S. Newman, M. Chimney, R. Bearzotti, and G. Goforth. 1995. Large-scale constructed wetlands for nutrient removal from stormwater runoff: an Everglades restoration project. Environmental Management 19:879–889.

Howard-Williams, C. 1985. Cycling and nutrient retention of nitrogen and phosphorus in wetlands: a theoretical and applied perspective. Fresh Water Biology 15:391–431.

Jammal and Associates. 1991. Geotechnical services, SFWMD Everglades Nutrient Removal Project. Final report submitted to Burns and McDonnell, Overland Park, KS.

Knight, R.L., R.W. Ruble, R.H. Kadlec, R.H., and S. Reed. 1993. Wetlands for wastewater treatment: performance database. Pages 35–58. In: G.A. Moshiri (ed.). Constructed Wetlands for Water Quality Improvement. Lewis Publishers, Boca Raton, FL.

Kadlec, R.H. and S. Newman. 1992. Phosphorus removal in wetland treatment areas, principles and data. DOR 106. Support document prepared for South Florida Water Management District, West Palm Beach, FL.

Koch, M.S. 1991. Soil and surface water nutrients in the Everglades Nutrient Removal Project Tech. Pub. 9104. South Florida Water Management District, West Palm Beach, FL.

Koch, M.S. and K.R. Reddy. 1992. Distribution of soil and plant nutrients along a trophic gradient in the Florida Everglades. Soil Science Society of America Journal 56:1492–1499.

Light, S.S. and J. W. Dineen. 1994. Water control in the Everglades: a historical perspective. Pages 74–84. In: S.M Davis and J.C, Ogden (eds.). Everglades – the Ecosystem and its Restoration. St. Lucie Press, Delray Beach, FL.

Lodge, T.E. 1994. The Everglades Handbook – Understanding the Ecosystem. St. Lucie Press, Delray Beach, FL.

McCormick, P.V. and M.B. O'Dell. 1996. Quantifying periphyton responses to phosphorus in the Florida Everglades: a synoptic experimental approach. Journal of North American Benthological Society 15:450–468.

Moustafa, M.Z., M.J. Chimney, T.D. Fontaine, G. Shih, and S. Davis. 1996. The response of a freshwater wetland to long-term "low level" nutrient loads—marsh efficiency. Ecological Engineering 7:15–33.

Newman, S., K. C. Pietro, and K.R. Reddy. 1997. Nutrient release in response to flooding: implications for Everglades stormwater treatment areas. In review.

Odgen, J.C. 1994. A comparison of wading bird nesting colony dynamics (1931–1946 and 1974–1989) as an indication of ecosystem conditions in the southern Everglades. Pages 533–570. In: S.M Davis and J.C, Ogden (eds.). Everglades – the Ecosystem and its Restoration. St. Lucie Press, Delray Beach, FL.

Prymas, A. A. 1997. Calibration of seepage from steady state simulation for water budget estimation. Master thesis. Florida Atlantic University, Boca Raton, FL.

Reddy, K.R. 1983. Soluble phosphorus release from organic soils. Agriculture, Ecosystems and Environment 9:373–382.

Reddy, K. R., and D. A. Graetz. 1991. Phosphorus dynamics in the Everglades Nutrient Removal system. Final report for Contract C90-1168 submitted to the South Florida Water Management District, West Palm Beach, FL.

Reddy, K.R., and P.S.C. Rao. 1983. Nitrogen and phosphorus flux from a flooded organic soil. Soil Science 136:300–307.

Rutchey K. and L. Vilcheck. 1994. Development of an Everglades vegetation map using a SPOT image and the global positioning system. Photogrammetric Engineering and Remote Sensing 6:767–775.

South Florida Water Management District. 1997. Everglades Nutrient Removal Project, 1966 Monitoring Report. South Florida Water Management District, West Palm Beach, FL.

South Florida Water Management District. 1996. Everglades Nutrient Removal Project, 1995 Monitoring Report. South Florida Water Management District, West Palm Beach, FL.

South Florida Water Management District. 1990. Surface water improvement and management plan for the Everglades. South Florida Water Management District, West Palm Beach, FL.

Walker, W.W. Jr., 1995. Design basis for Everglades stormwater treatment areas. Water Resources Bulletin 31: 671–685.

Walker, W.W. 1989. Software and documentation updates BATHTUB and FLUX, prepared for Environmental Laboratory, Waterways Experiment Station, U.S. Army Corps of Engineers, Vicksburg, MS.

22 The Role of Phosphorus Reduction and Export in the Restoration of Lake Apopka, Florida

Lawrence E. Battoe, Michael F. Coveney, Edgar F. Lowe, and David L. Stites

22.1 ABSTRACT

Lake Apopka is a large (125 km^2), shallow ($\overline{Z} = 1.6$ m) lake in central Florida made hypereutrophic by 50 years of agricultural stormwater discharges from farms on 80 km^2 of drained littoral marshes. The lake is characterized by high nutrient levels, high turbidity caused by algae and resuspended sediments, and almost no remaining submersed or emergent macrophytic vegetation. Phosphorus loading to Lake Apopka is being reduced through the purchase of the riparian farms and restoration to aquatic habitat. Additional management activities to accelerate recovery of the lake are creation of a treatment wetland to remove nutrients and suspended solids from lake water, removal of gizzard shad (*Dorosoma cepedianum*), and replanting of littoral vegetation. Because of concerns that reduction in phosphorus (P) loading will be ineffective in restoration of Lake Apopka, we have reexamined the empirical and theoretical basis for P load reduction as a lake restoration technique.

Case studies of control of P loading show that a proportional improvement in lake trophic condition often is obtained when a significant reduction in P loading is effected. Restoration of hypereutrophic, shallow, turbid lakes requires reduction in P loading to lower the stability of phytoplankton dominance and increase the stability

1-56670-331-X/99/$0.00+$.50

of macrophyte dominance. Several characteristics of Lake Apopka increase the probability that the lake will respond to reduction in P loading from adjacent farms. First, P loading (approx. 0.55 g P m^{-2} yr^{-1}) to the lake during the last 30 years has been elevated about seven-fold compared to prefarming levels, and P loading is dominated by farm discharges. Second, biogeochemical processes will dampen internal loading from P-rich sediments after external loading is reduced. The majority of P enters the lake as soluble reactive P. However, as a result of chemical and biological processes, almost 80% of the P in surficial sediments is in Ca-Mg-bound (33%) or organic (46%) forms resistant to rapid biological uptake.

Since summer 1995, trophic indicators (TP, TSS, Chl, Secchi depth) in Lake Apopka have improved significantly based on an 11-year data set. These changes are consistent with modest reductions in P loading achieved since 1993 through regulatory actions. Recently, patches of submersed vegetation (e.g., *Vallisneria*, *Chara*) have established naturally at more than 20 sites around the lake.

22.2 INTRODUCTION

Lake Apopka is the fourth largest lake in Florida. Once a nationally famous fishing lake, Lake Apopka suffers from hypereutrophication as a result of receiving agricultural drainage water rich in phosphorus (P) for the past 50 years. Based on diagnostic research and critical evaluations of restoration techniques (Conrow et al., 1993), we recommended phosphorus load reduction as the primary restoration strategy.

Phosphorus loading to Lake Apopka will be drastically reduced through the purchase of almost all of the 80 km² of riparian farms and restoration of these areas to aquatic habitat. Phosphorus stores in the lake will be reduced by filtration of algae and resuspended sediments in a recirculating treatment wetland and by the mass removal of gizzard shad. Gizzard shad removal may further contribute to improved water quality through biological control mechanisms (biomanipulation) (Moss et al., 1996). Finally, habitat restoration is starting through planting of native vegetation in the littoral zone.

Recently, some limnologists have argued that, for many hypereutrophic lakes, release of internal stores of P can be so substantial that a reduction in P loading may not cause a decline in limnetic P concentration, or that the decline may require many decades (Chapra and Canale, 1991; Welch and Cooke, 1995; Carvalho et al., 1995). Moreover, some workers have taken recent theories of alternative stable states in shallow lakes (Scheffer et al., 1993) to mean that P load reduction is unnecessary or ineffective for shallow lakes (Canfield et al., 1996). In the case of Lake Apopka, several arguments have been made against P load reduction as a means for restoration.

1. The present state of the lake did not result from increased P loading but did result from a hurricane in the late 1940s that destroyed submersed plant beds, and P load reduction will not reverse these effects.
2. Frequent sediment resuspension releases enough P from the sediments that the effectiveness of P load reduction will not reduce the P concentration.

3. A decrease in the P concentration will be ineffective, because resuspended sediments will maintain turbid conditions and prevent reestablishment of macrophytes (Bachmann and Canfield, 1996; Canfield et al., 1996; Bachmann et al., 1997, 1998).

These arguments indicated to us a need to reexamine the empirical and theoretical basis for P load reduction as a lake restoration technique for lakes in general, and for Lake Apopka specifically.

22.3 LOCATION AND DESCRIPTION

Lake Apopka is a large (125 km^2), shallow (mean depth = 1.6 m) lake located centrally on the Florida peninsula in Lake and Orange Counties and approximately 11 km northwest of Orlando (Fig. 22.1). It is the headwater lake of the Harris chain of lakes and the Ocklawaha River that flows into the St. Johns River. A spring is located at the southern end of Lake Apopka, and a few small streams flow into the lake, but the major source of water is rainfall on the lake surface. A single, artificial outflow, the Apopka-Beauclair Canal, was dug for navigation and completed in the late 1890s. The canal apparently lowered the lake's high water level approximately 1 m. Topography indicates that prior to construction of the canal, the lake drained through a swamp to the northwest (Double Run Swamp) and into Little Lake Harris.

Lake Apopka is hypereutrophic (mean TP = 0.204 mg L^{-1}, mean chlorophyll-*a* = 0.092 mg L^{-1}, mean Secchi depth = 0.23 m) (Table 22.1). Almost the entire lake bottom is covered by organic sediments, with high water and P content deposited since the 1940s. These sediments of algal origin are underlain by a more-consolidated organic sediment of macrophytic origin and by peat, sand, or marl (Schneider and Little, 1969; Reddy and Graetz, 1991; Schelske, 1997). Dense phytoplankton and suspended surficial sediments cause high turbidity. Diffuse extinction coefficients for photosynthetically active radiation range from 3.1 to 12.8 m^{-1} (Schelske et al., 1992). In enrichment bioassays, nitrogen (N) was most frequently the limiting nutrient for production of phytoplankton (Aldridge et al., 1993). This situation is expected given the high P loading.

22.4 BRIEF HISTORY OF LAKE APOPKA

Through 1946, the lake had clear water and submersed vegetation dominated by *Potamogeton illinoensis* growing to a depth of 2.4 m (Clugston, 1963). The lake was nationally recognized as a trophy bass fishery, and more than 24 fish camps ringed the lake. The original lake/marsh system had an area of approximately 214 km^2 (Lowe, et al., 1998). During the 1940s, portions of the lake bottom and most of the approx. 80 km^2 saw grass (*Cladium jamaicense*) marsh across the north end of Lake Apopka were diked, ditched, and drained for agriculture.

The first recorded lake-wide algal bloom occurred in 1947. Some reports attribute a sudden shift from macrophyte to phytoplankton dominance to uprooting of vegetation by a hurricane in the fall of 1947 (USEPA, 1978; Schelske and Brezonik,

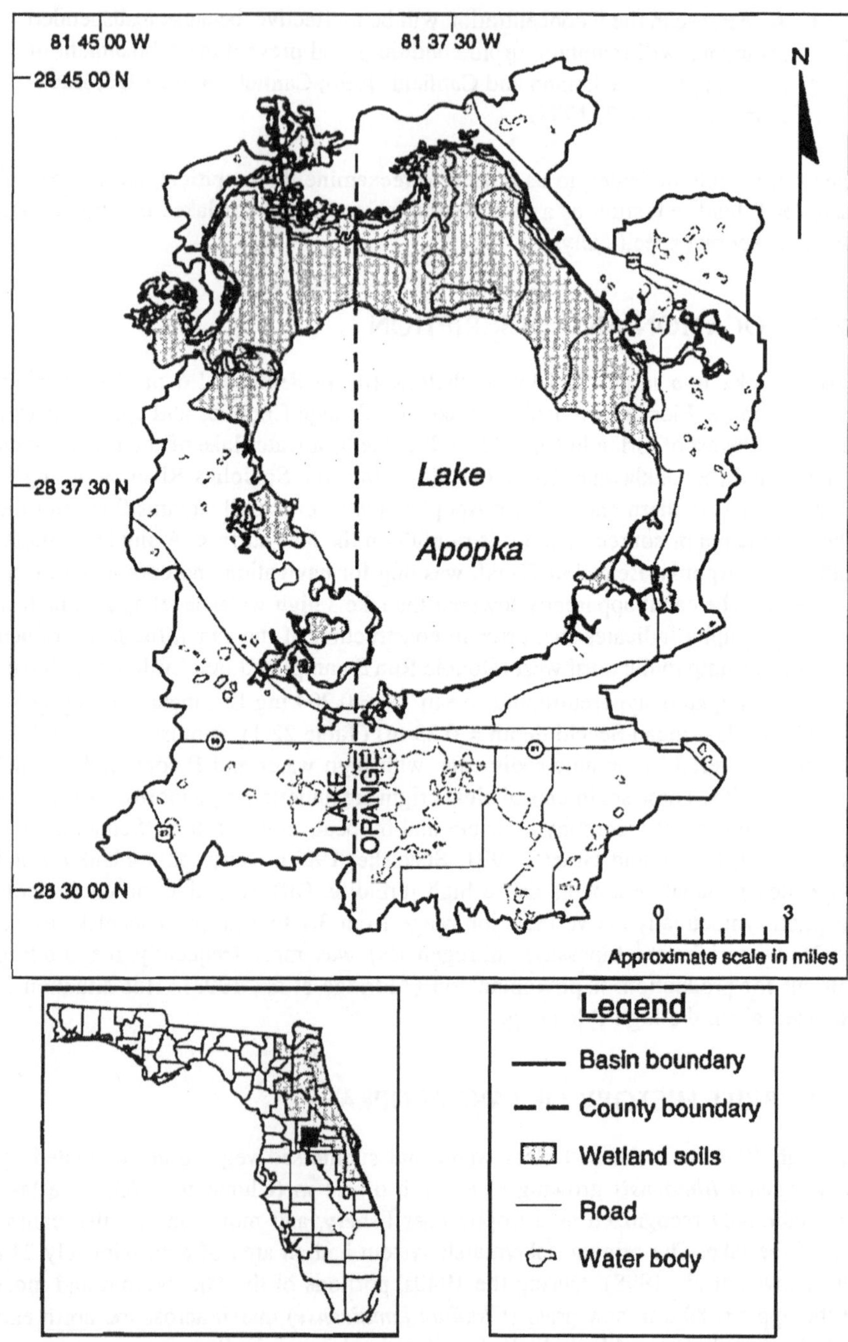

FIGURE 22.1 Location of Lake Apopka, Florida. The agricultural area is indicated approximately by the shading showing wetland soils (contours within the shaded area show different wetland soil types).

TABLE 22.1
Characteristics of Lake Apopka, Florida

Variable	Mean	SD	Method
Mean depth, m	1.63	NA	NA
Secchi depth, m	0.23	0.07	25-cm diameter disk
Chlorophyll-*a*, mg L^{-1}	0.092	0.033	EPA 10200 H (not corrected for pheopigments)
Total phosphorus, mg L^{-1}	0.204	0.063	EPA 365.4
Total nitrogen, mg L^{-1}	5.14	1.27	Sum of Kjeldahl N (EPA 351.2) and NO_3/NO_2 N (EPA 353.2)
Total suspended solids, mg L^{-1}	79.0	31.2	EPA 160.2

Notes: Mean values and standard deviations (SD) were calculated for January 1987–February 1997. Mean depth was calculated at lake surface elevation 66.5 ft NGVD. Sampling frequency was variable but was at least monthly. EPA: Kopp and McKee (1983). NA = not applicable.

1992; Bachmann and Canfield,1996). However, no hurricane passed within 160 km of Lake Apopka in 1947. Photographic evidence and historic accounts suggest that the increase in phytoplankton and decline in macrophytes actually occurred over a several-year period from 1947 to 1951 (Lowe et al.,1998). Since the 1950s, the lake has had high levels of P and N, high turbidity caused by algae and resuspended sediments, almost no remaining submersed macrophytes, and only a thin fringe of emergent macrophytic vegetation.

In 1985, with the passage of the Lake Apopka Restoration Act (LARA), the Florida State Legislature directed that an "environmentally sound and economically feasible" means be found to restore Lake Apopka to Class III water quality standards and identified the St. Johns River Water Management District (SJRWMD) as the lead agency. In Florida law, Class III refers to a lake suitable for recreational use and for the propagation of fish and wildlife. In 1987, the Surface Water Improvement and Management (SWIM) Act augmented the LARA and provided further emphasis and funding for Lake Apopka as one of several priority waterbodies requiring restoration work.

In the ensuing years, eight diagnostic projects were conducted to better understand the problems and the character of Lake Apopka, and ten alternate restoration methods were reviewed and evaluated (Conrow et al., 1993). Several restoration techniques originally proposed for Lake Apopka were found on examination to be either ineffective or too costly. These included large-scale dredging, phosphorus inactivation in the lake sediments, enhanced microbial decomposition, and the introduction of exotic fish species. A pilot-scale wetland filtration system was operated for six years, and several littoral zone revegetation demonstration projects were completed.

22.5 SOURCES AND EXTENT OF POLLUTION

For the last 50 years, Lake Apopka has been a source for irrigation water for the approx. 80 km^2 area of farms established on the former littoral marsh. In addition, fallow fields periodically have been flooded with lake water to prevent subsidence of the organic soils and to help control nematode plant pests. Because surface elevations of the farms are below lake level, excess storm water and flood water, enriched with P and N, have been pumped into the lake. Farm pumping contributed an average P loading of more than 50×10^6 g P yr^{-1} in a recent six-year P budget (Stites et al., 1997).

Lake Apopka also received waste water from citrus processing plants and effluent from the Winter Garden sewage treatment plant. Phosphorus loading from these sources combined was about 4×10^6 g P yr^{-1} in the 1940s and had increased to about 7×10^6 g P yr^{-1} when these sources were controlled in 1977 (Lowe, et al., 1998). These point-source loads were appreciable when compared with pristine loading from precipitation, spring flow, seepage, and runoff (about 14.7×10^6 g P yr^{-1} for the original 214 km^2 lake/marsh system) (Lowe et al., 1998), but were much less than loading from the farms.

Areal P loading (approx. 0.55 g P m^{-2} yr^{-1}) during at least the last 30 years (Coveney, 1997) has been elevated about seven-fold compared to prefarming levels (approx. 0.077 g P m^{-2} yr^{-1}). Farm nutrient loading has decreased in recent years because of a consent agreement between the farmers and the SJRWMD, but further decreases are necessary for the lake to meet Class III standards.

22.6 WHY PHOSPHORUS CONTROL? THE EMPIRICAL AND THEORETICAL BASIS

As discussed above, it has been argued that P load reduction will be ineffective for restoration of Lake Apopka (Canfield et al., 1996; Bachmann et al., 1997, 1998). These arguments stemmed from two basic conclusions: (1) the P concentration will not fall following P load reduction because of internal P loading, and (2) the lake is in an algal-dominated state that will not change as a result of declining P concentration. It was suggested that the theory of alternative stable states (Scheffer et al.,1993) supports the conclusion that P load reduction will be ineffective. We find no support for these views either in new or classical theories of lake eutrophication, in data from lake surveys, in case studies of P-load reduction, or in studies of Lake Apopka.

It has long been recognized that the P concentration of lakes stems from the balance between P loading and P losses through flushing and sedimentation. Many simple mass balance models have been developed on this premise and employed to predict lake responses to P-load reductions (Reckhow and Chapra, 1983). Since these early models were developed, it has been recognized that internal recycling of P between the sediments and water column render the simple models inadequate for predicting the time-course of decline in the P concentration following P-load reduction because, for a time, the sediments can be a net source of P to the water column (e.g., Chapra and Canale, 1991).

Most concede, however, that P-load reduction eventually will lower the P concentration even in lakes with large sedimentary P stores (Marsden, 1989; Chapra and Canale, 1991; Cooke et al., 1993; Welch and Cooke, 1995). Case studies support this view. For example, Sas (1989) reviews the responses of 18 lakes to P-load reduction and concludes that the period of net P release from the sediments is a transient state and that a new steady state is "gradually approached as the old stock of phosphorus in the water-phase is flushed out and net annual release from the phosphorus pool in the sediment decreases and stops." Lake Sammamish, Washington, has been used as an example of the reduced response from a shallow lake because of internal P loading, compared to Lake Washington's rapid recovery (Henderson-Sellers and Markland, 1987). However, 14 years after the diversion of P loading from Lake Sammamish, a 36% decrease in P load had resulted in a 45% decrease in lake P concentration (Cooke et al., 1993). Shagawa Lake, Minnesota, responded more rapidly to reduced P loading (Cooke at al., 1993) than predicted by an internal P loading model (Chapra and Canale, 1991). Reductions in P concentration following P-load reduction also occurred in Lake Tohopekaliga, Florida (James et al., 1994) and Lake Thonotosassa, Florida (SWFWMD, 1992; Brenner et al., 1996). For most lakes in this diverse set, the percent reduction in P concentration is roughly equivalent to or greater than the percent reduction in the P load (Fig. 22.2).

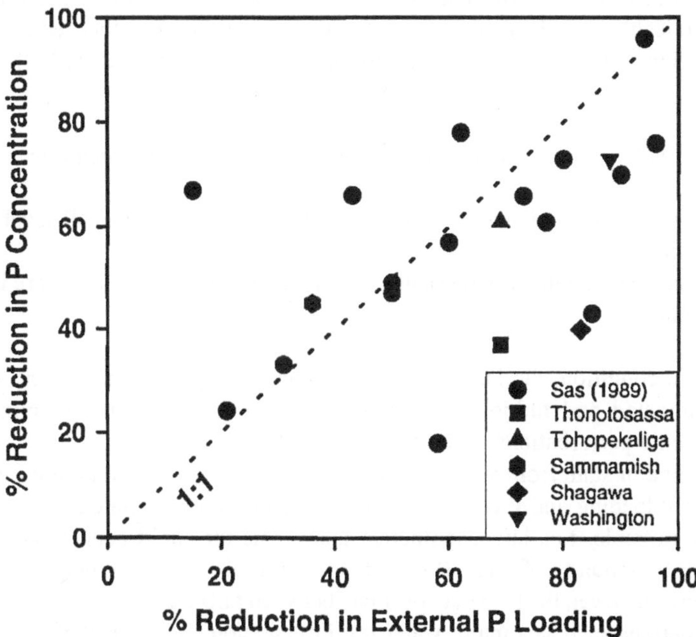

FIGURE 22.2 Percent change in lakewater TP concentration related to percent reduction in external P loading for a number of lake restoration projects. Dotted line indicates 1:1 relationship. Changes in loading and concentration were calculated from the extreme high and low annual values in each case study by Sas (1989) and for Lake Thonotosassa. For the other lakes, multiyear mean values were used to calculate changes in loading and concentration. All case studies analyzed in Sas (1989) are shown here except for Lough Neagh, where no systematic decrease in loading was demonstrated.

In parallel to the work that related the P concentration to the P load, a large body of evidence was developed that related the P concentration to other indicators of trophic state. In the 1970s, Schindler (1974) demonstrated that P, rather than C or N, elicited the changes associated with cultural eutrophication. Many cross-system studies showed a strong, positive correlation between P concentration and algal abundance as indicated by chlorophyll-*a* (Reckhow and Chapra, 1983). This log-linear relationship also holds for Florida lakes (Canfield, 1983; Huber et al., 1982). Thus, both experimental and correlational evidence indicates that P concentration controls algal abundance. Algal abundance, as measured by chlorophyll-*a*, in turn, is negatively correlated with water clarity, as measured by Secchi disc depth across a broad range of lakes (Reckhow and Chapra, 1983) including Florida lakes (Huber et al., 1982; Canfield and Hodgson, 1983).

Improvements in the trophic condition of lakes following reductions in P loading may not always meet a specific restoration goal, such as a shift from a eutrophic to a mesotrophic classification. These situations can be caused by insufficient reduction in overall P loading or insufficient elapsed time to achieve a new equilibrium. However, these situations do not indicate that lakes fail to improve in response to lower external P loading (Cooke et al., 1993).

Recent theories that envision two alternative states in shallow lakes are particularly applicable to Lake Apopka and do not support the view that reducing the P concentration will be ineffective. As discussed by Scheffer et al. (1993), shallow lakes may exist in three conditions determined by P levels.

1. A stable, macrophyte-dominated, clear-water condition when P levels are low
2. A stable phytoplankton-dominated, turbid condition when P levels are high
3. A condition with two alternate, quasistable states when P levels are intermediate

In the third condition, the lake may be either macrophyte dominated and clear or algae dominated and turbid, and the state may shift in response to climatic or anthropogenic perturbations (Scheffer et al., 1993).

A change in state from a stable, turbid condition to a clear condition requires a decline in P loading and P concentrations at least to a level where two quasistable states are possible. For large, temperate lakes, this threshold concentration for P appears to be around 100 mg TP m^{-3} (Jeppesen et al., 1990; Klinge et al., 1995). At or below this level, the lake may oscillate between turbid and clear phases initiated by perturbations such as a major water level fluctuation (Blindow et al., 1993), ice-scouring of the bottom, or reduction in planktivorous fish (Scheffer et al., 1993). At a lower range in P concentration, about 25 to 50 mg TP m^{-3}, the transition to a stable, macrophyte-dominated, clear-water state can occur (Moss et al., 1996).

In this theory, the restoration of shallow, hypereutrophic lakes requires reduction of P loading to lower the stability of phytoplankton dominance and increase the stability of macrophyte dominance. Once the P concentration falls below the threshold for stable algal dominance, other strategies may hasten the shift from phytoplankton to macrophyte dominance. These include biomanipulation of the fish community (removal of planktivores or increase in piscivores) and planting littoral vegetation (Moss et al., 1996).

Lake Apopka was apparently in a stable, macrophyte-dominated state prior to the 1940s. Evidence for this conclusion is found in historical accounts of macrophyte growth and water quality, and in the fact that extreme perturbations, for example numerous hurricanes, did not elicit a change to a turbid state. After P loading was greatly increased through the development of agriculture, the lake entered a turbid, phytoplankton-dominated state. This condition has been stable since the early 1950s and was unaffected by periods of extremely low water in 1956 and 1971.

22.7 RESPONSE OF LAKE APOPKA TO P LOAD REDUCTION

Several characteristics of Lake Apopka indicate that a large reduction in farm P loading will result in a large reduction in the P budget and that lake P concentrations will respond. First, because other important anthropogenic sources of P were mitigated in the 1970s, P loading to the lake is dominated by farm discharges. In a recent six-year P budget study (Stites et al., 1997), farm discharges (0.42 g P m^{-2} yr^{-1}) averaged 85% of total P loading. Once this massive farm loading is curtailed, the overall P budget will fall precipitously. By comparison, areal P loading from uncontrollable sources (seepage, direct basin runoff, spring flow, atmospheric deposition) from 1989 to 94 averaged about 0.06 g P m^{-2} yr^{-1} (Stites et al., 1997) which was equal to estimated areal loading to Lake Apopka in its original, pristine condition (Lowe et al., 1998).

Second, biogeochemical P cycling in Lake Apopka will dampen internal loading from P-rich sediments. Redox potential had minimal effects on soluble reactive P (SRP) levels in sediment porewater from Lake Apopka (Olila and Reddy, 1997). The lake water is calcium rich (mean Ca = 46 mg L^{-1}). Based on chemical extraction, high levels of Ca-Mg-bound P were present in the sediments (Olila et al., 1995). These compounds are poorly soluble in lake water and should be relatively unavailable for algal growth. In addition, high year-round algal productivity, algal sedimentation, and bacterial decomposition create refractory forms of organic P. The majority of P enters the lake as soluble reactive P. However, as a result of chemical and biological processes, almost 80% of the P in surficial sediments was in mineral (33%) or organic (46%) forms resistant to rapid biological uptake (Reddy and Graetz, 1991; Olila et al., 1995). These processes should continue to convert available P in the water column and sediments to refractory P after external loading is reduced.

Recent changes in water chemistry in Lake Apopka provide strong support that our understanding of the functioning of Lake Apopka is accurate, and that reduction in P loading is the correct restoration strategy. Starting in summer 1995, trophic state indicators (TP, TN, TSS, Secchi depth, Chl-*a*) in Lake Apopka significantly improved compared to previous years in an 11-year data set. For example, total P, chlorophyll-*a*, and suspended solids averaged about 30% lower after mid-1995, and Secchi depth averaged about 23% greater (Figs. 22.3 and 22.4). Evaluated with a

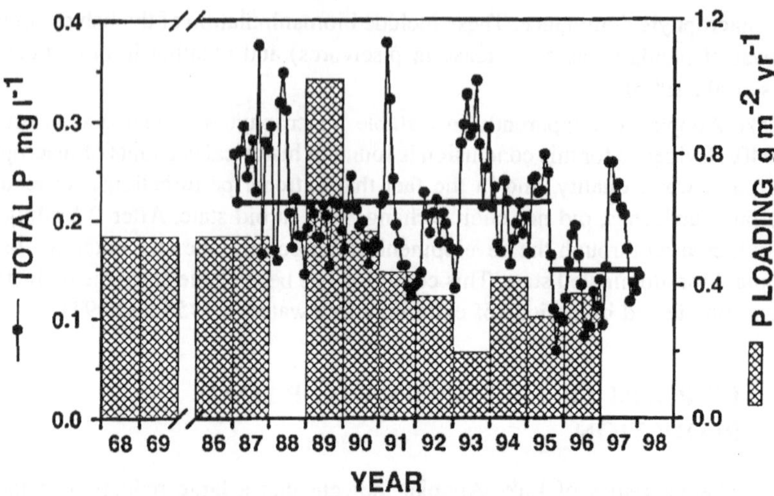

FIGURE 22.3 Annual P loading and mean monthly TP concentrations for Lake Apopka. A mean value for annual loading for the period 1968–87 was derived from sediment stratigraphy (Coveney, 1997), and annual loading for each year 1989–94 was measured (Stites et al., 1997). Annual loading was not measured in 1988 and has not yet been evaluated after 1996. TP concentrations are shown for the SJRWMD period of record January 1987 to the present. Horizontal lines indicate mean TP concentrations for the periods January 1987 to July 1995 and August 1995 to January 1998. Mean values for these two periods differ significantly ($p < 0.005$).

distribution-free resampling technique, mean values for each variable were significantly different before and after mid 1995 ($p < 0.005$).

The most likely explanation for these changes was a decrease in P loading that began around 1992 as a result of improved water management required by regulatory agreements between the SJRWMD and the farmers (Fig. 22.3). In 1993, improved management practices and low summer rainfall resulted in the lowest annual P load to the lake on record (0.20 g P m^{-2} yr^{-1}). Phosphorus loading in subsequent years was higher, but mean annual loading in 1993 to 1996 (0.32 g P m^{-2} yr^{-1}) was still about 40% less than long-term annual loading for 1968 to 1987 (0.55 g P m^{-2} yr^{-1}) derived from sediment studies (Coveney, 1997). A two-year delay between the lowest P loading (1993) and reduced levels of TP in lake water (1995) is reasonable, given internal loading and the mean 2.2-yr hydraulic residence time for 1993 to 1995.

The period after mid 1995 also has seen the spontaneous development of macrophyte (*Vallisneria americana, Chara sp.*) beds at more than 20 sites around the lake. Regrowth of submersed macrophytes in response to increased transparency is the response that we predict to occur when the P concentration in Lake Apopka is reduced below the threshold where the macrophyte-dominated state is favored. The modest improvement in water quality apparent in 1995 likely will not be permanent, since farm loading typically will vary until final discharge limits are met. However, the improved conditions demonstrated that the concentration of P in Lake Apopka will decline following load reduction. With lower P levels, beneficial ecological

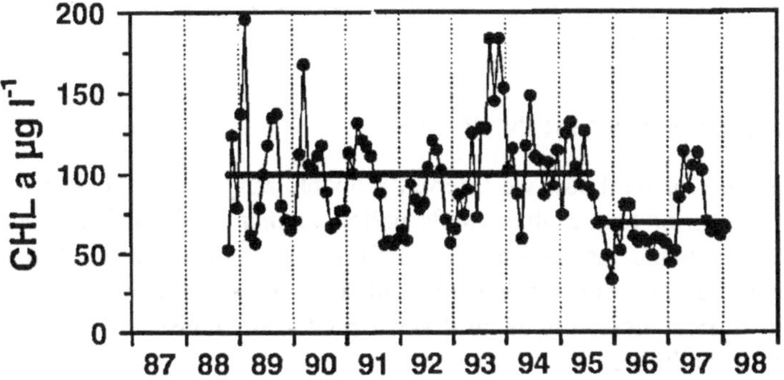

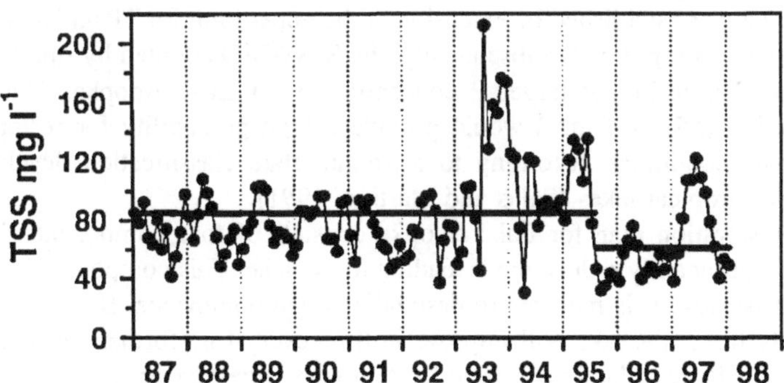

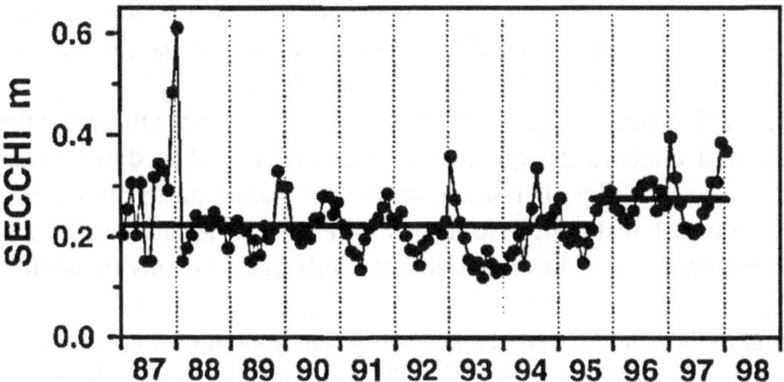

FIGURE 22.4 Mean monthly concentrations of chlorophyll-*a*, total suspended solids, and Secchi depth for Lake Apopka. Data are shown for the SJRWMD period of record January 1987 to the present (chlorophyll data start October 1988). Horizontal lines indicate mean values for the periods January 1987 through July 1995 and August 1995 through January 1998. Mean values for the two periods differ significantly ($p < 0.005$). Secchi depth values in December 1987 and January 1988 likely are erroneous but were not excluded from the analyses.

changes such as lowered algal biomass, increased transparency, and increased growth of macrophytes will occur.

22.8 THE RESTORATION PLAN

Using a variety of methods, we inferred probable ranges for trophic variables for Lake Apopka prior to the after 1940s period of increased nutrient loading from agriculture. Probable ranges were 32 to 51 mg m^{-3} for TP, 8 to 38 mg m^{-3} for chlorophyll-*a*, and 1.39 to 0.76 m for Secchi depth (Lowe et al., 1998). These ranges and the historical descriptions of Lake Apopka are consistent with an earlier mesotrophic condition.

The restoration goal for Lake Apopka is Florida Class III water quality (suitable for recreation and fish and wildlife). Class III standards allow some degradation in water quality from natural background conditions. Therefore, the restoration target range for TP was established by extension of the upper limit for TP under antecedent conditions to 55 mg P m^{-3}. Subsequently, the SJRWMD adopted by rule this upper limit as a P criterion or target P concentration for Lake Apopka. This total P concentration (55 mg P m^{-3}) would provide a high probability for restoration of mesotrophic conditions according to a trophic state classification developed for warm-water tropical lakes (Salas and Martino, 1991).

The restoration plan for Lake Apopka consists of four components. The first step is to significantly reduce the P loading to the lake. This component is the most important because it is required to destabilize algal dominance. If the reduction is sufficiently large, the lake will improve in time even if no further steps are taken. To achieve a large reduction in farm loading (the largest external source of P) the SJRWMD has purchased most of the riparian farms using state and federal funds. Restoration of the approximately 80 km^2 of agricultural lands to wetlands and other aquatic habitat and regulation of any remaining farm P load will decrease overall P loading to about 0.13 g P m^{-2} yr^{-1}. Input-output modeling with a sedimentation coefficient for P developed for Lake Apopka predicts that the resultant equilibrium TP will meet the 55 mg P m^{-3} target (Coveney, 1997).

The remaining three components of the restoration plan will hasten the recovery of the lake and improve the quality of water discharged to downstream lakes. If Lake Apopka follows the alternative stable-state paradigm (Scheffer et al., 1993), then the additional restoration strategies will help to accelerate the shift toward macrophyte dominance as lowered nutrient levels make the phytoplankton state less stable.

The first of these additional components is a treatment wetland to filter suspended solids from the lake water (Lowe et al., 1992). Algae and resuspended lake sediments will be removed by sedimentation in the 14 km^2 wetland. The projected storage of P in the wetland of about 30 × 10^6 g P yr^{-1} is significant compared with measured long-term mean sedimentation of P in Lake Apopka (about 46 × 10^6 g P yr^{-1}; Coveney, 1997; Schelske, 1997). The wetland will have a flow capacity sufficient to treat two lake volumes per year. The performance for six years of a 2 km^2 demonstration wetland has substantiated the nutrient removal ability of the larger wetland (Coveney et al., 1998).

Another component of the restoration plan is the continued removal of gizzard shad (*Dorosoma cepedianum*) from the lake by commercial fisherman. In addition to being a relatively inexpensive method for P removal (less than \$0.05 g^{-1} P, depending on market conditions), the trophic effects of shad removal can include a decrease in the rate at which P is recycled from particulate to soluble forms, a reduction in sediment resuspension, and an increase in zooplankton populations (Schaus et al., 1997). These effects combine to cause a decrease in algal abundance. Gizzard shad dominate the fish community in Lake Apopka and account for 96% to 99% of all fish caught in commercial gill nets. From 1993 through 1997, annual removal of gizzard shad averaged 4.4×10^5 kg (fresh weight), equivalent to 3.1×10^6 g P (Crumpton and Godwin, 1997). A pilot gizzard shad removal project conducted by the SJRWMD in hypereutrophic Lake Denham was followed by reduced P and chlorophyll levels and greater water transparency (SJRWMD, unpublished). There was also increased growth of submersed macrophytes, and increased gamefish recruitment (W. F. Godwin, personal communication). Reduction of planktivorous fish populations, either through their removal or through stocking of predatory fish, is a biomanipulation technique that has been used successfully to promote shifts from phytoplankton to macrophytes in the restoration of shallow European lakes (Moss et al., 1996). For more than a decade, however, it has been recognized that biomanipulation will not be effective in hypereutrophic lakes without reduction of P loading (Benndorf, 1987).

The final component of the restoration plan calls for replanting littoral macrophyte beds around the lake, and 25 sites have been planted with varying success in a pilot-scale effort. Along with submersed macrophyte beds that spontaneously develop, these plantings will help to stabilize the sediments and provide habitat for spawning of gamefish. Theoretical and empirical evidence shows that once a threshold water clarity is achieved in shallow lakes, then growth of submersed macrophytes results in direct and indirect feedback mechanisms to further reduce nutrient levels, algal biomass, and turbidity (Scheffer, 1993). Macrophytes reduce wind-driven resuspension of sediments by stabilizing sediments and damping wind-generated water movements. Less resuspension decreases the flux of nutrients from sediments to water and increases water clarity. The large fetch and shallow depth of Lake Apopka make this step very important. Macrophyte beds provide habitat for piscivorous fish and for macro- and microinvertebrates that prey on phytoplankton, and macrophytes compete with phytoplankton for nutrients. We expect that initial establishment of macrophyte beds will, through these feedback mechanisms, further improve local water clarity so that expansion of macrophytes occurs. Improvement in water quality and littoral habitat in Lake Apopka will be a progressive process.

22.9 CONCLUSIONS

The preponderance of lake management theory and experience indicates that the long-term trophic condition of lakes is determined primarily by P loading, sedimentation, and flushing. Decreased P loading to lakes results in an approximately proportionate decrease in P concentration (Marsden, 1989). Moreover, availability of P in lake water typically controls algal abundance (Schindler, 1974; Dillon and Rigler,

1974; and others). Shallow lakes can occupy different stable states at the extremes of the range of P concentrations. At intermediate P concentrations shallow lakes may be either macrophyte-dominated and clear or phytoplankton-dominated and turbid and can switch between these alternate quasistable states (Scheffer et al., 1993; Klinge et al.,1995).

The large body of experience and theory leads to the conclusion that P load reduction should be the primary tool for restoration of Lake Apopka and other large hypereutrophic lakes. The response to reduced P loading may be small if the reduction in loading is insufficient to limit phytoplankton growth. The response may be slow until internal loading declines. The response may be strongly nonlinear in shallow lakes where multiple feedback loops stabilize the turbid state. These factors should be considered in prediction of the temporal response to P load reduction or of the equilibrium condition. However, these factors do not mean that initial control of P loading is ineffective or unnecessary.

REFERENCES

Aldridge, F. J., C. L. Schelske, and H. J. Carrick. 1993. Nutrient limitation in a hypereutrophic Florida lake. Arch. Hydrobiol. 127:21–37.

Bachmann, R. W. and D. E. Canfield, Jr. 1996. "The big switch: From macrophytes to algae in Lake Apopka, Florida". Abstract published in the Proceedings of NALMS 16th Annual International Symposium on Lake, Reservoir, and Watershed Management, Minneapolis/St. Paul, MN, Nov. 13–16, 1996. p. 59.

Bachmann, R. W., M. V. Hoyer and D. E. Canfield, Jr. 1997. "Can watershed management restore shallow, Florida lakes?". Abstract published in the Proceedings of the Annual Meeting of the Florida Lake Management Society, "New Perspectives and Tools for Lake and Watershed Management", May 7–9, 1997, Palm Beach Shores, FL. p.115.

Bachmann, R. W., M. V. Hoyer and D. E. Canfield, Jr. 1998. "Fluid mud, and the restoration of Lake Apopka". Abstract published in the Proceedings of the Seventh Annual Southeast Lakes Management Conference, "Integrating Water Resources and Growth into the 21st Century", April 15–18, 1998, Orlando, FL. p. 175.

Benndorf, J. 1987. Food web manipulation without nutrient control: A useful strategy in lake restoration. Schweiz. Z. Hydrol. 49(2):238–247.

Blindow, I., G. Andersson, A. Hargeby, and S. Johansson. 1993. Long-term pattern of alternative stable states in two shallow eutrophic lakes. Freshw. Biol. 30:159–167.

Brenner, M, T. J. Whitmore, and C. L. Schelske. Paleolimnological evaluation of historical trophic state conditions in hypereutrophic Lake Thonotosassa, Florida, USA. Hydrbiologia 331:143–152.

Canfield, D. E. Jr. 1983. Prediction of chlorophyll-*a* concentrations in Florida lakes: The importance of phosphorus and nitrogen. Water Resources Bulletin 19(2):255–262.

Canfield, D. E. Jr., M. Hoyer, and R. W. Bachmann. 1996. "Why nutrient control won't restore Lake Apopka". Abstract published in the Proceedings of NALMS 16th Annual International Symposium on Lake, Reservoir, and Watershed Management, Minneapolis/St. Paul, MN, Nov. 13–16, 1996. p. 65.

Canfield, D. E. and L. Hodgson. 1983. Prediction of Secchi disc depths in Florida lakes: impact of algal biomass and organic color. Hydrobiologia 99:51–60.

Carvalho, L., M. Beklioglu, and B. Moss, 1995. Changes in a deep lake following sewage diversion—A challenge to the orthodoxy of external phosphorus control as a restoration strategy? Freshwater Biology, 34:399–410.

Chapra, S. and R. Canale. 1991. Long-term phenomenological model of phosphorus and oxygen for stratified lakes. Water Res., 25(6):707–715.

Clugston, J. P. 1963. Lake Apopka, Florida, A changing lake and its vegetation. Florida Acad. Sci., 26(2):168–174.

Conrow, R. C., W. F. Godwin, M. F. Coveney, and L. E. Battoe. 1993. Surface water improvement and management plan for Lake Apopka. St. Johns River Water Management District, Palatka, Florida. 163 p.

Cooke, G., E. Welch, S. A. Peterson, and P. R. Newroth, 1993. Restoration and Management of Lakes and Reservoirs. 2nd edition. Lewis Publishers, Boca Raton, FL. 548 p.

Coveney, M. F. 1997. Sedimentary phosphorus stores, accumulation rates, and sedimentation coefficients in Lake Apopka: Prediction of the allowable phosphorus loading rate. Draft Technical Memorandum. St. Johns River Water Management District, Palatka, Florida.

Coveney, M. F., D. L. Stites, E. F. Lowe, L. E. Battoe, and R. Conrow. 1998. Nutrient removal from eutrophic lake water by wetland filtration. In preparation.

Crumpton, J. E. and W. F. Godwin. 1997. Rough fish harvesting in Lake Apopka, summary report, 1993–1997. St. Johns River Water Management District Special Publication SJ97-SP23, 24 p.

Dillon, P. J. and F. H. Rigler. 1974. The phosphorus-chlorophyll relationship in lakes. Limnol. Oceanogr.19:767–773.

Henderson-Sellers, B. and H. R. Markland. 1987. *Decaying Lakes*. John Wiley & Sons, Ltd., Great Britain. 254 p.

Huber, W., P. Brezonik, J. Heaney, R. Dickinson, S. Preston, D. Dwornik, and M. DeMaio. 1982. A classification of Florida Lakes. Publ. No. 72, Water Resources Research Center, Dept. Env. Eng. Sci., Univ. Fl. Gainesville. 529 p.

James, R. T., K. O'Dell, and V. H. Smith. 1994. Water quality trends in Lake Tohopekaliga, Florida, USA: Responses to watershed management. Water Resources Bulletin 30(3):531–546.

Jeppesen, E., Jensen, J. P., Kristensen, P., Søndergaard, M., Mortensen, E., Sortkjaer, O., and Olrik K. 1990. Fish manipulation as a lake restoration tool in shallow, eutrophic, temperate lakes. 2: Threshold levels, long-term stability and conclusions. Hydrobiologia 200/201:219–227.

Klinge, M., M. P. Grimm, and S. H. Hosper. 1995. Eutrophication and ecological rehabilitation of Dutch lakes: Presentation of a new conceptual framework. Wat. Sci. Tech. 31(8):207–218.

Kopp, J. F. and G. D. McKee. 1983. Methods for chemical analysis of water and wastes. United States Environmental Protection Agency Report No. EPA-600/4-79-020. 521 p.

Lowe, E. F., L. E. Battoe, M. F. Coveney, and D. L. Stites. 1998. Setting water quality goals for restoration of Lake Apopka: Inferring past conditions. Submitted to Lake and Reservoir Management.

Lowe, E. F., L. E. Battoe, D. L. Stites, and M. F. Coveney. 1992. Particulate phosphorus removal via wetland filtration: an examination of potential for hypertrophic lake restoration. Environ. Management 16:67–74.

Marsden, M. W. 1989. Lake restoration by reducing external phosphorus loading: the influence of sediment phosphorus release. Freshwater Biol.21:139–162.

Moss, B., J. Madgwick and G. Phillips. 1996. *A Guide to the restoration of nutrient-enriched shallow lakes*. W.W. Hawes, UK. 179 p.

Olila, O. G., and K. R. Reddy. 1997. Influence of redox potential on phosphate-uptake by sediments in two subtropical eutrophic lakes. Hydrobiologia 345:45–57.

Olila, O. G., K. R. Reddy, and W. G. Harris, Jr. 1995. Forms and distribution of inorganic phosphorus in sediments of two shallow eutrophic lakes in Florida. Hydrobiologia 302:147–161.

Reckhow, K. H., and S. C. Chapra. 1983. Empirical models for lake trophic state evaluation. In: Engineering Approaches for Lake Management, Vol. 1. Data analysis and empirical modeling. Butterworth Publishers, Boston. 340 p.

Reddy, K. R., and D. A. Graetz. 1991. Internal nutrient budget for Lake Apopka. Special Publication SJ 91-SP-6, plus Addendum. St. Johns River Water Management District. Palatka, Florida.

Salas, H. J., and P. Martino. 1991. A simplified phosphorus trophic state model for warm-water tropical lakes. Wat. Res. 25:341–350.

Sas, H. 1989. *Lake restoration by reduction of nutrient loading: Expectations, experiences, extrapolations*. Academia Verlag Richarz. Sankt Augustin. 497 p.

Schaus, M. H., M. J. Vanni, T. E. Wissing, M. T. Bremigan, J. E. Garvey, and R. A. Stein. 1997. Nitrogen and phosphorus excretion by detritivorous gizzard shad in a reservoir ecosystem. Limnol. Oceanogr. 42:1386–1397.

Scheffer, M., S. H. Hosper, M.-L. Meijer, B. Moss, and E. Jeppesen. 1993. Alternative equilibria in shallow lakes. TREE 8:275–279.

Schelske, C. L. 1997. Sediment and phosphorus deposition in Lake Apopka. Special Publication SJ97-SP21. St. Johns River Water Management District. Palatka, Florida. 208 p.

Schelske, C. L., F. J. Aldridge, and H. J. Carrick. 1992. Phytoplankton-nutrient interactions in Lake Apopka. Special Publication SJ92-SP9. St. Johns River Water Management District. Palatka, Florida. 181 p.

Schelske, C. L., and P. Brezonik. 1992. Restoration case studies. Can Lake Apopka be restored? In: *Restoration of Aquatic Ecosystems: Science, Technology, and Public Policy.* Report of Committee on Restoration of Aquatic Ecosystems, National Research Council, National Academy Press, Washington, D.C. p. 393–398.

Schindler, D. W. 1974. Eutrophication and recovery in experimental lakes: Implications for lake management. Science 184:897–899.

Schneider, R. F., and J. A. Little. 1969. Characterization of bottom sediments and selected nitrogen and phosphorus sources in Lake Apopka, Florida. Report for the U. S. Dept. of the Interior, Federal Water Pollution Control Administration, Southeast Water Laboratory, Technical Programs, Athens, GA. 35+ p.

Stites, D. L., M. F. Coveney, L. E. Battoe, and E. F. Lowe. 1997. An external phosphorus budget for Lake Apopka. Draft Technical Memorandum. St. Johns River Water Management District. Palatka, Florida.

Southwest Florida Water Management District (SWFWMD). 1992. Final Report: Lake Thonotosassa diagnostic feasibility study—Southwest Florida Water Management District, SWIM Department. Produced under contract with the Dynamac Corporation.

United States Environmental Protection Agency (USEPA), 1978. Draft Environmental Impact Statement. Lake Apopka Restoration Project, Lake and Orange Counties, Florida. USEPA, Region 4, Atlanta, Georgia.

Welsh, E. and G. Cooke. 1995. Internal phosphorus loading in shallow lakes: Importance and control. Lake and Reservoir Management, 11(3):273–281.

23 Phosphorus in Lake Okeechobee: Sources, Sinks, and Strategies

Alan D. Steinman, Karl E. Havens, Nicholas G. Aumen, R. Thomas James, Kang-Ren Jin, Joyce Zhang, and Barry H. Rosen

23.1 ABSTRACT

Increased phosphorus loads to Lake Okeechobee, the largest lake in Florida, have resulted in higher total phosphorus concentrations in the sediments and water column of the lake's pelagic region since the mid-1900s. In this chapter, we examine the role of phosphorus (P) in this dynamic and heterogeneous ecosystem. The chapter is subdivided into six sections: (1) lake and watershed description, (2) phosphorus in lake sediments, (3) pelagic-littoral zone interactions, (4) phosphorus content in different pools, (5) modeling lake responses to watershed management of phosphorus, and (6) strategies for reducing phosphorus in the future.

Given the large amount of phosphorus stored in the lake's sediments (2.87×10^7 kg) and its frequent resuspension, it may take a considerable period of time to measure noticeable reductions in P concentration in the lake's water column. However, the sooner P loads are reduced from sources in the watershed, the sooner recovery may begin. Progress has been made, in cooperation with the farmers and ranchers in the watershed, to reduce loads over the past two decades, but further reductions still are needed. In-lake P removal strategies, such as dredging, do not appear to be viable in Lake Okeechobee; both ecological and economic uncertainties restrict their utility and, at present, we discourage their implementation.

1-56670-331-X/99/$0.00+$.50

23.2 INTRODUCTION

Lake Okeechobee, Florida (USA) is a multipurpose regional water body, serving the following functions:

- flood control
- irrigation
- municipal and industrial water supply
- enhancement of fish and wildlife resources
- navigation
- prevention of saltwater intrusion for coastal well fields
- recreation and ecotourism
- water supply to Everglades National Park (Aumen 1995)

As a consequence, the environmental quality of this lake is of major concern to the residents of the region.

Phosphorus loading from agricultural activities in the Lake Okeechobee watershed has resulted in increased total phosphorus (TP) concentrations in the lake since the early 1970s (Flaig and Havens, 1995; James et al., 1995b). Increased frequency of algal blooms, and associated deterioration of water quality, has warranted increased attention into the sources, sinks, and cycling of phosphorus in this water body. Although increased P concentration is widely recognized as a potential problem in freshwater ecosystems (Wetzel, 1983), developing and implementing solutions to shallow lake eutrophication often is an elusive task (Hofstra and Van Liere, 1992; Nixdorf and Deneke, 1997).

In this chapter, we present an overview of P dynamics in Lake Okeechobee in sections covering (1) lake and watershed description, (2) phosphorus in lake sediments, (3) pelagic-littoral zone interactions, (4) phosphorus content in different pools, (5) modeling lake responses to watershed management of phosphorus, and (6) strategies for reducing phosphorus in the future. We focus primarily on in-lake processes, but we recognize the importance of understanding watershed dynamics as well. We agree with the statement of Flaig and Reddy (1995), that "effective P control strategies can only be implemented if we understand the fate and transport of P in the uplands, wetlands, and streams of the watershed."

23.3 LAKE AND WATERSHED DESCRIPTION

Lake Okeechobee has a surface area of 1732 km^2, making it the third largest lake located entirely within the United States (Herdendorf, 1982). It is extremely shallow, with a mean depth of 2.7 m, resulting in a large surface area-to-volume ratio and making the lake very susceptible to the influence of wind-induced sediment resuspension (Maceina and Soballe, 1990). An earthen levee, initiated in the early 1900s and completed in 1967, now encircles the lake, and 32 water control structures serve to regulate inflows and outflows (Fig. 23.1). Prior to construction of the levee, the lake was much larger, had a vast littoral marsh to the south and west, and outflow occurred as a several kilometer wide sheet flow directly into the Everglades (Havens et al., 1996a).

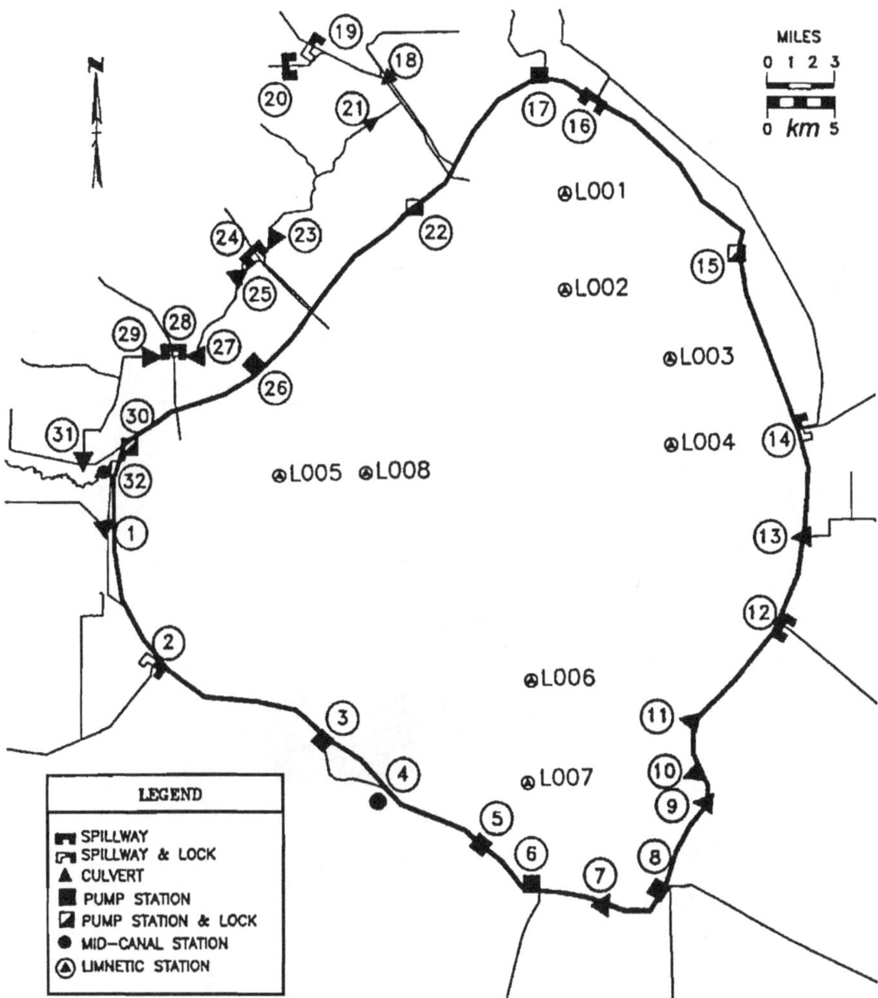

FIGURE 23.1 Map of Lake Okeechobee showing the monitored inflow/outflow structures (circles 1–32) and the long-term monitoring stations (L001–L008) (after James et al., 1995).

The lake consists of two adjacent but hydrologically uncoupled parts: a 1332 km^2 pelagic zone, with a mean TP concentration of approximately 95 μg L^{-1}, and a 400 km^2 littoral zone, with a mean TP concentration of 5 to 10 μg L^{-1} (Steinman et al., 1998). The pelagic zone is underlain by a mosaic of sediment types, which correspond loosely to different ecological zones (Olila and Reddy, 1993; Phlips et al., 1995). Mud is the dominant sediment in the central and north regions of the lake (Fig. 23.2) where easily resuspended sediments often result in light limitation of phytoplankton growth (Phlips et al., 1997). Sand sediments dominate in a transitional region between the central area and the pelagic/littoral ecotone. Sand, rock, and peat underlie the ecotone area. The sediments in the littoral region are heterogeneous in

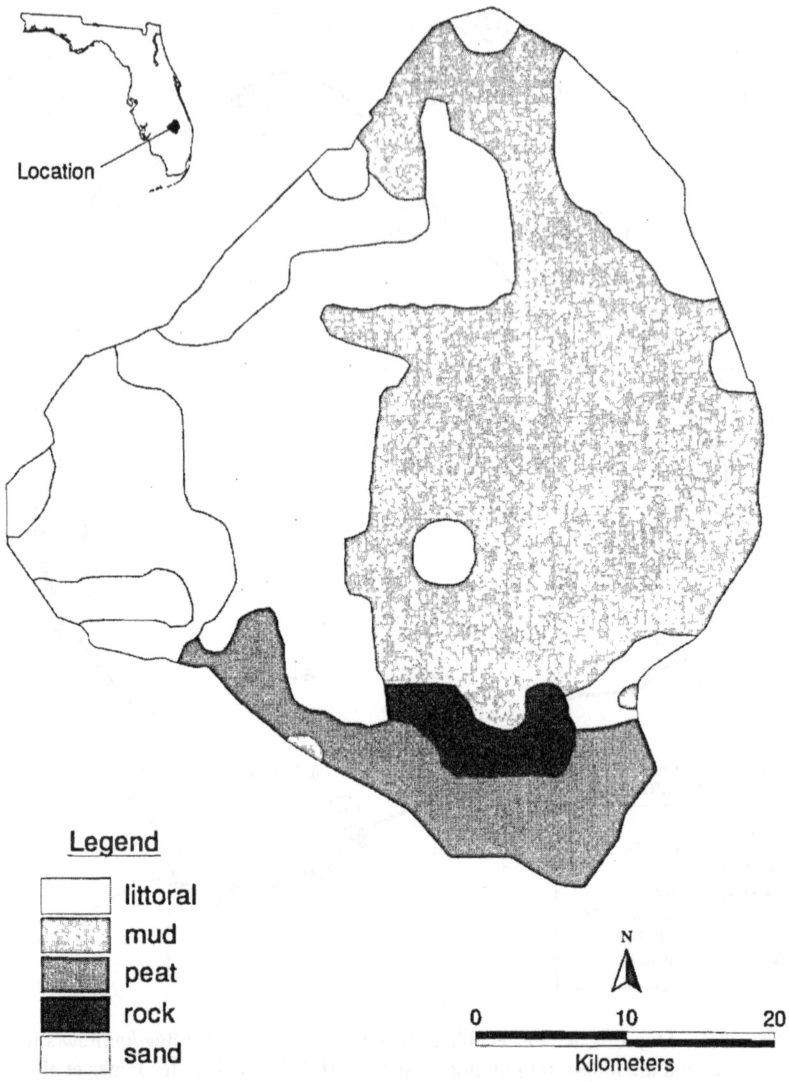

FIGURE 23.2 Distribution of sediment types in Lake Okeechobee (after Reddy et al., 1995). The spatial coverage of each sediment type, in terms of relative abundance, is: muds, 44.3%; sand, 24.5%; rocky reef, 3.7%; peat, 8.9%; and littoral, 18.6%.

composition (Olila and Reddy, 1993). Dating of sediments from the mud zone of Lake Okeechobee using ^{210}Pb has revealed that sediment accumulation rates have increased by an average of two-fold during this century (Brezonik and Engstrom, 1998). Phosphorus accumulation rates during this century have increased approximately four-fold, from approx. 250 mg P m^{-2} yr^{-1} pre-1910 to approx. 1000 mg P m^{-2} yr^{-1} in the 1980s (Reddy et al., 1995). This compares to an average rate of 360 mg P m^{-2} yr^{-1} over the past 50 years in Lake Apopka, FL (Schelske, 1997).

Enclosure of the lake by the levee reduced the area of the historical littoral zone, although it still covers almost 400 km^2. The littoral vegetation is a diverse mosaic of emergent, floating-leaved, and submersed macrophytes (Richardson and Harris, 1995). However, rapid expansion of the invasive exotic plant, torpedo grass (*Panicum repens*), is threatening the diversity of the plant community in the littoral zone (Hanlon, C., personal communication, 1997).

The drainage basin of Lake Okeechobee covers approximately 22,500 km^2. Land use is largely agricultural, with beef cattle ranching and dairy farming accounting for over 50% of land use in the watershed (Flaig and Havens, 1995). For the north Okeechobee watershed, total net P imports were estimated at 2380 metric tons (t) P yr^{-1}, of which approximately 300 t P yr^{-1} reach the lake (Boggess et al., 1995). Phosphorus imports in the form of dairy cow feed and pasture fertilizer were the primary sources of P in the northern watershed (Boggess et al., 1995).

The primary sources of water to the lake are direct precipitation (39%) and surface inflows from the Kissimmee River (31%). The primary losses of water from the lake are evapotranspiration (66%) and discharges to canals draining south to the Everglades region (22%) (James et al., 1995a). A total phosphorus budget, based on a 20-yr period of record (1973 to 1992) revealed that inputs and outputs average 518 and 137 t per year, respectively, and that the lake acted as a net sink during 19 of those 20 years (James et al., 1995a). The mean net sedimentation coefficient for phosphorus was 1.25 yr^{-1} (James et al., 1995a).

23.4 PHOSPHORUS IN LAKE SEDIMENTS

Lake Okeechobee is comprised of several sediment types (Fig. 23.2), each of which is characterized by different P storage, uptake, exchange, and assimilative capacities. For example, of the 2.87×10^7 kg of TP stored in lake sediments, the TP distribution is 42% in muds, 41% in sands, 14% in littoral sediments, and 3% in peat (Reddy et al., 1995). Concentrations of readily available P (NH_4Cl-extractable) are 2% of TP in mud sediments, and 9.7% and 17.4% of TP in littoral and peat sediments, respectively (Olila et al., 1995). In contrast, Ca- and Mg-bound P account for 65% of TP in the mud sediments, and 28% and 41% of TP in the littoral and peat sediments, respectively (Olila et al., 1995).

The phosphate retention capacity of the sediments is: mud > littoral > peat > sand (Olila and Reddy, 1993). Batch incubation experiments were conducted to determine P adsorption coefficients and equilibrium P concentrations (EPC) for different sediment types (Olila and Reddy, 1993). The EPC is the concentration at which P sorption by solid phase is equal to desorption; these values are used to determine the direction of P movement between sediment and the overlying water column. EPC values in the 0 to 5 cm sediment layer averaged approximately 38 μg L^{-1} in the peat and sand sediments, 25 μg L^{-1} in the mud sediments, and 3 μg L^{-1} in littoral sediments (Table 23.1); if the water column P concentrations are lower than these sediment EPC values, then the sediments at that location will function as a source of P to the water column. These data suggest that the capacity of the sediments to retain P is spatially highly variable, but given the current TP concentrations in the lake, the sediments should generally act as a net sink for P.

TABLE 23.1
Mean Phosphorus Flux (mg P m^{-2} d^{-1}) and Range from Different Sediment Types to the Overlying Water Column in Lake Okeechobee[a] and Equilibrium Phosphorus Concentration (EPC, µg L^{-1}). Values for 0–5 cm Sediment Depth[b]

Sediment type	P flux	Range	EPC
Littoral	1.09	0.64–1.54	3.1
Peat	0.91	0.16–2.22	40.3
Mud	0.70	0.14–1.89	24.9
Sand	0.29	0.11–0.52	37.2

[a]Data from Reddy et al., 1995
[b]Data from Moore and Reddy, 1994

However, under certain conditions, the sediments also can act as a source of P (see below).

Long-term batch incubation experiments indicated that Lake Okeechobee sediments are capable of assimilating large quantities of P under high loading conditions (Reddy, 1991). Bulk surface sediment samples were thoroughly mixed with filtered lake water containing known amounts of soluble reactive phosphorus (SRP). In the 10 mg P L^{-1} treatments, porewater SRP decreased to below 0.1 mg P L^{-1} in the mud, littoral, and sand sediments after six months, but was approx. 1.5 mg L^{-1} in the peat sediments even after one year. Reddy (1991) concluded that the removal or sorption of P from solution by the solid phase of the sediments in Lake Okeechobee is regulated largely by amorphous oxides of iron and aluminum, and that P release from most sediments is highly dependent on redox conditions (as Fe controls P geochemistry). This does not apply to peat sediments, however; they apparently do not contain sufficient Fe to influence P behavior.

Stirred mesocosm experiments were used to examine the effect of redox potential (Eh) and pH on P mobility in the sediments. Dissolved P concentrations were about an order of magnitude lower under oxidized compared to reduced conditions, and these declines were coupled with decreases in dissolved Fe (Moore and Reddy, 1994). These results indicate that precipitation of ferric phosphate or phosphate adsorption by Fe oxides was the removal mechanism. Mineral equilibria calculations, and P fractionation data that showed Fe-bound P increased under oxidized conditions at the expense of Ca-bound P, suggest that Fe controls P behavior under oxidizing conditions in Lake Okeechobee sediments, but that Ca-phosphate precipitation governs P solubility under reducing conditions (Moore and Reddy, 1994).

Organic phosphorus comprises 20 to 75% of the total P in Lake Okeechobee sediments, and decomposition studies revealed that mineralization of this organic P can contribute significantly to the overall P flux from sediments (Reddy, 1991). These studies showed that this flux was particularly important in mud and peat sediments, was not important in sand sediments, and was variable both in space and time in littoral sediments.

Phosphorus flux from Lake Okeechobee sediments to the overlying water column also varies with sediment type, according to the following order: littoral > peat > mud > sand, with an overall range in P flux being 0.14 to 2.22 mg P m^{-2} d^{-1} (Table 23.1; Reddy et al., 1995). Phosphorus flux was several orders of magnitude greater when the overlying water was anaerobic vs. aerobic (Reddy, 1991). Another source of P flux to and from sediments is bioturbation. Laboratory experiments using cores of sand, mud, sand, and littoral sediments revealed that diffusion coefficient values, which represent the pooled contributions of molecular diffusion and solute mixing from benthic activity, were 1.6 to 15 times greater in sediments with macrobenthic populations than sediments without benthic activity (Van Rees et al., 1996). Finally, it was estimated that rates of internal P loading from bottom sediments are approximately equal to the rates of external P loading (approx. 1 mg P m^{-2} d^{-1}), indicating the critical role of internal loading in the P dynamics of Lake Okeechobee (Reddy, 1991).

Laboratory incubations indicated that the influence of sediment resuspension on inorganic P release into the water column was complex. Once suspended in the water column, soluble P release rates under aerobic conditions were higher in sand and peat sediments than in mud and littoral sediments (Reddy, 1991). However, the total capacity of sand and peat to release P is low because of their relative inability to resuspend into the overlying water column. The lower release rates in mud and littoral sediments were likely attributable to their high concentrations of extractable Fe; mixing of anaerobic sediments with the aerobic water column can oxidize Fe^{2+} to Fe^{3+} and precipitate soluble P as ferric phosphate. Under low total suspended solid (TSS) concentrations (< 2 g L^{-1}) and low DO levels (< 1 mg L^{-1}), the P flux from resuspension was about 6 to 18 times the diffusive flux measured from the same sediments. Under similarly low TSS concentrations but oxygenated water column conditions, the suspended sediment particles served as a sink, and soluble P concentrations declined in the water column because adsorption and precipitation (Reddy, 1991). These results suggest that soluble P release may not occur during every sediment resuspension event. Factors such as water column and sediment pH, as well as SRP concentration, sediment redox values, and chemical properties of the sediment will influence whether the suspended sediments function as a source or a sink (Bostrom et al., 1982, Reddy, 1991). Research currently is being conducted *in situ* to assess diel and seasonal changes in redox and dissolved oxygen in order to validate the laboratory-generated estimates of internal loading in the lake.

23.5 PELAGIC-LITTORAL ZONE INTERACTIONS

The pelagic and littoral regions are functionally distinct, despite being adjacent to each other. One of the reasons for this is the hydrologic uncoupling between the two regions. Sheng (1991) noted that because circulation in the open water zone tends to flow parallel (rather than perpendicular) to the vegetation ecotone, exchange of P between the regions generally is weak. Advection of P from the pelagic into the littoral zone occurs only during high lake stage periods, when perpendicular flow into the littoral zone is enhanced. Mesocosm studies, conducted deep within the marsh area, have shown that the vegetation in this region is a major sink for phosphorus. P additions of up to 700 μg L^{-1} are virtually undetectable within days of addition, with most of the nutrient being tied up in the surface algal mat (Havens, K., unpublished data, 1996).

It also is possible that the littoral zone might act periodically as a source of P to the pelagic region, given the amount of phosphorus stored in the aquatic vegetation. However, calculations by Dierberg (1992), assuming no immobilization or recycling of P within the littoral zone and complete transport from the littoral to the pelagic zone, revealed that release and transport of P from the vegetation could increase TP in the open water by a maximum of 11 μg L^{-1}. Once sediment immobilization was taken into account, the increase in TP was only 1 μg L^{-1}. However, it is possible that remineralization of P may have localized effects, particularly in the ecotone region, where competition for nutrients between phytoplankton and the attached algae/macrophyte complex can be strong (Phlips et al., 1993; Havens et al., 1996b).

Competitive interactions for P may vary spatially within the littoral zone, as well. Recent studies conducted in several locations in the littoral zone revealed that phytoplankton and attached algae were P-limited during summer months only at sites along the pelagic-littoral ecotone, but that these communities were P limited year round at a site deep within the littoral marsh (Hwang, S.-J., Havens, K.E., and Steinman, A.D., unpublished data, 1997). Phosphorus kinetics data were consistent with this pattern, as the ratios of maximum uptake velocity (V_m) to half saturation constant (K_s) for both plankton and periphyton were usually greater at the marsh site than the ecotone sites. In addition, the P maximum uptake rates of plankton increased over 1500% during the P-limited summer months compared to the P-replete winter months at the ecotone sites. Interestingly, the marsh site was capable of sustaining high periphyton biomass levels, despite much lower ambient P concentrations, presumably because of efficient cycling of P and the absence of light limitation.

23.6 PHOSPHORUS POOLS

Using available data, we have calculated the amounts of phosphorus stored in the major compartments of Lake Okeechobee. Given that the littoral and pelagic zones represent two very different regions of the lake, separate estimates were made for each of these zones. In both regions, the vast majority (> 94%) of P is stored in the sediments (Table 23.2), calculated based on a sediment depth of 10 cm (Reddy et al., 1995). In the pelagic zone, the only other compartment with > 1% of P is the abiotic particulate fraction; the total quantity of P in this fraction is dynamic and related to periodic resuspension by wind. In the littoral zone, the fish and macrophytes account for 2.7 and 2.5% of P, respectively (Table 23.2). Some of the data seem counterintuitive; for example, one would expect absolute P storage in macrophytes to be greater in the littoral region than in the pelagic region. However, the pelagic data are based on information in Zimba et al. (1995), which reflect the very high biomass of submerged aquatic vegetation (SAV) that existed in the pelagic region (adjacent to the littoral zone) during a postdrought period in the early 1990s. High water levels in recent years have resulted in a marked decline in these com-

TABLE 23.2
Estimates of Phosphorus Mass (kg) in Different Compartments of the Open Water and Littoral Zones of Lake Okeechobee

	Open-water region			Littoral region		
Compartment	**Mass**	**Percent**	**Note**	**Mass**	**Percent**	**Note**
Water column						
Dissolved P	8.40×10^4	0.3	1	2.33×10^3	0.1	9
Abiotic PP	3.36×10^5	1.3	2	2.02×10^3	<0.1	10
Sediments						
Abiotic	2.47×10^7	97.4	3	4.02×10^6	94.2	3
Biotic	0.52×10^3	<0.1	4	0.66×10^3	<0.1	4
Biota						
Phytoplankton	4.61×10^4	0.2	5	3.00×10^3	0.1	11
Zooplankton	1.55×10^3	<0.1	6	2.70×10^1	<0.1	12
Macrophytes	1.25×10^5	0.5	7	1.08×10^5	2.5	13
Periphyton	6.10×10^4	0.2	8	1.90×10^4	0.4	11
Invertebrates	ND			ND		
Fish	ND			1.14×10^5	2.7	14

[1]SFWMD water quality data for dissolved PO_4, concentration multiplied by volume of water in littoral zone.

[2]SFWMD water quality data for TP, minus P content of 5 and 6 (below), concentration multiplied by volume of water in littoral zone.

[3]Estimate from Reddy et al. (1995), based on sediment depth of 10 cm.

[4]Unpublished data (A. D. Steinman et al., SFWMD), core depths averaged 5 cm.

[5]Havens and East (1997) plankton carbon data, C:P ratio of 50:1 from WASP model, concentration multiplied by volume of water in littoral zone.

[6]Havens and East (1997) plankton carbon data, Anderson and Hessen (1991) copepod C:P ratio of 212:1, concentration multiplied by volume of water in littoral zone.

[7]Estimate from Zimba et al. (1995).

[8]Estimate from Zimba et al. (1995).

[9]SFWMD water quality data for dissolved PO_4, concentration multiplied by volume of water in pelagic zone.

[10]SFWMD water quality data for TP, minus P content of 11 and 12 (below), concentration multiplied by volume of water in pelagic zone.

[11]Unpublished data (S.-J. Hwang et al., SFWMD).

[12]Unpublished data (K. E. Havens et al., SFWMD).

[13]Based on FDEP estimates for area (R. Kipker, pers. comm.), Harris et al. (1995) for P standing stock of *Eleocharis*, *Panicum*, *Typha*, and *Scirpus* (39% of total coverage), and mean P standing stock (Harris et al. 1995) applied to other spp. (61% of coverage). SAV estimates within littoral zone from Zimba et al. (1995).

[14]P content of selected fish taxa provided by D. Fox (FGFWFC), multiplied by areal biomass data from Bachmann et al. (1996), and littoral regions from Richardson and Harris (1995).

munities (Steinman et al., 1997), and presumably much lower P storage. In addition, the greater absolute levels of P in pelagic periphyton compared to littoral periphyton seem anomalous but again reflect the high epiphyton load associated with SAV in the early 1990s. We report no data for P storage in fish in the open water region, because available fisheries data are based on catch per unit effort, which cannot be converted reliably to densities. In the littoral zone, a relatively large portion of biotic P is stored in the fish. Although there is considerable uncertainty inherent in these extrapolated values, the biotic P fractions are clearly substantially lower than the abiotic P fractions, again reflecting the important role of sediments in the storage of P in Lake Okeechobee.

23.7 MODELING LAKE RESPONSES TO WATERSHED P MANAGEMENT

In the last decade, watershed and in-lake models have been developed that permit us to estimate how changes in land use in the Lake Okeechobee watershed affect P runoff rates, P loads to the lake, and in-lake water quality conditions. The watershed model, known as the Lake Okeechobee Agricultural Decision Support System (LOADSS) is described in detail in Negahban et al. (1994); it is a GIS-based decision support system that evaluates the effectiveness of different land use and phosphorus control practice (PCP) combinations in the basin for reducing P loads to the lake. The in-lake model, known as the Lake Okeechobee Water Quality Model (LOWQM), is used to relate external loading rates to water column conditions. It uses an enhanced version of the USEPA's Water Quality Analysis Simulation Program (WASP) framework. The details of LOWQM are given in James et al. (1997); LOWQM simulates the transformations of organic and inorganic materials in the lake based on input data regarding nutrient and solute loads, water budgets, and solar insolation. The present version of the model considers the pelagic region of the lake as a single mixed unit, but a revised version of the model (under development) partitions the lake into five discrete "ecological zones" to more accurately reflect the lake's natural heterogeneity.

We used the two models in combination to predict how specified changes in phosphorus control practice might ultimately impact the following ecologically relevant attributes within the lake: water column TP, SRP, and chlorophyll-*a*. The latter variable is most directly related to the lake's ecological and societal values because high levels of chlorophyll-*a* correspond with a greater frequency of noxious cyanobacteria blooms (Walker and Havens, 1995). These blooms, in turn, can harm the natural biota, cause taste and odor problems in drinking water supplies, and degrade the quality of the lake for recreational uses.

We considered three combinations of changes in land use and PCPs in the Lake Okeechobee watershed, and evaluated their impacts on total P loading to the lake using LOADSS. The predicted load (Table 23.3) was modified from a base plan, developed in a previous exercise using the LOADSS model (Negahban et al., 1994), and deals with dairy and beef cattle operations in the watershed's S-154, S-191, S-65D, and S-65E basins (Fig. 23.3). These regions presently are responsible for much

TABLE 23.3
Measured and Predicted Phosphorus Loads to Lake Okeechobee

Basin	Predicted Load (t)	Plan I (t)	Plan II (t)	Plan III (t)
S-154	20.6	13.0	11.9	7.1
S-191	92.3	67.4	55.4	37.5
S-65D	27.3	18.7	16.7	11.1
S-65E	14.3	10.2	9.0	6.0
Total	155.0	110.0	93.0	62.0
(Reduction)		(44)	(61)	(93)

of the excessive P loading to the lake and are considered prime targets for additional management actions in the Lake Okeechobee Surface Water Improvement and Management (SWIM) Plan (SFWMD, 1997). Of the various land uses in the watershed, beef cattle pastures are considered to be one of the largest remaining uncontrolled sources of P (Fluck et al., 1992). The predicted load has dairies operating according to their existing (1991) design, and beef cattle pastures being fertilized at a rate of 19 kg P ha^{-1} y^{-1}.

The first PCP (Plan I) reduces beef pasture fertilization rates from 19 to 12 kg P ha^{-1} y^{-1} (Table 23.3). According to the LOADSS output, a total lake P load reduction of 44 t y^{-1} could be expected with this practice. Plan II includes the same reduced P fertilization rate for beef pastures and hypothetical changes to dairy practices, including a confinement collection system, a chemical treatment with spray field application, and compost and sale for all dairy solid wastes. A total lake P load reduction of 61 t y^{-1} is predicted under this scenario. Plan III includes all of the dairy PCPs listed for plan II, and a no P-fertilization practice for beef cattle. The maximal total P load reduction, 93 t y^{-1}, is predicted under this final scenario (Table 23.4). This amount of P matches, in general, the "over target" loading to the lake, which is the difference between actual loads and the loading target specified in the SWIM Act. However, a complete elimination of P fertilizer use for beef cattle pastures may be an unrealistic PCP. An ongoing experimental research project (see below), designed to optimize beef cattle operations but minimize nutrient runoff in surface waters, is being conducted in collaboration with the South Florida Water Management District, University of Florida's Institute of Food and Agricultural Science, Florida Cattlemen's Association, and the Natural Resource Conservation Service. Results from this research may provide new approaches to meet the SWIM-mandated P loading target.

We used the maximal load reduction (93 t) obtained from LOADSS, and existing loading rates from other regions of the watershed, to generate an estimate of average inflow P concentration to the lake. This value (along with a baseline value reflecting loads without the PCP and landuse changes) was used as input for the WASP model, which gave output data for the in-lake water quality attributes listed above. The maximal load reduction at S-154, S-191, S-65D, and S65-E corresponded to a 20%

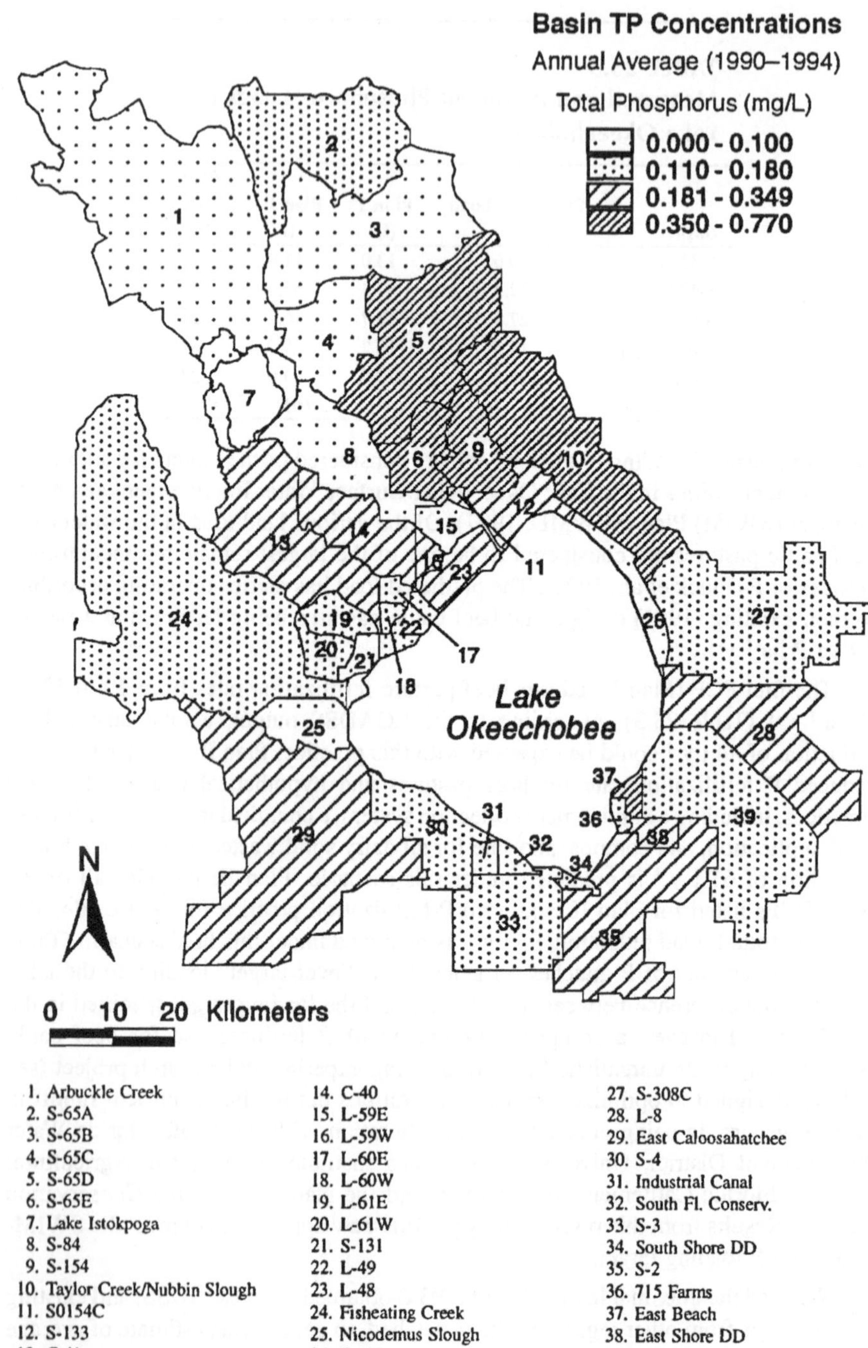

FIGURE 23.3 Contributions of phosphorus from individual basins in the Lake Okeechobee watershed (after SFWMD, 1997).

TABLE 23.4
Comparison of Model Results from Two Alternative Scenarios of the Lake Okeechobee Water Quality Model: Base (inflow concentration = 154 μg P L^{-1}) and Reduction (inflow concentration = 120 μg P L^{-1})

Parameter	Base	Reduction	Charge (%)
Four-basin load reduction (metric tons yr^{-1})	381	305	−76 (−20%)
Total load to lake: 24 yr (kg)	1.07×10^7	8.88×10^6	-1.82×10^6 (−17.0%)
Average water column TP (μg L^{-1})	82	78	−4 (−5.6%)
Average water column SRP (μg L^{-1})	30	26	−4 (−12.1%)
Average water column chl *a* (μg L^{-1})	25.9	25.4	−0.5 (−2.3%)

surface load reduction for the entire lake, from 381 to 305 t y^{-1}, and a total loading reduction (including rainfall inputs) of 17% (Table 23.4). The corresponding reduction in surface inflow total P concentration was from 154 to 123 μg L^{-1}. When considered in the context of a 24-year period of record (the 1973 to 96 baseline period used in the model runs), this gave a 1,620 t reduction in P accumulated in the lake ecosystem (water and sediments). However, because the load reduction also affected the relative rate of P losses to lake sediments, the water column total P concentration was reduced, on average, by only 5.6% (from 82 to 78 μg L^{-1}) over those 24 years. The corresponding reduction in SRP averaged over the 24 year period of record was from 30 to 26 μg L^{-1} (12%) and chlorophyll-*a* was reduced from 25.9 to 25.4 μg L^{-1} (2%), an insignificant change given the uncertainties associated with the predictions.

Model results support the hypothesis that sediment-water interactions lead to a high degree of resistance to change in this and other shallow eutrophic lakes in response to external load reductions (Sas, 1989; James et al., 1995b).

23.8 STRATEGIES

Strategies to reduce P impacts in Lake Okeechobee can be classified, for convenience, into watershed and in-lake approaches. Several programs already have been implemented to reduce P in runoff from the watershed, and several other new programs are being proposed. The programs already implemented include (1) the Taylor Creek Headwaters Project and the Taylor Creek/Nubbin Slough Rural Clean Waters Program, which involved the implementation of Best Management Practices (BMPs), including fencing cows away from streams, (2) the Florida Department of Environmental Regulation's Dairy Rule, which required that all dairy operations within the Lake Okeechobee watershed and its tributaries implement BMPs for the purpose of reducing P inputs into the lake, (3) the Dairy Buy-Out Program, which facilitated the removal of milking cows from those dairy farmers unable to comply with the Dairy Rule, and (4) the Works of the District rule, a regulatory program that established a numeric P concentration limit for runoff from nondairy land uses. Each of these programs is described in greater detail in the Lake Okeechobee SWIM Plan Update (SFWMD, 1997).

The management programs described above have been successful in reducing P loading to Lake Okeechobee (Havens et al., 1996c) but have not brought loads down to the SWIM-mandated target values determined from the modified Vollenweider model (Vollenweider, 1976; Federico et al., 1981). This target value is a function of both the amount of water entering the lake and the water's residence time in the lake. The five-year rolling average (1992 to 1996) of loading of TP to Lake Okeechobee was approximately 470 tons, which was approximately 100 tons over target. As a consequence, new programs/strategies have been proposed to reduce loads even further.

The first program focuses on the four over-target basins identified in the SWIM Plan update and then modeled in the LOADSS simulation (Table 23.3). These basins contain most of the dairies, as well as the majority of Works of the District sites that are out of compliance (Zhang and Essex, 1997). The strategy involves implementing nonregulatory, landowner-based initiatives to restore isolated and riverine wetlands in the watershed. The current system results in rapid runoff of water off the watershed into the canal and tributary system that drains into Lake Okeechobee. Restoration of these drained wetlands in the northern watershed of the Lake, as well as the creation of new ones to retain/detain the water in the watershed, is expected to slow the runoff and attenuate peak flows.

The second program is to dredge tributary sediments that are rich in P. These sediments may represent a large source of P to Lake Okeechobee. An ongoing study indicates that approximately 900 t of P are located in a thin, surface layer throughout the watershed (Stuck, J.D., Bottcher, A.B., Rosen, B.H., Hiscock, J.G., Mehta, A.J., Flaig, E., unpublished data, 1997). The sediments are rich in organic material and are highly mobile.

The third program/strategy is to develop BMPs for beef cattle operations. Although animal densities and runoff P concentrations associated with beef cattle operations are relatively low, almost 200,000 ha of land are dedicated to beef cattle, making them a major contributor of P to the lake. It is essential that new or innovative BMPs be developed or implemented to reduce runoff from this land use. These practices may include better management of stocking rates, improved grazing approaches such as winter/summer range rotation, and rotational grazing to ensure biomass and nutrient management.

An optimization project has been initiated at the Buck Island Ranch at the MacArthur Agro-Ecology Research Center in Lake Placid, FL. The infrastructure for this project consists of 16, field-scale experimental pastures: eight 20-ha plots on summer pasture land (improved pastures consisting of planted bahiagrass, occasionally fertilized) and eight 32-ha plots on winter pasture land (native range not typically fertilized or planted). Each pasture is individually fenced and ditched, and instrumented, so that all surface water runoff can be captured and analyzed. It is anticipated that this program will last for ten years. Three primary management practices will be investigated in successive phases: optimization of beef cattle stocking densities; optimization of pasture fertilization practices, and optimization of grazing and cattle rotation schemes. There are three overall objectives to this project: (1) to optimize beef cattle BMPs that ensure both environmentally and economically sustainable beef cattle practices in Florida, (2) to communicate these optimized

BMPs to beef cattle ranchers through extension publications or other appropriate mechanisms, and (3) to generate the data to enhance a Beef Cattle Management Decision Support System, which along with information provided by extension publications, will allow ranchers to make more informed management decisions.

In-lake options to reduce P impacts deal with the lake's sediments; it is clear that the vast majority of P in Lake Okeechobee is tied up in the sediments. Although it is essential that we reduce P runoff from the watershed, our modeling efforts indicated that even with drastic changes in PCPs, phosphorus levels in the lake will be slow to respond, given the huge reservoir of P in the sediments. This is not unexpected or unusual for a lake and watershed of this size, but public pressure for improved water quality requires examining the viability of alternative measures to restore lake health. For example, alternative engineering solutions may accelerate recovery rates. Sediment dredging, or treatment of surface sediment layers with alum or other P-binding chemicals, has been shown to greatly reduce rates of internal P loading in other shallow eutrophic lakes (Welch and Cooke, 1995). However, the sheer size of Lake Okeechobee may preclude their utility. Cooke et al. (1993) estimated that the average cost for sediment removal was $20,000 per ha. Lake Okeechobee contains approximately 81,000 ha of mud, resulting in an estimated cost of $1.6 billion. Restoration of shallow Banana Lake, Florida, cost approximately $1.30 per cubic meter, and required one full year of dredging to remove 300 ha of mud (Kelly, M., personal communication, 1997). By extrapolation, it would cost $249 million and require 270 years for a similar effort at Lake Okeechobee. Finally, one needs to consider the risks associated with these in-lake measures, including detrimental impacts on fish and wildlife because of high turbidity and nutrient release during the process. Alternatively, we can increase our efforts to implement nonregulatory land use changes in the watershed, which ultimately will result in reduced P concentrations in the lake's water column and sediments. However, the response time may be on the order of decades. In the interim, implementing BMPs will reduce P loads to the lake, and hasten lake recovery, and therefore should be encouraged wherever possible.

23.9 ACKNOWLEDGMENTS

The authors are grateful to D. Fox, Florida Game and Fresh Water Fish Commission, for providing estimates of tissue P concentrations in littoral fish taxa. Comments from Garth Redfield, Paul McCormick, Zhenquan Chen, Sue Newman, Mike Chimney, Mike Coveney, and Claire Schelske improved the manuscript.

REFERENCES

Andersen, T., and D.O. Hessen. 1991. Carbon, nitrogen, and phosphorus content of freshwater zooplankton. Limnol. Oceanogr. 36:807–813.

Aumen, N.G. 1995. The history of human impacts, lake management, and limnological research on Lake Okeechobee, Florida (USA). Arch. Hydrobiol. Beih. Ergebn. Limnol. 45:1–16.

Bachmann, R.W., B.L. Jones, D.D. Fox, M. Hoyer, L.A. Bull, and D.E. Canfield. 1996. Relations between trophic state indicators and fish in Florida (U.S.A.) lakes. Can. J. Fish. Aquat. Sci. 53:842–855.

Boggess, C.F., E.G. Flaig, and R.C. Fluck. 1995. Phosphorus budget-basin relationships for Lake Okeechobee tributary basins. Ecol. Eng. 5:143–162.

Bostrom, B., M. Janson, and C. Forsberg. 1982. Phosphorus release from lake sediments. Arch. Hydrobiol. 18:5–59.

Brezonik, P.L., and D.R. Engstrom. 1998. Modern and historic accumulation rates of phosphorus in Lake Okeechobee, Florida. J. Paleolimnology 20:31–46.

Cooke, G.D., E.B. Welch, S.A. Peterson, and P.R. Newroth. 1993. Restoration and Management of Lakes and Reservoirs. 2nd Ed. Lewis Publ., Boca Raton, FL, USA.

Dierberg, F.E. 1992. The littoral zone of Lake Okeechobee as a source of phosphorus after drawdown. Environ. Manage. 13:729–742.

Federico, A., K. Dickson, C. Kratzer, and F. Davis. 1981. Lake Okeechobee water quality studies and eutrophication assessment. Technical Publication 81-2. South Florida Water Management District, West Palm Beach, FL.

Flaig, E.G., and K.E. Havens. 1995. Historical trends in the Lake Okeechobee ecosystem. I. Land use and nutrient loading. Arch. Hydrobiol. Suppl. Monogr. Beitr. 107:1–24.

Flaig, E.G., and K.R. Reddy. 1995. Fate of phosphorus in the Lake Okeechobee watershed, Florida, USA: overview and recommendations. Ecol. Eng. 5:127–142.

Fluck, R.C., C. Fonyo, and E.G. Flaig. 1992. Land use based phosphorus balances for Lake Okeechobee, Florida, drainage basins. Appl. Eng. Agr. 8:813–820.

Harris, T.T., K.A. Williges, and P.V. Zimba. 1995. Primary productivity and decomposition of five emergent macrophyte communities in the Lake Okeechobee marsh ecosystem. Arch. Hydrobiol. Beih. Ergebn. Limnol. 45:63–78.

Havens, K.E., N.G. Aumen, R.T. James, and V.H. Smith. 1996a. Rapid ecological changes in a large subtropical lake undergoing cultural eutrophication. Ambio 25:150–155.

Havens, K.E., T.L. East, R.H. Meeker, W.P. Davis, and A.D. Steinman. 1996b. Phytoplankton and periphyton responses to in situ experimental nutrient enrichment in a shallow subtropical lake. J. Plankton Res. 18:551–566.

Havens, K.E., E.G. Flaig, R.T. James, S. Lostal, and D. Muszick. 1996c. Results of a program to control phosphorus discharges from dairy operations in south-central Florida, USA. Environ. Manage. 21:585–593.

Havens, K.E., and T.L. East. 1997. Carbon dynamics in the "grazing food chain" of a large subtropical lake. J. Plankton Res. 19:1687–1711.

Herdendorf, C.E. 1982. Large lakes of the world. J. Great Lakes Res. 8:379–412.

Hofstra, J.J., and L. Van Liere. 1992. The state of the environment of the Loosdrecht lakes. p. 11–20. *In*: L. Van Liere and R.D. Gulati (ed.) Restoration and recovery of shallow eutrophic lake ecosystems in The Netherlands. Kluwer Acad. Publ., Dordrecht, the Netherlands.

James, R.T., B.L. Jones, and V.H. Smith. 1995a. Historical trends in the Lake Okeechobee ecosystem. II. Nutrient budgets. Arch. Hydrobiol. Suppl. Monogr. Beitr. 107:25–47.

James, R.T., V.H. Smith, and B.L. Jones. 1995b. Historical trends in the Lake Okeechobee ecosystem. III. Water quality. Arch. Hydrobiol. Suppl. Monogr. Beitr. 107:49–69.

James, R.T., J. Martin, T. Wool and P.F. Wang. 1997. A sediment resuspension and water quality model of Lake Okeechobee. J. Amer. Water Res. Assoc. 33:661–680.

Maceina, M.J., and D.M. Soballe. 1990. Wind-related limnological variation in Lake Okeechobee, Florida. Lake Reserv. Manage. 6:93–100.

Moore, P.A., and K.R. Reddy. 1994. Role of Eh and pH on phosphorus geochemistry in sediments of Lake Okeechobee, Florida. J. Environ. Qual. 23:955–964.

Negahban, B., C.B. Moss, J.W. Jones, J. Zhang, W.D. Boggess and K.L. Campbell. 1994. Optimal field management for regional water quality planning. ASAE Paper No. 94-3553. St. Joseph, MI.

Nixdorf, B., and R. Deneke. 1997. Why very shallow lakes are more successful opposing reduced nutrient loads. Hydrobiologia 342/343:269–284.

Olila, O.G., and K.R. Reddy. 1993. Phosphorus sorption characteristics of sediments in shallow eutrophic lakes of Florida. Arch. Hydrobiol. 129:45–65.

Olila, O.G., K.R. Reddy, and W.G. Harris. 1995. Forms and distribution of inorganic phosphorus in sediments of two shallow eutrophic lakes in Florida. Hydrobiologia 302:147–161.

Phlips, E.J., F.J. Aldridge, P. Hansen, P., Zimba, P.V., Inhat, J., Conroy, M., and Ritter, P. 1993. Spatial and temporal variability of trophic state parameters in a shallow subtropical lake (Lake Okeechobee, Florida, USA). Arch. Hydrobiol. 128:437–458.

Phlips, E.J., F.J. Aldridge, and C. Hanlon. 1995. Potential limiting factors for phytoplankton biomass in a shallow subtropical lake (Lake Okeechobee, Florida, USA). Arch. Hydrobiol. Beih. Ergebn. Limnol. 45:137–155.

Phlips, E.J., M. Cichra, K.E. Havens, C. Hanlon, S. Badylak, B. Rueter, M. Randall, and P. Hansen. 1997. Relationships between phytoplankton dynamics and the availability of light and nutrients in a shallow subtropical lake. J. Plankton Res. 19:319–342.

Reddy, K.R. 1991. Lake Okeechobee Phosphorus Dynamics Study, Volume IV, Biogeochemical Processes in the Sediments. South Florida Water Management District, West Palm Beach, FL.

Reddy, K.R., Y.P. Sheng, and B.L. Jones. 1995. Lake Okeechobee Phosphorus Dynamics Study, Volume 1, Summary. Report, South Florida Water Management District, West Palm Beach, FL.

Richardson, J.R., and T.T. Harris. 1995. Vegetation mapping and change detection in the Lake Okeechobee marsh ecosystem. Arch. Hydrobiol. Beih. Ergebn. Limnol. 45:41–61.

Sas, H. 1989. Lake Restoration by Reduction of Nutrient Loading: Expectations, Experiences, and Extrapolations. Academia Verlag Richarz, Germany.

Schelske, C.L. 1997. Sediment and phosphorus deposition in Lake Apopka. Special Publication SJ97-SP21, St. Johns River Water Management District, Palata, FL.

SFWMD. 1997. Surface Water Improvement and Management (SWIM) Plan update for Lake Okeechobee. South Florida Water Management District, West Palm Beach, FL.

Sheng, Y.P. 1991. Lake Okeechobee Phosphorus Dynamics Study, Volume VII, Hydrodynamics and Sediment Dynamics—A Field and Modeling Study. South Florida Water Management District, West Palm Beach, FL.

Steinman, A.D., R.H. Meeker, A.J. Rodusky, W.P. Davis, and S-J. Hwang. 1997. Ecological properties of charophytes in a large subtropical lake. J. No. Am. Benthol. Soc. 16:781–793.

Steinman, A.D., K.E. Havens, J.W. Louda, N.M. Winfree, and E.W. Baker. 1998. Characterization of the photoautotrophic algal and bacterial communities in a large, shallow subtropical lake using HPLC-PDA based pigment analysis. Can. J. Fish. Aquat. Sci. 55: xxx–xxx.

Van Rees, K.C.J., K.R. Reddy, and P.S.C. Rao. 1996. Influence of benthic organisms on solute transport in lake sediments. Hydrobiologia 317:31–40.

Vollenweider, R.A. 1976. Advances in defining critical loading levels for phosphorus in lake eutrophication. Mem. Ist. Ital. Idrobiol. 33:53–83.

Walker, W.W., Jr., and K.E. Havens. 1995. Relating algal bloom frequencies to phosphorus concentrations in Lake Okeechobee. Lake Reserv. Manage. 11:77–83.

Welch, E.B., and G.D. Cooke. 1995. Internal phosphorus loading in shallow lakes: importance and control. Lake Reserv. Manage. 11:273–281.

Wetzel, R.G. 1983. Limnology. 2nd Edition, Saunders, Philadelphia.

Zhang, J., and A. Essex. 1997. Phosphorus load reductions from out-of-compliance sites in the Lake Okeechobee watershed. Appl. Eng. Agr. 13:193–198.

Zimba, P.V., M.S. Hopson, J.P. Smith, D.E. Colle, and J.V. Shireman. 1995. Chemical composition and distribution of submersed aquatic vegetation in Lake Okeechobee, Florida (1989–1991). Arch. Hydrobiol. Beih. Ergebn. Limnol. 45:241–246.

24 Effects of Freshwater Inputs and Loading of Phosphorus and Nitrogen on the Water Quality of Eastern Florida Bay

Joseph N. Boyer and Ronald D. Jones

24.1 ABSTRACT

The eastern portion of Florida Bay was chosen as the area most affected by freshwater input and, therefore, nutrient loading. Freshwater flows to eastern Florida Bay via Taylor Slough (S175 and S332) and the coastal basin (S18C) fluctuated widely during 1984 to 1996 and were not significantly related to precipitation. However, after 1990, a seven-fold increase in freshwater input was observed. At the same time, the contribution of the S332 to the total flow increased from 47 to 256 × 10^6 m^3 yr^{-1} (17 to 32%). We believe part of this effect was because of a management strategy of increased pumping at the S332 and that these management efforts were effective in increasing water to the Taylor Slough. Average annual salinity in eastern Florida Bay during 1989 to 1996 declined by 18 g kg^{-1} and was significantly related to

1-56670-331-X/99/$0.00+$.50

increased flow ($r^2 = 0.75$) but not to precipitation. Total phosphorus (TP) loading ranged from 0.9 to 7.7×10^6 g yr^{-1} and was significantly related to flow ($r^2 = 0.82$) with the average concentration being 0.32 μM (10 ppb). While TP loading increased, ambient TP concentration in eastern Florida Bay decreased from 0.4 to 0.2 μM. Total nitrogen (TN) loading ranged from 98.7 to 635×10^6 g yr^{-1} and was also highly correlated with flow ($r^2 = 0.92$). Average input TN concentration was 62 μM while ambient TN concentration in eastern Florida Bay averaged 55 μM. Ambient TN:TP ratios in eastern Florida Bay were maintained at or near the high input TN:TP ratios (mean = 224), implying that both the input ecosystem (Everglades) and the receiving waters are P limited. There was no relationship between either TP or TN loading and chlorophyll-*a* in eastern Florida Bay; instead, there was a significant regression with ambient TP concentration ($r^2 = 0.32$). This indicates that phytoplankton are responding to internally cycled P and not to external loading. We argue that, to maintain the oligotrophic status of eastern Florida Bay, we must be concerned with the critical loading concentration as well as the annual total mass load.

24.2 INTRODUCTION

For many years, estuaries and coastal environments have been subjected to close scrutiny by scientists, managers, politicians, and the public for signs of stress and damage. The ecological status of an estuary is directly tied to its geomorphology, hydrology, and watershed characteristics (Day et al., 1989). The shape of an estuary can be modified rapidly by dredge and fill activities and slowly by sea level rise. Estuarine hydrology is most affected by modifications and diversions of freshwater inputs, which induce changes in the quantity and timing of freshwater flows (Kennish, 1986). The watershed itself may be subject to large changes in land-use patterns, which can affect both the hydrologic regime and the water quality of freshwater inputs (Correll et al., 1992). All of these modifications can occur as a direct result of human intervention resulting in eutrophication of many estuaries worldwide (Hutchinson, 1969; Neilson and Cronin, 1981; Vollenweider et al., 1992; Nixon, 1995).

The symptoms of estuarine eutrophication have stimulated the formation of monitoring and research programs whose purpose is to characterize existing conditions, detect trends over both space and time, and provide insight for further management activity (NRC, 1990). In the case of Florida Bay, the impetus of such a program was a combination of events during the mid-1980s that included the invasion of exotic species in the freshwater wetlands and uplands (Bodle et al., 1994), periods of prolonged hypersalinity of coastal embayments (Fourqurean et al., 1993), a poorly understood seagrass die-off (Robblee et al., 1991), sponge mortality events (Butler et al., 1995), and elevated phytoplankton abundance (Phlips and Badylak, 1996). In response, a monitoring network operated by the Southeast Environmental Research Program at FIU was established in 1991 so that trends in water quality might be addressed (Boyer and Jones, 1998). Herein we attempt to relate freshwater inputs and the associated nutrient loading to salinity, phosphorus, and nitrogen concentrations in surface waters of the eastern portion of Florida Bay. We also provide a comparison of ambient nutrient characteristics of eastern Florida Bay with those from other estuarine systems, gaining new perspectives on their similarities and differences in the process.

24.3 METHODS

24.3.1 Site Description

Florida Bay is a wedge-shaped, shallow lagoonal estuary located off the southern tip of the Florida peninsula bounded by the Everglades to the north and open to the Gulf of Mexico along its western margin. The Florida Keys, an almost continuous Pleistocene reef along the southern boundary, separates Florida Bay from the Atlantic Ocean and restricts tidal exchange to a few channels. Shallow mud banks divide Florida Bay into relatively discrete basins that restrict water mixing between basins and attenuate both tidal range and current speed.

A previous analysis of six years of monthly collected data by the FIU monitoring program resulted in the delineation of three distinct zones (Fig. 24.1) that had robust

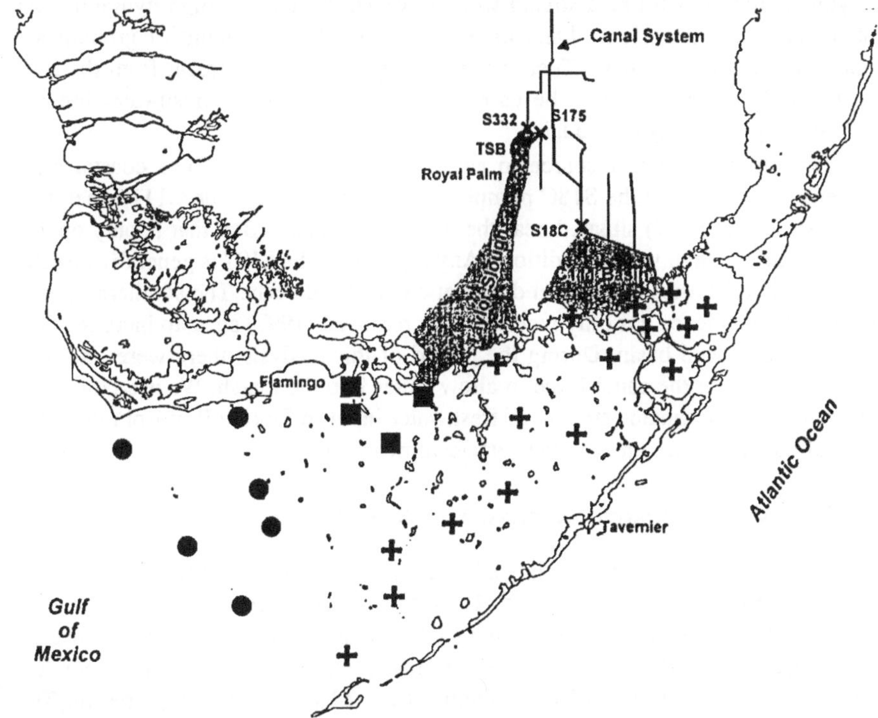

FIGURE 24.1 Map of Florida Bay station clusters (Boyer et al. 1997). Stations as labeled are: eastern Florida Bay (✚), central Florida Bay (■), and western Florida Bay (●). Gauging stations of the water distribution network (✖) include the Taylor Slough (S175 and S332), Taylor Slough bridge (TSB), and C111 coastal basin (S18C). Locations of precipitation gauges are indicated as O.

similarities in water quality (Boyer et al., 1997). Eastern Florida Bay was defined as being the zone most like a river dominated estuary in that it has a semilongitudinal salinity gradient. Central Florida Bay is a physically isolated area with low freshwater input and high evaporative potential. Western Florida Bay is the zone most influenced by winds and tides from the Gulf of Mexico. See Fourqurean et al. (1998) for a more detailed description of the Florida Bay ecosystem. Because the hydrology of Florida Bay is so complex and the advection among zones fairly restricted, we limited our analysis to the eastern Florida Bay, as this is the area that receives the bulk of freshwater input.

Eastern Florida Bay receives freshwater at its northeast end in the form of overland flow from Taylor Slough and the C111 coastal basin of the Everglades and from direct precipitation to the Bay surface. The three main sources of freshwater input to Taylor Slough are the S175 and S332 pumping stations as well as direct precipitation in the watershed itself (Fig. 24.1). A levee road maintained by ENP bisects Taylor Slough downstream of these structures and restricts sheet flow to the south along its length except for a bridge over the slough. The flow gauge at the Taylor Slough bridge (TSB, Fig. 24.1) integrates all overland flow to the southern Taylor Slough, much like a stream weir. However, because of high hydraulic conductivity of the underlying karst limestone, a significant amount of groundwater may traverse this boundary. There are no other gauges downstream from the road, so the TSB gauge may only be used as a rough estimate of freshwater input to Florida Bay from Taylor Slough.

The other freshwater source to eastern Florida Bay, the C111 coastal basin system, is supplied by the S18C pumping station (Fig. 24.1). The C111 canal itself has a structure (S197) situated near the mouth at Manatee Bay that is only opened during extreme high water conditions. Any input from the S197 is generally directed northwards to Barnes Sound and does not enter Florida Bay. The southern edge of the C111 canal levee has only recently been removed (1997) so as to increase sheet flow to the coastal basin. During the period of this analysis, there were only segmented breaks in the canal levee to allow water into the marsh. We used the S18C gauge as a approximate estimate of freshwater input to Florida Bay from the C111 coastal basin as there are no other gauges downstream.

24.3.2 Water Quality Sampling and Analysis

A total of 28 water quality sampling sites are located throughout Florida Bay as part of the FIU monitoring program (Fig. 24.1). Most of the sites were situated near the center of the basins, others were evenly spaced so as to give as complete a coverage of the area as possible. Stations were originally sampled semimonthly from July 1989 to December 1990 and then monthly from March 1991 to the present. Two consecutive days were required to complete each sampling event. The period of record used in this analysis is from March 1991 to December 1997.

Water quality parameters measured at each station included salinity (g kg^{-1}), temperature (°C), dissolved oxygen (DO, mg l^{-1}), nitrate (NO_3^-), nitrite (NO_2^-), ammonium (NH_4^+), total nitrogen (TN), total organic nitrogen (TON), total phosphorus (TP), soluble reactive phosphorus (SRP), total organic carbon (TOC), chlorophyll-*a* (Chl-*a*, $\mu g\ l^{-1}$), alkaline phosphatase activity (APA, $\mu M\ hr^{-1}$), and turbidity (NTU). Analytical details are provided in Boyer et al. (1997); all nutrient concentrations are reported in μM unless otherwise noted.

24.4 FRESHWATER INPUT AND NUTRIENT LOADING CALCULATIONS

Freshwater flow data ($m^3\ d^{-1}$) from 1984 to 1996 was provided at gauging stations in Everglades National Park (ENP) as operated by the South Florida Water Management District (SFWMD) and US Geological Service (USGS). Daily flow data at each structure were summed to get monthly and annual flow ($10^6\ m^3\ yr^{-1}$). Daily precipitation ($cm\ d^{-1}$) at the ENP Royal Palm monitoring site (Fig. 24.1) was summed to get monthly and annual inputs in $cm\ yr^{-1}$. We did not have any estimates of groundwater inputs to ENP or Florida Bay at this time.

Monthly nutrient concentrations at the structures were collected and analyzed by SFWMD. Annual loading of TP and TN (10^6 g yr^{-1} = metric tons yr^{-1}) were calculated at the S18C, S175, and S332 gauging stations by William Walker (pers. comm., 1997) and kindly provided to us. No attempt was made to derive N or P loading from the western boundary of Florida Bay; neither was loading from precipitation nor groundwater included in the calculations. These are serious omissions to the nutrient budget but none of these data existed when preparing this synthesis.

A seasonal Kendall-τ analysis was performed on monthly flow and precipitation data to test for monotonic trends (Hirsch et al., 1991). The seasonal Kendall-τ test is a nonparametric regression analysis that determines: direction of trend (+ or –), goodness of fit(τ), statistical significance of fit (P) not accounting for serial correlation, and trend slope estimate over the period of record (TSE in units yr^{-1}). The seasonal Kendall-τ test requires the dataset to be contiguous and cannot detect reversals in trend direction, such as might be seen in the case of interannual oscillations.

24.5 RESULTS AND DISCUSSION

24.5.1 Freshwater Inputs

The C111 coastal basin (S18C) accounted for the largest proportion of overland freshwater flow to the system, ranging from 70 to 435 × $10^6\ m^3\ yr^{-1}$ (Fig. 24.2). Losses to groundwater from this structure were unknown as was the amount of evapotranspiration occurring along the flow path, therefore, the actual amount of freshwater entering eastern Florida Bay from this system was probably significantly attenuated. However, because of its proximity to the coast, actual output to the Bay is probably closest to the measured freshwater flow through the S18C gauge than for any of the other structures. There was no significant trend in S18C input for the 1984 to 96 period of record ($\tau = 0.33$, $P = 0.22$), however for the period 1990 to 96 (post-1989 drought), flow increased significantly by 38 × $10^6\ m^3\ yr^{-1}$ ($\tau = 0.69$, $P = 0.02$).

Flows through the S175 fluctuated from lows of 8.3 and 3.6 × 10^6 m^3 yr^{-1} in 1989 to 1991 to a high of 152 × 10^6 m^3 yr^{-1} in 1993 (Fig. 24.2). No trend in S175 flow was evident ($\tau = 0.09$, $P = 0.72$). The S332 flows prior to 1992 were ~40 × 10^6 m^3 yr^{-1}; from 1992 onward the flows increased dramatically up to 256 × 10^6 m^3 yr^{-1} in 1995 (Fig. 24.2). In addition, there was a change in the relative flows at S332 and S18C. Prior to 1992, the S332 made up only 20% of the combined flow; from 1992 onward, the S332 constituted 33% of the total. The increased contribution of S332 was because of a change from tail water height-limited pumping schedule to one of continuous pumping, which was instituted to increase the hydroperiod of Taylor Slough. This change to continuous pumping was also reflected as a concurrent increase in flows through TSB, downstream of the S332 (Fig. 24.2). Therefore, as a result of management activity, the annual TSB flows increased from an average of 20 × 10^6 m^3 yr^{-1} prior to 1992 to 73 × 10^6 m^3 yr^{-1} afterward.

Precipitation at the Royal Palm station (ENP) was used as a proxy of rainfall input to the lower Everglades. Annual precipitation during 1984 to 1996 ranged from 97 to 220 cm yr^{-1}, averaging 136 cm yr^{-1} (Fig. 24.3). A slight increasing trend in monthly precipitation for the 1984 to 1996 period of record was evident ($\tau = 0.25$, $P = 0.12$), but no trend during 1990 to 1996 was observed ($\tau = 0.05$, $P = 0.78$). Regressions of detrended monthly flows at S18C and TSB (1 month lag) and precipitation were significant but explained <2% of the variance, supplying very little predictive power. This means that increased inputs from S332 were because of the increased pumping schedule and not to local precipitation events.

This uncoupling of precipitation from flow raises an important point concerning the relationship between combined S175 + S332 canal inputs and the Taylor Slough flows through the TSB (Fig. 24.2). The differences in gauge measurements indicated that there was significant attenuation in overland flow in the Taylor Slough watershed over a relatively short distance. The flow ratio of TSB to S332 + S175 was low (mean = 0.22) and relatively constant for the period of record, meaning that only 22% of the total canal input to Taylor Slough made it through the bridge. This low flow ratio was because of two unmeasured processes: evapotranspiration (ET) and

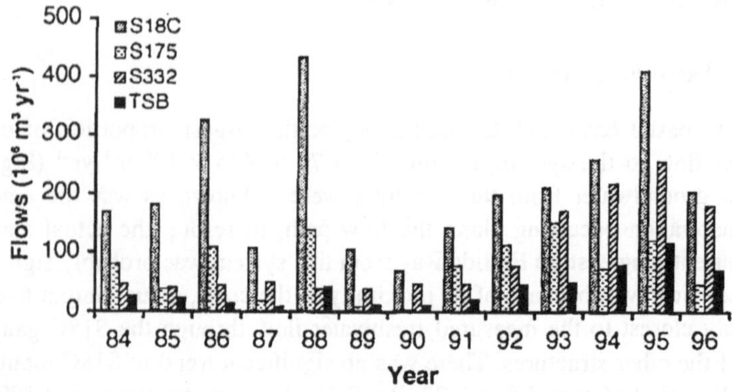

FIGURE 24.2 Annual potential freshwater inputs to eastern Florida Bay (10^6 m^3 yr^{-1}) via the Taylor Slough (S175, S332, TSB) and coastal basin (S18C).

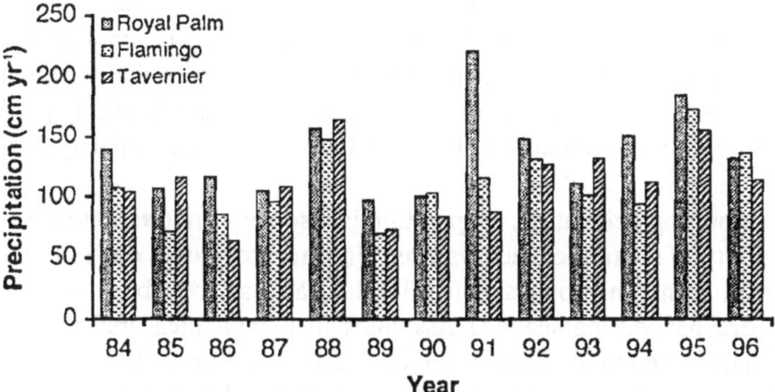

FIGURE 24.3 Annual precipitation at the Royal Palm station (ENP) in cm yr^{-1}.

groundwater recharge. Since TSB flow also integrates the direct precipitation to the upstream watershed as well, the actual flow ratio might be lower than estimated. That the TSB:S332 + S175 flow ratio was relatively constant implies that the output through the bridge is directly proportional to some combination of precipitation, ET, and groundwater recharge.

Another possible explanation for the low TSB:S332 + S175 flow ratio might be groundwater recycling. Because of high hydraulic conductivity of the underlying karst limestone, it is possible that a significant percentage of the water being pumped through the structures drains back into the canals and is returned to once again to a pumping station. Any recycling of Taylor Slough water via groundwater return probably occurs downstream of the TSB gauge where Taylor Slough abuts the L-31W canal (Fig. 24.1). If groundwater recycling were large, then pumping rates would go up without increasing flows through the TSB. This is obviously not the case for the Taylor Slough.

As mentioned previously, we have no estimates of groundwater flow to eastern Florida Bay but can only speculate that it might be substantial, as any increase in hydraulic head above the level of Florida Bay will result in an input to the system. It is clear then that increased pumping at the S332 is directly translated to increased TSB flows, but just how much of this water actually gets to Florida Bay is unknown.

24.5.2 Salinity Relationships

The annual median salinity in eastern Florida Bay declined from 40 g kg^{-1} to 22 g kg^{-1} during 1989 to 1996 (Boyer et al. in press). As precipitation during this period did not increase significantly, increased overland and groundwater flows must have accounted for the observed salinity decline. However, we must stress that the volume of direct precipitation to eastern Florida Bay dwarfs the maximum possible freshwater input from land. If we assume the area of eastern Florida Bay to be 1000 km^2 (1×10^9 m^2) and the average depth to be 1 m, then its volume is 10^9 m^3. For 1995 the annual precipitation was 184 cm yr^{-1} or a total volume of precipitation of 1.84×10^9 m^3, which is almost twice the volume of the estuary itself. The overland flow estimate (S18C + TSB) was 0.53×10^9 m^3 or <20% the precipitation volume. Therefore, a small error in precipitation measurement will be translated to a large change in salinity. However, since precipitation is highest in the north central bay (W. Nuttle, personal communication) and significant landward salinity gradients are routinely observed, it is unlikely that these patterns can be explained by precipitation alone.

In addition to precipitation, evaporation is also very important process in determining salinity of subtropical ecosystems. Unfortunately, only one evaporation rate measurement from only one site in Florida Bay has ever been performed (M. Robblee, personal communication). Assuming that single measured rate (4 mm d^{-1}) to be constant results in a potential evaporative loss of 1.46×10^9 m^3 yr^{-1}, a volume comparable to the average annual precipitation input of 1.36×10^9 m^3 yr^{-1}. In eastern Florida Bay, a simple annual net water flux volume calculation for 1989 to 1996 (sensu Smith and Atkinson, 1994), where:

$$\text{Net Water Flux} = \text{Evaporation} - \text{Flow} - \text{Precipitation}$$

resulted in a net export of water from the system and predicted decreases in salinity. On average, actual freshwater inputs to eastern Florida Bay exceeded evaporation by only 0.26×10^9 m^3 yr^{-1} that is lower than for most estuaries (Smith and Atkinson, 1994). Only in 1989 and 1990, the driest years of the study, would eastern Florida Bay be considered evaporative (negative estuary). The net water flux volume is underestimated by omission of groundwater inputs and overestimated by pan evaporation measurements. Regardless of these errors, the net water flux is driven by measured freshwater inputs during this time period and can account for much of the change in salinity during this period of record.

Our analysis shows that increased freshwater inputs to Taylor Slough and the C111 coastal basin were more important in alleviating the hypersalinity in eastern Florida Bay than precipitation. Measured flows during this time period increased seven-fold ($\tau = 0.38$, $P = 0.008$) while precipitation remained relatively constant. The regression of salinity with freshwater input (Fig. 24.4) was highly significant ($r^2 = 0.75$, $P = 0.001$). In addition, time series analyses of well stage measurements in ENP (Nuttle, 1997) were shown to be statistically significant and predictive of surface salinities in selected regions of northeastern Florida Bay.

24.5.3 Phosphorus Loading

Annual TP loading estimates for 1984 to 1996 ranged from 0.9 to 7.7×10^6 g yr^{-1} with the mean being 3.8×10^6 g yr^{-1} (Fig. 24.5). Unlike the freshwater inputs, the contribution of TP loading from the S332 structure did not change with increased pumping rate but stayed relatively constant at 24.5% of the total load. TP loading did not increase significantly over the long period of record ($r^2 = 0.106$, $P = 0.48$), but during the time of increasing flow, 1990 to 1995, load increased by 1.4×10^6 g yr^{-1} ($r^2 = 0.95$). Interestingly, during this same time period, median TP concentrations declined from 0.4 μM (12 ppb) to 0.2 μM (6 ppb) in eastern Florida Bay (Boyer et al. in press). Chl-*a* concentrations decreased as well, indicating that increased phy-

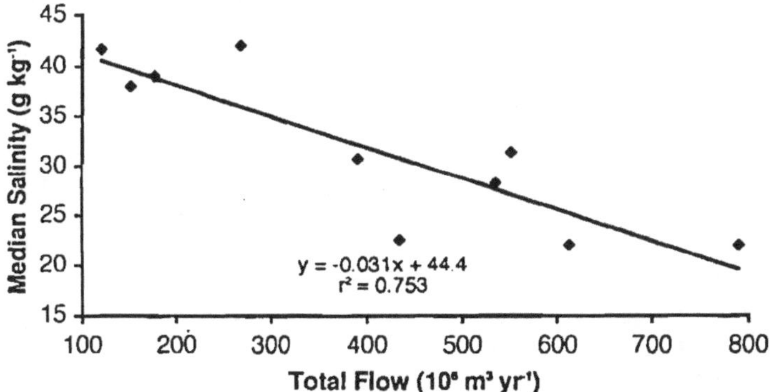

FIGURE 24.4 Linear regression of annual median salinity (g kg^{-1}) in eastern Florida Bay with annual potential freshwater input (10^6 m^3 yr^{-1}) from S175, S332, and S18C.

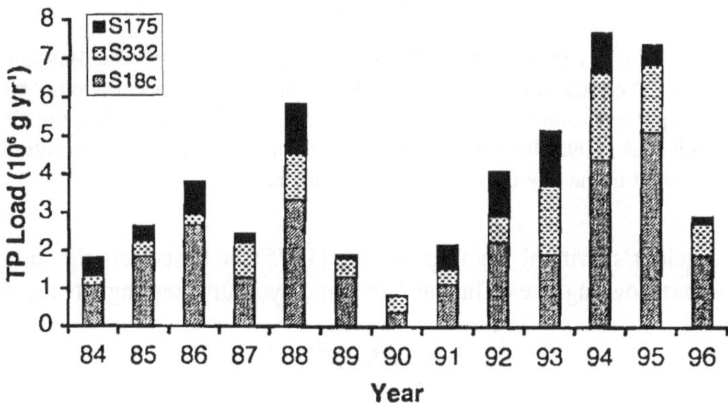

FIGURE 24.5 Annual TP loading (10^6 g yr^{-1}) for the Taylor Slough (S175 andS332) and C111 coastal basin (S18C).

toplankton uptake did not account for the decline in TP. An increase in seagrass cover or increased binding to sediments with lower salinity could possibly account for the TP decline but has not been quantified.

Although TP loading increased with flow, TP loading was highly correlated with total flow ($r^2 = 0.82$, $P = 0.002$) indicating that the TP concentration of the incoming water remained relatively constant (mean 0.32 μM or 10 ppb). Remarkably, the input TP concentrations in the canals was about equal to the ambient TP concentrations found in eastern Florida Bay (Fig. 24.6). This is notable as the freshwater input TP concentrations of other estuaries are generally much higher than the ambient waters (see Chesapeake Bay example in Fig. 24.6). In addition, >80% of the TP in ambient waters was in the form of organic P with the inorganic, SRP fraction being 0.039 μM (1.2 ppb). Assuming a similar split between organic and inorganic P for input waters results in SRP of concentrations of 0.045 μM (1.4 ppb). These values are so low as to be at or below the threshold (K_s) of typical phytoplankton uptake kinetics.

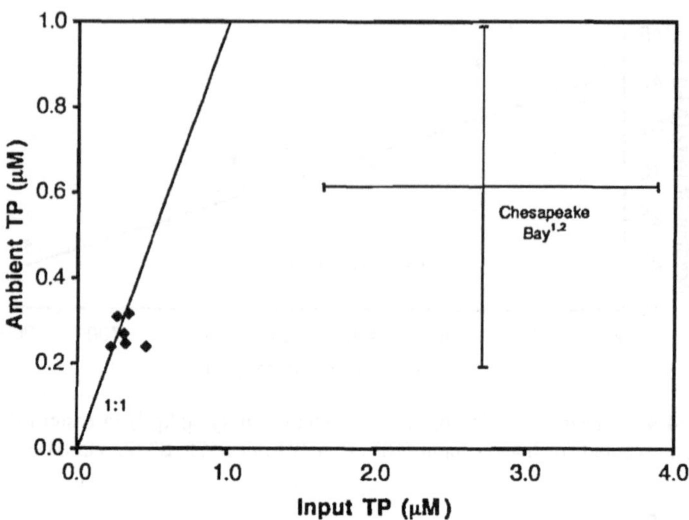

FIGURE 24.6 Plot of annual average TP concentration in eastern Florida Bay (µM) against annual average TP concentration of loading. The Florida Bay concentrations (◆) are very low for both input and ambient and fall at or near the 1:1 ratio line while the Chesapeake Bay ([1]Magnien et al., 1992; [2]Boynton et al., 1995) has much greater TP concentration in the input than is found in the bay (cross denotes the data range).

Vollenweider's critical loading model (1975) is based on the relationship between P mass loading rate (L in g m^{-2} yr^{-1}) and hydraulic loading rate (q_s in m yr^{-1}).

$$q_s = Z/(Q/V)$$

where Z = mean depth (m), Q = annual flow (m^3 yr^{-1}), and V = volume (m^3).

The P mass loading rate for eastern Florida Bay is L = 0.0038 g m^{-2} yr^{-1} while q_s = 1.887 m yr^{-1}. At $q_s < 12$, the critical load (L_c) is ~1.2 g m^{-2} yr^{-1}, therefore eastern Florida Bay is almost three orders of magnitude below the Vollenweider definition of an oligotrophic ecosystem. The expected ambient TP concentration (L/Q) using this model is 0.007 g m^{-3} (7 ppb) that is very close to the measured ambient TP concentrations ~10 ppb (0.3 µM).

24.5.4 Nitrogen Loading

Annual TN loading estimates for 1984 to 1996 ranged from 98.7 to 635 × 10^6 g yr^{-1} with the average being 365 × 10^6 g yr^{-1} (Fig. 24.7). The trend in TN loading at the S332 structure did not follow TP in that TN loading increased after 1992 from 19.4 to 33.4% of the total. TN loading was also highly correlated with total flow (r^2 = 0.92) with the average concentration of the incoming water being 62 µM (~0.9 ppm). TN loading did not increase significantly over the period of record (r^2 = 0.01, P = 0.88), but during 1990 to 1995 it increased by 104 × 10^6 g yr^{-1} (r^2 = 0.94), leading to a five-fold increase in load. TN concentration of the freshwater inputs varied

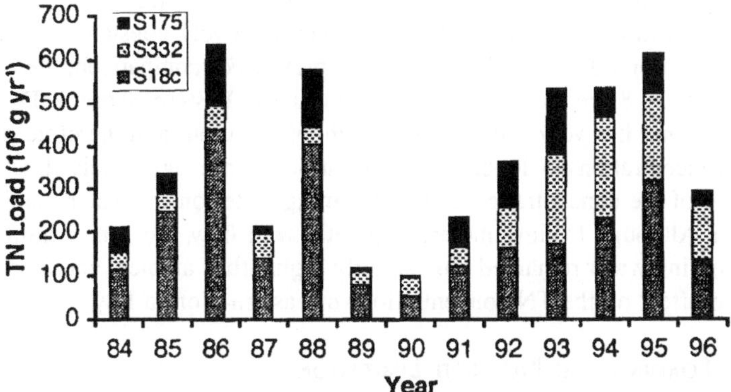

FIGURE 24.7 Annual TN loading (10^6 g yr^{-1}) for the Taylor Slough (S175 and S332) and C111 coastal basin (S18C).

somewhat over the period of record (48 to 91 μM, Fig. 24.8) but was not more variable than input TP concentrations (0.20 to 0.45 μM). Most of the TN as loading and in ambient water was in the form of organic N (data not shown). This is in contrast with other estuaries such as the Chesapeake Bay (Magnien et al., 1992; Boynton et al., 1995) and the Neuse River estuary (Boyer et al., 1988), for which inorganic sources such as NO_3^- and NH_4^+ constitute a large component of the loading.

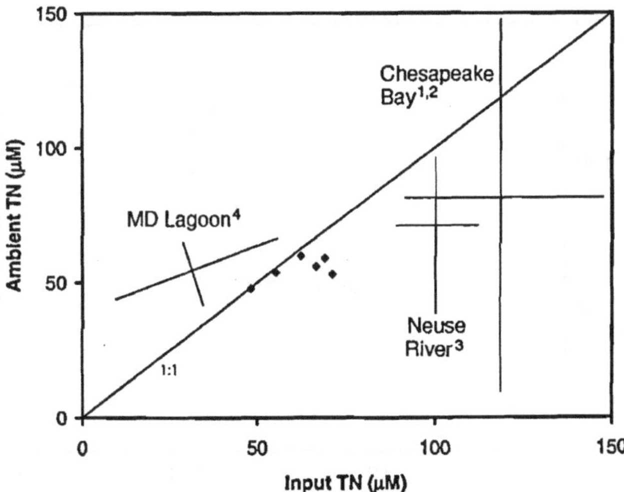

FIGURE 24.8 Plot of annual average TN concentration in eastern Florida Bay (μM) against annual average TN concentration of loading. The Florida Bay concentrations (◆) are low for both input and ambient and fall at or near the 1:1 ratio line. Both the Chesapeake Bay ([1]Magnien et al., 1992; [2]Boynton et al., 1995) and Neuse River estuary ([3]Boyer et al., 1988) have much higher input TN concentrations than ambient. The Eastern Shore, MD data ([4]Boynton et al., 1996) are representative of a lagoonal system with very little freshwater input and loading (cross denotes the data range).

While TN loading increased greatly during 1990 through 1995, average TN concentrations in eastern Florida Bay remained relatively constant at 55 μM (0.8 ppm) with no trend detected (Boyer et al. in press). For comparison, the other data points in Fig. 24.8 are from a lagoon system on the Eastern Shore, MD (Boynton et al., 1996) that has very little freshwater input or output and therefore maintains ambient concentrations > loading concentrations. Once again, this illustrates the importance of the concentration of the incoming water on the water quality of the ecosystem. Although loading increased greatly with flow, the concentration of TN in the incoming water remained only slightly higher than ambient and therefore had little or no effect on the TN concentrations of eastern Florida Bay.

24.5.5 Loading and Nutrient Limitation

The TN:TP molar ratio of loading input ranged from 153 to 367 (mean = 224) that is very high compared to other estuaries (Fig. 24.9). The input TN:TP ratio has implications in the potential for selective nutrient limitation of in situ primary production. Ambient TN:TP ratios in eastern Florida Bay are maintained at or near input TN:TP ratios, implying that both the input ecosystem (Everglades) and the receiving waters are P limited. In contrast, Chesapeake Bay and other estuaries display much lower input and ambient TN:TP ratios, most probably as a result of the low TN:TP of nonpoint source inputs (Nixon and Pilson, 1984; Boynton et al., 1995).

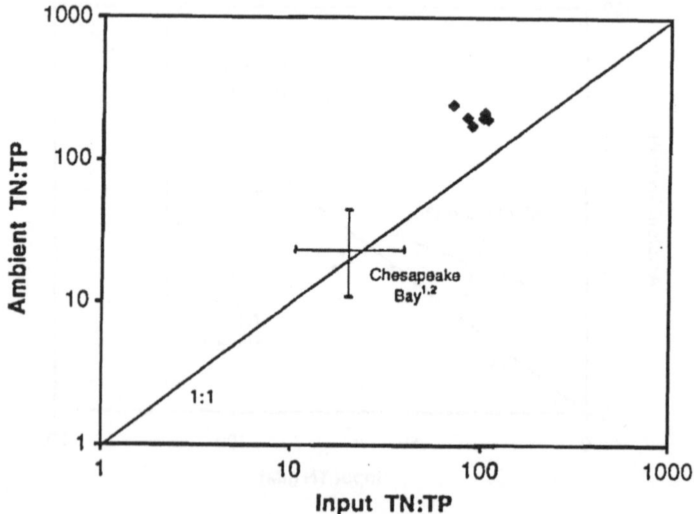

FIGURE 24.9 Plot of annual average molar TN:TP ratio of loading input and ambient water showing the potential for eastern Florida Bay to be highly P limited. Both input and ambient TN:TP (◆) fall at or near the 1:1 ratio line with input TN:TP being most variable. TN:TP of Chesapeake Bay ([1]Magnien et al. 1992; [2]Boynton et al. 1995) are much closer to the Redfield ratio (16:1) with ambient TN:TP being generally higher than input TN:TP (cross denotes the data range).

Schelske et al. (in press) have suggested that the DIN:SRP ratio may be a better statistic for determining ecosystem nutrient limitation status than the TN:TP ratio. The DIN:SRP ratio for eastern Florida Bay ranged from 3.3 to 8415 with a median of 133. Only 44 of the 1287 observations (3.5%) had a DIN:SRP ratio of 16 or less indicating the prevalence of potential P limitation in this region. The very broad range in this ratio points out a problem of interpreting nutrient ratios in oligotrophic systems; high DIN:SRP ratios are because of extremely low SRP concentrations and not to high DIN. Schelske et al. (in press) also concluded that these ratios have little utility in oligotrophic systems. We did not have loading DIN:SRP ratios, but we assumed they were similar to ambient as with TN:TP. Unlike the eutrophic Florida lakes example (Schelske et al. in press), the small difference between DIN:SRP and TN:TP ratios implies that the ratios of inorganic to organic species are similar for both N and P pools. We do not have any information as to direct bioavailability of the organic pools, but our APA values for eastern Florida Bay are relatively high (~0.5 μM hr^{-1}), implying significant microbially-mediated enzyme conversion of the TP pool.

Unlike many estuaries that may fluctuate seasonally between N and P limitation, eastern Florida Bay is a more consistently P limited ecosystem (Fourqurean et al., 1993; Boyer et al., in press). We did not expect nor did we observe a significant relationship between TN loading and phytoplankton biomass (Chl-*a*) as is generally seen in many other estuaries (Boynton et al., 1982). Nor was there a significant relationship between TP loading and Chl-*a* ($P = 0.89$) as might be expected in oligotrophic lakes (Vollenweider, 1976). Instead, there was a significant regression ($P = 0.07$) between Chl-*a* and ambient TP concentration of eastern Florida Bay. However, the weak relationship ($r^2 = 0.32$) as well as the low slope of the regression line ($y = 0.80x + 0.75$) made this association ineffective for predictions. The characteristics of these TP to Chl-*a* relationships indicate that phytoplankton respond only weakly to internally cycled P but not at all to external loading.

Recently, Boynton et al. (1995) compiled loading rates for some 20 estuaries in an effort to compare trophic status among ecosystems. Figure 24.10 shows the inclusion of our loading estimates in relation to these other estuaries. We also include an estimate from the Peel-Harvey estuary in southwest Australia (McComb and Humphries, 1992) for comparison. The Peel-Harvey is a eutrophic lagoonal system set in a Mediterranean climate with low tidal amplitude that also experiences periods of hypersalinity. It is clear that eastern Florida Bay has the lowest range of TP loading of any other estuary reported, being two to three orders of magnitude lower. It is also one of the lowest in TN loading as well (but see Fig. 24.8). In addition, loading rates in eastern Florida Bay deviate the most from Redfield ratio, being very high (>200) compared to other estuaries.

The geomorphology of Florida Bay with its shallow basins and mud banks in conjunction with the almost unbroken boundary of the Florida Keys has a very significant impact on advective P and N loading from both the Gulf of Mexico and the Atlantic Ocean. We will not have any estimates of these loadings until a working hydrodynamic model is completed by the Army Corps of Engineers. Addition of advective P and N loading will probably affect mostly Western Bay, as there is little mixing to the east on an annual cycle. However, precipitation inputs may be signif-

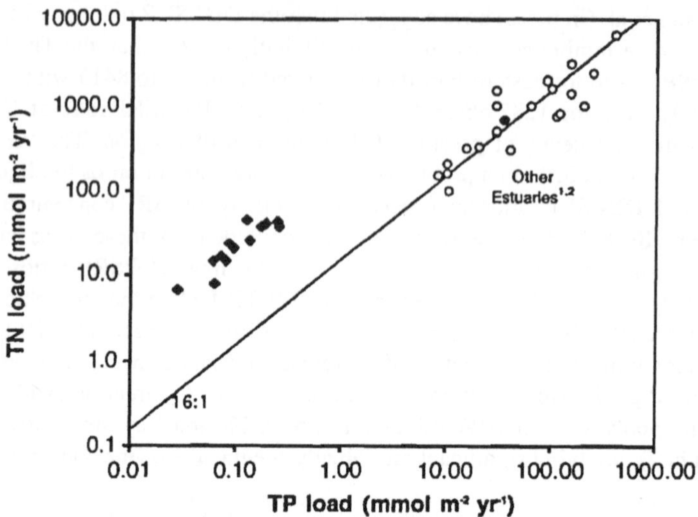

FIGURE 24.10 Areal annual loads of TN and TP (mmol m^{-2} yr^{-1}) assuming eastern Florida Bay to cover 1000 km^2. TP loads (◆) are 2-4 orders of magnitude lower than those for 20 other estuaries (○) while TN loads are only 1-2 orders of magnitude less ([1]Boynton et al. 1995). Also shown is areal loading from Peel-Harvey estuary (●), a subtropical lagoon ([2]McComb and Humphries 1992). The 16:1 line accentuates the deviation of eastern Florida Bay from standard Redfield ratio.

icant, especially for N. It has been estimated that for many estuaries, N loading by direct precipitation can account for up to a third of the total N load (Fisher et al., 1988; Hinga et al., 1991; Fu and Winchester, 1994). This being the case, we would expect the input TN:TP ratio to be even higher than 200, making eastern Florida Bay even more P limited. Finally, the groundwater loading component may also be quite significant but, because of difficulty in measurement, remains unknown at this time.

24.6 CONCLUSIONS

Increased freshwater inputs to Taylor Slough and the C111 coastal basin by canal inputs were more important in alleviating the hypersaline condition in eastern Florida Bay than was precipitation alone. Granted that climactic cycles have much to do with the absolute amount of water in South Florida, it was the input of water from outside the Taylor Slough watershed that probably made the difference. Continuous pumping at the S332 most probably acts to recharge the local groundwater field in Taylor Slough, alleviating the deficit and allowing natural precipitation events to become immediate runoff.

For an estuary to exhibit increased P loading while P concentration declines usually requires an increase in either storage or advection. We believe the answer lies in the fact that the TP concentration of the incoming freshwater is approximately equal to ambient TP concentrations in the Bay. We propose that management activities to increase input of freshwater from Taylor Slough and the coastal basin may have effected a significant decline in both salinity and TP concentrations in eastern Florida Bay via simple dilution.

Our calculations point out two important concerns in the interpretation of nutrient loading estimates. The first is the importance of reporting nutrient loading concentrations in addition to total mass load. Reporting only mass loading of P may be misleading when, during certain times of the year, increased freshwater flows actually dilute the nutrient concentrations in the receiving waters. The second important aspect of nutrient loading is that of chemical speciation. Almost all the P input is in organic form of unknown biological availability. The concentration of inorganic P is so low that it may be effectively unavailable to promote the primary production of phytoplankton and macrophytes.

Finally, what is the impact of nutrient loading on an ecosystem? We know that for oligotrophic ecosystems such as lakes and the Everglades, increased concentrations of P can have deleterious effect on the natural communities allowing invasion of mesotrophic and nonnative species. One line of reasoning is that any nutrient loading may be detrimental as it becomes entrained in internal nutrient cycles and is delayed from being exported. But what if the nutrient concentration of the inflowing water is at or below the ambient concentration? We argue that the total mass load is important but it is the concentration of the loading that becomes critical in maintaining an oligotrophic ecosystem. For a given mass load, it is less damaging for the nutrients to be delivered at a dilute concentration over the period of a year than it is for all of it to be added during a single event.

The above results point to the simple conclusion that Florida Bay is not like other estuaries of the eastern USA. Many people continue to believe the dogma that Florida Bay is a eutrophic ecosystem polluted by nutrient runoff from agriculture. Were it not for the Everglades marshes, this might indeed be true. We have shown that the actuality is one in which the loading inputs of the limiting nutrient, P, are at or below the ambient concentrations found in eastern Florida Bay. We believe this idea of critical loading concentration should be incorporated into the future planning of South Florida restoration activities.

24.7 ACKNOWLEDGMENTS

We especially want to thank Bill Walker for providing the Everglades loading estimates. In addition we thank all the field and laboratory technicians involved with this project including: Jeff Absten, Omar Beceiro, Sylvia Bolanos, Tom Frankovich, Bill Gilhooly, Scott Kaczynski, Elaine Kotler, Cristina Menendez, Jennifer Mohammed, Susy Perez, Pierre Sterling, and especially Pete Lorenzo. This project was possible because of the continued funding of the South Florida Water Management District through the Everglades National Park (SFWMD/NPS Cooperative Agreement C-7919 and NPS/SERP Cooperative Agreement 5280-2-9017).

This is a contribution of the Southeast Environmental Research Program at Florida International University.

REFERENCES

Bodle, M.J., A.P. Ferriter, and D.D. Thayer. 1994. The biology, distribution, and ecological consequences of *Melaleuca quinqueneria* in the Everglades. p. 341–356. *In* S. M. Davis and J. C. Ogden (ed.) Everglades, the ecosystem and its restoration. St. Lucie Press, Delray Beach, FL.

Boyer, J.N., D.W. Stanley, R.R. Christian, and W.M. Rizzo. 1988. Modulation of nitrogen loading impacts within an estuary. p. 165–176. *In* W.L. Lyke and T.J. Hoban (ed.) Proceedings of the American Water Resources Association, Symposium on Coastal Water Resources. AWRA Technical Publication Series TPS-88-1. AWRA, Bethesda, MD.

Boyer, J.N., J.W. Fourqurean, and R.D. Jones. 1997. Spatial characterization of water quality in Florida Bay and Whitewater Bay by multivariate analysis: Zones of similar influence (ZSI). Estuaries 20:743–758.

Boyer, J.N., J.W. Fourqurean, and R.D. Jones. 1998. Seasonal and long term trends in the water quality of Florida Bay (1989–1997). Estuaries. In press.

Boynton, W.R., W.M. Kemp, and C.W. Keefe. 1982. A comparative analysis of nutrients and other factors influencing estuarine phytoplankton production. p. 69–90. *In* V. S. Kennedy (ed.) Estuarine Comparisons. Academic Press, New York.

Boynton, W.R., J.H. Garber, R. Summers, and W.M. Kemp. 1995. Inputs, transformations, and transport of nitrogen and phosphorus in Chesapeake Bay and selected tributaries. Estuaries 18:285–314.

Boynton, W.R., J.D. Hagy, L. Murray, C. Stokes, and W.M. Kemp. 1996. A comparative analysis of eutrophication patterns in a temperate coastal lagoon. Estuaries 19:408–421.

Butler, M.J. IV, J.V. Hunt, W.F. Herrnkind, M.J. Childress, R. Bertelson, W. Sharp, T. Matthews, J.M. Field, and H.G. Marshall. 1995. Cascading disturbances in Florida Bay, USA: cyanobacteria blooms, sponge mortality, and implications for juvenile spiny lobsters *Panulirus argus*. Mar. Ecol. Prog. Ser. 129:119–125.

Correll, D.L., T.E. Jordan, and D.E. Weller. 1992. Nutrient flux in a landscape: effects of coastal land use and terrestrial community mosaic on nutrient transport to coastal waters. Estuaries 15:431–442.

Day, J.W. Jr., C.A.S. Hall, W.M. Kemp, and A. Yanez-Arancibia. 1989. Estuarine ecology. John Wiley, New York.

Fisher, D.J., J. Cerasco, T. Matthew, and M. Oppenheimer. 1988. Polluted coastal waters: the role of acid rain. Environmental Defense Fund, New York.

Fourqurean, J.W., R.D. Jones, and J.C. Zieman. 1993. Processes influencing water column nutrient characteristics and phosphorus limitation of phytoplankton biomass in Florida Bay, FL, USA: Inferences from spatial distributions. Estuar., Coast. Shelf Sci. 36:295–314.

Fourqurean, J.W., M.B. Robblee, and L. Deegan. 1998. Florida Bay: a brief history of recent ecological changes. Estuaries. In press.

Fu, J.-M. and J.W. Winchester. 1994. Sources of nitrogen in three watersheds of northern Florida, USA: mainly atmospheric deposition. Geochim. Cosmochim. Acta 58:1581–1590.

Hinga, K.R., A.A. Keller, and C.A. Oviatt. 1991. Atmospheric deposition and nitrogen inputs to coastal waters. Ambio 20:256–260.

Hirsch, R.M., R.B. Alexander, and R.A. Smith. 1991. Selection of methods for the detection and estimation of trends in water quality. Water Resour. Res. 27:803–813.

Hutchinson, G. E. 1969. Eutrophication, past and present. p. 17–26. In Eutrophication: causes, consequences, and correctives. National Academy of Sciences, Washington, DC.

Kennish, M.J. 1986. Ecology of estuaries. Vol. 1: Physical and chemical aspects. CRC Press, Boca Raton, Florida.

Magnien, R.E., R.M. Summers, and K.G. Sellner. 1992. External nutrient sources, internal nutrient pools, and phytoplankton production in Chesapeake Bay. Estuaries 15:497–516.

McComb, A.J., and R. Humphries. 1992. Loss of nutrients from catchments and their ecological impacts in the Peel-Harvey estuarine system, western Australia. Estuaries 15:529–537.

National Research Council. 1990. Managing troubled waters: the role of marine environmental monitoring. National Academy Press, Washington, DC.

Neilson, B.J., and L.E. Cronin. 1981. Estuaries and nutrients. Humana Press, New Jersey.

Nixon, S. W. 1995. Coastal marine eutrophication: a definition, social causes, and future concerns. Ophelia 41:199–219.

Nixon, S.W. and M.E.Q. Pilson. 1984. Estuarine total system metabolism and organic exchange calculated from nutrient ratios: an example from Narragansett Bay. p. 261–290. *In* V.S. Kennedy (ed.) The Estuary as a Filter. Academic Press, New York.

Nuttle, W.K. 1997. Salinity transfer functions for Florida Bay and west coast estuaries. Report to Everglades National Park.

Phlips, E.J. and S. Badylak. 1996. Spatial variability in phytoplankton standing stock and composition in a shallow inner-shelf lagoon, Florida Bay, Florida. Bull. Mar. Sci. 58:203–216.

Robblee, M.B., T.B. Barber, P.R. Carlson Jr., M.J. Durako, J.W. Fourqurean, L.M. Muehlstein, D. Porter, L.A. Yabro, R.T. Zieman, and J.C. Zieman. 1991. Mass mortality of the tropical seagrass *Thalassia testudinum* in Florida Bay (USA). Mar. Ecol. Prog. Ser. 71:297–299.

Schelske, C.L., F.J. Aldridge, and W.F. Kenney. (in press) Assessing nutrient limitations and trophic state in Florida lakes. *In* K.R. Reddy (ed.), Phosphorus biogeochemistry in subtropical ecosystems. CRC Press.

Smith, S.V., and M.J. Atkinson. 1994. Mass balance of nutrient fluxes in coastal lagoons. p. 133–155. *In* B. Kjerfve (ed.) Coastal Lagoon Processes. Elsevier.

Vollenweider, R.A. 1975. Input-output models with special reference to the phosphorus loading concept in ecology. Schwei. Zeit. Hydrol. 37:53–84.

Vollenweider, R.A. 1976. Advances in defining critical loading levels for phosphorus in lake eutrophication. Mem. Instit. Ital. Idrobiol. 33:53–83.

Section VII

Synthesis and Modeling

25 Management Models To Evaluate Phosphorus Transport from Watersheds

Kenneth L. Campbell

25.1 ABSTRACT

Spatial decision support systems developed using integrated geographic information systems (GIS) and hydrologic/water quality models have great potential to provide the information needed by local decision makers and policy makers, who often must act without the benefit of detailed information on the impacts of their actions. Hydrologic/water quality models (CREAMS-WT, FHANTM, FHANTM 2.0) have been developed and modified for use in Florida's flat, sandy, high water table (flatwoods) soils. GIS-based decision support system tools (LOADSS, IDM, BRADSS) for farm and watershed-scale water quality planning have been developed based on these hydrologic/water quality models. These tools can be used for analyzing the effects of phosphorus control practices at different geographic scales. LOADSS was designed to allow regional planners to alter land uses and management practices in the Lake Okeechobee watershed and then view the environmental and economic effects resulting from the changes on a regional scale. IDM was constructed specifically for application on dairies in the Lake Okeechobee watershed for evaluation of the water quality effects of specific combinations of dairy management systems. BRADSS was recently developed for use as a tool for evaluating water, nitrogen, and phosphorus losses from individual ranches in the Lake Okeechobee watershed when various management practices are implemented on individual fields of the ranch.

There is great public concern about the quality of Florida's environment and the potential adverse environmental impacts from agriculture if appropriate management

1-56670-331-X/99/$0.00+$.50

practices are not implemented. The search for solutions to the many problems related to nutrient management that affect water resources implies a continued demand for the development of modeling systems that can be used to analyze, in a holistic approach, the impact of alternative management policies. These tools have great potential to help determine the best combinations of management to minimize adverse nutrient impacts on the environment while generating reasonable socioeconomic outcomes for the people who depend on these lands and serve as the direct stewards of these natural systems.

25.2 INTRODUCTION

Assessment and mitigation of adverse effects of agricultural practices on water quality are necessary to protect our natural resources and environment. Agricultural production operations are highly variable in their physical facilities, management systems and the soil, drainage, and climatic conditions that affect the risk of water pollution from nutrients. The modeling of hydrology and water quality using computer simulation models is an efficient and cost-effective way of determining the effects of related agricultural management practices. Linkage between geographic information systems (GIS) and hydrologic models offers an excellent way for representing the spatial features of the fields being simulated and improving results. In addition, a GIS containing a relational database is a functional way of storing, retrieving, and formatting the various types of spatial and tabular data required to run a simulation model. Use of these hydrologic/water quality models and GIS tools in a decision support system framework results in a user-friendly system that can provide resource managers and decision makers an integrated view of large amounts of spatially distributed information on which to base important resource management decisions. Tim (1996) addresses this and other technologies as they may affect the future of hydrologic/water quality modeling in greater detail.

Capece et al. (1987, 1988) analyzed hydrologic data collected from Kissimmee River basin small watersheds and developed improved rainfall-runoff relationships. Heatwole et al. (1987a, 1987b, 1988) subsequently incorporated these findings into and developed the CREAMS-WT and BASIN models. This software package and database allows evaluation of best management practice (BMP) water quality impacts upon the region. In brief, modifications to the hydrologic component of CREAMS restrict deep seepage out of the surficial water table and introduce a recession curve model to handle water-table dynamics below the root zone (Heatwole et al., 1987b). Within the root zone, an evapotranspiration-precipitation accounting procedure budgets soil moisture. The high phosphorus (P) buffering capacity of soils found in the CREAMS soil P model was also drastically reduced to better reflect flatwoods soil conditions (Heatwole et al., 1988). This was accomplished by increasing the depth of the active layer within which soluble P is assumed to be available for extraction into runoff and for leaching into the root zone. This has the same effect as reducing the P extraction coefficients. Outputs from the modified CREAMS model serve as inputs to the BASIN model. BASIN takes the CREAMS-WT generated "edge-of-stream" nutrient loads and introduces factors to reflect nutrient loads, BMPs, and stream and wetlands attenuation, as well as nonagricultural background levels, to arrive at estimates of water quality impacts at the basin outlet (Heatwole et al., 1987a).

These efforts answered some questions regarding P sources and transport but also identified other important questions. Among the additional needed research topics identified was that of the region's unique runoff flowpaths and the related P transport mechanisms. Specifically identified was the need for clarification of appropriate hydrologic parameters relating watershed characteristics, land use, and management practices to the runoff parameters used by CREAMS-WT. The need for more detailed information regarding hydrologic events and associated P transport led to development of FHANTM (Campbell et al., 1995) and its recent enhancements in FHANTM 2.0 (Fraisse and Campbell, 1997a).

Effective P control strategies depend upon an understanding of the fate and transport of P in the watershed. FHANTM (Field Hydrologic And Nutrient Transport Model) is a field-scale model developed to simulate water and P movement from individual fields (Campbell and Tremwel, 1992; Campbell et al., 1995). The model is based on DRAINMOD (Skaggs, 1980) with modifications to include simulation of P movement and routing of overland flow. It was developed as part of an effort by the South Florida Water Management District (SFWMD) to develop P control practices aimed at reducing the levels of P in agricultural runoff and subsurface lateral flow in the Lake Okeechobee watershed. The DRAINMOD model developed to simulate water-table management (Skaggs, 1980) proved to be a useful platform because of its focus on continuous storage calculations. FHANTM was developed from DRAINMOD for continuous simulation of the phreatic zone moisture balance, thereby allowing the prediction of runoff volumes, peaks, and timing. The following functions were added to DRAINMOD:

1. An algorithm for overland flow routing
2. A dynamic deep seepage boundary
3. Algorithms to describe the fate of soluble P input, mass balance, and transport

Seepage volumes are predicted in accordance with water table fluctuations. FHANTM also simulates P (or any conservative solute) concentration in the phreatic zone and P loads in runoff. The results were an improved ability to model P movement when compared to previous field-scale hydrologic models used in this setting.

FHANTM assumes that P is not adsorbed to the soil and therefore moves freely with water through the soil profile and in the surface runoff. This assumption is adequate under some high loading conditions associated with dairies and other intensive uses where large amounts of P in forms that move freely through the system dominate. In beef pasture and other low-intensity systems, however, the amounts of P are much smaller and soil adsorption and other sinks and sources, such as decaying plant residues and soil P transformations, become a more important part of the P cycle and transport processes. Although spodic soils have a low retention capacity in the upper horizons, the above assumption may lead to an overestimation of the amount of P moving in the soil system. This may be why FHANTM simulated the intensive dairy pasture much better than the low-intensity beef pastures in its verification testing (Campbell et al., 1995).

FHANTM 2.0 is a modification of the previous version of the FHANTM model developed to simulate water and P movement from individual fields. FHANTM 2.0 replaced the P component developed for the previous version of the model and incorporated a nitrogen (N) component, providing improved performance in pasture and rangeland areas (Fraisse and Campbell, 1997a). FHANTM 2.0 further enhances the representation of nutrient processes and transformations in the model by incorporating the N and P components of the GLEAMS (Groundwater Loading Effects of Agricultural Management Systems) model (Knisel et al., 1993). The existing P component was completely replaced by the GLEAMS P-handling component. All major processes and transformations are currently represented in the model, including the exchange of soluble with adsorbed P, simulation of P release from decaying plant material and the possibility of simulating P loads resulting from intensive spreading or incorporation of solid and liquid animal waste. An option to add P present in rainfall to the surface P pools, that is not available in GLEAMS, also was incorporated into the FHANTM model. The current nutrient component includes mineralization from crop residue, from soil organic matter, and from animal waste, immobilization to crop residue, solution and adsorbed phases for transport and routing, and crop uptake. Nutrient specific processes such as N_2 fixation by legumes, denitrification and ammonia volatilization from animal waste are also taken into consideration by the model. The model simulates land application of animal wastes by creating appropriate N and P pools for mineralization. Initial testing of FHANTM 2.0 shows that it gives consistently better results for P transport than the previous version of FHANTM.

25.3 HYDROLOGIC MODELS AND GIS

By using models, we can better understand or explain natural phenomena and under some conditions make predictions in a deterministic or probabilistic sense (Woolhiser and Brakensiek, 1982). A hydrologic model can be defined as a mathematical representation of the transport of water and its constituents on some part of the land surface or subsurface environment. Hydrologic models can be used as planning tools for determining management practices that minimize nutrient loadings from an agricultural activity to water resources. The results obtained depend on a good representation of the environment through which water flows and of the spatial distribution of rainfall characteristics. These models have been quite successful in dealing with time but they are often spatially aggregated or lumped parameter models. Recently, hydrologists have turned their attention to GIS for assistance in studying the movement of water and its constituents in the hydrologic cycle. GIS are computer-based tools to capture, manipulate, process, and display spatial or georeferenced data. They contain both geometry data (coordinates and topological information) and attribute data; that is, information describing the properties of geometrical objects such as points, lines, and areas (Fedra, 1996). A GIS can represent the spatial variation of a given field property by means of a cell grid structure in which the area is partitioned into regular grid cells (raster GIS) or using a set of points, lines, and polygons (vector GIS).

There is obviously a close connection between GIS and hydrologic models and tremendous benefits in integrating them. Parameter determination is currently one of the most active areas in GIS related to hydrology. Parameters such as land surface slope, channel length, land use, and soil properties of a watershed are being extracted from both raster and vector GIS, with most work up to this time in raster-based systems. The spatial nature of a GIS also provides an ideal structure for modeling. A GIS can be a substantial time saver that allows different modeling approaches to be tried, sparing manual encoding of parameters. Furthermore, it can provide the tool for examining the spatial information from various user-defined perspectives (Tim et al., 1992). It enables the user to selectively analyze the data pertinent to the situation and try alternative approaches toward the analysis. GIS has been particularly successful in addressing environmental problems.

25.3.1 Approaches for Integrating GIS and Models

There has been a significant amount of work integrating both raster and vector GIS with hydrologic/water quality models (Maidment, 1993; Fedra, 1996; Zhang et al., 1990; Hallam et al., 1996; Goodchild et al., 1996; Jain et al., 1995; Tim and Jolly, 1994). Several strategies and approaches for the integration have been tried. Early work tended to use simpler models such as DRASTIC (Whittemore et al., 1987) and the Agricultural Pollution Potential Index (Petersen et al., 1991). In these cases, the models were implemented within the GIS. These studies aimed at the development of GIS-based screening methods to rank the nonpoint pollution potential. The use of more complex models requires that the GIS be used to retrieve, and possibly format, the model data. The model itself is implemented separately and communicates with the GIS via data files. This mode is referred to by Goodchild (1993) as "loose coupling," implying that the GIS and modeling software are coupled sufficiently to allow the transfer of data, and perhaps also of results in the reverse direction. Fedra (1996) refers to this level of integration as "shallow coupling" (Fig.25.1). Only the file formats and the corresponding input and output routines, usually of the model, have to be adapted. An application of this type is described by Liao and Tim (1992), in which an interface was developed to automatically generate topographic data and simplify the data input process for the Agricultural Non-Point-Source Pollution Model (AGNPS) (Young et al., 1989) water quality model.

In higher forms of connection, a common interface and transparent file or information sharing and transfer between the respective components is provided (Fig. 25.2). The BRADSS model, discussed in the next section, is an application of this kind. It links the FHANTM 2.0 model and GIS for the purpose of evaluating the potential water, N and P losses from individual ranches. LOADSS is an extension of this type of application, since it includes an optimization module that will enable the system to select the best P control practices at the regional scale based on goals and constraints defined by the user. Both applications use ARC/INFO's AML, which is a higher-level application language built into the GIS. A subset of functions of a

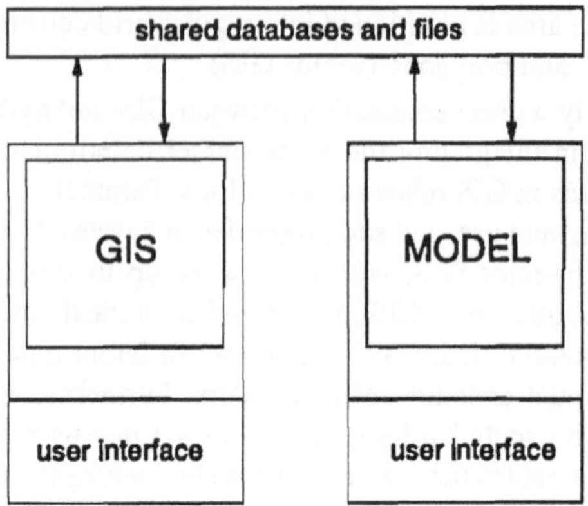

FIGURE 25.1 Loose or shallow coupling through common files (Fedra, 1996).

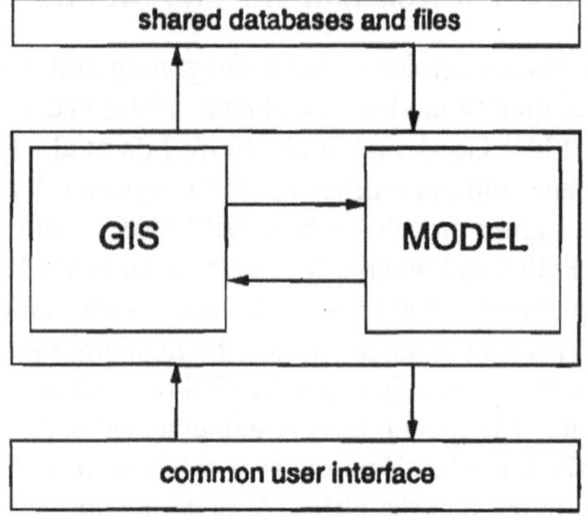

FIGURE 25.2 Deep coupling in a common framework (Fedra, 1996).

full-featured GIS, such as creation of maps (including model output) and tabular reports as well as model-related analysis, are embedded in the applications, giving the system great flexibility and performance. A deeper level of integration described by Fedra (1996) would merge the two previous approaches, such that the model becomes one of the analytical functions of a GIS, or the GIS becomes yet another option to generate and manipulate parameters, input and state variables, and model output, and to provide additional display options. In this case, software components would share memory rather than files.

The choice between integrating a water quality model with a raster or vector GIS depends on the importance of spatial interactions in the process being studied and the nature of the model itself. Some water quality models such as FHANTM are field-scale models that provide edge-of-field values for hydrologic and water quality parameters. In this case, spatial interactions between adjacent fields are ignored, and a vector GIS can be used to describe the system. Moreover, important factors in the simulation process, such as land use and management practices, are normally field attributes and are thus better represented in a vector structure. However, other factors playing an important role in the hydrologic process, such as field slope, aspect, and specific catchment area are hard to estimate in vector systems. Watershed models in which the process of routing is important and spatial interactions are considered are better handled by raster-based GIS. Several algorithms for estimating important terrain attributes are often incorporated in commercially available raster-based GIS packages.

25.4 DECISION SUPPORT SYSTEMS

Spatial decision support systems developed using integrated GIS and hydrologic/water quality models have great potential to provide the information needed by local decision makers and policy makers who often must act without the benefit of detailed information on the impacts of their actions. An integrated decision support system should be capable of providing resource analysts with timely information on the relative cost-effectiveness of alternative agricultural management strategies for a given watershed (Lovejoy et al., 1997). Output from the decision support system could be used by farmers and local, state, and federal agency personnel to identify specific geographic areas that have high nutrient losses, soil erosion, or sedimentation. Lovejoy et al. (1997) recommend the development and use of such systems to increase the quantity and quality of information available for use by resource managers to make more informed decisions on the environmental and economic tradeoffs of alternative management systems.

Hydrologic/water quality models developed and modified for Florida conditions (flat, sandy, high-water-table) including CREAMS-WT, FHANTM, and FHANTM 2.0, are being used as planning tools for implementation of management systems to minimize nutrient loads to lakes and streams. Engineering workstations and GIS software are being applied in a decision support system framework for regional water quality planning. The GIS manages spatial databases and provides data to a hydrologic/water quality model. This model supplies output data to the database so that water and nutrient loads can be viewed spatially as an aid to decision-making. The search for solutions to the many problems related to nutrient management that affect water resources implies a continued demand for the development of modeling systems that can be used to analyze, in a holistic approach, the impact of alternative management policies.

The current research program has resulted in a set of GIS-based decision support system tools for farm and basin-scale water quality planning including:

- LOADSS (Lake Okeechobee Agricultural Decision Support System)—a regional planning decision support system structured around a geographic information system (ARC/INFO) (Negahban et al., 1995; Negahban et al., 1996a). Its purpose is to aid in determining the P loads reaching Lake Okeechobee from each subbasin of its watershed when specific management practices are applied to each land use in the basin. CREAMS-WT

(USDA-ARS CREAMS model modified for high-water-table) and FHANTM are used within LOADSS to provide field-scale water and P loads for use in regional planning.

- IDM (Interactive Dairy Model)—a direct linkage of FHANTM with a geographic information system (ARC/INFO) for use in evaluating water and P losses from individual dairies in the Lake Okeechobee basin when various management practices are implemented on individual fields of the dairy (Negahban et al., 1993). This model can be used within the LOADSS structure or independent from it.
- BRADSS (Beef Ranch Decision Support System)—a direct linkage of FHANTM 2.0 with a geographic information system (ARC/INFO) for use in evaluating water, N and P losses from individual ranches in the Lake Okeechobee basin when various management practices are implemented on individual fields of the ranch (Fraisse and Campbell, 1997b). It was designed for flexibility and ease of use in a generalized beef ranch type of application. BRADSS uses a GIS-based interface to develop field-level management plans for beef ranches, runs the FHANTM 2.0 simulation model, and displays the results obtained in results tables, spatial maps, and time-series graphs.

The development of LOADSS is an example of how the integration of hydrologic models and GIS can be used for analyzing nutrient control practices at different scales. The addition of optimization algorithms further enhances the ability of policy and decision makers for analyzing the impact of alternative management practices and land uses at the regional level. LOADSS (version 3.1), including the CREAMS-WT regional-scale model, the IDM components and the optimization component, is fully functional and currently available at the South Florida Water Management District. Results show that LOADSS behaves consistently with measured data at the lake basin scale. The optimization component generates the best combination of management practices for the basin(s) of concern given the goals and constraints as established by the user.

IDM is a direct linkage of FHANTM with ARC/INFO for use in evaluating water and P losses from individual dairies in the Lake Okeechobee basin when various management practices are implemented on individual fields of the dairy. Details of the technical design and implementation of IDM are available in previous reports (Fonyo et al., 1993; Negahban et al., 1993). IDM was constructed specifically for application on dairies in the Lake Okeechobee basin for evaluation of the water quality effects of specific combinations of dairy management systems.

The development of BRADSS represents a different approach in integrating water quality models and GIS in the sense that it is designed to be generic and focused mainly on the ranch or farm level. It is primarily designed to help decision and policy makers to analyze the effects of alternative water quality management practices at the ranch or farm level. The framework can easily be adapted to handle different types of animal wastes (such as beef, dairy, and poultry) as well as to simulate the impact of other crop management practices such as pesticide applications.

25.4.1 LOADSS

LOADSS was developed to help address problems created by P runoff into Lake Okeechobee. It was designed to allow regional planners to alter land uses and management practices in the Lake Okeechobee basin and then view the environmental and economic effects resulting from the changes. The Lake Okeechobee basin coverage incorporates information about land uses, soil associations, weather regions, management practices, hydrologic features, and political boundaries for approximately 1.5 million acres of land and consists of close to 8000 polygons.

The SFWMD, responsible for managing Lake Okeechobee, has initiated numerous projects to develop effective control practices to reduce the level of P in agricultural runoff as part of the Lake Okeechobee SWIM (Surface Water Improvement and Management) Plan (SFWMD, 1989). These projects, numbering over 30, have been designed to develop information on the control and management of P within the lake basin and to determine the costs and effectiveness of selected management options. There are three types of control options being studied: (1) nonpoint source controls, such as pasture management, (2) point source controls, such as sewage treatment, and (3) basin-scale controls, such as aquifer storage and retrieval. With the completion of the majority of these research efforts, the need arose for a comprehensive management tool that could integrate the results for all three classes of P control practices. In response to these needs, design and implementation of a decision support system was initiated with the following objectives (Negahban et al., 1995):

- Organize spatial and nonspatial knowledge about soils, weather, land use, hydrography of the lake basin and P control practices (PCPs) under a GIS environment.
- Develop and implement algorithms for modeling nonpoint source, point source and basin-scale PCPs.
- Develop and implement mechanisms for evaluating the performance of the entire Lake Okeechobee basin under different combinations of PCPs applied to the basin.
- Design and develop a user interface that would facilitate the use of the system by noncomputer experts.

The goal in developing LOADSS was to create an information system that would integrate available information for regional planners to make decisions. It is able to generate reports and maps concerning regional land attributes, call external hydrologic simulation models, and display actual water quality and quantity sampling station data. LOADSS is a collection of different components:

- The regional-scale GIS-based model used to develop and manipulate regional plans aimed at reducing P loading to Lake Okeechobee.
- The IDM used to develop field-level management plans for dairies and run the FHANTM simulation model for nutrient transport modeling.
- An optimization module that enables the system to select the best P control practices at the regional-scale given an objective function and a set of constraints.

Although these components can run independently, they are fully integrated in the LOADSS package and are able to exchange information where necessary. A design schematic of LOADSS is given in Figure 25.3.

Regional-Scale GIS-Based Model

LOADSS serves both as a decision support system for regional planning and as a graphical user interface for controlling the different components. One consideration in the design of LOADSS was the size of the database that was being manipulated. Since the land use database consisted of nearly 8000 polygons, it was decided that running the simulation models interactively would not be a feasible option. Thus, the CREAMS-WT (Heatwole et al., 1987b, 1988) hydrologic/water quality model was prerun for different levels of inputs and management for each land use, soil association, and weather region (Kiker et al., 1992). Depending on the land use and its relative importance as a contributor of P to the lake, anywhere from 1 (background levels of inputs to land uses like barren land) to 25 (dairies, beef pastures) levels of inputs were selected. Each set of inputs to a particular land use was given a separate PCP identification code. A CREAMS-WT simulation was performed for each PCP, on each soil association and weather region. This resulted in about 2,600 simulation runs. Annual average results were computed from the 20-year simulations for use in LOADSS. CREAMS-WT provides an average annual estimate of P runoff from each polygon. Phosphorus assimilation along flow paths to Lake Okeechobee are estimated as an exponential decay function of distance traveled through canals and wetlands (SFWMD, 1989).

The imports, exports, and economics of each PCP are based on a per production unit basis. Depending on the type of polygon, the production unit can be acres (pastures, forests, etc.), number of cows (dairies), or millions of gallons of effluent (waste treatment plants and sugar mills). Developing a regional plan in LOADSS is a process of assigning a PCP identification code to each one of the polygons in the Lake Okeechobee basin. Assessing the results of a regional plan is a process of multiplying the production unit of each polygon by its appropriate database import, export, or economic attribute and summing the resulting values over all polygons in the Lake Okeechobee basin. LOADSS runs in the ARC/INFO version 6.1.1 GIS software on SUN SPARC® stations.

Interactive Dairy Model

Although the LOADSS level of detail is adequate for regional planning, a more detailed model was necessary to analyze individual dairies in the Lake Okeechobee basin, as dairies were one of the large and concentrated sources of P runoff into the lake. Thus, the IDM was developed and incorporated into LOADSS. The IDM uses FHANTM to simulate P movement in dairy fields. FHANTM is a modification of DRAINMOD (Skaggs, 1980) with added functions to handle overland flow routing, dynamic seepage boundary, and soluble P algorithms for P input, mass balance, and transport (Tremwel and Campbell, 1992). Unlike in LOADSS, FHANTM is run interactively, as IDM requires. Furthermore, in LOADSS the user can only select between a number of

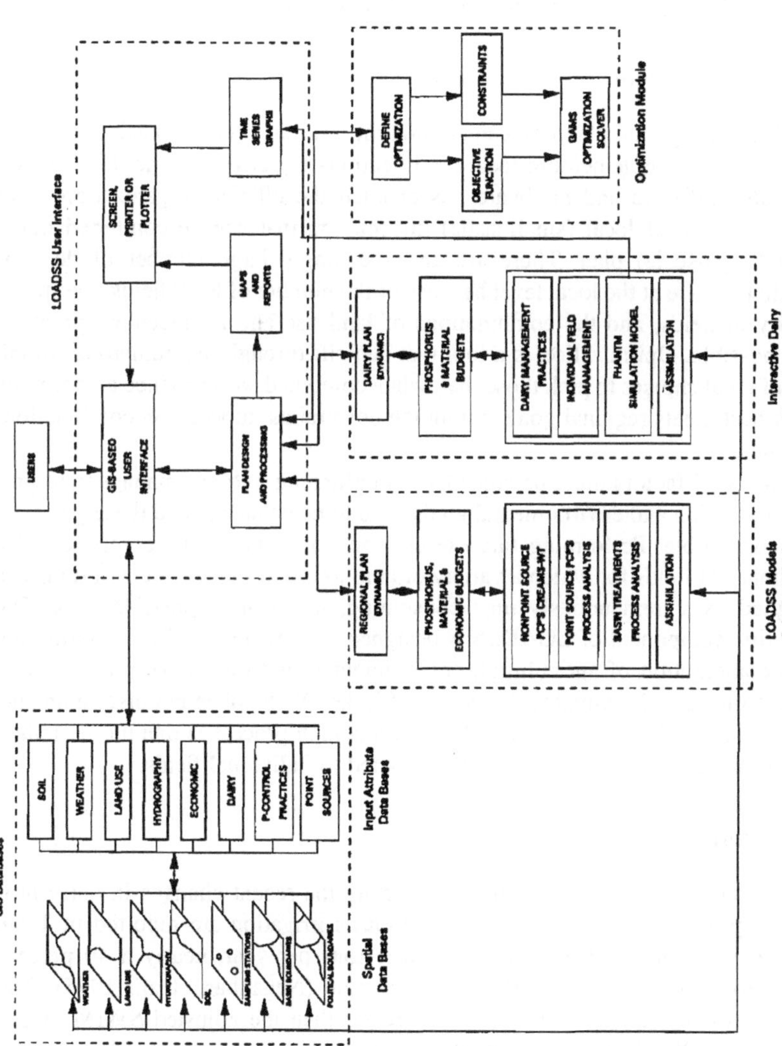

FIGURE 25.3 LOADSS design schematic.

predefined PCPs, while in IDM the user has access to more than 100 input and management variables, all of which can take a range of values. This allows for the development and evaluation of detailed dairy management plans that would be otherwise impossible at a regional scale. While LOADSS only provides average annual results, IDM is able to display daily time series simulation results to the user. IDM uses the same assimilation algorithm and can produce the same P budget maps and reports as LOADSS. A design schematic of IDM is shown in Figure 25.4.

Optimization Module

The process of regional planning can be thought of as an attempt to achieve certain regional goals by implementing local modifications in activities. The development of computer software and hardware has enabled the efficient organization of an extensive amount of local (small-scale) information that can be combined into a regional (large-scale) plan. There are an exceedingly large number of decision choices to be made at the local level because of the number of local fields, the spatial variability in fields, and the combinations of land use and management practices that are possible. Optimization is a tool that can sift through the numerous combinations of local choices to pick those that, when combined, will produce an optimum plan that best meets regional goals within the constraints imposed on combinations of activities.

A variety of factors must be considered in planning nutrient management programs. Production and environmental goals need to be balanced, and these goals are often incompatible. Performing this exercise on a regional scale, composed of a large number of fields for which a variety of land uses and management options can be assigned, is a tremendously time consuming, if not an impossible, task. The optimization component of LOADSS is designed to provide a tool for determining the best combination of agricultural, environmental, and regulatory practices that not only protects and maintains the health of Lake Okeechobee but also maintains the economic viability of the region. The optimization process is one more method by which the user can modify the PCPs assigned to individual fields.

LOADSS Results

LOADSS estimates of 1993 lake loads reflecting the recent changes in dairy production, improved pasture fertilization, and citrus irrigation are reported in Table 25.1. Total loads are estimated at 213 metric tons per year, nearly a 40 percent reduction from the adjusted SWIM load estimate (Negahban et al., 1995). The estimate of 1993 loads is 57 metric tons greater than the adjusted SWIM target. However, 14 of the 22 basins meet or exceed the target reductions. Three basins (i.e., S-191, S-154, and S-65D) account for 78 percent of the target shortfall, despite the fact that these basins are estimated to have achieved a 50 percent average reduction in load. These three basins are all important dairy producing areas.

The optimization module in LOADSS was tested on a pilot basin (Negahban et al., 1996b). Of the six optimal solutions obtained from eight optimization formulations specified for the basin, all but one were able to improve on the LOADSS base

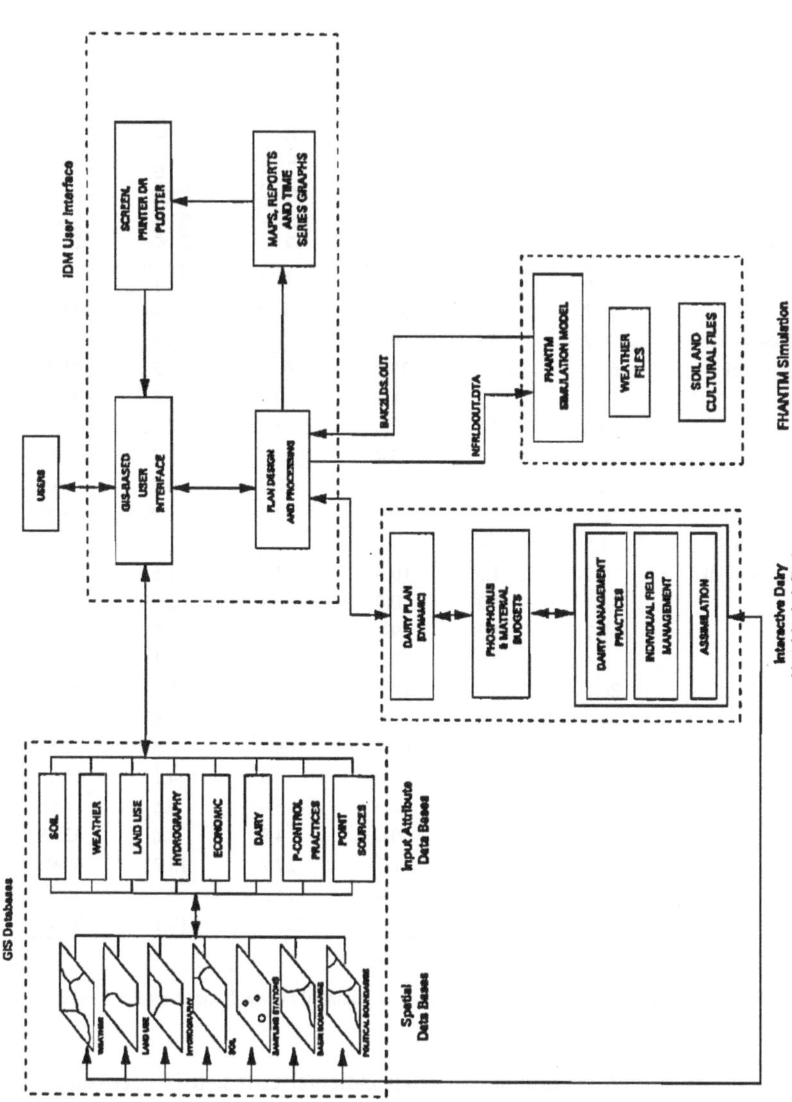

FIGURE 25.4 IDM design schematic (Negahban et al., 1993).

TABLE 25.1
Average Annual Adjusted SWIM Loads, SWIM Target Loads, LOADSS Calibrated Predairy Rule Loads, and LOADSS 1993 Estimates (Negahban et al., 1995)

Basin name	Adjusted SWIM P load, 1973–1987 avg. (ton/yr)	Adjusted SWIM target P load (ton/yr)	Target reduction (%)	LOADSS predairy rule (ton/yr)	LOADSS current conditions (ton/yr)	Estimated reduction (%)
C-40	15.6	6.7	57.0	15.4	11.2	27.1
C-41	41.2	19.0	54.0	40.0	31.4	21.5
C-41A (S-84)	6.2	6.4	0.0	6.3	4.6	26.1[a]
Fisheating Creek	60.3	45.0	25.4	60.7	45.3	25.4[a]
L-48	10.1	3.2	68.5	9.9	6.7	32.1
L-49	3.2	2.8	12.9	3.3	2.42	5.0[a]
L-59E, L-59W	3.5	3.4	5.4	3.4	2.23	6.8[a]
L-60E, L-60W	0.9	1.2	0.0	0.9	0.7	20.0[a]
L-61E, L-61W	2.5	2.7	0.0	2.5	1.92	5.0[a]
Lake Istokpoga	1.5	—	—	1.5	1.1	25.0
Nicodemus Slough	1.7	1.5	10.5	1.7	1.5	10.5[a]
S-131	1.1	1.2	0.0	1.1	0.8	25.0[a]
S-133	8.7	5.3	39.8	8.5	4.8	43.6[a]
S-135	2.4	2.4	0.0	2.4	2.1	11.5[a]
S-154	26.5	5.5	79.1	26.8	16.0	40.5
S-154C	0.5	0.4	19.0	0.5	0.4	33.3[a]
S-191	109.7	20.7	81.2	108.7	45.8	57.8
S-65A	4.8	5.0	0.0	4.8	4.1	15.1[a]
S-65B	6.1	6.5	0.0	5.9	5.6	4.6[a]
S-65C	4.7	5.0	0.0	4.9	2.8	42.6[a]
S-65D	28.3	6.4	77.3	27.7	15.2	44.9
S-65E	14.4	5.6	61.0	14.5	6.3	56.3
Total	354	156	56.0	351.4	213.1	39.4

[a]Indicates target load reduction reached.

Note: SWIM P loads were adjusted for differences in average annual flow volumes between LOADSS results and SWIM plan data.

plan for the basin in both economic and environmental terms. It is necessary for water resource planners to carefully analyze optimization formulation as well as suggested solutions, looking for logical loopholes that may have been left open. Although solutions may look good as far as the bottom line is concerned (for example, P loading to the lake), care should be taken to make sure that suggested activities (PCPs) on each field make practical sense. Optimization can be a very powerful tool provided that meaningful instructions are given to the algorithm.

Alternative means for obtaining additional reductions in lake loads can be evaluated using the components of the LOADSS system. The IDM can be used in consultation with district specialists and dairy operators to examine alternative P management practices for particular fields on individual dairies. The regional LOADSS planning model with the optimization module can be used to evaluate the cost effectiveness of additional nonpoint, point and basin-scale control practices for individual basins or for the entire drainage area.

25.4.2 BRADSS

BRADSS was recently developed by linking the FHANTM 2.0 model with ARC/INFO for use as a tool for evaluating water, N, and P losses from individual ranches in the Lake Okeechobee basin when various management practices are implemented on individual fields of the ranch. BRADSS is designed to be an additional tool for answering questions related to the environmental impacts of beef operations. A design schematic of BRADSS is given in Fig. 25.5. It operates on an

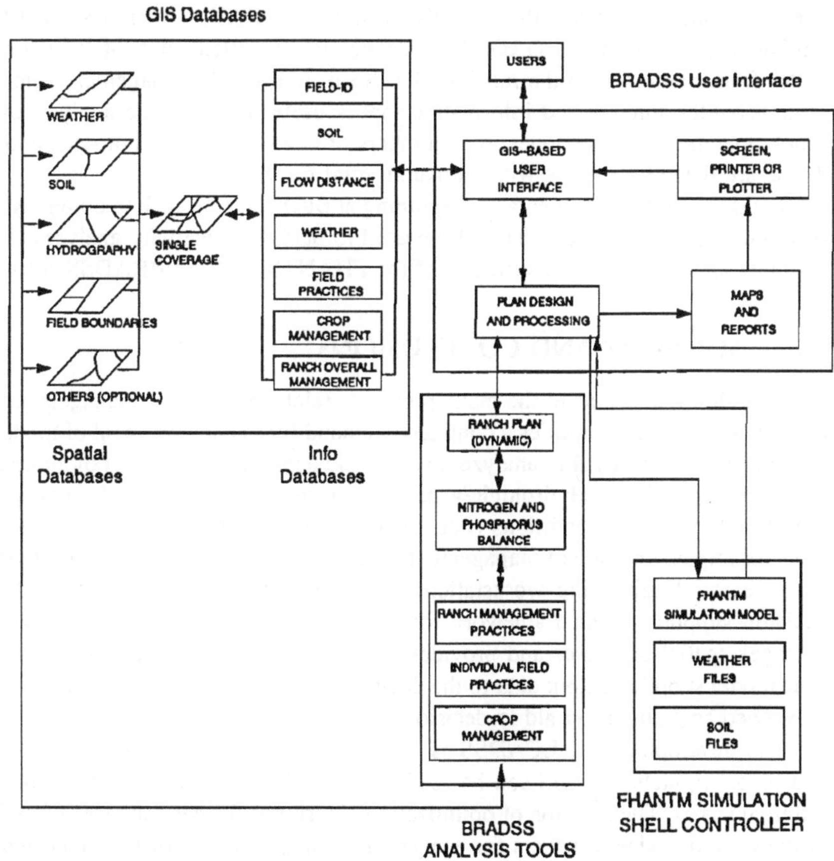

FIGURE 25.5 BRADSS design schematic (Fraisse and Campbell, 1997b).

individual ranch basis and incorporates the FHANTM 2.0 water quality model for simulating nutrient transport of both N and P for specific fields of a ranch. BRADSS was tailored to be used by researchers at the Buck Island Ranch, which is a 4,170 ha ranch located in south Florida and currently used for a number of research projects in the region. However, its design is generic, so that any ranch represented by a coverage for which relevant data such as soil characteristics, weather, and field boundaries are available can be simulated.

BRADSS was designed for flexibility and ease of use in a generalized beef ranch type of application. All model development was conducted on Unix workstations using ARC/INFO, Fortran, and C software. In developing BRADSS, the user interface menus were configured to fit a varied type of management on each individual field of the ranch. Experience gained with development of other decision support systems described above was used to design an improved functional framework for this new-generation decision support system. A time-series graphing capability is included that allows the user to view selected input and output variables in paired graphs. This provides a visual display of daily water and nutrient responses as a function of time, in addition to the more customary maps and tabular reports of average annual responses that are also available. BRADSS includes algorithms similar to those used in LOADSS to account for the attenuation of P in flowing water from field-edge to the ranch outlet. An overall ranch management summary table provides summarized information that is useful in economic analysis of the ranch operation. This includes total numbers of cows, acres, and nutrients applied. These are listed by pasture with an overall total for the ranch.

Field research experiments are beginning at Buck Island Ranch to obtain hydrology and water quality data that will be used to perform sensitivity analyses and to calibrate and verify the performance of the FHANTM 2.0 and BRADSS software.

25.5 SUMMARY AND CONCLUSIONS

The search for solutions to the many problems related to nutrient management that affect water resources implies a continued demand for the development of modeling systems that can be used to analyze, in a holistic approach, the impact of alternative management policies. Hydrologic/water quality models have been developed and modified for Florida conditions (flat, sandy, high water table) to use as planning tools for implementation of management systems to minimize nutrient loads to lakes and streams. Engineering workstations and GIS software are being applied in a decision support system framework for regional water quality planning. The GIS manages spatial databases and provides data to a hydrologic/water quality model. This model supplies output data to the database so that water and nutrient loads can be viewed spatially as an aid to decision-making.

The development of LOADSS is a good example of how the integration of hydrologic models and GIS can be used for analyzing nutrient control practices at different scales. The addition of optimization algorithms further enhances the ability of policy and decision makers for analyzing the impact of alternative management practices and land uses at the regional level. LOADSS (version 3.1), including the CREAMS-WT regional-scale model, the IDM components, and the optimization component is fully functional and currently available at the South Florida Water Management District. Results show that LOADSS behaves consistently with measured data at the lake basin scale. The optimization component generates the best combination of management practices for the basin(s) of concern given the goals and constraints as established by the user. Initial results from LOADSS indicate that current regulations and P control practices have achieved approximately three-fourths of the target load reduction. The remaining 25 percent reduction is likely to be more difficult and expensive. The LOADSS system provides a means for "what if" evaluations of the cost-effectiveness of additional P control practices applied at various scales within the drainage area. In a more general sense, the LOADSS system provides a versatile set of components useful in regional environmental planning. The system was designed to be as flexible as possible in order to allow for additional modifications to the capabilities of the system as well as to facilitate application to other regions and problems.

The development of BRADSS represents a different approach in integrating water quality models and GIS in the sense that it is designed to be generic and focused mainly on the ranch level. BRADSS has been developed as a tool for helping decision and policy makers to analyze the effects of alternative management practices at a ranch level. BRADSS integrates the FHANTM 2.0 water quality model with ARC/INFO GIS software and uses a GIS-based interface to develop field-level management plans for a ranch, run the FHANTM 2.0 water quality model, and analyze the results obtained. BRADSS is designed to allow the user a site-specific investigation of the water quality impacts resulting from field management practices. Although, currently, the management practices are uniform for an entire field, BRADSS is a first step toward the development of a tool that could analyze the impacts in distinct areas of a field and also assign site-specific management practices, further reducing the overall nutrient loadings to surface and ground waters.

There is great public concern about the quality of Florida's environment and the potential adverse environmental impacts from agriculture if appropriate management practices are not implemented. These hydrologic/water quality models and decision support system tools have great potential to help determine the best combinations of management to minimize adverse nutrient impacts on the environment while generating the best economic return possible for agricultural producers.

25.6 ACKNOWLEDGMENTS

This chapter is a contribution from the Institute of Food and Agricultural Sciences, University of Florida, as part of Southern Region Project S-273 of the USDA-CSREES.

REFERENCES

Campbell, K.L., J.C. Capece, and T.K. Tremwel. 1995. Surface/subsurface hydrology and phosphorus transport in the Kissimmee River Basin, Florida. Ecological Engineering 5:301–330.

Campbell, K.L. and T.K. Tremwel. 1992. Biogeochemical behavior and transport of phosphorus in the Lake Okeechobee basin: FHANTM users manual. Deliverable 2.4.4. South Florida Water Management District. West Palm Beach, Florida. 65 p.

Capece, J.C., K.L. Campbell, and L.B. Baldwin. 1988. Estimating runoff peak rates from flat, high-water-table watersheds. Transactions of the ASAE 31:74–81.

Capece, J.C., K.L. Campbell, L.B. Baldwin, and K.D. Konyha. 1987. Estimating runoff volumes from flat, high-water-table watersheds. Transactions of the ASAE 30:1397–1402.

Fedra, K. 1996. Distributed models and embedded GIS: Integration strategies and case studies. In GIS and environmental modeling: Progress and research issues. Edited by Goodchild, M.F., L.T. Steyaert, B.O. Parks, C. Johnston, D. Maidment, M. Crane, and S. Glendinning. GIS World, Inc., Fort Collins, Colorado. pp. 413–417.

Fonyo, C., B. Negahban, W. Boggess, K. Campbell, and J. Jones. 1993. LOADSS version 2.2 design document. Agricultural Engineering Research Report No. 93-L3, Department of Agricultural Engineering, University of Florida. Gainesville, Florida. 320 p.

Fraisse, C.W. and K.L. Campbell. 1997a. FHANTM (Field Hydrologic And Nutrient Transport Model) version 2.0 user's manual. Research Report. Agricultural and Biological Engineering Department, University of Florida. Gainesville, Florida. 185 p.

Fraisse, C.W., and K.L. Campbell. 1997b. BRADSS (Beef Ranch Decision Support System) version 1.0 user's and developer's manual. Research Report. Agricultural and Biological Engineering Department, University of Florida. Gainesville, Florida. 99 p.

Goodchild, M. 1993. The state of GIS for environmental problem-solving. In Environmental Modeling with GIS. Edited by Goodchild, M., B. Parks, and L. Steyaert. Oxford University Press, New York. pp. 8–15.

Goodchild, M.F., L.T. Steyaert, B.O. Parks, C. Johnston, D. Maidment, M. Crane, and S. Glendinning, eds. 1996. GIS and environmental modeling: Progress and research issues. GIS World, Inc. Fort Collins, Colorado. 486 p.

Hallam, C.A., J.M. Salisbury, K.J. Lanfear, and W.A. Battaglin, eds. 1996. Proceedings of the AWRA annual symposium, GIS and water resources. American Water Resources Association. Herndon, Virginia. TPS-96-3. 482 p.

Heatwole, C.D., A.B. Bottcher, and K.L. Campbell. 1987a. Basin scale water quality model for Coastal Plain flatwoods. Transactions of the ASAE 30:1023–1030.

Heatwole, C.D., K.L. Campbell, and A.B. Bottcher. 1987b. Modified CREAMS hydrology model for Coastal Plain flatwoods. Transactions of the ASAE 30:1014–1022.

Heatwole, C.D., K.L. Campbell, and A.B. Bottcher. 1988. Modified CREAMS nutrient model for Coastal Plain watersheds. Transactions of the ASAE 31:154–160.

Jain, D.K., U.S. Tim, and R.W. Jolly. 1995. A spatial decision support system for livestock production planning and environmental management. Applied Engineering in Agriculture 11:711–719.

Kiker, G.A., K.L. Campbell, and J. Zhang. 1992. CREAMS-WT linked with GIS to simulate phosphorus loading. ASAE Paper No. 92-9016. American Society of Agricultural Engineers, St. Joseph, Michigan. 18 p.

Knisel, W.G., R.A. Leonard, and F.M. Davis. 1993. GLEAMS version 2.10, part I: Nutrient component documentation. Unpublished, USDA/ARS. Tifton, Georgia. 49 p.

Liao, H.H., and U.S. Tim. 1992. Integration of geographic information system (GIS) and hydrologic/water quality modeling: An interface. ASAE Paper No. 92-3612. American Society of Agricultural Engineers. St. Joseph, Michigan. 16 p.

Lovejoy, S.B., J.G. Lee, T.O. Randhir, and B.A. Engel. 1997. Research needs for water quality management in the 21st century: A spatial decision support system. Soil and Water Conservation 52:18–22.

Maidment, D.R. 1993. GIS and hydrologic modeling. In Environmental Modeling with GIS. Edited by Goodchild, M., B. Parks, and L. Steyaert. Oxford University Press, New York. pp. 147–167.

Negahban, B., C. Fonyo, W.G. Boggess, J.W. Jones, K.L. Campbell, G. Kiker, E. Flaig, and H. Lal. 1995. LOADSS: A GIS-based decision support system for regional environmental planning. Ecological Engineering 5:391–404.

Negahban, B., C. Fonyo, K.L. Campbell, J.W. Jones, W.G. Boggess, G. Kiker, E. Hamouda, E. Flaig, and H. Lal. 1996a. LOADSS: A GIS-based decision support system for regional environmental planning. In GIS and Environmental Modeling: Progress and Research Issues. M. F. Goodchild, L. T. Steyaert, B. O. Parks, C. Johnston, D. Maidment, M. Crane, and S. Glendinning (eds.). GIS World, Inc.: Fort Collins, CO. pp. 277–282.

Negahban, B., G. Kiker, K. Campbell, J. Jones, W. Boggess, C. Fonyo, and E. Flaig. 1993. GIS-based hydrologic modeling for dairy runoff phosphorus management. In Proceedings of ASAE International Symposium on Integrated Resource Management & Landscape Modification for Environmental Protection. ASAE Publication 13-93. St. Joseph, Michigan. pp. 330–339.

Negahban, B., C.B. Moss, J.W. Jones, J. Zhang, W.G. Boggess, and K.L. Campbell. 1996b. Integrating optimization into a regional planning model using GIS. In Security and Sustainability in a Mature Water Economy: A Global Perspective. Water and Resource Economics Consortium, Centre for Water Policy Research, University of New England, Armidale NSW 2351, Australia. pp. 347–361.

Petersen, G.W., J.M. Hamlett, G.M. Baumer, D.A. Miller, R.L. Day, and J.M. Russo. 1991. Evaluation of agricultural nonpoint pollution potential in Pennsylvania using a geographic information system. Environmental Resources Research Institute—ER9105. University Park, Pennsylvania. 60 p.

SFWMD. 1989. Interim surface water improvement and management (SWIM) plan for Lake Okeechobee. South Florida Water Management District, West Palm Beach, Florida. 212 p.

Skaggs, R.W. 1980. DRAINMOD reference report: Methods for design and evaluation of drainage-water management systems for soils with high water tables. USDA-SCS, South National Technical Center, Fort Worth, Texas. 329 p.

Tim, U.S. 1996. Emerging technologies for hydrologic and water quality modeling research. Transactions of the ASAE 39:465–476.

Tim, U.S., and R. Jolly. 1994. Evaluating agricultural nonpoint-source pollution using integrated geographic information systems and hydrologic/water quality model. Journal of Environmental Quality 23:25–35.

Tim, U.S., M. Milner, and J. Majure. 1992. Geographic information systems / simulation model linkage: Processes, problems and opportunities. ASAE Paper No. 92-3610. American Society of Agricultural Engineers, St. Joseph, Michigan. 20 p.

Tremwel, T.K., and K.L. Campbell. 1992. FHANTM, a modified DRAINMOD: Sensitivity and verification results. ASAE Paper No. 92-2045. American Society of Agricultural Engineers, St. Joseph, Michigan. 23 p.

Whittemore, D.O., J.W. Merchant, J. Whistler, C.E. McElwee, and J.J. Woods. 1987. Groundwater protection planning using the ERDAS geographic information system: Automation of DRASTIC and time-related capture zones. Proceedings of the NWWA (National Water Well Association) FOCUS Conference on Midwestern Ground Water Issues. Dublin, Ohio. pp. 359–374.

Woolhiser, D.A., and D.L. Brakensiek. 1982. Hydrologic system synthesis. In Hydrologic Modeling of Small Watersheds. Edited by C.T. Haan, H.P. Johnson, and D.L. Brakensiek. ASAE Monograph No. 5. St. Joseph, Michigan. pp. 3–16.

Young, R.A., C.A. Onstad, D.D. Bosch, and W.P. Anderson. 1989. AGNPS: A nonpoint source pollution model for evaluation of agricultural watersheds. Journal of Soil and Water Conservation 44:168–173.

Zhang, H., C.T. Haan, and D.L. Nofziger. 1990. Hydrologic modeling with GIS: An overview. Applied Engineering in Agriculture 6:453–458.

26 Ecosystem Analysis of Phosphorus Impacts and Altered Hydrology in the Everglades: A Landscape Modeling Approach

H. Carl Fitz and Fred H. Sklar

1-56670-331-X/99/$0.00+$.50

26.1 ABSTRACT

The Everglades has undergone significant change in response to altered hydrology and water quality, which is why natural resource managers are now evaluating alternative water and nutrient management strategies for the region. Simulation models are an integral part of the process of understanding complex ecological systems, providing a means to evaluate potential ecosystem response to changes in management.

To evaluate various management alternatives on the Everglades ecosystems, we developed a spatially explicit ecosystem model for Water Conservation Area 2A (the Conservation Area Landscape Model, or CALM). The CALM simulates interactions among hydrology, chemistry, and biology of the marsh systems across the landscape, synthesizing ecosystem behavior in response to changing environmental inputs. Calibration results for 1980 to 1996 indicated good hydrologic and ecological agreement with observed data. Observed and simulated water stage were well correlated ($r^2 = 0.70$). North-south gradients of simulated dissolved inorganic phosphorus in the surface waters (50 – 4 μg P L^{-1}) and in the pore waters (950 – 10 μg P L^{-1}) were spatially and temporally realistic. Likewise, simulated and observed data along the gradient had similar values of peat accretion (5.5 – 3.1 mm yr^{-1}, respectively), macrophyte biomass (1100 – 300 g C m^{-2}, respectively), calcareous periphyton biomass (0 – 52 g C m^{-2}, respectively), and community shifts in periphyton and macrophytes.

The model captures the feedbacks among plants, hydrology, and biogeochemistry through process-based algorithms, including the influence of spatial patterns on these processes. For example, in a 17-year (1980 to 96) simulation sensitivity analysis of increased evaporative losses, there were more pronounced regions of increased soil P remineralization associated with reduced hydroperiods, which in turn increased macrophyte growth. If future management of the area lowers water levels below those that occurred in the 1980s, the simulation under drier conditions indicates that changes in the system's biogeochemistry are likely to occur beyond those directly linked to reduced allochthonous nutrient loads.

In another simulation scenario in which external phosphorus loads were reduced, the CALM indicated that some landscape measures, such as sorbed phosphorus and macrophyte biomass, would not be immediately affected (improved) in all currently-impacted areas. That scenario indicated that internal cycling of phosphorus would likely continue to maintain a eutrophic state in some impacted areas for years, with soil porewater P increasing for a decade, albeit at reduced rates compared to a simulation with actual, observed loads. In that reduced-load scenario run, appearance of calcareous periphyton within the currently-impacted zone reflected the reduction in surface water nutrients compared to the nominal run. However, macrophyte biomass (and thus shading) was not greatly reduced in much of that zone, thus preventing calcareous periphyton from attaining high densities that are found in pristine areas.

We are currently working on refinements and further model verification in anticipation of applying the model framework toward evaluating Everglades restoration alternatives in the entire Everglades/Big Cypress region.

26.2 INTRODUCTION

During the early decades of the 20th century, large regions of south Florida were drained by construction of canals. In 1948, the U.S. Army Corps of Engineers (USACOE) initiated the Central and South Florida (C&SF) Project Study, an extensive engineering project that ultimately led to the compartmentalization of large regions of the Everglades. These impounded Water Conservation Areas (WCAs) are now part of a network of some 2500 km of canals and levees that provide flood control and water supply for the urban and agricultural sectors of south Florida, with about one half of the original Everglades converted to agricultural and urban land uses. However, with recognition that the natural system of the region has been impacted by these engineering works, the USACOE, authorized by the U.S. House of Representatives and Section 309(1) of the Water Resources Development Act of 1992, has implemented a "Restudy" of the management network with the primary goal of restoring the remnant Everglades back toward historical attributes while still providing adequate water supply and flood control. A large part of this effort involves restoring historical water depths and hydroperiods in the WCAs and Everglades National Park (ENP). Critical to successful Everglades restoration are evaluations of the water quality and biology associated with Restudy flow patterns and sources.

Many of the Restudy evaluations use models to analyze possible results of Everglades restoration scenarios. One of the principal tools used in evaluation of alternate management scenarios is the South Florida Water Management Model (SFWMM) (MacVicar et al., 1984; HSM, 1997), a grid-based hydrologic model that simulates regional water management. A companion model to the SFWMM is the Natural System Model (NSM) (Fennema et al., 1994; Bales et al., 1997), which simulates predrainage hydrology in the region by removing all water management engineering works and using best estimates of historical land elevation and vegetation type. One of the fundamental Restudy assumptions is that these models can be used to develop predrainage hydrologic targets for Everglades restoration. However, restoration strategies also need to evaluate how the Everglades system will respond to hydrology plus other variables, such as those that are associated with phosphorus (McCormick and O'Dell, in press; Miao and Sklar, in press; Rutchey and Vilchek, submitted), heavy metals (Ogden et al., 1974), and sea level rise (Meeder et al., 1996). Therefore, other models are being developed to furnish a more comprehensive analysis of ecosystem change resulting from restoration activities. For example, the Everglades Water Quality Model (EWQM) (Chen et al., 1997), using output from the SFWMM, applies statistically derived, first-order settling rates of total phosphorus to determine transport of phosphorus through the Everglades. Another modeling tool is a tracking model (Walker, 1999) that employs regression analyses of observed data to predict total phosphorus at various water control structures in the Everglades.

An important distinction between the above models and the more process-oriented model we present is the extent to which underlying mechanisms of ecosystem dynamics are explicitly incorporated. The process-oriented model may have statistically derived parameters in its equations. However, the model structure incorporates dynamic feedbacks among the variables, and it responds effectively to a broad range of inputs and conditions. A purely statistical model should not be used to predict the response of a dependent variable when the independent variable goes outside the range of observed values used to construct the model. A process-based ecological model can characterize ecosystem dynamics under a wide variety of conditions, using known characteristics associated with the processes such as nutrient uptake and remineralization.

The importance of dynamic interaction is further accentuated when spatial heterogeneity of the model space is considered. Spatially explicit process-based models account for this heterogeneity. The pattern of these processes is a function of the connectivity of different subregions and the exchange of matter or information among them. As spatial patterns change, they alter flow of material and information and thus influence local processes. Conversely, changing processes alter ecosystem characteristics within a region and thus alter its spatial pattern. It is the confluence of spatial patterns and system processes that is our focus of ecological modeling of the Everglades. We present here a process-based, spatially explicit ecological model framework in order to understand the long-term, large-scale consequences of management changes. The Everglades Landscape Model (ELM), designed to evaluate the ecosystem responses to modified water and nutrient management policies (Fitz et al., 1993), is the framework for this paper. The ELM covers most of the natural system of the Everglades and Big Cypress region and incorporates the direct and indirect interactions associated with hydrology, phosphorus cycling, detrital decomposition, primary production, and habitat succession. In this manuscript, we present results of a rescaled ELM that was applied to Water Conservation Area 2A (WCA-2A) in the northern Everglades. WCA-2A has well documented ecological gradients (Koch and Reddy, 1992; Jensen et al., 1995; McCormick and O'Dell, in press; McCormick et al., in press; McCormick et al., in press[a]; Miao and Sklar, in press) in response to altered nutrient and water flows associated with water and agricultural management in the Everglades Agricultural Area (EAA). This rescaled ELM is called the Conservation Area Landscape Model (CALM), and it was developed to (1) address questions concerning the processes underlying specific landscape changes in WCA-2A and (2) calibrate and debug the ELM code in an area with significantly fewer vegetation types and less water management engineering complexity compared to the entire Everglades/Big Cypress region. We will discuss the characteristics of WCA-2A, the CALM structure, its calibration, and ecosystem properties that emerge from CALM simulation sensitivity analyses.

26.3 LANDSCAPE CHARACTERISTICS

26.3.1 WATER MANAGEMENT

In the northern Everglades, WCA-2A is a 433 km^2 wetland that was impounded in its present form in 1961 (Plate 26.1*). The Everglades Agricultural Area (EAA) is on its northwest boundary, with urban development along its eastern boundary. Entirely surrounded by levee systems with water control structures along its perimeter, WCA-2A has undergone a number of operational changes during its history

* Plate 26.1 appears as a color plate following p. 590.

(Light and Dineen, 1994). WCA-2A water stage has been regulated for various combinations of water storage (relatively deep inundation) and environmental protection (lower, varying stages) of the marshes and tree islands in the area.

Major water discharges into WCA-2A are along the L-39 levee through the S-10 A, C, and D gated spillways (Plate 26.1), each of which has a design discharge capacity of 132 m^3 s^{-1} (Cooper and Roy, 1991). This water enters an interior borrow canal that extends along portions of levees of the northeast and eastern WCA-2A borders, respectively, exchanging water with the adjacent interior marshes. The S-10E inflow structure at the northern tip of L-39 is relatively small in capacity, with a design discharge capacity of 12 m^3 s^{-1}. Another major inflow is via S-7 (70 m^3 s^{-1} capacity) at the western tip of the WCA, where water enters the North New River Canal along the WCA interior and exchanges with the surrounding marshes. Major discharges from the WCA are via the three S-11 structures (each with 158 m^3 s^{-1} capacity) from the North New River Canal in the southwest, with comparatively minor discharge volumes from an interior borrow canal along the southeastern levee via S-143 (14 m^3 s^{-1} capacity), S-144, 145, and 146 (each with 6 m^3 s^{-1} capacity), and S-38 (14 m^3 s^{-1} capacity).

Water flow through these structures has varied dramatically within and among years. Seasonal and interannual changes in rainfall intensity alter the inflows to the WCA, water management regulation schedules have varied over the years, and deviations from those targets occurred based on overriding water supply and flood control needs elsewhere. Interannual variations in structure discharges are large, with a pattern that generally follows the trends in annual rainfall (Fig. 26.2).

26.3.2 Rainfall

South Florida rainfall is generally described as having a wet season from June through September and a dry season during the remainder of the water year. Annual rainfall sums for WCA-2A (measured at S-7) averaged 132 cm yr^{-1} between (calendar years) 1965 to 1996 (Fig. 26.2), ranging from 107 to 210 cm yr^{-1}. During several years in the mid-1990s, there were extreme flood conditions resulting from very heavy rains. The late 1960s and early 1980s also were conspicuous for unusually high rainfall periods. Droughts of varying magnitudes occurred on a number of occasions during this record.

26.3.3 Nutrient Loading

External nutrient loads to WCA-2A come from several sources: wet and dry atmospheric deposition, the S-10 structures, and the S-7 structure. The S-10A, C, D, and E structures receive water and dissolved/suspended nutrient constituents from the Hillsboro Canal along the southern portion of WCA1, whose principal source is the S-6 structure discharging from the EAA. In general, the highest P loads from the Hillsboro Canal into WCA-2A have been from S-10D (Fig. 26.3a), with decreases downstream at S-10 C followed by S-10A. The most upstream S-10E structure has high nutrient concentrations, but loads are low relative to S-10D because of S-10E's lower volume capacity and operational rules. EAA discharges through S-7 in the west is the other significant contributor to the WCA-2A P loads.

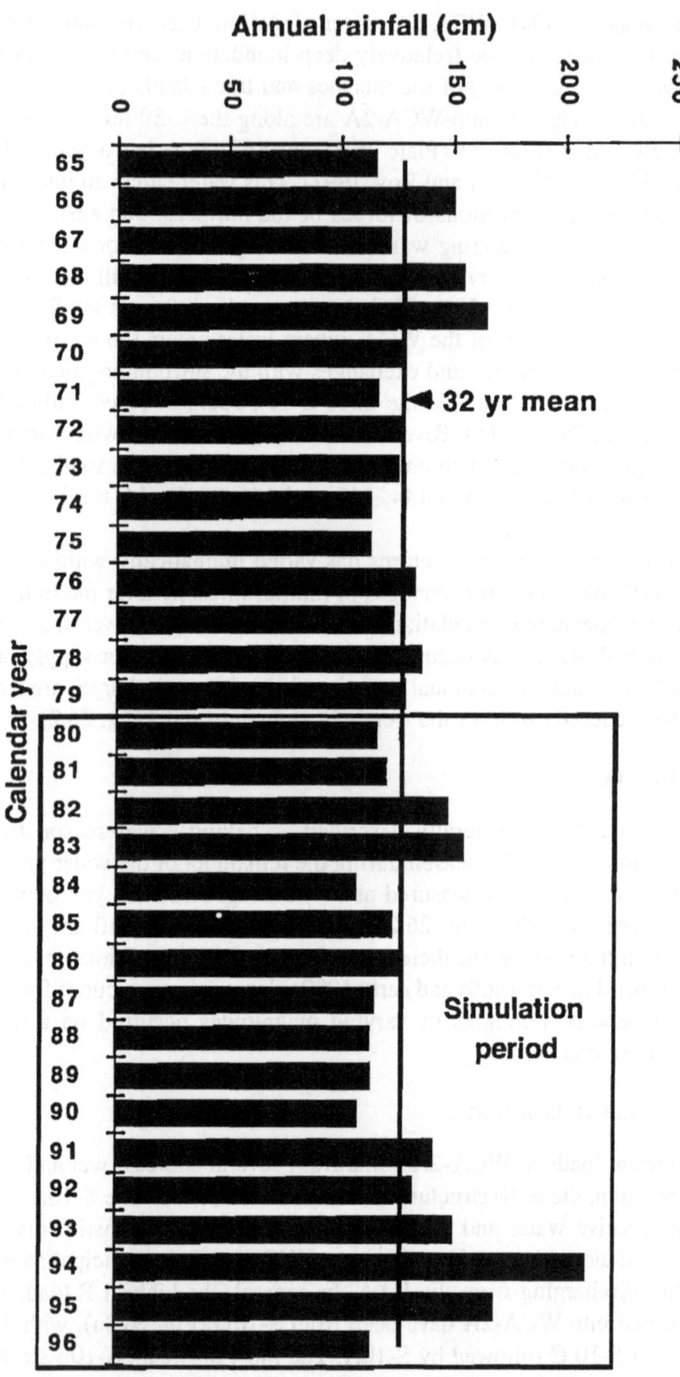

FIGURE 26.2 Observed annual rainfall for WCA-2A. The simulation period encompasses 1980–1996, inclusive of extremely low- and high-rainfall years.

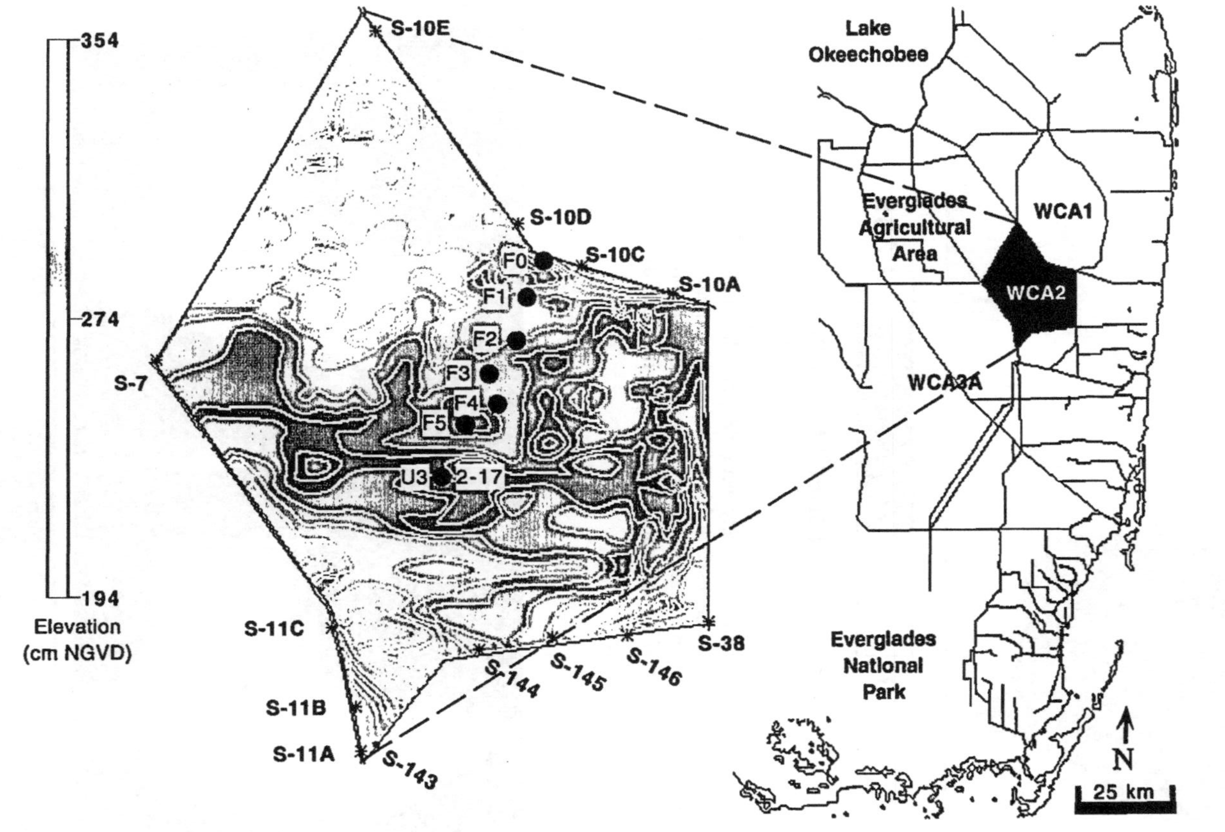

Plate 26.1
The location and attributes of Water Conservation Area 2A in South Florida. Elevation (cm 1929 NGVD) indicates the presence of the major topographic gradients. Levees bound all sides of the impoundment. With the names of water control structures indicated around the periphery. Seven of the monitoring sites along a WCA-2A gradient transect are labeled FO-F5 and U3. Hydrologic monitoring gage 2-17 is adjacent to site U3.

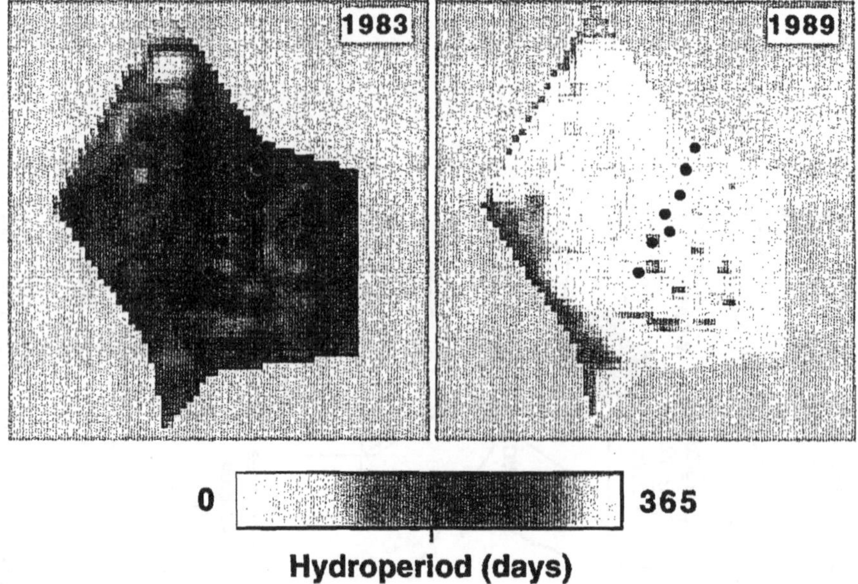

Plate 26.7
Simulated annual hydroperiod for a relatively wet (1983) and dry (1989) year. The effect of the ridge and slough elevation pattern can be discerned in the central region, and the areas near the inflow structures and canals tend to have slightly longer hydroperiods than would otherwise be the case. The monitoring sites along a WCA-2A gradient transect are indicated by unlabeled black circles.

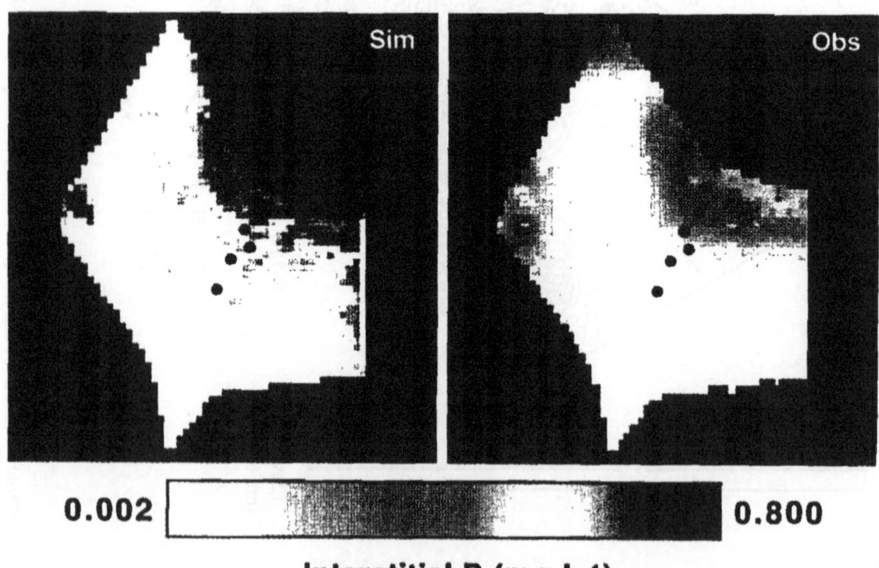

Plate 26.9
Simulated and observed concentration of P in the interstitial porewater in July 1990. The monitoring sites along a WCA-2A gradient transect are indicated by unlabeled black circles.

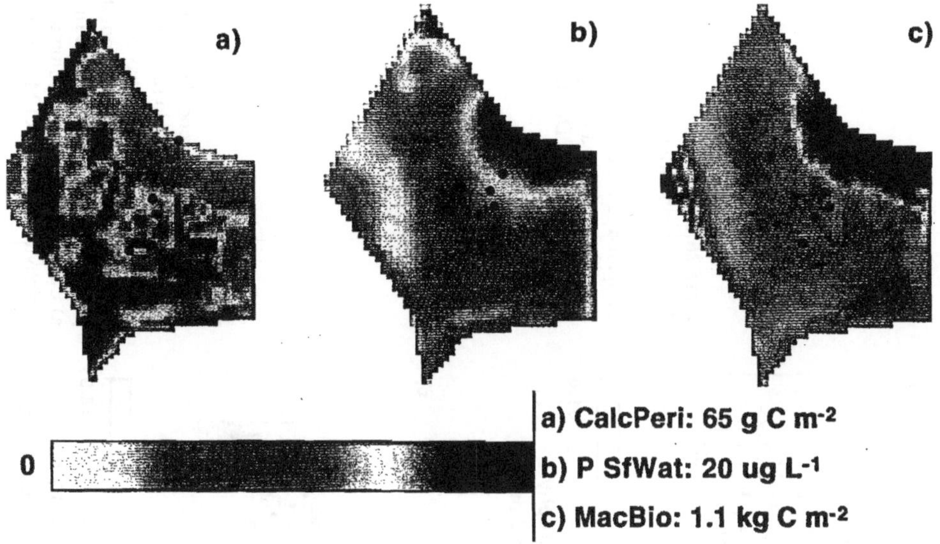

late 26.12

lean values of selected variables over the 17-yr simulation showing, a) calcareous periphyton iomass, b) surface water P concentration, and c) macrophyte total (photosynthetic and non photo-ynthetic) biomass.

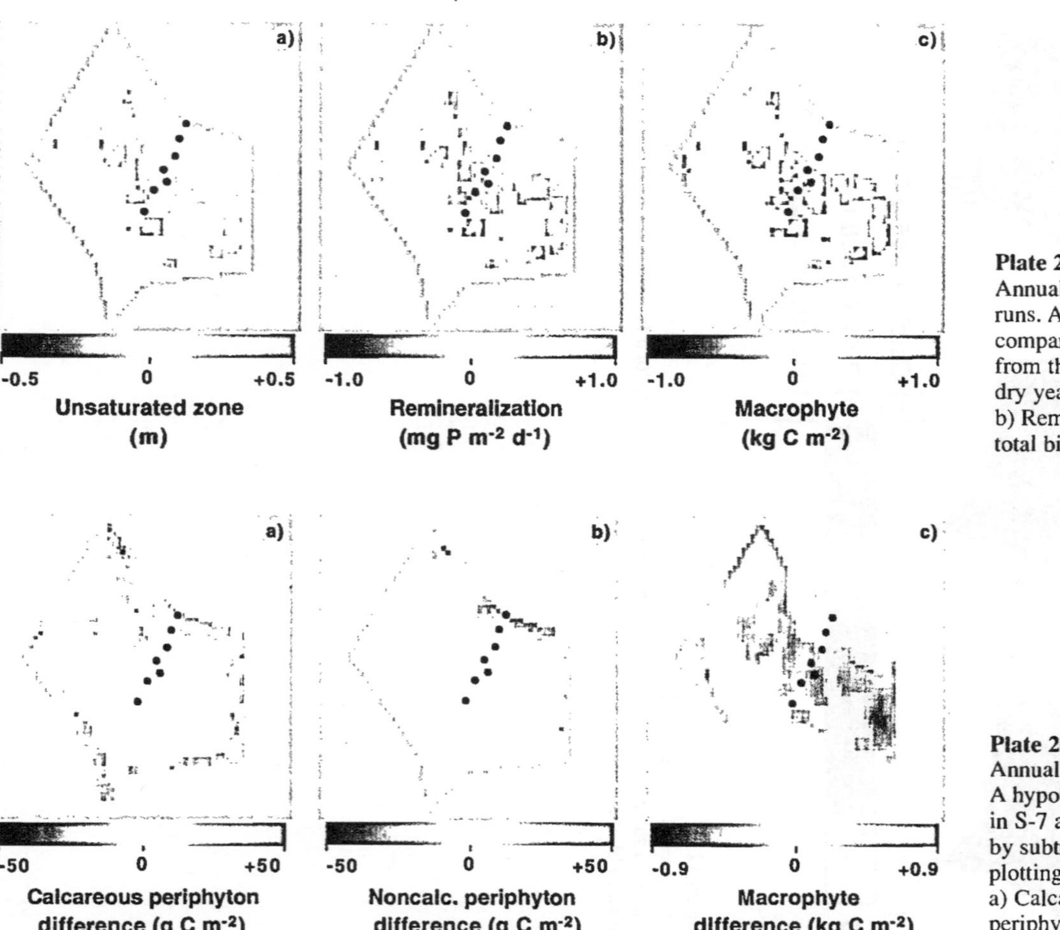

Plate 26.13
Annual mean difference between two 17-yr simulation runs. A test run with increased evapotranspiration was compared to the nominal run by subtracting the nominal from the test run values, plotting the differences for one dry year (1989). a) Depth of the zone of unsaturated water. b) Remineralization rate of P in the soil. c) Macrophyte total biomass.

Plate 26.14
Annual mean difference between two 17-yr simulation runs. A hypothetical scenario run with reduced P concentration in S-7 and S-10 inflows was compared to the nominal run by subtracting the nominal from the scenario run values, plotting the differences for the final simulation year (1986). a) Calcareous periphyton biomass. b) Noncalcareous periphyton biomass. c) Macrophyte total biomass.

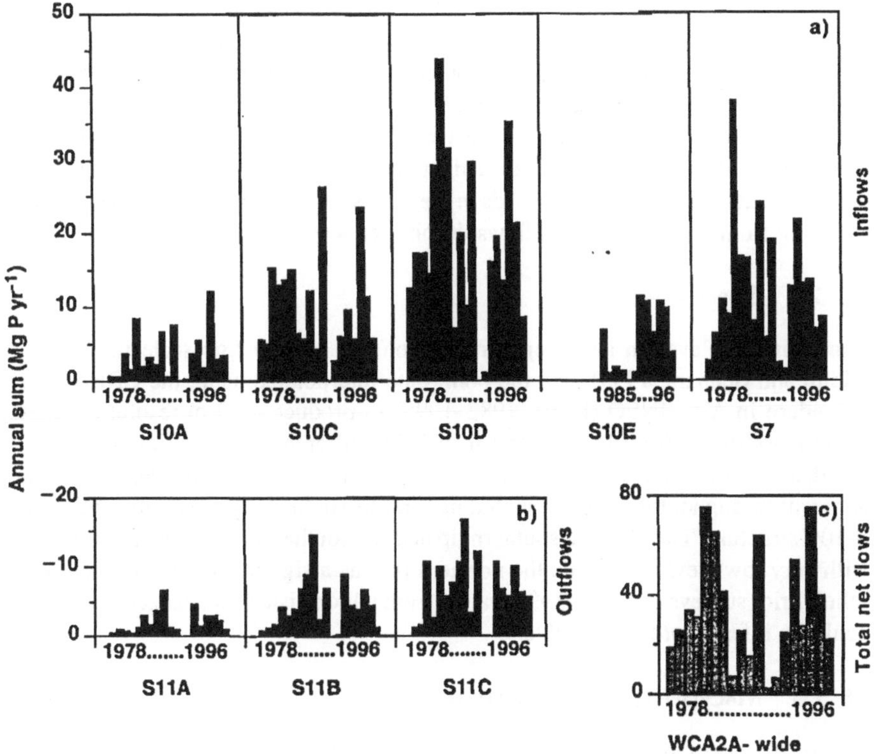

FIGURE 26.3 Observed annual sum of P (Mg = metric ton) flows through major water control structures of WCA-2A: (a) inflows (S-7, S-10A, S-10C, S-10D), (b) outflows (S-11A, S-11B, S-11C), and net flows for those structures (Σinflow-Σoutflow). S-10E began operations in 1985.

Annual load from atmospheric sources is uncertain because of the difficulties in measuring wet and dry deposition. For the CALM, we assumed a constant median (reducing bias from possible contamination resulting from birds and insects) of 0.010 mg L^{-1} dissolved inorganic P in rainfall based on a two-year period from collectors managed by the South Florida Water Management District. This equals a mean wet deposition rate on the order of 20 mg P m^{-2} yr^{-1} between 1978 to 1996. There are few available dry deposition measurements, but several sources at a conference on atmospheric deposition in south Florida indicate that the rate may be approximately half that of wet deposition in the vicinity of WCA-2A. Thus, we assumed a total median atmospheric deposition on the order of 30 mg P m^{-2} yr^{-1} to WCA-2A, or approximately 12.9 Mg (metric ton) P yr^{-1} to the entire WCA. This *WCA-wide* atmospheric load of 12.9 Mg is only 21% of the 19-year annual average total water control structure and atmospheric P load. Input through the S-7 plus S-10 structures (47.7 mg P yr^{-1}), compared with atmospheric deposition, has a more localized distribution within WCA-2A.

The largest structure outflows are via the S-11 structures (Fig. 26.3b), which transport water and nutrients from the North New River Canal to WCA-3A. While summed S-11 dissolved inorganic P outflows are similar in magnitude to that of the corresponding S-7 inflow (12.6 and 13.2 mg P yr^{-1}, respectively), P introduced through the S-10 structures is largely routed through the marshes. During the period of record for nutrient measurements (1978 to 1996) at the major water control structures, there has been a large, positive net P load to WCA-2A. The temporal distribution of this load exhibited distinct peaks in 1982, 1983, 1988, 1992, and 1994 (Fig. 26.3c), with the peak loads in the 1990s (> 75 mg P yr^{-1} in 1994) primarily resulting from very large water flows despite reduced P concentrations.

26.3.4 Topography

Land surface elevation has been surveyed on a relatively fine scale in WCA-2A (Keith and Schnars, 1993), and the point data interpolated using the CREATETIN procedure in ARC/INFO (ESRI, 1982–1995) to produce a 20 m resolution digital raster map (Plate 26.1). There is less than 1.5 m difference in elevation over the 30-km distance between the north and south extremes. Sloughs and depressions are apparent throughout the region, including a north-south slough starting south of the S-10C structure. The extreme southern tip and the southeast corner are two locations with very low elevation, while the northern tip has a significant rise in topographic relief. Prior surveys of WCA-2A elevation from which historical changes in elevation could be inferred are unavailable.

26.3.5 Macrophytes and Periphyton

WCA-2A is a complex mosaic of habitats such as tree islands, saw grass (*Cladium*) marshes, spikerush (*Eliocharis*) sloughs, open water, and cattail (*Typha)* marshes (Rutchey and Vilchek, 1992), with mixtures, patches, and gradients of these macrophytes throughout the region. Rutchey and Vilchek (1992) found significant areas dominated by *Typha*, which did not appear to be found in abundance within the historical Everglades (Davis, 1943). Whereas satellite image classification techniques were employed to estimate changes in *Typha* distribution between 1973 and 1991 (Jensen et al., 1995), more accurate recent photointerpretations (Rutchey and Vilchek, submitted) found a 70% increase in cattail coverage in WCA-2A between 1991 and 1995.

26.3.6 WCA-2A Chemical and Biological Gradient

As evidenced by the changes in *Typha* distribution, a number of biogeochemical characteristics have significantly changed in the region south of the S-10 structures. Downstream gradients of soil nutrients, surface water nutrients, periphyton, and macrophyte community structure along a ~10 km distance have been well documented (Swift, 1984; Swift and Nicholas, 1987; Koch and Reddy, 1992; Urban et al., 1993; Jensen et al, 1995; McCormick and O'Dell, in press; McCormick et al., in press [a]; Miao and Sklar, in press). The causal mechanisms for these biological and chemical gradients have been the focus of extensive research and debate in south Florida. Reddy et al. (1991) measured the spatial distribution of a variety of biogeochemical variables and accumulation rates, and found that soil nutrient concentrations at the S-10 structures were an order of magnitude higher than reference sites in the center of WCA-2A. Koch-Rose et al. (1994) provided evidence that microbial activity and P remineralization were P-limited in pristine areas of WCA-2A. McCormick and Scinto (1999) found a periphyton community shift downstream of the S-10 structures where total (organic and inorganic) phosphorus (TP) concentrations dropped below 12 μg L^{-1}. Miao and Sklar (in press) found physiological differences between cattail populations immediately downstream of the S-10 structures and those at interior reference sites. Richardson et al. (1995) determined the response of a variety of biological indicators to changes in phosphorus and hydroperiod in WCA-2A. The ELM framework was used to synthesize this type of gradient information for calibration and validation of the CALM.

26.4 MODEL CHARACTERISTICS

26.4.1 Objectives

The Everglades Landscape Model (ELM) was designed to be a regional landscape simulation model to address the effects of different management scenarios on the ecosystems in the entire Everglades. The Conservation Area Landscape Model (CALM) was an implementation of the same code applied to a smaller, simpler region. This application was used to evaluate and calibrate some of the model's ecological dynamics within the intensively studied landscape of WCA-2A. The ELM and CALM are intended to provide a tool to

1. Estimate the water demands of the Everglades in terms of adequacy of water flow and water levels to achieve user-defined landscape/ecosystem characteristics.
2. Predict vegetation changes that result from specific hydrology and associated water quality regimes, simulating the interrelationships among water quality, hydrology, and vegetation, and the influence of these relationships on habitat quality.
3. Provide a focus to research programs and models on other scales.

26.4.2 General Structure

The CALM simulates interactions among fundamental ecological processes (Fig. 26.4), with the objective of quantifying the response of vegetation. Simulations can be several decades in duration, or as many as 50 years depending on the available information. Growth of macrophyte and periphyton communities responds to available nutrients, water, sunlight and temperature. Hydrology in the model responds directly in turn to the vegetation via linkages such as Manning's roughness coefficient and transpiration losses (Fitz et al., 1993). Phosphorus and nitrogen cycles include uptake, remineralization, sorption, diffusion, and organic soil loss/gain. The magnitude and duration of nutrient availability and water depths alters habitat patterns via a habitat transition algorithm.

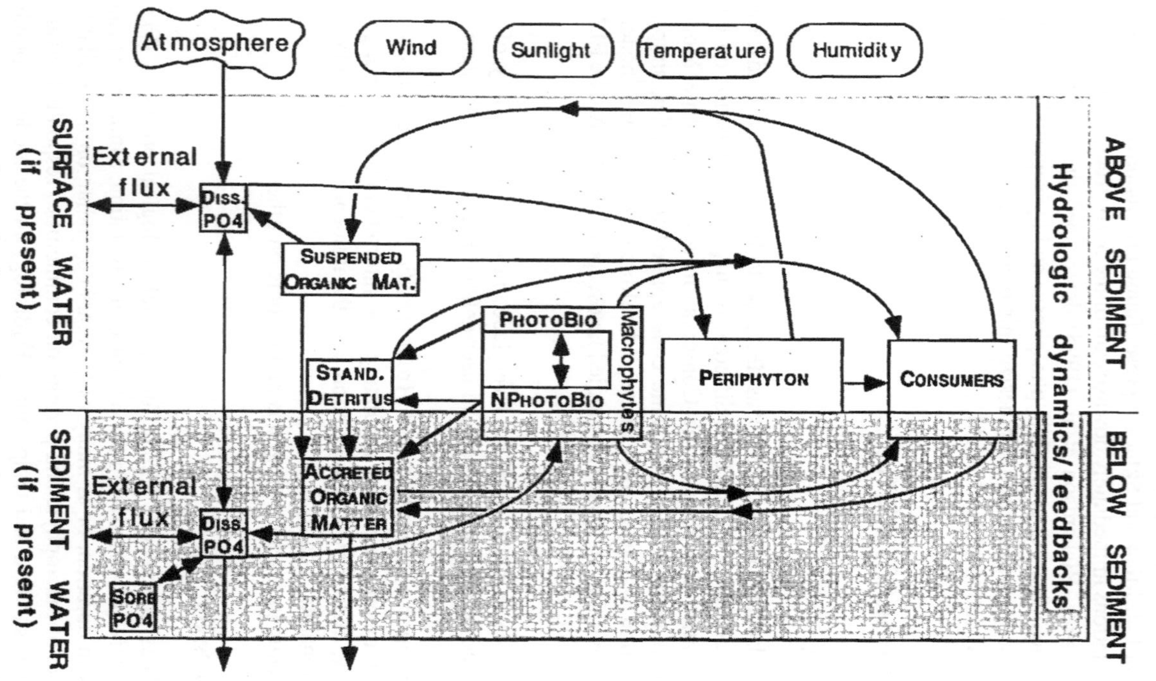

FIGURE 26.4 The state variables and within-cell flows of the CALM's unit model, excluding hydrology that is also a component of the model dynamics (Fig. 26.6). State variables are enclosed within rectangles in SMALL CAPS. Environmental forcing functions are in ovals. Phosphorus is in the form of dissolved P (DISS. P) in the surface water and soil porewater, P sorbed to soils (SORB. P), and incorporated in living and nonliving organic carbon stocks. The latter include suspended organic matter in the surface water (SUSPENDED ORGANIC MAT.), organic matter accreted in soils (ACCRETED ORGANIC MATTER), photosynthetic (PHOTOBIO) and nonphotosynthetic (NPHOTOBIO) components of macrophytes, standing dead detritus (STAND. DETRITUS), PERIPHYTON, and generalized CONSUMERS.

Central to the CALM structure is division of the landscape into square grid cells to represent the landscape in digital form. Superimposed on this grid network are canal/levee vectors along the boundaries of the region. Within this spatial structure are four fundamental components of the simulation.

1. *The unit model* (Fitz et al., 1996). This is the most basic building block of the model, simulating the temporal dynamics of important biological and physical processes within a grid-cell. Different vegetative habitats have unique parameter values, but all habitats run the same unit model.
2. *The management component.* Canals and associated levees are represented by a set of vector objects that interact with a specific set of raster landscape cells (Fitz et al., 1993; Voinov et al., submitted). This allows for rapid flux of water and dissolved nutrients over long distances.
3. *The Spatial Modeling Environment* (Maxwell and Costanza, 1995). This provides for the translation of the unit model into a spatially explicit framework, integrates all of the spatial and nonspatial fluxes, and coordinates input/output.
4. *The data component.* Spatially explicit data such as habitat type, elevation, and canal vectors are maintained in GIS layers, and other databases store time series inputs (e.g., rainfall) and parameters that vary with habitat (e.g., growth rates). The data structure organizes the information and alleviates the need to recompile the model code when evaluating the effects of different management scenarios.

The resulting spatially explicit model (Fig. 26.5) contains a unit model that calculates within-cell dynamics depending on each cell's parameter set, with the interactions among cells via hydrologic fluxes propagating state changes across a heterogeneous space. It is this simulation of the influence of the landscape heterogeneity on ecosystem (cell) processes, and the concomitant influence of the ecosystem (cell) processes on shaping local pattern, that allows such spatially explicit process models to explore ecosystem properties at relatively large scales.

26.4.3 Scales and Boundary Conditions

The CALM boundaries encompass the entire 433 km^2 wetland within the levees that define WCA-2A. We used 1,734 cells of 0.25 km^2 fixed grid size and assumed that each individual cell is homogenous in all of its characteristics. The model was run with a one-hour time step for all horizontal flows of water and dissolved nutrients. All other flux calculations were made at a 12-hr time step. Boundary conditions for water and nutrient inflows and outflows were associated with the atmosphere and water management along the WCAs levees. Daily rainfall measured at the S-7 structure (Plate 26.1) was applied to all model cells. As described earlier, dry deposition was not explicitly simulated. Atmospheric deposition was assumed to have a constant concentration of 0.015 mg P L^{-1}. The total deposition using these rainfall-driven inputs averaged approximately 30 mg P m^{-2} yr^{-1}.

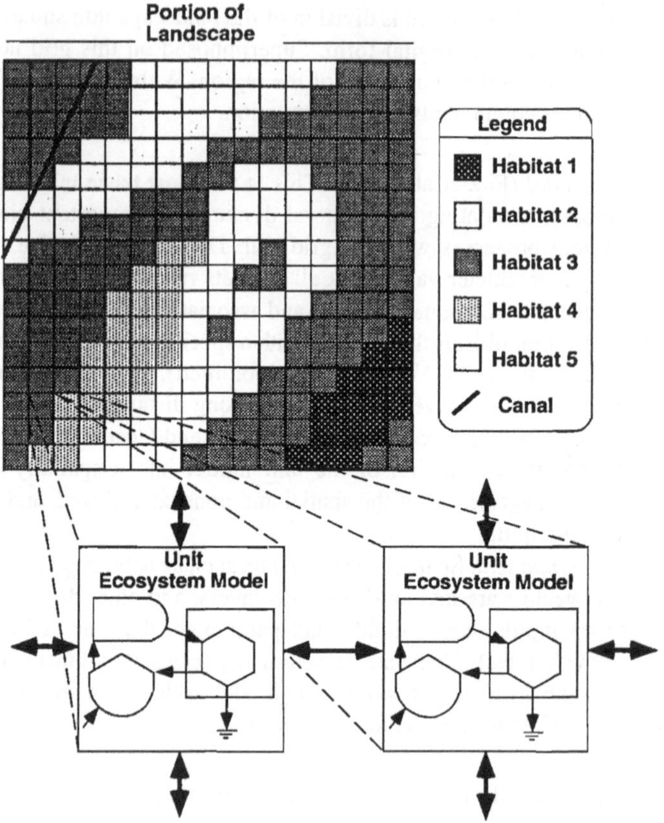

FIGURE 26.5 The spatial integration of a general unit model within the raster grid cell and canal/levee vector landscape. Each homogeneous cell in the modeled landscape is assigned a habitat type (shown here as Habitats 1–5), which is used to parameterize the model for that cell. Water and dissolved nutrients are fluxed spatially among grid cells and canal vectors. Within-cell dynamics and those spatial fluxes can alter the attributes (e.g., plant biomass, soil nutrient concentration) of the grid cells, including the habitat type.

Daily observed measurements of water volume and associated P concentration were applied to the S-7, S-10A, S-10C, S-10D, and S-10E inflow water control structures. Because nutrient concentrations were measured at a minimum of biweekly intervals, we used a linear interpolation across dates with missing values (Walker, 1998). For outflow of nutrients, observed measurements of water outflows through S-11A, S-11B, and S-11C were multiplied by nutrient concentrations in adjacent donor cells/canals from the simulation. Outflows through the S-38, S-143-146 structures were driven by management rules, in which the targeted stage was compared at each time step with simulated stage at the 2-17 gauge (see Plate 26.1). Regulatory water (and nutrient) releases were applied through these structures when the stage was greater than the target. Seepage across boundary levees was calculated using a fixed hydraulic head in the borrow canals bordering WCA-2A.

26.4.4 Process Simulation

Feedbacks among physical, geochemical, and biological components of a system have long been recognized as critical in determining ecosystem properties. Because of the nature of the myriad interactions within the Everglades and the CALM, processes that have an influence on, and are affected by, phosphorus can virtually include every ecosystem component. The CALM does not attempt to simulate them all. Here we present a summary of the model structure and provide details of some of the modules and equations that most directly relate to transport and fate of phosphorus. Other details of all state variables, algorithms, and equations are published elsewhere.

26.4.5 Hydrology

The CALM hydrology module simulates ponded surface water, water in unsaturated soil, and saturated water below the water table, coupling the surface and ground water components. Fluxes among those variables (Fig. 26.6) involve rainfall, evaporation, infiltration, percolation, saturated/unsaturated transpiration, and horizontal movement of surface and saturated water. Many of the hydrologic algorithms are based on those developed for the SFWMM (MacVicar et al., 1984; HSM, 1997). Differences are associated with accommodating fine-scales and direct linkages with vegetation. Evaporation and transpiration in CALM are driven by the calculated air saturation deficit, leaf area exposed above any ponded surface water, and canopy

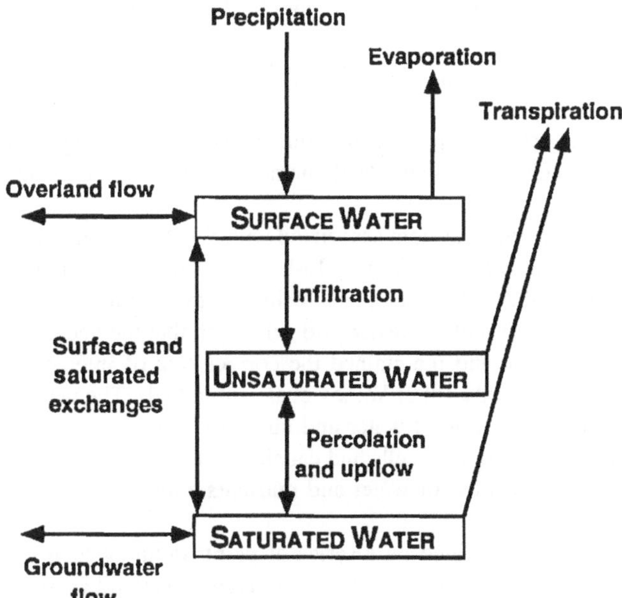

FIGURE 26.6 Water storages and flows for the hydrologic module. The depths associated with water in surface, unsaturated, and saturated storages all vary dynamically.

morphology (Fitz et al., 1996), in relation to a simple evaporative model (Christiansen, 1968). Transpiration losses further depend upon the availability of water relative to the root zone depth. As will be indicated in the model results section, transpiration in a flooded wetland can transport P as a result of the associated surface to soil porewater advective flux of water and dissolved constituents.

Horizontal transport of water and nutrients is a fundamental component of the water quality component of the model. Flows of water among grid cells were calculated using a mass balance, finite difference algorithm solving two dimensional horizontal diffusion equations. We used an alternating direction, explicit method to solve the Manning's equation for overland flow, which for equal, square grid cells can be discretized into:

$$Q = \frac{h^{\frac{5}{3}} L^{\frac{1}{2}} \Delta H^{\frac{1}{2}}}{n} \tag{1}$$

where Q = volumetric flow velocity ($m^3\ d^{-1}$), h = water depth (hydraulic radius, m) above ground elevation in the source cell, L = length of a grid cell (m), H =difference (m) in water stage between the source and destination cells, and n = empirically derived Manning's roughness coefficient.

The equation was constrained such that the volume flux could not result in a reversal of the sign of the H within any iteration of a cell-cell flux. A four-way, alternating direction algorithm (Voinov et al., in press) was used to solve this flow velocity equation. The Manning's n value was a function of the dynamic height of the particular vegetation type relative to water depth (Fitz et al., 1996) according to:

$$n = n_{max} - \left|(n_{max} - n_{min})(2^{(1-(h/\ )mac)} - 1)\right| \tag{2}$$

where n_{min} and n_{max} are the respective minimum and maximum roughness coefficients associated with a cell's macrophyte/soil characteristics, and mac is the macrophyte height.

The water management system associated with WCA-2A, while not as extensive and complex as found within the larger ELM region, has significant impact on water distribution in WCA-2A. We developed a technique to dynamically exchange water among the canal vectors and the raster grid cells such that the vectors overlaid cells in their true orientation and maintained the correct area of interaction among the two object types (Fitz et al., 1993; Voinov et al., submitted). The water management module provided for (a) flow of water and nutrients along canals, (b) exchange of water and nutrients among grid cells and canal vectors via overland or groundwater flow, and (c) controlled flow of water and nutrients among canals and/or cells via water control structures.

Canal reaches were defined by a series of points defining their exact location. A geometry algorithm determined the length and type (overland vs. seepage) of each segment's cell interaction using a technique that can be generalized to other raster and vector objects for simulation (Voinov et al., submitted). Water and nutrients were distributed along the length of a canal, with exchanges through control structures and among cells and canal reaches. Stages in the canals and in all of the interacting cells were updated using an iterative relaxation routine (used in the SFWMM) (HSM, 1997) to equilibrate the water levels until converging to a minimum error. All flow calculations contained constraints to ensure mass balance. When the Manning's equation (1) was applied to overland flows between a canal and a cell, the hydraulic radius was that of the canal (hr_{can}) if flux was from the canal into a cell, or that of the cell (hr_{cell}) if flow was from the cell into the canal:

$$hr_{can} = \frac{l_{can} w_{can} d_{can}}{A}, \qquad hr_{cell} = \frac{(A_{cell} - l_{can} w_{can})h}{A_{cell}} \tag{3}$$

where l_{can} = canal length (m), w_{can} = canal width (m), d_{can} = canal depth (m), and A_{cell} = area of the cell (m^2).

While the algorithm allows levees to be absent, on both sides of a canal reach, or on either side, the latter was the only case needed in the CALM. Because we used fixed external stages for boundary conditions, seepage flows across levees and subsurface groundwater flows exiting the system depended only on changing water stage within WCA-2A. This simplification was justified under the assumption that water management generally targets a maximum stage in the urban and agricultural areas adjacent to WCA-2A.

The WCA-2A stage regulation targets determined whether the S-143-146 and SS-38 structures (Plate 26.1) were open to release water out of the WCA, the rate of which was calculated using a simplified weir flow equation:

$$Q_{out} = Q(h_{can} - h_{ext})^{1.5} \tag{4}$$

where Q_{out} = calculated flow through the structure (m^3 d^{-1}), Q = weir flow rate coefficient using a common 10 m weir length, and h_{ext} = fixed external water stage height (m).

Structures (S-7, S-10s, and S-11s) driven by observed fluxes had all flows determined by the time series data.

26.4.6 Biogeochemistry

The principal components of the P biogeochemical dynamics included (a) P storage in suspended organic matter (SOM), accreted organic matter (AOM) in soils, standing dead detritus (SDD), macrophytes, periphyton, and consumers, (b) remineralization of SOM and AOM into inorganic P dissolved in surface and soil water, (c) reversible sorption of P to soils, (d) uptake of soil and surface water inorganic P by macrophytes and periphyton, respectively, and (e) horizontal (among cells and canals) and vertical (within-cell) flows of P dissolved in water (Fitz et al., 1996). (A similar nitrogen module exists but was not fully implemented for this version of CALM). To facilitate calibration, the live organic matter (macrophytes, periphyton, and consumers) and SDD stocks were expressed in organic carbon units (kg C m^{-2}), and the AOM and SOM stocks were expressed in ash free dry mass (kg OM m^{-2}, kg OM m^{-3}, respectively). The P storage in these stocks was determined by fixed, habitat-specific P:C (and C:OM) ratios; the exception was that of the soil AOM, which had variable P:C stoichiometry for mass balance accounting.

Organic carbon from mortality of photosynthetic and nonphotosynthetic stocks of macrophytes was partitioned between the AOM and SDD stocks, according to the macrophyte type. AOM dynamics were:

$$AOM_t = AOM_{(t-dt)} + (AOM_{set} - AOM_{dec}) \cdot dt \qquad (5)$$

where AOM_t and $AOM_{(t-dt)}$ = mass densities of accreted organic matter (kg OM m^{-2}) at time t and (t–1); AOM_{set} = the additive ("settling") input from the macrophyte, periphyton, consumer, SDD, and SOM organic stocks (kg OM m^{-2} d^{-1}); and AOM_{dec} = the loss resulting from decomposition (kg OM m^{-2} d^{-1}).

Implicit microbial decomposition of the AOM carbon stocks was controlled by temperature, available nutrients, moisture availability, and a classification of aerobic vs. anaerobic conditions as shown in the following equations. This AOM decomposition was described by:

$$AOM_{dec} = AOM \cdot k_{ar} T_{cf} P_{cf} \left[Usat_{cf} \frac{H_{ar}}{H_{AOM}} + k_{an} \frac{H_{an}}{H_{AOM}} \right] \qquad (6)$$

where k_{ar} = maximum specific rate of aerobic decomposition (d^{-1}), k_{an} = modifier for anaerobic decomposition, $Usat_{cf}$ = proportion (0–1) of soil moisture in the unsaturated zone, and T_{cf} = the 0–1, dimensionless temperature control function (Lassiter, 1975):

$$T_{cf} = e^{0.2(T - T_{opt})} \left(\frac{T_{max} - T}{T_{max} - T_{opt}} \right)^{0.2(T_{max} - T_{opt})} \qquad (7)$$

where T = soil water temperature, T_{max} = maximum temperature of acclimation (°C), and T_{opt} = optimal temperature at which the decomposition rate is maximal (°C).

P_{cf} is the available phosphorus control function (dimensionless, 0 – 1), in the form of:

$$P_{cf} = \frac{P_{pwat}}{(P_{pwat} + K_{AOM})} \frac{PC_{AOM}}{(PC_{AOM} + PC_{ref})} \qquad (8)$$

where K_{AOM} = half saturation coefficient (mg P L^{-1}) for microbial uptake, P_{pwat} = concentration of P in the soil pore water, PC_{AOM} = P:C ratio of the AOM, and PC_{ref} = reference P:C ratio for that habitat type (Reddy et al., 1991). H_{ar} and H_{an} were the thicknesses (m) of the aerobic and anaerobic soil zones, respectively:

$$H_{ar} = \min(H_{us} + H_{tar}, H_{AOM}), \qquad H_{an} = H_{AOM} - H_{ar} \qquad (9)$$

where H_{us} = variable height of the unsaturated water zone (m), H_{tar} = fixed height (m) of the thin aerobic layer for each habitat type in the model, and H_{AOM} = fixed height (m) of the active AOM layer for each habitat type. No plant uptake or microbial dynamics were considered below this (30 to 50 cm) active layer.

Phosphorus and organic carbon from periphyton mortality were routed into the SOM compartment at the same P:C ratio as live periphyton. Decomposition of SOM followed the same general form as that of *AOM* [Eq. (6)], except we assumed only aerobic conditions and fixed stoichiometry. Similarly, the SDD stock assumed the same P:C ratio as its macrophyte source.

Losses of SOM and AOM phosphorus from carbon decomposition were determined through the respective SOM and AOM P:C stoichiometry. The P:C ratio of AOM was calculated from the AOM carbon and AOM phosphorus (AOM_P, kg m^{-2}) variables. AOM_P dynamics were determined by:

$$AOM\_P_t = AOM\_P_{t-1} + (AOM\_P_{set} - AOM\_P_{dec}) \cdot dt \tag{10}$$

where AOM_P_t and $AOM\_P_{(t-1)}$ = AOM_P at time t and (t − 1), respectively; AOM_P_{set} = "settling" input associated with the source (fixed stoichiometry) AOM_{set}; and AOM_P_{dec} = P loss associated with the AOM_{dec} decomposition such that:

$$AOM\_P_{dec} = AOM_{dec} PC_{AOM} \max\left(1 - \frac{PC_{ref}}{PC_{AOM}}, 0\right) \tag{11}$$

where PC_{AOM} converges to PC_{ref} as new organic matter decomposes.

In addition to total P storages associated with AOM in soil, we considered the dynamics of labile inorganic P that is loosely sorbed to the soils in the active AOM layer (*P_sorb*). The active *AOM* layer of the soils in the model was assumed to be homogeneous in vertical profile, with a fixed porosity and no change in the P:C ratio or *P_sorb* with depth. We used a modified Freundlich adsorption model (Richardson and Vaithiyanathan, 1995), accounting for antecedent phosphate sorbed to the soils:

$$P\_sorb(t) = P\_sorb(t-1) + (k_{sb} P_{pwat}{}^{0.8} - P\_sorb[t-1])dt \tag{12}$$

where k_{sb} = adsorption coefficient (L kg^{-1}) and P_{pwat} = P concentration in the soil pore water (mg L^{-1}). Using data from Richardson and Vaithiyanathan (1995), k_{sb} varied linearly with P_{pwat} within the observed range of Everglades soil pore water concentrations.

26.4.7 Biology

Growth of macrophytes and periphyton was influenced by available nutrients, light, water, temperature, and density dependence. While these relations are described in Fitz et al. (1996), below we describe some of the details related to P uptake by plants and algae. Plant growth via carbon fixation was responsible for the removal of *P* from the soil or surface water. This removal was based upon a fixed stoichiometric (P:C) ratio for each habitat type. Growth was limited by inorganic P according to Monod kinetics:

$$MAC\_P_{cf} = \frac{P_{pwat}}{(P_{pwat} + MAC\_K_s)} \tag{13}$$

where MAC_P_{cf} = dimensionless P-limiting function for macrophytes and MAC_K_s = half saturation coefficient (mg P L^{-1}) for macrophyte uptake. The same relationship, parameterized with a K_s value and compared to P in surface water, was used to constrain growth of noncalcareous periphyton communities. Calcareous periphyton, most abundant in unenriched regions of lower phosphorus, rapidly ceased growth as an upper surface water P concentration (McCormick et al., in press [a]) was approached:

$$CPer\_P_{cf} = \frac{P_{swat}}{(P_{swat} + K_{CPer})} \min\left(\frac{ThrP_{CPer} - \min(P_{swat}, ThrP_{CPer})}{ThrP_{CPer} - K_{CPer}}, 1\right) \tag{14}$$

where $CPer\_P_{cf}$ = dimensionless nutrient control function for calcareous periphyton, K_{CPer} = half saturation coefficient (mg P L^{-1}) for calcareous periphyton uptake, P_{swat} = P concentration in surface water, and $ThrP_{CPer}$ = threshold P concentration above which growth ceases.

While shifts in periphyton community type (calcareous vs. noncalcareous) occurred because of their relative growth characteristics in the presence of varying water column *P* concentrations, macrophyte community shifts were based on the cumulative impacts of both available P and water depth. For each cell we evaluated the number of weeks that contained conditions favorable for each possible habitat type, switching to the new habitat type when conditions merited. Each model cell was evaluated on a daily basis for each of the two conditions for all possible habitat types:

$$PLo_i \le P_{pwat} \le PH_i, \qquad HLo_i \le h \le HH_i \tag{15}$$

where PLo_i and PHi_i = respective lower and upper thresholds of porewater P concentrations (P_{pwat}) for the ith habitat type (i = 0,1,2, for saw grass-dominant, cattail-dominant, and cattail-present), and HLo_i and HHi_i were the respective lower and upper thresholds of surface water depth (h) for the ith habitat type.

If a cell met either criteria for habitat i, a counter was incremented for that habitat type, regardless of the cell's current habitat type designation. When counters for water depth and for P conditions in a cell exceeded the elapsed (potentially discontinuous) number-of-weeks criterion, the cell's habitat type classification switched to the new type and counters were set to 0.

26.5 MODEL INITIALIZATION

The CALM was initialized using a series of maps for the conditions expected to have been present in the 1980 landscape. Elevation was aggregated from 20 m to 500 m resolution (Plate 26.1) using cell averages, and 3 cm was uniformly subtracted to account for soil accretion (Reddy et al., 1993) during the 13 years between 1980 and the elevation measurements. The 1982 cattail and saw grass distribution map was aggregated from 30 m to 500 m resolution using modal frequencies within the model grid cells, with a total of 26 km² coverage by cattail and a saw grass/cattail mixture.

An estimate of the spatial distribution of soil P was made from recent data collected by Reddy et al. (1991). Their 1990 data of total phosphorus in the 10 to 20 cm soil depth layer was used to estimate the total P in the 0 to 30 cm soil profile for 1980. This was accomplished with a linear, inverse-weighted nearest neighbor algorithm and the assumption that similar soil equilibrium processes existed in 1980. We also assumed that this 1980 pattern of soil P influenced the initial distribution of macrophyte biomass. A scaling factor based upon the soil P was used to calculate the initial 1980 standing stock of live and dead macrophyte biomass for each vegetation type. Periphyton were initialized to a constant throughout the region. Initial water stage for each cell was developed by running the CALM for one year with a constant "management" target (controlling outflows) that was equal to the observed, January 1980, stage at the 2 to 17 gauge.

26.6 MODEL RESULTS: CALIBRATION

We tested the model over a 17-year time span (1980 to 1996) that encompassed the hydrologic spectrum ranging from extreme drought years to extremely wet years (Fig. 26.2). Water quality monitoring data at water control structures used to drive the model were generally unavailable before 1980.

26.6.1 Hydrology

A useful summary of hydrologic conditions is the annual hydroperiod, or the number of days during a water year (starting in October) with positive water depths. Hydroperiod patterns (Plate 26.7*) were indicative of the elevation gradients in the ridge and slough regions interspersed through the area and the influence of nearby canal inflows and outflows. Simulated water stage had a substantial degree of concordance with observed stage (m, 1929 NGVD) at the 2 to 17 gauge for the 17 yr simulation period ($R^2 = 0.70$, RMSE = 0.21 m). As Fig. 26.8 indicates, there was good agreement between simulated and observed data for the majority of the simulation record during both low and high water events. However, several substantial deviations occurred during the later years of the simulation, and in particular during the 1995 drawdown of WCA-2A following an extreme flood year. We feel that, because the CALM v1.0 was driven by observed flows through only the major structures (i.e., S-7, S-10, and S-11), these discrepancies are the result of historical managerial deviations from water control structure operational criteria for S-143, S-144, S-145, S-146, and S-38 (which were driven by management rules in the simulation).

26.6.2 Phosphorus

Simulated P dissolved in the interstitial porewater gradually increased in the regions near the S-7 and S-10 structures and produced a spatial pattern by mid-1990 similar

* Plate 26.7 appears as a color plate following p. 590.

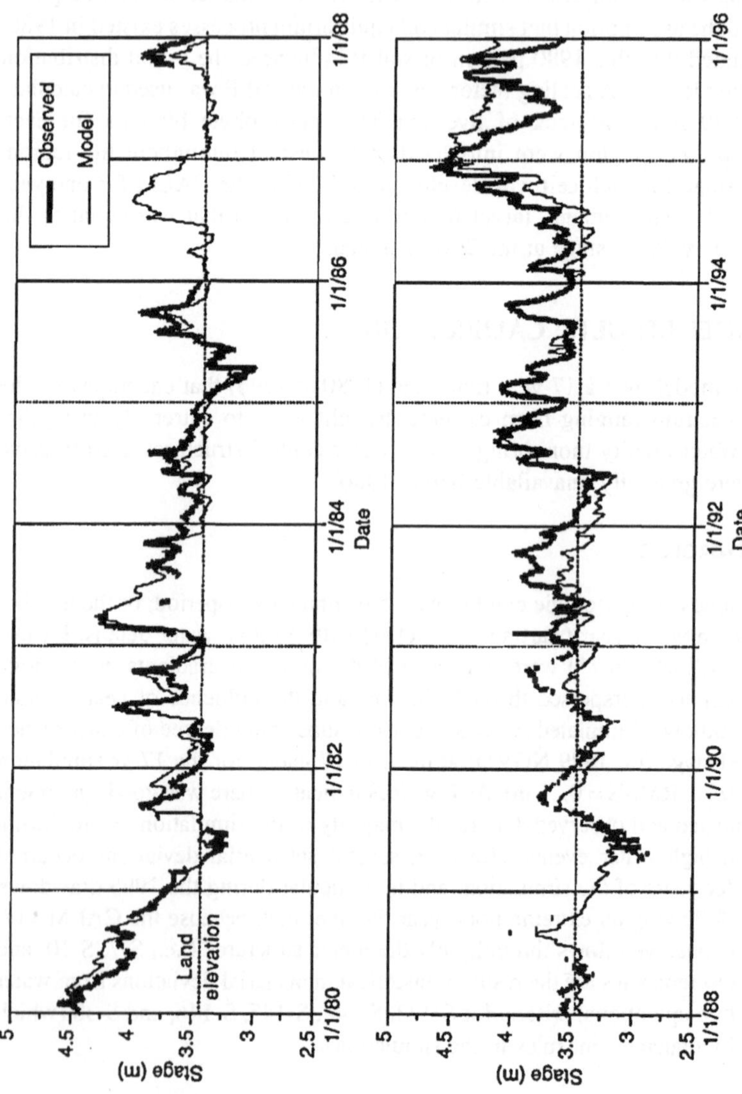

FIGURE 26.8 Simulated and observed water stage at the 2-17 gauge in the central region of WCA-2A (close to site U3 in Plate 26.1). The overall correlation coefficient (R^2) of the model fit to observed data is 0.70. There are a number of periods (e.g., 1986) for which we could not obtain observed data.

to that found by Reddy et al. (1991) and DeBusk et al. (1994) (Plate 26.9*). Mean porewater P concentrations of the simulated and observed data along the transects (Fig. 26.10a) were similar at all but the most eutrophic (F1) sites. The field observations are mean values from a 1.5-year (August 1995 to 1997) unpublished study

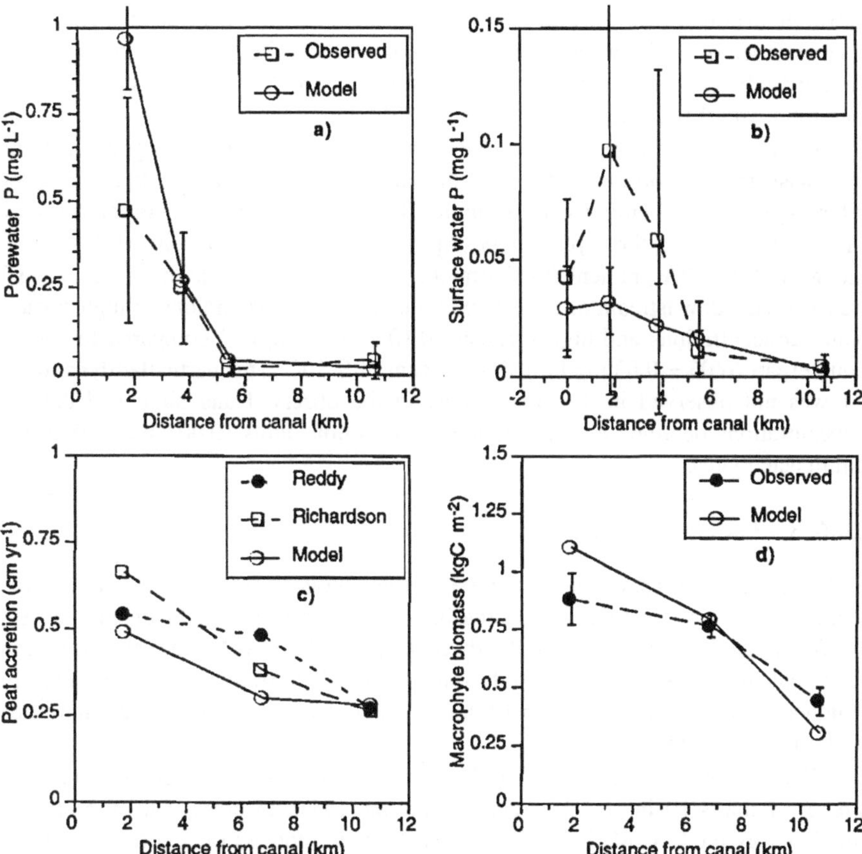

FIGURE 26.10 Simulated and observed data along a WCA-2A gradient transect. Site locations are shown in Plate 26.1. (a) Porewater P concentration, comparing observed mean values from a 1.5-year (1995–1997) study with quarterly samples to the simulation means during that same period. (b) Surface water P concentration, comparing observed mean values from a three-year (1994–1996) study with biweekly samples to the corresponding 1994–1996 simulation means. (c) Simulated and observed mean annual peat accumulation rates, with the observed rates from Cesium-137 dating (1964–1990) of soil cores, and simulated rates are the mean rate during the 1980–1996 simulation. (d) Total (photosynthetic and nonphotosynthetic) biomass of macrophytes. The observed data are the mean total biomass of a number of samples at each site in the 1994 spring/summer growing season, and the corresponding simulated data are total biomass for the corresponding cells in June 1994.

* Plate 26.9 appears as a color plate following p. 594.

(Newman, personal communication) with quarterly samples along the WCA-2A gradient. The simulated and observed P concentrations declined rapidly along the transect to near 0.01 mg L^{-1} in the unimpacted sites. The model overestimated mean porewater P at the enriched F1 site (0.95 ± 0.2 mg L^{-1} vs. 0.5 ± 0.3 mg L^{-1}), but closely simulated the observed data at the enriched F2 site, only 2 km further downstream.

The simulated final concentrations of phosphorus that was sorbed to the peat substrate declined from 81.2 to 3.6 mg kg^{-1} from F1 to U3, respectively, similar to that observed by Richardson and Vaithiyanathan (1995) along three 10 km transects. In that study, the maximum concentration was 95.6 mg kg^{-1} near the canal and the minimum was 1.0 mg kg^{-1} at a distance of 10.5 km downstream.

The P dissolved in surface water likewise demonstrated a realistic gradient along the transects from north to south, with simulated and observed concentrations during a three-year period being generally similar (Fig. 26.10b). The field observations are mean values from a three-year study (April 1994 to 1996) with biweekly samples along the WCA-2A gradient (McCormick, personal communication). Observed F1 transect data, downstream of S-10D, exhibited numerous excursions of high surface water concentrations and high variance (0.10 ± 0.11 mg L^{-1}) compared to the E1 counterpart (0.05 ± 0.03 mg L^{-1}), downstream of S-10C. This particular dynamic at F1 was not observed in the simulations. In the observed and simulated data, P concentrations declined to near laboratory detection limits (0.004 mg L^{-1}) in the unimpacted sites.

26.6.3 Organic Matter

Soil peat has been estimated to accrete at rates between 0.5 to 0.7 cm yr^{-2} in the vicinity of the enriched F1 site, decreasing to 0.25 cm yr^{-1} near the unenriched U3 site (Craft and Richardson, 1993; Reddy et al, 1993). The CALM simulated accretion rates, calculated as the average rate during the 17 yr simulation period, closely matched these estimates (Fig. 26.10c), slightly underestimating the accretion rates in the impacted area.

Observed biomass of *Cladium* and *Typha* communities steadily decreased downstream of the S-10 inflow structures (Miao and Sklar, in press) and was closely simulated by the CALM (Fig. 26.10d). Macrophyte biomass ranged from approximately 1.0 kg C m^{-2} at impacted sites where *Typha* was dominant, to 0.3 kg C m^{-2} at unimpacted sites where *Cladium* was dominant. There were no consistent seasonal biomass changes of *Typha* or *Cladium* observed in WCA-2A (Davis, 1989) or the simulation. Interannual variations were common during the period, with simulated *Cladium* biomass at the U3 site exhibiting short term (ca. one-season) increases up to densities of 0.5 to 0.6 kg C m^{-2} (e.g., in 1988), in response to combinations of relatively low water levels (Fig. 26.8) and high nutrient loads (Fig. 26.3c). Much of the pattern of the documented (Rutchey and Vilchek, submitted) spread of *Typha* southward along the transect was captured by the model (Fig. 26.11), with 44 km^2 cattail cover by 1991 (53 km^2 observed) and increasing to 117 km^2 coverage by 1995 (95 km^2 observed). The overestimates of *Typha* in 1995 were due to the simulation's increased occurrence of low density *Typha* mixed with *Cladium* in the ridge and slough regions south of the S-10 structures.

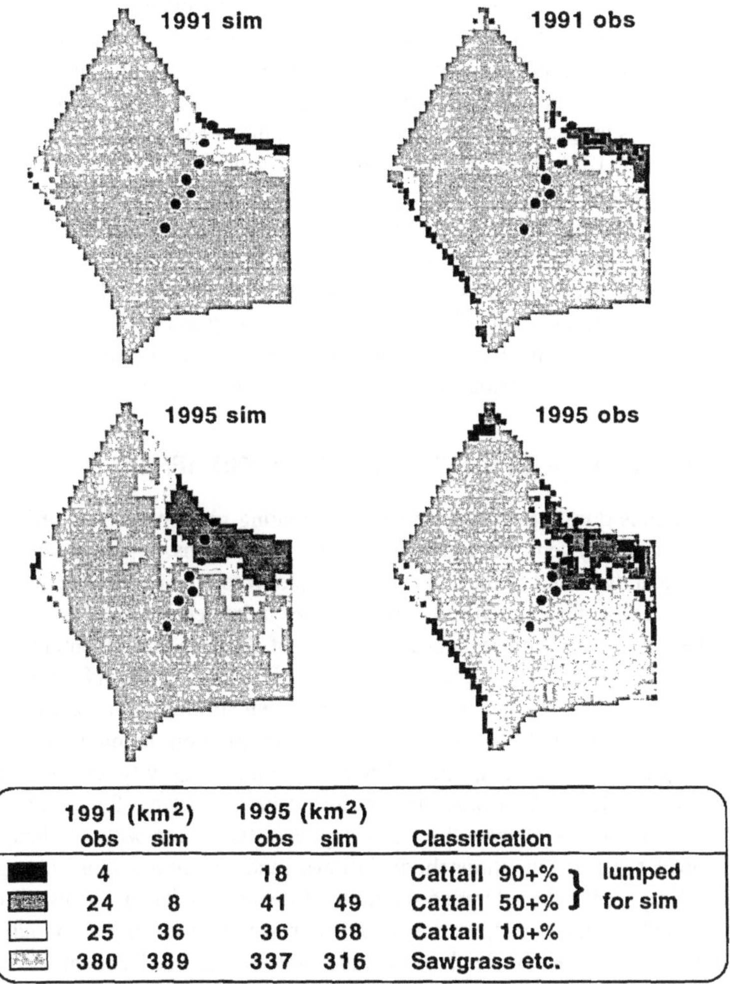

	1991 (km²) obs	1991 (km²) sim	1995 (km²) obs	1995 (km²) sim	Classification	
■	4		18		Cattail 90+%	lumped for sim
▒	24	8	41	49	Cattail 50+%	
□	25	36	36	68	Cattail 10+%	
░	380	389	337	316	Sawgrass etc.	

FIGURE 26.11 Simulated (sim) and observed (obs) transitions in habitat types in WCA-2A. The observed data from classified aerial photography were aggregated into the 0.25 km^2 grid cell resolution of the CALM, and the simulation did not distinguish between the cattail and cattail dominant mix (90+% and 50+%) classifications used in the photograph classifications.

Simulated periphyton communities showed more temporal and spatial heterogeneity than macrophytes, as periphyton respond quickly to changes in water availability (depth) and nutrient concentration (McCormick, personal communication). The CALM produced realistic periphyton responses to water and phosphorus in addition to the longer temporal response to shading by macrophytes. McCormick et al. (in press [b]) estimated total periphyton biomass in oligotrophic sloughs or sparsely vegetated habitats of WCA-2A to be ~1000 g AFDM m^{-2}, or approximately 150 g (organic) C m^{-2} depending on the ash content and the ratio of carbon to organic matter in those values reported by McCormick et al. Simulated maxima of 75 g C m^{-2} were underestimates of those observed values because the model underestimates the amount of slough, and thus overestimates macrophyte shading of periphyton. It should be noted that the vegetation mapping techniques used for WCA-2A did not separate sparse saw grass habitats from slough habitats (Rutchey and Vilchek, submitted). Plate 26.12a* shows the magnitude and spatial distribution of calcareous periphyton biomass, with the 17-year mean increasing along the gradient from the eutrophic to oligotrophic areas (F0 to U3). This was a response to both the influence of "excess" P in the surface water near the S-10 inflows (Plate 26.12b) and simulated macrophyte biomass changes along the gradient (Plate 26.12c). Much of the heterogeneity in the simulated pattern within the interior region of WCA-2A was due to elevation differences, with habitats along slough ridges that dried out periodically, as reflected by the hydroperiod distribution (Plate 26.7).

26.7 MODEL RESULTS: SENSITIVITY EXPERIMENTS

Of the numerous questions that arise when evaluating phosphorus fate and transport in the Everglades system, several relate to the interactions among hydrologic and biogeochemical dynamics in the soils. For example, when the water table drops below the soil surface, the unsaturated soil zone becomes more aerobic and thus more favorable for peat oxidation. Subsequent rehydration of the area can potentially increase levels of P in the porewaters. Of course, many other factors can control P concentrations such as sorption, plant uptake, and advection between surface and pore waters.

To determine the effects of altered evapotranspiration on the model P dynamics, we increased the potential evapotranspiration within a sensitivity analysis. We considered increased losses by evapotranspiration to be a surrogate for lower managed water levels, effectively altering the water depths in the region without changing the water control structure or atmospheric inflows (and the concomitant change in P loading). Compared to the reference simulation (the nominal case), the test simulation with increased potential evaporation had lower water levels during much of the simulation. In particular, the depth of the unsaturated zone was frequently greater for more prolonged periods. This produced a feedback that increased decomposition and soil remineralization to inorganic P by as much as 1 mg P m^{-2} d^{-1}, resulting in a greater accumulation of porewater phosphorus in particular regions. Plate 26.13* shows difference (test run minus nominal run) maps of the mean depth of the unsaturated zone during a dry year (1989), along with the corresponding difference in the mean of the remineralization rate. Associated with the drier zones in the central region was increased P release compared to the nominal run and an increase in macrophyte biomass in response to increased available nutrients during the simulation. In the eutrophic zone near the S-10 inflows, the area was drier in the test run compared to nominal, but there was little change in decomposition rates, which were high in both the nominal and test runs because of high nutrient availability for microbial processes.

Not all of the increased porewater P in the above analysis was necessarily because of increased peat decomposition. In this model, higher transpiration from

* Plates 26.12 and 26.13 appear as color plates following p. 590.

the macrophytes can advect surface water and its dissolved nutrients into the porewater zone. Davis (1982) showed that more than half of radio-labeled phosphorus introduced into surface water of marsh enclosures moved into the subsurface soils within ten days. With molecular diffusion being a relatively minor flux, this is indicative that downward advection of surface water appears to be a likely flow pathway for this phosphorus. Water losses by transpiration from the saturated zone are replaced by downflow of ponded surface water (Fig. 26.6), along with any constituents dissolved in that water. How significant is this process in making P available to the root zone? To answer this question, we can conduct a simulation sensitivity analysis to investigate the effects of a difference between a densely vegetated and a sparsely vegetated region. We adjusted the leaf area index of saw grass and cattail, and we introduced a conservative tracer into the flows from the S-10 and S-7 control structures (Table 26.1). Because the model's total evapotranspiration (ET) is apportioned between evaporation and plant transpiration, we can qualitatively demonstrate the relative importance of the advective flow of P into the porewater zone because of transpiration. For this analysis, we halved the leaf area index associated with each cell's macrophyte biomass (mimicking less dense canopy cover) for a 1980 test run compared to a 1980 nominal run. We used 1980 because it was a period when the region was continuously flooded (mean depth of 0.7 m), without large differences in plant biomass between runs. Table 26.1 shows that in the central saw grass region (site U3, total biomass approximately 250 g C m^{-2}), average total ET was approximately the same for both the nominal- and reduced-canopy runs. In the northern cattail region (F1, total biomass approximately 1000 g C m^{-2}), total ET in the reduced-canopy condition was similar to that of the saw grass site under both conditions (3.5 to 3.6 mm d^{-1}), but the denser canopy, nominal-condition at this site had somewhat higher ET (4.1 mm d^{-1}) than all others.

TABLE 26.1
Transpiration and Advection: Simulated Response of Year-1 (1980) Mean Total ET, Surface Water Evaporation, and Porewater Transpiration to Varying Macrophyte Canopy Conditions

Variable	F1 nominal	F1 low canopy	U3 nominal	U3 low canopy
Total ET (mm d^{-1})	4.1	3.6	3.6	3.5
Evaporation (mm d^{-1})	1.4	2.5	3.0	3.4
Porewater tracer conc. (μg L^{-1})	2.7	1.1	0.6	0.2
Root zone turnover (d)	420	190	90	30
	89	218	400	1200

Note: The Nominal condition used the nominal, best estimates of the leaf area index associated with the macrophytes at different sites (F1, U3) along the transect. The run under the Low Canopy condition used leaf area indices that were 0.5 × Nominal values. The final concentration of a conservative tracer determined the extent to which dissolved constituents were advected from the surface to the pore waters. Root zone turnover is the time required for transpiration and downward advection to completely replace the water volume in the upper 30 cm of soil of a continuously flooded area. (Actual root zone is 35 cm and 30 cm for saw grass and cattail, respectively).

The principal differences emerged when we examined the relative contributions of transpiration vs. surface water evaporation. Transpiration increased consistently with increasing macrophyte canopy cover (from 0.1 to 2.7 mm d^{-1}), while surface water evaporation decreased with less radiation reaching the water surface. With the higher transpiration and its advection of surface waters into the porewater, the conservative tracer concentration at least doubled at both the saw grass and cattail sites. In terms of hydrologic exchange resulting from transpiration and downward advection, the time required for the upper 30 cm of the soil porewater to be completely replaced was as little as 3 months in the cattail site and several years in the reduced-canopy condition at the saw grass site. Thus, depending on the concentration of P in the surface water relative to porewater and sorbed concentrations in the soil, this transport mechanism can potentially be a significant source of phosphorus to the plant root zone.

26.8 MODEL RESULTS: SCENARIOS

The present-day WCA-2A eutrophication gradient includes noticeable changes in surface water and sediment phosphorus, macrophyte and periphyton biomass, and community types. What may have occurred if the nutrient loads had been significantly reduced in the early 1980s, when eutrophication impacts were apparent in WCA-2A? For a hypothetical scenario, we curtailed the simulation's (1980 to 1996) inorganic P inflow concentrations to a maximum of 6.0 μg L^{-1} (compared to the observed mean of 62 μg L^{-1}).

Average differences between the nominal and reduced-load runs across the whole region were generally not large (Table 26.2). The expanse of marsh that is unimpacted by external phosphorus loads under either scenario heavily weights the regional mean values. However, subregions near the water control structure inflows were substantially affected by reduced loads. While mean surface water P concentration for the entire region was 6 μg L^{-1} greater in the nominal run, more than 40% of the region had >20 μg L^{-1} higher P concentration compared to the reduced-load scenario in 1996. The periphyton response to the reduced P load was not apparent when viewing the regional mean difference (Table 26.2), but spatial differences were pronounced as discerned from maps of calcareous periphyton biomass. A mean difference map (Plate 26.14a*) indicated a region of much higher biomass under the reduced-load scenario in the region immediately downstream of the S-10 A-E structures. The noncalcareous periphyton assemblage, while generally at a lower biomass in the region, tended to reflect an opposing trend (Plate 26.14b), but differences between runs were restricted to the area immediately downstream of the S-10 structures. These responses were primarily resulting from significantly reduced surface water nutrient concentrations in that region (Fig. 26.15a), as there was little difference in shading from macrophyte biomass in the immediate vicinity of the S-10 structures (Plate 26.14c).

Simulation of soil nutrients produced a similar trend of more pronounced localized differences compared to region-wide changes under reduced-inflow loads.

* Plate 26.14 appears as a color plate following p. 590.

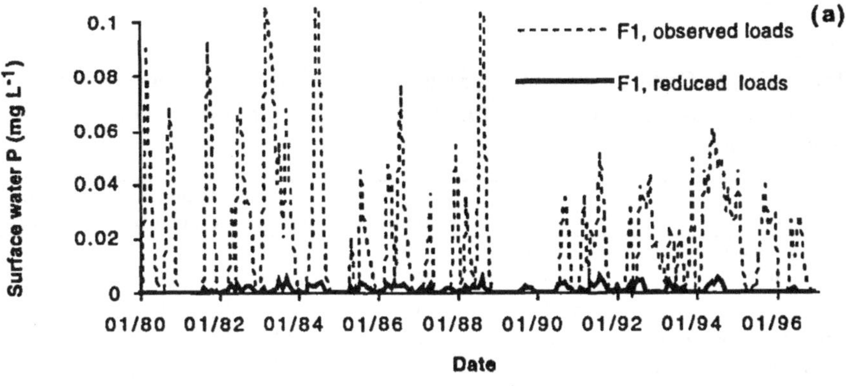

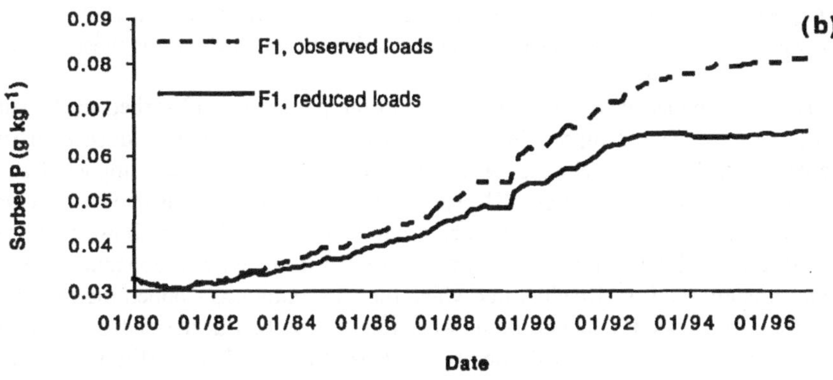

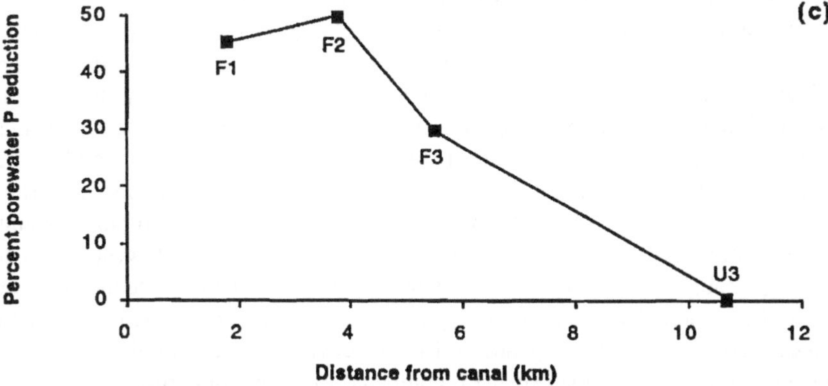

FIGURE 26.15 Simulated surface water and soil nutrient concentrations under the nominal condition of observed P loads and under the reduced-load condition. (a) Surface water P concentration at the F1 (eutrophic) site. (b) Sorbed P concentrations at the F1 site. (c) Reduction, relative to the nominal simulation, in porewater P concentration with distance from the S-10 inflow structures.

TABLE 26.2
Region-Wide Differences Among Scenarios: End-of-Simulation (1996) Summary of Selected Variables for the Nominal vs. Reduced-Load (1980–1996) Scenario

Variable	Units	Nominal	Difference	Minimum difference	Maximum difference
P surface water	μg P L^{-1}	6	6	– 6	58
P porewater	μg P L^{-1}	109	28	–92	863
P sorbed	mg P kg^{-1}	12	3	– 3	37
Periphyton—calcareous	g C m^{-2}	52	–0.1	–60	30
Periphyton—noncalcareous	g C m^{-2}	4	0.6	– 8	57
Macrophytes	g C m^{-2}	520	240	–69	904

Note: The Nominal run is the 12-month, WCA-wide mean of each variable. The Difference is the WCA-wide mean difference (nominal minus reduced load) for that year. The Minimum and Maximum differences are the cell-specific minimum and maximum differences for each variable during that year.

While region-wide mean P concentrations in the porewater and sorbed to the substrate were somewhat higher in the nominal compared to the reduced-load run (Table 26.2), the largest changes were observed in the current-day impacted zone near the S-10 structures. At site F1 near the inflow source, sorbed phosphorus was reduced 20% by 1996 under reduced external loads. Figure 26.15b shows that even in that reduced-load condition, the area continued to increase in concentration for approximately the first decade, after which the concentrations stopped their gradual increase. Porewater P concentrations were reduced by half at the end of the 17 yr simulation under reduced-load conditions at the F1 and F2 sites (Figure 26.15c). Sites farther downstream of the S-10 structures had smaller differences, with no apparent effect at the unimpacted U3 site 10 km downstream.

While this scenario analysis indicated that the magnitude and spatial extent of an elevated nutrient "front" was reduced with decreased loads, the simulated macrophyte biomass response was less pronounced in what is currently the impacted region. Under both scenarios, biomass of *Cladium* and *Typha* were near their maximum within the first ~4 km south of the S-10 A-D structures (e.g., south to F2), with little difference between the two scenarios in that particular subregion (Plate 26.14c). Differences on the order of 500 to 800 g C m^{-2} higher total biomass under the nominal vs. reduced load runs were evident in parts of the northwest region and the ridge/slough areas further south of the S-10 A-D structures. Reduced inflows also shifted macrophyte community types. There was a 33 km^2 reduction of the saw grass/cattail mixture category and a 14 km^2 cover reduction of the pure cattail category in the reduced-load scenario compared to the nominal run.

26.9 DISCUSSION

Spatial and temporal characteristics of the Everglades are well described for some ecological processes, but there remain gaps in our quantitative understanding of all of the system's dynamics. For example, whereas some of the graminoid plant communities are reasonably well quantified with respect to standing biomass and tissue nutrient stoichiometry (Toth, 1987; Toth, 1988; Davis, 1991; and Miao and Sklar, in press), changes in these characteristics are less well known across time and space for all species. In particular, the interaction of multiple factors (such as hydroperiod, nutrient availability, and fire regime) that are driving the changes in plant stock are not well understood, but relatively recent studies are providing better insight (Urban et al., 1993, Newman et al., 1996). Similarly, only recently have fine-scaled data been available on some of the processes behind changes in periphyton communities.

Hydrology drives the Everglades, and hydrology was a priority in our modeling and data compilation efforts. Because a hydrology model exists (MacVicar et al., 1984; HSM, 1997), a reasonably detailed understanding of the system hydrology was available. However, a significant effort remained to obtain the relevant data for the purposes of our model that is of a different spatial scale and areal extent, and which also incorporates dynamic feedbacks among hydrology and plant biology. Water losses through evapotranspiration and friction effects of vegetation on overland flow are two very sensitive and important processes in determining regional and/or local water budgets. While macrophyte biomass dynamics over short time scales and limited areas may not significantly influence the total water budget for a region as large as WCA-2A, there is evidence (e.g., Table 26.1) that altered evapotranspiration associated with plant biomass and community shifts has the potential to alter water storages within particular locations. The linkages among hydrology, biogeochemistry, and plant biology can alter other ecosystem characteristics involving nutrient budgets and plant biomass (Plate 26.13 and Table 26.1).

Despite uncertainties associated with fully characterizing all of the dynamics of the complex Everglades system, we were able to calibrate and verify the CALM at a number of levels. The most rudimentary level of determining the extent to which a model is capturing reality is what we term a Level 1 calibration, where the model behavior is judged to exhibit "reasonable" behavior. Level 1 calibration for the CALM did not indicate any variable performance that was inconsistent with ecological principles or observations. Most of our calibration was at the intermediate Level 2, where we had useful spatial and temporal resolution, but which was only complete enough to characterize the system variables over subsets of the entire spatial and temporal domains. We demonstrated the extent to which simulated hydrologic and ecological variations in space and time were consistent with observed data (Fig. 26.8 through Plate 26.12).

Presentation of a formal verification of model behavior is lacking for the current implementation of the CALM. However, the general model fit to observed water stages over widely varying conditions is an informal verification that the model's hydrologic behavior is accurate. (In actuality, the initial model calibration was conducted for the years 1980 through 1984, and subsequent extension of that time domain provided an implicit verification that we do not present here). An ideal level of calibration and verification for the CALM would be to have spatially distributed measurements of all important state variables at temporal scales sufficient to capture the spatial and temporal heterogeneity of the system, one set for calibration and an independent set for verification. Because of the difficulty of making all of these measurements, and because many variables of interest (such as vegetation biomass and community type) have not been measured historically over the appropriate spatiotemporal scales, this Level 3 calibration/verification was not attained. Nevertheless, we achieved realistic results and obtained good levels of fit of model and observations. Thus, we believe that for significant changes in environmental forcings, the model will provide realistic behavior.

26.9.1 Ecological Implications

One of the fundamental aspects of model development is recognizing the degree of process complexity needed for the stated objectives and goals. We have arrived at what we believe to be an appropriate level of process-detail after extensive data analyses and ad hoc unit model experiments to analyze the effects of aggregation according to the principles described by Rastetter et al (1992). The CALM/ELM is designed to run at varying spatial scales and for widely varying habitat types and environmental conditions. For habitats as diverse as fresh marshes and upland forests, we have evaluated the degree of sensitivity of the parameter set (Fitz et al., 1995). Using a modular framework, we have a modeling tool that can easily be aggregated or disaggregated. For example, the model's simulation of organic material decomposition and remineralization of nutrients currently is controlled by several factors, including temperature, moisture and substrate quality. We implicitly incorporate the redox potential in the sediments using a simple water depth, aerobic zone relation, and generalized rate parameters for aerobic and anaerobic environments. This appears to be adequate for our current objectives but would require more detailed relationships for finer scale modeling of nutrient availability in different layers of the root zone.

After evaluating the ecosystem's properties and the model's structure and dynamics, there emerge some aspects of the model that we would like to improve. While we are able to simulate the eutrophication gradient associated with the inflows from the water management structures, additional realism could be incorporated by introducing and tracking total phosphorus, in addition to P, as input from the structures. Currently, we assume that biological processes in surface and pore waters are driven by dissolved inorganic P, and the organic phosphorus contribution to the sediment is via periphyton and macrophyte detritus, but not by direct deposition of allochthonous organic phosphorus. In addition, it could be useful to implement variable C:P stoichiometry dynamics of the plant biomass. However, all of these processes in combination will significantly increase the model's complexity. Given the uncertainties associated with many of the process rates for these complex interactions, it is "uncertain" that there will be a relative benefit associated with better predictability or effectiveness (Costanza and Sklar, 1985; Costanza and Maxwell, 1994) for our objectives of evaluating ecosystem responses on a regional scale. We anticipate future efforts to evaluate how these changes could affect simulation results.

Simulation results must be interpreted in a realistic and conservative manner. In some instances, the CALM produced spatial output for some variables that did not match measured data. For example, the model overestimated porewater concentration at the most eutrophic site (F1) (Fig. 26.10b), while it underestimated peat accretion (and P storage) at that site (Fig. 26.10c). The C:P stoichiometry and homogenous soil profile are simplified compared to the real system dynamics. Therefore, there is some uncertainty in our estimates of the timing and magnitude of the propagation of the nutrient "front" along the gradient associated with the S-10 structures. Similarly, it is likely that there is some uncertainty in model estimates of the time that internal P loading will continue in the highly impacted areas. Nevertheless, the model captured realistic dynamics of a complex system across large temporal and spatial extents, allowing us to further investigate the potential relative changes to be expected under a variety of altered external forcing functions. It is apparent that parts of the WCA-2A ecosystem will continue in a eutrophic state for a significant time period after reducing external inputs, and some ecological indicators (such as macrophytes) point to some continued expansion of the zone of impact from external forcing functions. The utility of the CALM lies not in producing exact spatial results, but rather in evaluating the types of interactions that occur in the ecosystem that are difficult to predict *a priori*, and are not captured in models of either hydrology or surface water quality alone. An example is the result showing that internal cycling of nutrients within the soils can continue to impose a load on the ecosystem long after external sources of nutrient inputs are curtailed (Fig. 26.15). Alterations in hydroperiod, such as those that may occur with new water management scenarios, can significantly alter the pattern and magnitude of this internal cycling.

The simulated difference in remineralization rates (Plate 26.13) was as much as 2.5 to 3 times greater in drier regions, contributing to internal loading of available phosphorus. Koch-Rose et al. (1994) found peak porewater SRP at eutrophic sites following a summer drydown with high temperatures, and data to indicate that P may be a limiting factor in microbially mediated nutrient mineralization along the WCA-2A gradient. Following agricultural drainage of the northern Everglades, soil elevations decreased dramatically because of subsidence and peat oxidation (Snyder and Davidson, 1994). The CALM incorporates both these feedbacks of hydroperiod and nutrient availability on decomposition rates. The temporal and spatial extent of the interactions that lead to such processes is fundamental to a spatially explicit ecological model such as CALM. Capturing these dynamic interactions in simulations allows an evaluation of the potential for the system to self-organize and change.

It is difficult, if not impossible, to use just one or two variables to predict total ecosystem response to altered management scenarios, given the multitude of interactions that occur in the real ecosystem. Water levels alone do not fully characterize the ecosystem response. Water, nutrients, and plants are important to understanding the system structure and function. For example, vegetation biomass and community types may change over relatively long time scales. This affects transpiration, nutrient uptake, and detrital accumulation, further influencing the ecosystem through continued feedbacks among hydrology, geochemistry, and ecosystem structure. These aspects of the ecosystem are important to how it functions: with changes to surface water phosphorus loading, a component of the system such as periphyton is indirectly altered by macrophyte shading in addition to the direct effect of new surface water chemistry (Plate 26.14). The results of the model sensitivity and scenario analyses suitably demonstrate these interrelationships, and provide an improved understanding of the ecological response to water management.

26.9.2 Restoration Implications

Significant reductions in nutrient loads to WCA-2A are anticipated within the next decade. However, our results suggest that reductions in nutrient loads through the S-10 structures will not have the immediate and dramatic effects that many are hoping to observe within the currently impacted regions of WCA-2A. This may be relatively surprising to some since nutrient processing by wetlands can be relatively fast (in the range of days). One might expect to see rapid restoration to impacted areas once nutrient loads are reduced. This does not occur in the CALM, because some regions of these wetland soils have significant nutrient storage that maintains the system in its present state, and can potentially continue to accumulate available nutrients, while at the same time spreading out spatially. Although surface water nutrients decline to low levels under reduced external load, the macrophyte biomass (and cattail areal coverage) associated with the currently impacted areas is not likely to decline in the near term. In fact, even after 17 years of reduced loads, the total plant biomass for much of the impacted zone remains similar to that under observed loads (Plate 26.14c). This indicates that another component of WCA-2A habitats, that of calcareous periphyton communities, will not likely undergo restoration in the near term because of light limitation under the dense vegetation canopy in the impacted zone.

There are a variety of tools, including models and other techniques, that will be used to aid in identifying optimal scenarios for improved water management in south Florida. In conjunction with boundary condition inputs from the SFWMM that simulates the entire system of water management, the ELM/CALM tools provide an ecological basis for evaluating indirect effects of altered water and nutrient flows as they propagate across the landscape mosaic of the Everglades wetlands. Without considering feedbacks among nutrient cycling/availability, hydrology, and plant community dynamics, simple predictions of the response to a change in one factor alone would be problematic. Changes in water delivery to the Water Conservation Areas can alter plant growth resulting from water limitation or excess. Different water levels also alter the decomposition rates of organic material in the soil, and thus affect the relative availability of nutrients. Thus, the interactions of plants and nutrient cycling could produce a different state of plant biomass and composition, one that was not expected from hydrologic considerations alone. The CALM has served as the testing platform for development of the larger ELM, which encompasses the entire "natural" area of the Everglades/Big Cypress system. Using a process-based landscape model to depict ecosystem dynamics in the Everglades will assist in developing more informed recommendations concerning management strategies. In this first version of a landscape modeling program for WCA-2A, we have demonstrated the nature of these ecosystem responses to a variety of environmental changes. We will continue to refine the model to incorporate new empirical studies, and are implementing the modeling framework within the entire Everglades system.

26.10 ACKNOWLEDGMENTS

This work was initially supported by funding from the SFWMD (C-3123) to the University of Maryland, and continues to be supported at the SFWMD. R. Costanza was the Principal Investigator on the initial project, providing support and valuable input along the course of the work. A. Voinov was instrumental in developing and implementing many aspects of the hydrologic algorithms, particularly the raster-vector interaction modules. K. Rutchey and L. Vilchek were valuable contributors in many aspects of spatial characterization of WCA-2A, including the elevation data processing. P. McCormick and S. Newman kindly provided us with recent unpublished data from their field studies in WCA-2A, and on many occasions R. Reddy provided us with early results from his laboratory. Y. Wu and E. Reyes provided useful data summaries and advice during the development of the project. We would like to thank D. Worth and T. Fontaine for motivating and supporting the initiation of this project. Special thanks go to T. Maxwell for his Spatial Modeling Environment (http://kabir.cbl.umces.edu/SME3/) used in developing the CALM and ELM. We appreciate the thoughtful reviews of T. Fontaine, T. James, C. Madden, Z. Moustafa, and Y. Wu. Further descriptions of the model structure and implementation, along with recent animations of model output, can be found at:

http://sfwmd.gov/org/erd/esr/projects/ELM.html

REFERENCES

Bales, J. D., J. M. Fulford, and E. Swain. 1997. Review of selected features of the Natural Systems Model, and suggestions for application in South Florida. Report 97-4039. Water Resources Investigations, U.S. Geological Survey, Raleigh, NC.

Browder, J. A. 1982. Biomass and primary production of microphytes and macrophytes in periphyton habitats of the southern Everglades. South Florida Research Center. Homestead, FL. 49 pp.

Chen, Z., T. D. Fontaine, and R. Z. Xue. 1997. Phosphorus transport simulation in the Everglades Protection Area: a GIS-modeling approach. *In:* Proceedings of the 22nd Annual Conference of the National Association of Environmental Professionals, Orlando, FL.

Christiansen, J. E. 1968. Pan evaporation and evapotranspiration from climatic data. J. of Irrig. and Drain. Div. 94:243–265.

Cooper, R. M. and J. Roy. 1991. An atlas of surface water management basins in the Everglades: the Water Conservation Areas and Everglades National Park. South Florida Water Management District, West Palm Beach, FL. Sept. 1991.

Costanza, R. and T. Maxwell. 1994. Resolution and predictability: an approach to the scaling problem. Landscape Ecology. 9:47–57.

Costanza, R. and F. H. Sklar. 1985. Articulation, accuracy and effectiveness of mathematical models: a review of freshwater wetland applications. Ecological Modeling. 27:45–68.

Craft, C. B. and C. J. Richardson. 1993. Peat accretion and N, P, and organic C accumulation in nutrient-enriched and unenriched Everglades peatlands. Ecological Applications. 3(3):446–458.

Davis, J. H. 1943. The Natural Features of Southern Florida, Especially the Vegetation, and the Everglades. Florida Geological Survey, Tallahassee, FL.

Davis, S. M. 1982. Patterns of radiophosphorus accumulation in the Everglades after its introduction into surface water. Report 82-2. South Florida Water Management District, West Palm Beach, FL. February, 1982.

Davis, S. M. 1989. Sawgrass and cattail production in relation to nutrient supply in the Everglades. pp. 325–341 *In:* Sharitz, R. R. and J. W. Gibbons (eds.). Freshwater Wetlands and Wildlife, DOE Symposium Series No. 61. USDOE Office of Scientific and Technical Information, Oak Ridge, Tennessee.

Davis, S. M. 1991. Growth, Decomposition and Nutrient Retention of *Cladium jamaicense* Crantz and *Typha domingensis* Pers. in the Florida Everglades. Aquatic Botany. 40:203–224.

DeBusk, W. F., K. R. Reddy, M. S. Koch, and Y. Wang. 1994. Spatial distribution of soil nutrients in a northern Everglades marsh—Water Conservation Area 2A. Soil Science Society of America Journal. 58(2):543–552.

Fennema, R. J., C. J. Neidrauer, R. A. Johnson, T. K. MacVicar, and W. A. Perkins. 1994. A computer model to simulate natural Everglades hydrology. pp. 249–290 *In* Davis, S. M. and J. C. Ogden (eds.). Everglades: the Ecosystem and its Restoration. St. Lucie Press, Delray Beach, FL.

Fitz, H. C., R. Costanza, and E. Reyes. 1993. The Everglades Landscape Model (ELM): Summary Report of Task 2, Model Development. Report to South Florida Water Management District, West Palm Beach, FL. 109 pp.

Fitz, H. C., E. B. DeBellevue, R. Costanza, R. Boumans, T. Maxwell, L. Wainger, and F. H. Sklar. 1996. Development of a general ecosystem model for a range of scales and ecosystems. Ecological Modeling. 88:263–295.

Fitz, H. C., A. Voinov, and R. Costanza. 1995. The Everglades Landscape Model: multiscale sensitivity analysis. Report to South Florida Water Management District, Everglades Systems Research Division. 88 pp.

HSM. 1997. Documentation update for the South Florida Water Management Model. Hydrologic Systems Modeling Division, South Florida Water Management District. West Palm Beach, FL. 239 pp.

Jensen, J. R., K. Rutchey, M. S. Koch, and S. Narumalani. 1995. Inland wetland change detection in the Everglades Water Conservation Area 2A using a time series of normalized remotely sensed data. Photogrammetric Engineering & Remote Sensing. 61:199–209.

Keith and Schnars. 1993. G.P.S. Geodetic Survey in the Everglades WCA-2A, and WCA-3A. Keith and Schnars, Lakeland Division, Lakeland, FL.

Koch, M. S. and K. R. Reddy. 1992. Distribution of soil and plant nutrients along a trophic gradient in the Florida Everglades. Soil Science Society of America Journal. 56(5):1492–1499.

Koch-Rose, M. S., K. R. Reddy, and J. P. Chanton. 1994. Factors controlling seasonal nutrient profiles in a subtropical peatland of the Florida Everglades. J Environ Qual. 23(3):526–533.

Lassiter, R. 1975. Modeling dynamics of biological and chemical components of aquatic ecosystems. Report EPA-660/3-75-012. Southeast Environmental Research Laboratory, U.S. Environmental Protection Agency.

Light, S. S. and J. W. Dineen. 1994. Water control in the Everglades: a historical perspective. pp. 47–84 *In* Davis, S. M. and J. C. Ogden (eds.). Everglades: The Ecosystem and its Restoration. St. Lucie Press, Delray Beach, FL.

MacVicar, T. K., T. VanLent, and A. Castro. 1984. South Florida Water Management Model: documentation report. Report 84-3. South Florida Water Management District. West Palm Beach, FL. December, 1983.

Maxwell, T. and R. Costanza. 1995. Distributed modular spatial ecosystem modeling. International Journal of Computer Simulation: Special Issue on Advanced Simulation Methodologies. 5(3):247–262.

McCormick, P. V. *pers. comm.* South Florida Water Management District, West Palm Beach, FL.

McCormick, P. V. and M. B. O'Dell. *in press*. Quantifying periphyton responses to phosphorus in the Florida Everglades: a synoptic-experimental approach. Journal of the North American Benthological Society.

McCormick, P. V., P. S. Rawlick, K. Lurding, E. P. Smith, and F. H. Sklar. *in press-a.* Periphyton-water quality relationships along a nutrient gradient in the Florida Everglades. Journal of the North American Benthological Society.

McCormick, P. V. and L. Scinto. 1998. Influence of phosphorus loading on periphyton communities in wetlands. p.? *In* Reddy, K. R. (ed.) Phosphorus Biogeochemistry in subTropical Ecosystems.

McCormick, P. V., R. B. E. Shuford, J. G. Backus, and W. C. Kennedy. *in press-b.* Spatial and seasonal patterns of periphyton biomass and productivity in the northern Everglades, Florida, USA. Hydrobiologia.

Meeder, J. F., M. S. Ross, G. J. Telesnicki, P. L. Ruiz, and J. P. Sah. 1996. Vegetation analysis in the C-111aylor Slough Basin. Report to the South Florida Water Management District. Southeast Environmental Research Program, Florida International University. Miami, FL.

Miao, S. L. and F. H. Sklar. *in press*. Biomass and nutrient allocation of sawgrass and cattail along a nutrient gradient in the Florida Everglades. Journal of Wetland Ecology and Management.

Newman, S. *pers. comm.* South Florida Water Management District, West Palm Beach, FL.

Newman, S., J. B. Grace, and J. W. Koebel. 1996. Effects of nutrients and hydroperiod on mixtures of *Typha*, *Cladium*, and *Eleocharis*: implications for Everglades restoration. Ecological Applications. 6:774–783.

Ogden, J. C., Robertson, W.B., G. E. Davis, and T. W. Schmidt. 1974. Pesticides, polychlorinated biphenyls and heavy metals in upper food chain levels, Everglades National Park and vicinity. U.S. Department of the Interior Report.

Rastetter, E. B., A. W. King, B. J. Cosby, G. M. Hornberger, R. V. O'Neill, and J. E. Hobbie. 1992. Aggregating fine-scale ecological knowledge to model coarser-scale attributes of ecosystems. Ecological Applications. 2:55–70.

Reddy, K., W. DeBusk, Y. Wang, D. R, and M. Koch. 1991. Physico-chemical properties of soils in the Water Conservation Area 2 of the Everglades. University of Florida. Gainesville, FL. 214 pp.

Reddy, K. R., R. D. Delaune, W. F. Debusk, and M. S. Koch. 1993. Long-term nutrient accumulation rates in the Everglades. Soil Science Society of America Journal. 57(4):1147–1155.

Richardson, C. J., C. B. Craft, R. G. Qualls, J. Stevenson, and P. Vaithiyanathan. 1995. Effects of Phosphorus and Hydroperiod Alterations on Ecosystem Structure and Function in the Everglades. Duke Wetland Center, Durham, NC. 371 p.

Richardson, C. J. and P. Vaithiyanathan. 1995. Phosphorus sorption characteristics of Everglades soils along a eutrophication gradient. Soil Science Society of America Journal. 59:1782–1788.

Rutchey, K. and L. Vilchek. 1992. Development of an Everglades vegetation map using a SPOT image and the global positioning system. Photogrammetric Engineering & Remote Sensing. 60:767–775.

Rutchey, K. and L. Vilchek. *submitted.* Air photo-interpretation and satellite imagery analysis techniques for mapping cattail coverage in a northern Everglades impoundment. Journal of Photogrammetric Engineering and Remote Sensing.

SFWMD. 1997. Proceedings of the Conference on: Atmospheric Deposition into South Florida: Measuring Net Atmospheric Inputs of Nutrients. South Florida Water Management District. West Palm Beach, FL.

Snyder, G. H. and J. M. Davidson. 1994. Everglades Agriculture—Past, Present, and Future. pp. 85–115 *In* Davis, S. M. and J. C. Ogden (eds.). Everglades: the Ecosystem and its Restoration. St. Lucie Press, Delray Beach, FL.

Swift, D. R. 1984. Periphyton and water quality relationships in the Everglades Water Conservation Areas. pp. 97–117 *In* Gleason, P. J. (ed.) Environments of South Florida: Present and Past. Miami Geological Society, Coral Gables, FL.

Swift, D. R. and R. B. Nicholas. 1987. Periphyton and water quality relationships in the Everglades Water Conservation Areas 1978–1982. Report 87-2. South Florida Water Management District, West Palm Beach, FL. March, 1987.

Toth, L. A. 1987. Effects of hydrologic regimes on lifetime production and nutrient dynamics of sawgrass. Report 87-6. South Florida Water Management District, West Palm Beach, FL.

Toth, L. A. 1988. Effects of hydrologic regimes on lifetime production and nutrient dynamics of cattail. Report 88-6. South Florida Water Management District, West Palm Beach, FL.

Urban, N. H., S. M. Davis, and N. G. Aumen. 1993. Fluctuations in sawgrass and cattail densities in Everglades Water Conservation Area 2A under varying nutrient, hydrologic and fire regimes. Aquatic Botany. 46:203–223.

Voinov, A., C. Fitz, and R. Costanza. *in press*. Surface water flow in landscape models: 1. Everglades case study. Ecological Modeling.

Voinov, A., H. C. Fitz, and R. Costanza. *submitted*. Everglades landscape hydrology: coupling raster and vector based models. Landscape Ecology.

Walker, W. W. 1998. Long term water quality trends in the Everglades. *In* Reddy, K. R. (ed.) Phosphorus Biogeochemistry in subTropical Ecosystems.

27 Management Models to Evaluate Phosphorus Impacts on Wetlands

Robert H. Kadlec and W.W. Walker

27.1 ABSTRACT

Additions of phosphorus (P) and water can alter the status of many types of receiving wetlands. Typical responses involve alterations of sediments and soils, as well as micro and macro flora, and the associated faunal uses. Oligotrophic ecosystems, dominated by sedges (*Cladium* or *Carex*) typically respond to P fertilization by increasing productivity. In the longer term, opportunistic species (such as *Typha*) may replace the original vegetation. Less obvious changes occur sooner in soils, algae and microbes.

Wetland soils in general, and Everglades peats in particular, have sorption capacity for phosphorus. That storage is typically considered to be reversible, and the sorbed P to be available. New additions of P-containing waters may either increase or decrease this temporary P storage on a time scale of weeks. On approximately the same time scale, algal and microbial communities may undergo expansion and shift to a new species composition. Existing wetland flora respond to an increase in P availability by increasing biomass, and by reapportioning that biomass

1-56670-331-X/99/$0.00+$.50

from below to above ground parts. This is termed *fertilizer response* and proceeds over several turnover times to a new state characteristic of the new nutrient availability.

If the magnitude of the P load increase is sufficiently large, the macrophyte community (if any) may undergo alteration, as new wetland species are able to assert dominance, and replace some or all of the antecedent species. This change in community species composition is also to some degree dependent on seed banks and vegetative reproduction, and adjacent community structure. Physical processes, such as shading out and the accretion of new rooting media, are long-term influences that contribute to the structure of the replacement ecosystem. The speed of community shifts is typically slower than the fertilizer response.

Under continuously wet conditions, the several large and small components of the wetland biogeochemical cycle produce accretion of new organic sediments. A first-generation model of the carbon and P cycles produces estimates of the alteration of the rhizosphere, which may then be used as an indicator of potential macrophyte colonization. The components of the model include P removal rates linked to surface water concentrations, and the relation between P accretion and soil accretion, as evidenced in soil column P profiles. Calculations of time trends in root zone average P content are then possible, which may then be used as a trigger variable in the prediction of community changes. Stochastic processes are not included in this deterministic model and create the need for probability density overlays.

Calibration of the model to existing Florida wetland data shows reasonable representations of field phenomena.

27.2 INTRODUCTION

Wetlands often change character when subjected to new phosphorus and water loadings. However, wetlands also act to incorporate new phosphorus (P) into the biosphere and soils, and thus protect downstream portions of the ecosystem. Consequently, new P loadings create zonation within the receiving wetland, with stronger effects near the point of discharge, diminishing in the direction of water flow as P is stripped from the water (Fig. 27.1). In general terms, the zone nearest the new discharge may undergo species alteration; zones farther away may retain their species under nutrient enrichment, and at long distances the background ecosystem will continue to prevail (Lowe and Keenan, 1997).

Wetland responses are keyed to P concentrations in the water and to concentrations in the root zone of the soil column (rhizosphere). The aquatic components, such as periphyton, interact directly with dissolved reactive P, which stimulates growth and induces changes in community species composition and relative abundance. These changes appear to occur over a continuum of P concentrations, ranging from essentially zero phosphorus up to moderately high concentrations (McCormick, 1996). Rooted macrophytes respond to increasing soil P concentrations, which are the result of increasing water concentrations. The first response is often biomass increase for the antecedent assemblage of macrophytes, possibly later followed by changes in species composition and abundance. In addition, the magnitude of the

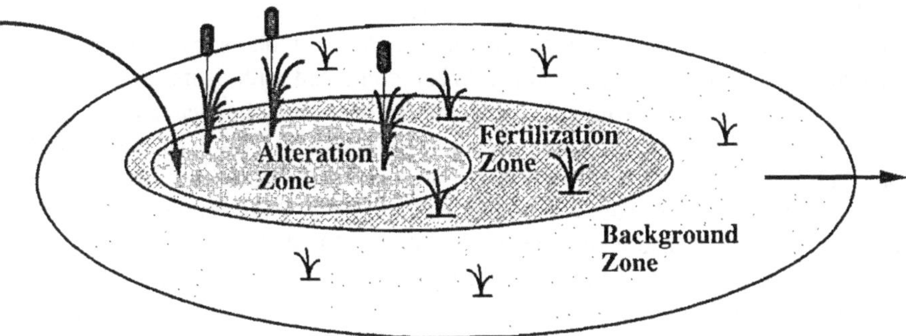

FIGURE 27.1 New phosphorus inputs tend to create zonation in the vicinity of the addition.

biogeochemical cycle increases, with the production of larger amounts of new sediments. Such a sequence of responses is also found along the gradient from the new source (Fig. 27.2). Sudden changes in ecosystem characteristics do not occur along either the spatial or temporal gradients. Rather, these are blurred changes, with new communities gradually interspersing with the old.

It is common for cattails (*Typha* spp.) to invade wetland areas receiving new inputs of water and phosphorus. This phenomenon has been reported for both northern peatlands (Kadlec and Bevis, 1990) and southern peatlands (Davis, 1994).

Phosphorus is not the only factor that may determine wetland changes. In most instances, new P inputs are accompanied by new water inputs, creating a wetter hydrologic regime. Changes in hydrologic regime can cause ecosystem effects in the absence of any additional P. The antecedent soil condition is also of importance, especially if there are prior alterations because of dryout, fire or mechanical disturbance.

A first step toward prediction of possible ecosystem effects is a prediction of the altered surface water total P concentrations, and the spatial and temporal allocation of the new cumulative P load. In the following sections, a P removal model

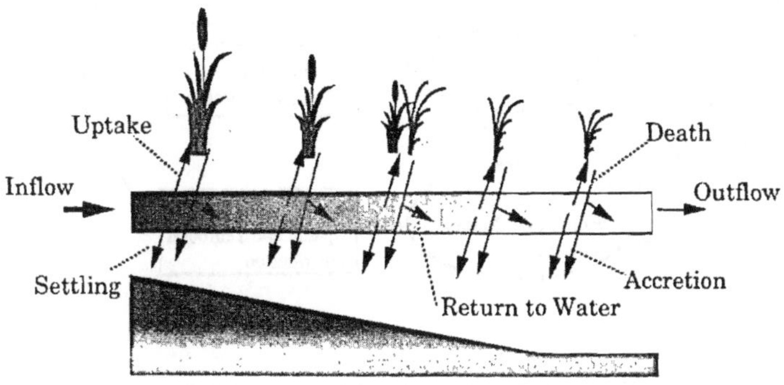

FIGURE 27.2 Gradients in surface water P lead to enhanced biomass, a larger biogeochemical cycle, and larger accretion rates for recalcitrant residuals. Arrows indicate the movement of phosphorus.

for the water is combined with a P mass balance on the soil to determine those allocations. A second step to improving the predictive tools is including the mass balance on the biomass in the wetland. When combined with estimates of wetland change criteria, potential wetland biological responses may be evaluated.

This chapter is based on three precursor publications: Walker (1995), Walker and Kadlec (1996), and Kadlec (1997). Walker (1995) describes a first-order P removal model that reproduces stationary soil and water P concentrations along the gradient in Water Conservation Area 2A (WCA-2A). Walker and Kadlec (1996) predicts the accretion of soil P with time in several wetlands. Kadlec (1997) shows that the first-order model is a close approximation to the response of the wetland biogeochemical cycle and its development.

This suite of models embodies water and phosphorus mass conservation and contains only a few relevant features of the wetland. It is intended to provide a tool for design and management of P in wetlands. A flowchart of the calculation strategy shows the integration of the several mass balances, and the input information needs (Fig. 27.3).

27.3 SURFACE WATER REMOVAL MODEL: ONE PARAMETER

27.3.1 Model Development

Under conditions of increased P inputs, concentrations typically decline within a wetland, along the flow direction (Fig. 27.4). The simplest fit to these observed

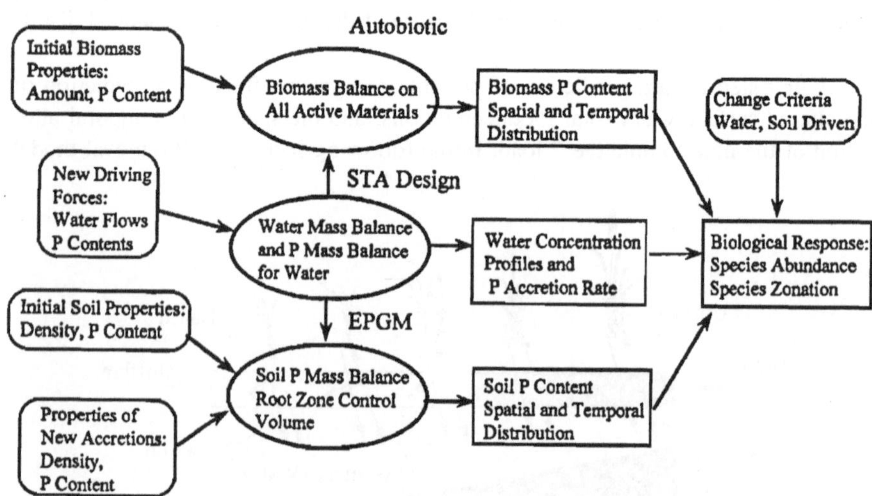

FIGURE 27.3 Interactions and data needs for the impact model network. The Stormwater Treatment Area (STA) design model (Walker, 1995) is represented by the center oval. The bottom two ovals are the Everglades Phosphorus Gradient Model (EPGM) (Walker and Kadlec, 1996). The top two ovals are the biomachine model (Kadlec, 1997).

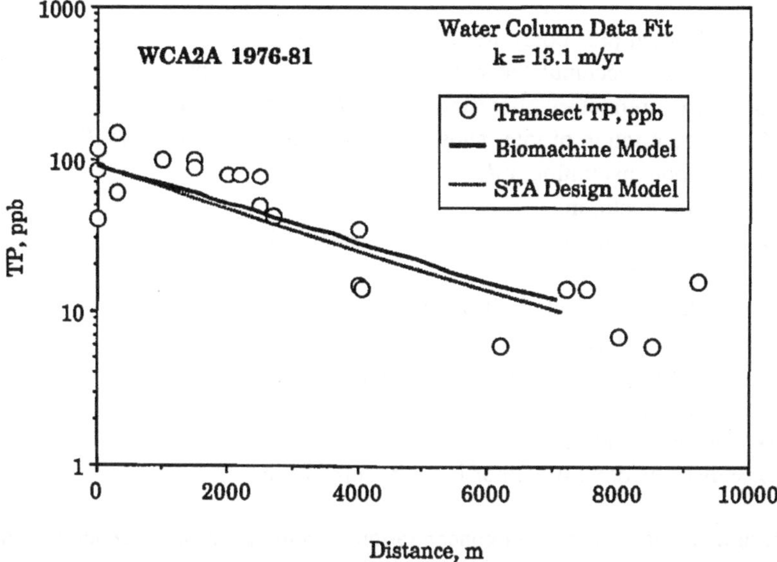

FIGURE 27.4 STA design and biomachine models fit the transect data for WCA-2A.

concentration gradients is a global, one parameter model. Several studies have shown that a first-order areal model provides a reasonable description (i.e., $R^2 \approx 0.8$) of the long-term sustainable phosphorus removal in emergent marshes (Kadlec, 1993, Walker, 1995, Mitsch et al., 1995). This model is an extension of the traditional lake model for phosphorus (Vollenweider, 1975; Kirchner and Dillon, 1975).

Phosphorus is presumed to be removed from water at an areal rate that is proportional to the P concentration in the water at the location in question. This removal rate is incorporated into a nondispersive (plug flow) mass balance. This model is here written for a one-dimensional, time averaged situation. The component parts are shown below.

- Dynamic, spatial water mass balance:

$$\frac{\partial h}{\partial t} + \frac{\partial Q}{\partial \mathcal{A}} = P - ET \tag{1}$$

- Dynamic, spatial P mass balance on water plus biomass and active soils:

$$\frac{\partial (hC)}{\partial t} + \frac{\partial M}{\partial t} + \frac{\partial A}{\partial t} + \frac{\partial (QC)}{\partial \mathcal{A}} = PC_p - S \tag{2}$$

- Dynamic, spatial P mass balance on inactive soils:

$$\frac{\partial B}{\partial t} = S \tag{3}$$

where

A = adsorbed plus porewater mobile phosphorus, g P/m^2
$\mathcal{A}$ = W • x = accumulated wetland area downstream, m^2
B = buried refractory phosphorus in soil/sediment, g P/m^2
C = concentration of total phosphorus (TP) in water, g/m^3 = mg/L
ET = evapotranspiration, m/yr
h = water depth, m
M = biomass temporary phosphorus content, g P/m^2
P = rainfall rate, m/yr
PC_p = rainfall plus dryfall phosphorus deposition, g P/m^2/yr
Q = water flow rate, m3/yr
S = net phosphorus removal rate, g P/m^2/yr
t = time, yr
W = wetland width, m
x = distance downstream of P addition point, m

It is assumed that surface water concentration is an indicator of the general chemical and biological activity at any given location, and that phosphorus deposition to the sediment follows in direct proportion to that activity. Justification is found in the transect profiles of TP concentrations in several studies (Kadlec and Knight, 1996). It is also presumed that chemical and biological activity, the macrophyte biomass and sediment-water interface are locally proportional to land surface area. The equation that quantifies these statements is:

$$S = kC \tag{4}$$

where k = removal rate constant, m/yr.

Terminology has evolved to designate the constant of proportionality k, which is a first-order areal rate constant, as the net apparent removal rate constant. There is a fraction of the phosphorus entering a wetland that is particulate and physically settles to the bottom, but the removal rate also includes the particulates generated within the wetland, including those formed underground because of root death and decomposition. The net phosphorus deposition rate, S, has come to be designated as the phosphorus removal rate. Care must be taken to properly interpret these terms. The removal rate is for all undecomposed particulate matter, which includes incoming suspended particulate matter and the detritus from carbon cycling in the wetland, and any precipitates that may form because of chemical reactions.

Surface water may also be considered as the control volume, or enclosed system for the mass balance. The transfers to this compartment are water flows, soil leaching, and atmospheric deposition. The transfers from this compartment are water flows, biomass accumulation, sorption and soil/sediment accretion. The dynamic, spatially distributed phosphorus budget is:

$$\frac{\partial(hC)}{\partial t} + \frac{\partial(QC)}{\partial \mathcal{A}} = PC_p - U \tag{5}$$

where U = net phosphorus uptake rate, g/m²/yr.

The net uptake is the difference between removals from the water to biological and soil compartments, and additions to the water by releases from biological and soil compartments. These transfers include leaching, sediment accretion, and uptake into growing biomass. Therefore, by comparison to Eqs. (2) and (5), the net uptake is:

$$U = S + \frac{\partial M}{\partial t} + \frac{\partial A}{\partial t} \tag{6}$$

Analysis of data from existing wetlands logically follows from application of Eq. (6) for determination of the net uptake. In the limit of a stable ecosystem, in which biomass phosphorus is not increasing and sorption sites are saturated, $dM/\partial t = 0$, $dA/\partial t = 0$, and $U = S$. Therefore, the form of Eq. (4) may also be applied to net uptake:

$$U = k_u C \tag{7}$$

where k_u = uptake rate constant, m/yr.

In a wetland that is leaching phosphorus from the soils because of antecedent conditions, U, and therefore k_u, may be small or even negative. In the limit of a stable ecosystem, $k_u = k$, the removal rate constant. In a developing wetland, large amounts of biomass are accumulating and dM/dt is large, leading to values of k_u that are larger than k. U is much larger than S in that situation. If hydroperiod (percentage of days wet) is less than 100%, then, as an ad hoc procedure, the value of k_u should be reduced by a factor not less than the hydroperiod, because P removal from water cannot occur when there is not water present.

Solution

Equations (5) and (7) may be solved under the assumption of negligible storage change. It will also be assumed that there is no groundwater recharge or discharge. Averaging over several water displacements, or over several flow events, is also assumed. Solution yields:

$$C = C_\infty + (C_i - C_\infty)\left(\frac{Q_i + a\alpha}{Q_i}\right)^{-(1+k/a)} \qquad a \neq 0 \tag{8a}$$

$$C = C_\infty + (C_i - C_\infty)\exp\left[-\frac{k\alpha}{Q_i}\right] \qquad a = 0 \tag{8b}$$

where C_i = inlet concentration, mg/L; C_∞ = background concentration, mg/L; and Q_i = inlet volumetric flow rate, m³/yr; and where

$$a = P - ET \tag{9}$$

$$C_\infty = \frac{PC_p}{(P - ET + k)} \tag{10}$$

The wetland background concentration C_∞ is that which would exist very far from a P addition point, where the only source of phosphorus would be from the atmosphere. Because k is an order of magnitude greater than P or ET, Eq. (10) predicts a very low limiting value of C_∞, about a factor of ten lower than rainfall phosphorus.

27.3.2 Calibration

The flow and water-column mass balances have been calibrated to data from WCA-2A (Walker, 1995). Long-term average, steady-state flow, and phosphorus concentration profiles downstream of the S10 structures are predicted. Flow input terms include entering flows and rainfall. Flow output terms include downstream discharge and evapotranspiration. Phosphorus input terms include incoming phosphorus loads and atmospheric deposition (uniform over simulated area). Phosphorus output terms include downstream discharge and net deposition to soils. That calibration produced k =10.2 m/yr for WCA-2A. Similar values have been determined for many other wetlands, including other Florida wetlands (Kadlec, 1994).

27.4 SURFACE WATER REMOVAL MODEL: THE BIOMACHINE MODEL

An alternate approach for modeling the removal of P from surface water involves calculation of the total biomass at a point within the wetland. Growth is driven by the surface water concentration, and thus this approach acknowledges the presence of the biological mediation of the transfer of P from the water to accreting solids. As a premium, this model describes the spatial and temporal variations in total biomass; as a penalty, it requires input parameters that increase the tasks of calibration. This model has been presented and calibrated to extensive data from a wastewater impacted wetland (Kadlec, 1997), and here for WCA-2A. The model is autobiotic, because more P stimulates more cycling and more P removal.

27.4.1 Model Development

The biomass cycle is the prime driving force for the creation of refractory, P-containing residuals that add to the sediments. The size of the return flux from the various biomass compartments is clearly most directly related to the size of the biomass pool, here termed the biomachine. It is indirectly related to the concentration of P in the water, because more nutrients stimulate more growth and higher standing crops. In this section, the local accretion flux is presumed to be proportional to the lumped biomass at that given location in the wetland. This approach has been described for rivers, in which the attached plant biomass is assumed to be the primary determinant of P uptake (Thomann and Mueller, 1987).

The rate of P burial is therefore written as:

$$S = k_N N = [x_N m(1 - \beta)]N \tag{11}$$

where k_N = burial rate constant, g P/g/yr; N = total biomass, g/m^2; m = biomass loss rate constant, yr^{-1}; x_N = P concentration in biomass, g P/g; and β = fraction of biomass P returned to the water.

This redefinition of the accretion rate has the advantage of tying the P removal calculation directly to the size and speed of the removal "engine" or "biomachine." It suffers from a disadvantage in that biomass is not frequently measured in wetland nutrient studies.

The size of the total, lumped biomass pool may be calculated from a growth-death model:

$$\frac{\partial N}{\partial t} = \frac{m}{N_{max}}(L - N)N \tag{12}$$

Equation (12) is the inhibited form of Malthus' law, long used to describe population growth (Bailey and Ollis, 1986). It postulates that growth will occur in proportion to the biomass present, but only up to a limiting density determined by the local surface water P concentration and physical space limitations. The form of the relation for the growth limit is presumed to be:

$$L = N_{max}\left(\frac{C}{C + s}\right) \tag{13}$$

where L = upper limit of biomass that can exist at a given C, g/m^2; N_{max} = maximum biomass that can exist per unit area, g/m^2; and s = biomass half saturation concentration, g/m^3.

27.4.2 CALIBRATION

The biomachine model was calibrated to data from WCA-2A. The parameters required are β, s, N_{max}, m, and x_N. The following requirements were imposed as objectives of model calibration: reasonable descriptions of (1) P concentration profiles in surface waters under long-term stationary conditions, both in the shape of the concentration gradient and in the global removal as characterized by the predicted first-order areal removal rate constant (k), (2) the increase in areal extent with time of the affected biomass zone, (3) the gradient in biomass from the discharge out to the background area, and (4) the amount of solids accretion in the wetland. The selected values were required to meet these conditions simultaneously. Both k and N are quite sensitive to the fit parameters, which implies that the optimal set occurs within a narrow range of acceptably close values. In other words, there is not much leeway in selecting β, N_{max}, m, and x_N to provide the observed uptake rates and biomass (Kadlec, 1997).

Data for calibration was available from the sources used by Walker (1995), for accretion and water phase P concentrations along the gradient. Information on standing crop along the gradient, and for implied turnover times was from Davis (1989); for tissue P concentrations in live and dead leaves, along the WCA-2A gradient from Davis (1991); and for P concentrations in a variety of plant parts in north and south WCA-2A from Toth (1988). Numerical values for the combined biomass pool were:

- $x_N = 0.0012$ g P/gm
- m = 5 turnovers per year
- $\beta = 93\%$ biomass P returned to the water
- $N_{max} = 12{,}200$ g/m^2, corresponding to an inlet aboveground crop of 584 g/m^2
- s = 320 μgP/L, which limits the inlet total crop to 2694 g/m^2 for the sum of above and below ground macrophytes, litter, and live and dead periphyton

The fit of this model to water P concentrations and P accretion is comparable to that of the one parameter settling rate model (Fig. 27.4).

27.5 BENTHIC AND ROOT ZONE PROCESSES

A number of processes typically occur in and on the wetland rhizosphere. The simple first-order removal model presumes that P is accreted directly into new sediments and soils. The contributing processes layer some new material on top of older, but add accretions to lower horizons as well (Fig. 27.5). The overall biogeochemical

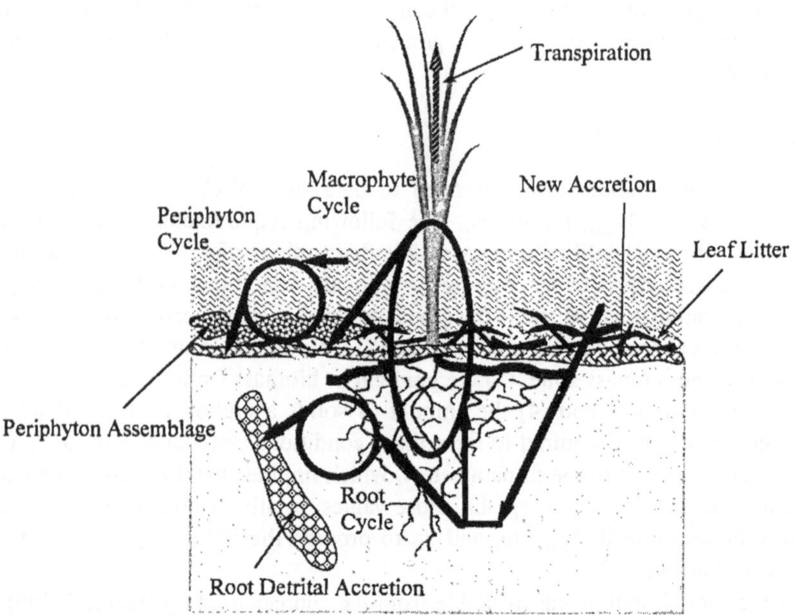

FIGURE 27.5 The total biogeochemical cycle may be broken into loops representing roots, shoots, and periphyton. The two macrophyte cycles interact via translocation and transpiration flows. The periphyton cycle draws phosphorus directly from the water column.

cycle can be broken down into three distinct loops, based on location. The aboveground macrophyte parts draw nutrients and water from the root zone, and return a large proportion of the nutrients back to the water, as a result of litter decomposition. In addition, nutrients can be translocated from aboveground tissues to rhizomes. The residuals of aboveground tissue decay are incorporated into the top layer of the soil column. At the top of the soil column, there is often a periphyton benthic assemblage that also undergoes a growth/death/decomposition cycle. That assemblage can obtain nutrients directly from the water, and also deposits its detrital residue on the top of the soil column. The plant roots grow and die, with an undecomposing residual that becomes an inert part of the rhizosphere. The roots draw and use nutrients from adjacent porewaters, and supply aboveground plant parts.

Shoot turnover times in the Everglades may be as short as two to three months (Davis, 1994). Root turnover times are largely unstudied, but have been reported to be on the order of 1.5 to 3.0 years (Prentki, 1978). Periphyton growth is fast, and lifespan relatively short, with 2 to 10 turnovers per year. In an emergent macrophyte stand, root and shoot biomass often dominate the biomass standing crop.

Diffusion, infiltration, and transpiration pumping carry phosphorus into the root zone. In a "sealed" wetland, infiltration is blocked, and the other two mechanisms must transport the required P. Calculations show that diffusion alone cannot supply the needs of the macrophytes, rates are an order of magnitude too slow. However, vertical water flow from the litter benthic zone can carry sufficient P to meet growth needs. For conditions of nearly equal periphyton and macrophyte cycling, most of the accretion (90%) is into a top soil layer (Table 27.1). This is caused by the presumed low turnover rate of root biomass compared to aboveground plant parts and periphyton.

TABLE 27.1
Hypothetical Multicycle Apportionment of Phosphorus Removal

	Standing crop, gmDW/m^2	Turnover time, 1/yr	P content, g P/gmDW	Gross P req'd, g P/m^2 • yr	Cycle burial efficiency, %	Net P req'd., g P/m^2 • yr	P standing crop, g P/m^2
Below-ground macrophytes	500	1.0	0.0010	0.50	10	0.05	0.50
Above-ground macrophytes	500	4.0	0.0010	2.00	10	0.20	0.50
Periphyton mat	250	8.0	0.0015	3.00	10	0.30	0.38
Total above ground	750	5.7		5.00		0.50	0.88
Total or avg.	1250	4.0		5.50		0.55	1.38

At 50 μg/L TP in the water, the settling rate constant for the example in Table 27.1 would be $0.55 \div 0.050 = 11$ m/yr. If the plant transpired 0.5 cm/d [75th percentile for the monthly ET means for the Everglades Nutrient Removal (ENR) project

(Abtew, 1996)], the infiltrating concentration would be (2.00 + 0.05) g P/m2 • yr ÷ (0.005 • 365) m/yr = 1.12 g P/m3 (1,123 μg/L). This is typical of the top layer porewater P measured by Richardson et al (1992) and Reddy et al (1991).

27.6 INTEGRATION OF PHOSPHORUS LOADS

27.6.1 Model Development

Phosphorus removed from the water column is deposited in the soil sediment compartment, primarily in new top layers. A control volume is selected that occupies a fixed vertical distance from the sediment-water interface. That control volume moves slowly upward as material accretes, with older solids passing out through the bottom surface (Fig. 27.6). Rates of accretion in WCA-2A were measured to be 0.27 to 1.13 cm/yr (Reddy et al, 1991) and 0.003 to 0.66 cm/yr (Craft and Richardson, 1993), depending on position along the nutrient gradient. These accretions are predicted by the water phase mass balance on phosphorus [Eqs. (4) and (8)], coupled with information on soil P content and bulk density.

The upward movement of the soil surface varies along the gradient in WCA-2A so that predischarge vertical soil P profiles are covered to different depths along the gradient, ranging from about 30 cm near the S10s to about 5 cm toward the center of WCA-2A, as a result of accretion over 30 years. Therefore, when results are presented in terms of the vertical depth below the current soil surface, there exists a variable vertical displacement along the WCA-2A gradient. These P profiles can be adjusted to a common datum by use of the depth of the Cesium-137 peak at each station along the gradient. When this is done, there is good conformity of all vertical profiles below the Cesium-137 datum (Fig. 27.7).

Mass balances on soil and soil phosphorus may be constructed for two time intervals:

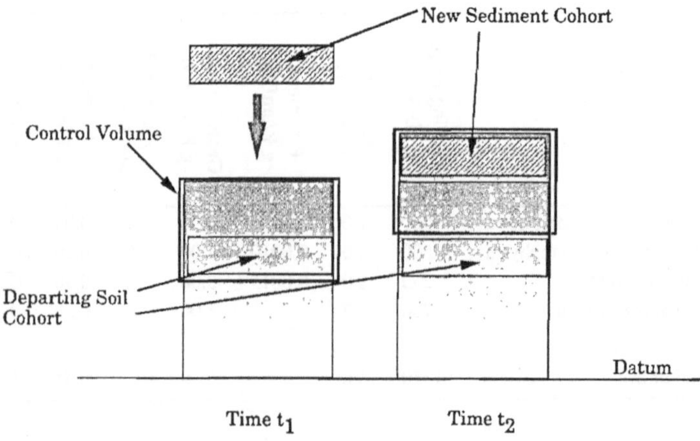

FIGURE 27.6 Soil phase mass balances use a control volume that moves upward with the soil surface.

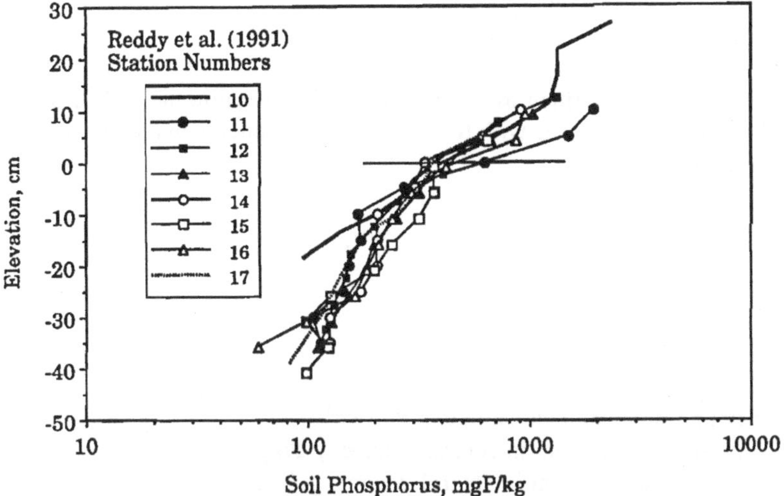

FIGURE 27.7 Soil phosphorus profiles in WCA-2A adjusted to the Cesium-137 datum.

1. *Initial phase.* New soil accumulates on top of the initial soil at a fixed rate. Output concentrations at the bottom of the control volume (fixed depth) reflect vertical gradients in the initial soil profile.
2. *Final phase.* Steady-state: starts when the depth of new soil equals the depth of the control volume. Soil properties in control volume equal properties of new soil.

Advective and diffusive transport of phosphorus across the bottom of the soil control volume (in pore waters) are ignored. There is no evidence to suggest that such mechanisms are important. If they do exist, their influences are implicit in the calibration of the settling rate.

The following differential equation describes the soil mass balance on the control volume:

$$\frac{\partial M}{\partial t} = T - \rho_z V \tag{14}$$

$$M(0) = M_i = \rho_i Z$$

where

M = soil mass in control volume, kg/m²
M_i = initial soil mass in control volume, kg/m²
ρ_i = initial soil mean density, kg/m³
ρ_z = soil density at depth Z, kg/m³
t = time, yr
T = new soil accretion rate, kg/m²/yr

V = soil volume accretion rate, m/yr
Z = control volume depth, m

The time derivative is written as a partial derivative, because the soil mass varies along the gradient as well as with time. The soil phosphorus mass balance on the control volume is:

$$\frac{\partial(MY)}{\partial t} = S - p_z V Y_z \tag{15}$$

$$M(0) = M_i Y_i$$

where

S = net phosphorus removal rate, g P/m²/yr
Y = mean soil P content in control volume, g P/kg
Y_i = initial soil P content in control volume, g P/kg
Y_z = bottom layer soil P content in control volume, g P/kg

An estimate of the soil mass accretion rate (T, kg/m²/yr) is required in order to solve above equations. This estimate is derived from an empirical model relating the average phosphorus content of soil above the Cesium-137 peak to the average phosphorus accretion rate:

$$Y_s = a + bS \tag{16}$$

where Y_s = mean soil P content in control volume at steady state, g P/kg.

From the definition of soil P content and Eq. (16), the required elation is:

$$T = \frac{S}{Y_s} = \frac{S}{a + bS} \tag{17}$$

27.6.2 Calibration

Data are available on the mean soil P content (Y_s) and settling rate (S), from Cesium-137 studies in WCA-1, WCA-2A, and WCA-3. The relationship between soil phosphorus content and phosphorus accretion rate [Eq. (16)] has been calibrated by Walker and Kadlec (1996) to data from WCA-2A (Reddy et al., 1991, Richardson et al., 1992; Craft & Richardson, 1993), WCA-3A (Reddy et al., 1994b, Robbins et al., 1996), and WCA-1 (Reddy et al., 1994a, Robbins et al., 1996). Observed and predicted values for soil phosphorus content and mass accretion rate are shown in Fig. 27.8. The calibrated parameters and equations are as follows:

$$Y_s = a + bS \quad (R^2 = 0.78, \text{SE} = 171 \text{ mgP/kg})$$

$$T = \frac{S}{a + bS} \quad (R^2 = 0.88, \text{SE} = 0.05 \text{ kg/m}^2 \bullet \text{yr})$$

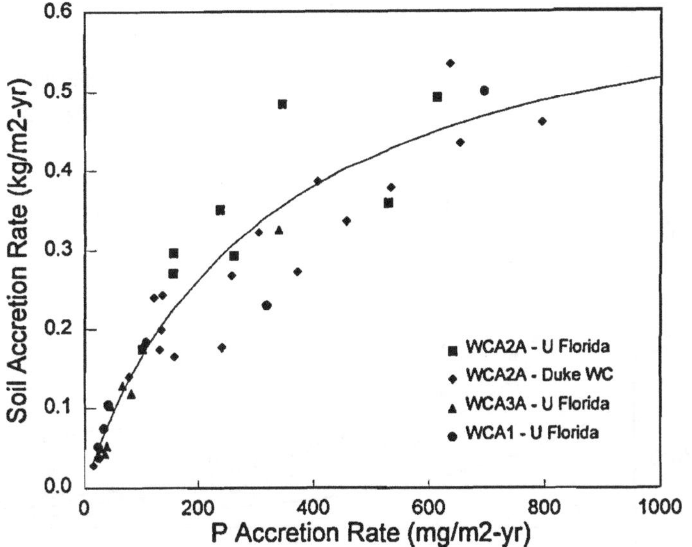

FIGURE 27.8 The fit of the regression for soil accretion and P accretion.

$$a = 463 \pm 27$$

$$b = 1.467 \pm 0.124$$

The calibration reflects the average soil response integrated over 26 to 29 years of peat accretion. Model parameter values may deviate from these average values during the start-up or transition period when the system is responding to a change in phosphorus loading. A measure of the goodness of fit is provided by comparing the data on soil phosphorus in the top 20 cm along the gradient in WCA-2A with the model calculations based on Eqs. (8), (14), and (15) (Fig. 27.9).

Solution

The soil phosphorus mass balance equations may be solved analytically for any specified settling rate (S), provided that the initial soil P content vertical profile is known. In the case of data from WCA-2A, those profiles are approximately linear. The soil bulk density for any specific wetland is nearly independent of depth, and may be considered constant (Walker and Kadlec, 1996).

The appropriate control volume depth (Z) is arbitrary but should contain the majority of the root zone if influences on vegetation are to be inferred. Roots are typically located in the upper 30 cm of the soil column, and are distributed approximately linearly decreasing with depth (Tanner, 1996)

The model returns the mean soil P content for the control volume as a function of time and surface water P concentration. The water P mass balance provides values of surface water P. Consequently, in combination these mass balances allocate historic P loads along a gradient from the source. Impacts to the ecosystem may

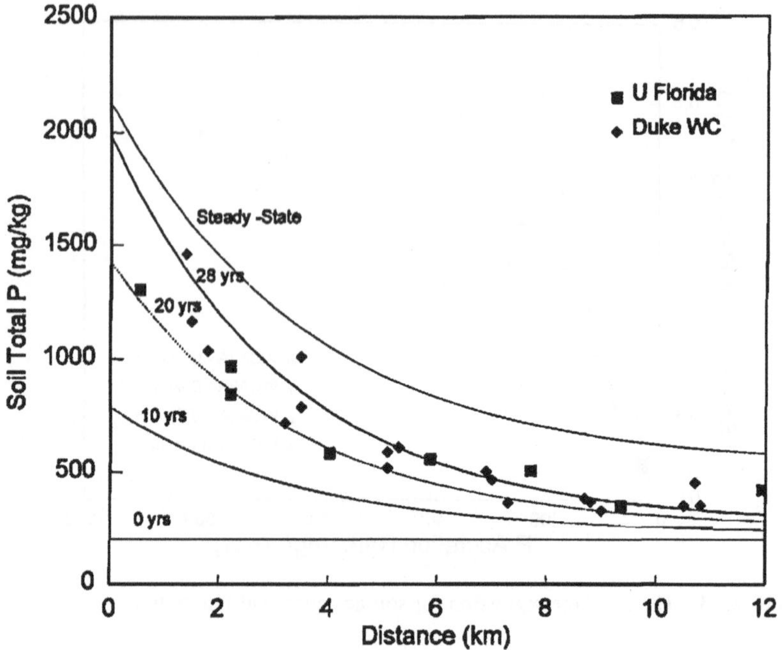

FIGURE 27.9 The time progression of soil P in the top 30 cm for WCA-2A.

then be inferred, as occurring in response to either the new surface water concentrations or the new soil P concentrations.

27.7 TRANSITION TRIGGERS

As illustrated in Fig. 27.3, the last step in model linkage is to predict biological responses, based on predicted changes in vegetation, and in water-column and soil P concentrations. Biological responses may be expressed in the following terms:

1. Marsh areas with long-term average water-column concentrations exceeding a specified threshold criterion
2. Marsh areas with soil P concentrations exceeding criteria or thresholds for changes in species composition or relative abundance
3. Total cattail area, estimated from a logistic equation relating cattail density (% coverage) to soil P concentration
4. Marsh areas exhibiting a growth response above a specified criterion

Item 1 is a surrogate for impacts on ecosystem components that respond to water-column concentrations (e.g., periphyton, algae). Items 2 and 3 are surrogates for impacts on ecosystem components that respond to soil P concentrations (e.g., cattails and other rooted vegetation). Item 4 is a direct indication of a more productive suite of biota (e.g., bigger plants and more algae).

Soil threshold values were calibrated to data from WCA-2A. The model estimates changes in cattail areas and densities potentially resulting from changes in external phosphorus loads. Changes resulting from other factors (water depths, fire, etc.) may occur but are not considered here. Although macrophyte changes may be driven by available phosphorus (vs. total), a much more complex model would be required to predict individual phosphorus fractions. Available P (as measured by bicarbonate extractable P) averages less than 2% of total P, but is highly correlated with total P in WCA-2A soils (Reddy et al., 1991).

Previous studies have correlated spatial variations in dominant vegetation with soil P levels in WCA-2A. Data summarized by Richardson, et al. (1995) indicate that increases in soil P levels are spatially correlated with declines in native slough macrophyte species (e.g., *Eleocharis, Utricularia, Cladium*). These species are replaced by cattail and other macrophytes characteristic of eutrophic Everglades. In discussing these results, Richardson (1996) noted that shifts in dominant vegetation from oligotrophic to eutrophic species generally occurred at surface soil P levels above 500 to 700 mg/kg. DeBusk et al. (1994) reported average soil phosphorus concentrations (0 to 10 cm) in three WCA-2A plant communities: saw grass 473 ± 134; mixed 802 ± 444; cattail 1338 ± 381 mgP/kg.

To provide a basis for comparing model with results with those reported by SFWMD (1996), predictions of "total cattail area" are developed by mapping the spatial distribution of soil P levels predicted for a given year onto a logistic function relating cattail density (% of area) to soil P (Fig. 27.10). The model is similar in form to that used by Wu et al. (1996) for predicting annual vegetation transition probabilities as a function of soil P levels. Details may be found in Walker and Kadlec (1996).

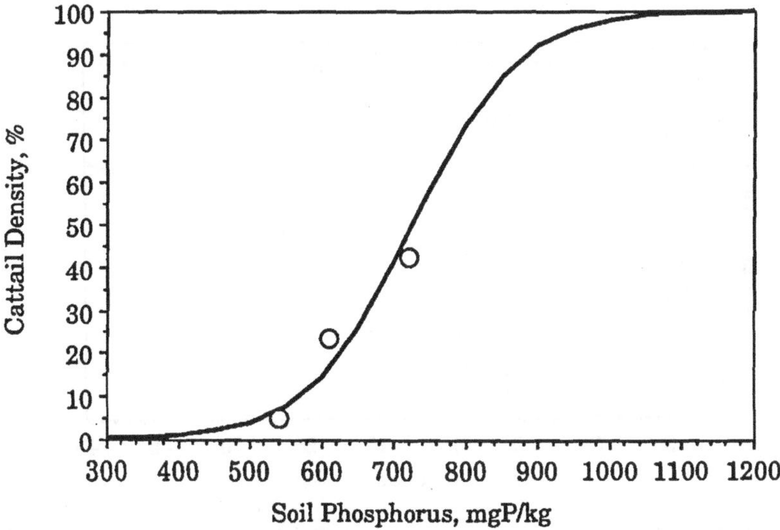

FIGURE 27.10 A logistic curve fit to data on cattail density and soil phosphorus.

27.8 RESULTS AND DISCUSSION

The methods above have been applied to the several water conservation areas and the projected Stormwater Treatment Areas (STAs) (Walker and Kadlec, 1996). Some results for WCA-2A are presented here.

The average P loading to WCA-2A was 42.36 metric tons per year, at an incoming concentration of 122 µgP/L. The areas required to reduce surface water TP concentrations to various levels were projected (and calibrated) to be:

- 5,723 ha to 30 µgP/L
- 7,613 ha to 20 µgP/L
- 13,393 ha to 10 µgP/L

The area requirement to reach 30 µgP/L was thus 135 ha/mt. This area requirement varies with the inlet P concentration as well as the inlet P load.

The cattail area, determined as the areal total over a spectrum of densities predicted by the logistic criterion, was found to increase with time (Fig. 27.11). Model predictions are in generally good agreement with measurements by SFWMD, although that data was based on total area with more than a fixed low percentage of cattail cover.

The biomachine model forecasts increases in total biomass, including roots, shoots, litter and periphyton, over a time period of about 15 years (Fig. 27.12). This period is shorter than that predicted for responses of the soil P content, and the resultant changes in species composition. A 20-cm soil column takes 20 years to "flush" to a new steady state for a high-end accretion rate of one centimeter per year, that represents the high end of the WCA-2A gradient. Further downgradient, those rates diminish to 10 to 20% of those near the inlet, and consequently the 20

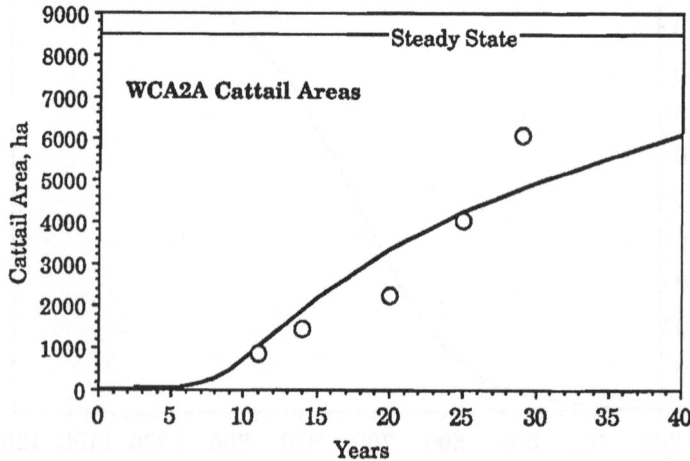

FIGURE 27.11 EPGM model prediction and SFWMD data on cattail expansion in WCA-2A.

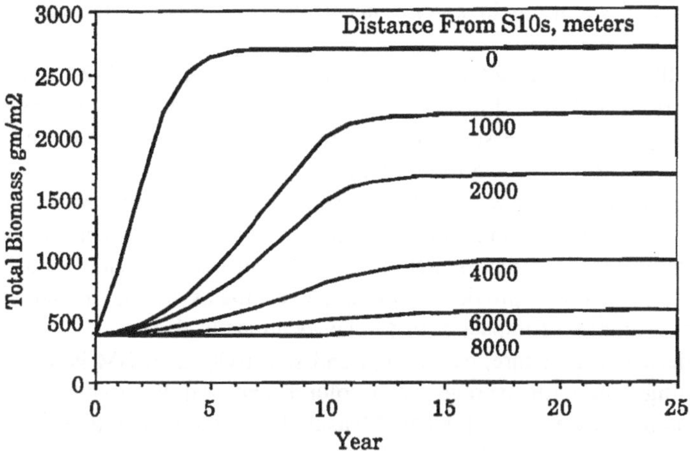

FIGURE 27.12 Autobiotic model predictions for active biomass in WCA-2A.

cm control volume takes many decades to replace. The growth response is greatest in the inlet region, where both soil P and water P are highest.

27.9 CONCLUSIONS

The surface water quality model used as a basis for STA design has been extended to include mass balances on the water-column and surface soils in marsh areas downstream of STA discharges. The revised model (labeled EPGM = Everglades Phosphorus Gradient Model) has been used to project impacts of discharges into Everglades ecosystems. Impacts are expressed in terms of areas exceeding threshold criteria for water-column and soil phosphorus concentrations, increases in total cattail area, and increases in cattail density.

The model used as a basis for STA design has been modified to include mass balances on the water-column and surface soils in marsh areas downstream of STA discharges. The revised model (labeled EPGM = Everglades Phosphorus Gradient Model) has been used to project impacts of discharges into Everglades ecosystems. Impacts are expressed in terms of areas exceeding threshold criteria for water-column and soil phosphorus concentrations, increases in total cattail area, and increases in cattail density.

A further model extension replaces the first-order removal rule, based on water concentration, with a first-order removal rule based on total lumped biomass. This autobiotic model requires a model for biomass growth in response to phosphorus concentration, and hence increases both the input data requirements and the number of predicted attributes of the ecosystem. Impact assessment may then be extended to include the spatial and temporal distributions of total active biomass.

The EPGM model successfully predicts observed spatial variations in water concentrations and soil phosphorus below the S10s, averaged over a depth 20 cm after 28 years of loading (1962 to 1990). The autobiotic model successfully predicts observed spatial variations in water concentrations, accretion rates, and total biomass.

Soil P thresholds for cattail expansion estimated from WCA-2A and WCA-1 data range from 540 to 720 mg/kg for a 20 cm soil depth. Predicted increases in cattail density and area are surrogates for impacts on any ecosystem components that respond to soil P levels in these ranges.

Simulations are for *average* hydrologic conditions. Actual responses will deviate from the predictions, depending on actual hydrologic conditions and system sensitivity. Since an idealized representation of flow distribution is employed (uniform sheet flow), simulations provide approximate estimates of the spatial scales of impact, not estimates of impact at particular locations or dates.

Considering its structure, calibration, and sensitivities, EPGM is most reliable for predicting long-term average water-column and soil P concentrations along gradients induced by external P loads. Measured initial soil conditions have strong influences on predicted soil P and cattail responses within four- to eight-year time frames. Refinements to the model structure are needed to improve model performance over short time scales in response to variations in hydrology (flow, hydroperiod, drought), P loading, biomass P storage, and start-up phenomena. Compilation of other data sets will support future refinement, calibration, and testing of the model.

REFERENCES

Abtew, W., 1996. Evapotranspiration measurements and modeling for three wetland systems in south Florida. Water Resources Bulletin, Vol. 32, No. 3, pp. 465–473.

Bailey, J.E. and D.F. Ollis, 1986. Biochemical Engineering Fundamentals, McGraw-Hill, New York, NY.

Craft, C. B. and C. J. Richardson. 1993. Peat accretion and phosphorus accumulation along a eutrophication gradient in the northern Everglades. *Biogeochemistry,* Vol. 22:133–156.

Davis, S.M., 1989. Sawgrass and cattail production in relation to nutrient supply in the Everglades. in: Freshwater Wetlands and Wildlife, R. R. Sharitz and J.W. Gibbons, eds., US Dept. of Energy, DE90005384, NTIS, Springfield, VA, pp. 325–341.

Davis, S.M., 1991. Growth, decomposition, and nutrient retention of *Cladium jamaicense* Crantz and *Typha domingensis* Pers. in the Florida Everglades. Aquatic Botany, Vol. 40, pp. 203–224.

Davis, S.M., 1994. Phosphorus inputs and vegetation sensitivity in the Everglades. in: Everglades: The Ecosystem and Its Restoration, S.M. Davis and J.C. Ogden, Eds., St. Lucie Press, Delray Beach, FL, 1994, pp. 357–378.

DeBusk, W. F., K.R. Reddy, M. S. Koch and Y. Wang, 1994. Spatial distribution of soil nutrients in a northern Everglades marsh: Water Conservation Area 2A. Soil Sci. Soc. Am. J., Vol. 58, pp. 543–552.

Kadlec, R. H., 1993. "Natural Wetland Treatment at Houghton Lake: The First Fifteen Years," in: Proc. WEF 66th Annual Conf., Anaheim, CA, WEF, Alexandria, VA, pp. 73–84.

Kadlec, R. H., 1994. Phosphorus Uptake in Florida marshes. Water Science and Technology, Vol. 30, No. 8, pp. 225–234.

Kadlec, R. H., 1997. An autobiotic wetland phosphorus model. Ecological Engineering, Vol. 8, No. 2, pp. 145–172.

Kadlec, R. H., and F. B. Bevis, 1990. Wetlands and wastewater: Kinross, Michigan. Wetlands, J. Society Wetland Scientists, Vol. 10, No. 1, pp. 77–92.

Kadlec, R.H., and R.L. Knight, 1996. Treatment Wetlands, CRC Press, Boca Raton, FL, 893 pp.

Kirchner, W.B and P.J. Dillon, 1975. "An Empirical Method of Estimating the Retention of Phosphorus in Lakes," Water Resources Research, Vol. 11, pp. 182–183.

Lowe, E. F. and L. W. Keenan, 1997. "Managing phosphorus-based cultural eutrophication in wetlands: a conceptual approach," Ecological Engineering, Vol. 9, Nos. 1,2, pp. 109–118.

McCormick, P. V., 1996. Effects of phosphorus and hydrology on the Everglades. Presentation to the Florida Environmental Regulatory Commission, October 25, 1996.

Mitsch, W.J, J.K. Cronk, X. Wu, R. W. Nairn and D. L. Hey, 1995. "Phosphorus Retention in Constructed Freshwater Riparian Marshes," Ecological Applications, Vol. 5, No. 3, pp. 830–845.

Prentki, R.T., T. D. Gustafson, and M. S. Adams, 1978. "Nutrient movements in lakeshore marshes," in: R. E. Good, D.F. Whigham and R.L. Simpson, (Eds.) Freshwater wetlands: Ecological processes and Management Potential, Academic press, New York, pp. 307–323.

Reddy, K. R., DeBusk, W. F., Wang, Y., DeLaune, R. and M. Koch, 1991. Physico-Chemical Properties of Soils in the Water Conservation Area 2 of the Everglades, Report to the South Florida Water Management District, West Palm Beach Florida.

Reddy, K.R., W.F. DeBusk, Y. Wang, and S. Newman, *Physico-Chemical Properties of Soils in the Water Conservation Area 1 (WCA-1) of the Everglades*, prepared for South Florida Water Management District, Soil and Water Science Department, University of Florida, Contract No. C90-1168, 1994a.

Reddy, K.R., W.F. DeBusk, Y. Wang, and S. Newman, *Physico-Chemical Properties of Soils in the Water Conservation Area 3 (WCA-3) of the Everglades*, prepared for South Florida Water Management District, Soil and Water Science Department, University of Florida, Contract No. C90-1168, 1994b.

Richardson, C. J., 1996. Presentation to the Florida Environmental Regulatory Commission, June, 1996.

Richardson, C. J., C. B. Craft, R.G. Qualls, J. Stevenson, P. Vaithiyanathan, M. Bush and J. Zahina, 1995. *Effects of Phosphorus and Hydroperiod Alterations on Ecosystem Structure and Function in the Everglades*. Duke Wetland Center publication 95-05. Nicholas School of the Environment, Duke University, Durham, NC. 372p.

Richardson, C. J., C. B. Craft, R.G. Qualls, R.B. Rader and R. R. Johnson, 1992. *Effects of Nutrient Loadings and Hydroperiod Alterations on Control of Cattail Expansion, Community Structure and Nutrient Retention in the Water Conservation Areas of South Florid*a, Publication 92-11, Duke Wetland Center, Duke University, Durham, NC.

Robbins, JA, X. Wang, and R.W. Rood, *Sediment Core Dating*, Semi-Annual Report, prepared for South Florida Water Management District, Contract Number C-5324, Great Lakes Environmental Research Laboratory, National Oceanic and Atmospheric Administration, U.S. Department of Commerce, January 1996.

SFWMD (South Florida Water Management District), 1996. *Evaluation of Benefits and Impacts of the Hydropattern Restoration Components of the Everglades Construction Project*, 87 pp. + Appendices.

Tanner, C. C., 1996. Plants for constructed wetland treatment systems—a comparison of the growth and nutrient uptake of eight emergent species. *Ecological Engineering*, Vol. 7, No. 1, pp. 59–83.

Thomann, R. V. and J. A. Mueller, 1987. *Principles of Surface Water Quality Modeling and Control*, Harper and Row, New York, NY.

Toth, L. A., 1988. *Effects of Hydrologic Regimes on Lifetime Production and Nutrient Dynamics of Cattail*, Technical Publication 88-6, South Florida Water management District, West Palm Beach, FL, 26 pp.

Vollenweider, R.A., 1975. "Input-Output Models with Special reference to the Phosphorus Loading Concept in Limnology," *Schweiz. Zeit. Hydrol.*, Vol. 37, 53–84.

Walker, W.W and R. H. Kadlec, 1996. *A Model for Simulating Phosphorus Concentrations in Waters and Soils Downstream of Everglades Stormwater Treatment Areas*, Report to U. S. Dept. of Interior, Everglades National Park, August, 1996. Also included in: *Florida Everglades Program, Everglades Construction Project: Final Programmatic Environmental Impact Statement*, Appendix Vol. III, U. S. Army Corps of Engineers, South Atlantic Division, Jacksonville, FL.

Walker, W. W., 1995. Design basis for Everglades stormwater treatment areas. *Water Resources Bulletin*, Vol. 31, No. 4, pp. 671–685.

Wu, Y., F. H. Sklar and K. Rutchey, 1996. Analysis and simulations of fragmentation patterns in the Everglades. *Ecological Applications*, accepted.

28 Management Models To Evaluate Phosphorus Loads in Lakes

Steven C. Chapra and Martin T. Auer

28.1 ABSTRACT

Present capabilities for modeling the impact of phosphorus loads on lake water quality are reviewed. The historical development of phosphorus models is examined within the context of the overall evolution of the discipline of surface water quality modeling. Trade-offs in model reliability, model complexity, and cost/ease of application are considered for three broad classes of model frameworks: empirical models, simple budget models, and more complex nutrient food-web models. Questions of spatial, temporal, and kinetic resolution important to the development of a model framework are addressed. The capabilities and limitations of two readily available water quality management models, WASP and CE-QUAL-W2, are described. Finally, focus is placed on the shortcomings of the existing nutrient-food web models, those tools from which a new generation of phosphorus management models will evolve. Research gaps, as well as cases where science has evolved, but incorporation into model frameworks has lapsed, are identified. A plea is made to

1-56670-331-X/99/$0.00+$.50

integrate field monitoring programs and experimental studies at all stages of model development. Researchers are encouraged to reexamine phosphorus models developed in the 1970s and 1980s, incorporating recent scientific advances and bringing a fresh perspective on phosphorus modeling to bear on issues of concern to water quality managers.

28.2 INTRODUCTION

Mathematical models for water quality have been available to the scientific and engineering community for almost 75 years (Fig. 28.1). Great strides were made in model development in the 1970s, stimulated in large part by societal concern over the eutrophication of lakes and impoundments. This manuscript describes present capabilities for modeling the impact of nutrient loads on lake water quality. Since the model frameworks developed in the 1970s continue to represent extant capabilities, this review will focus on these vintage algorithms. Of course, in the ensuing years, scientific research progressed at a vigorous pace. Unfortunately, much of this research has not been integrated into our modeling arsenal, and hence current models lag the state of science. Consequently, after our review, the latter part of this chapter is devoted to key innovations that should be integrated to upgrade model performance in management contexts.

28.3 REVIEW OF LAKE/IMPOUNDMENT EUTROPHICATION MODELS

Three general approaches have been applied for simulating the impact of nutrients (especially phosphorus) on the quality of standing waters:

- Empirical models
- Simple budget models
- Nutrient good-web models

These approaches are listed in order of increasing complexity and higher mechanistic definition. It should be stressed that higher complexity does not necessarily connote inherent superiority. As in Fig. 28.2, there is always a balance that must be struck among model reliability, complexity, and cost that suggests an optimum choice rather than a "more is better" philosophy.

28.3.1 Empirical Models

Empirical models are data-based, often graphical, approaches drawing on measurements from many lakes and reservoirs. The pioneering work in this area was performed by Vollenweider (1968, 1969, 1975, 1976) and Dillon and Rigler (1974, 1975). Other investigators (notably, Rast and Lee, 1978; Reckhow, 1977, 1979; Chapra, 1979; Reckhow and Chapra, 1979; and Chapra and Reckhow, 1983) extended and broadened the approach. Empirical models can be loosely divided into two categories: (1) phosphorus loading plots and (2) trophic parameter corre-

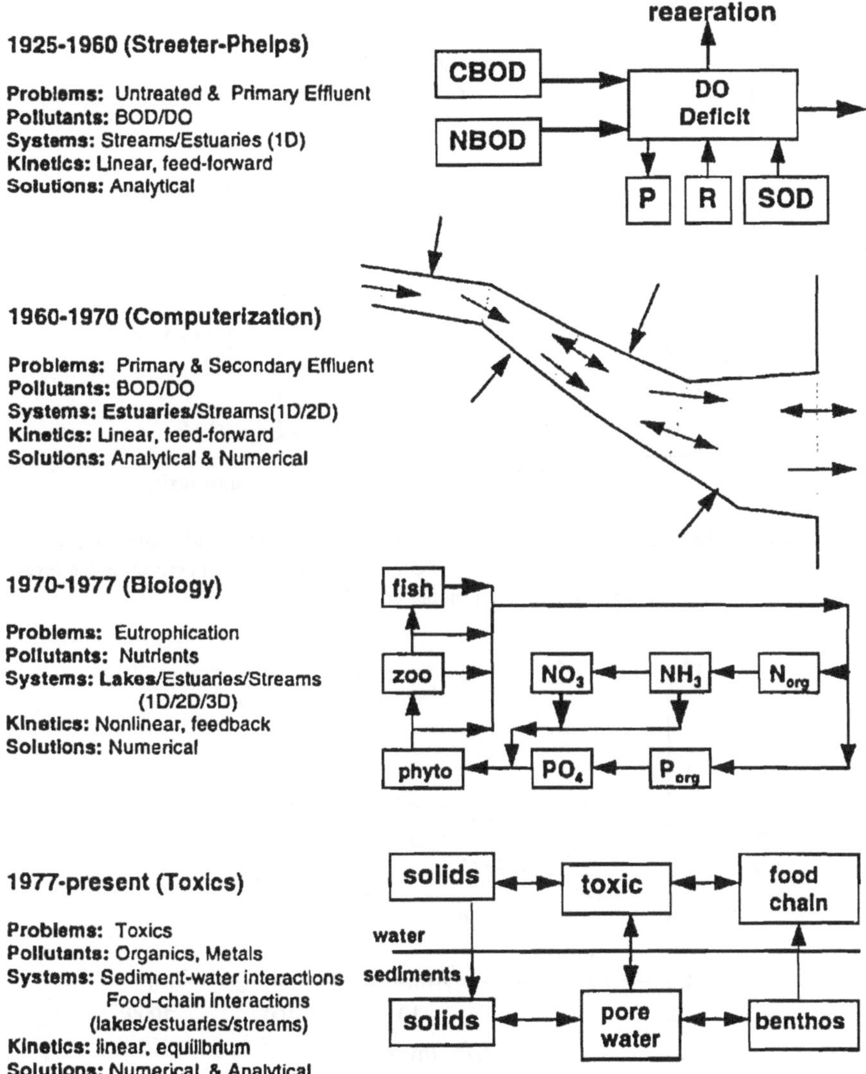

FIGURE 28.1 Four periods in the development of water-quality modeling (reprinted from Chapra, 1997).

lations. As depicted in Fig. 28.3, phosphorus loading plots typically graph lakes on a two-dimensional space with the log of the areal phosphorus loading on the ordinate and the log of hydrogeometric parameters on the abscissa. For example, Fig. 28.3 has the log of ratio of the lake's mean depth to its residence time as the abscissa. Lines are then superimposed to demarcate different trophic states. The plots can then be used to predict the trophic state of a lake based on its loading and hydrogeometry. It should be noted that a number of investigators (e.g., Chapra and Tarapchak, 1976; Vollenweider, 1976; Thomann, 1977) have illustrated how

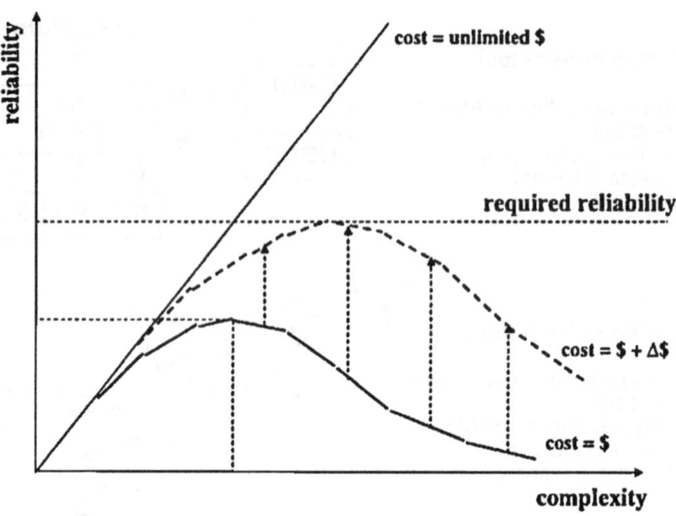

FIGURE 28.2 The trade-offs between model reliability, complexity, and budget (reprinted from Chapra, 1997). Note that the availability of economic resources can determine the level of model reliability associated with a given model complexity.

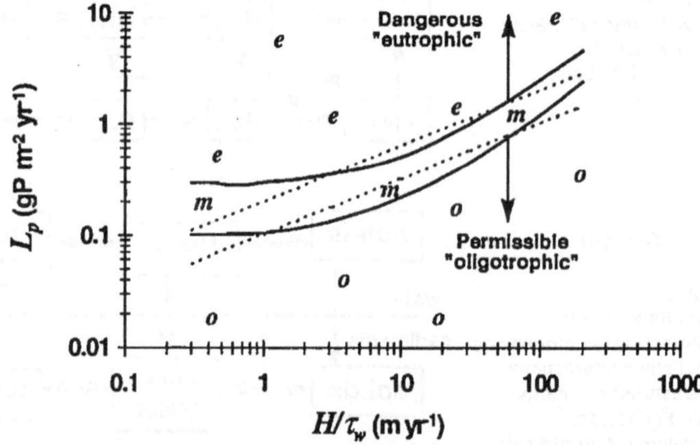

FIGURE 28.3 Vollenweider's (1975) loading plot. The location of lakes on the space are designated by the letters *o* (oligotrophic), *m* (mesotrophic), and *e* (eutrophic).

such plots can be related to and derived from the simple phosphorus budget models to be described in the next section. Thus, aside from predicting trophic state, the plots can be structured to predict in-lake total phosphorus (TP) concentration as a function of loads. Trophic parameter correlations are usually log-log plots relating two trophic parameters. For example, Fig. 28.4 shows a correlation between chlorophyll-*a* and TP concentration. As depicted in Fig. 28.5, the loading plots (or simple mass balances) and the correlations can be used in tandem as originally suggested by Dillon and Rigler (1974). Together they form a causal chain starting

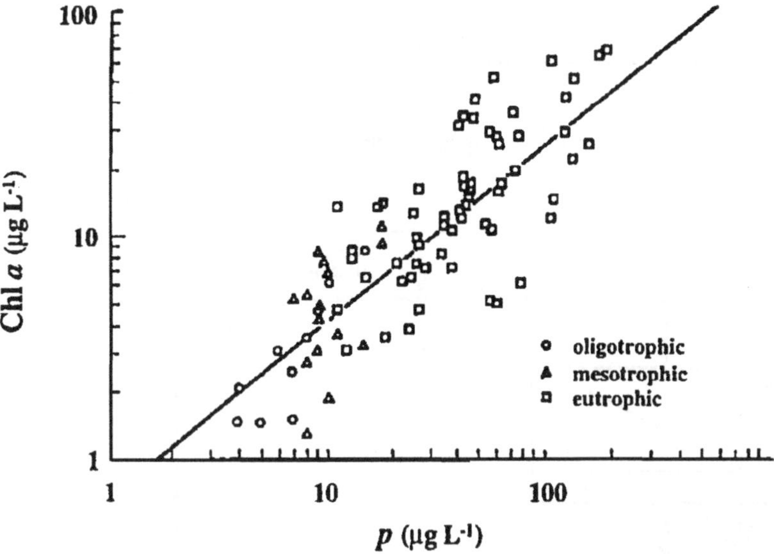

FIGURE 28.4 The relationship between chlorophyll and phosphorus in some U.S. lakes and reservoirs (from Bartsch and Gakstatter, 1978).

with phosphorus loading and ending with trophic parameters such as Secchi disk depth and hypolimnetic oxygen demand. In addition, some investigators have developed hybrid approaches (notably Rast and Lee, 1978) that link trophic state variables directly back to loadings.

The empirical models have several strengths and weaknesses. Their strengths are that they are extremely easy to use; that they provide a quick means to identify "outlier" lakes and that, if based on regional or local data representing relatively homogeneous populations (e.g., hardwater lakes in northern Michigan), they are capable of producing predictions considered adequate for management purposes. Their primary weakness relates to the fact that, if based on global data (e.g., "north temperate lakes"), they tend to have very large standard errors of prediction. Unfortunately, the plots are often presented in a manner that does not make this uncertainty explicit. Hence, naive users can develop predictions, unaware that their result may have a very substantial error. A number of investigators, notably Reckhow (1979) and Walker (1977, 1980), have sought to have uncertainty estimates accompany empirical model predictions. Figure 28.6 shows a recently developed empirical correlation (Chapra et al. 1997) that attempts to quantify the prediction error in the form of 95% prediction intervals. This plot indicates that a lake with a TP of 100 $\mu g \cdot L^{-1}$ should have a total organic carbon (TOC) concentration of approximately 9.5 $mg \cdot L^{-1}$. However, the range encompassing 95% of the predictions goes from 6.2 to 17 $mg \cdot L^{-1}$. Hence, the relative prediction error is over ±50%.

In summary, although they have some utility, empirical models (and particularly those based on global data) do not usually have the required precision on which high-cost decisions can be predicated. As such, they should be relegated to broad

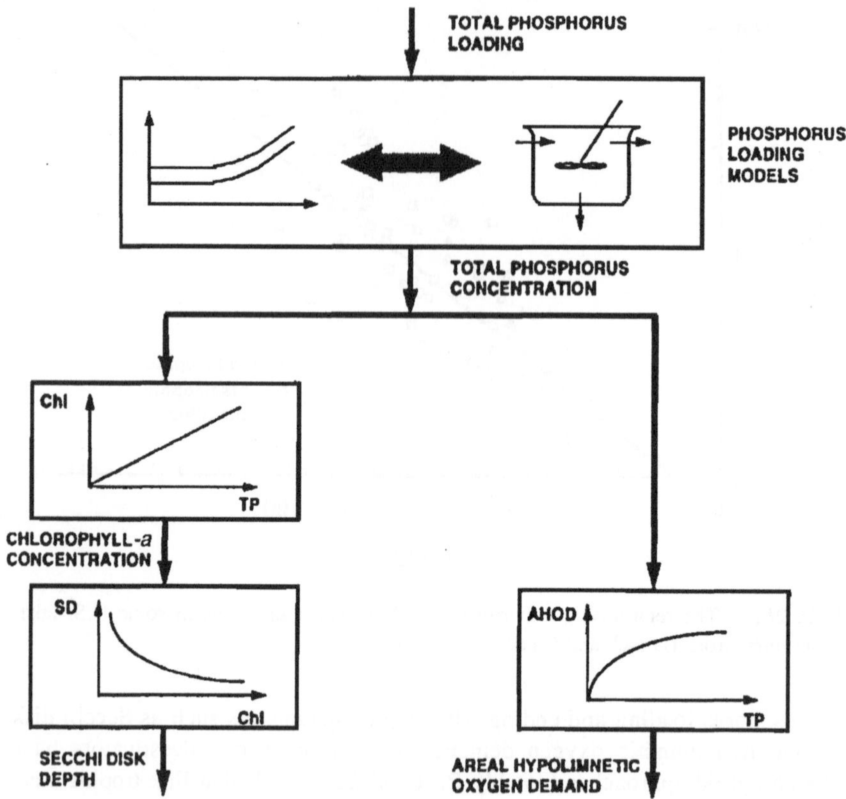

FIGURE 28.5 Schematic of approach used by Chapra (1980) to predict trophic state variables based on phosphorus loading model predictions. The approach consists of a number of submodels and correlations that form a hypothesized causal chain that starts with total P concentrations predictions based on budget models or loading plots. This concentration is used in conjunction with a series of correlation plots to estimate symptoms of eutrophication such as chlorophyll-*a* concentration, Secchi disk depth, and hypolimnetic oxygen demand.

screening applications and for identifying atypical lakes. However, they may have sufficient precision if developed and applied for regional populations of lakes and reservoirs.

28.3.2 Simple Budget Models

Early on (Vollenweider, 1969), it was recognized that simple mass-balance models could provide similar predictions to phosphorus loading plots. These models do not attempt a detailed characterization of the division of phosphorus within the water column. Rather, they focus of characterizing major inputs and outputs in order to predict the long-term trends of a lake's response to loading changes. The simplest example of a TP budget model was developed by Vollenweider (1969, modified by Chapra, 1975)

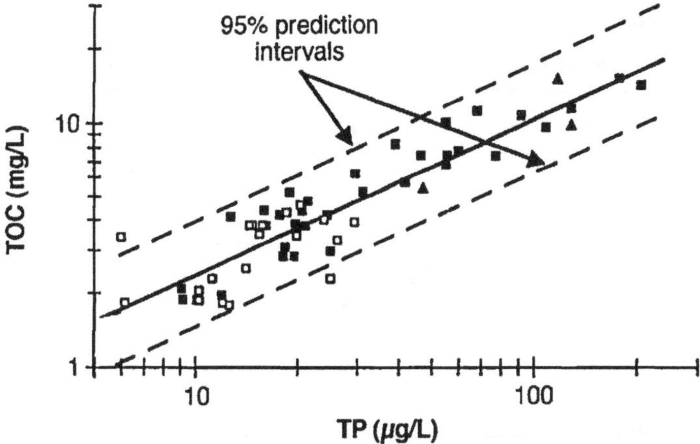

FIGURE 28.6 Relationship between total phosphorus and total organic carbon in lakes and reservoirs (Chapra et al., 1997). The best-fit line is shown along with 95% prediction intervals.

$$V\left(\frac{dP}{dt}\right) = W - QP - vAP$$

where V = volume, P = TP concentration, t = time, W = loading, Q = outflow, v = an apparent settling velocity, and A = surface area.

As in Fig. 28.7a, the key feature of this model is the simple way in which it characterizes the input-output terms for TP. In particular, it attempts to characterize sedimentation losses as a simple one-way settling of TP.

As with loading plots, steady-state solutions can be developed by setting the derivative to zero and solving for $P = W/(Q + vA)$. If levels of TP can be associated with trophic state, the model can be used to determine the loading required to maintain a particular lake at a desired quality in a fashion similar to the loading plots. The model also provides a framework to determine the temporal response of a lake to loading changes. Thus, it has the advantage over loading plots, that system dynamics can be characterized. These models can and have been improved in several ways. For incompletely mixed systems, the lake can be divided into a system of interconnected well mixed systems. This can be done horizontally or vertically. For example, Chapra (1979) used two mass balances to characterize a lake with a major embayment. Vertically, O'Melia (1972) and others have divided the water column of thermally stratified lakes into surface and bottom layers. Efforts have also been made to better characterize sediment-water interactions. Chapra and Canale (1991) represented a lake and its underlying sediments as a two-layer system (Fig. 28.7b). Along with phosphorus settling, this model also allows sediment feedback. A simple oxygen model is used to simulate hypolimnetic anoxia that triggers sediment release of TP into the overlying waters. This mechanism is significant, because sediment feedback can retard the recovery of lakes after TP load reductions. Hybrid models have been developed that use mass balance and multiple segments to characterize

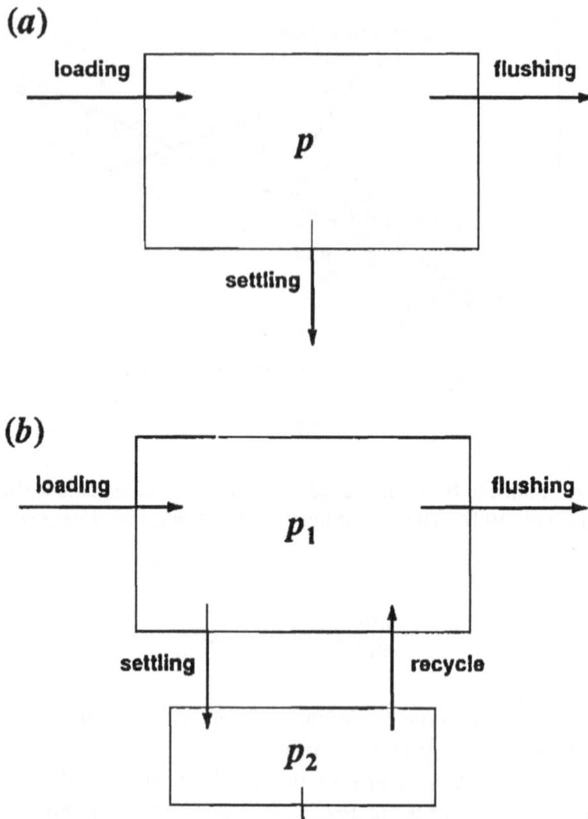

FIGURE 28.7 Two phosphorus budget models: (a) characterizes sedimentation as a simple one-way loss to the sediments, and (b) includes sediment feedback to more accurately assess cases where the lake's bottom waters go anoxic.

transport but use empirically derived relationships to quantify kinetics. Walker's (1996) "bathtub" model for impoundments is a good representative of this type.

In summary, the simple budget models use a mass balance to characterize how phosphorus levels change in lakes in response to load modifications. The performance of these tools is, however, highly sensitive to the quality of input information (e.g., loads, settling velocities). For example, the trophic state prediction generated by the Vollenweider TP budget approach can vary from oligotrophic to eutrophic over the range of settling velocities observed in lakes. Furthermore, the assumption is made that "as goes phosphorus, so goes eutrophication." Water quality managers are acutely aware that it is not phosphorus, but the water quality manifestations of this nutrient, that are of importance to those who use the resource. It can be concluded, therefore, that simple budget models can be useful for simulating long-term trends in the quality of P-limited lakes and impoundments, when applied with appropriate knowledge and guidance.

28.3.3 Nutrient/Food Web Models

In contrast to the simple budget models described in the previous section, nutrient/food-web models attempt to mechanistically characterize the partitioning of matter within the lake over time and space. Based on the pioneering work of investigators like Riley (1946) and Steele (1962), these models were first developed in the 1970s to expressly address the impact of nutrients on natural waters (e.g., Chen, 1970; Chen and Orlob, 1975; DiToro et al., 1971; Canale et al., 1974). They typically have a number of common characteristics as described below.

Transport

An effort is made to characterize the internal physics of a lake or impoundment. Thus, rather than representing the lake as a single well mixed entity, multiple segments are typically used to model the internal physics. The most common approach is to use two vertical layers to characterize thermal stratification. More refined vertical segmentation is sometimes used to resolve hypolimnetic gradients, particularly near the sediment-water interface. In addition, multiple horizontal segments are employed for incompletely mixed systems such as elongated impoundments. There are two ways in which the magnitude of mixing and interflow between segments is modeled. First, it can be treated as a model input. This is done by specifying turbulent diffusion coefficients and inter-segment flows. In many such applications, the temperature distribution is also treated as a model input. Second, water motion can be calculated internally using energy and momentum balances. Thus, a separate hydrodynamic model is used to supply the physics. In some cases, temperature is calculated as a part of the hydrodynamic simulation.

Kinetic Characterization

Matter in the lake is divided into several forms of nutrients and a food web. A typical example of how this is done is shown in Fig. 28.8. Several nutrients are usually included. Hence, the model is capable of simulating multiple nutrient limitation. As in this figure, phosphorus and nitrogen are the common choices. These are usually divided into available and unavailable components. The latter can be broken down further; for example, into dissolved and particulate fractions. The food web shown in the figure consists of a single algal compartment, along with two zooplankton compartments. Algal growth is calculated as a function of temperature, light, and available nutrient concentrations. All other rates are temperature dependent. All three organism groups experience respiration/excretion losses, contributing material to either the available or unavailable nutrient pools. Grazing is inefficient, with a fraction of the material grazed egested to the unavailable pools. This framework can be simplified by dropping a nutrient (usually nitrogen). It is more likely to be made more complicated by adding nutrients (e.g., silicon), or making them more refined (e.g., breaking the unavailable components into dissolved and particulate fractions). The food web can be made more complex by breaking the single algal compartments into components (e.g., diatoms, greens, and cyanobacteria). Similar refinements can be made to the zooplankton. When this is done, feeding preferences are usually

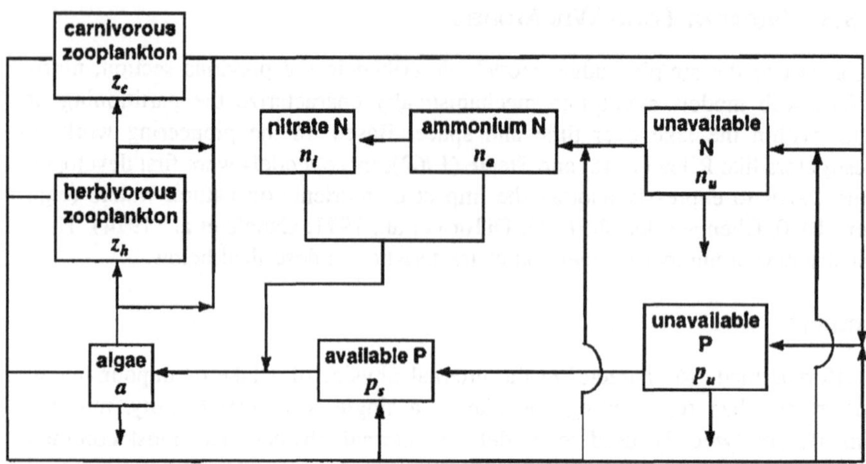

FIGURE 28.8 Kinetic segmentation.

specified. It should be noted that other variables such as oxygen and pH can be integrated into these frameworks. In these cases, it is usually necessary to simulate organic carbon.

28.3.4 Available Models

Major modeling advances and some excellent research models have been developed for specific nutrient/food-web applications over the past 20 years. Recent contributions include a lake and reservoir framework developed by Hamilton and Schladow (1997; also Schladow and Hamilton, 1997), food web studies related to biomanipulation (Bakema et al., 1990; Jayaweera and Asaeda, 1996) and modeling of energy flow in lacustrine food webs (Halfon et al., 1996). Jorgensen (1995) has provided a review the state of the art in ecological research modeling in limnology. Here, we focus on management models that are generally available (open code, nonproprietary) and useful for management applications. Several such public-domain computer programs are available to implement nutrient-food web models. The two most widely known are WASP5 and CE-QUAL-W2.

WASP5 (Ambrose et al. 1993) is a general-purpose water-quality model developed by U.S. EPA. It is designed as a flexible tool capable of simulating pollutant dynamics in 1, 2, or 3 dimensions in lakes, estuaries, and rivers. The software contains a hydrodynamic module and simulates conventional pollutants (DO and BOD), eutrophication and toxics. WASP5 has been criticized for seeking to do too much. General-purpose transport codes may not be particularly well suited for simulation of certain features of stream (e.g., dead zones, drop structures) or impoundment (e.g., withdrawal structures) hydrodynamics, requiring specialized schemes to incorporate their impacts on water quality. For eutrophication, WASP5 uses the kinetic framework of Thomann and Fitzpatrick (1982), an approach similar to that depicted in Fig. 28.8. As with many generally available water-quality models, no major developments have been made since the early 1980s (with the exception of the addition of a toxic substance module). Thus, as we will describe in the following section, WASP5 does not have the benefit of scientific advances that have occurred in the interim. In some cases, this limitation has been overcome by adding kinetic enhancements (e.g., multiple algal groups, additional nutrients, sediment linkages) to the WASP code (cf. James and Bierman, 1995; James et al., 1997; Jin et al., 1998). However, the software does not easily accommodate user-specified modification of its kinetic framework, a shortfall shared by most general-purpose models.

CE-QUAL-W2 (Cole and Buchak, 1995), developed under the auspices of U.S. Army Corps of Engineers, is a water-quality model specifically designed for the analysis of impoundment eutrophication. Its hydrodynamic module is set up to simulate water motion and temperature in the vertical and longitudinal dimensions of elongated and dendritic reservoirs. It also has a detailed representation of the outflow regime for such systems. Thus, it has a very sophisticated physical characterization. The model's kinetic structure is its primary deficiency. The kinetics are not highly developed but generally follow a scheme that is similar in spirit but simpler than Fig. 28.8 (e.g., there are no zooplankton). Sediment-water interactions are handled in a primitive fashion. Improved kinetic and sediment modules are expected to accompany the next release.

Thus, nutrient/food web models, at least theoretically, provide a means to address detailed management questions related to lake eutrophication. Unfortunately, the readily available manifestations of this class of model are either not designed for the analysis of lakes and impoundments (do not simulate stratification or the location of outlet structures) and hence are deficient from a hydrodynamic perspective, or are designed for impoundments and lakes and hence well developed hydrodynamically, but are deficient kinetically. Furthermore, although they have the potential for more refined predictions, they also require more data and user sophistication than simpler models. Finally, as mentioned previously, although the science available to support water-quality modeling has progressed greatly over the past 20 years, the nutrient/food-web models that are available for widespread use are "frozen in time" back in the late 1970s and early 1980s. The lack of progress in the field may be attributed to a reduced support from granting agencies, the misperception that extant models are adequate, and redirection of research priorities to other venues (toxics, acid rain, global climate change). Thus, it is now necessary that these generally available models be refined in light of recent research on nutrient/food-web dynamics in order that they have optimal value for management applications. In addition, a number of research gaps must be filled for the models to be generally effective.

28.4 SUMMARY: TOWARD IMPROVEMENT OF AVAILABLE MODEL FRAMEWORKS

A wide range of models have been developed to make predictions regarding lake and reservoir eutrophication. On the one hand, simple empirical loading plots are available to answer broad questions with low precision. On the other hand, complex mechanistic nutrient-food web models offer the potential to answer more detailed questions with greater precision. However, it should be noted that a price must be paid to obtain adequately precise predictions with the mechanistic models. This can be understood by recognizing that the information embodied in mechanistic models comes in two forms, as described below.

28.4.1 Mechanism and Structural Information

When mass-balance equations of nutrients or elements of the food web are written, we are embedding scientific understanding into a model. By formulating that the herbivores consume algae and carnivores consume herbivores, we are constraining the model to do something very specific with the autochthonous carbon produced through photosynthesis. Similarly, when we divide a lake vertically into surface and bottom layers or horizontally into a bay and a main lake, the very act of division has information content. Assuming that our choices are not frivolous and are based on sound science, models that make such distinctions are intrinsically superior (in the sense of having more information content) to simpler representations.

28.4.2 Parameter Information

Refinement of mechanistic resolution incurs additional cost. Mechanisms beget parameters and, in all cases, parameter estimation costs something. For example, for the lake/embayment mechanism, a turbulent diffusion coefficient must be estimated. For zooplankton/phytoplankton interactions, grazing rates and efficiencies must be determined experimentally. The general-purpose models described above, which are not kinetically complex by most standards, require specification of 50 to 60 kinetic coefficients.

Thus, as depicted in Fig. 28.2, there are trade-offs between model complexity and reliability that turn on the issue of imbedded information and parameter estimation. The straight line in Fig. 28.2 reflects the underlying assumption that, if we have an unlimited budget, a more complex model will be more reliable. In essence, as we add complexity to the model (that is, more mechanisms with more parameters), we assume that there are sufficient funds to accomplish the field and laboratory studies required to specify the additional parameters. In fact, this assumption itself may not be true, because there are limits to our ability to mathematically characterize the complexity of nature (sort of an "uncertainty principle" of ecology). However, even though we may not be able to totally characterize a natural water system, we generally function under the notion that if we have more and better information, our models will be more reliable predictors. The real rub comes from the fact that we almost never have a large enough budget to even come close to approaching that limit. Rather, we are usually in the position that we must make due with a limited budget for sampling and laboratory analysis. In such cases, there are two extreme outcomes. At one extreme, a very simple model will be so unrealistic that it will not yield reliable predictions. At the other, a very complex model will be so detailed that it outpaces available data. As a consequence, it can be equally unreliable because of the uncertainty of the parameters. As in Fig. 28.2, there is an intermediate point where the model is consistent with the available level of information. At this point, a third dimension must be interjected: the reliability required to solve the problem. While you might be at the optimum for your data, you also might have a model that gives predictions that are inadequate for the problem being addressed. Consequently, additional information must be collected to bring the model up to the required level. Therefore, different problem contexts require different reliability, and hence require different models; there is no one "best" model for all problems.

How, then, can the modeling frameworks described above be improved? The simple empirical frameworks can be improved in three ways. First, wherever possible, empirical models should be developed on a regional basis, i.e.,for areas having similar environmental (geologic, climatic, etc.) influences. These should include explicit error analyses to easily include probabilistic information in the resulting predictions. Second, efforts should be made to reduce the variability inherent in these correlations. Light limitation, secondary nutrient limitation, and differences in particulate phosphorus bioavailability have all been suggested as reasons for the order-of-magnitude differences in chlorophyll levels predicted by many correlations for a given TP concentration. Finally, the models should be implemented as a part of a decision-support system. Reckhow et al. (1992) have made a first cut at such an interface with the Eutromod framework. This software prompts the user for information regarding the lake (e.g., land use, depth, etc.) and develops trophic-state predictions along with error bounds. If developed properly, such a tool would allow nonmodelers to effectively apply such models to their system. In addition, if the framework included artificial intelligence and help facilities, it could provide a normative function by, for example, constraining the user to employ realistic parameter values.

The simple budget models could be improved in three ways as well. First, mechanisms such as sediment-water exchange and burial could be better characterized. Second, site-specific determination of all kinetic coefficients is a worthy goal. Auer et al. (1997) describe a "zero-degree-of-freedom" approach to the development of a TP model where all model inputs, including kinetic coefficients, were determined on a site-specific basis through field measurement and laboratory experimentation. Last, the models could be improved by developing better user interfaces. Chapra (1992) has done this to an extent by implementing the Chapra and Canale (1991) phosphorus/oxygen model as a software package called PHOSMOD. This interface, although now dated, allows the user to input data via convenient tables and view graphical output immediately. Programming tools such as Visual BASIC® and Excel® now provide a means to conveniently develop such user interfaces. Doerr et al. (1997) provide an excellent example of the application of Visual BASIC in developing user-friendly graphic interfaces. Implementation of simple lake models in this way could stimulate and facilitate their use.

28.5 ISSUES DRIVING THE NEXT GENERATION OF MODELS

While existing nutrient food-web models could certainly be improved through the development of user-friendly interfaces and a capability for user-specified kinetic constructs, it is the interface with advances in science that requires the most significant upgrade. Hence, we devote the remainder of this chapter to that issue. This involves both research gaps as well as cases where the science has been developed but not incorporated in models. The discussion is divided into three categories: hydrodynamics, kinetics, and bottom interactions.

28.5.1 Hydrodynamics

Scientific and computing advances over the past 20 years have improved our ability to simulate water movement in lakes and impoundments. From a theoretical perspective, turbulence closure schemes have allowed a more mechanistic characterization of turbulence to be attained. From a computing perspective, the inclusion of sound hydrodynamics into water-quality simulations is not as economically prohibitive as it once was. Sheng (1999) provides a review of advances in hydrodynamic modeling as applied to aquatic systems in Florida.

The benefit of accommodating hydrodynamics in water-quality modeling is that, if the physics is characterized properly, errors in kinetic formulations (i.e., chemical and biological transformations) cannot be masked through poorly defined mass transport. An example is provided by the issue of settling velocities. It is well known among modelers that measured settling velocities fail to properly characterize phenomena such as the deep chlorophyll and benthic nepheloid layers within the crude physical characterization of two-layer lake models (Fig. 28.9). Matter may be rapidly delivered to these boundaries at a rate consistent with measured settling. However, transient mixing events could reintroduce these into the overlying waters before they are subsumed into the lower layers. The net result is that the settling velocity used

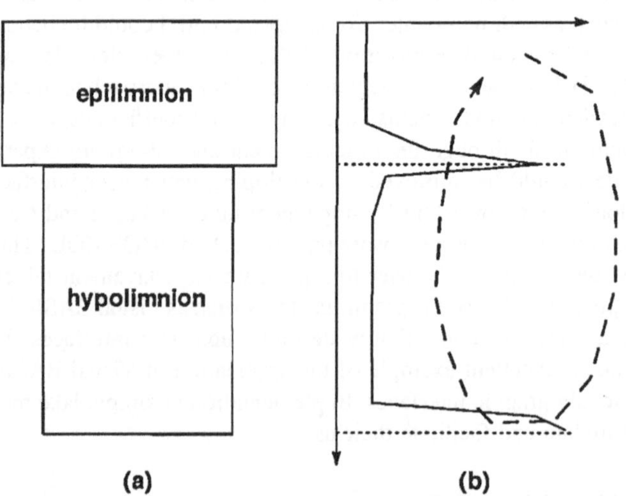

FIGURE 28.9 (a) A two-layer characterization of a lake. (b) The vertical structure of matter due to the buildup of matter at interfaces such as the thermocline and the sediment-water boundary. Such interfaces result in layers (the metalimnion and the nepheloid layer, respectively) where matter builds up during quiescent conditions. Episodic turbulent mixing can reintroduce this matter into the upper layers.

in a two-layer model may have to be smaller than measured to account for the upward transport. A simple two-layer segmentation cannot be expected to resolve the interfacial build-up of organic matter and nutrients. A multilayer model that adequately accounted for transient mixing and hindered settling in boundary layers would not have to make such compensation.

Although fine-scale physical resolution is certainly not justified for all problems, it provides a means to reduce the uncertainty of nutrient-food web models. The one situation where this may be routinely necessary is the simulation of impoundment eutrophication. Because of their elongated and dendritic morphometry and shorter residence times, the standard well mixed horizontal and two-layer vertical representation is rarely appropriate. Reservoir water quality can also be impacted by short-circuiting of inflows and selective withdrawals of outflows, thus demanding a higher level of physical characterization than for lakes (and particularly larger, more circular lakes). Advances in hydrodynamic modeling such as those contained within CE-QUAL-W2 are therefore critical to the effectiveness of that model.

28.5.2 Kinetics

The kinetic characterizations used in today's management models have changed little since they were originally developed in the 1970s. Two areas must be addressed to improve their predictive capabilities. The first involves a research question: Is there a minimal food-web representation that will adjust properly across the oligotrophic-to-eutrophic continuum? The second involves the explicit simulation of organic carbon.

Food-Web Definition

Standard water-quality models are based on a kinetic framework similar to that outlined in Fig. 28.8, limiting the biological characterization to a single phytoplankton compartment and a food web consisting of an herbivore and a carnivore. Thus, management models are not presently able to answer one of the most important questions related to eutrophication: At what point do increased nutrient loadings lead to a shift to less desirable algal species?

An alternative scheme that begins to address this question is shown in Fig. 28.10. The phytoplankton are divided into three functional groups: diatoms, green algae, and cyanobacteria. Notice that these groups require different nutrients. A silicon compartment has been added to the nutrient group to accommodate the requirements of diatoms. All groups depend on phosphorus, whereas, in this framework, the cyanobacteria are assumed to be able to fix nitrogen and hence are independent of N. There are three consumers: an herbivore, which feeds exclusively on greens and diatoms; a carnivore, which feeds on herbivores; and an omnivore, which feeds on diatoms, greens, and the herbivores. Notice that, in this conceptual framework, none of the grazers impact cyanobacteria. Together with appropriate kinetic information (growth and respiration rates, formulations describing light and nutrient limitation, feeding preferences, etc.) such a framework could provide a means to predict species shifts in lakes and reservoirs.

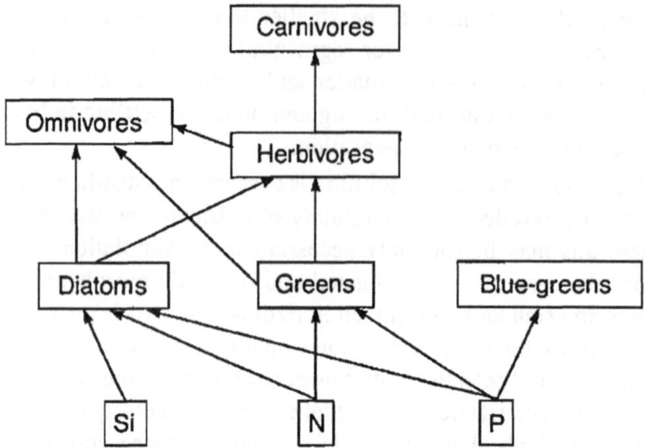

FIGURE 28.10 A more refined conceptual framework (as compared with Fig. 28.8) that attempts to include enough food web detail to effectively predict the shift between desirable (diatoms, greens) and undesirable (blue-greens) species of phytoplankton.

We do not suggest that the particular scheme in Fig. 28.10 is the proper one to make such a determination. For example, the susceptibility of certain of the cyanobacteria to grazing losses and the role of the microbial loop in cycling of organic carbon and nutrients suggests additional "arrows" presently absent from the framework presented as Fig. 28.10. Rather, we would like to suggest that these new research findings be synthesized and new research be conducted to identify (1) the proper level of food-web resolution and (2) values for the kinetic coefficients that support such a framework. Without such research, present models are generally inadequate to address questions related to species shifts and other management issues such as top-down control of water quality.

Organic Carbon

Although organic carbon production and decomposition are at the heart of the eutrophication problem, they are not explicitly simulated in most standard nutrient/food-web frameworks. In fact, this is a characteristic of each of the three issues that have dominated water-quality modeling (Fig. 28.11). In all these cases, the omission was in large part due to the historical difficulty in measuring organic carbon.

How we have avoided organic carbon

Management Issue	Organic C Surrogate
Dissolved oxygen	BOD
Eutrophication	Chlorophyll-*a*
Toxics	Fraction organic carbon

FIGURE 28.11 In the three problems that have dominated water-quality modeling, organic carbon was at the heart of the problem, yet carbon was never directly simulated in management models.

For the case of eutrophication, it is also due to the fact that chlorophyll provides the only convenient and economic means to directly measure the fraction of the organic carbon associated with plants.

Today, there are several reasons for recasting nutrient/food-web models on an organic carbon basis. First, organic carbon is now measured more easily and precisely than in the past. Next, although chlorophyll remains the only economic means to directly measure phytoplankton biomass, it is well known that the chlorophyll-to-carbon ratio in phytoplankton varies by about a factor of five, depending on ambient light and nutrient levels (Laws and Chalup, 1990). Consequently, paired determinations of particulate organic carbon and chlorophyll would provide a better measure of the amount of autochthonous carbon present, particularly during bloom conditions. Models could also be developed that simulate algae as carbon and use calculated chlorophyll-to-carbon levels to express these results as chlorophyll (i.e., for comparison with measurements). In addition, the use of eutrophication models as the basis for examining the transport and fate of toxic substances (e.g., organics, metals, disinfection by-products) in lakes and impoundments requires specification of the amount and forms of organic carbon present. Direct modeling of organic carbon becomes especially significant for systems where both allochthonous and autochthonous carbon sources are important. Finally, the state of the lake's bottom sediments is inextricably tied to the amount of carbon it receives from the overlying waters. This has ramifications for the exchange of oxygen, nutrients, reduced chemical species, and toxics at the sediment-water interface.

28.5.3 Bottom Interactions

It is important that bottom sediments be explicitly included in nutrient/food web models. Today it is recognized that sediments represent the lake's "memory." Nutrient release from the sediment reservoir may persist for many years following external load reductions. Thus, inclusion of the sediments is critical for predicting the long-term recovery of lakes and impoundments. Some specific areas of research related to the sediments are as follows.

Sediment-Water Exchange

Standard nutrient/food-web models have traditionally handled sediment-water exchange (sediment oxygen demand, SOD, and nutrient release) in a primitive fashion. For deeper, unproductive lakes, these processes were usually disregarded completely. This made sense, because sediment-water exchange are usually negligible for such systems. For shallower lakes, where exchange at the sediment-water interface is typically quite important, it is usually handled in two ways. First, sediment exchange is treated as a zero-order term, i.e., a constant prescribed flux, usually based on field or laboratory measurements. Alternatively, chemical exchange is sometimes modeled linearly, i.e., exchange (e.g., SOD) is directly proportional to the downward flux (e.g., of organic carbon). Both of these approaches are inadequate. As depicted in Fig. 28.12, the relationship between SOD and chemical oxygen demand (COD) and that between areal hypolimnetic oxygen deficit (AHOD) and

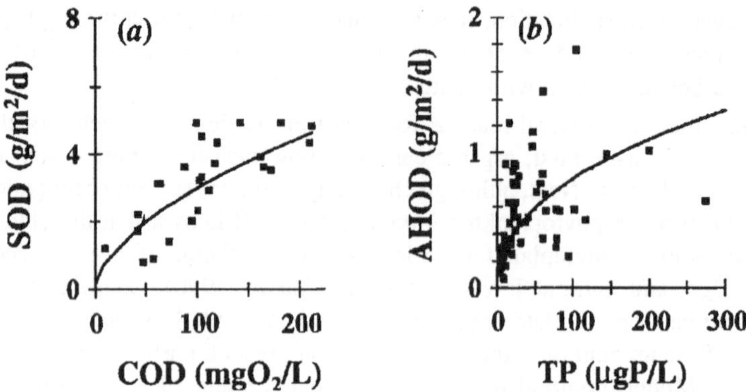

FIGURE 28.12 Observations that indicate a square-root relationship between SOD and organic content of sediments: (a) SOD vs. surface sediment COD (Gardiner et al., 1984) and (b) AHOD versus total phosphorus concentration (from Rast and Lee, 1978; Chapra and Canale, 1991).

TP are curvilinear. The problem with not capturing this nonlinearity (overestimate of SOD) is depicted in Fig. 28.13. Over the past decade, great advances have been made in our ability to simulate the sediment diagenesis process. In particular, Di Toro et al. (1990), Di Toro and Fitzpatrick (1993), Penn et al. (1995), and Wickman (1996) have developed computational frameworks to calculate sediment oxygen demand and phosphorus and nitrogen release as functions of the downward flux of carbon, nitrogen, and phosphorus from the water column. These approaches, well founded in diagenetic theory and supported by field and laboratory measurements, are important advances in the field of sediment-water interactions and must be integrated into nutrient/food-web models if those tools are to be adequate predictors of lake trophic state.

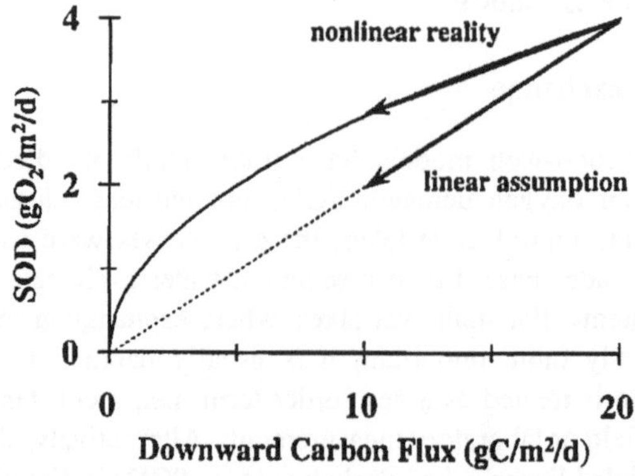

FIGURE 28.13 A graphical depiction of the type of discrepancy that can result if SOD is assumed to behave linearly with changes in organic carbon flux. The result depicted would overestimate the reduction of SOD for a given reduction in carbon flux.

Rooted and Attached Plants

From a management perspective, there are two primary reasons why rooted and attached plants are important and should be included in general nutrient/food web frameworks. First, in some lakes and impoundments, they are the primary way in which eutrophication interferes with beneficial uses. This is particularly true because the problem occurs in shallower water (e.g., near shore) where human use is typically intense. Most kinetic structures developed for large bodies of water in the 1970s (e.g., Great Lakes water quality models) do not accommodate rooted and attached plants. Canale and Auer (1982) describe an approach for modeling attached algal growth in the near-shore waters of the Great Lakes. More recently, Asaeda and VanBon (1997) have developed a coupled macrophyte-phytoplankton model and applied that framework for simulation of algal blooms in shallow, eutrophic lakes. Additionally, advances in modeling rooted aquatic plants in wetland systems (cf. Kadlec and Walker, 1997) may have some utility for application in lakes and reservoirs.

Secondly, where they are abundant, rooted and aquatic plants can significantly impact a system's light environment and nutrient budget. Scheffer et al. (1993) present an alternate state hypothesis that explains shifts between phytoplankton and macrophyte dominance in lakes. Schelske (1999) has suggested that this approach may explain observations of the dramatic change in the ecological state of Lake Apopka (Florida). Inclusion of rooted and attached aquatic plants in a generally available management model for nutrient/food web interactions would represent an important advance in water quality modeling.

Sediment Transport

The shallower the lake, the more its dynamics can be dictated by the sediments. Nowhere is this as striking as with regard to the issue of sediment resuspension. High-intensity, short-duration wind events such as storms can resuspend large quantities of bottom sediments into the overlying waters of shallow lakes. Aside from direct impacts on light extinction, such events can also introduce nutrients, organic matter, bacteria, algae, and toxic substances. Coupled with the advances in hydrodynamics and sediment kinetic processes described earlier, the basic science needed to incorporate such phenomena into nutrient/food-web frameworks is well developed (cf. James et al., 1997). It is now time to incorporate those advances in generally available model frameworks.

28.6 DISCUSSION

In summary, mathematical models ranging from simple empirical correlations to sophisticated computer packages are available to simulate the impact of nutrients on lakes. These models (most of which stopped evolving in the early 1980s because of curtailment of funding) provide a preliminary framework for assessing the impact of load reductions on lake and reservoir eutrophication. Unfortunately, because modeling in the 1980s focused on important systems such as Lake Ontario and Lake

Michigan, the nutrient/food-web models that resulted (and which are the basis for models such as WASP5) are oriented toward deep, long residence-time lakes where phytoplankton dominate and bottom processes are not critical. These two features allow great simplifications to be invoked (e.g., well mixed hypolimnion, omitting sediment processes, etc.). Although such lakes are certainly important, many of the smaller systems taking center stage in the management arena today cannot be simplified in this way.

In the documentation for CE-QUAL-W2, Cole and Buchak (1995, p. 19) offer some modeling philosophy that provides a particularly appropriate closure for this review:

> Ideally, a model should be used as a starting point for limnological investigations of a waterbody, with the data and formulations continuously refined to reflect the increased understanding of the system and processes gained over time. Unfortunately, this approach is rarely taken in practice due in large part to the inability of aquatic biologists/limnologists and engineers to collaborate together. This cooperative approach between experimentalists and theoreticians is the main impetus behind the tremendous advances in physics, chemistry, and, to some extent, biology (e.g., genetic research) during the last century, but is seldom seen in the field of water quality modeling.

This level of cooperation, what Auer and Canale (1986) termed an "integrated approach," has been achieved in research models, selected examples of which have been cited here. An ability to follow the lead of those scientists/modelers will be fundamental to the development of the next generation of generally available water quality management models.

REFERENCES

Ambrose, R.B., T.A. Wool, and J.L. Martin. 1993. The Water Quality Analysis Simulation Program, WASP5. Part A: Model Documentation. U.S. EPA, Environmental Research Laboratory, Athens, Georgia, 209 pp.

Asaeda, T. and T. VanBon. 1997. Modeling the effects of macrophytes on algal blooming in eutrophic shallow lakes. Ecological Modeling, 104: 261–287.

Auer, M.T. and R.P. Canale. 1986. Mathematical modeling of primary production in Green Bay (Lake Michigan, USA): A phosphorus and lightlimited system. Hydrobiological Bulletin, 20: 195211.

Auer, M.T., S.M. Doerr, S.W. Effler, and E.M. Owens. 1997. A zero degree of freedom total phosphorus model: Development for Onondaga Lake. Lake and Reservoir Management, 13(2): 118–130.

Bakema, A.H., W.J. Rip, M.W. deHaan, and F.J. Los. 1990. Quantifying the food webs of Lake Bleiswijkse Zoom and Lake Zwemlust. Hydrobiologia, 200/201: 487–495.

Bartsch, A. F. and J.H. Gakstatter. 1978. Management Decisions for Lake Systems on a Survey of Trophic Status, Limiting Nutrients, and Nutrient Loadings. pp. 372–394, In: AmericanSoviet Symposium on Use of Mathematical Models to Optimize Water Quality Management, 1975. U.S. Environmental Protection Agency Office of Research and Development, Environmental Research Laboratory, Gulf Breeze, FL, EPA600/978024.

Canale, R.P. and M.T. Auer. 1982. Ecological studies and mathematical modeling of *Cladophora* in Lake Huron: 5. Model development and calibration. Journal of Great Lakes Research, 8: 112–126.

Canale, R.P., D.F. Hinemann, and S. Nachippan. 1974. A Biological Production Model for Grand Traverse Bay. University of Michigan Sea Grant Program, Technical Report No. 37, Ann Arbor, Michigan, 116 pp.

Chapra, S.C. 1975. Comment on "An empirical method of estimating the retention of phosphorus in lakes" by W.B. Kirchner and P.J. Dillon. Water Resources Research, 11:1033–1034.

Chapra, S.C. 1979. Applying phosphorus loading models to embayments. Limnology and Oceanography, 24: 163–168.

Chapra, S.C. 1980. Application of the phosphorus loading concept to the Great Lakes. pp. 135–152, In: R.C. Loehr et al. (eds.), Phosphorus Management Strategies for Lakes, Ann Arbor Science Publishers, Ann Arbor, MI.

Chapra, S.C. 1992. PHOSMOD 1.0: Software to Model Seasonal and Long-Term Trends of Total Phosphorus and Oxygen in Stratified Lakes. CADWES Working Paper No. 14, Software Package No. 7, North American Lake Management Society, Madison, WI, 20 pp.

Chapra, S.C. 1997. Surface Water-Quality Modeling, McGraw-Hill, New York, New York, 844 pp.

Chapra, S.C. and R.P. Canale. 1991. Long-term phenomenological model of phosphorus and oxygen in stratified lakes. Water Research, 25:707–715.

Chapra, S.C. and K.H. Reckhow. 1983. Engineering Approaches for Lake Management, Vol. 2: Mechanistic Modeling. Ann Arbor Science/Butterworth, Woburn, MA, 492 pp.

Chapra, S.C. and S.J. Tarapchak. 1976. A chlorophyll-*a* model and its relationship to phosphorus loading plots for Lakes. Water Resources Research, 12: 1260–1264.

Chapra, S.C., R.P. Canale and G.L. Amy. 1997. Empirical models for disinfection by-products in lakes and reservoirs. Journal of Environmental Engineering, 123: 714–715.

Chen, C.W. 1970. Concepts and utilities of ecological models," Journal of the Sanitary Engineering Division, ASCE, 96:1085–1086.

Chen, C.W. and G.T. Orlob. 1975. Ecological Simulation for Aquatic Environments. pp. 475–588, In: B.C. Patton (ed.), Systems Analysis and Simulation in Ecology, Vol. III, Academic Press, New York.

Cole, T.M. and E.M. Buchak. 1995. CE-QUAL-W2: A Two-Dimensional, Laterally Averaged, Hydrodynamic and Water Quality Model, Version 2.0. User Manual. Instruction Report EL-95, U.S. Army Corps of Engineers, Waterways Experiment Station, Vicksburg, Mississippi, 57 pp.

Dillon, P.J. and F.H. Rigler. 1974. The phosphorus-chlorophyll relationship for lakes. Limnology and Oceanography, 19:767–773.

Dillon, P.J., and F.H. Rigler. 1975. A simple method for predicting the capacity of a lake for development based on lake trophic status. Journal of the Fisheries Research Board of Canada, 31:15191531.

Di Toro, D.M. and J.J. Fitzpatrick. 1993. Chesapeake Bay Sediment Flux Model. Tech. Report EL-93-2, U.S. Army Corps of Engineers, Waterways Experiment Station, Vicksburg, Mississippi, 316 pp.

Di Toro, D.M., R.V. Thomann and D.J. O'Connor. 1971. A dynamic model of phytoplankton population in the Sacramento-San Joaquin Delta. pp. 131–180, In R.F. Gould (ed.), Advances in Chemistry Series 106: Nonequilibrium Systems in Natural Water Chemistry, American Chemical Society, Washington, D.C.

Di Toro, D.M., P.R. Paquin, K. Subburamu, and D.A. Gruber. 1990. Sediment oxygen demand model: Methane and ammonia oxidation. Journal of Environmental Engineering, 116: 945–986.

Doerr, S.M., E.M. Owens, R.K. Gelda, M.T. Auer and S.W. Effler. 1997. Development and testing of a nutrient-phytoplankton model for Cannonsville Reservoir. In Press. Lake and Reservoir Management.

Gardiner, R.D., M.T. Auer, and R.P. Canale. 1984. Sediment oxygen demand in Green Bay (Lake Michigan), pp. 514519, In: M. Pirbazari and J. S. Devinny, (eds.), Proceedings of the 1984 Specialty Conference on Environmental Engineering, American Society of Civil Engineers, New York, NY.

Halfon, E., N. Schito, and R.E. Ulanowicz. 1996. Energy flow through the Lake Ontario food web: conceptual model and attempt at mass balance. Ecological Modeling, 86: 1–36.

Hamilton, D.P. and S.G. Schladow. 1997. Prediction of water quality in lakes and reservoirs. Part 1 – Model description. Ecological Modeling, 96: 91–110.

James, R.T. and V.J. Bierman, Jr. 1995. A preliminary modeling analysis of water quality in Lake Okeechobee, Florida: Calibration results. Water Research, 29(12): 2755–2766.

James, R.T., J. Martin, T. Wool, and P.F. Wang. 1997. A sediment resuspension and water quality model of Lake Okeechobee. Journal of the American Water Resources Association, 33(3): 661–680.

Jayaweera, M. and T. Asaeda. 1996. Modeling of biomanipulation in shallow, eutrophic lakes: An application to Lake Bleiswijkse Zoom, the Netherlands. Ecological Modeling, 85: 113–127.

Jin, K-R., James, R.T., W-S. Lung, D.P. Loucks, R.A. Park, and T.S. Tisdale. 1998. Assessing Lake Okeechobee eutrophication with water quality models. Journal of Water Resources Planning and Management, 124(1): 22–30.

Jorgensen, S.E. 1995. State of the art the art of ecological modeling in limnology. Ecological Modeling, 78: 101–115.

Kadlec, R.H. and W.W. Walker, Jr. 1997. Management models to evaluate phosphorus impacts on wetlands. This volume.

Laws, E.A. and M.S. Chalup. 1990. A microalgal growth model. Limnology and Oceanography, 35: 597–608.

O'Melia, C.R. 1972. An approach to the modeling of lakes. Schweizerische Zeitschrift fur Hydrologie, 34: 1–34.

Penn, M.R., M.T. Auer, E.L. VanOrman, and J.J. Korienek. 1995. Phosphorus diagenesis in lake sediments: Investigations using fractionation techniques. Marine and Freshwater Research, 46: 89–99.

Rast, W. and G.F. Lee. 1978. Summary Analysis of the North American Project (US portion) OECD Eutrophication Project: Nutrient Loading-Lake Response Relationships and Trophic State Indices, USEPA Corvallis Environmental Research Laboratory, Corvallis, OR, EPA-600/3-78-008, 455 pp.

Reckhow, K.H. 1977. Phosphorus Models for Lake Management. PhD Dissertation, Harvard University, Cambridge, MA.

Reckhow, K.H. 1979. Empirical lake models for phosphorus development: Applications, limitations, and uncertainty. pp. 183–222, In: D. Scavia and A. Robertson (eds.), Perspectives on Lake Ecosystem Modeling, Ann Arbor Science, Ann Arbor, MI.

Reckhow, K.H and S.C. Chapra.1979. Error analysis for a phosphorus retention model. Water Resources Research, 15:1643–1646.

Reckhow, K.H., S.C. Coffey, M.H. Henning, K. Smith, and R. Banting. 1992. EUTROMOD: Technical Guidance and Spreadsheet Models for Nutrient Loading and Lake Eutrophication. North American Lake Management Society, Madison, WI, 95 pp.

Riley, G.A. 1946. Factors controlling phytoplankton population on Georges Bank. Journal of Marine Research, 6:104–113.

Schladow, S.G. and D.P. Hamilton. 1997. Prediction of water quality in lakes and reservoirs. Part 2 – Model calibration, sensitivity analysis and application. Ecological Modeling, 96: 111–123.

Scheffer, M., S.H. Hosper, M-L. Meijer, B. Moss, and E. Jeppesen. 1993. Alternative Equilibria in Shallow Lakes. Trends Ecol. Evol., 8: 275–279.

Schelske, C.L. 1999. Assessing nutrient limitation and trophic state in Florida lakes. Chapter 13, this book.

Sheng, P. 1999. Effect of hydrodynamic processes on phosphorus distribution in aquatic ecosystems. Chapter 16, this book.

Steele, J.H. 1962. Environmental control of photosynthesis in the sea. Limnology and Oceanography, 7:137–150.

Thomann, R.V. 1977. Comparison of lake phytoplankton models and loading plots. Limnology and Oceanography, 22:370–373.

Thomann, R.V. and J.F. Fitzpatrick. 1982. Calibration and Verification of a Mathematical Model of the Eutrophication of the Potomac Estuary. Report to District of Columbia, Department of Environmental Services, HydroQual, Inc. Mahwah, New Jersey, 500 pp.

Vollenweider, R.A. 1968. Scientific Fundamentals of the Eutrophication of Lakes and Flowing Waters, with Particular Reference to Phosphorus and Nitrogen as Factors in Eutrophication. Technical Report DA5/SCI/68.27, Organization for Economic Cooperation and Development, Paris, France, 250 pp.

Vollenweider, R.A. 1969. Possibilities and limits of elementary models concerning the budget of substances in lakes. Archive fur Hydrobiology, 66: 1–36.

Vollenweider, R.A. 1975. Input-output models with special reference to the phosphorus loading concept in limnology. Schweizerische Zeitschrift fur Hydrologie, 37:53–84.

Vollenweider, R.A. 1976. Advances in defining critical loading levels for phosphorus in lake eutrophication. Mem. Ist. Ital. Idrobiol., 33:53–83.

Walker, W.W., Jr. 1977. Some Analytical Methods Applied to Lake Water Quality Problems. PhD Dissertation, Harvard University, Cambridge, MA.

Walker, W.W., Jr. 1980. Variability of Trophic State Indicators in Reservoirs. pp. 344–348, In: Restoration of Lakes and Reservoirs, U.S. Environmental Protection Agency, Office of Water Regulations and Standards, Washington, DC, EPA 440/5-81-010.

Walker, W.W., Jr. 1996. Simplified Procedures for Eutrophication Assessment and Prediction: User Manual. Instructional Report W-96-2, U.S. Army Corps of Engineers, Waterways Experiment Station, Vicksburg, MS.

Wickman, T.R. 1996. Modeling and Measurement of Nitrogen Diagenesis and Ammonia Release from the Sediments of a Hypereutrophic Lake. M.S. Thesis, Department of Civil and Environmental Engineering, Michigan Technological University, Houghton, Michigan, 124 pp.

29 Policy Implications to Phosphorus Management in Florida Ecosystems

Jerry Brooks

29.1 ABSTRACT

Nutrient loading to surface waters is a complex water resource management issue facing state policy makers. Management efforts to control the effects of nutrient loading are often focused on phosphorus. Point source discharges, both domestic and industrial, commonly contain phosphorus concentrations at levels exceeding background, thus representing a potential threat to receiving surface waters. Likewise nonpoint sources, both urban and agricultural, are also significant contributors of phosphorus loading. Because phosphorus is so ubiquitous, it is frequently an issue of concern for surface water management programs. These programs include both regulatory and restoration activities. Central to the successful implementation of

1-56670-331-X/99/$0.00+$.50

these activities are the surface water quality standards that reflect federal and state policy.

A phosphorus criterion, as a component of the water quality standards, establishes the threshold level that, when exceeded, results in unacceptable risk to aquatic biota. Establishment of this threshold is often complicated, because the ecosystem response to phosphorus is governed by a number of physical, chemical and biological variables that are often water body specific. The influence of these other variables leads to scientific uncertainty in defining a phosphorus threshold. Because of this uncertainty, policy makers in establishing phosphorus criteria have maintained the opportunity for flexibility. While this flexibility is appropriate to account for individual water body variability, it does place considerable responsibility on ecologists to understand cause and effect relationships. The weight of that responsibility is often proportional to the level of societal and economic interest associated with individual bodies of water.

Florida's water quality standards for total phosphorus are broad and provide flexibility. This flexibility allows for the consideration of those environmental variables that influence the cause and effect relationships found within specific water bodies. As a narrative standard the limit is established as a concentration which, when exceeded, causes "an imbalance in the natural populations of aquatic flora and fauna." In establishing the allowable concentration, policy makers will weigh ecological risk against societal and economic costs. Insight into this decision making process can be gained through a review of standards and associated permit limits both in Florida and across the nation.

29.2 INTRODUCTION

Water quality standards represent the foundation of management strategies for the protection and restoration of waters. As water quality standards are adopted through a public process, they ultimately represent a balance of public interest that considers social, economic, and scientific variables. While societal and economic impacts are typically easily qualified and quantified, the level of scientific understanding desired for the development of standards is often incomplete. As a result, the establishment of a standard not only represents a policy of balanced societal, economic, and environmental factors but also a policy defining the level of risk deemed acceptable in light of scientific uncertainty.

Nutrient loading to surface waters, originating through both point and nonpoint sources, represents a water quality variable having considerable environmental consequence. While this is indisputable, the level of loading and the relationship of that loading to the health of a specific aquatic ecosystem is often the subject of great scientific uncertainty. This scientific uncertainty is the most contentious issue facing policy makers in the adoption of phosphorus standards. As research of the ecological dynamics of phosphorus continues, it is useful to consider how policy makers have dealt with this scientific uncertainty in the past. Through this understanding one can perhaps gain insight to those factors that have most influenced past policy decisions.

29.3 OVERVIEW OF PHOSPHORUS STANDARDS NATIONWIDE

State policies and/or standards relevant to phosphorus are guided by national policy and regulations. Understanding current phosphorus related policies and standards is necessary to help anticipate future directions. At the national level, Section 303(3) of the Clean Water Act provides the statutory basis for the water quality standards program. The requirements and procedures for developing, reviewing, revising and approving water quality standards by the States are defined in 40 CFR 131. Pursuant to this regulation a water quality standard must consist of three primary components: (1) water quality goals that establish the designated use of that water body, (2) water quality criteria necessary to protect the use, and (3) an antidegradation policy to maintain and protect existing uses and water quality, to provide protection for higher quality waters, and to provide protection for outstanding national resource waters.

For all water bodies, each state must specify appropriate water uses (goals) to be achieved and protected. The classification of the waters must take into consideration the use and value of water for public water supplies; protection and propagation of fish, shellfish and wildlife; recreation in and on the water; and agricultural, industrial, and other purposes including navigation. Consistency with the Clean Water Act requires that wherever attainable states must adopt water quality standards that minimally provide for the protection and propagation of fish, shellfish, and wildlife. Waters that have been designated for uses that do not support propagation of fish, shellfish, and wildlife are required to be reassessed on a reoccurring schedule every three years.

To ensure the protection of surface waters and their designated uses, surface water criteria are established. These criteria may be either numerical or narrative. Numerical criteria are associated with specific contaminants for which an acute or chronic end-point can be derived. Derivation of numerical criteria is unambiguous as the measurement of acute and chronic toxicity responses can be easily measured for multiple aquatic organisms. To protect waters from contaminates that do not readily lend themselves to numerical criteria, narrative criteria are used. Narrative criteria are designed to prevent unacceptable environmental responses that might be associated with specific contaminates or interactions of multiple variables. Many states employ the use of narrative criteria for phosphorus because its affect is governed by a multitude of site specific chemical, physical, and biological conditions. Because of the complex interaction of these variables controlling the phosphorus-related response of the biological community, it is the narrative criteria that is often subject of great scientific uncertainty.

29.4 NATIONAL OVERVIEW OF STATE IMPLEMENTATION

Nutrients represent a group of water quality constituents that have well recognized environmental implications. In the absence of regulatory controls on nutrient discharges to all water bodies, ecological degradation will result. While the ecological implications of nutrient loading to surface waters is widely recognized, the nutrient dynamics of surface waters are varied and often poorly understood. In reviewing nutrient standards across the nation, policy makers have responded to this complex issue in various ways. For phosphorus, a number of different management strategies have been implemented. Narrative criteria are the most common for the control of phosphorus discharges. However, in many states where site specific data have been collected, the narrative criteria have served as the foundation for development of more definitive criteria. These more definitive criteria are typically associated with specific water bodies or classes of water bodies where the phosphorus dynamics have been defined. In these waters, the more definitive criteria or limits include (1) numerical criteria, (2) required use of best available treatment technology, (3) prohibition of phosphorus loading, and (4) various combinations of the above.

29.4.1 States with Narrative Limits

Narrative limits on the discharge of phosphorus is the most common management strategy employed by states. In some states, the narrative limit is used in combination with other more narrowly defined limits. A common element of the narrative limit is a restriction on loading that encourages accelerated eutrophication. Many narrative limits make specific reference to the control of algae and/or macrophyte growth. However, some states have not restricted their phosphorus controls to primary production. Vermont and Michigan have additionally prohibited the stimulation of growth in fungi and bacteria to the extent that it adversely affects designated uses. Florida has a very broad standard that prohibits phosphorus concentrations that cause an imbalance in the natural populations of aquatic flora and fauna. Florida's water quality standards are discussed later in this paper.

29.4.2 States with Established Numerical Limits

While a number of states have established numerical limits, categorically those states falling within EPA Region 9 have the most well defined numerical criteria for the control of phosphorus discharges. States within Region 9 include Hawaii, California, Arizona, and Nevada. Of those states, Hawaii is perhaps the most complex. Hawaii has categorized its surface waters ranging from fresh water streams to oceanic waters. Within each of six categories, it has established mean and single-value concentration limits for both wet and dry seasons. Limits expressed as geometric means range from 10 μg/L for oceanic waters to 60 μg/L for the Pearl Harbor Estuary. It is worth noting that the criteria for Pearl Harbor Estuary is 2.4 times higher than the limit for other estuaries, a reflection of societal and economic factors.

29.4.3 States Requiring the Use of Best Available Technology

Required use of best available technology for removal of phosphorus is a management strategy used in some states. In Maine, new discharges of phosphorus to any lake or pond or tributary thereto are prohibited if best available technology for removal of phosphorus has not been employed. For existing discharges, phosphorus treatment is required to the maximum extent technically feasible. These technology-based requirements are complementary to established numerical limits for specified classes of surface water. This reflects a recognition of the scientific uncertainty surrounding a numerical limit and provides for the highest level of treatment feasible regardless of compliance with the numerical criteria.

29.4.4 States with Prohibitions of Phosphorus Discharge

While no state has a blanket prohibition of phosphorus discharge to all waters, several states have designated classes of waters to which phosphorus discharges are prohibited or require stringent demonstrations of no effect. These are most often waters that fall into the category of *special waters*. Special waters are those water bodies designated by the state that receive a higher level of protection than those generally classed for recreation and fish and wildlife propagation. Special waters are typically water bodies of unusual value ecologically resulting from the presence of unusual species, productivity, diversity, or other unique ecological characteristics.

29.5 FLORIDA'S WATER QUALITY STANDARDS

29.5.1 Summary of Applicable Standards

Florida's water quality standards are consistent with federal law. All waters are classified according to their designated use, and water quality criteria are established to protect those uses. The classifications are arranged in order, according to the degree of protection required with Class I water having the most stringent water quality criteria and Class V the least. However, consistent with federal law, Class I, II, and III surface waters share water quality criteria established to protect recreation and the propagation and maintenance of a healthy, well balanced population of fish and wildlife. A combination of both numerical and narrative criteria are used. For the regulation of phosphorus the most apparent criterion is a narrative criterion governing the discharge of nutrients. It states: "In no case shall nutrient concentrations of a body of water be altered so as to cause an imbalance in natural populations of aquatic flora and fauna."

Chapter 62.302.300 (13), F.A.C. provides additional policy guidance relevant to the control of nutrients. It states:

> The Department finds that excessive nutrients (total nitrogen and total phosphorus) constitute one of the most severe water quality problems facing the state. It shall be the Department's policy to limit the introduction of man-induced nutrients into waters of the State. Particular consideration shall be given to the protection from further nutrient enrichment of waters which is presently high in nutrient concentrations or sensitive to further nutrient concentrations and sensitive to further nutrient loading. Also, particular consideration shall be given to the protection from nutrient enrichment of those presently containing very low nutrient concentrations: less than 0.3 milligrams per liter total nitrogen or less than 0.04 milligrams per liter total phosphorus.

While the narrative criterion is directed specifically at controlling the discharge of phosphorus, other numerical criteria are relevant. For example the transparency criterion states, "the depth of the compensation point for photosynthetic activity shall not be reduced by more than 10% compared to the natural background value." Application of this and other numerical criteria for the control of phosphorus does require an understanding of the cause and effect relationships unique to each body of water.

In addition to the classifications mentioned above, the state has also designated certain waters as "Outstanding Florida Waters" (OFW). OFWs are waters designated as worthy of special protection because of their natural attributes. Within these waters, no degradation of water quality is allowed, and degradation is evaluated relative to a defined ambient condition.

29.5.2 Application of Florida's Water Quality Standards

Development of nutrient limitations for a surface discharge is typically based on a review of historical water quality information and the water quality criteria applicable to the water body. Defining the phosphorus threshold related to an imbalance in flora and fauna generally requires the use of water quality models that consider nutrient-algal-dissolved oxygen interaction, various biometrics and specific assays, statistical comparisons with similar systems, and best professional judgment. Intensive studies that involve the collection of physical, biological, and chemical information under specific spatial and temporal conditions are used for the calibration and validation of models.

In reviewing how Florida's narrative nutrient criterion has been applied to specific water bodies, many similarities can be found to those states that have adopted more definitive criteria. A common point of reference is algal production and chlorophyll concentrations. Application of both numerical and narrative criteria often revolve around a biological response of the algal community. Most common among those states with narrative phosphorus criteria are limits on the loading that result in unacceptable levels of algal production. Where numerical limits have been established, either as part of water quality standards or as permit limits, algal growth or chlorophyll production is often the operative variable. In many instances, the allowable threshold for algal production is linked to other water quality characteristics governed by primary production. These linked water quality variables include dissolved oxygen and transparency.

29.5.3 Defining Regulatory Thresholds

Defining the phosphorus concentration that cannot be exceeded is a subject of considerable scientific inquiry. Cause-and-effect relationships are often very complex because of the interaction of numerous controlling variables. For open-water bodies, such as streams, lakes, and estuaries, inferences can be drawn from other state standards and specific permit limits that have been imposed in Florida. Clearly, phosphorus discharges that result in the violation of any numerical water quality criterion can and have been considered in establishing water body specific limits. This relationship is appropriate in that the numerical limits associated with a designated use have been determined necessary to protect those uses. Consequently, any phosphorus discharge that leads to, for example, a lowering of dissolved oxygen below 5 ppm is considered unacceptable for protecting the designated use of that water. Any relief provisions applicable to the numerical criteria must also be considered in establishing this limit. In Florida and other states where phosphorus limits and or rule criteria have been established, it is typically based on the demonstrated relationship of phosphorus concentrations to other numerical limits. Typically, this relationship is linked to algal production and associated chlorophyll levels.

29.5.4 Defining Imbalance to Natural Populations of Flora and Fauna

While most numerical criteria or regulatory limits on phosphorus have been based on established links to other water quality variables, concerns for imbalance to flora and fauna must consider other factors.

This is particularly true where discharges are to surface waters with nutrient sensitive biota. Defining phosphorus limits based on an imbalance to natural populations of flora and fauna is most often a contentious process. Scientific uncertainty is often the focus of this contentiousness when there are conflicting social, economic, and environmental interest.

29.5.5 Environmental Issues

Scientific uncertainty is almost always an issue to be considered in the field of ecology. In natural systems, there exist numerous interactions among the physical, chemical, and biological components of the system that are both complex and dynamic. Because of this complex and dynamic relationship, predicting an ecosystem response to the alteration of one variable is difficult. The precision of such a prediction in most instances is subject to considerable uncertainty. Unlike laboratory experiments where all variables can be controlled, ecological studies must attempt to consider the influence of other controlling factors. The fact that most phosphorus limits have been established based on nutrient/chlorophyll relationships is a reflection of the level of research focused on this issue. Even this relationship can be difficult to predict and subject to question in waters with complex hydrologic characteristics.

Within wetland systems, the relationship of phosphorus to other water quality criteria and ecosystem health is less well understood. Derivation of phosphorus limits for discharges to wetlands is an emerging process. While much research has been focused on understanding the fate and transport of phosphorus in wetlands, the understanding of cause and effect relationships is far from complete. As research in this area progresses it is anticipated that the use of biological criteria will become a more visible component of the state's water quality standards. Biological metrics that consider structure and/or function will likely become the focus of establishing phosphorus limits.

29.5.6 Balancing Social, Economic, and Environmental Interest

Policy makers are routinely faced with the responsibility of balancing social, economic, and environmental interest. Pivotal to that task is a clear understanding of the conflicting factors. Where the development of standards or specific regulatory criterion have clear economic implications, scientific uncertainty immediately becomes the issue. As economic pressures increase, so too does the level of scientific understanding required for regulatory management. In fact, this need to clearly define the science is often driven by the parties most affected economically. To protect against regulations that might be more stringent than necessary to achieve resource protection, economic forces frequently raise questions regarding scientific uncertainty.

29.6 CONCLUSION

Florida's rules and policies reflect a recognition of the threat nutrients represent to the state's surface water resources. Embodied within these rules and policies are provisions that allow for management flexibility to account for site specific conditions. These site specific conditions account not only for ecosystem diversity but also social and economic interest. In the balancing of these often conflicting areas of interest, the scientific uncertainty of ecosystem functions is often the focal point of conflict resolution. Because ecosystems are so complex and dynamic, precise prediction of their response to perturbation is difficult. Where the social and economic interest are strong, the greater the burden becomes for the ecologist to make these predictions. The level of risk determined acceptable in predicting ecosystem response is often a reflection of the strength of social and economic interest. It is for this reason that a universal definition of the standard, "...imbalance to natural populations of fish and wildlife" is intangible. Policy makers must acknowledge this uncertainty and accept that establishment of a phosphorus limit will often be a matter of establishing public policy. To ensure protection of nutrient sensitive waters, policy makers cannot wait for absolute certainty but must make judgments based on the best information available.

Section VIII

Panel Discussion

30 Phosphorus Biochemistry In Subtropical Ecosystems

Panel Summary

30.1 PANEL SUMMARY

A scientific panel consisting of six members was convened to discuss the technical issues related to phosphorus biogeochemistry of subtropical ecosystems, as related to surface water quality. This chapter summarizes the results of the panel discussion on critical issues and future research needs using Florida as a case example. The panel members were:

1. G.A. O'Connor, Panel Chairman, University of Florida
2. N.G. Aumen, South Florida Water Management District
3. S.R. Humphrey, University of Florida
4. T.J. Logan, Ohio State University

1-56670-331-X/99/$0.00+$.50

5. T. MacVicar, MacVicar, Federico and Lamb, Inc.
6. R.G. Wetzel, University of Alabama

30.2 GEOCHEMICAL PROCESSES: G.A. O'CONNER

30.2.1 Chemical Fractionation Schemes

These schemes are widely used to determine P forms and/or transformations in soil/sediments. Don Graetz did a nice job reviewing the historical basis for some schemes. Sue Newman appropriately listed attributes and limitations of some schemes. In particular, Newman and Graetz cautioned us that the schemes yield only operationally defined information about P forms. Each scheme includes assumptions about the "forms" and bioavailability, involving major leaps of faith. This is especially true when we attempt to apply schemes developed for one system to another, e.g., methods developed for inorganically dominated, upland, nonbiological systems extended to biologically dominated, wetland systems. Similar dangers are involved in using soil tests (usually developed to identify nutrient deficient soils) on highly polluted systems. Greater levels of extractable ion do not necessarily mean greater bioavailability—only that the extract's capacity has not yet been exceeded.

Thus, several authors appropriately reminded us of the need to relate (calibrate) extractant levels or "forms" to organism response: plants to soil tests, microbes to extractant fractions, rules to actual bioavailable (effective) concentrations. Surprisingly, I heard little discussion of how to determine truly soluble P and its forms' correlation to biological response. I refer specifically to consideration of filtration techniques (0.1 vs. 0.45 μm), speciation determinations, etc. Only when we really understand the forms of P and have validated, highly correlated results can we hope to measure the right parameters, and to devise appropriate remediation techniques.

30.2.2 Models

Dean Rhue reviewed the various reactions and equilibrium and kinetic models of P sorption in soils and sediments. He and Jim Davidson also reminded us that desorption cannot be ignored if we are to truly understand/predict P transport. Most chemical models assume equilibrium and complete reversibility. Neither assumption typically characterizes natural systems. Similarly, lab studies, and even some field studies involving large spikes of solute, are only snapshots of possible long-term effects. Given the dynamic nature of most natural systems, we may never be able to simulate what goes on very well, especially if we insist on using equilibrium vs. kinetic models. (See Wetzel's remarks that follow.) We must research system characteristics and use models to set limits of what can and cannot be attained. Models, of course, have their own limitations (assumptions) that must be understood. Models, however, are useful because they allow data integration to set boundaries of consideration.

We also don't want to ignore possible changes in a system's P sorbing capacity associated with waste streams, nature, and amendments.

30.2.3 Wastes

Effluents, manures, biosolids, etc. can add significantly to a soils/sediments P sorption through additions of Ca^{++} (as a precipitating cation with P) or through additions of sorbing solids such as Fe and Al oxides, $CaCO_3$, etc.

30.2.4 Nature

Soils/sediments are constantly rejuvenating themselves via weathering (encouraged by alternately wetting and drying) or accretion. Thus, adsorption "maxima" determined in the laboratory (days) may well underestimate real world retention capacities (months, years).

30.2.5 Amendments

We may also want to consider intentionally increasing a system's sorption capacity through the addition of Fe, Al (alum), Ca (lime), etc. at various points along the P trail to reduce P load and/or availability. There are several "wastes" (by-products) available in Florida, e.g., fly ash, biosolids, water treatment sludges, "high-clay alumina," etc. that may be beneficially used. It may not be feasible to try wholesale ecosystem amendment, but strategic use to reduce major inputs may be effective.

30.3 BIOGEOCHEMICAL PROCESSES: ROBERT G. WETZEL

The objectives of this symposium focused on effective management of eutrophication processes. The fundamental roles of phosphorus in increasing and altering productivity of wetland and lake ecosystems were addressed, particularly in relation to changes in species and community composition to less desirable biota or functional capacities of these ecosystems.

Management of eutrophication problems has been dominated by controlling P loadings. Reduction in loadings has often resulted, for many well known physiological and biogeochemical reasons, in a decline in primary production and shifts to more desirable species. Problems also often emerged in which conditions of oligotrophication were not as predicted or desired. These conditions did not follow models and predictions based on generalized data bases. Clearly, our understanding of functional processes controlling metabolism, competitive interactions, and rates of growth under differing time and spatial scales is incomplete. Repeatedly, I see interpretative problems focused on controls of static concentrations, both in chemical or mass of biota, rather than on rate functions.

Management has the responsibility to understand the control mechanisms regulating these rate functions in aquatic ecosystems. Phosphorus cycling is just one of the many such functions. The physiological understanding of the dynamic factors regulating the rate functions of biogeochemical cycling would not only be "nice," as one modeler stated in this symposium, but is essential. Until we know and understand the physiology and control mechanisms regulating the biotic dynamics, we cannot effectively manage these ecosystems. It is particularly not possible to manage them economically. That continued ignorance is prohibitively expensive, particularly in the long term.

Aquatic ecosystems are not physical structures, such as a bridge, in which the engineers can compensate for unknown stresses by adding a safety factor of 10 because the dynamics of the environment are so poorly known. As understanding of molecular adhesion and flexing of construction materials and the dynamics of the environmental conditions improved, the excessive compensation for uncertainty (safety factor) could be reduced markedly at great functional and economic savings. A more relevant analogy is in human medicine. Anatomical knowledge of the structure of the human body did not lead to effective management and improvement of human health. Only with detailed experimentally founded understanding of the physiology of rate functions and factors controlling those functions has effective human health management and remediation occurred. Managers of aquatic ecosystems have a responsibility to understand the physiological control of aquatic ecosystem health.

Bridges are not the same, and humans are not the same. Each one is uniquely different. I frequently hear the same tired song that ecosystems are too heterogeneous to study adequately, as I insist is essential. Just as in human physiology, functional controls of ecosystems are generic and supersede that heterogeneity. It is the specific range of fluctuations of parameters and rate functions that must be known in specific cases. The actors may differ, but the play is the same.

If one concedes the point that functional understanding improves the efficacy and economics of management, how do we obtain such knowledge? The present largely trial-and-error operation, followed by remediation to alter or correct bad directions, is much too slow and expensive. It is much more prudent to underwrite with time and expenses the experimental approaches in which controlled rate functions are assayed to quantitatively delineate emergent regulating processes. Just as in human medicine, these ecosystems are complex, and experimentation requires adequate funding and interdisciplinary expertise to gain the essential operational and controlling insights.

Modeling can be a powerful tool, but it is not the panacea often proclaimed. Rate functions within natural ecosystems are highly dynamic, interactive, and interdependent. Effective modeling depends on accurate input-response data and an appreciation and incorporation of nonlinear real-world kinetics of those data. Because of the present acute limitations of input-response data, even the best of mechanistic models can do little more than point to narrowed probability of system responses under a set of input circumstances. Under no circumstances will modeling supplant the need for the physiological, experimentally derived input-response data essential for development of powerful generic models. Models can be effective, however, in addressing scaling problems, particularly spatial scales, along gradients within ecosystems and integration of spatial and temporal differences among functional components of ecosystems.

In conclusion, I wish to emphasize markedly neglected areas of understanding in biogeochemical cycling of phosphorus in ecosystems, particularly aquatic ecosystems. Study of geochemical factors alone cannot possibly yield significant understanding of P dynamics in ecosystems. The storage pools and changing environmental conditions that regulate the dynamics of P mobility and transfer rates are almost entirely microbial. Because most aquatic ecosystems of relevant interest and use to humankind are shallow, most of the activities regulating the redox conditions, solubilities, and fluxes among different storage pools are by microbial communities attached to surfaces, as is similarly the case in terrestrial ecosystems. The regulation of fluxes is highly dynamic and often at short time scales. A majority of the biogeochemical regulation of nutrient, particularly P, dynamics in ecosystems is at the microbial level. Most of our attention has been directed to other ecosystem components even though easily over 70% of the control of P fluxes is mediated by microbial metabolism. To effectively manage nutrient fluxes at any level much beyond the input-output approach to a "block box" ecosystem, we must understand more effectively the microbial retention and recycling capacities within those ecosystems.

I also note the importance of gradients in ecosystems. Nutrients move along gradients, along and within a series of storage sites (e.g., biota, sorption to inert surfaces, etc.), release to the environment, movement, and redeposition into a storage site. The resulting series of metabolic coupling between organisms and environment is largely at the microbial level and is often rapid. The rapidity of recycling increases as the availability of essential nutrients decreases. Although this nutrient "spiraling" is complex spatially and temporally, it is essential to quantify pools and fluxes among pools and factors regulating these transfer fluxes.

30.4 WATERSHED PROCESS: TERRY J. LOGAN

30.4.1 General Observations

1. Phosphorus biogeochemistry in Florida is unique because of its soils (generally sandy or organic, low in total P and low in P binding capacity), hydrology (shallow water tables, interconnected riverine, lacustrine and wetland systems), and land use (intensive livestock and crop production, rapidly urbanizing geography). Soils are easily overloaded with added P, and P is easily transferred to the aquatic ecosystems. Management will require not only source control to avoid future problems, but restoration of P-impacted aquatic systems. The ultimate storage for P appears to be in lake or wetland biota or sediments.
2. Even though the Florida ecosystems are relatively unique, the issues I have heard discussed for the last three days are the same ones I heard in the Great Lakes 20 years ago. What is different is the rapid rate of change in Florida, particularly with respect to the adoption of agricultural BMPs (because of strong water management districts and court decisions) and the response of the water bodies to increased P loading (e.g., Everglades). How the water systems will respond to P load reductions is variable (Lake Apopka vs. Lake Okeechobee), and the role of biota and accreted sediment P in buffering the systems against P reduction is not known.
3. We must give the public reasonable and rational choices on P management (drastically reduce the sources and then let the lakes restore themselves,

or spend money to remediate the impacted water bodies) and the time scales for expected results (years or decades).

30.4.2 Watershed Processes

1. Phosphorus transport in Florida ecosystems is primarily as dissolved P. We are unsure as to the exact speciation of the dissolved P (orthophosphate, polyphosphate, organic, colloidal), making it difficult to interpret plant response to P loadings and how dissolved P interacts with soils and sediments.
2. Manure (and to a lesser extent biosolids) are major sources of concentrated P. Initial dissolved P concentrations from concentrated manure sources are orders of magnitude larger than concentrations from crop or pasture fertilization.
3. Once P enters the watershed transport system, it has very little attenuation: sandy soils, low P retention, shallow water tables, little eroded suspended sediment. Runoff or seepage P moves rapidly to the stream systems and final receptors.
4. The ultimate P sinks are the wetlands and lakes where P attenuation by sediments varies with sediment chemistry (calcareous or organic), and biotic processes like periphyton assimilation and storage (perhaps as calcite precipitates) are important. The possible coprecipitation of P in calcite precipitated by the periphyton is an important mechanism for wetland P removal because, unlike P adsorption of sediments, indefinite P removal is possible provided that adequate Ca is available.

30.4.3 Management Implications

1. Phosphorus sources, particularly manure, must be controlled at the source. BMPs must emphasize source control over hydrologic control; in hydrologically active watersheds as are found in Florida, changing watershed hydrology (except in pumped drainage areas) is difficult.
2. Lake or wetland restoration, after P inputs have been controlled, is a function of the bioavailability of stored sediment P. We need to know the mechanisms for P buffering after P gradients are reversed, and we must model these mechanisms to determine how long recovery is likely to take. Chemical additions (e.g., alum or like materials) may accelerate restoration by immobilizing stored P.

30.4.4 Scientific and Sociopolitical Implications

1. The large wetland and lake reclamation projects underway are a unique opportunity to study the reversal of P-induced eutrophication. They will tell us how well buffered these systems are when nutrient loadings are reduced, and where the buffering lies: in the sediments, or in the biota.

2. The effect of rapidly urbanizing Florida landscapes has been overlooked in this conference. The perspective of the conference has been on the effect of agriculture on Florida water systems; this may be today's problem, but not tomorrow's.

30.5 ECOSYSTEM MANAGEMENT: NICHOLAS G. AUMEN

30.5.1 Comments on the Development of Synthesis Tools

The symposium presentations highlighted the increasing complexity of hydrological, water quality, and ecological models and their enhanced value as synthesis tools for phosphorus biogeochemistry research and natural resource management. Advances in Geographical Information Systems (GIS) software and user-friendly interfaces have further increased the applicability of these models (e.g., Negahban et al., 1995). The impetus for development of these models has been to increase basic understanding of the role of phosphorus in aquatic and terrestrial ecosystems, and to develop a predictive understanding of biogeochemical processes across space and time scales. The symposium provided examples of model uses such as evaluation of phosphorus transport in watersheds, prediction of phosphorus impacts on wetlands and lakes, and landscape-scale evaluation of the role of phosphorus in ecological processes. All of these models use and build upon a foundation of physical, chemical, and biological process data gained through experimental research, process measurement, and observational studies. Model calibration, validation, and application can help identify future ecosystem trends, highlight information needs, and serve as an organizational framework for future research.

Care must be taken however, to ensure a balance of effort between further model complexity and model validation. As models become easier (and more fun) to develop and enhance, model development can quickly outpace field and laboratory research needed to improve understanding of basic ecosystem processes. Data from field and laboratory research are necessary for model calibration and validation. At worst, a poorly calibrated (although complex) model might generate a false impression of understanding among those unfamiliar with the tools. A reasonable balance of resource expenditure can be assured if models are appropriately calibrated and validated, and are applied as they are being refined, to identify further data and research needs.

30.5.2 Research Recommendations

To understand phosphorus biogeochemistry better, more emphasis should be given to well designed experiments and process measurements, as compared to strictly observational studies (e.g., water quality data collection). Observational data, while essential to determine long-term trends and to track ecosystem health, provide only correlative information and alone cannot reveal cause/effect relationships. Properly designed and relevant experiments, both in the laboratory and in the field, will help us move away from the "black boxes" of phosphorus biogeochemistry toward a better understanding of ecosystem-level mechanisms and processes.

With respect to research topics related to phosphorus biogeochemistry, more effort is needed to understand microbial processes in sediments and the water column. Microbial communities play an important role in the transformations of phosphorus in aquatic systems, and a perusal of the symposium topics suggests a lack of this type of information in Florida ecosystems. In addition, a better understanding is needed of the sources and transformations of various phosphorus species, especially organic forms, and their relative bioavailability. This information would be extremely useful in the design and implementation of restoration programs in the greater Everglades ecosystem.

Finally, several presenters suggested the value of applying nutrient spiraling theory to wetlands to estimate the retentiveness of these systems. The spiraling concept has been applied most effectively to lotic ecosystems, where it has been used to describe the simultaneous processes of nutrient cycling and downstream transport (Newbold et al., 1983). In theory, nutrient spiraling is as applicable to wetlands as it is to lotic ecosystems; however, application of the concept in wetlands may not be practical. Determinations of nutrient spiraling lengths require experimental releases of nutrient and conservative tracers, and estimation of dilution and dispersion. Limitations imposed by very slow water velocities, the relatively large contributions of lateral and subterranean hydrologic inputs, and vegetative roughness may make estimation of these parameters impossible in the field. One alternative for estimating spiraling or uptake lengths in wetlands is to use flumes to isolate a known area, and then conduct releases within the flume. In this manner, one can control for the potential problems of dilution and dispersion. However, caution is needed when scaling up the results from flume studies to the entire ecosystem (Lamberti and Steinman, 1993).

30.5.3 Phosphorus Control and Ecosystem Restoration Recommendations

Phase I of south Florida's Everglades restoration consists of Best Management Practices (BMPs) implemented on farms, and construction of approximately 16,000 ha of wetlands to reduce phosphorus in agricultural runoff. Because the phosphorus concentration goal of Phase I (50 µg L^{-1} TP) still would cause down stream changes in the phosphorus-limited Everglades, a Phase II restoration program likely will be needed. Phase II will consist of supplemental technologies implemented in conjunction with the BMPs and constructed wetlands to achieve a phosphorus concentration target that may be as low as 10 µg L^{-1} TP. Potential supplemental technologies to accomplish this additional reduction range from chemical addition/filtration systems, to more passive technologies that take advantage of natural chemical and biological processes.

As restoration research progresses, it is important to consider that national and international experience in nonpoint source pollution shows that controls implemented close to nutrient sources may be more cost effective than controls only implemented downstream (National Research Council 1992). Therefore, phosphorus reduction approaches such as on-site BMPs are likely to be more effective, less costly, and longer lived than highly engineered, downstream solutions. The goal should be development of agricultural practices that are both environmentally and economically sustainable.

Research examining phosphorus reduction and restoration programs in the Everglades should look for unifying concepts and lessons learned from other locations. For example, research on the ecology of Lake Okeechobee's phosphorus-limited littoral zone has wide applicability to phosphorus biogeochemistry research in the Everglades (Steinman et al., this volume).

Finally, scientists should be more proactive in ensuring the best possible use of research results. Instead of ending their involvement at data analysis and publication, it is imperative that they become more involved in the guidance of how their information is used, and reduce the possibility of misuse. This involvement can occur in an unbiased and nonadvocacy fashion so that the objectivity of science can be maintained.

30.6 POLICY IMPLICATIONS: S.R. HUMPHREY AND T. MAC VICAR

Policy makers always must make decisions in the face of uncertainty, and failure to make a decision by an uncertain policy-maker is itself a decision. The adaptive-management solution to this problem is to make decisions in a timely manner (regardless of uncertainty), design the policy decision to yield lessons, and modify the policy as needed, based on the lessons learned. The scientific process is designed, of course, to produce reliable knowledge and thus reduce uncertainty. Science rarely supplies all the answers we desire, but sometimes the limitations are not as severe as reluctant policy makers believe. Consider the issue of numerical limits of P required to protect the environmental quality of Florida's waters. A limit for fresh waters was first hypothesized and then supported with empirical data almost 50 years ago by Sawyer. Are such data not useful today? Are data available for other freshwater systems not applicable to freshwater wetlands?

Even given hard scientific evidence, should societal and economic interests be allowed to outweigh normal standards for environmental protection? Under what circumstances, and to what extent, should such degradation be permissible in Florida?

Two difficult standard-setting issues, especially important in south Florida, were not adequately addressed in the conference. One is the development of a definition of "natural population of flora and fauna" for a man-made system. How should one interpret that phrase when setting nutrient limits for flood control canals that are, by nature and design, "unnatural"? The other issue is the process of defining "natural populations" for areas that are both natural and artificial. This is the Everglades case. Parts of the Everglades were dedicated to receiving storm water decades before the Clean Water Act. The policy issues of whether the standards should be imposed after the fact as a means of ecosystem restoration far outweigh the risk and uncertainty issue that received most attention. That is why the Everglades case was approached through a "P management plan," i.e., Storm water Treatment Areas and BMPs, rather than standard setting and discharge permits.

More attention should be given to the "Outstanding Florida Water" concept, especially given the prominence of the Everglades at the conference. The limitations imposed by the federal court and the Everglades Forever Act (EFA) are based on a technical interpretation of Florida's nondegradation requirement, not on a standard derived from an algae- or chlorophyll-based evaluation. The standard setting exercise is still years away, but in the meantime, $750 million is being spent on "P management."

REFERENCES

Lamberti, G.A., and A.D. Steinman (editors). 1993. Research in artificial streams: applications, uses, and abuses. Journal of the North American Benthological Society 12:313–384.

National Research Council, Committee on Restoration of Aquatic Ecosystems. 1992. Restoration of aquatic ecosystems: science, technology, and public policy. National Academy Press, Washington D.C. 552 pp.

Negahban, B., C. Fonyo, W.G. Boggess, J.W. jones, K.L. Campbell, G. Kiker, E. Flaig, and H. Lal. 1995. LOADSS: a GIS-based decision support system for regional environmental planning. Ecological Engineering 5:391–404.

Newbold, J.D. J.W. Elwood, R.V. ONeill, and A.L. Sheldon. 1983. Phosphorus dynamics in a woodland stream ecosystem: a study of nutrient spiraling. Ecology 64:1249–1265.

Steinman, A.D., K.E. Havens, N.G. Aumen, R.T. James, K.-R. Jin, J. Zhang, and B.H. Rosen. 1998. Phosphorus in Lake Okeechobee: sources, sinks and strategies. Ch. 23, this volume.

Index

Index

F

K

L

M

Q

R

S

W

Z

T - #0218 - 101024 - C0 - 234/156/34 [36] - CB - 9781566703314 - Gloss Lamination